Strongly Coupled Plasma Physics

NATO ASI Series

Advanced Science Institutes Series

A series presenting the results of activities sponsored by the NATO Science Committee, which aims at the dissemination of advanced scientific and technological knowledge, with a view to strengthening links between scientific communities.

The series is published by an international board of publishers in conjunction with the NATO Scientific Affairs Division

A	**Life Sciences**	Plenum Publishing Corporation
B	**Physics**	New York and London
C	**Mathematical and Physical Sciences**	D. Reidel Publishing Company Dordrecht, Boston, and Lancaster
D	**Behavioral and Social Sciences**	Martinus Nijhoff Publishers
E	**Engineering and Materials Sciences**	The Hague, Boston, Dordrecht, and Lancaster
F	**Computer and Systems Sciences**	Springer-Verlag
G	**Ecological Sciences**	Berlin, Heidelberg, New York, London,
H	**Cell Biology**	Paris, and Tokyo

Recent Volumes in this Series

Volume 150—Particle Physics: *Cargèse 1985*
edited by Maurice Lévy, Jean-Louis Basdevant, Maurice Jacob, David Speiser, Jacques Weyers, and Raymond Gastmans

Volume 151—Giant Resonances in Atoms, Molecules, and Solids
edited by J. P. Connerade, J. M. Esteva, and R. C. Karnatak

Volume 152—Optical Properties of Narrow-Gap Low-Dimensional Structures
edited by C. M. Sotomayor Torres, J. C. Portal, J. C. Mann, and R. A. Stradling

Volume 153—Physics of Strong Fields
edited by W. Greiner

Volume 154—Strongly Coupled Plasma Physics
edited by Forrest J. Rogers and Hugh E. Dewitt

Volume 155—Low-Dimensional Conductors and Superconductors
edited by D. Jérome and L. G. Caron

Volume 156—Gravitation in Astrophysics: *Cargèse 1986*
edited by B. Carter and J. B. Hartle

Series B: Physics

Strongly Coupled Plasma Physics

Edited by

Forrest J. Rogers

and

Hugh E. Dewitt

Lawrence Livermore National Laboratory
Livermore, California

Plenum Press
New York and London
Published in cooperation with NATO Scientific Affairs Division

Proceedings of a NATO Advanced Research Workshop on
Strongly Coupled Plasma Physics,
held August 4-9, 1986,
in Santa Cruz, California

Library of Congress Cataloging in Publication Data

NATO Advanced Research Workshop on Strongly Coupled Plasma Physics (1986: Santa Cruz, Calif.)
Strongly coupled plasma physics.

(NATO ASI series. Series B, Physics; v. 154)
"Proceedings of a NATO Advanced Research Workshop on Strongly Coupled Plasma Physics, held August 4-9, 1986, in Santa Cruz, California"—T.p. verso.
"Published in cooperation with NATO Scientific Affairs Division."
Includes bibliographical references and index.
1. Plasma (ionized gases)—Congresses. 2. Plasma density—Congresses. I. Rogers, Forrest J. II. Dewitt, Hugh E. III. Title. IV. Title: Coupled plasma physics. V. Series.
QC717.6.N366 1986 530.4'4 87-7972
ISBN-13: 978-1-4612-9053-7 e-ISBN-13: 978-1-4613-1891-0
DOI: 10.1007/ 978-1-4613-1891-0

Softcover reprint of the hardcover 1st edition 1987
A Division of Plenum Publishing Corporation
233 Spring Street, New York, N.Y. 10013

PREFACE

A NATO Advanced Research Workshop on Strongly Coupled Plasma Physics was held on the Santa Cruz Campus of the University of California, from August 4 through August 9, 1986. It was attended by 80 participants from 13 countries, 45 of whom were invited speakers. The present volume contains the texts of the invited talks and many of the contributed papers. The relative length of each text is roughly proportional to the length of the workshop presentation.

The aim of the workshop was to bring together leading researchers from a number of related disciplines in which strong Coulomb interactions play a dominant role. Compared to the 1977 meeting in Orleans-la-Source, France and the 1982 meeting in Les-Houches, France, it is apparent that the field of strongly coupled plasmas has expanded greatly and has become a very significant field of physics with a wide range of applications.

This workshop had a far greater participation of experimental researchers than did the previous two, and some confrontations of real experiments with theoretical calculations occurred. In the two earlier meetings the theoretical presentations were dominated by numerical simulations of static and dynamic properties of various strongly coupled plasmas. The dearth of experiments in the 1970's is now replaced by some very good experimental efforts. At the University of California San Diego a device for magnetically confining electrons cryogenically has made it possible to produce stable strongly coupled electron plasmas that are essentially the same as the one component classical plasma (OCP) that has been so extensively studied with computer simulations. Similarly the group at the National Bureau of Standards in Boulder, Colorado, have developed a successful arrangement for cryogenically trapping heavy ions to a coupling constant as high as $\Gamma = 100$. They anticipate seeing the fluid-solid phase transition predicted by simulation studies. These experiments give the promise in the next few years of observing in a laboratory strongly coupled Coulombic effects that are normally found in extremely high density astrophysical objects such as white dwarf stars and neutron stars.

Other experimental groups from West Germany, Yugoslavia, France, and the United States reported on recent work on liquid metals in the vicinity of the critical point, electrical conductivity at intermediate coupling, light absorption in cesium plasmas, and high compression measurements on liquid metals. Some of these experimental results could be directly compared with earlier theoretical calculations, and have already provided theorists with suggestions for important future work.

The theoretical calculations continue to include computer simulations of thermodynamic and transport properties from groups in Japan, France, and the United States. It is important to note that "numerical experiments" have led to a much better understanding of how to calculate the fluid state properties of dense plasmas from liquid-state-integral equations which require only a small fraction of the computer time needed for numerical simulation. The Japanese and the U.S. groups reported on successful use of coupled sets of hypernetted chain equations (HNC) that can largely reproduce most of the known numerical simulation results, and which have been extended to deal with electron screening and partial ionization.

The density functional theory has emerged as a powerful theoretical method for dealing with most strongly coupled plasma problems. Significant applications were reported for a first principles calculation of the fluid-solid phase transition, treatment of high Z ions in laboratory dense plasmas, and to a general treatment of quantum effects.

Just as with the two earlier meetings, strongly coupled plasmas in astrophysics was an important topic with results reported for white dwarf star interiors, Jovian planetary interiors, transport properties of dense stellar interiors, and plasma effects on neutrino emission.

As directors of the workshop, we would like to thank the North Atlantic Treaty Organization for its generous sponsorship of the workshop. We also wish to thank the National Science Foundation (U.S.A.) and the Lawrence Livermore National Laboratory (U.S.A.) for their supplementary sponsorship.

Special thanks are due to many individuals who assisted in various aspects of the organization of the workshop and the preparation of this volume:

- to the International Organizing Committee, (J.-P. Hansen, S. Ichimaru, G. Kalman and H. Van Horn) for suggesting subjects and speakers;
- to D. B. Boercker, B. G. Wilson and J. K. Nash for help with local arrangements;
- to H. C. Graboske for making Lawrence Livermore National Laboratory staff and facilities available;
- to Nancy Willard and Judy Gomez for help with the extensive correspondence and recordkeeping associated with organizing and advertizing the workshop and especially for making the workshop itself function so smoothly; and
- to Donna McWilliams for overseeing and carrying out the difficult task of typing and retyping many of the manuscripts in this volume.

F. J. Rogers
H. E. Dewitt

CONTENTS

CHAPTER I. CLASSICAL STRONG COUPLING

CHAPTER II. PLASMA EXPERIMENTS

CHAPTER III. MOLECULAR DYNAMICS AND KINETIC THEORY

CHAPTER IV. ASTROPHYSICS

CHAPTER V. QUANTUM PLASMAS

CHAPTER VI. DENSITY FUNCTIONAL THEORY

CHAPTER VII. 2-D PLASMAS

CHAPTER VIII. METALLIC FLUIDS

CHAPTER IX. RESPONSE FUNCTIONS AND ION STOPPING POWER

CHAPTER X. ELECTRIC MICROFIELD AND OPTICAL PROPERTIES

CHAPTER XI. INTEGRAL EQUATIONS

CHAPTER I

CLASSICAL STRONG COUPLING

THERMODYNAMIC FUNCTIONS, TRANSPORT COEFFICIENTS

AND DYNAMIC CORRELATIONS IN DENSE PLASMAS

Setsuo Ichimaru, Hiroshi Iyetomi and Shigenori Tanaka

Department of Physics
University of Tokyo
Bunkyo, Tokyo 113, Japan

I. INTRODUCTION

Since one of the present authors wrote a previous review on strongly coupled plasmas [Ichimaru, 1982], remarkable progress has been achieved in the field of statistical physics of dense plasmas. It is the purpose of this article to present a coherent review on the recent progress in our understanding of the thermodynamic properties, transport coefficients and dynamic correlations in dense plasmas and related plasmalike materials. This paper takes a form of an extended abstract on those various topics in dense plasma physics; a detailed account of the review will be published elsewhere [Ichimaru, Iyetomi and Tanaka, 1987].

II. PARAMETERS OF DENSE PLASMAS

We begin by introducing several of dimensionless parameters characterizing the dense plasmas; those will facilitate classifying the physical problems involved in each case of the plasma under consideration. Let us for the moment assume a plasma consisting of a species of ions (with the electric charge Ze, the mass M and the number density n_i) and the electrons (with the electric charge -e, the mass m and the number density $n_e = Zn_i$), a system referred to as a two-component plasma (TCP). The one-component plasma (OCP), on the other hand, consists of a single species of charged particles embedded in a uniform background of neutralizing charges.

For the ion system, the Wigner-Seitz radius or the ion-sphere radius, defined as

$$a = (3/4\pi n_i)^{1/3}, \qquad (1)$$

measures the average distance between neighboring ions. A comparison between the ion-sphere radius and the thermal de Broglie wavelength yields

$$\frac{a}{\hbar(Mk_BT)^{-1/2}} \simeq 2 \times 10^5\ A^{1/2}\left(\frac{n_i}{10^{14}\ \mathrm{cm}^{-3}}\right)^{-1/3}\left(\frac{T}{10^6\mathrm{K}}\right)^{1/2}$$

$$\simeq 12\left(\frac{A}{12}\right)^{5/6}\left(\frac{\rho_m}{10^6\ \mathrm{g/cm^3}}\right)^{-1/3}\left(\frac{T}{10^7 \mathrm{K}}\right)^{1/2} \tag{2}$$

where T is the temperature and k_B(= 1.3807 × 10^{-16} erg/deg) denotes the Boltzmann constant. In Eq. (2) as well as in analogous expressions occurring later, we use the mass number A and the mass density $\rho_m = Mn_i$ of the ions, for convenience in application to examples of real plasmas. When a >> $\hbar(Mk_BT)^{-1/2}$, one may ignore the wave nature of the ions and treat them as a system of particles obeying the classical dynamics and statistics.

The Coulomb coupling constant of such a classical ion system is defined as [Baus and Hansen, 1980; Ichimaru, 1982]

$$\Gamma \equiv \frac{(Ze)^2}{ak_BT} \simeq 10^{-4}Z^2\left(\frac{n_i}{10^{14}\ \mathrm{cm^{-3}}}\right)^{1/3}\left(\frac{T}{10^6 \mathrm{K}}\right)^{-1}$$

$$\simeq 36\left(\frac{Z}{6}\right)^2\left(\frac{A}{12}\right)^{-1/3}\left(\frac{\rho_m}{10^6\ \mathrm{g/cm^3}}\right)^{1/3}\left(\frac{T}{10^7 \mathrm{K}}\right)^{-1} \quad . \tag{3}$$

A weakly coupled plasma corresponds to the case with $\Gamma \ll 1$, where the Coulomb interaction can be treated perturbation-theoretically. A strongly coupled plasma refers to the case with $\Gamma \gtrsim 1$, where a perturbation theory is no longer valid and the system begins to exhibit features qualitatively different from those in a weakly coupled plasma. The statistical physics of dense plasmas involves the charged liquid (or solid) problems where the strong Coulomb-coupling effects play a major part.

A typical dimensionless parameter characterizing the system of electrons is [Pines and Nozières, 1966]

$$r_s \equiv \left(\frac{3}{4\pi n_e}\right)^{1/3}\frac{me^2}{\hbar^2} \simeq \left(\frac{n_e}{1.6\times 10^{24}\ \mathrm{cm^{-3}}}\right)^{-1/3} \quad . \tag{4}$$

It is the Wigner-Seitz radius of the electrons in units of the Bohr radius and depends only on the electron density. The Fermi energy of the electrons is then given by

$$E_F = mc^2\{[1 + 1.96\times 10^{-4}\ r_s^{-2}]^{1/2} - 1\} \tag{5}$$

with inclusion of the relativistic effect. The electrons can be treated nonrelativistically in the low-density regime such that $r_s \gg 10^{-2}$.

The degree of the Fermi degeneracy is described by the parameter,

$$\theta \equiv \frac{k_BT}{E_F} = 2\left(\frac{4}{9\pi}\right)^{2/3} Z^{5/3}\frac{r_s}{\Gamma} \quad . \tag{6}$$

In the final expression of Eq. (6), $r_s \gg 10^{-2}$ has been assumed. When $\theta \ll 1$, the electrons are in the state of complete Fermi degeneracy; $\theta \simeq 1$ corresponds to a state of intermediate degeneracy; when $\theta \gg 1$, we may regard the system of electrons as in the nondegenerate, classical state. The Coulomb coupling constant of the completely degenerate electrons is given by r_s of Eq. (4), rather than by Γ of Eq. (3). A remarkable feature in dense-plasma problems is involvement of the varied degrees of Fermi degeneracy in the treatment of the electrons.

The condition that the atomic nuclei are all stripped of their orbital electrons may be derived roughly from the requirement that the Fermi energy be greater than the binding energy of an orbital electron, $E_F > 13.6\ Z^2$[eV], that is

$$\rho_m > 0.38\ AZ^2\ [g/cm^3] \quad . \tag{7}$$

When this condition is not satisfied, the atomic nuclei may retain some of the orbital electrons.

Although Eq. (7) is known to provide qualitatively a correct criterion for the pressure ionization, its quantitative accuracy remains to be ascertained. It is in fact related directly to those frontal problems in condensed-matter physics such as the metal-insulator transition and the localization of electrons in random fields. Those involve strong interplay between atomic physics and statistical physics in dense plasmalike materials.

III. DENSE PLASMAS IN NATURE

The neutron star [Shapiro and Teukolsky, 1983], one of the final stages of the stellar evolution, is a highly condensed material corresponding approximately to a compression of a solar mass ($\simeq 2 \times 10^{33}$g) into a radius of ~10km. According to theoretical model calculations, it has a crust with a thickness of several hundred meters and a mass density in the range of $10^4 - 10^7$ g/cm^3, consisting mostly of iron. The condition that $\rho_m > 10^4$ g/cm^3 corresponds to Eq. (7), so that we may assume each iron atom contributing 26 conduction electrons. When $T = 10^7 - 10^8$K, the ratio (2) takes on a magnitude greater than 20, so that we may regard the iron nuclei as forming a classical ion system. The Γ value varies in the range of $10 - 10^3$. It is thus an essential problem to analyze the phase properties of the system, with inclusion of the possibilities of Wigner crystallization [Slattery, Doolen and DeWitt, 1980 and 1982] and the glass transition [Ichimaru, Iyetomi, Mitake and Itoh, 1983; Ichimaru and Tanaka, 1986].

The electron system, with the r_s value ranging $10^{-2} - 10^{-1}$, satisfies the condition for the complete Fermi degeneracy. At $r_s \simeq 10^{-2}$, the Fermi energy $E_F \simeq mc^2$ ($\simeq$ 0.5 MeV) is much greater than the typical value $mZ^2e^4/2\hbar^2$ ($\simeq$ 9 keV) of the electron-ion interaction energy. Hence, the Coulomb field associated with the iron nucleus does not significantly disturb the distribution of the conduction electrons; the polarization (screening) effect of the electrons can thus be ignored. Consequently, the system of electrons acts as a uniform background of negative charges neutralizing the average space charge of the positive ions. It is in this sense that we may treat the outer crustal matter of a neutron star as an OCP of iron. The state of matter and the transport properties in the outer crust are considered to form those physical elements which crucially control the cooling rate of a neutron star [Gudmundsson, Pethick and Epstein, 1982].

The interior of a white dwarf [Shapiro and Teukolsky, 1983], another final stage of stellar evolution, consists of dense material with ρ_m and T comparable to those of the neutron-star crust. In connection with the supernova explosion, one may extend the range of the parameters and consider the cases up to $\rho_m \simeq 10^{10}$ g/cm^3. For the progenitor of the type I supernova, one often assumes a white dwarf with interior consisting of carbon-oxygen mixture, a kind of the binary-ionic mixture (BIM). Physical problems in BIM include assessment of the possibilities of phase separation and formation of eutectic alloys [Stevenson, 1980]; those are related to the cooling rate [Mochkovitch,

1983] and a detailed mechanism of the supernova explosion [Canal, Isern and Labay, 1982; Isern, Labay and Canal, 1984].

The material inside a Jovian planet offers an important subject of study in the dense plasma physics [Stevenson, 1982]. Here one considers a hydrogenic plasma with a few percent admixture of helium at $\rho_m = 1 - 10$ g/cm^3 and $T \simeq 10^4$ K. It is thus a strongly coupled plasma with mixed ionic species at $\Gamma = 20 - 50$ and $r_s = 0.6 - 1$. The electron density of Jovian interior being smaller substantially than that of a white dwarf, new electronic problems emerge in the treatment of dense Jovian matter, such as the polarization (and screening) effect of the electrons and a possibility of electrons forming bound states with the helium nuclei. Jupiter, for example, is known to emit radiation energy in the infrared range, 2 - 3 times as much as that which it receives from the sun [Hubbard, 1980]. To account for the source of this excess energy as well as the internal structure of a Jovian planet, thermodynamic and transport properties of dense Jovian matter need to be clarified.

The interiors of main sequence stars such as the sun are plasmas constituting mostly of the hydrogen. The central part of the sun has the pressure of approximately 10^5 Mbar and the temperature of approximately 10^7 K. Since $\Gamma \simeq 0.05$, $\theta \simeq 4$ and $r_s \simeq 0.4$, the plasma may not be said strictly in the strongly coupled state; the polarization and quantum effects of the electrons play significant parts in determining the plasma properties, however. The dense plasma effects are crucial also to the analyses of atomic states for those "impurities" starting with helium. In the calculation of miscibilities for high-Z elements such as iron, the strong coupling effects need to be carefully taken into account [Alder, Pollock and Hansen, 1980; Iyetomi and Ichimaru, 1986b].

The states of those plasmas aimed at in the inertial confinement fusion (ICF) researches [Brueckner and Jorna, 1974] are similar to those in the solar interior mentioned above. The projected temperatures in the ICF plasmas need to be on the order of 10^8 K, so that the Γ values of the "fuel" material (isotopes of the hydrogen) may remain smaller than unity. Those materials which drive implosion of the fuel, however, consist of high-Z elements, such as C, Al, Fe, Au, Pb •••, which after ionization form plasmas with $\Gamma > 1$. Atomic physics of those high-Z elements is influenced strongly by the correlated behaviors of charged particles in dense plasmas.

The conduction electrons in metals and in liquid metals form strongly coupled, quantum plasmas, where the wave nature of the electrons as fermions plays an essential part. The metallic electrons at room temperatures have $r_s = 2 - 6$, and may be regarded as in a state of complete Fermi degeneracy ($\theta \ll 1$). Owing to the presence of the core electrons, the ion-ion and electron-ion interactions are described by the pseudopotentials, deviating away from the pure Coulombic form. The strong coupling effect between the conduction electrons has a strong influence in the determination of those pseudopotentials [Singwi and Tosi, 1981].

Some of the strongly coupled plasmas in the laboratory setting have the spatial degrees of freedom in the particle motion less than three. For example, those electrons (or holes) trapped in the surface states of liquid helium [Grimes, 1978; Ando, Fowler and Stern, 1982] or in the interfaces of the metal-oxide-semiconductor system [Ando, Fowler and Stern, 1982] form a pseudo-two-dimensional system. The electrons on the liquid-helium surface are characterized by the densities and temperatures in the ranges of $10^7 - 2 \times 10^9$ cm^{-2} and 0.1 -1 K; they thus form a classical two-dimensional OCP. Grimes and Adams [1979] found a crystallization of such a system at $(\pi n)^{1/2} e^2/k_BT \simeq 137$.

In addition to those two-dimensional systems mentioned above, there exists a second class of strongly coupled charged systems in two dimensions, where the "particles" interact via a logarithmic potential in the x-y plane. This system corresponds physically to a collection of line charges in the z direction and has been adopted as an approximate model to those electrons in strong magnetic field.

Important examples of the three-dimensional strongly coupled plasmas in the laboratory include those plasmas produced in the shock tubes [Fortov, 1982] and the pure electron [Malmberg and O'Neil, 1977; Driscoll and Malmberg, 1983] or ion [Bollinger and Wineland, 1984], Penning trapped plasmas at cryogenic temperatures (10^{-2} - 10^{0} K). The latter plasmas rotate around the magnetic axis due in part to the space-charge field in the radial direction. In the frame corotating with the bulk of the plasma, such a system of charged particles may be regarded effectively as an OCP. The pure ion Penning-trapped plasmas have been stably maintained [Bollinger and Wineland, 1984] for many hours at a Γ value on the order of 10.

IV. THERMODYNAMIC AND CORRELATIONAL PROPERTIES OF FINITE-TEMPERATURE ELECTRON LIQUIDS IN THE SINGWI-TOSI-LAND-SJÖLANDER APPROXIMATION

The electron liquid is a strongly coupled OCP of the electrons embedded in a uniform neutralizing background of positive charges. The static correlations in such electron liquids at finite temperatures were studied in the dielectric formulation mostly with the random-phase approximation (RPA) [see e.g., Fetter and Walecka, 1971], where the local-field correction (LFC) [Ichimaru, 1982] is set equal to zero. The properties of the free-electron polarizability at finite temperatures have been analyzed extensively [Khanna and Glyde, 1976; Gouedard and Deutsch, 1978; Arista and Brandt, 1984].

The strong exchange and Coulomb coupling effects beyond the RPA may be taken into account through the static LFC; the static correlations and the thermodynamic properties are thereby analyzed. On the basis of the Singwi-Tosi-Land-Sjölander (STLS) [1968] approximation, Tanaka, Mitake and Ichimaru [1985; see also Tanaka and Ichimaru, 1986a] calculated the static correlation functions and the interaction energies of the finite-temperature electron liquids for 70 combinations of the density and temperature parameters in the range of $r_s \leq 73.66$ and $\theta = 0.1$, 1 and 5. Tanaka and Ichimaru [1987a] then used the computed results to construct an analytic expression for the interaction energy in the form:

$$-\frac{1}{\Gamma}\frac{E_{int}}{Nk_BT} = \frac{a(\theta) + b(\theta)\Gamma^{1/2} + c(\theta)\Gamma}{1 + d(\theta)\Gamma^{1/2} + e(\theta)\Gamma} \quad . \tag{8}$$

Here

$$a(\theta) = \left(\frac{3}{2\pi}\right)^{2/3} \frac{0.75 + 3.04363\theta^2 - 0.09227\theta^3 + 1.7035\theta^4}{1 + 8.31051\theta^2 + 5.1105\theta^4} \tanh\left(\frac{1}{\theta}\right) \tag{9}$$

represents the Hartree-Fock contribution derived originally by Perrot and Dharma-wardana [1984],

$$b(\theta) = \frac{0.341308 + 12.070873\theta^2 + 1.148889\theta^4}{1 + 10.495346\theta^2 + 1.326623\theta^4}\,\theta^{1/2} \tanh\left(\frac{1}{\theta^{1/2}}\right) \quad , \tag{10}$$

$$c(\theta) = [0.872496 + 0.025248 \exp(-\frac{1}{\theta})]e(\theta) \quad , \tag{11}$$

$$d(\theta) = \frac{0.614925 + 16.996055\theta^2 + 1.489056\theta^4}{1 + 10.109350\theta^2 + 1.221840\theta^4} \theta^{1/2}\tanh(\frac{1}{\theta^{1/2}}) \quad , \tag{12}$$

$$e(\theta) = \frac{0.539409 + 2.522206\theta^2 + 0.178484\theta^4}{1 + 2.555501\theta^2 + 0.146319\theta^4} \theta\tanh(\frac{1}{\theta}) \quad . \tag{13}$$

In the classical limit ($\theta >> 1$), the ratio $c(\theta)/e(\theta)$ approaches 0.897744, the coefficient a in the liquid internal energy formula of Slattery et al. [1982] derived from their Monte Carlo (MC) simulation data. The formulas (8) - (13) in fact reproduce the hypernetted chain (HNC) values for $\Gamma \leq 1$ with digressions of less than 1% and agree with the liquid internal-energy formula within 0.5% for $1 \leq \Gamma \leq 200$ in the classical limit.

The functions $b(\theta)$, $c(\theta)$, $d(\theta)$ and $e(\theta)$ vanish at $\theta = 0$ in such a way that Eq. (8) becomes a function of r_s. The formulas (8) - (13) are therefore applicable to the electron liquids in the ground state as well, and the interaction energy (8) agrees with the results of Green's function Monte Carlo (GFMC) calculations [Ceperley and Alder, 1980] for $r_s \leq 100$ within 0.4%.

It has been well known [e.g., Ichimaru, 1982] that the STLS values of the internal energy exhibit systematic departures from the exact MC or GFMC values as the Coulomb coupling constant Γ or r_s increases in the classical ($\theta >> 1$) or degenerate ($\theta \to 0$) limit. In the derivation of Eqs. (8) - (13), this feature has been taken into consideration by anticipating similar deviations in the 70 STLS values computed at $\theta = 0.1$, 1 and 5: Those formulas reproduce the 70 STLS values so corrected with digressions of less than 0.6%.

The expression for the excess free energy F_{ex} is then obtained by performing the Γ integration [e.g., Ichimaru. 1982] of the interaction energy. Figure 1 compares the values of $f_{ex} \equiv F_{ex}/Nk_BT$ on the basis of Eq. (8) with those in other theoretical schemes at $\theta = 1$. As one would expect, the RPA values show a trend of systematic underestimation of f_{ex} as compared with the present evaluation; the deviations between those two values become remarkable for $\Gamma \gtrsim 1$ at $\theta = 1$.

The formula proposed by Richert and Ebeling [1984] results from a Padé-approximant fitting by the use of only the information obtained from the GFMC values at $\theta = 0$ and an expansion of Debye- Hückel type with quantum corrections. It fails to account for the exchange effects appropriately in the weak coupling regime, and thereby predicts the values of f_{ex} even lower than the RPA values over a significant domain of Γ, as Fig. 1 illustrates. Since no reliable information was included at $\theta \simeq 1$ or for $\Gamma \gtrsim 1$ at $\theta >> 1$, their formula appears applicable only in the domain $\theta >> 1$ and $\Gamma << 1$.

Pokrant [1977] evaluated f_{ex} by a method in which the quantum pair potential was obtained with the aid of a finite-temperature variational principle and the correlation functions were calculated in the HNC approximation. His results appear to contain slight but systematic overestimation of f_{ex} as Fig. 1 illustrates.

V. HYPERNETTED CHAIN ANALYSES OF DENSE PLASMAS

It has been known empirically that the HNC approximation provides an accurate description of correlations in the classical plasmas [e.g., Ichimaru,

1982]. The HNC internal energy reproduces the exact MC data [Slattery, Doolen and DeWitt, 1980 and 1982] within errors of 1% over the whole fluid region; the HNC scheme correctly accounts for the qualitative features of the correlation functions in the OCP. This situation presents a sharp contrast to the cases of a short-ranged hard-core system, for example, where the Percus-Yevick equation is known to be superior to the HNC equation. A question then arises as to why the HNC equation works so well for the Coulombic system.

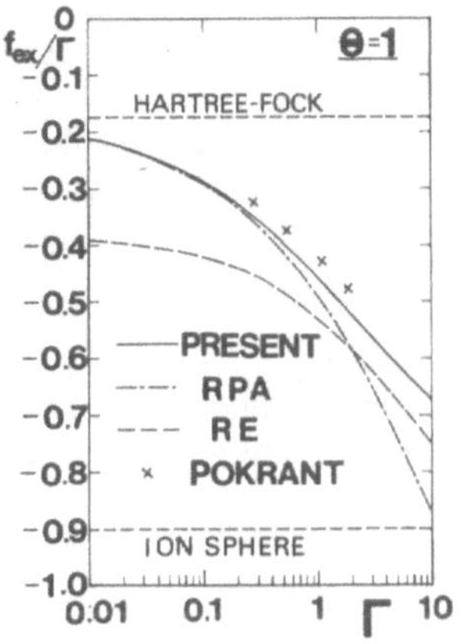

Fig. 1. Excess free energy f_{ex} divided by Γ calculated in various schemes at $\theta = 1$. "PRESENT" refers to the values based on Eq. (8); RPA, the RPA calculations; RE, the formula by Richert and Ebeling [1984]; "POKPANT", the calculation by Pokrant [1977]. Two horizontal dashed lines represent the evaluations based on the Hartree-Fock approximation ($f_{ex} = -0.174\Gamma$) and the ion-sphere model ($f_{ex} = -0.9\Gamma$).

This question has been answered by Iyetomi [1984] through diagrammatic analyses of the bridge functions, which are the neglected terms in the HNC approximation. For the long-ranged Coulombic system, it has been recognized essential to maintain the sequential relations or the charge neutrality conditions at each stage of the higher-order correlation functions. The charge neutrality conditions then guarantee the short-rangedness of the bridge functions to all orders, and may thereby be interpreted as conditions ensuring perfect screening in the Coulombic system. It is shown that the multiparticle correlation functions constructed in the convolution approximation exactly satisfy the sequential relations and lead to the HNC equation.

Iyetomi and Ichimaru [1986a] have derived free energy formulas applicable to the electron-screened ion plasmas in the HNC approximation. The formulas, expressed in terms of the correlation functions, enable one to avoid the more cumbersome and less accurate calculations involving the thermodynamic integrations.

As an application of the generalized HNC free-energy formula, Iyetomi and Ichimaru [1986b] then revisited the miscibility problem of iron atoms in hydrogen plasmas under the solar interior conditions, with a hope of shedding light on solution to the solar neutrino problem [Alder and Pollock, 1978]. The temperature and the pressure of the solar plasma were assumed to be in the vicinity of 1.5×10^7 K and 10^5 Mbar. The relative concentration of the irons near the thermonuclear burn region was assumed to take on a value close to the cosmic abundance (2.5×10^{-5} ionic mole fraction).

The calculation [Iyetomi and Ichimaru, 1986b] have been carried out with special emphasis on the role of the screening effect arising from semiclassical electrons in the solar interior; such a semiclassical electron gas acts to screen the ion-ion interaction quite efficiently, and hence modifies the thermodynamic properties of the plasma substantially. The calculations thus improve over those of Alder, Pollock and Hansen [1980] in two ways: (i) a proper account of the electronic polarization through the static screening function of the electrons, and (ii) a corresponding account of the exchange and correlation contributions to the thermodynamic functions for the electron system. Since the electrons are weakly coupled in the solar interior, the RPA is applicable for the description of the correlational properties of the electrons. The strong coupling effects between ions are treated accurately in the HNC scheme, as Alder et at. [1980] have done.

The Gibbs free energy of mixing is expressed as the sum of the electronic, ideal-gas, and excess contributions. Qualitatively, the electronic and ideal-gas terms favor phase mixing, whereas the excess term promotes phase demixing. Phase separation of the plasma mixtures takes place as a consequence of delicate balance between those physically distinct contributions.

Figure 2 shows the phase diagram for the hydrogen-iron mixture calculated in the present scheme at $P = 0.5 \times 10^5$ Mbar. The critical point for demixing takes place at $T_c \simeq 5.5 \times 10^6$ K and $x_c \simeq 2.4 \times 10^{-2}$. Comparing the present results with those of Alder et al, we find an increase of T_c by 15% arising from the electronic screening effect. The increase, however, is not sufficient so as to resolve the solar neutrino dilemma through the idea of a limited solubility of the iron atoms in the solar interior plasma. Figure 2 also exhibits the substantial influence of the adopted electronic equation of state exerted on the phase diagram calculations.

The calculations presented in the preceding paragraphs have been successful because the plasma density in the solar interior is not so high as to require an improvement over the HNC approximation. In many other examples of astrophysical dense plasmas, such as the interiors of Jovian planets and white dwarfs, the relevant density and temperature parameters are such that it becomes essential to develop a theoretical scheme which significantly improves over the HNC approximation.

Numerous schemes have been proposed thus far, aiming at such an improvement over the HNC approximation. Particularly notable among them is the semi-empirical scheme developed by Rosenfeld and Ashcroft [1979] on the basis of the universality ansatz for the bridge function; they assumed the OCP bridge function B(r) as given in effect by that of an equivalent hard-sphere reference system and modified the HNC scheme by adopting the effective HNC potential

$$v_{eff}(r) = (Ze)^2/r - k_B T B(r). \qquad (14)$$

Iyetomi and Ichimaru [1982 and 1983] proposed a scheme of improvement over the HNC approximation, on the basis of the density-functional analysis of the multiparticle correlations. It has been noted that the convolution approximation on which the HNC scheme is based takes accurate account of the long-range correlations and that the bridge functions, which are neglected in the HNC approximation, represent basically a short-range effect. An approximate expression for B(r), to be substituted in Eq. (14), is thus obtained in accord with the ion-sphere model by paying a special attention to the correlations in the short-range domain. The improved HNC scheme has reproduced almost exactly the existing MC data [Slattery, Doolen and DeWitt, 1980 and 1982] of the radial distribution function for $\Gamma \leq 160$.

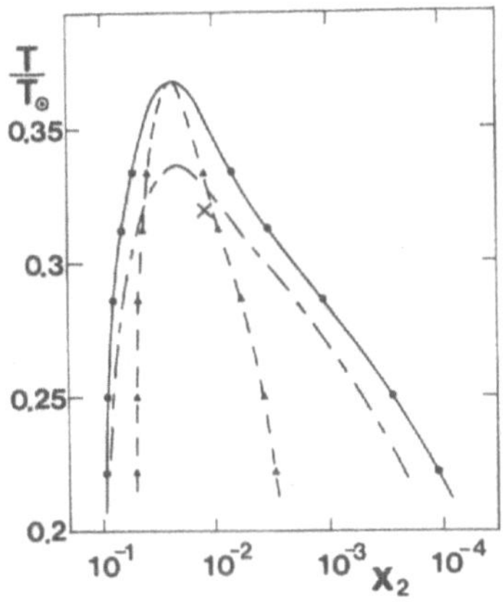

Fig. 2. Phase diagrams for the $H^+ - Fe^{24+}$ mixture with the electronic screening at $P = 0.5 \times 10^5$ Mbar. The temperature T in the ordinate is normalized with $T_\odot = 1.5 \times 10^7$ K, the interior temperature of the sun. The solid and dashed curves are the coexistence and spinodal curves, interpolated by the spline method with the third order polynomials; the closed circles and triangles represent the calculated points. The cross refers to the critical point obtained without the electronic screening by Alder, Pollock, and Hansen [1980]. The chain curve is the coexistence curve calculated by retaining only the ideal-gas term in the equation of state for the uniform electron gas.

Solution to the improved HNC scheme has been extended to cover the supercooled fluid regime up to $\Gamma = 1000$ [Ichimaru and Tanaka, 1986; Tanaka and Ichimaru, 1987b]. Figure 3 exhibits the graphs of the radial distribution function calculated in this scheme. We clearly observe splitting of the second peak and structural developments around the third peak in g(r) as Γ increases to and beyond $\Gamma = 500$.

Noting that an exact summation of all the bridge diagrams can be carried out in the density-functional formalism [e.g., Evans, 1979], Iyetomi and Ichimaru [1987] have derived new formulas for the bridge function with the aid of a nonlocal density-functional approximation to the direct correlation function. Consequences of those new formulas in the improvement of the HNC scheme have thereby been numerically examined.

VI. DYNAMIC THEORY OF THE GLASS TRANSITION IN DENSE CLASSICAL PLASMAS

A new theory of dynamic correlations in a strongly coupled, classical OCP is developed within the generalized viscoelastic formalism [Ichimaru and Tanaka, 1986; Tanaka and Ichimaru, 1987b]. Fully convergent kinetic equations for the strongly coupled OCP are thereby derived with the aid of a fluctuation-theoretic formulation of the collision integrals [e.g., Ichimaru, 1986].

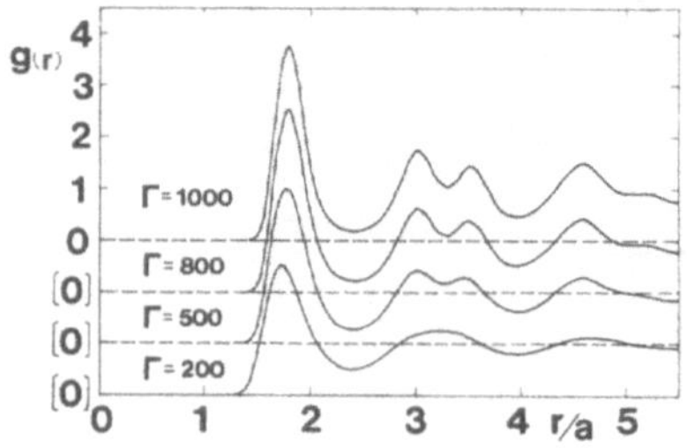

Fig. 3. Radial distribution functions of the supercooled OCP computed in the improved HNC scheme at various values of Γ.

The dynamic structure factor $S(k,\omega)$ and the coefficient η of shear viscosity are calculated both in the ordinary fluid state and the metastable supercooled state through a self-consistent solution to the kinetic equation [Ichimaru and Tanaka, 1986; Tanaka and Ichimaru, 1987b]. The numerical results for $S(k,\omega)$ in the ordinary fluid state are shown to agree well with other theoretical [e.g., Sjödin and Mitra, 1977; Bosse and Kubo, 1978; Cauble and Duderstadt, 1981] and molecular-dynamics (MD) simulation [Hansen, Pollock and McDonald, 1974; Hansen, McDonald and Pollock, 1975] results. The computed values of η in the fluid state also agree with other theoretical [Wallenborn and Baus, 1978] and MD simulation [Bernu, Vieillefosse and Hansen, 1977; Bernu and Vieillefosse, 1978] data, as Fig. 4 illustrates.

A possibility of the dynamic glass transition is predicted in the supercooled OCP at Γ = 900 - 1000 through the analyses of the variation in η (see Fig. 4), the quasielastic peak in $S(k,\omega)$ and the behavior of the self-diffusion coefficient. Relevance to a laboratory experiment [Bollinger and Wineland, 1984] is examined in terms of the metastable-state lifetimes against homogeneous nucleation of the crystalline state.

VII. THERMODYNAMIC AND TRANSPORT PROPERTIES OF DENSE, HIGH-TEMPERATURE HYDROGENIC PLASMAS APPROPRIATE TO THE INERTIAL-CONFINEMENT-FUSION EXPERIMENTS AND INTERIORS OF THE MAIN-SEQUENCE STARS

In a series of papers we have presented the results of a systematic study of multiparticle correlation effects in those dense ($n \leq 10^{28}$ cm^{-3}), high-temperature ($T = 10^6 - 10^9$ K) hydrogenic plasmas appropriate to the ICF experiments and the interior of the main-sequence stars. The Coulomb coupling constant takes on a value $\Gamma \lesssim 3$, while the degree of Fermi degeneracy θ varies widely.

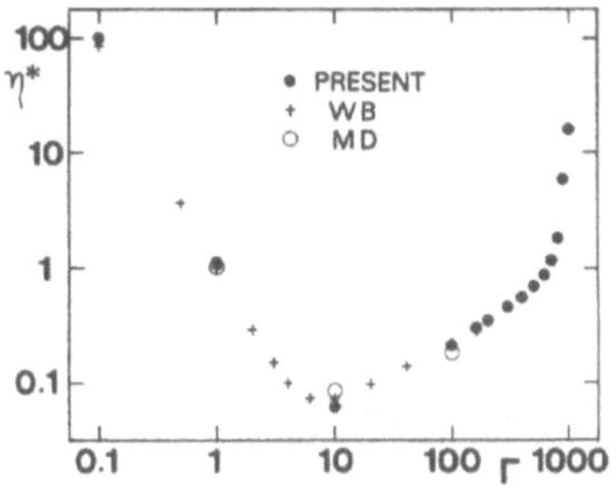

Fig. 4. The reduced shear viscosity $\eta^* \equiv \eta/Mn\omega_p a^2$ calculated in the generalized viscoelastic theory [Ichimaru and Tanaka, 1986] (solid circles). The crosses refer to the calculation by Wallenborn and Baus [1978] the open circles, the MD simulation result [Bernu, Vieillefosse and Hansen, 1977].

A general density-response formalism has been developed with inclusion of the varied degrees of the electron degeneracy and the LFC's describing the strong Coulomb-coupling effects [Ichimaru, Mitake, Tanaka and Yan, 1985]. An explicit theoretical scheme of calculating the static LFC's has been advanced on the basis of the HNC approximation.

Interparticle correlations in dense plasmas have been investigated quantitatively and the physical implications are clarified [Tanaka and Ichimaru, 1984; Mitake, Tanaka, Yan and Ichimaru, 1985].

On the basis of the general formalism and the calculations of the correlation functions mentioned above, various thermodynamic quantities have been evaluated explicitly for the dense, high-temperature plasmas [Tanaka, Mitake, Yan and Ichimaru, 1985]. The numerical data for the interaction and excess free energies have been parametrized accurately, so that the resulting analytic formulas exactly satisfy the known boundary conditions at complete degeneracy as well as in the weak- and strong-coupling regimes [Tanaka and Ichimaru, 1985].

The stopping power of a dense TCP has been calculated in the dielectric formulation, where the static and dynamic LFC's are explicitly taken into account [Yan, Tanaka, Mitake and Ichimaru, 1985]. The extent to which the LFC's and the presence of ions act to modify the rate of inelastic scattering has been clarified.

Ichimaru and Tanaka [1985] calculated the electric and thermal conductivities, ρ^{-1} and κ, of the dense, high-temperature hydrogen plasmas over the domain $\Gamma \leq 2$ and $0.1 \leq \theta \leq 10$ on the basis of the aforementioned correlation analyses. They used the correlation functions obtained by Mitake, Tanaka, Yan and Ichimaru [1985] and parametrized the numerical results for the generalized Coulomb logarithms L_E and L_T, introduced via

$$\rho = 4\left(\frac{2\pi}{3}\right)^{1/2} \frac{\Gamma^{3/2}}{\omega_{pe}} L_E \quad , \tag{15}$$

$$\frac{1}{\kappa} = \frac{52(6\pi)^{1/2}}{75} \left(\frac{e^2}{k_B{}^2 T}\right) \frac{\Gamma^{3/2}}{\omega_{pe}} L_T \quad , \tag{16}$$

where $\omega_{pe} = (4\pi n e^2/m)^{1/2}$. Their results are expressed as

$$L_E(\Gamma,\theta) = \frac{a(\theta)\ln\Gamma + b(\theta) + c(\theta)\Gamma}{1 + d(\theta)\Gamma^3} \quad , \tag{17}$$

$$L_T(\Gamma,\theta) = \frac{p(\theta)\ln\Gamma + q(\theta) + r(\theta)\Gamma}{1 + s(\theta)\Gamma^3} \quad , \tag{18}$$

where

$$a(\theta) = \frac{-\theta^{3/2}}{2\theta^{3/2} - 0.57923\theta + 0.23272\theta^{1/2} + 1.4853} \quad ,$$

$$b(\theta) = \frac{\theta^3(\frac{1}{2}\ln\theta - 0.18603) + 1.2704\theta^{3/2}}{\theta^3 + 1.8993\theta^2 + 4.3243\theta + 1} \quad , \tag{19}$$

$$c(\theta) = \theta^{3/2} \frac{0.69460\theta^{1/2} + 0.24228}{\theta^2 + 1.7768} \quad ,$$

$$d(\theta) = \theta^{3/2} \frac{0.13550\theta^{1/2} + 0.083521}{\theta^2 + 0.36797} \quad ,$$

$$p(\theta) = \frac{-\theta^{3/2}}{2\theta^{3/2} + 0.029220\theta - 1.4661\theta^{1/2} + 2.6858} \quad ,$$

$$q(\theta) = \frac{\theta^3(\frac{1}{2}\ln\theta - 0.18603) - 0.98787\theta^2 + 0.87422\theta^{3/2}}{\theta^3 + 4.9312\theta^2 + \theta + 1} \quad , \tag{20}$$

$$r(\theta) = \theta^{3/2}\,\frac{0.636070\theta^{1/2} + 0.033439}{\theta^2 - 0.361860\theta + 1} ,$$

$$s(\theta) = \theta^{3/2}\,\frac{0.031856\theta^{1/2} + 0.42460}{\theta^2 - 0.29933\theta + 0.5} .$$

The analytic forms of those formulas retain the following features:
(i) In the classical ($\theta \gg 1$) and weak-coupling ($\Gamma \ll 1$) limit, both L_E and L_T approach the same value,

$$L_0 = -\frac{1}{2}\ell n\zeta - \frac{1}{2}(\gamma + \ell n2) + O(\zeta, \zeta\ell n\zeta) , \tag{21}$$

where $\zeta = \hbar^2 k_D{}^2/4mk_BT$, $k_D = (4\pi ne^2/k_BT)^{1/2}$ and $\gamma = 0.57721 \cdots$ is Euler's constant.
(ii) In the limit of complete Fermi degeneracy ($\theta \to 0$), L_E and L_T behave proportionally to $\theta^{3/2}$.
(iii) In the strong-coupling regime ($\Gamma \gg 1$), both Eqs. (17) and (18) behave proportionally to Γ^{-2}. This is a consequence of the ion-sphere scaling in the interparticle correlations for a strongly coupled plasma.

In a remarkable experiment, Ivanov, Mintsev, Fortov and Dremin [1976] measured the Coulomb conductivity of non-ideal plasmas which were produced by a dynamic method based on compression and irreversible heating of gases in the front of high-power ionizing shock waves. Gases used were argon, xenon, neon and air; those were regarded as forming singly ionized ($Z = 1$) plasmas. Each of the experimental values σ_{exp} for the Coulomb conductivity derives from an average of five to ten independent measurements and is attached to a 10-50% error bar.

We compare those experimental values with the present theoretical predictions. Since the classical statistics applies to the electrons for all the cases of the experiment, we take account of the electron-scattering factor 1.97 and write the electric conductivity as

$$\sigma = 1.97\left(\frac{3\pi}{2}\right)^{1/2} \frac{\omega_{pe}}{4\pi\Gamma^{3/2}L} . \tag{22}$$

When L_E given by Eq. (17) is substituted in place of L, we denote the resulting value of Eq. (22) as σ_{th}. When the first two terms on the right-hand side of Eq. (21) is used for L in Eq. (22), the resulting value of σ is called σ_0.

In the weak-coupling domain $\Gamma < 1$, we find that σ_{exp} is fairly well represented by σ_0. In the four strong-coupling cases ($\Gamma > 1$) of Xe, however, σ_0 shows a large departure from σ_{exp}, which increases systematically with Γ.

In the comparison between σ_{exp} and σ_{th}, such a systematic discrepancy is completely erased, and we now find that the values of

$$\delta = |\sigma_{exp} - \sigma_{th}|/\sigma_{exp} \tag{23}$$

are confined within 0.31 for all the 15 cases of the experiment. In view of the large error bars associated with the experimental data, we find such an overall agreement to be rather remarkable. We emphasize in this connection that the generalized Coulomb logarithms are functions of two parameters

Γ and θ, rather than of a single parameter Γ, even for those plasmas where the electrons may obey the classical statistics.

The coefficient of the ionic shear viscosity has been calculated through a solution to the kinetic equation for dense hydrogenic TCP with a fully convergent collision integral [Tanaka and Ichimaru, 1986b]. An analytic expression for the generalized Coulomb logarithm of the shear viscosity has been obtained through parametrization of those numerical results, in a way analogous to the derivation of Eq. (17) or (18).

VIII. ELECTRICAL AND THERMAL CONDUCTIVITIES OF DENSE MATTER IN THE LIQUID METAL PHASE, APPROPRIATE TO THE INTERIORS OF WHITE DWARFS AND THE CRUST OF NEUTRON STARS

Electrical and thermal conductivities have been calculated for the dense matter ($\Gamma \gg 1$ and $\theta \ll 1$) in the liquid metal phase for various elemental compositions of astrophysical importance [Itoh, Mitake, Iyetomi and Ichimaru, 1983]. The calculation based on the Ziman formula takes into account the dielectric screening due to the relativistic degenerate electrons [Jancovici, 1962] and uses the ionic structure factors obtained in the improved HNC scheme. The low-temperature quantum corrections to the transport coefficients arising from the quantum nature of the semiclassical ions have been evaluated by using the frequency-moment sum rules and the Wigner expansion in powers of $\hbar$ for the ionic correlation [Mitake, Ichimaru and Itoh, 1984].

IX. SUMMARY

In the main text we have reviewed the present status of the theoretical understanding, concerning the static and dynamic properties as well as the transport and elementary processes in dense plasmas and plasmalike materials. It may fairly be said that we now have reliable theoretical devices, supported by the computer-simulation results, by which to analyse the strong Coulomb-coupling effects in classical and quantum, OCP systems; their static and dynamic properties have been elucidated.

Good progress has been achieved also in the understanding of the ion-electron TCP systems, where one takes account of the strong Coulomb-coupling effects between ions and the varied degrees of Fermi degeneracy in the electrons. Strong Coulomb-coupling effects between the ions and the electrons, including the possibility of formation of the bound states, have been investigated to an extent [see e.g., Yan and Ichimaru, 1986a, b], but it still appears that much more work remains to be done in this area.

In this connection we remark that the atomic and molecular processes in dense plasmalike materials, involving electrons in bound or localized states, offer varieties of outstanding, unsolved problems that deserve further study in the future. Here it is necessary to solve a self-consistent problem in which the states of atoms and localized electrons are influenced in an essential way by the correlations between the atoms and the charged particles in free states, while the correlated states of the plasma particles depend strongly on the states that the atoms and the localized electrons assume.

We have seen some significant progress in the study of the properties of multi-ionic plasmas including the effects of electronic polarization. Construction of phase diagrams for realistic plasmas, describing the possibilities of demixing and solidification, requires an extremely accurate assessment of the relevant thermodynamic functions. We will see further progress in these directions in the coming years.

REFERENCES

Alder, B.J. and E.L. Pollock, 1978, Nature 275, 41.
Alder, B.J., E.L. Pollock and J.-P. Hansen, 1980, Proc. Natl. Acad. Sci. USA 77, 6272.
Ando, T., A.B. Fowler and F. Stern, 1982, Rev. Mod. Phys. 54, 437.
Arista, N.R. and W. Brandt, 1984, Phys. Rev. A 29, 1471.
Baus, M. and J.-P. Hansen, 1980, Phys. Rep. 59, 1.
Bernu, B. and P. Vieillefosse, 1978, Phys. Rev. A 18, 2345.
Bernu, B., P. Vieillefosse and J.-P. Hansen, 1977, Phys. Lett. 63A, 301.
Bollinger, J.J. and D.J. Wineland, 1984, Phys. Rev. Lett. 53, 348.
Bosse, J. and K. Kubo, 1978, Phys. Rev. A 18, 2337.
Brueckner, K.A. and S. Jorna, 1974, Rev. Mod. Phys. 46, 325.
Canal, R., J. Isern and J. Labay, 1982, Nature 296, 225.
Cauble, R. and J.J. Duderstadt, 1981, Phys. Rev. A 23, 3182.
Ceperley, D.M. and B.J. Alder, 1980, Phys. Rev. Lett. 45, 566.
Driscoll, C.F. and J.H. Malmberg, 1983, Phys. Rev. Lett. 50, 167.
Evans, R., 1979, Adv. Phys. 28, 143.
Fetter, A.L. and J.D. Walecka, 1971, Quantum Theory of Many-Particle Systems (McGraw-Hill, New York).
Fortov, V.E., 1982, Usp. Fiz. Nauk 138, 361 [Sov. Phys.-Usp. 25, 781 (1983)].
Gouedard, C. and C. Deutsch, 1978, J. Math. Phys. 19, 32.
Grimes, C.C., 1978, Surf. Sci. 73, 397.
Grimes, C.C. and G. Adams, 1979, Phys. Rev. Lett. 42, 795.
Gudmundsson, E.H., C.J. Pethick and R.I. Epstein, 1982, Astrophys. J. Lett. 259, L19.
Hansen. J.-P., I.R. McDonald and E.L. Pollock, 1975, Phys. Rev. A 11, 1025.
Hansen, J.-P., E.L. Pollock and I.R. McDonald, 1974, Phys. Rev. Lett. 32, 277.
Hubbard, W.B., 1980, Rev. Geophys. Space Phys. 18, 1.
Ichimaru, S., 1982, Rev. Mod. Phys. 54, 1017.
Ichimaru, S., 1986, Plasma Physics: An Introduction to Statistical Physics of Charged Particles (Benjamin/Cummings, Menlo Park, Calif.).
Ichimaru, S., H. Iyetomi, S. Mitake and N. Itoh, 1983, Astrophys. J. Lett. 265, L83.
Ichimaru, S., H. Iyetomi and S. Tanaka, 1987, Phys. Rep., to be published.
Ichimaru, S., S. Mitake, S. Tanaka and X.-Z. Yan, 1985, Phys. Rev. A 32, 1768.
Ichimaru, S. and S. Tanaka, 1985, Phys. Rev. A 32, 1790.
Ichimaru, S. and S. Tanaka, 1986, Phys. Rev. Lett. 56, 2815.
Isern, J., J. Labay and R. Canal, 1984, Nature 309, 431.
Itoh, N., S. Mitake, H. Iyetomi and S. Ichimaru, 1983, Astrophys. J. 273, 774.
Ivanov, Yu.V., V.B. Mintsev, V.E. Fortov and A.N. Dremin, 1976, Zh. Eksp. Teor. Fiz. 71, 216 [Sov. Phys. -JETP 44, 112(1976)].
Iyetomi, H., 1984, Prog. Theor. Phys. 71, 427.
Iyetomi, H. and S. Ichimaru, 1982, Phys. Rev. A 25, 2434.
Iyetomi, H. and S. Ichimaru, 1983, Phys. Rev. A 27, 3241.
Iyetomi, H. and S. Ichimaru, 1986a, Phys. Rev. A 34, 433.
Iyetomi, H. and S. Ichimaru, 1986b, Phys. Rev. A 34, 3203.
Iyetomi, H. and S. Ichimaru, 1987, in Proc. Intern. Meeting on Condensed Matter Theories, edited by R.F. Bishop, R.K. Kalia and P. Vashishta (Plenum, London) Vol. 2, to be published.
Jancovici, B., 1962, Nuovo Cimento 25, 428.
Khanna, F.C. and H.R. Glyde, 1976, Can. J. Phys. 54, 648.
Malmberg, J.H. and T.M. O'Neil, 1977, Phys. Rev. Lett. 39, 1333.
Mitake, S., S. Ichimaru and N. Itoh, 1984, Astrophys. J. 277, 375.
Mitake, S., S. Tanaka, X.-Z. Yan and S. Ichimaru, 1985, Phys. Rev. A 32, 1775.
Mochkovitch, R., 1983, Astron. Astrophys. 122, 212.
Perrot, F. and M.W.C. Dharma-wardana, 1984, Phys. Rev. A 30, 2619.
Pines, D. and P. Nozières, 1966, The Theory of Quantum Liquids (Benjamin,

New York), Vol. I.
Pokrant, M.A., 1977, Phys. Rev. A 16, 413.
Richert, W. and W. Ebeling, 1984, Phys. Status Solidi (b) 121, 633.
Rosenfeld, Y. and N.W. Ashcroft, 1979, Phys. Rev. A 20, 1208.
Shapiro, S.L. and S.A. Teukolsky, 1983, Black Holes, White Dwarfs and Neutron Stars (Wiley, New York).
Singwi, K.S. and M.P. Tosi, 1981, in Solid State Physics, edited by H. Ehrenreich, F. Seitz and D. Turnbull (Academic, New York), Vol. 36, p. 177.
Singwi, K.S., M.P. Tosi, R.H. Land and A. Sjölander, 1968, Phys. Rev. 176, 589.
Sjödin, S. and S.K. Mitra, 1977, J. Phys. A 10, L163.
Slattery, W.L., G.D. Doolen and H.E. DeWitt, 1980, Phys. Rev. A 21, 2087.
Slattery, W.L., G.D. Doolen and H.E. DeWitt, 1982, Phys. Rev. A 26, 2255.
Stevenson, D.J., 1980, J. de Physique 41, C2-61.
Stevenson, D.J., 1982, Ann. Rev. Earth Planet. Sci. 10, 257.
Tanaka, S. and S. Ichimaru, 1984, J. Phys. Soc. Jpn. 53, 2039.
Tanaka, S. and S. Ichimaru, 1985, Phys. Rev. A 32, 3756.
Tanaka, S. and S. Ichimaru, 1986a, J. Phys. Soc. Jpn. 55, 2278.
Tanaka, S. and S. Ichimaru, 1986b, Phys. Rev. A 34, 4163.
Tanaka, S. and S. Ichimaru, 1987a, to be published.
Tanaka, S. and S. Ichimaru, 1987b, to be published.
Tanaka, S., S. Mitake and S. Ichimaru, 1985, Phys. Rev. A 32, 1896.
Tanaka, S., S. Mitake, Yan, X.-Z. and S. Ichimaru, 1985, Phys. Rev. A 32, 1779.
Wallenborn, J. and M. Baus, 1978, Phys. Rev. A 18, 1737.
Yan, X.-Z. and S. Ichimaru, 1986a, Phys. Rev. A 34, 2167.
Yan, X.-Z. and S. Ichimaru, 1986b, Phys. Rev. A 34, 2173.
Yan, X.-Z., S. Tanaka, S. Mitake and S. Ichimaru, 1985, Phys. Rev. A 32, 1785.

STATIC AND DYNAMIC PROPERTIES OF STRONGLY-COUPLED CLASSICAL ONE-COMPONENT PLASMAS: NUMERICAL EXPERIMENTS ON SUPERCOOLED LIQUID STATE AND SIMULATION OF ION PLASMA IN THE PENNING TRAP

Hiroo Totsuji

Department of Electronics, Okayama University

Tsushimanaka, Okayama 700, Japan

I. INTRODUCTION

Among various strongly coupled Coulombic systems, the classical one-component plasma (OCP), the classical system of charged particles of one species in the uniform background of opposite charges, is the simplest one which nevertheless manifests fundamental characteristics of Coulomb system. The OCP may also be one of most-thoroughly-investigated classical systems from statistical mechanical point of view. Since most of its static and dynamic properties in thermal equilibrium are known accurately (e.g., Baus and Hansen, 1980; Ichimaru, 1982), it works as a useful reference system for other more complicated Coulombic systems.

Since OCP is a classical system with unique dimensionless coupling parameter, its state in thermodynamical equilibrium is either liquid (or fluid, including gaseous state) or solid. Static and dynamic properties of OCP in the domain of liquid has been clarified through various theoretical approaches and extensive numerical experiments and recent investigation is focused on the domain of liquid with extremely strong coupling near or beyond the liquid-solid phase transition. Here OCP may possibly become the metastable supercooled liquid associated with this first order phase transition or even the amorphous glass as in the case of rapidly quenched metals or simple liquids. The interparticle potential in OCP, however, is very soft and has different nature from the short-ranged ones such as the Lennard-Jones potential. It is therefore of interest to follow the behavior of rapidly quenched OCP and observe the properties of these states, if they exist.

The first part of this paper is concerned with the static and dynamic properties of these strongly coupled OCP liquids. We analyze the results of molecular dynamics numerical experiments in comparison with those obtained earlier for liquids with smaller coupling parameters.

In investigating the properties of OCP, results of numerical experiments of both Monte Carlo and molecular dynamics have been quite useful as a guide for theoretical approaches giving various thermodynamic quantities and transport coefficients: Real experiments on OCP in laboratories have been possible only for the two-dimensional electron system on the surface of liquid He and other material.

In these circumstances, recent experiments indicating the possibility of realizing the strongly coupled OCP of ions in the Penning trap (Bollinger and Wineland, 1984) are of interest and simulation of this plasma may be useful for observation of strong coupling effect by laboratory experiments. In the second part of this paper, we present the results of numerical simulations of ion plasmas in the Penning trap and show some examples of strong coupling effects appearing in these experiments.

II. OCP IN SUPERCOOLED LIQUID STATE

A. Method

Our system is composed of charged particles of one species in a uniform background interacting through the Coulomb interaction. The nondimensional coupling constant Γ characterizing our system is defined by

$$\Gamma = e^2/ak_BT, \tag{1}$$

where e is the charge of a particle, a the mean distance between charges given by $a=(3/4\pi n)^{1/3}$, n the density, T the temperature, and k_B the Boltzmann constant.

In order to analyze both static and dynamic properties, we apply the method of microcanonical molecular dynamics to our system (Hansen et al., 1975; Totsuji et al., 1980). We put N particles in the cubic cell and impose periodic boundary conditions. We use $N=432=2\cdot6^3$ or $1024=2\cdot8^3$ independent particles and main results presented here are based on experiments with 432 particles. The force acting on each particle is computed by the Ewald method and the equations of motion are integrated by the fourth order Runge-Kutta method. The time step is taken to be as large as possible to minimize the computational time satisfying the condition that the total energy and total momentum are conserved with sufficient accuracy during the whole experiment.

The supercooled liquid state is obtained by quenching liquid OCP in thermodynamic equilibrium into the domain where the coupling parameter is larger than the critical value of solidification $\Gamma_m=178$ (Slattery et al., 1982). To realize the rapid cooling of the system, we simultaneously scale the velocity of all particles by a factor between 0.7 and 0.8. By this scaling, the total kinetic energy is instantaneously reduced in proportion to the square of the scaling factor. In the course of subsequent microcanonical evolution, the kinetic energy partly recovers from this reduction. This recovery takes place in a relatively short time less than $10\omega_p^{-1}$. The pair correlation function relaxes to a new stationary value much more slowly than energy. The relaxation time for the pair correlation function, however, is less than $500\omega_p^{-1}$. We monitor the behaviors of kinetic and potential energies and the pair distribution function, and regard the system to be in a new stationary state when the pair distribution function becomes stationary. Our analyses of new stationary states are made for durations of more than $3000\omega_p^{-1}$ with N=432 and $1500\omega_p^{-1}$ with N=1024.

In contrast to the Monte Carlo (MC) numerical experiments, the value of the coupling parameter Γ cannot be specified in advance but have to be determined by the average value of the kinetic energy per particle $K=\langle mv^2/2\rangle$ as

$$\Gamma = e^2/[a(2/3)K]. \tag{2}$$

Here we assume that velocities are distributed by the Maxwellian. The parameters of our liquid states and supercooled liquid states are summarized in Table 1.

Table 1. Value of Γ Estimated by Kinetic Energy and Thermal Part of Correlation Energy.

	N=432			
Γ	100.2±0.2	175.0±0.6	229.2±1.7	294.6±1.5
$(e_c)_{th}/k_BT$	2.07±.007	2.39±.014	2.56±.02	2.67±.02

	N=1024		
Γ	202.5±0.7	313.6±2.1	394.4±2.1
$(e_c)_{th}/k_BT$	2.43±.01	2.61±.02	2.74±.02

B. Correlation Energy

The correlation energy e_c (per particle) of OCP in thermal equilibrium has been known very accurately from the Monte Carlo numerical experiments and fitting formulae have been given for both the liquid and solid states (e.g. Slattery et al., 1980 and 1982). When the coupling parameter is sufficiently large, the correlation energy is dominated by the contribution of the Madelung-energy-like term proportional to Γ. To observe the behavior of the correlation energy more closely, we define the thermal part of the correlation energy $(e_c)_{th}$ subtracting the Madelung energy of the bcc lattice $(e_c)_{bcc}/k_BT=0.895929\Gamma$ as

$$(e_c)_{th} = e_c - (e_c)_{bcc}: \tag{3}$$

The bcc lattice has the lowest energy among simple lattices of OCP.

In Fig.1 we show the values of the thermal part of the correlation energy obtained by our numerical experiments in comparison with those for liquid and solid in thermal equilibrium (Slattery et al., 1982). We see that the thermal part of the correlation energy of the supercooled metastable state is clearly larger than that of solid state with the same value of Γ.

In the domain of supercooled liquid, there has previously been reported one result for Γ=200 obtained by Monte Carlo method by Slattery et al. (1982). As is shown in Fig.1, this results is consistent with our results. We also plot the values given by extrapolating the interpolation formula for liquid (Slattery et al., 1982) into the supercooled domain and see that our experimental results are close to those extrapolation. It should be kept in mind that this domain is beyond the original applicability of the formula.

C. Pair Distribution Function

The values of the pair distribution function (PDF) for Γ=175, 229, and 295 obtained by experiments with 432 particles and those for Γ=203 and 314 obtained with 1024 particles are shown in Fig.2. We also plot the pair distribution function obtained by the Monte Carlo numerical experiments by Slattery et al. for Γ=180 (1980) and 200 (DeWitt, 1982) to show that our results are consistent with MC experiments.

With the increase of Γ, the height of the first peak increases. We note, however, that, in the domain of Γ of our experiments, there seems to be no remarkable change in the structure at the second peak of the pair distribution function such as the splitting which characterizes the so-

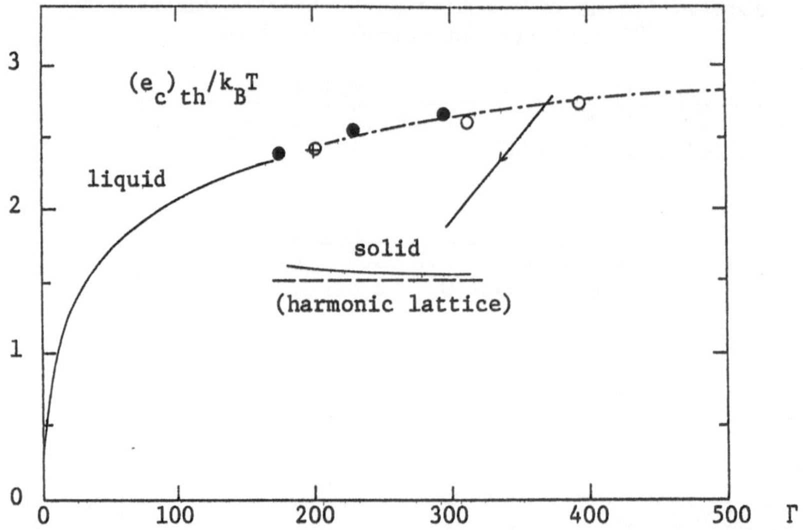

Fig.1. Thermal part of correlation energy is shown by filled (N=432) and open (N=1024) circles. MC values are plotted by solid lines (liquid and solid in thermal equilibrium) and cross (supercooled liquid). Solid line with arrow follows an example of solidification.

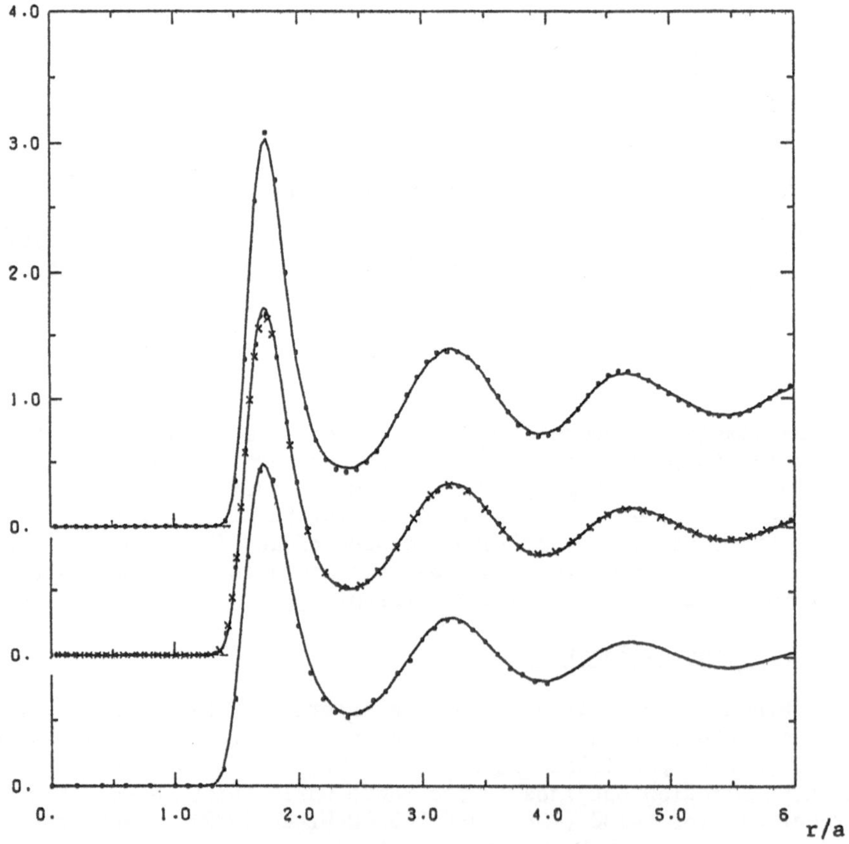

Fig.2. Pair distribution function.
(a) Solid line: Γ=175(N=432). Dot: MC Γ=180.
(b) Solid line: Γ=229(N=432). Dot: Γ=203(N=1024). Cross: MC Γ=200.
(c) Solid line: Γ=295(N=432). Dot: Γ=314(N=1024).

called amorphous state (e.g., Kimura and Yonezawa, 1983; Ichimaru and Tanaka, 1986).

D. Velocity Autocorrelation Function and its Spectrum

The velocity autocorrelation function (VAF) Z(t) is defined by

$$Z(t) = \langle \vec{v}_i(t + t')\cdot\vec{v}_i(t')\rangle/3, \tag{4}$$

where $\vec{v}_i(t)$ is the velocity of i-th particle and < > denotes the average with respect to t' and particles.

The behavior of VAF in the domain of liquid (Hansen et al., 1975) may be summarized as follows. For sufficiently small values of Γ, the VAF decays monotonically as a function of time. When $\Gamma \sim 10$ the tail begins to have the component which oscillates with the plasma frequency and the tail can be regarded as the damped oscillation with the plasma frequency for $100 \lesssim \Gamma \lesssim 150$.

The values obtained by our experiments are shown in Fig.3. We see that the values of the first and second peaks and the second dip decrease with further increase of Γ and the first peak becomes negative for $300 \lesssim \Gamma$. At the same time the oscillation with the plasma frequency becomes less significant. The velocity autocorrelation function may thus be considered as a superposition of damped oscillation with the plasma frequency ω_p and the overdamped oscillation which is observed in simple liquids, for example, of inert atoms (e.g., Hansen and McDonald, 1976). With the increase of Γ, the relative importance of the ω_p-component first increases for $100 \lesssim \Gamma \lesssim 180$ and then decreases for $200 \lesssim \Gamma$.

In Fig.4 we show the spectrum of the velocity autocorrelation function Z(ω) defined by

$$Z(\omega) = (1/2\pi)\int dt\, \exp(i\omega t) Z(t). \tag{5}$$

In Z(ω) we observe the above mentioned change of the relative importance of two components. For Γ=175, the spectrum is not so different from the one obtained previously for liquid with Γ=152 (Hansen et al., 1975). For Γ=229 and 295, the relative importance of the peak corresponding to the plasma oscillation is decreased. This change is consistent with the behavior of VAF in time space.

E. Diffusion Constant

The self-diffusion constant D is defined by

$$D = \lim_{t\to\infty} \langle [\Delta\vec{r}_i(t)]^2\rangle/6t, \tag{6}$$

where

$$\Delta\vec{r}_i(t) = \vec{r}_i(t + t') - \vec{r}_i(t'). \tag{7}$$

We plot the numerator of (6), the mean square displacement (MSD), in Fig.5 as a function of t. For Γ=175 and 229, the MSD increases almost linearly with time. For Γ=295, however, it first seems to increase rapidly and then the phase of slow increase appears. The latter behavior of the mean square displacement has been observed in various rapidly quenched systems (e.g., Kimura and Yonezawa, 1983).

The diffusion constant is evaluated by the slope of MSD plotted as a function of time. The results are shown in Fig.6 where the diffusion

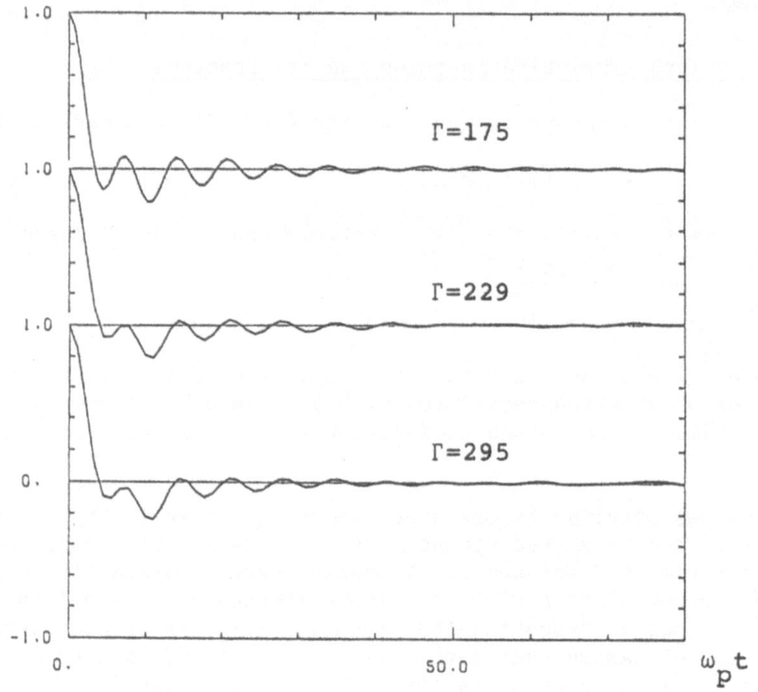

Fig.3. Velocity autocorrelation function normalized by Z(t=0).

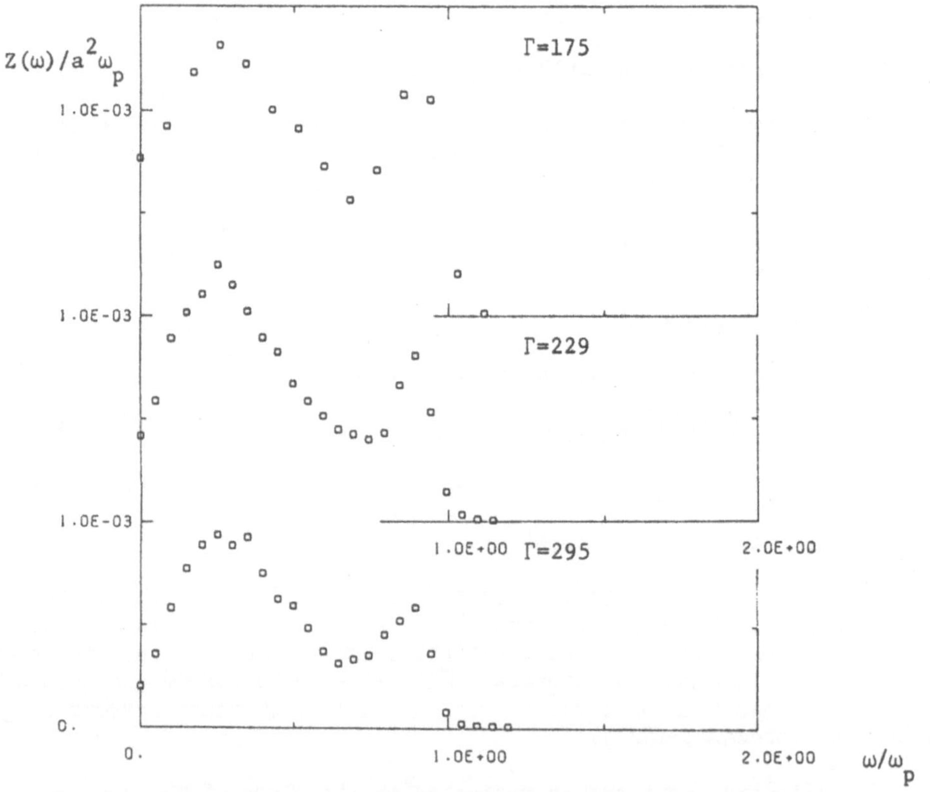

Fig.4. Spectrum of velocity autocorrelation function Z(ω).

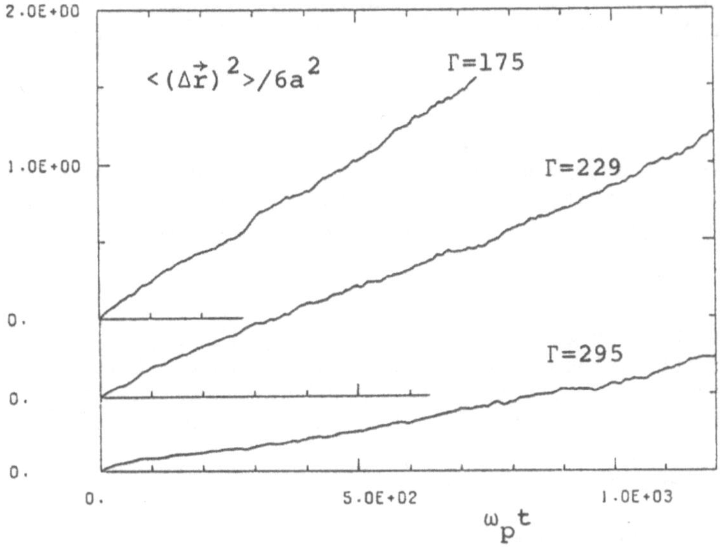

Fig.5. Mean square displacement vs. time.

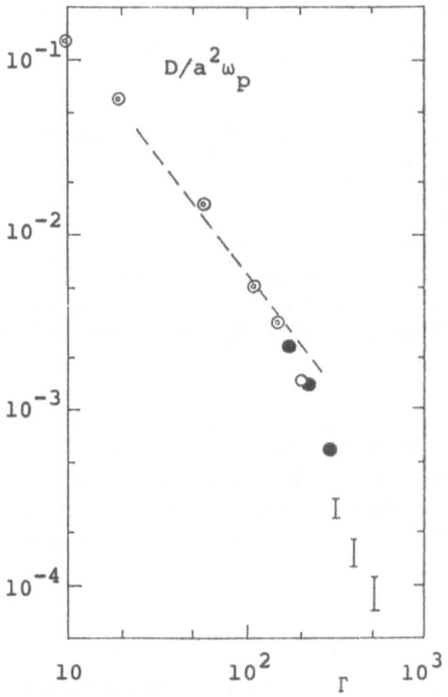

Fig.6. Self-diffusion coefficient.
Filled circles: N=432.
Open circle and bars (upper and lower bounds): N=1024.
Double circles and broken line: Hansen et al. (1975).

constant obtained by experiments with 1024 particles are also shown. We note that the self diffusion constant D is related to $Z(\omega=0)$ as

$$D = \pi Z(\omega = 0). \tag{8}$$

Values of the diffusion constant given by (8) are not exactly the same as those obtained from MSD but agreement is satisfactory. In Fig.6 we also show the values for liquid state obtained by Hansen et al. (1975). They have interpolated their results by the fitting formula

$$D/a^2\omega_p = 2.59\Gamma^{-1.34}. \tag{9}$$

We see that the diffusion constant decreases rapidly and becomes much smaller than the values extrapolated from the values for liquid state (9). It is, however, difficult to discuss whether the diffusion constant shows some qualitative change or not in the domain of supercooled liquid.

F. Shear Viscosity

The shear viscosity coefficient η is evaluated from the long-wavelength limit of the autocorrelation function of the transverse part of the stress tensor $\sigma_{k,\alpha\beta}$ as

$$\eta = \int_0^\infty dt\ \eta(t), \tag{10}$$

where

$$\eta(t) = \lim_{k\to 0} \langle\sigma_{k,\alpha\beta}(t)\sigma_{k,\alpha\beta}(0)\rangle, \quad \alpha \neq \beta. \tag{11}$$

The values of the transverse stress autocorrelation function are shown in Fig.7 and resultant values of shear viscosity are plotted in Fig.8. The correlation function is obtained by dividing the whole stationary state into several shorter parts and taking the average of the correlation functions obtained in each division.

The most remarkable change of the stress correlation function is the increase of the relaxation time with the increase of Γ. The estimated values of relaxation time (in the unit of ω_p^{-1}) for Γ=175, 229, and 295 are 6, 9, and 11, respectively, when simple exponential decay is assumed for the first part of the autocorrelation function. For Γ=229 and 295, however, there appears the tail which decays much more slowly: The relaxation time of the tail for Γ=295 is about 40. These values are much larger than that in the domain of liquid obtained by Bernu, Vieillefosse, and Hansen for Γ=100 (1977, 1978).

The increase of relaxation time naturally leads to the increase of the shear viscosity as is shown in Fig.8. Combined with earlier results, we see that the viscosity increases with the increase of Γ after attaining its minimum around Γ=10. It seems, however, to be impossible to draw definite conclusion about the existence of drastic change related to transition to amorphous state.

G. Spectrum of Density Fluctuation

The density fluctuation spectrum is expressed by the dynamic form factor defined by

$$S(k,\omega) = (1/2\pi)\int dt\ \exp(i\omega t)\langle\rho_k(t)\rho_{-k}(0)\rangle, \tag{12}$$

where

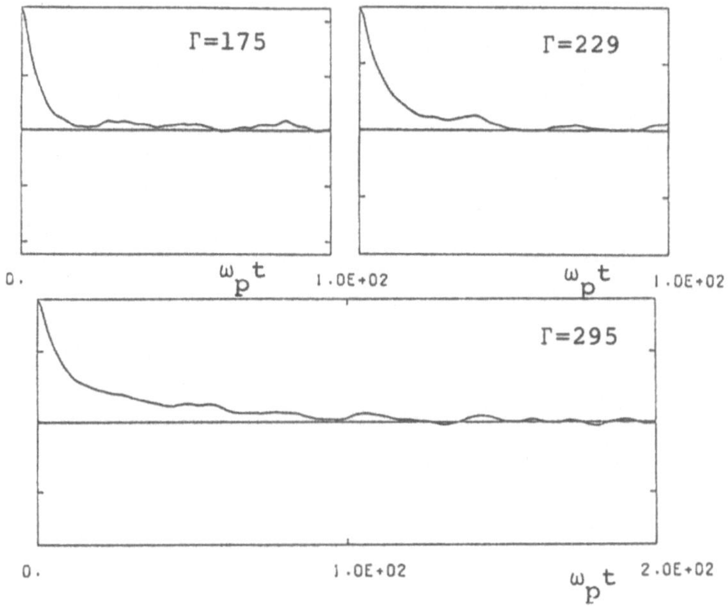

Fig.7. Autocorrelation function of transverse part of the stress tensor.

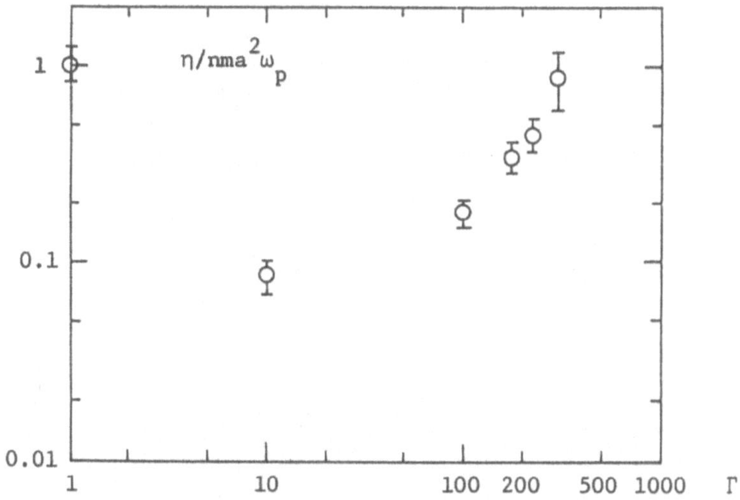

Fig.8. Shear viscosity. Values for Γ<100 are due to Bernu et al. (1977, 1978).

$$\rho_k(t) = \sum_i \exp[-i\vec{k}\cdot\vec{r}_i(t)]. \tag{13}$$

The results for $S(k,\omega)$ are shown in Fig.9.

Compared with the values in the domain of liquids (Hansen et al., 1975), we observe that the peak structure representing the well-defined plasma oscillation extends to larger values of the wave number. At the same time, we observe the concentration of the spectrum at zero frequency with the increase of Γ.

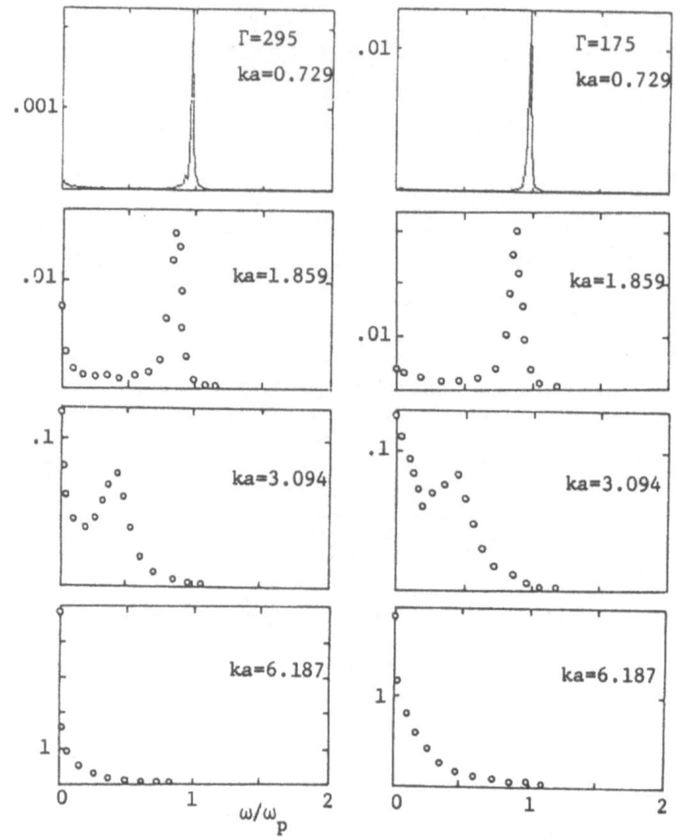

Fig.9. Dynamic form factor $(\omega_p/N)S(k,\omega)$.

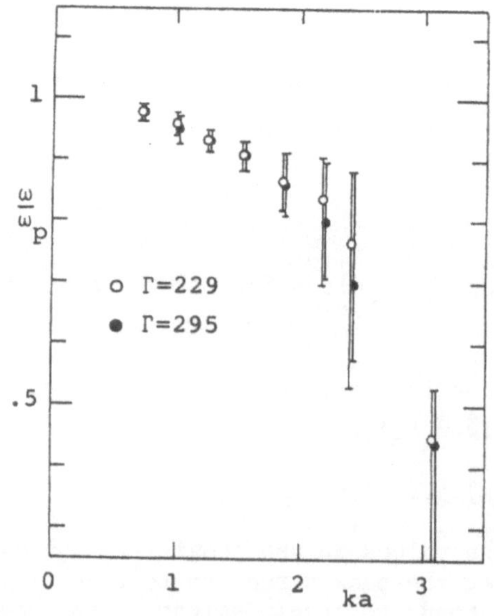

Fig.10. Behavior of plasmon.
Circles: Position of peak.
Bars: Frequencies at half maximum.

The behavior of the plasmon peak is summarized in Fig.10 where the central frequency and the width at the half maximum are shown. The central frequencies for Γ=229 and 295 are fitted by the dispersion relation as

$$\omega/\omega_p = 1 - 0.04(ka)^2. \tag{14}$$

H. Transverse Current Fluctuation Spectrum

The spectrum of current fluctuation is defined by

$$C(\vec{k},\omega) = (1/2\pi)\int dt\, \exp(i\omega t)\langle\vec{g}_k(t)\vec{g}_{-k}(0)\rangle, \tag{15}$$

where

$$\vec{g}_k(t) = \sum_i \vec{v}_i(t)\exp[-i\vec{k}\cdot\vec{r}_i(t)]. \tag{16}$$

The spectrum is divided into the longitudinal and the transverse parts as

$$C(\vec{k},\omega) = (\vec{k}\vec{k}/k^2)C_\ell(k,\omega) + (I - \vec{k}\vec{k}/k^2)C_t(k,\omega), \tag{17}$$

and the former is related to the dynamic form factor by the equation of continuity. We have confirmed that this relation is satisfied by our results.

The results for the transverse part are shown in Fig.11. We observe that the shear mode extends to smaller wave numbers with the increase of Γ. We also observe the small shoulder which has been observed earlier by Hansen et al. (1975). Compared with the longitudinal or density fluctuation spectrum, the increase of Γ does not cause significant change in the transverse current fluctuations.

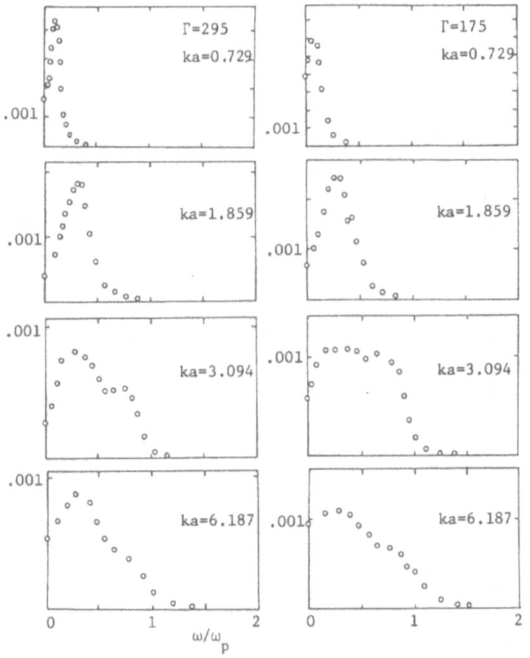

Fig.11. Transverse part of current fluctuation spectrum $C_t(k,\omega)/a^2\omega_p N$.

I. Example of Solidification

In order to show the difference in the pair distribution functions in supercooled liquid and solid states, we present an example which undergoes the solidification. We have prepared this system by quenching the system at $\Gamma=295$ which, after microcanonical evolution of more than $3000\omega_p^{-1}$, can be regarded as in stationary state from the behaviors of the pair distribution function and thermodynamic quantities.

After quenched into the supercooled liquid state with $\Gamma\cong380$, the system makes the transition to the solid with dislocations. The orbit of particles projected onto planes and the change in the pair distribution function in the course of solidification are shown in Fig.12. We see that the PDF changes into that of bcc solid. (The analysis of the distribution of Voronoi polyhedra also confirms that the lattice is bcc.)

Since the latent heat of liquid-solid transition is released, the value of Γ estimated by (2) decreases when the system solidifies. The behavior of estimated value of Γ is shown in Fig.1. The mean square displacement needed for solidification is about $3a^2$.

III. STRONGLY-COUPLED ION PLASMA IN THE PENNING TRAP

A. Method

The Penning trap is composed of the uniform magnetic field (in the z-direction) which prevents ions from escaping in the x- or y-direction and the electric field due to electrode which suppresses the motion of ions along the z-axis (e.g., Brown et al., 1986).

In these magnetic and electric fields, charged particles rotate as a whole and, in the rotating coordinate system, ions behave as if they are in the neutralizing background or OCP (Malmberg and O'Neil, 1977). In the experiment by Bollinger et al. (1984) the narrow laser beam is used to cool the ion plasma. In the process of fluorescence scattering, ions lose their kinetic energy and the total angular momentum.

In order to observe the behavior of strongly coupled ion plasma realized in these experiments and examine the possibility to observe strong coupling effects by these experiments, we perform numerical simulation.

We numerically integrate the equations of motion for N ions

$$\begin{aligned} m d\vec{v}_i/dt &= \vec{F}_i, \\ \vec{F}_i &= \sum_j e^2(\vec{r}_i - \vec{r}_j)/|\vec{r}_i - \vec{r}_j|^3 + e\vec{E} + (e/c)\vec{v}_i\times\vec{B}, \\ \vec{E} &= -\mathrm{grad}[m\omega_z{}^2(2z^2 - x^2 - y^2)/4e], \\ \vec{B} &= B\hat{z}. \end{aligned} \tag{18}$$

Here $\vec{E}$ is the electric field due to electrode and the magnetic field is uniform and in z-direction.

We take the origin of the coordinates at the center of the plasma and assume that the laser beam in the positive x-direction is irradiating the plasma in the domain with y>0. In order to simulate the cooling by laser beam, we scale the velocity by a ratio α as

$$v_x \to \alpha v_x, \quad \text{when } y>0 \text{ and } v_x<0, \tag{19}$$

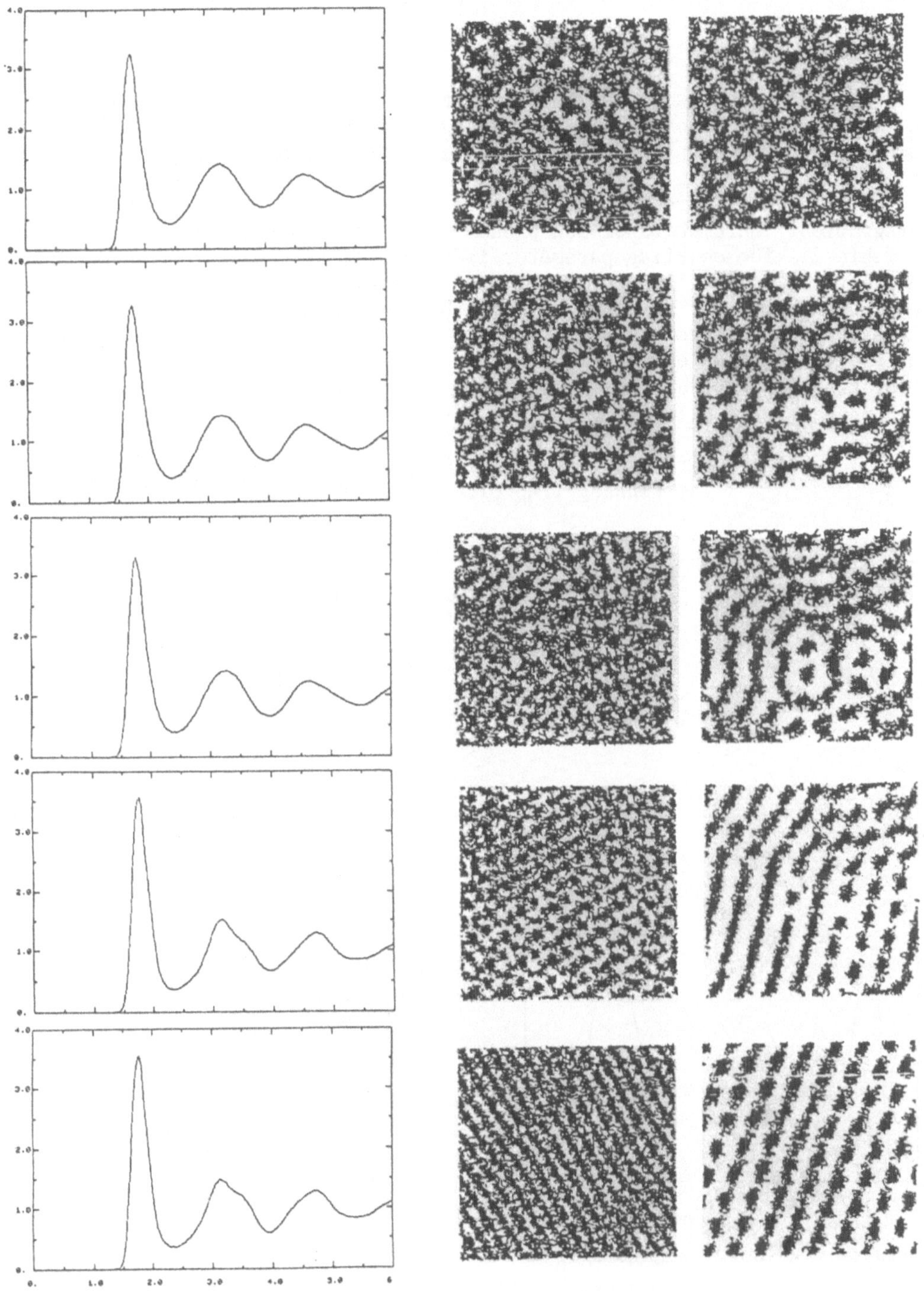

Fig.12. Solidification of supercooled liquid OCP into bcc lattice. Pair distribution function and orbits of particles projected onto x-y and y-z planes. Orbits are followed for 250 ω_p^{-1} after (from top to bottom) 130, 640, 1140, 1650, and 2160 ω_p^{-1} from quenching.

after each step of integration by the Runge-Kutta method of fourth order. The scaling ratio α is chosen between 0.995 and 0.9: This cooling is about 7 to 25 times faster than the value calculated from the rate of scattering phenomena (e.g., 10^6 per second per ion, as in Bollinger et al., 1984).

B. Result

Starting from the random configuration in a sphere of radius $R_0=10^{-2}$cm, we integrate equations (18). The average angular frequency of rotation is computed and the temperature is evaluated as the kinetic energy in the frame of coordinate rotating with average angular frequency. The total energy and the total angular momentum are monitored to check the accuracy of numerical integration. Parameters of ion plasma obtained by our simulations are shown in Table 2. The coupling parameter is estimated by the central density and the temperature parallel to the magnetic field. We have used larger cooling rate than experiment in order to save the computational time. Resultant plasmas, however, seem to be similar to those obtained by real experiments: For example, ratios of the parallel and perpendicular temperatures are consistent with extrapolated value from experiments.

Table 2. Parameters of Plasma Composed of 150 ^{9}Be Ions in the Penning Trap with B=0.819T and $\omega_z/2\pi$=200kHz.

n(center)	a	$T_\perp$	$T_{//}$	$T_\perp/T_{//}$	Γ
$2.8\cdot10^7$cm^{-3}	$2.0\cdot10^{-3}$cm	1.2mK	24mK	0.05	35
$3.3\cdot10^7$cm^{-3}	$1.9\cdot10^{-3}$cm	2.9mK	35mK	0.08	25

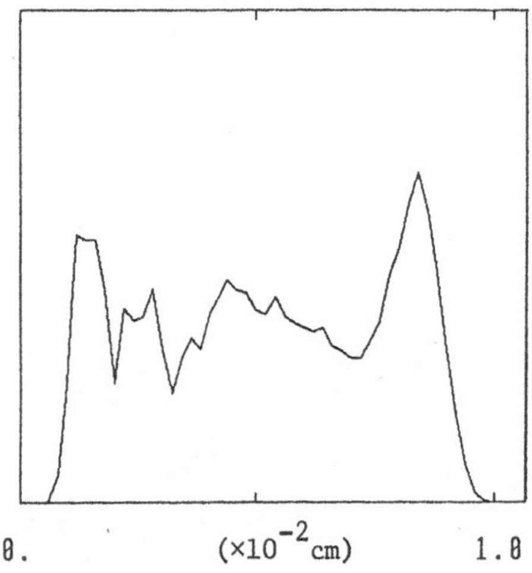

Fig.13. Distribution of ions in the Penning trap with Γ=35 as a function of distance from the center.

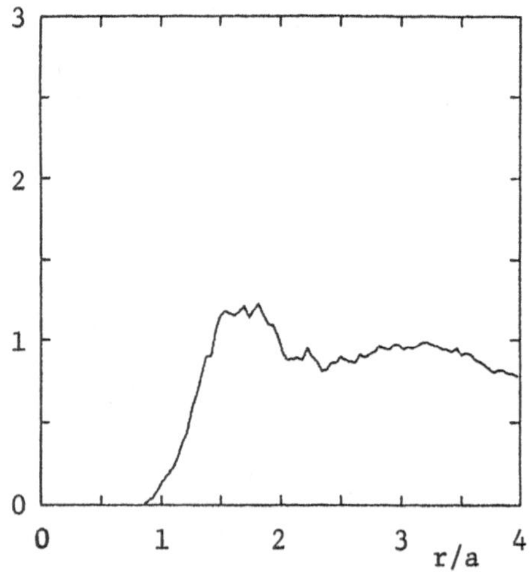

Fig.14. Pair distribution function in the central part of ion plasma in the Penning trap with Γ=35.

The distribution of ions computed assuming the spherical symmetry is shown in Fig.13. We clearly see that the oscillation of density occurs for these intermediate values of the coupling parameter.

The oscillation of the density profile has been observed in numerical experiments where the background charge density is distributed uniformly in a sphere (Badiali et al., 1983; Levesque and Weis, 1983) or in a slab (Totsuji, 1986). Fig.13 shows that this oscillation can be observed by real experiments in the Penning trap.

The pair distribution function is computed for ions in the central domain where the density is approximately uniform. An example is plotted in Fig.14. Our results are similar to the pair distribution function of OCP with the same value of the coupling parameter.

These results of numerical simulations indicate that ion plasmas in the Penning trap may be quite useful to observe various strong-coupling effects in OCP by real experiments in laboratories.

This work has been partially supported by the Grant-in-Aid for Scientific Researches from the Ministry of Education, Science, and Culture of Japan and by the Yamada Science Foundation. The author also thanks Dr. H. E. DeWitt for his kindness in providing unpublished MC data.

REFERENCES

Badiali, J.-P., Rosinberg, M.-L., Levesque, D., and Weis, J. J., 1983, J. Phys. C: Solid State Phys. 16: 2183.
Baus, M. and Hansen, J.-P., 1980, Phys. Rep. 59: 1.
Bernu, B.,and Vieillefosse, P., 1978, Phys. Rev. A18: 2345.
Bernu, B., Vieillefosse, P., and Hansen, J.-P., 1977, Phys. Lett. 63A: 301.
Bollinger, J. J., and Wineland, D. J., 1984, Phys. Rev. Lett., 53: 348.
Brown, L. S., and Gabrielse, G., 1986, Rev. Mod. Phys. 58: 233.
DeWitt, H. E., 1982, unpublished.
Hansen, J.-P., and McDonald, I. R., 1976, "Theory of Simple Liquids," Academic Press, London, New York, and San Francisco.
Hansen, J.-P., McDonald, I. R., and Pollock, E. L., 1975, Phys. Rev. A11: 1025.
Ichimaru, S., 1982, Rev. Mod. Phys. 54: 1017.
Ichimaru, S. and Tanaka, S., 1986, Phys. Rev. Lett. 56: 2815.
Kimura, M, and Yonezawa, F., 1983, in "Topological Disorder in Condensed Matter," T. Ninomiya and F. Yonezawa, eds., Springer, Berlin, Heidelberg, New York, and Tokyo.
Levesque, D. and Weis, J. J., J. Stat. Phys. 33: 549.
Malmberg, J. H., and O'Neil, T. M., 1977, Phys. Rev. Lett. 39: 1333.
Slattery, W. L., Doolen, G. D., and DeWitt, H. E., 1980, Phys. Rev. A21: 2087.
Slattery, W. L., Doolen, G. D., and DeWitt, H. E., 1982, Phys. Rev. A26: 2255.
Totsuji, H., 1986, J. Phys. C: Solid State Phys. 19: L573.
Totsuji, H., and Kakeya, H., 1980, Phys. Rev. 22: 1220.

EXTRACTION OF THE ONE-COMPONENT PLASMA BRIDGE FUNCTION FROM COMPUTER SIMULATION DATA

P. D. Poll and N. W. Ashcroft

Laboratory of Atomic and Solid State Physics
Cornell University
Ithaca, New York 14853-2501

INTRODUCTION

What follows is motivated by the classical inverse problem for homogeneous highly correlated systems: Given complete structural information [at the pair level this will be the static structure factor S(k) for all k], find the effective pair-potential $\phi(r)$ that will generate that structure. There is a unique functional relationship $\beta\phi(r) = F[S(k)]$ [with β the inverse temperature $1/(kT)$], between the pair-potential and the static structure factor. However, in practice this is a very difficult problem to solve, in part, because of the extreme sensitivity of the resultant $\beta\phi(r)$ to the input functions S(k).

Consider the diagrammatic resummation result for homogeneous classical fluids within the pair-potential approximation

$$\beta\phi(r) = -\ln[g(r)] + g(r) - 1 - c(r) + E(r), \tag{1}$$

where g(r) is the radial distribution function and c(r) is the Ornstein-Zernike direct correlation function defined, again for homogeneous classical fluids by

$$h(r) = c(r) + \rho\int d\vec{r}'\ h(|\vec{r} - \vec{r}'|)c(r'). \tag{2}$$

Here $h(r) = g(r) - 1$, ρ is the average one-particle number density, and E(r) is the bridge function. Now in general, E(r) is also a unique functional of the total correlation function h(r), as is known from its expansion in highly connected h-bond diagrams. Thus the classical inverse problem is solved in principle through complete knowledge of h(r), or of S(k) [$= 1 + \rho h(k)$]. The diagrammatic expansion, however, is slowly convergent and is not suitable for practical calculations. To perform a practical inversion of structural data, theories for the bridge function E(r) are needed. The simplest such theory sets the bridge function to zero: this is the hypernetted-chain (HNC) approximation. The results of this approximation in the context of the inverse problem are generally very poor for highly correlated systems. The appropriate energy scale is incorrectly determined and spurious structure appears in the resultant pair-potential. A more refined theory, the modified hypernetted-chain approximation,[1] replaces the bridge function by its value for some hard-sphere reference system. Even though this equation yields fluid structure to within 1% of the best computer simulation results in the forward direction, where the pair-potential is known, the work of Levesque, Weis, and Reatto[2,3] has shown that the theory is actually inadequate for a detailed direct inversion of known structural data.

There has been progress in determining even more accurate representations of the bridge function for an eventual application to the classical inverse problem. However, it is already clear from (1) that we need a complete set of functions $\{\beta\phi(r), g(r), c(r), E(r)\}$ for various reference systems in order to test any proposed inversion scheme. The purpose of this article is to provide this reference data. Two sets, one for the classical Lennard-Jones (LJ) system and one for the classical one-component plasma (OCP), will be presented. The latter involves a new extension of finite range computer simulation data.

EXTENSION OF COMPUTER SIMULATION RESULTS

General Remarks

Computer simulation methods, either Monte Carlo or Molecular Dynamics, normally start with an assumed pair-potential (one body, or three and higher body, potentials may also, in principle, be considered). The simulation, for three dimensional systems, is performed in a box with volume L^3 and with periodic boundary conditions imposed. The result is generally a radial distribution function for radii up to a certain cutoff r_c, with r_c typically less than L/2. Now, to obtain the radial distribution function for all r we need a procedure for dealing with the region $r>r_c$. The subsequent evaluation of the radial distribution function for all r, along with all the other correlation functions, for a given procedure, is then referred to as an "extension" of computer simulation data. The key to meaningful extensions is a corresponding physically meaningful and accurate procedure for the behavior of the system under consideration in this unsimulated region.

In early work, Ceperley and Chester[4] assumed that the radial distribution function takes the form

$$g(r) = 1 + \mathrm{Re}\Big[\sum_{j=1}^{n} A_j \frac{\exp(-z_j r)}{r}\Big], \qquad r>r_c. \tag{3a}$$

The parameters A_j and z_j are complex numbers, and are simply chosen to fit the simulation results for $r<r_c$; usually 3 or 4 terms are sufficient for a reasonable extension. In an alternative approach, Verlet[5] has worked instead with the direct correlation function, and has invoked the Percus-Yevick approximation for the unsimulated region, i.e.,

$$c(r) = c_{PY}(r) = g(r)\{1 - \exp[\beta\phi(r)]\}, \qquad r>r_c. \tag{3b}$$

By using the simulation results for g(r) for r less than r_c, both g(r) and c(r) are then obtained via a standard iterative solution. In a variation of this method, Galam and Hansen[6] have proposed the mean-spherical approximation, i.e.,

$$c(r) = c_{MSA}(r) = -\beta\phi(r), \qquad r>r_c. \tag{3c}$$

This form for c(r) might be expected to yield a more meaningful extension of the computer simulation data for the OCP, a system that is very poorly described by the Percus-Yevick approximation.

The methods just described are not exhaustive, they merely demonstrate that different approaches exist and a good choice for the extension procedure is by no means unique. As noted, to test an inversion scheme we need very accurate determinations of the various correlation functions, typically better than 0.5%. To this end accurate simulation data as well as accurate extensions of the simulation data are both essential.

The Lennard-Jones Potential

The (6-12) pair-potential for the classical Lennard-Jones system is given by

$$\phi(r) = 4\epsilon[(\frac{\sigma}{r})^{12} - (\frac{\sigma}{r})^{6}], \tag{4}$$

where ϵ is the well depth and σ is the diameter. For this system Molecular Dynamics results have been obtained by Levesque, Weis, and Reatto.[2,3] These correspond to a reduced density $\rho\sigma^3 = 0.84$ and a reduced temperature $kT/\epsilon = 0.747$. Some 6800 steps were used with 864 particles, and a cutoff in the pair-potential (4) of 4σ was introduced. Typical fluctuations in the resultant g(r) were at most 0.3% in the region $r \sim \sigma$. This is considered to be a very accurate simulation, and it has, in fact, been used by the original authors as a test of their own predictor-corrector approach to the classical inverse problem. We now report our own extension of this data using the MSA extension method (3c). Since g(r) from the simulation is given to the limit of the pair-potential cutoff of 4σ, we have chosen a cutoff r_c in the data at the last node prior to the pair-potential cutoff. The extended g(r) shown in Figure 1a is in units of r/σ (the arrow indicates the position of r_c/σ), and the corresponding bridge function E(r), determined by use of the diagrammatic resummation result together with knowledge of the initial pair-potential, is presented in Figure 1b, again in units of r/σ.

A test of the accuracy of any extension procedure, including those to be described below, can be obtained by investigating the change in the extracted pair-potential as r_c is increased towards the actual value used. This can be done by using any of the theories mentioned in the introduction; most simply the HNC approximation. An accurate extension requires that the effective pair-potential should stabilize before the actual data cutoff used is reached. The Molecular Dynamics simulation data used here passes this test; this success is due to the fact that simulation data is supplied right up to the pair-potential cutoff of 4σ (most earlier work using just 2.5σ).

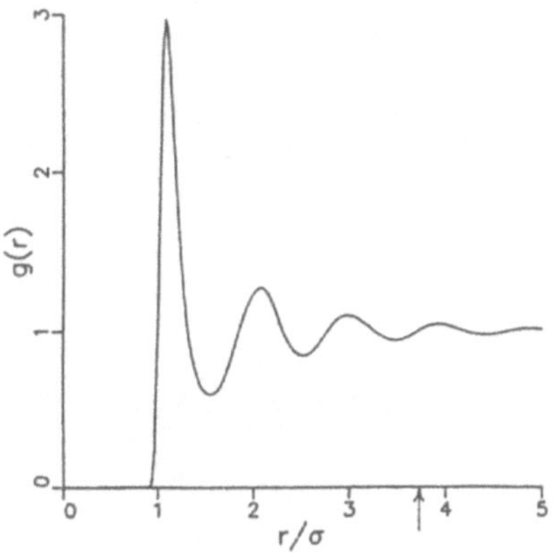

Figure 1a

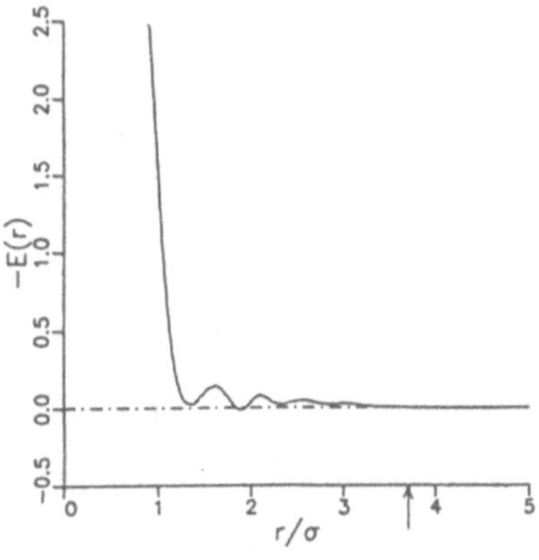

Figure 1b

The Classical One-Component Plasma

The classical one-component plasma consists of N particles of charge Q, interacting through a Coulomb pair-potential

$$\phi(r) = \frac{Q^2}{r}, \qquad (5)$$

and immersed in a uniform, rigid, compensating background. We now report our analysis of the Monte Carlo results of Slattery, Doolen, and DeWitt[7,8] at $\Gamma = 100$, where $\Gamma = \beta Q^2 r_{WS}$ is the Coulomb coupling parameter and $r_{WS} = (4\pi\rho/3)^{-1/3}$ is the Wigner-Seitz or ion-sphere radius. This particular simulation was carried out for 1024 particles, and involved 21.3 million configurations.

The long-range nature of the Coulomb interaction complicates the issue of the choice of an appropriate extension procedure for the one-component plasma correlation functions beyond r_c. As a first attempt, we may extend the $\Gamma = 100$ Monte Carlo data with the MSA extension method of Galam and Hansen,[6] again with r_c taken at the last node in the supplied g(r). This procedure leads to a failure of the accuracy criterion discussed above and to a discontinuity in all of the real space correlation functions, the reason for this being well understood. From the work of Ng[9] we know that when seeking numerical iterative solutions with Coulomb like pair-potentials, the pair-potential must usually be formally rewritten as a long-range piece $\phi_{lr}(r)$, and a short-range piece $\phi_{sr}(r)$, with $\phi(r) = S_{lr}(r) + \phi_{sr}(r)$. This division then allows the use of Fast Fourier transform numerical techniques. We then have a short-range Ornstein-Sernike direct correlation function $c_{sr}(r) = c(r) + \phi\beta_{lr}(r)$, and a new diagrammatic resummation result with c(r) and $\beta\phi(r)$ replaced by their short-range values. As a consequence the Ornstein-Zernike relation is modified; if this were <u>not</u> the case, all systems could be trivially transformed into an ideal gas! Although $-\beta\phi(r)$ gives a very accurate determination of c(r) relative to $-\beta\phi(r)$ beyond r_c, the short-range part $-\beta\phi_{sr}(r)$ does <u>not</u> give an accurate determination of $c_{sr}(r)$. This discontinuity in $c_{sr}(r)$ enters, by continued iteration of the extension algorithm, into the extended g(r) and then into the extracted bridge function E(r).

To overcome this difficulty we propose a new procedure which, by construction, corrects for any discontinuity in $c_{sr}(r)$. It is

$$c_{sr}(r) = -\beta\phi(r) + \frac{A\exp(-\lambda r)}{r}, \qquad r>r_c, \qquad (6)$$

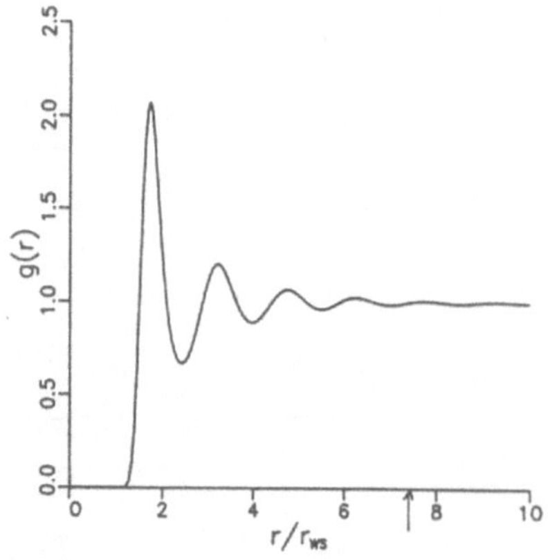

Figure 2a

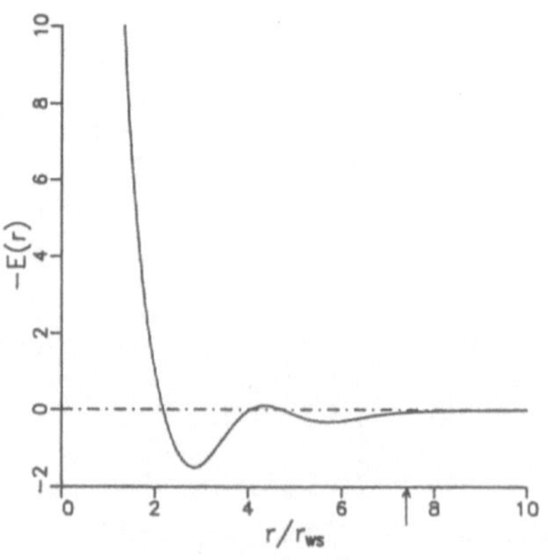

Figure 2b

with A and λ now chosen to enforce continuity in $c_{sr}(r)$ and its first derivative. Equation 6 is physically motivated by the appearance of Debye-Huckel like functions in a variety of diagrammatic approximations to the plasma problem. In Figure 2a we summarize the extended g(r) in this new extension procedure in units of r/r_{WS}. As before, the arrow indicates the position of r_c/e_{WS}. In Figure 2b we summarize the extracted bridge function for this extended data, again in units of r/r_{WS}.

Further Remarks

A comparison of Figures 1b and 2b show that the basic features of the bridge function are identical. This is in agreement with the short-range universality hypothesis of Rosenfeld and Ashcroft;[1] both bridge functions continue to be well represented at short-range by the bridge function of an appropriately chosen hard-sphere reference system. But beyond this universal core behavior the OCP bridge function becomes more highly "attractive" than its LJ counterpart. In addition, the OCP bridge function does not display the characteristic $-Dh^2(r)$ behavior at long-range; this is a direct consequence of the long-range nature of the Coulomb pair-potential itself.

We believe that our extended results of the OCP at moderate coupling represent an accurate extension of the most precise computer simulation data to date. There remain, however, deficiencies in the extension procedure. In particular, the isothermal compressibility does not agree with the equation of state results of Slattery, Doolen, and DeWitt.[7,8] The extension estimate for the isothermal compressibility moves toward the desired value, but subsequently drifts away as the number of iterations increases. This can, in principle, be corrected by enforcing agreement with the equation of state result. Work on the consequence of imposing this additional constraint is progressing.

ACKNOWLEDGEMENT

This work was supported by the National Science Foundation through Grant No. DMR-83-14764. Computations supporting this research were performed on the Cornell Production Supercomputer Facility, which is supported in part by the National Science Foundation, New York State, and IBM Corporation. One of us (PDP) gratefully acknowledges the support of the Natural Sciences and Engineering Research Council of Canada. The authors gratefully acknowledge the assistance of H. E. DeWitt, J. J. Weiss, and L. Reatto in providing advice and additional data.

REFERENCES

1. Y. Rosenfeld and N. W. Ashcroft, Phys. Rev. A 20:1208 (1979).
2. D. Levesque, J. J. Weis, and L. Reatto, Phys. Rev. Lett. 54:451 (1985).
3. L. Reatto, D. Levesque, and J. J. Weiss, Phys. Rev. A 33:3451 (1986).
4. D. M. Çeperley and G. V. Chester, Phys. Rev. A 15:755 (1977).
5. L. Verlet, Phys. Rev. 165:201 (1968).
6. S. Galam and J. P. Hansen, Phys. Rev. A 14:816 (1976).
7. W. L. Slattery, G. D. Doolen, and H. E. DeWitt, Phys. Rev. A 21:2087 (1980).
8. W. L. Slattery, G. D. Doolen, and H. E. DeWitt, Phys. Rev. A 26:2255 (1982).
9. K. Ng, J. Chem. Phys. 61:2680 (1974).

SOME PROPERTIES OF A POLARIZED OCP

H. L. Helfer and R. L. McCrory

Laboratory for Laser Energetics
University of Rochester
250 East River Road
Rochester, New York 14623-1299

This paper reports on Monte Carlo calculations of properties of polarizable one-component plasmas (OCP). The purpose of this research is to supplement the classical OCP investigations by determining the thermodynamic properties of dense plasmas that are only partially degenerate. This work extends similar calculations by DeWitt and Hubbard (1976) and Totsuji and Takami (1984). An additional purpose is to determine radial distribution functions for partially degenerate plasmas; these can be used for testing theoretical methods of calculating plasma properties (cf. Rogers et. al., 1983).

To evaluate $U = -(\partial \ln \hat{Z}/\partial\beta)_V$, $\beta P = (\partial \ln \hat{Z}/\partial V)_T$, etc., we start with N pointlike ions and the expression for the partition function:

$$\hat{Z} = e^{-\beta F_{ions}} \bullet \int \frac{d^{3N}R}{V^N} e^{-\beta U_{ii}} Tr(e^{-\beta(K_e+U_{ee}+U_{ie})}) .$$

Following a procedure discussed by Ashcroft and Stroud (1978), the trace may be evaluated when the electron density fluctuations are linear in the electric potential, say, $\delta\tilde{\rho}(k) = q^2\eta(k)\tilde{\varphi}(k)$ (where the tilde signifies the Fourier transform). The Helmholtz free energy, F, can be evaluated for a given Hamiltonian, by the prescription:

$$H = H_0 + H_1 \Rightarrow F = F_0 + \int_0^1 \frac{d\lambda}{\lambda} \langle\lambda H_1\rangle_{H_0+\lambda H_1}$$

The form of the dielectric function used was calculated using the linear form of a density matrix procedure developed by March and Murray (1961); this gives:

$$q^2\eta(k) = \frac{4}{\pi a_{Bohr} k} \int_0^\infty \frac{k'}{1+\exp\{\beta[E(k')-\mu]\}} \ln \left|\frac{k+2k'}{k-2k'}\right| dk' .$$

Here μ is the chemical potential, and $k^2[\epsilon(k)-1] = q^2\eta(k)$.

This form is in agreement with the RPA dielectric function used by Totsuji and Takami. The potential closely approximates exp(-qr)/r, where q is the Thomas-Fermi wavenumber. There are additional oscillatory terms, $\propto r^{-3}$, which amount to a few percent at values of r of interest. These minor Friedel oscillation terms can provide long-term coupling of the plasma at distances greater than a few ion-sphere radii.

One gets $\hat{Z}$ in a form useful for performing a Monte Carlo calculation:

$$\hat{Z} = e^{-\beta[F_{(ideal\ ions)}+F_{(free\ electrons)}+F_{pol}]} \cdot \int \frac{d^{3N}R}{V^N} e^{-\beta U_{eff}}$$

where

$$\beta U_{eff} = \frac{1}{2}\beta \{\sum_{I\neq J}\sum Z_I Z_J \phi(|R_I - R_J|) + \sum_I Z_I^2 \lim_{\xi\to 0} [\phi(\xi) - \frac{1}{\xi}]\} ,$$

and

$$\tilde{\phi}(k) = \frac{4\pi}{k^2 + q^2\ \eta(k)} .$$

For the plasmas being considered, F_{pol} is a small second order term which can be ignored. The calculations use an Ewald sum technique for calculating $<\beta U_{eff}>$ and evaluating the pair distribution function. In effect, the plasma is represented as a cubic lattice, with N (=128) ions per cell. (For details, see Helfer et. al., 1984).

In addition to the internal energy and pressure terms one associates with non-interacting electron and ion gases, one finds excess energy and pressure terms attributable to the interactions; these are:

$$\frac{\beta U_{excess}}{N} \simeq < \frac{\beta U_{eff}}{N} > ,$$

$$\frac{\beta U_{excess}}{N} \simeq \frac{1}{3}(1 + \frac{1}{2}<qr>) < \frac{\beta U_{eff}}{N} > + \frac{1}{12} Z^2 \Gamma q^* a$$

where $<\beta U^{eff}/N>$ and qr, are quantities resulting from the Monte Carlo calculation, and q* is calculated from the limiting value of the effective two body potential when $r \Rightarrow 0$. Here, a is the ion sphere radius and $\Gamma = e^2/akT$.

Figure 1 shows the excess energy term. It has been divided into two parts: (1) the energy per ion of a reference rigid BCC lattice; and (2) the difference in energy per ion between the plasma and the BCC lattice. At large Γ the BCC lattice energy dominates. The BCC excess energy decreases from the OCP value as the density decreases. This reflects the binding energy of the electrons as they cluster around individual ions. For the plasmas studied, the difference in excess energy, plasma - lattice, also decreases with density, amounting to ~10.5 kT per ion at ion densities of 2×10^{23} cm^{-3} (for Z=1). The difference in energies is not a strong function of Γ at low densities. The very low density high-Γ models may be quite unphysical because de-ionization is not taken into account; for these models the plasma excess energy is less than that of the BCC lattice.

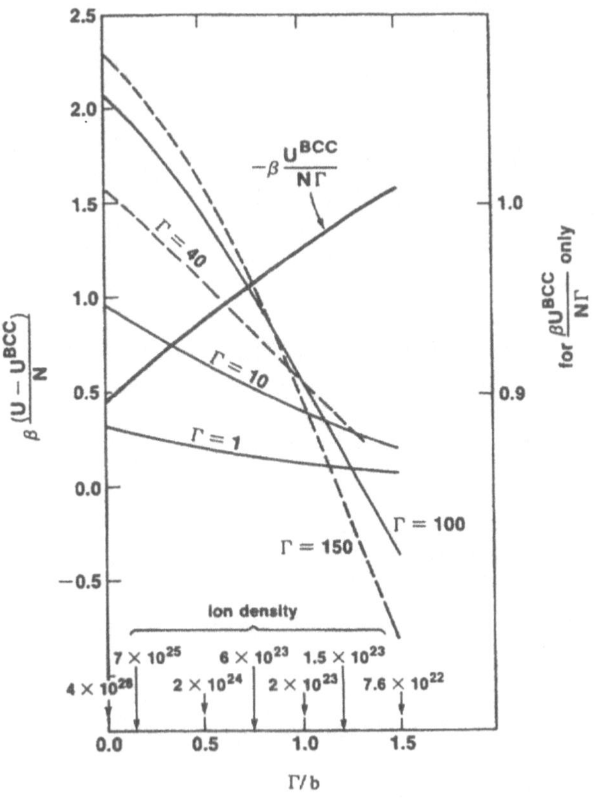

Fig. 1. Excess energies vs Γ/b ($=\Gamma kt/\mu \simeq 0.5r_s$). The heavy solid line refers to a reference BCC lattice and the scale on the right. The other curves, for constant Γ, use the scale on the left.

The calculations show the following crude approximations may be used:

$$< \frac{\beta U_{eff}}{N} > \sim - (0.9 + 0.033r_s)\Gamma Z^2 \text{ where } r_s = a/a_{Bohr} ,$$

and

$$<qr> \sim \min (\tfrac{2}{3}qa, 2) \text{ and } q^* \sim q .$$

For the OCP, the terms involving q and q* are absent in the expression for the excess pressure. These extra terms can cause the excess pressure to be more negative by up to ~20% more than in the OCP case.

The pair correlation functions show some unusual features (cf. Fig. 2). The minor oscillations beyond the first maximum first decrease in amplitude as one goes to lower densities and then increase in strength (with a phase shift) as one goes to the lowest densities studied. This behavior is seen for all the models, down to the lowest value of Γ(>10) for which the oscillations can be studied.

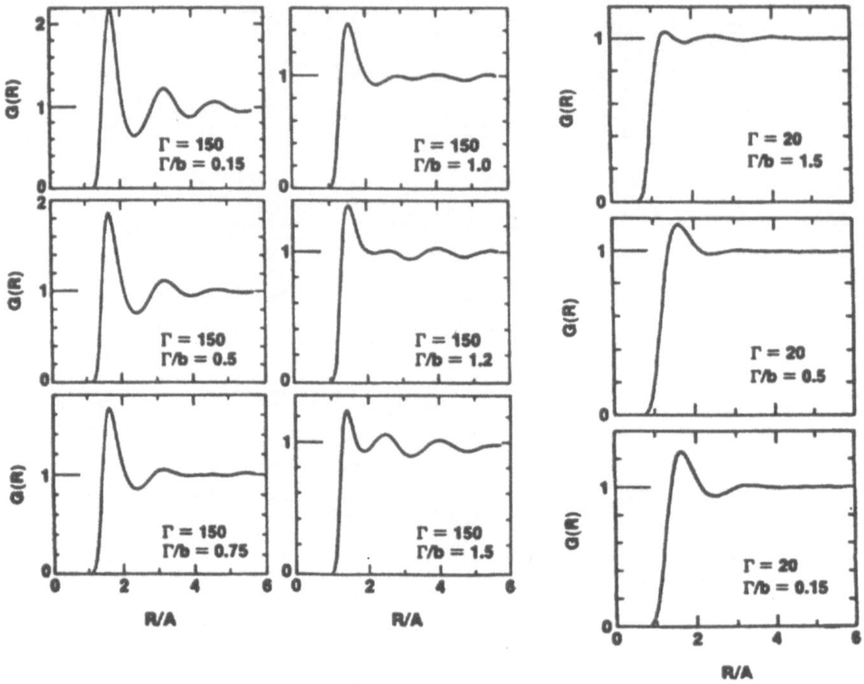

Fig. 2. Pair correlation functions vs r/a, where a is the ion sphere radius. The various curves are labeled by a density parameter (see Fig. 1).

ACKNOWLEDGMENT

This work was supported by the Laser Fusion Feasibility Project at the Laboratory for Laser Energetics which has the following sponsors: Empire State Electric Energy Research Corporation, General Electric Company, New York State Energy Research and Development Authority, Ontario Hydro, Southern California Edison Company, and the University of Rochester. Such support does not imply endorsement of the content by any of the above parties.

REFERENCES

Ashcroft, N. W., and Stroud, D., Solid State Phys., 33:2 (1978).
DeWitt, H. E., and Hubbard, W. B., Astrophys. J., 168:131 (1976)
Helfer, H. L., McCrory, R. L., and Van Horn, H. M., J. Stat. Phys., 37:577 1984).
March, N., and Murray, A., Phys. Rev. A, 120:830; Proc. Roy. Soc. Lon. Ser. A, 261:119 (1961).
Rogers, F. J., Young, D. A., DeWitt, H. E., and Ross, M., Phys. Rev. A, 28:2990 (1983).
Totsuji, H., and Tokami, K., Phys. Rev. A 30:3175 (1984).

PERTURBATION THEORY OF THE MISCIBILITY GAP IN METAL SALT SOLUTIONS

G. Chabrier

Dpt. Physique des Materiaux
Universite Claude Bernard Lyon I
69622 Villeurbanne Cedex, France

REFERENCE SYSTEM: IONIC FLUID IN A RIGID NEUTRALIZING BACKGROUND

The metal-salt solutions are regarded as being composed of N_1 positive ions of charge Z_1e and N_2 negative ions of charge Z_2e in a volume Ω; the corresponding number densities and concentrations are defined as $\rho_\alpha = N_\alpha/\Omega$ and $x_\alpha=N_\alpha/N$ $(\alpha=1,2)$, with $N=N_1+N_2$. The excess of positive charge is compensated by the conduction electrons which are assumed to provide a rigid, uniform background of charge density $e\rho_o$, ensuring overall charge neutrality:

$$\rho_1 Z_1 + \rho_2 Z_2 + \rho_o = 0 \tag{1}$$

This reference system will serve as a starting point for our perturbation expansion. In that case the ions are assumed to interact via the following potential:

$$U_{\alpha\beta}(r) = \frac{Z_\alpha Z_\beta e^2}{r} + (1-\delta_{\alpha\beta})V_o(r) \tag{2a}$$

where the short range repulsion, which acts only between oppositely charged ions, is taken to be of exponential form:

$$V_o(r) = A_{12}\exp(-\alpha_{12}r) \tag{2b}$$

The potential (2) is a simplified version of the usual Born-Huggins-Mayer potential, retaining only its essential features. We have dropped the Van der Waals dispersion terms, as well as the short range repulsion between equally charged ions, since the Coulomb repulsion is sufficiently strong to keep them apart. The limiting situations of the potential (2) are: i) x=1, i.e. the pure metal, for which we recover the one component plasma (O.C.P.) model[3,4]; ii) x=0, i.e. the pure salt, reasonably well described by a Born-Huggins-Mayer rigid ion potential, provided ion polarizability effects are not too important.[5] The pair correlation functions are calculated through the closed set of equations composed of the Ornstein-Zernicke relation and the hypernetted-chain (H.N.C.) equations.[2] The reduced excess (non ideal) internal energy $\frac{U_i^{EX}}{Nk_BT}$ and excess pressure $\frac{P_i^{EX}}{\rho k_BT}$

are calculated in terms of integrals over the static correlation functions, via the standard energy and virial equations. In the HNC approximation the excess chemical potentials μ_α^{EX} of both ionic species are also expressible in terms of the pair correlation functions.[6] Hence the excess free enthalpy and free energy per ion follow directly from:

$$\frac{G_i^{EX}}{Nk_BT} = \sum_{\alpha=1,2} x_\alpha \mu_\alpha^{EX}; \quad \frac{F_i^{EX}}{Nk_BT} = \frac{G_i^{EX}}{Nk_BT} - \frac{P_i^{EX}}{\rho k_BT} \tag{3}$$

PERTURBATION EXPANSION: RESPONSIVE BACKGROUND[7]

The total hamiltonian of the system is written now as the sum of three terms:

$$H = H_{ii} + H_{ee} + V_{ie} \tag{4}$$

H_{ii} is the hamiltonian for the ions in a neutralizing background and has been detailed in the previous section. H_{ee} is the familiar "jellium" hamiltonian for the electrons in a uniform background which exactly cancels the previous one. V_{ie} describes the ion-electron interaction minus the interaction energy of the ions with their associated background.[7,8] This term can be split in two terms:

$$V_{ie} = \frac{1}{\Omega} \sum_{\alpha=1,2} \sum_{\vec{k}}{}' v_{o\alpha}(k) \rho_{\vec{k}\alpha} \rho^*_{\vec{k}_o} + U_o \tag{5a}$$

where U_o describes the non-coulombic structure independent term:

$$U_o = \frac{1}{\Omega} \lim_{k\to o} \left[\hat{v}_{o\alpha}(k) + \frac{4\pi Z_\alpha e^2}{k_BT} \right] \rho_{\vec{k}_\alpha} \rho^*_{\vec{k}_o} = \rho_o \sum_{\alpha=1}^{2} N_\alpha \int_\Omega \left[v_{o\alpha}(r) + \frac{Z_\alpha e^2}{r} \right] d\vec{r} \tag{5b}$$

The $\rho_{\vec{k}i}$ (i=0,1,2) denote the Fourier components of the microscopic densities (0 for the electrons) and $v_{o\alpha}(k)$ is the dimensionless Fourier transform of the ion-electron pseudopotential $v_{o\alpha}(r)$. The hamiltonian specified by Eqs. (2), (4), and (5) is in fact quite general and describes a number of coulombic systems besides metal-salt solutions M_xM_{1-x}: molten salt (x=0), liquid metals (x=1), binary ionic mixture ($Z_1Z_2>0$) in which case the short range repulsive term may be omitted, binary metallic alloys which differ from the case of BIM in that the ions have a finite core radius so that the ion-electron interaction is no longer purely conlombic but must be described by a psuedopotential. For the ion electron psuedopotentials occurring in Eqns. (5a) and (5b), we have chosen the Ashcroft "empty core" form[8] for the cation, with a core diameter r_c determined at melting, from compressibility data. Much less is known about the electron-anion pseudopotential which we have chosen to be an interpolation between the Ashcroft and Shaw[9] forms,[10] i.e.:

$$\begin{aligned} v_{o2}(r) &= -\xi Z_2 e^2/r_{c_2}; \quad r<r_{c_2} \\ &= -Z_2 e^2/r; \quad r>r_{c_2} \end{aligned} \tag{6}$$

For r_{c_2} we have chosen the ionic Pauling radius. ξ will in fact be the only adjustable parameter in our perturbation theory. The special case $\xi=0$ corresponds to the "empty core" model. To this order the ionic and electronic components, neutralized by their respective uniform backgrounds, are assumed to be completely decoupled so that the Helmholtz free energy is then simply the sum of the two independent contributions:

$$F^{(0)} = F_i + F_e \tag{7}$$

F_i is given by Eqn. (3) whereas for F_e we have used the equation of state given by Richert and Ebeling[11] calculated in a wide range of density, even in the strongly coupled regime ($r_s \gg 1$). Since the two components are completely decoupled in the reference system, $\langle \rho_{\vec{k}\alpha} \rho^*_{\vec{k}o} \rangle = \langle \rho_{\vec{k}\alpha} \rangle_o \langle \rho^*_{\vec{k}o} \rangle = 0$ due to translational invariance, it follows immediately that to first order in perturbation theory the ion-electron coupling (5a) contributes the term[7]:

$$F^{(1)} = \langle v_{ie} \rangle_o = U_o = \frac{3}{r_s^3} \quad N_1 Z_1 r_c^2 + N_2 Z_2 r_{c_2}^2 \ (1 - \tfrac{2}{3}\xi) \text{ Ryd} \tag{8}$$

the second order ion-electron contribution to the free energy is calculated via linear response theory with the result

$$F^{(2)} = \int_0^1 \langle v_{ie} \rangle_\lambda d\lambda = \frac{1}{2} \sum_{\alpha\beta} (N_\alpha N_\beta)^{1/2} \frac{1}{(2\pi)^3}$$

$$\int \hat{v}_{o\alpha}(K) \hat{v}_{o\beta}(K) S_{\alpha\beta}(K) \frac{K^2}{4\pi e^2} \quad \frac{1}{\epsilon_e(k)} - 1 \quad d\vec{K}. \tag{9}$$

For the dielectric function $\epsilon_e(k)$ we have chosen the zero-temperature form proposed by Ichimaru and Utsumi, adapated to the highly correlated regime ($r_s \gg 1$). By truncating the perturbation expansion of the free energy after second order, we restrict ourselves to linear screening in the description of ion-electron coupling. This is a priori inapplicable in the regime of low metallic concentration. However, since the weight of the electronic contributions to the thermodynamic properties of a metal salt solution decreases with decreasing metal concentration x, we have used the results of second order perturbation theory throughout the entire range of concentration.

THE PHASE DIAGRAM

The total free energy of the system is now taken to be the sum:

$$F = F_i + F_e + F^{(1)} + F^{(2)} \tag{10}$$

The volume derivative yields the total pressure P, and the molar Gibbs free energy $G_m(T,P,x)$ follows from $G_m = F_m + P\Omega_m$ where F_m and Ω_m are the Helmholtz free energy and volume per mole of the solution. The corresponding excess free energy of mixing is defined in the ususal way as:

$$\Delta G_m(P,T,x) = G_m(P,T,x) - xG_m(P,T,x{=}1) - (1-x)G_m(P,T,x{=}0) \tag{11}$$

Table 1. Critical Coordinates for K_xKcl_{1-x} for several values of ξ.

ξ	Experiment	0.01	0.05	0.2	0.333
T_c(k)	1073	1250	1650	2300	2500
x_c	0.35	0.25	0.25	0.35	0.4

In practice the various contributions to F_m or G_m are calculated as functions of the variables T, x and Ω, and for fixed T and x the latter is then varied to yield zero total pressure. The corresponding excess molar volume of mixing:

$$\Delta\Omega_m(P=o,T,x) = \Omega_m(P=o,T,x)-x\Omega_m(P=o,T,x=1)-(1-x)\Omega_m(P=0,T,x=o) \quad (12)$$

is shown to be surprisingly large and negative for all x (see Ref. 7).

To zeroth order, if only the ionic and electronic contributions were included, ΔG_m should be a convex function of $x((\partial^2\Delta G_m/\partial^2x^2)_{P,T}<o)$ so that the solution would be thermodynamically unstable at all concentrations. This behavior contrasts with the case of BIM or metallic alloys where the ionic contribution always tend to stabilize the mixture.[6,12] But when the first and second order corrections due to the ion-electron coupling are added, ΔG_m gradually builds up a convex portion on the salt rich side, signaling phase separation. The concentrations of the coexisting liquid phases are determined by the usual double tangent construction. The critical coordinates T_c and x_c depend sensitively on ξ and are compared to the experimental values in the case of K_xKcl_{1-x} in Table 1, for several values of ξ. While the calculated critical temperature can be brought into agreement by an adequate choice of ξ, the corresponding critical concentration x_c is too small; this is probably a consequence of the inadequacy of linear screening theory on the salt rich side of the diagram. Typical phase diagrams calculated for K_xKcl_{1-x} and Rb_xRbBr_{1-x} are shown in Ref. 7.

REFERENCES

1. M. A. Bredig, "Molten Salt Chemistry," M. Blauder ed., Wiley Interscience, (1964).
2. G. Chabrier, J. P. Hansen, Mol. Phys. 50 5:901 (1983).
3. H. Minoo, C. Deutsch, J. P. Hansen, J. Phys. Lett., Paris 38:L191 (1977).
4. D. K. Chaturvedi, M. Rovere, G. Senatre, M. P. Tosi, Physica B 111:11 (1981).
5. D. J. Adams, J. Chem. Soc. Faraday Trans. II 72:1372 (1976).
6. J. P. Hansen, G. M. Torrie, P. Vieillefosse, Phys. Rev. A 16:2153, (1977).
7. G. Chabrier, J. P. Hansen, J. Phys. C. L751 (1985); G. Chabrier, J. P. Hansen, in press in J. Phys. C. (1986).
8. N. W. Ashcroft, D. Stroud, "Solid State Physics, Vol. 33," F. Seitz and D. Turnbull ed., Academic Press, NY (1978).
9. R. W. Shaw, Phys. Rev. 174:769 (1968).

10. G. Chabrier, G. Senatore, M. P. Tosi, Nuovo Cin., 3D:4 (1984).
11. W. Richert, W. Ebeling, Phys. Stat. Solidi B 121:633 (1984).
12. R. L. Henderson, N. W. Ashcroft, Phys. Rev. A 13:859 (1976).

CHAPTER II

PLASMA EXPERIMENTS

A HIGH-Γ, STRONGLY-COUPLED, NON-NEUTRAL ION PLASMA†

L.R. Brewer, J.D. Prestage*, J.J. Bollinger, and D.J. Wineland

Time and Frequency Division
National Bureau of Standards
Boulder, Co. 80303

INTRODUCTION

We have produced a strongly coupled non-neutral $^9Be^+$ ion plasma with a coupling parameter of approximately 100 or greater. The ions were spatially confined by a Penning trap [Penning 1936, Dehmelt 1967, 1969, Wineland et al. 1983] and cooled and compressed using laser cooling [Itano and Wineland 1982]. Measurements were made of the plasma shape, rotation frequency, density and temperature. In this paper we describe the experimental confinement geometry, the laser cooling of ions and the experimental data which are compared with theoretical predictions. Future experiments to measure the plasma static structure function, measure ion diffusion, and improve the temperature measurement are discussed.

CONFINEMENT GEOMETRY

The Penning trap, shown in Fig. 1, is composed of two "endcap" electrodes and a "ring" electrode which are biased with respect to each other by a d.c. electric potential. The electrode surfaces are approximate hyperboloids of revolution. The symmetry axis of the trap is parallel with a static magnetic field. The configuration is similar to that used by the group at the University of California at San Diego (UCSD) [Malmberg and deGrassie 1975]. The hyperboloidal shaped electrodes give rise to an applied trap potential

$$\varphi_T = \frac{m\omega_z^2}{4q}(2z^2 - r^2) , \tag{1}$$

where m is the ion mass and the axial frequency ω_z is defined by the equation

$$\omega_z^2 = \frac{4qV_0}{m(r_0{}^2 + 2z_0{}^2)} . \tag{2}$$

V_0 is the electric potential applied between the ring and endcaps and r_0 and z_0 are the characteristic trap dimensions as shown in Fig. 1. There are three principal motions of a single ion in this trap. The potential φ_T is quadratic in z, and this gives rise to simple harmonic motion of the ion at frequency ω_z along the z axis of the trap. In the radial direction the ions are confined by a magnetic field, and the ion motion is a superposition of two circular motions, the cyclotron and magnetron motions. The cyclotron motion is shifted somewhat in frequency from its value in a pure magnetic field by the radial electric field [Dehmelt 1967, 1969, Wineland et al. 1983]. The magnetron or rotation motion is a circular drift of the guiding center of the cyclotron motion due to the $\mathbf{E} \times \mathbf{B}$ forces of the trap. These motions are shown in Fig. 2.

For the work discussed in this paper, typical trap parameters are an electric potential V_0 of 2 volts, a B field of 1.4 tesla and trap dimensions of $r_0 = 0.417$ cm and $z_0 = 0.254$ cm. For these parameters the magnetron frequency for a single $^9Be^+$ ion is 15.1 kHz, the cyclotron frequency is 2.38 MHz and the axial frequency is 267 kHz.

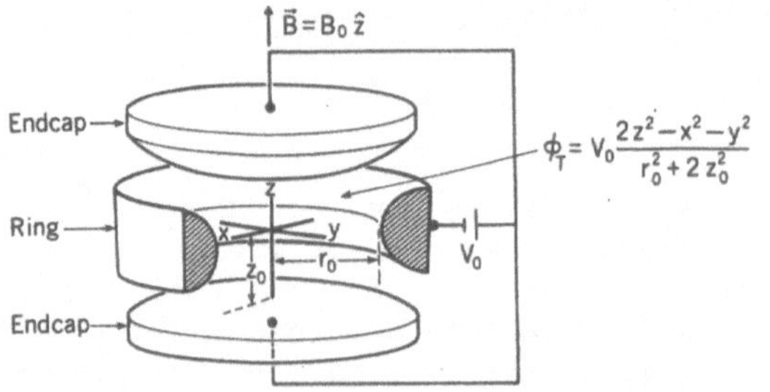

Fig. 1. The Penning trap consists of two endcaps and one ring electrode which are biased with respect to each other by a potential V_0. The hyperbolic surfaces of the electrodes produce a quadrupole potential which confines the ions in the z direction. The static B field confines the ions in the radial direction.

THERMAL EQUILIBRIUM STATE

For a collection of ions in the trap, the resulting single species plasma is assumed to be in thermal equilibrium because of Coulomb collisions. If the plasma is in thermal equilibrium there is no shear in the plasma and the plasma rotates as a rigid body. The density is constant up to the edge of the plasma where it drops off in a distance that is characterized by the Debye length [Prasad and O'Neil 1979].

The single particle distribution function for a magnetically confined non-neutral ion plasma has been given by Davidson [Davidson 1974] and by Malmberg and O'Neil [Malmberg and O'Neil 1977]. For positive ions we have

$$f(r,z,\mathbf{v}) = n(r,z)[m/(2\pi k_B T)]^{3/2}\exp\{-(m/2k_B T)(\mathbf{v} + \omega r\hat{\theta})^2\} \qquad (3)$$

$$n(r,z) = n_0\exp\{-(1/k_B T)[q\varphi(r,z) + (m\omega/2)(\Omega - \omega)r^2]\}. \qquad (4)$$

Here φ is the total electrostatic potential, ω can be identified as the rotation frequency of the plasma, $\Omega = qB_0/mc$ is the cyclotron frequency, and n_0 is the density of the ions at the center of the trap.

In the T = 0 limit, in order that f and n remain finite we find that

$$q\varphi(r,z) + (m\omega/2)(\Omega - \omega)r^2 \to 0 \qquad (5)$$

for r,z inside the plasma. This equation tells us that the electrostatic potential is independent of z and along with Poisson's equation, that the density must be constant throughout the plasma and equal to n_0. From Poisson's equation and Eq. 5, n_0 is given by

$$n_0 = \frac{m\omega(\Omega-\omega)}{2\pi q^2} \quad . \qquad (6)$$

The distribution function predicts simple shapes for the plasma in the limits that T = 0 and the trap dimensions are large compared to the plasma dimensions. The potential of the plasma is given by the expression

$$\varphi = \varphi_I + \varphi_T + \varphi_{ind}, \qquad (7)$$

where φ_I is the Coulomb potential of the ions in the absence of the trap walls and φ_{ind} is the potential due to the charges induced on the trap electrodes. If the electrode spacing is large compared to the dimensions of the plasma we can neglect φ_{ind} and solve for the ion potential. From Eqs. 1, 5, and 7 we find that

$$\varphi_I = \frac{-m}{2q}\{\omega(\Omega - \omega) - \omega_z{}^2/2\}r^2 - \frac{m\omega_z{}^2z^2}{2q} \qquad (8)$$

$$= -2/3\ \pi q n_0(\alpha r^2 + \beta z^2) \qquad (9)$$

where Eq. 9 is used to define α and β. In general Eq. 9 represents the potential inside a uniformly charged spheroid. For example for α=β=1, φ_I is the potential inside a uniformly charged sphere.

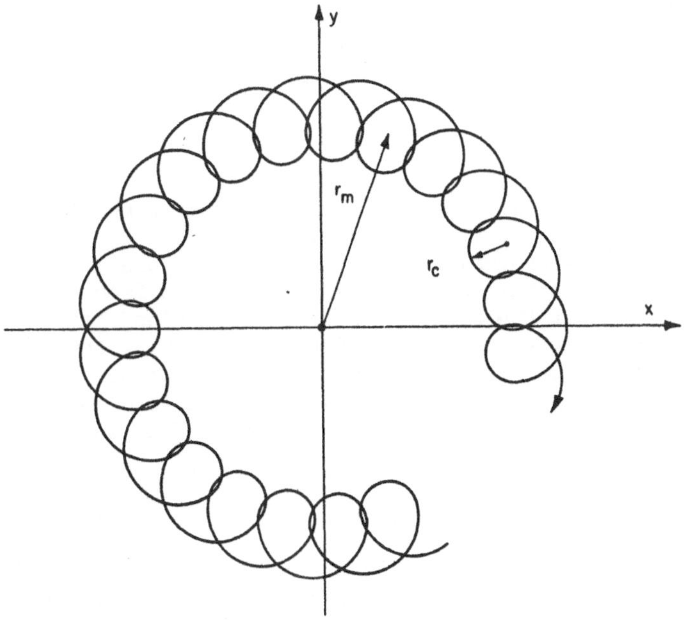

Fig. 2. The orbit of a single ion in the x-y plane consists of two circular motions. r_c is the radius of the cyclotron motion and r_m is the radius of the magnetron motion. The figure sketches the ion orbit for the case $r_c<r_m$.

ONE COMPONENT PLASMA

In the frame of reference rotating about the trap axis with frequency ω, the ion plasma behaves like a neutral one component plasma. That is, the positively charged ions behave as if thay are moving in a uniform negatively charged background. In particular, Malmberg and O'Neil [Malmberg and O'Neil 1977] have shown that the static properties of magnetically confined non-neutral plasma are the same as those for a one component plasma.

A one component plasma [Ichimaru 1982] is composed of a single species of charge imbedded in a uniform background of neutralizing charge. The one component plasma is characterized by a coupling constant

$$\Gamma = q^2/ak_BT \qquad (10)$$

which is the ratio of the nearest neighbor Coulomb energy to the thermal energy of a particle. The Wigner-Seitz radius a is defined by

$$a^3 = 3/(4\pi n) \qquad (11)$$

where n is the particle density. When $\Gamma > 1$ the plasma is said to be strongly coupled. When $\Gamma > 2$ the plasma should exhibit liquid like behavior and the particles should exhibit short range order. Slatterly, Doolen,and DeWitt [Slatterly et al. 1980] have predicted that at $\Gamma = 178$ the plasma undergoes a phase transition to become a crystal like structure.

Because the transition is of the first order, the plasma may remain in a metastable fluid like state when it is supercooled below the transition temperature. Ichimaru and Tanaka have investigated the supercooled one component plasma and presented evidence for a possible dynamic glass transition at a value of $\Gamma \approx 1{,}000$ [Ichimaru and Tanaka 1986]. (The correspondence between the magnetically confined non-neutral plasma and the one component plasma rigorously exists only for static properties. The possibility of a dynamic glass transition in a one component plasma is therefore a suggestion of what might happen in the magnetically confined non-neutral plasma.)

LASER COOLING

The technique of laser cooling utilizes the resonant scattering of laser light by atomic particles. By directing a laser beam at the plasma one can decrease the thermal velocity of the particle in a direction opposite to the laser beam. The laser is tuned in frequency to the red, or low frequency side of the atomic "cooling transition" (typically an electric dipole transition). Some of the ions moving toward the laser will be Doppler shifted into resonance and absorb a photon. Ions moving away from the laser will be Doppler shifted away from resonance and the absorption rate will correspondingly decrease. When an ion absorbs a photon its velocity is decreased by an amount

$$\Delta v = \hbar k/m \qquad (12)$$

due to momentum conservation. Here Δv is the change of the ion's velocity in the direction of the laser beam, $k = 2\pi/\lambda$ where λ is the wavelength of the cooling radiation, and m is the mass of the ion. The ion spontaneously re-emits the photon in a random direction (for low laser intensities where the stimulated emission rate is small), which when averaged over many scattering events does not change the momentum of the ion. The net effect then is that for each photon scattering event the ion's average velocity is reduced by the amount shown in Eq. 12. To cool an atom from 300 K to mK temperatures takes typically 10^4 scattering events. For this experiment the cooling limit, due to photon recoil effects [Wineland and Itano 1979], for a given transition is given by a temperature equal to $\hbar\gamma/2k_B$, where γ is the radiative linewidth of the atomic transition. For a linewidth γ of $2\pi\cdot 19.4$ MHz, which was the natural linewidth of the cooling transition in our experiment, the minimum obtainable temperature is 0.5 mK.

It is interesting to see how the laser affects the angular momentum of the plasma [Wineland et al. 1985]. The z component of the canonical angular momentum for a single particle is

$$l_z = mv_\Theta r + qBr^2/2c \tag{13}$$

where v_Θ is the ion's azimuthal component of velocity and r is the ion's radial cylindrical coordinate. The two terms represent the mechanical angular momentum and the field angular momentum. The total z component of the angular momentum is

$$L_z = \int dz \int 2\pi r dr \int d^3\mathbf{v} f(r,z,\mathbf{v}) l_z \tag{14}$$

$$= m(\Omega/2 - \omega)N\langle r^2\rangle \ . \tag{15}$$

Eq. 15 tells us that the total angular momentum about the z axis is proportional to the mean of the square of the radius of the plasma. Here N is the number of ions and f is the distribution function. For our experimental conditions Ω is usually much larger than ω [O'Neil 1980] so that

$$L_z \approx m(\Omega/2)N\langle r^2\rangle \geq 0 \ . \tag{16}$$

If the cooling laser is directed at the side of the plasma which is receding from the laser (due to the plasma rotation), angular momentum is removed from the plasma and the radius of the plasma must decrease. As the radius decreases, the density of the plasma increases. The limiting density, known as the Brillouin density [Davidson 1974], occurs when the rotation frequency $\omega = \Omega/2$. The Brillouin density is given by

$$n = \frac{m\Omega^2}{8\pi q^2} \ . \tag{17}$$

Collisions with background gas particles increase the angular momentum of the plasma. This is one of the effects that could limit the compression of the plasma. Axial asymmetry of the trap is also a limiting effect. The plasma group at UCSD has observed that the axial asymmetry of their cylindrical traps plays an important role in the electron confinement time [Driscoll et al. 1986]. It is also expected to be a limiting effect in the experiments reported here [Wineland et al. 1985].

At a magnetic field of 1.4 T, the Brillouin density for $^9Be^+$ is $5.9\cdot 10^8$ ions/cm^3. This density and the 0.5 mK temperature limit gives a theoretical limit on the coupling of $\Gamma \approx 4{,}500$. Consequently the possibility of obtaining couplings large enough to observe a liquid-solid phase transition looks promising. If the cooling or quench can be done rapidly enough, it appears possible to investigate the existence of a dynamic glass transition at $\Gamma \approx 1000$. Currently the laser cooling technique is capable of reducing the temperature of a cloud of ions from room temperature to less than 10 mK ($\Gamma \approx 100$) in a few seconds. This cooling rate, if continued to lower temperatures, compares favorably with the minimum nucleation times ranging from 80 to $8\cdot 10^5$ seconds as estimated by Ichimaru and Tanaka [Ichimaru and Tanaka 1986] for a $^9Be^+$ plasma with the above density.

The experimental configuration reported in this paper was similar to that reported by Bollinger and Wineland [Bollinger and Wineland 1984] where plasma coupling parameters as high as 10 were achieved. In their paper they suggested that by cooling the plasma in directions both perpendicular and parallel to the trap B field one could achieve lower temperatures and higher coupling. The results of this experiment are reported below.

Excitation Scheme

The excitation scheme is shown in Fig. 3. The cooling laser drives $^9Be^+$ ions between the $2s^2S_{1/2}$, $m_j = +1/2$ and the $2p^2P_{3/2}$, $m_j = +3/2$ states. Ions are optically pumped into the $2s^2S_{1/2}$ $m_j = +1/2$ state with 94 % efficiency [Wineland et al. 1984]. That is, for laser intensity below saturation, 94 % of the ions reside in the $2s^2S_{1/2}$ $m_j = 1/2$ state. The cooling laser has a wavelength of 313 nm and a power of approximately 50 μW. Resonance fluorescence (i.e. the scattered light) from this transition is detected in a photomultiplier tube. A second laser drives ion population from the $2s^2S_{1/2}$, $m_j = +1/2$ state to the $2p^2P_{3/2}$, $m_j = -1/2$ state where the ions decay with 2/3 probability to the $2s^2S_{1/2}$, $m_j = -1/2$ state causing a decrease in the observed fluorescence from the ions. The power of this "probe laser" is « 1 μW. Fig. 4 shows the resonance line shape when the probe laser is scanned through the transition. The probe laser is used to measure the shape of the plasma, its rotation frequency, density, number of ions, and temperature as described below.

Experimental Apparatus

The experimental apparatus is shown in Fig. 5. The cooling laser passes through a 50% beam splitter. Upon exiting the splitter one beam enters the plasma perpendicular to B and the other beam (diagonal cooling beam) enters between the ring and one endcap at a angle of 55 degrees with respect to B. The probe beam passes through a telescope which is used to precisely translate the beam spatially. Because the diagonal cooling beam scatters so much light from the steering mirrors it is chopped at 1 kHz and resonance fluorescence from the perpendicular cooling beam is detected only when the diagonal beam is off. The B field strength is 1.4 tesla. The vacuum in the Penning trap is approximately 10^{-8} Pa (133 Pa = 1 torr) allowing the ions to be trapped for many hours.

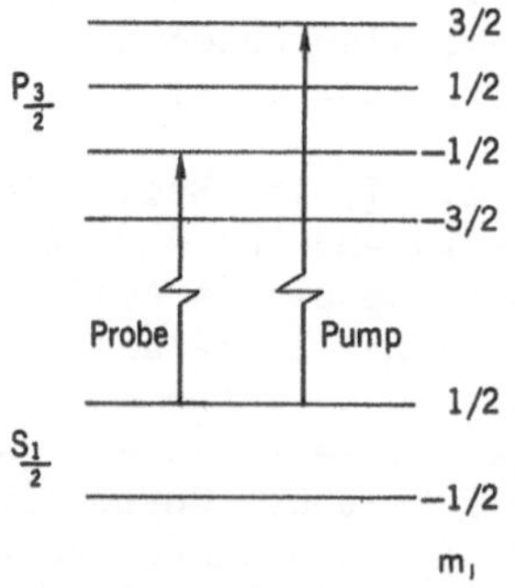

Fig. 3. The excitation scheme for the n=2 level of the $^9Be^+$ ion showing the laser cooling (pump) and depopulation (probe) transitions. Hyperfine structure has been neglected.

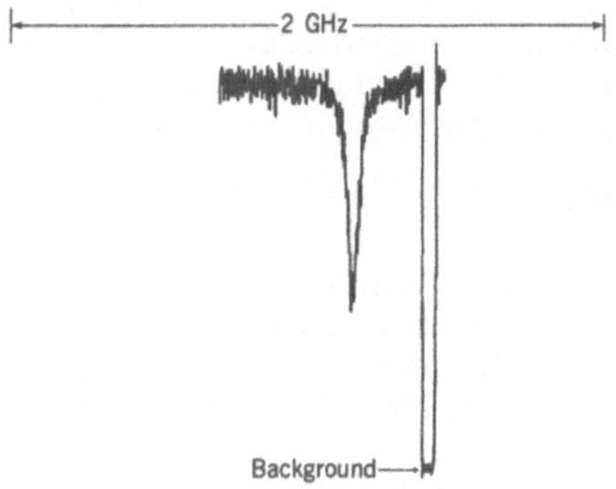

Fig. 4. The ion resonance fluorescence as a function of probe laser frequency. The bottom of the right most feature is the background signal.

Diagnostics

The depopulation signal is observable only when the probe beam intersects the plasma. This is used to determine the shape of the plasma. Spheroidal plasma shapes (with symmetry axis along z (Fig. 1)) with dimensions ranging from 200 μm to 500 μm were measured. A spheroid is the volume obtained when an ellipse is rotated about one of its axes.

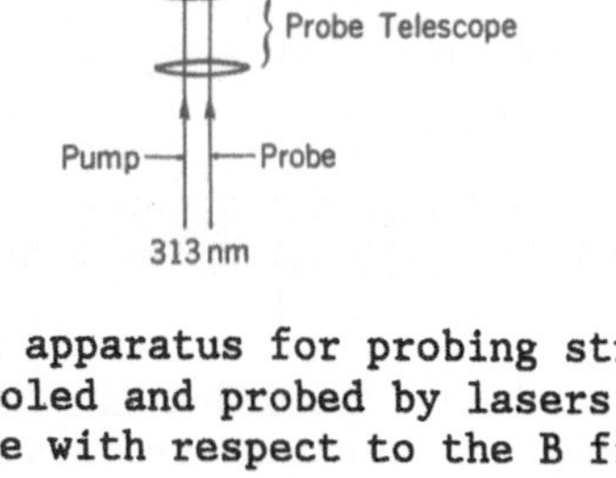

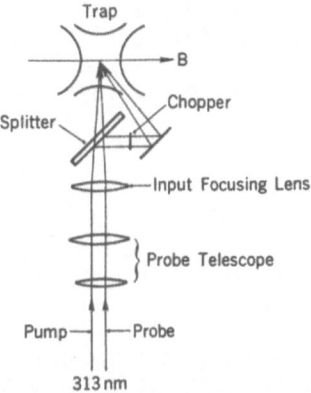

Fig. 5. The experimental apparatus for probing strongly coupled plasmas. The plasma is cooled and probed by lasers both perpendicular and at a 55 degree angle with respect to the B field.

The probe depopulation signal shifts in frequency as a function of the radial distance from the trap axis due to the Doppler shift caused by the rotation of the plasma. The rotation frequency

$$\omega/2\pi = (\Delta\nu/\Delta y)(\lambda/2\pi) \tag{18}$$

is calculated from the frequency shift $\Delta\nu$ when the probe laser position is moved by an amount Δy. The density was determined from the zero temperature formula Eq. 6. The number of ions is given by the volume of the spheroidal plasma times the density.

The temperature of the plasma can be measured in directions both perpendicular and parallel to the magnetic field. The temperature in the perpendicular direction is measured by pointing the probe laser at the plasma in the direction perpendicular to the B field. This laser is scanned in frequency and the full width half maximum of the unsaturated depopulation transition is measured. The lineshape is a Voigt profile whose width is composed of the natural linewidth, the Doppler width, and a width due to the convolution of the laser spot size with the plasma rotation. The Doppler width can be deconvoluted from the Voigt width giving the temperature of the plasma. The probe laser can also be pointed at the plasma at an angle of 55 degrees with respect to the B field. The full width at half maximum of this resonance contains Doppler widths from temperatures in both perpendicular and parallel directions to the magnetic field. With the perpendicular temperature from the previous measurement the parallel temperature can be determined. Table 1 summarizes the measurements on seven ion clouds.

Table 1. The experimental data. The error convention is as follows - 1.8(10) = 1.8 ± 1.0. V_0 is the trap voltage in V, 2Z and 2R are the axial and radial extent of the plasma in μm, n is the density of the plasma in units of $10^7/\text{cm}^3$, $\omega/2\pi$ is the plasma rotation in kHz, $T_\parallel$ and $T_\perp$ are the parallel and perpendicular temperatures of the plasma in mK, and Γ is the coupling parameter for the plasma based on $T_\parallel$.

V_0	2Z	2R	n	#IONS	$\omega/2\pi$	$T_\perp$	$T_\parallel$	Γ
2	130(30)	450(30)	2.4(6)	330(170)	25(6)	2.3(5)	10(5)	80
2	150 .	450 .	1.9(5)	300(150)	20(5)	1.8(10)	7.4(70)	100
2	130 .	480 .	2.2(6)	350(170)	23(6)	.9(15)	8.9(60)	90
2	160 .	450 .	2.6(7)	450(230)	27(7)	2.4(10)	6(6)	130
2	160 .	260 .	2.8(7)	150(80)	30(7)	2.7(30)	2.4(60)	340
4	80 .	390 .	3.9(10)	250(120)	40(10)	2.9(20)	20(12)	50
1	190 .	360 .	1.2(3)	150(80)	13(3)	2.9(10)	4.7(60)	130

DISCUSSION OF THE DATA AND CONCLUSIONS

The plasmas are oblate spheroids of revolution. The Brillouin density at a field strength of 1.4 tesla is 5.9×10^8 ions/cm^3. The measured densities were typically 2×10^7 ions/cm^3. This is a factor of 30 less than the theoretical limit. The mechanism which limits the density is not well understood but it may have to do with axial asymmetries in the Penning trap [O'Neil 1980, Wineland et al. 1985].

The theoretical cooling limit discussed earlier is 0.5 mK. While the uncertainty in the temperature measurement is large enough to include this limit in some cases, the temperatures measured were consistently higher by about an order of magnitude. The 0.5 mK limit was derived for the case of a single ion in a trap. Recently it has been shown [Itano 1986] that for a cloud of ions, the temperature limit depends on the distance the cooling laser is from the center of the cloud, the rotation frequency, and the saturation of the cooling transition. These factors could account for the temperatures we measured. With some small changes in the way the laser cooling is done we should be able to reach the 0.5 mK limit.

The largest coupling parameter measured was $\Gamma \approx 340$. The uncertainty in this measurement was large due to the uncertainty in the temperature measurement, which in this case was 2.4(60) mK. This temperature uncertainty results in a range of values for Γ of 100 to a maximum of 2,000 due to the theoretical cooling limit. This coupling may be in the range where we would expect the plasma to be crystalline.

The lowest temperatures were measured on plasmas of several hundred ions. This can not be truly called a three dimensional plasma. Since surface effects in the ion clouds may be important in our experiment the results are probably best compared to a theory which is somewhere between a plasma theory and a theory for ion clusters.

BRAGG SCATTERING FROM A STRONGLY COUPLED PLASMA

Slatterly, Doolen, and DeWitt [Slatterly et al. 1980] have derived expressions for the pair distribution function and the static structure function. The static structure function is the spatial Fourier transform of the pair distribution function and is what one expects to see in the diffraction pattern resulting from the scattering of coherent light from the ions. For low coupling parameter Γ the function is fairly flat but for $\Gamma \sim 100$ one sees sharp peaks in the amplitude of the structure function due to short range order. It should be possible to measure S(q) directly and compare this result to the calculations of Slatterly, Doolen and DeWitt.

An experimental apparatus to observe Bragg scattering which is currently under construction is shown in Fig. 6. Light from the cooling laser is scattered by the plasma and produces an interference pattern. This pattern is detected by a photon counting imaging tube.

For the densities we have measured the first interference fringe should occur at an angle of 0.6 degrees. We expect that the total count rate into the detector should be on the order of 100 counts/s. Therefore the suppression of background and scattered light into the detector will be of primary importance.

TAGGED ION DIFFUSION

A measurement of the ion diffusion may tell us whether the plasma is solid or liquid. Some experiments have observed crystallization in two dimensional [Grimes and Adams 1979] and solid state systems [Rosenbaum et al. 1985]. Wuerker et al. [Wuerker et al. 1959] observed crystallization of aluminum particles in a Paul trap. A possible technique for measuring ion diffusion in our experiment is as follows. The probe beam will be tuned on resonance and then pulsed on for a short period of time thereby "tagging" a group of ions. If the plasma is liquid, the tagged ions will diffuse between the spatially separated probe and cooling lasers and a depopulation signal will be observed at some time after the probe laser pulse. If the plasma is solid no tagged ion diffusion occurs and no signal will be observed.

IMPROVED TEMPERATURE MEASUREMENT

One difficulty with the present temperature measurement technique is that the natural linewidth of the probing transition ultimately limits the sensitivity of the measurement. Stimulated resonant Raman transitions, as for example studied by Thomas et al. [Thomas et al. 1982], avoid these difficulties. The natural linewidth of these nonlinear transitions is equal to the natural linewidth of the ground states which can be extremely small. If the angle between the two Raman beams is appropriately chosen the spectrum contains information about the velocity distribution of the plasma and is not affected by the upper state linewidth [Wineland 1984].

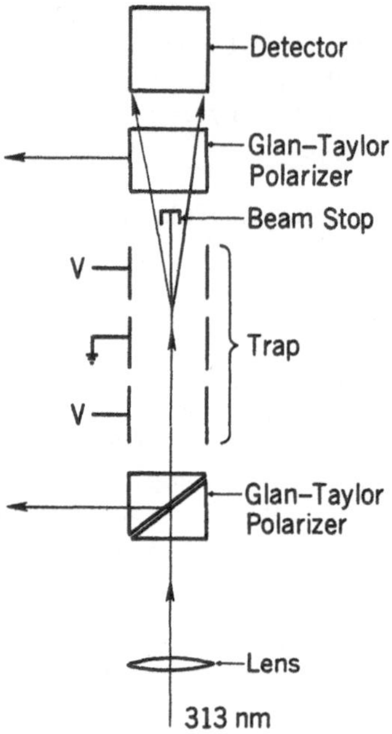

Fig. 6. The proposed apparatus for detecting the Bragg interference pattern. The probe beam is collinear with the B field along the symmetry axis of the trap. The Glan-Taylor polarizers are crossed to suppress light which does not come from the ions.

CONCLUSION

In this paper we have discussed the measurement of the temperature, density, rotation frequency, and shape of a $^9Be^+$ ion plasma. Temperatures as low as 2 mK were measured. This, along with a measured density of $3 \cdot 10^7$ cm^{-3} corresponds to a coupling parameter of $\Gamma = 340$. With an improved, highly axially symmetric trap operating at high magnetic fields we hope to be able to reach even lower temperatures and higher densities. This should result in values of the coupling parameter Γ that are much higher than the value predicted to observe crystallization in a one component plasma.

The technique of Bragg scattering resonant coherent laser light from the plasma makes possible a measurement of the static structure function. This measurement can be compared to the theoretically predicted value for a one component plasma. A measurement of the ion diffusion in the plasma may allow us to determine if the plasma is a liquid or a solid. Finally a measurement of the plasma velocity distribution using a stimulated Raman transition should improve the temperature measurement.

ACKNOWLEDGEMENTS

We gratefully acknowledge funding from the Office of Naval Research and the Air Force Office of Scientific Research. We thank W.M. Itano for his computer software and F. Walls, J. Cooper, W. Itano, and C. Weimer for carefully reading the manuscript.

*Present address: JPL, 298-104, 4800 Oak Grove Dr. Pasadena, CA 91109.

REFERENCES

Bollinger, J.J. and Wineland, D.J., 1984, Strongly coupled nonneutral ion plasma, Phys. Rev. Lett., 53:348.

Davidson, R.C. 1974, "Theory of Nonneutral Plasmas," W.A. Benjamin Inc., Reading, Massachusetts.

Dehmelt, H.G., 1967, "Radiofrequency spectroscopy of stored ions I: storage," in Advances in Atomic and Molecular Physics, 3:53.

Dehmelt, H.G., 1969, "Radiofrequency spectroscopy of stored ions II: spectroscopy," in Advances in Atomic and Molecular Physics, 5:109.

Driscoll, C.F., Fine, K.S., and Malmberg, J.H., 1986, Reduction of radial losses in a pure electron plasma, Phys. Fluids, 29:2015.

Grimes, G.C. and Adams, G., 1979, Evidence for a liquid-to-crystal phase transition in a classical, two-dimensional sheet of electrons, Phys. Rev. Lett., 42:795.

Ichimaru, S., 1982, Strongly coupled plasmas: high density classical plasmas and degenerate electron liquids, Rev. Mod. Phys., 54:1017.

Ichimaru, S. and Tanaka, S., 1986, Generalized viscoelastic theory of the glass transition for strongly coupled, classical, one-component plasmas, Phys. Rev. Lett., 56:2815.

Itano, W.M. and Wineland, D.J., 1982, Laser cooling of ions stored in harmonic and Penning traps, Phys. Rev. A, 25:35.

Itano, W.M, 1986, private communication.

Malmberg, J.H. and deGrassie, J.S., 1975, Properties of a nonneutral plasma, Phys. Rev. Lett., 35:577.

Malmberg, J.H., and O'Neil, T.M., 1977, Pure electron plasma, liquid, and crystal, Phys. Rev. Lett., 39:1333.

Malmberg, J.H., Driscoll, C.F., and White, W.D., 1982, Experiments with pure electron plasmas, Physica Scripta, T2:288.

O'Neil, T.M., 1980, Nonneutral plasmas have exceptional confinement properties, Comments Plasma Phys. Cont. Fusion, 5:231.

Penning, F.M., 1936, The glow discharge at low pressure between coaxial cylinders in a axial magnetic field, Physica, 3:873.

Prasad, S.A. and O'Neil, T.M., 1979, Finite length thermal equilibria of a pure electron plasma column, Phys. Fluids, 22:278.

Rosenbaum, T.F., Field, S. B., Nelson, D.A., and Littlewood, P.B., 1985, Magnetic field localization transition in HgCdTe, Phys. Rev. Lett., 54:241.

Slatterly, W.L., Doolen, G.D., DeWitt, H.E., 1980, Improved equation of state for the classical one-component plasma, Phys. Rev. A, 21:2087.

Thomas, J.E., Hemmer, P.R., Ezekiel, S., Leiby, C.C.Jr., Picard, R.H., Willis, C.R., 1982, Observation of Ramsey fringes using a stimulated, resonance Raman transition in a sodium atomic beam, Phys. Rev. Lett., 48:867.

Wineland, D.J. and Itano, W.M., 1979, Laser cooling of atoms, Phys. Rev. A, 20:1521.
Wineland, D.J., Itano, W.M., and Van Dyck, R.S. Jr., 1983, High resolution spectroscopy of stored ions, in: "Advances in Atomic and Molecular Physics," B. Bederson and D. Bates eds., Academic Press, New York.
Wineland, D.J., 1984, Trapped ions, laser cooling, and better clocks, Science, 226:395.
Wineland, D.J., Itano, W.M., Bergquist, J.C., Bollinger, J.J., and Prestage, J.D., 1984, Spectroscopy of Stored Atomic Ions, in "Atomic Physics 9," R.S. Van Dyck Jr. and E.N. Forstan eds., World Scientific, Singapore.
Wineland, D.J., Bollinger, J.J., Itano, W.M., and Prestage, J.D., 1985, Angular momentum of trapped atomic particles, J. Opt. Soc. Am. B, 2:1721.
Wuerker, R.F., Shelton, H., and Langmuir, R.V., 1959, Electrodynamic containment of charged particles, J. Appl. Phys., 30:342.

LASER SCATTERING MEASUREMENTS OF THERMAL ENTROPY AND ION-ACOUSTIC FLUCTUATIONS IN COLLISION-DOMINATED PLASMAS

Andrew N. Mostovych
U.S. Naval Research Laboratory

Yi Quang Zhang and Alan W. DeSilva
University of Maryland

Laser scattering is an ideal diagnostic tool for studying the kinetic properties of plasmas because the power spectrum of scattered light is directly proportional to the dynamic spectral density function $S(\tilde{k},\omega)$ of the plasma. Furthermore, $S(\tilde{k},\omega)$ is the Fourier transform of the spatial two-particle correlation function. As a result, scattering experiments, in addition to measuring the density fluctuation spectrum of the plasma, also obtain valuable information about pair correlations in the plasma.

In weakly coupled plasmas ($\Gamma \ll 1$) laser scattering has been routinely used to investigate collective plasma modes, to observe instabilities, and to measure electron temperatures[1]. While extremely useful for these purposes in the weakly coupled regime, laser scattering, also has the potential for producing the most direct experimental measurement of correlations and coupling properties in strongly coupled plasmas. In practice, this has been rather difficult because the strongly coupled plasmas produced, to date, in the laboratory have been either too optically thick to permit scattering diagnostics or so cold and dense that absorption of the probing laser would perturb the plasma.

In this paper we present laser scattering measurements of thermal ion-acoustic and entropy fluctuations from moderately coupled ($\Gamma \approx .05$), highly collisional Argon and Helium plasmas.† The plasma conditions ($n = 10^{17}cm^{-3}$, $T_e = 2eV$) were such that it was possible to circumvent the opacity problems of a strongly coupled plasma while still sampling a regime where the standard weakly coupled plasma approximations (Vlasov theory; $1/n\lambda_D^3 \ll 1$) are of dubious validity. The efforts by several authors [2-7] to generalize the kinetic theory for this parameter regime have produced differing predictions for which the current experimental data [8-10] are insufficiently precise to test their validity. In this experiment the measured spectra differed substantially from spectra observed in collisionless plasmas where $1/n\lambda_D^3 \ll 1$. Fluctuations at the ion-acoustic frequency were strongly enhanced and the width of the resonance was significantly narrowed in comparison to the colisionless case. Also, strong fluctuations around zero frequency, due to non-propagating entropy fluctuations, became visible for the first time in a plasma. Finally, the plasmas in this experiment were fully diagnosed by independent spectroscopic, interferometric, and probe

†The reader is refered to Refs. 14,15 for more details concerning this work.

measurements. This allowed a complete characterization of the plasma equilibrium, as well as faithful comparison to theoretical predictions without resorting to parameter fitting schemes.

Ion-acoustic fluctuations were selected because of their high sensitivity to the effects of collisions, given their relatively low frequency. In particular, the degree to which the scattering spectra can be expected to be altered by the presence of collisions is determined by the ratio (ν_{ii}/kC_s) of the ion-ion collision frequency to the ion-acoustic frequency [where $k=4\pi/\lambda_o \sin(\theta/2)$ is the fluctuation wave number and $C_s=(\gamma kT/m)^{1/2}$ is the ion-acoustic velocity, θ represents the scattering angle, and m is ion mass]. If, $\nu_{ii}/kC_s \geq 1$ collisional effects are important and for given plasma conditions they can be maximized by a choice of small wave numbers. This is best seen by examining a sample calculation (Fig. 1) of the ion-acoustic fluctuation spectra due to Debois and Gilinsky for several values of the normalized collision frequency (ν_{ii}/kC_s) . It is clear that the collision frequency must be at least several times the ion-acoustic frequency for a discernible effect. In this work collisional effects were maximized by the use of a long-wavelength (10.6 μm) CO_2 laser and small scattering angles (4° - 9°). This choice of scattering parameters was, in fact, ideally suited to test the validity of the various theories, since in this parameter regime the predicted ion-acoustic fluctuation spectra vary substantially from theory to theory, as is demonstrated by the calculations for an argon plasma in Fig. 2.

The curve due to Salpeter[11], the standard Vlasov result, was put in for comparison. In equilibrium where the electron and ion temperatures are equal it predicts a very broad quasi-resonance due to strong Landau damping. The theory of Dubois and Gilinsky corresponds to a solution of the Balescu-Lenard equation in the limit of a collision dominated plasma. It predicts an enhanced ion peak at the ion-acoustic frequency and a second peak at zero frequency corresponding to entropy fluctuations. It is interesting that the intensity, width, and position of these peaks are very different from the BGK calculations which also predict the same peaks. If the isothermal approximation is made in the BGK calculations then the entropy fluctuation peak totally disappears. Physically, the entropy fluctuations can be understood as non-propagating density fluctuations which have corresponding fluctuations in temperature, constrained to keep the pressure constant.

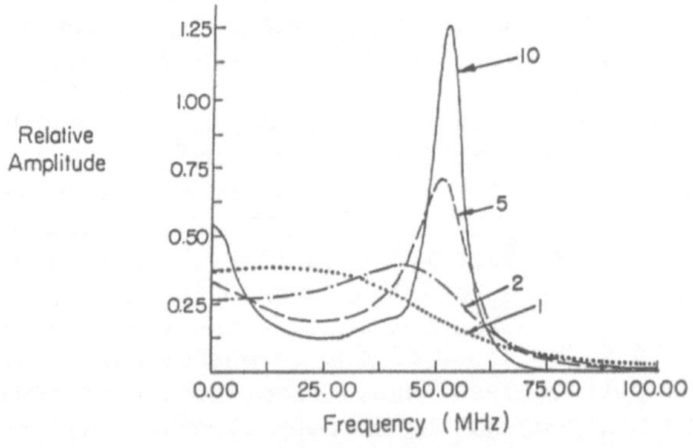

Fig. 1 Shape of the ion-acoustic resonance, as predicted by the Kivelson-Dubois solution of the Balescu-Lenard equation (Ref. 4) for various collision frequencies. Curves parametrized by ν_{ii}/kC_s .

Entropy fluctuations become visible in a collisional plasma because the thermal conductivity, the normal damping mechanism for temperature fluctuations, becomes relatively weak at high collision frequencies. Similarly, if the plasma is assumed isothermal then entropy fluctuations are not allowed because of the implied infinite thermal conductivity. The enhancement of the ion-acoustic resonance occurs because the otherwise strongly Landau damped electrostatic restoring force is

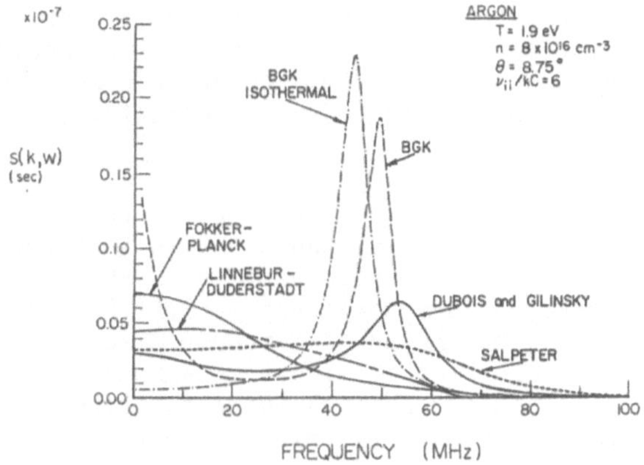

Fig. 2 Ion-acoustic spectra for a collisional agron plasma as calculated from the various collisional theories (Refs. 2-6) and compared to a collisionless calculation [Salpeter (Ref. 11)].

replaced with a much stronger ion-ion collisional restoring force. This is exactly the same mechanism as the one which supports ordinary sound waves in fluid media. Finally, if we look at the Linnebur-Duderstadt calculations, based on the generalized Langevin equation, and the Fokker-Planck equation, based on a Brownian collision term, we find that they predict an almost complete damping of the ion-acoustic resonance at high collision frequencies.

The testing of the various collisional theories previously discussed requires the use of a very dense low temperature plasma which is in thermodynamic equilibrium and for which the plasma density, temperature, and impurity content are accurately known. These requirements where satisfied by the use of a pulsed arc, designed to produce plasmas with $n = 10^{17}\,cm^{-3}$, T = 2eV and to last for about 100 μsec. The arc was made capable of high repetition rates (.5 Hz) in order to allow for integration of the scattering signals cumulatively over many shots.

The arc consisted of a quartz tube (r=2.3cm, ℓ=22cm) mounted between two copper-tungsten alloy electrodes which were fed by 1200 μF capacitor bank charged to 1-2 kV. To produce a relatively flat current pulse over the time of the discharge and hence a quiescent plasma, the capacitor bank used to drive the arc was constructed as a lumped delay line which has the characteristic of producing square discharge pulses.

In normal operation, the arc was filled with 1-7 Torr of helium or argon gas and discharge currents of about 10-25 kA produced plasmas with densities and temperatures of about 10^{16} - 10^{17} cm^{-3} and 2-4 eV. For scattering measurements, the plasma collisionality was maximized by operating the arc at the highest pressure (3 torr for Ar and 7 torr for He) and the lowest bank voltage (1500V) consistent with stable plasma discharges. This produced plasmas with 1.95 eV and $10^{17} cm^{-3}$ for argon and 2.3 eV and $10^{17} cm^{-3}$ for helium. The corresponding degree of ionization was 100% in argon and about 50% in helium.

The electron density and temperature of the arc plasma was determined from interferometric and spectroscopic measurements (Fig. 3). From these measurements, the electron density was found to be a smooth function of time, roughly following the profile of the current pulse. Similarly, measurements at various radial positions showed that the plasma density and temperature were relatively flat across the diameter of the discharge tube, indicating a very uniform arc discharge.

The stability properties of the plasma where checked by use of high speed photography, inductive probes, and electric Langmire probes. Framing camera photographs with exposure times as short as 5 nsec showed the plasma to be stable, uniform, and reproducible in both helium and argon discharges in the 1-7 torr and 1-3 torr pressure ranges, respectively. Magnetic and electric probe measurements showed that the plasma was very uniform and quiescent showing no signs of large-scale fluctuations or changes in the equilibrium state for times as long as 50 μsec.

Scattering measurements with long-wavelength lasers from dense low-temperature plasmas are generally very difficult to perform since whenever the input laser power is sufficiently intense to produce a discernible scattering signal, it is also sufficiently intense to perturb the plasma by heating it. In this experiment, the problem of heating the plasma was overcome by the use of a heterodyne technique[12,13] which is capable of boosting the detected signal many orders of magnitude; as a result, it permits the use of a relatively low-power laser (200 W) which does not perturb the plasma.

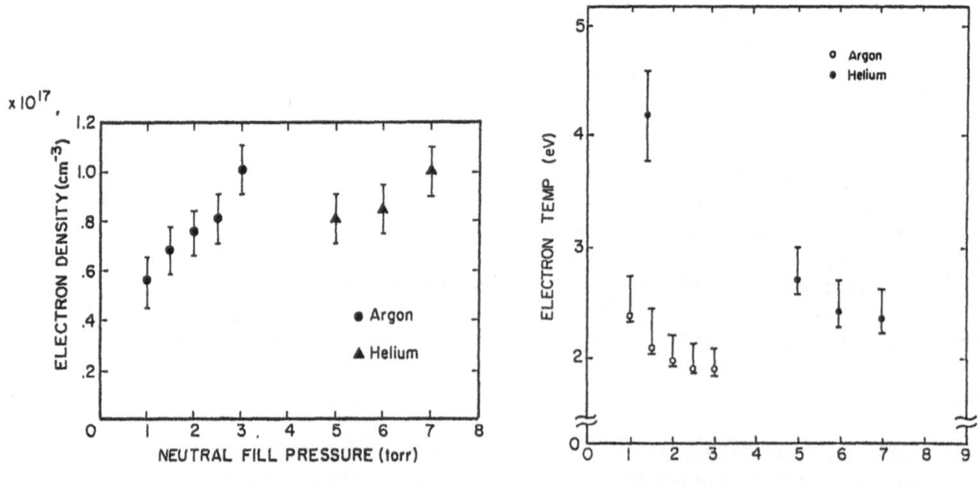

Fig. 3 Electron temperature and density of argon and helium plasmas as a function of fill pressure. Measured by interferometry and the ratio of spectral line intensities.

A schematic of the heterodyne scattering configuration is shown in Fig.4. First, a small fraction of the main laser beam (TEM_{00}, P = 200 watts, τ = 100μsec) is split off to form a local oscillator beam and, subsequently, the main and local oscillator beams are focused into the plasma with a 10-cm-focal-length lens. Any light that is scattered out of the main beam and into the solid angle subtended by the local oscillator beam is imaged onto a liquid-helium cooled Ge:Cu detector. The non-linear mixing of the scattered and local oscillator beams produce a detected photocurrent, which contains the plasma fluctuation spectra. In particular, the photocurrent can be expressed as:

$$I_0(t)=\frac{\eta Ge}{h\nu}\{P_S+P_{LO}+2\zeta[P_S(\omega)P_{LO}]^{1/2}\cos(\omega t)\},$$

where $P=c(EE^*/8\pi)$ is the average power of the individual fields, η and G are the detector quantum efficiency and gain, ω is the fluctuation frequency and ζ is the mixing efficiency which measures the extent to which the two radiation fields are in phase over the surface area of the detector. The form of this photocurrent shows that mixing the scattered light with a local oscillator beam produces a signal that is composed of two distinct components: a slowly varying (Freq. ≈ 1/laser pulse length) average power envelope that is due to the local oscillator and scattered light beams ($P_S + P_{LO}$); and a high frequency beat term (i.e., heterodyne) which contains the plasma fluctuation spectrum $[(P_SP_{LO})^{1/2}\cos(\omega t)]$. The heterodyne term is in the radio frequency range in this work. Consequently, it is easily differentiated from the low frequency average power terms and its spectrum is analyzed at high resolution (Δf = 6MH_z) by the use of a scannable electronic filter. In addition, the capability to separate the heterodyne term from the very large average power terms permits scattering measurements in the high stray light environments, as is the case for small angle scattering. The stray light not being frequency shifted by the scattering process only contributes to the average power terms which are easily filtered out.

In practice, the mean-square current analyzed by the data-acquisition unit is composed of not only the scattering signal, but also of signal currents from various noise sources in the detection system. As a result, good signal to noise ratios can only be obtained for long integration times. For the data presented in this paper, the integration time per shot was set at 40 μsec and 50 to 100 shots were averaged for each frequency value.

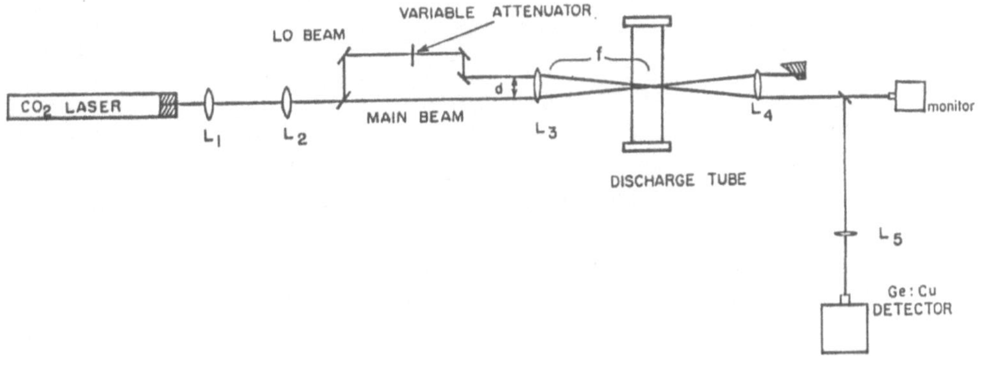

Fig. 4 Diagram of the optical system and scattering geometry.

Initial scattering measurements were made in argon plasmas for the two cases of $\tilde{k}$ parallel and perpendicular to the discharge current. The observed spectra are displayed in Fig. 5 and show substantial peaking near the ion-acoustic frequency (50 MHz at 2 eV). Each data point corresponds to a scattering signal averaged over 50 discharges of the arc and the error bars are derived from the standard deviations of the mean. Even though the observed enhancement of the ion-acoustic resonance is interpreted as evidence of collisional plasma behavior, it is important to rule out the possibility that such enhancement could also have been produced by collisionless mechanisms. Most importantly, reduction of Landau damping due to unequal electron and ion temperatures; and non-thermal excitations due to the discharge current could have both produced the observed spectra.

Reduction of Landau damping due to unequal electron and ion temperatures is very unlikely because the collisional equilibration time between ions and electrons (70nsec) is much shorter than the plasma lifetime (100 μsec). Similarly, the electron temperature would have to be about eight times that of the ion temperature in order for the collisionless result to qualitatively reproduce the measured spectra. Such a large temperature differential is sufficient to substantially reduce (30%) the ion-acoustic frequency in comparison to its adiabatic value. Since no such shift was observed, it is safe to assume that the observed enhancements of the ion-acoustic resonance were not caused by a reduction in Landau damping due to unequal electron and ion temperatures.

Enhancements of the spectra due to current-driven ion-acoustic waves were not expected to be very important because the electron drift velocity due to the current was only 20% of the ion-acoustic velocity. Nevertheless, measurements with $\tilde{k} \parallel \tilde{J}$ and $\tilde{k} \perp \tilde{J}$ showed that there was a 30% enhancement of the scattering spectrum parallel to the current. This relatively small level of enhancement suggested that the ion-acoustic waves parallel to the current are weakly driven, and are

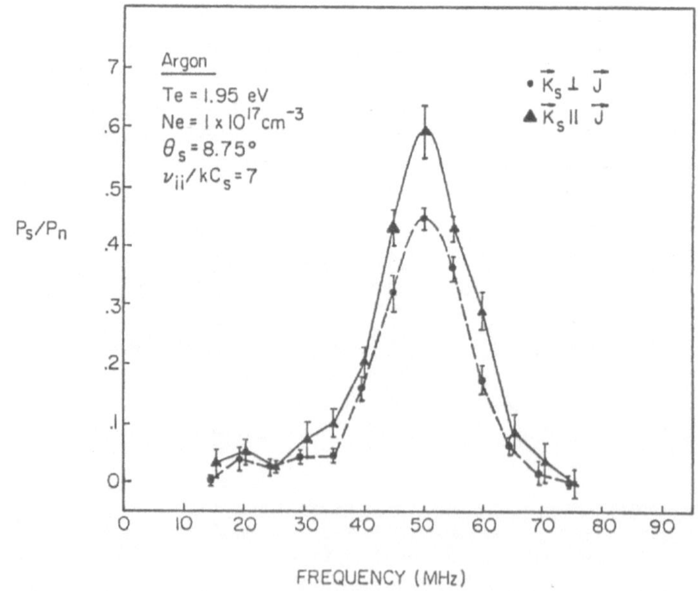

Fig. 5 Scattered ion-acoustic spectra for the two cases of $\tilde{k} \parallel \tilde{J}$ and $\tilde{k} \perp \tilde{J}$ which demonstrate the effect of the discharge current.

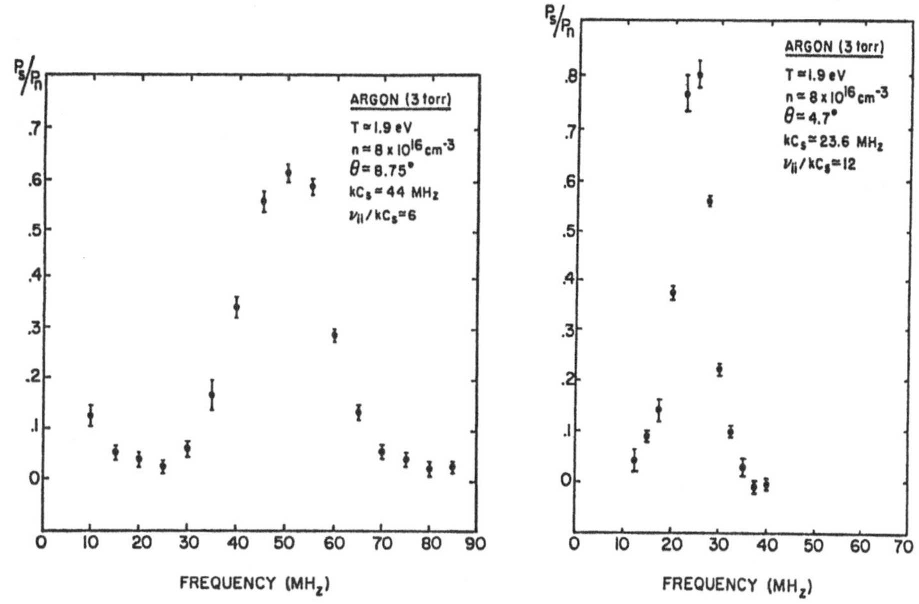

Fig. 6 Scattered light spectra in argon for 4.7° and 8.75° scattering angles.

thus uncoupled from fluctuations perpendicular to the current. As an additional test, the $\tilde{k} \parallel \tilde{J}$ scattering spectrum was also measured both during the discharge, when the current was about 17 kA, and during the afterglow, when the current had fallen to zero and was incapable of driving instabilities. Again, this measurement showed that the enhancement due to the discharge current was about 30%, thereby proving that the ion-acoustic waves perpendicular to the current were not coupled to the current, thus only exhibiting thermally excited fluctuation spectra.

A comparison of the theories displayed in Fig. 2 with the scattering results required that the data originate from thermal fluctuations. Final data were, therefore, taken with $\tilde{k} \perp \tilde{J}$. Specifically, scattering spectra from both helium (2.3 eV, 1×10^{17} cm^{-3}) and argon (1.95 eV, 1×10^{17} cm^{-3}) plasmas were each measured at scattering angles of 8.75° and 4.7°. These spectra are shown in Figs. 6 and 7; as expected, they show substantial enhancement and narrowing, with the peak amplitudes being about 10 times the predicted amplitudes of collisionless theory. For the spectra at 4.7° ν_{ii}/kC_s is about 2 times as large as for the one at 8.7°. This shows up as an increased narrowing of the ion-acoustic resonance at the lower ion-acoustic frequency. As before each data point corresponds to a scattering signal averaged over fifty discharges of the arc and the error bars are derived from the standard deviation of the mean.

From all of the theoretical models previously discussed, the BGK theory proved most successful in reproducing the data. Theoretical curves due to the BGK theory are presented alongside the argon data in Fig. 8a. These curves were calculated for the specific plasma temperatures and densities that were measured by independent diagnostics. No free parameters were used in the calculations other than a normalization of the calculated spectra to the peak amplitude of the data so as to allow for the uncertainty (factor of 2) in the absolute intensity calibration of the detection system. Even though the general shape and position of the ion-acoustic resonances is predicted fairly well, there exists a discrepancy between the measured and

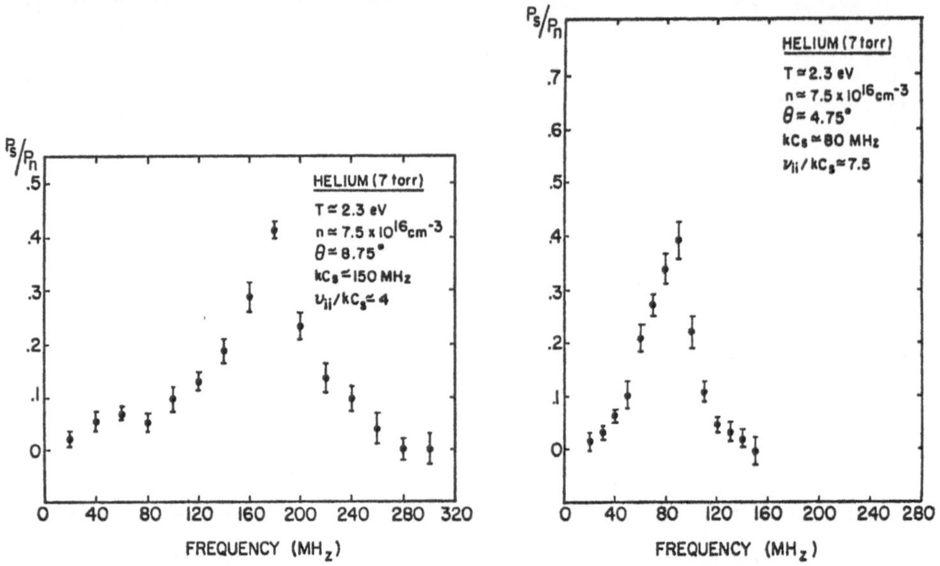

Fig. 7 Scattered light spectra in Helium for 4.7^{o} and 8.75^{o} scattering angles.

predicted widths of the resonances. This difference, except for the 4.7^{o} argon spectrum, cannot be accounted for by the finite resolution of the experiment (F=6Mhz, $\Delta k/k = 0.05$). A calculation where the finite wavelength and frequency resolution of the experiment are convoluted into the theory is displayed alongside the data (dashed curves) in Fig. 8a. The basis for this discrepancy probably lies in the fact that ion-electron collisions have been assumed to be unimportant in the theoretical formalism. For plasma conditions encountered in this experiment, however, such an assumption is most likely invalid because the collisional equilibration time between electrons and ions [ν_{ei}(He) ≈ 9nsec, ν_{ei}(Ar) ≈ 70 nsec] is comparable to the period of ion-acoustic oscillations. The main effect of collisional coupling between electrons and ions is to increase the effective thermal conductivity of the ions through contact with the highly conductive electrons. This increases the damping of ion-acoustic waves, thereby increasing the width of the observed resonances.

Fluctuation spectra predicted by the Kivelson-DuBois solution of the Balescu-Lenard equation were also compared to the scattering data (Fig. 8b). On the high frequency side of the ion-acoustic resonance the theory produces a very good fit; however, at lower frequencies this model reproduces the data fairly poorly. It appears that the approximations employed by Kivelson and Dubois, in solving the Balescue-Lenard equation, somehow overestimate the width of the entropy fluctuation contribution to the total spectrum. Finally, it is important to point out that the Fokker-Plank and Linnebur-Duderstadt models are in total disagreement with the data since they do not even predict the existence of an ion-acoustic resonance in highly collisional plasmas.

An issue still not addressed by this experiment is that of entropy fluctuations at zero frequency. Both the BGK and Debois-Gilinsky theories predict the enhancement of entropy fluctuations at zero frequency. In fact, the presence of these fluctuations is hinted by the data in Fig. 6. Complete verification of this effect could not be obtained with the initial experimental configuration because the finite pulse length of the laser and low frequency variations of the laser absorption in the plasma imposed a lower limit on the detection . frequency. To overcome this limitation the whole frequency spectrum of the scattered light was shifted to higher frequencies by a shift of the laser local oscillator frequency. The shift in the local oscillator was accomplished by scattering the local oscillator beam from ultrasonic waves in a germanium crystal (accousto-optic Bragg cell). The frequency shifted scattered beam was then used as the new local oscillator. With this modification to the experiment it now became possible to directly measure the low frequency entropy fluctuation contribution as well as the ion-acoustic resonance in a single spectrum. Initial measurements in an argon plasma at 4.7^{o} are shown in Fig. 9.† As expected from the BGK predictions, the entropy peak is strongly peaked and comparable to the ion-acoustic peak in intensity.

The simultaneous observation of the entropy and ion-acoustic peaks afforded a unique opportunity to measure some basic thermodynamic properties of the plasma. This was accomplished by assuming a fluid description of the plasma and comparing the scattered spectra to the general predictions of fluid theory. In particular, the spectrum of scattered light from a fluid in thermodynamic equilibrium is given by

$$\frac{\langle \rho(\tilde{k},\omega)\rho^{*}(\tilde{k},\omega)\rangle}{\langle \rho(\tilde{k})\rho^{*}(\tilde{k})\rangle} = \frac{C_p - C_v}{C_p}\,\frac{2\kappa k^2/\rho_o C_p}{(\kappa k^2/\rho_o C_p)^2+\omega^2} + \frac{C_v}{C_p}\,\frac{\Gamma k^2}{(\Gamma k^2)^2 + (\omega - C_s k)^2}$$

where C_S is the sound speed, C_P and C_V are the specific heats at constant pressure and volume, κ is the thermal conductivity, and Γ is the damping coefficient for sound waves:

$$\Gamma = \frac{1}{2}\left[\frac{4}{3}\,\frac{\eta_{sh} + \eta_b}{\rho_o} + \frac{\kappa}{\rho_o}\left(\frac{1}{C_V} - \frac{1}{C_P}\right)\right]$$

and η_{sh} and η_b are respectively the shear and bulk coefficients of viscosity.

An examination of this equation reveals that the fluctuation spectrum of a fluid plasma is composed of two distinct components: one at zero frequency due to entropy fluctuations and one at the ion-acoustic frequency. The width of the two resonances is determined, respectively, by the thermal conductivity and the ion-acoustic damping coefficient. Furthermore, the ratio of the peak intensities is related to the ratio of heat capacities $\gamma = C_\rho/C_v$ by $I(o)/I(kC_s) = 2(\gamma-1)$ and the width of the entropy peak is given by $\Delta\omega = \kappa k^2/\rho_o C_p$. The curve plotted alongside the data in Fig. 9 is a least squares fit to the fluid model. From the fit we find that $\gamma = 2.24$ and that the thermal conductivity is $\kappa = 3.84\times10^{3}$ ergs/deg-cm-sec . In comparison we find that a calculation of the Braginskii thermal conductivity gives $\kappa = 2.33 \times 10^{3}$ ergs/deg-cm-sec for the ions and $\kappa = 3.75 \times 10^{5}$ ergs/deg-cm-sec for the electrons. The fact that the experimental value of the conductivity is about 60% higher than calculated ion conductivity but about 2 orders of magnitude lower than the calculated electron conductivity suggests that there is a relative small amount of coupling between the ions and electrons. Nevertheless, the width of the resonances is increased by this interaction between the electrons and ions.

†The instrument width of this data has been reduced to $2MH_z$

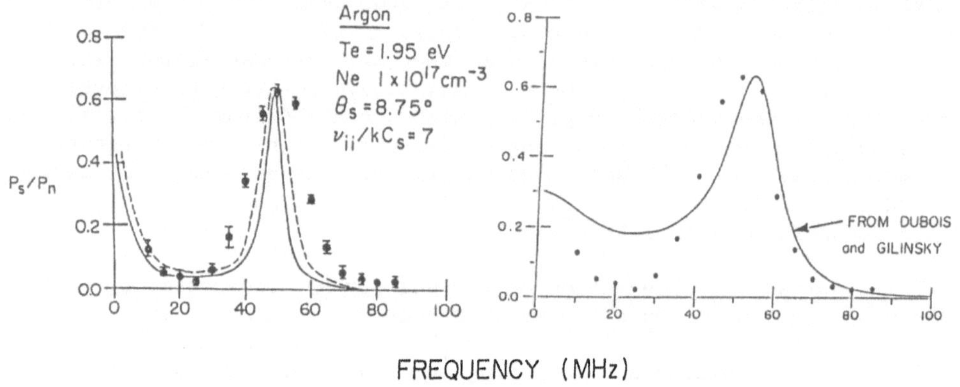

Fig. 8 a) Scattered ligh spectra in argon for 4.7° and 8.75° scattering angles. Calculated curves are due to the BGK theory (Refs. 2,6 and 15). The dashed curves are due to the BGK theory convoluted with the finite resolution of the experiment. b) Curve due to theoretical predictions from the solution of te Balescu-Lenard equation (Refs. 4, and 15).

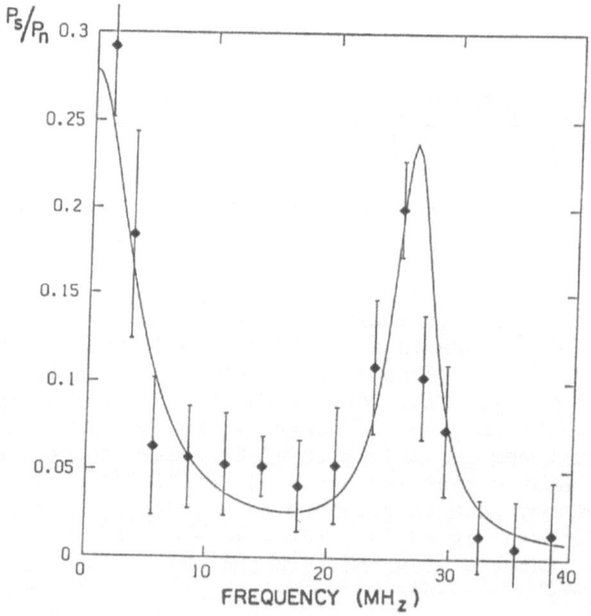

Fig. 9 Complete entropy and ion-acoustic spectrum is obtained with frequency-shifted local-oscillator beam. Curve due to least squares fit to fluid theory.

In conclusion, small-angle scattering measurements of thermal ion-acoustic fluctuations from highly collisional ($\nu_{ii}/kC_s \approx$ 5-13) argon and helium plasmas have been obtained. These measurements show substantial enhancements of the ion-acoustic and entropy fluctuation resonances due to collisional effects. A comparison to theoretical predictions calculated from independent measurements of temperature and density shows that the BGK theory is the most accurate in reproducing the data. It is demonstrated that fluid theory combined with laser scattering experiments can be used to measure plasma transport properties.

In addition to being very collisional, the plasma used in this experiment was characterized by a fairly large ($g = 1/m\lambda_D^3 \approx .3$) plasma parameter.
Thus, there was a good possibility that, in exception to the Linnebur-Duderstadt model, the theoretical models examined in this work were inapplicable, due to their formal dependence on g being small. Nevertheless, even with $g \approx .3$ the BGK and Balescu-Lenard models seem to be qualitatively correct.

REFERENCES

1. A. W. DeSilva and G. Goldenbaum, Methods of Experimental Physics (Academic, New York, 1970), Vol. 19A, Chap. 3.
   J. Sheffield, Plasma Scattering of Electromagnetic Radiation (Academic, New York, 1975).
2. P. L. Bhatnagar, E. P. Gross, and M. Krook, Phys. Rev. 94:511 (1954).
3. E. J. Linnebur and V. V. Duderstadt, Phys. Fluids 16:665 (1973).
4. D. F. Dubois and V. Gilinsky, Phys. Rev. 133:A1317 (1964).
5. M. S. Grewal, Phys. Rev. 134:A86 (1964).
6. O. Theimer and M. M. Theimer, Nuovo Cimento Soc. Ital. Fisc. B 65:207 (1981).
7. E. M. Leonard and R. K. Osborn, Phys. Rev. A 8:2021 (1973).
8. E. Holzhauer, Phys. Lett. 62A:495 (1977).
9. A. A. Offenberger and R. D. Kerr, Phys. Lett. 37A:435 (1971).
10. J. H. Massig, Phys. Lett. 66A:2097 (1978).
11. E. E. Salpeter, Phys. Rev. 120:1528 (1960).
12. R. E. Slusher and C. M. Surko, Phys. Fluids 23:472 (1980).
13. E. Holzhauer and J. H. Massig, Plasma Phys. 20:867 (1978).
14. A. N. Mostovych and A. W. DeSilva, Phys. Rev. Lett. 53:1563 (1984).
15. A. N. Mostovych and A. W. DeSilva, Phys. Rev. A 34:3238 (1986).

PRODUCTION AND DIAGNOSIS OF DENSE COOL ALKALI PLASMAS

O. L. Landen and R. J. Winfield

Lawrence Livermore National Laboratory, USA
Imperial College, Blackett Laboratory, London, UK

INTRODUCTION

In strongly-coupled plasmas, the electrostatic energy becomes comparable to, or larger than, the thermal energy, i.e.:

$$\frac{e^2}{4\pi E_o}\left(\frac{4\pi N_e}{3}\right)^{1/3} > kT_e$$

In addition, the parameter number of electrons/Debye sphere, $N_D = 1.7 \times 10^9\ T_e^{3/2}/N_e^{1/2}$ becomes less than 1 and hence the term non-Debye plasma. Under these conditions, deviations in electron and ion structure factors from the well-known weakly coupled limits(1) are expected. These deviations in plasma behavior should be observable spectroscopically from lineshape studies and Thomson scattering.

Non-Debye plasmas amenable to such diagnostics have been produced here by resonant and multiphoton excitation of sodium and cesium vapors.(2-5) Detailed time-resolved emission spectroscopy and Thomson scattering from the ensuing long-lived (> 100 ns) and optically thin plasmas was performed and compared with theory.

EXPERIMENTAL DETAILS

The experimental set-up consisted of a tunable dye laser pumped by a frequency-doubled Nd: glass laser, sodium and cesium ovens and a spectrometer/photomultiplier system for spatial, temporal and partial spatial resolution of plasma fluorescence and scattered laser light. The typical laser output was 25 mJ in 25 ns focussed to a few mm^2 in the oven and tunable between 5800 and 6900 Å by using a variety of laser dyes. Oven vapor densities were monitored by using saturated vapor pressure tables(6) and measured oven temperatures, and by curves of growth on sodium and cesium resonance lines. The plasma and laser light emitted at 90° to the laser beam within the ovens was focussed onto 50 to 100 µm wide horizontal slits of an f/4.2 30 cm grating spectrometer. Lineshapes were recorded on a shot-to-shot basis with 15 ns, 1.8-2.3 Å resolution for the 0.5 cm by 100 µm plasma area viewed. Detector sensitivities and spectral calibrations were performed using tungsten ribbon lamps and low pressure discharge lamps.

IONIZATION SCHEMES

Sodium plasmas were produced by saturating the resonance transition 3s→3p at λ = 5896 Å or by two-photon resonant (3s→3d) three-photon ionization at λ = 6854 Å. The ionization mechanism in the former case was believed to be associative ionization between two sodium atoms excited to the 3p state. Cesium plasmas were induced by two-photon ionization at λ = 5266 Å and λ = 6354 Å.

EMISSION SPECTROSCOPY: SODIUM

The column of laser excited sodium vapor emitted a white glow in which the following atomic doublet transitions were identified: 3p-ns for n=5-9, 3p-nd for n=3-10, 3p-nf (dipole forbidden) for n=4-6 and the D lines 3s-3p. Lineshapes were recorded for various times between 30 and 400 ns after the end of the laser pulse for the transitions 3p-3d, 3p-4d, 3p-5d, 3p-6d, 3p-5s, 3p-6s at various initial sodium densities, N_{Na}, between 2×10^{16} and 4×10^{17} cm^{-3}. A selection of doublet lineshapes with theoretical fits are shown in Figs. 1-3. The 3p-3d transition is optically thick, whereas the 3p-4d transition is optically thin but broadened and shifted by the ion microfield and electron collisions. The forbidden components 3p-4f (Fig. 3) appear on the blue wing of 3p-4d due to mixing of eigenstates by the ion microfield. The theoretical fits which include the instrument function, resonance broadening,(7) radiation transport and plasma-induced broadening(8) are good.

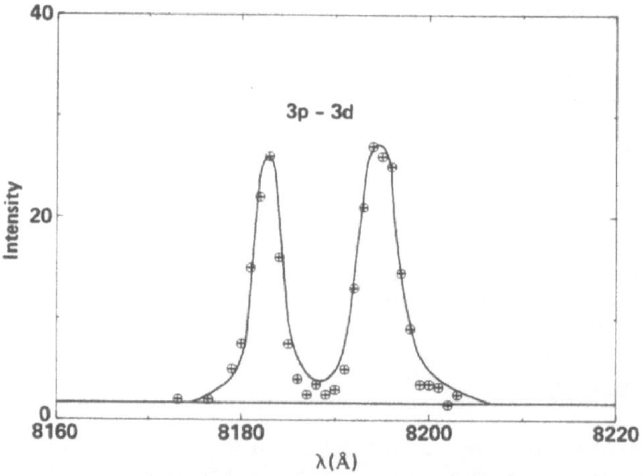

Fig. 1. 3p-3d emission line shape for $N_{Na} = 10^{17}$ cm^{-3}, t = 100 ns. Solid line is continuum level and solid curve is theoretical fit (Ref. 8) including radiation transport.

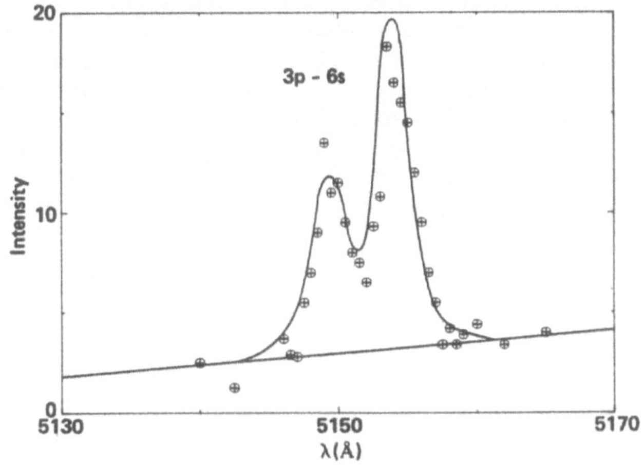

Fig. 2. 3p-6s emission line shape for N_{Na} = 8 x 10^{16} cm^{-3}, t = 50 ns. Solid line is continuum level and solid curve is theoretical fit (Ref. 8).

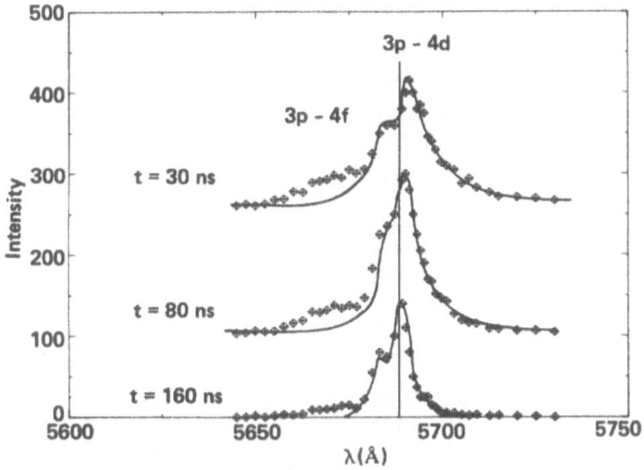

Fig. 3. Temporal history of 3p-4d emission line shape for N_{Na} = 4 x 10^{17} cm^{-3}. Solid curves are theoretical fits (Ref. 8) excluding forbidden components and including radiation transport. Line shapes are displaced vertically for ease of viewing. Vertical solid line is unperturbed $3p_{3/2}$-$4d_{5/2}$ line center.

For isolated lines, plasma-induced line shifts are predominantly due to the ion quadratic Stark effect and weak, distant elastic electron collisions. Since the shift depends on distant collisions, there is a greater discrepancy than for linewidths between theories including and ignoring Debye screening of collisions. Hence the ratio of shift-to-width can be used as a measure of the applicability of Debye shielding for dense cool plasmas. This ratio is plotted for various electron densities determined from the plasma-induced linewidths in Fig. 4, showing better agreement with theory including Debye screening.

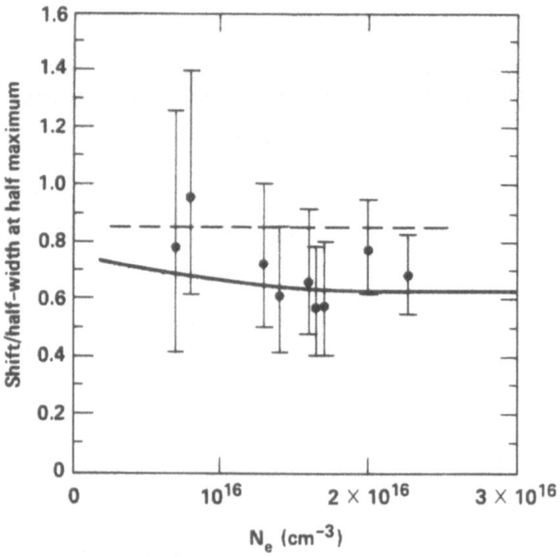

Fig. 4. 3p-4d shift-to-width ratio vs N_e measured from 3p-4d widths. Dashed line represents theory without shielding and solid line includes Debye shielding.

The spectrally integrated ratio of 3p-4f forbidden component intensity to 3p-4d allowed transition intensity is also an electron density diagnostic.[8] The ratio of electron densities determined by such forbidden component intensities and by the 3p-4d width is shown in Fig. 5 versus electron density determined from the 3p-4d width. Agreement is good at low densities ($N_e < 1.5 \times 10^{16}$ cm^{-3}). The discrepancy at higher densities is attributed to the breakdown of the perturbation theory used since the intensity ratio has reached 20% at $N_e = 1.5 \times 10^{16}$ cm^{-3}.

The electron temperature was measured from relative line intensities of high-lying transitions which are collisionally coupled to free electrons at the present densities. For both ionization schemes, T_e varied between 0.19 and 0.28 eV during the first 400 ns of the plasma recombination phase. Hence a minimum of 1 electron/Debye sphere was attained, limited by three-body recombination which acts to reduce the electron density and raise the electron temperature.

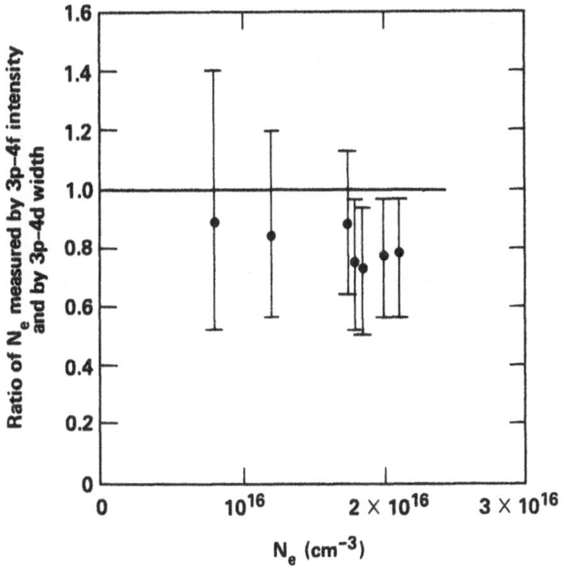

Fig. 5 Ratio of N_e measured by intensity of 3p-4f forbidden component and by 3p-4d width vs N_e measured from 3p-4d width. Solid line represents same N_e measured by both diagnostics.

EMISSION SPECTROSCOPY: CESIUM

A bluish-white column was formed for all Cs densities above 10^{16} cm^{-3} for both laser wavelengths used, λ = 5266 Å and λ = 6354 Å. Emission lineshapes from the 6p-6d, 6p-7d, 6s-7p, 6p-8d, 6p-9d, 6p-8s and 5d-5f transitions were recorded from the recombining plasmas for various times between 50 and 400 ns after the start of the plasma. Electron densities below 10^{17} cm^{-3} were inferred from the plasma-induced broadening of the 6p-7d, 6p-8d and 5d-5f transitions. Densities above 10^{17} cm^{-3} were deduced from the plasma-induced shift of the 6s-7p line.[8,9] Time resolved electron temperatures were deduced from relative line intensities of 6p-8d, 6p-7d, 6p-8s and 6p-6d, assuming partial LTE for all states above 5d. The results for various initial Cs densities, 50 - 100 ns into the plasma phase, are shown in Table 1 for λ = 6354 Å. The plasmas produced for λ = 5266 Å yielded electron densities between 10^{16} and 10^{17} cm^{-3} and hotter temperatures, 0.25 - 1 eV.

Table I. Results of emission spectroscopy on cesium plasmas produced by a dye laser at λ = 6354 Å, 50–100 nsec into the afterglow.

Value of N_{Cs}, cm^{-3}	N_e, cm^{-3}	T_e, eV	Ionization, %	N_D
1.5×10^{16}	$3.0 \pm 1 \times 10^{15}$	0.18 ± .02	20%	1.7 - 3.4
4.2×10^{16}	$1.5 \pm .3 \times 10^{16}$	0.3 ± .05	33%	1.6 - 3.2
10^{17}	$3.0 \pm .5 \times 10^{16}$	0.2 ± .05	30%	0.5 - 1.3
3×10^{17}	$1.0 \pm .3 \times 10^{17}$	0.25 ± .05	33%	0.4 - 1
5×10^{17}	$4.0 \pm 1 \times 10^{17}$	0.8 ± .1	60%	0.8 - 2

The Cs I 5d-5f lineshape was then recorded in detail at nearly constant density but varying temperature and hence varying degree of plasma nonideality. A typical optically thin allowed $5d_{5/2}$-$5f_{7/2}$ line and its associated forbidden component $5d_{5/2}$-$5g_{7/2,9/2}$ arising from the plasma microfield is shown in Fig. 6 for T_e = 0.5 eV. The electron density, N_e = 1.1 x 10^{16} cm^{-3}, determined from the linewidth and shift agree within experimental error (ΔN_e = ± 10^{15} cm^{-3}) and the lineshape fit is good. Figure 7 shows the same transition at a lower temperature (T_e = 0.2 eV). The experimental lineshape exhibits more asymmetry than predicted by theory. Moreover, the measured shift-to-width ratio, 0.13 ± .03, is smaller than the theoretical

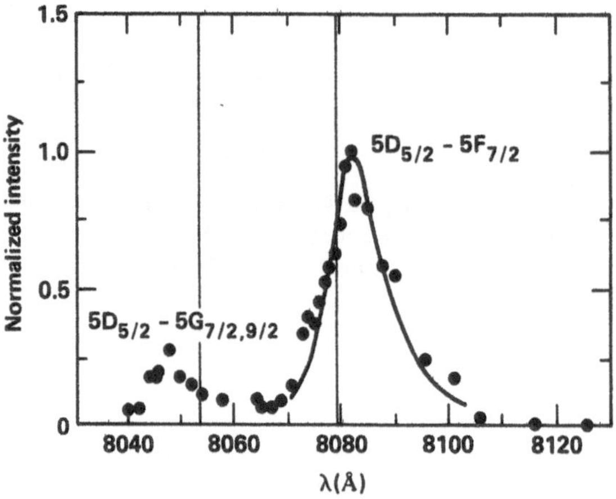

Fig. 6. Emission lineshape of the Cs I $5d_{5/2}$-$5f_{7/2}$ allowed transition and $5d_{5/2}$-$5g_{7/2,9/2}$ forbidden component at T_e=0.5 eV. The solid curve is a theoretical fit (Ref. 8) yielding N_e=1.1 ± 0.2 x 10^{16} cm^{-3}. The solid lines represent the unperturbed line centers.

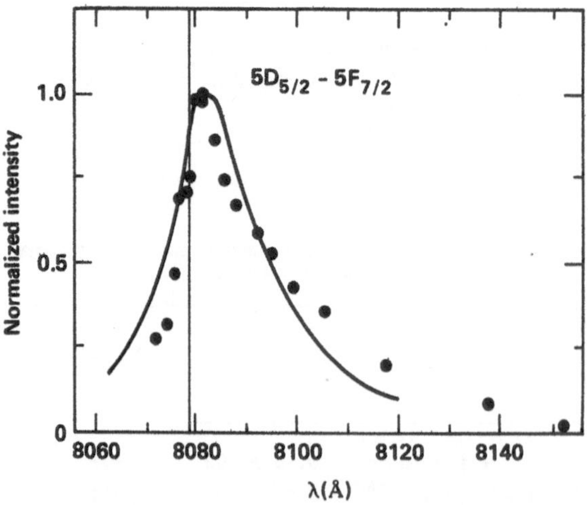

Fig.7 Emission lineshape of the Cs I $5d_{5/2}$-$5f_{7/2}$ allowed transition at T_e = 0.2 eV. The solid curve is a theoretical fit (Ref. 8) yielding N_e = 2.1 x 10^{16} cm^{-3}. The solid line represents the unperturbed line center.

values including Debye screening (0.18) and excluding screening (0.35).[8] Contributions to the lineshape from ion-quadrupole, Van der Waals and dipole-dipole interactions are calculated to be negligible. However, more exact potentials would be needed to evaluate the effects of close neutral-neutral collisions on the shape of the 5d-5f line wing.

LASER SCATTERING: CESIUM

A 0.1-1 MW, 30 ns laser beam at λ = 5266 Å was used to both ionize and scatter from the resultant Cs plasma. The results for the high frequency "electron feature" for an initial Cs density of $3.5 \pm 0.5 \times 10^{16}$ cm^{-3} at three progressively decreasing laser fluxes are shown in Fig. 8. The plasma parameters deduced are tabulated in Table II. The theoretical fits shown by the dashed lines which include the instrument function (FWHM = 2.5 Å) and laser bandwidth (FWHM = 2.5 Å) are derived from the usual collisionless theory.[10] For the nearly fully ionized plasmas of Table II laser flux inhomogeneities will create temperature rather than density variations. The wings of the spectra at large detunings (>20 Å) shown in Fig. 8 which deviate from the theoretical curves could then be explained by

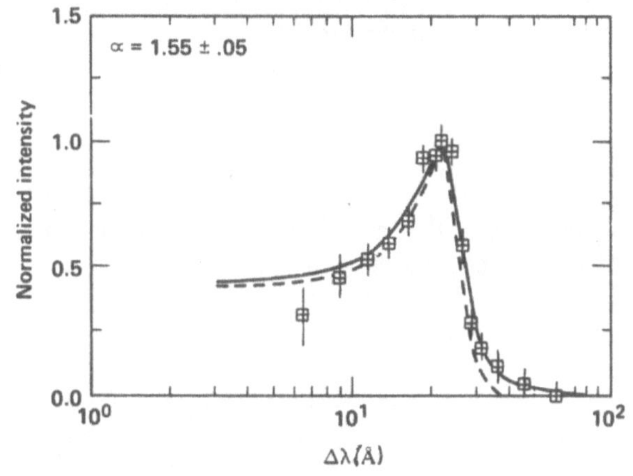

Figure 8a. (See page 84 for legend).

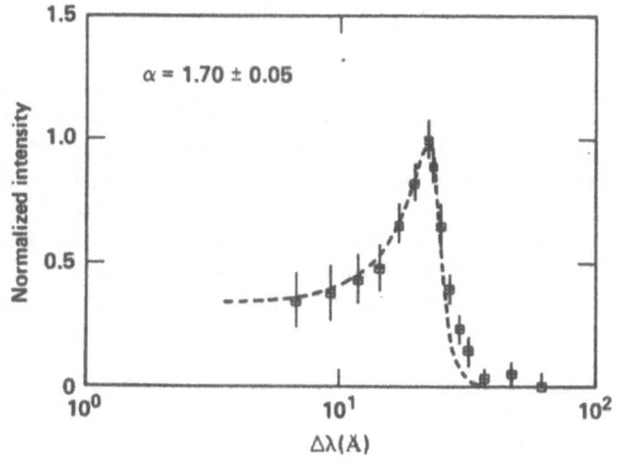

Figure 8b. (See page 84 for legend).

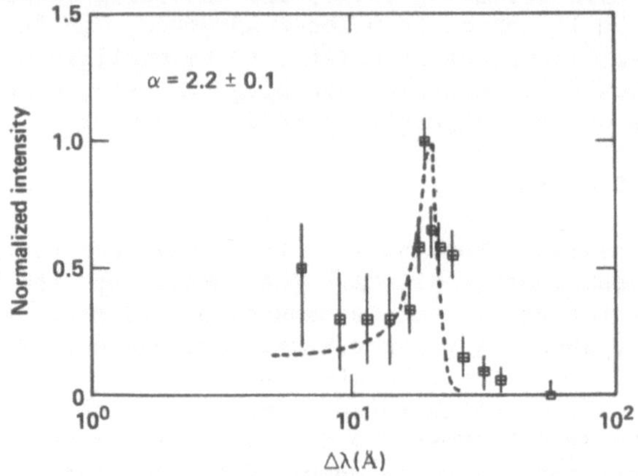

Fig. 8. (a) Electron feature of scattered spectrum fitted by $\alpha = 1.55 \pm 0.05$ of collisionless theory (Ref. 10), dashed line, and $\alpha = 1.55$, $c_e = 0.1$ of collisional theory (Ref. 13), solid line. (b) Electron feature of scattered spectrum fitted by $\alpha = 1.7 \pm 0.05$ of collisionless theory (Ref. 10). (c) Electron feature of scattered spectrum fitted by $\alpha = 2.2 \pm 0.1$ of collisionless theory (Ref. 10).

lower α scattering from higher temperature fully ionized regions, but only if these exceeded 2 eV. It is unlikely, however, that this explanation can account for all three wings of Fig. 8 since the laser flux was decreased by a factor of 7.5 between Figs. 8a and 8c.

TABLE II Results of laser scattering shown in Figs. 8(a)–8(c).

α	I (GW cm^{-3})	N_e (cm^{-3})	T_e (eV)	N_D
1.55 ± 0.05	1.5	$3.5 \pm 0.2 \times 10^{16}$	0.9 ± 0.05	7.8 ± 0.09
1.70 ± 0.05	0.7	$3.5 \pm 0.2 \times 10^{16}$	0.78 ± 0.04	6.3 ± 0.06
2.2 ± 0.1	0.3	$3.15 \pm 0.3 \times 10^{16}$	0.42 ± 0.04	2.6 ± 0.05

Moreover, collisions alter the scattered spectra as either the non-Debye limit is approached[11] or as α becomes large.[12] For $1 < \alpha < 3$, Lorentzian wings due to collisional damping become observable on the electron feature for $c_e = \nu_e\alpha/\omega_p > 0.05$, where c_e represents a ratio of collision frequency ν_e to electron plasma frequency ω_p.[13,14] The solid curve in Fig. 8a represents a better theoretical fit by including electron-ion collisions[13] for $\alpha = 1.55$ and $c_e = 0.1$. As expected theoretically,[11] it seems clear that the classical theory[10] does not breakdown substantially at the electron feature for N_D as low as 2.6 at α = 1–3.

SUMMARY

Long-lived (> 100 ns) alkali plasmas with < 1 electron/Debye sphere were produced by 1 MW, 25 ns laser beams focused to 10^7 - 10^9 W cm^{-2} in dense vapors. Good fits to emission spectra and scattered spectra were obtained after including the effects of electron-ion collisions and Debye shielding in all cases but the Cs I 5d-5f transition. Since the degree of nonideality achieved appeared to be limited by recombination and diffusion rates, experiments using shorter pulse lasers and faster diagnostics should be used to probe more non-Debye systems. Specifically, a 1 mJ, 1 ps dye laser at λ = 6150 Å focused to 10^{11} - 10^{12} W cm^{-2} in moderately dense (N_{Cs} = 10^{16} cm^{-3}) cesium vapor should produce a fully ionized (N_e = 10^{16} cm^{-3}) lower temperature (T_e = 0.1 eV) plasma amenable to pico-second Thomson scattering (α = 2-3) with 0.5 electron/Debye sphere.

ACKNOWLEDGMENT

Work performed under the auspices of the U.S. Department of Energy by Lawrence Livermore National Laboratory under contract #W-7405-Eng-48.

REFERENCES

1. D. B. Boercker, R. W. Lee and F. J. Rogers, J. Phys. B 16 (1983) p. 3279.
2. O. L. Landen, R. J. Winfield, D. D. Burgess, J. D. Kilkenny and R. W. Lee, Phys. Rev. A 32 (1985) p. 2963.
3. O. L. Landen and R. J. Winfield, Phys. Rev. A 32 (1985) p. 2972.
4. O. L. Landen and R. J. Winfield, J. Quant. Spectrosc. Radiat. Transfer 34 (1985) p. 177.
5. O. L. Landen and R. J. Winfield, Phys. Rev. Lett. 54 (1985) p. 1660.
6. JANAF Thermochemical Tables, edited by D. R. Stull (U.S. National Bureau of Standards, Washington, D.C., 1971).
7. I. I. Sobelman, V. A. Vainshtein and E. A. Yukov, "Excitation of Atoms and Broadening of Spectral Lines" (Springer, New York, 1981).
8. H. R. Griem, "Spectral Line Broadening by Plasmas" (Academic, New York, 1974).
9. P. M. Stone and L. Agnew, Phys. Rev. 127 (1962) p. 1157.
10. E. E. Salpeter, Phys. Rev. 120 (1960) p. 1528.
11. R. Cauble and D. B. Boercker, Phys. Rev. A 28 (1983) p. 944; E. J. Linnebur and J. J. Duderstadt, Phys. Fluids 16 (1973) p. 665.
12. A. N. Mostovych and A. W. DeSilva, Phys. Rev. Lett. 53 (1984) p. 1563.
13. J. Kopainsky and F. Vilsmeier, Appl. Phys. 8 (1975) p. 223.
14. I. Gorog, Phys. Fluids 12 (1969) p. 1702.

SHOCK WAVES AND THERMODYNAMICS OF STRONGLY COMPRESSED PLASMAS

V. E. Fortov and V. K. Gryaznov

Institute of High Temperatures
USSR Academy of Sciences
Moscow

INTRODUCTION

Nonideal plasmas, the most widely spread state of matter occurring in nature, have always attracted the attention of physicists due to a great variety of physical properties[1-3] and practical applications in some modern high energy installations and astrophysical projects.[4,5] The physical properties of plasmas are greatly simplified at extremely high pressures and temperatures, when the kinetic energy of particles considerably exceeds that of interparticle interaction, such that models of ideal homogeneous degenerate (or Boltzmann) plasmas can be applied with assurance. A weak interparticle interaction can then be taken into account with the perturbation theory methods in the framework of classical[6] (Debye-Hückel) or quasiclassical[7] (Thomas-Fermi) self-consistent field methods. In strongly compressed plasmas the interaction energy is comparable to or exceeds the kinetic energy of particle motion, which hinders the application of perturbation theory to such systems. Parameterless numerical simulation methods (Monte-Carlo, molecular dynamics)[8-10] provides comprehensive information about the simplest models beyond the framework of the perturbation theory, e.g., the one-component plasma[8,9] and the pseudopotential model of multicomponent plasma.[10] However, for the second model great difficulties arise when one tries to choose a qualitatively correct electron-ion pseudopotential, while it is difficult to apply the one-component plasma results to real plasmas. Therefore, for a qualitative analysis of the thermodynamical properties of strongly compressed plasmas there are heuristic models in use now, based on extrapolations of general ideas concerning the role of collective and quantum effects by the Coulomb interaction. These models predict physical effects that are new in principle, e.g., metalization and clusterization of plasma as well as the formation of yet unknown exotic plasma phases.[2,10-12] Naturally, all these theoretical predictions need verification in experiments with real plasmas at high pressure.

In spite of the fact that the major part of matter in the Universe is in the state of a strongly compressed plasma, our experimental knowledge about such plasma has been quite limited until now because of great difficulties in generation and diagnostics of the high pressure plasma under laboratory conditions.[2,12-14]

The main difficulties in the production of a nonideal plasma are to make considerable local energy concentrations, which produce high pressures and temperatures above the thermostrength limits of the devices structural materials. Consequently, it is necessary to carry out experiments in a forced pulsing regime at high power levels. In this case a serious problem is the diagnostics of a strongly compressed plasma which is opaque to the light.

In the experimental physics of strongly compressed plasmas the most widely used are dynamical methods[13,15] which employ intense shock wave techniques for the compression and irreversible heating of matter due to the viscous dissipation of energy in the shock front. In this way physical measurements have been carried out over a wide range unaccessible to traditional plasma experimental methods; in particular aluminum superdense plasma of extremely high energy concentration ~ 0.7 GJ/cc and pressure ~ 4 Gbar has been created.[16]

In this review we discuss experiments on the thermodynamics of nonideal plasmas and theoretical models for their interpretation. Data on radiative, electrophysical and gas-dynamic properties have been given.[1-4]

SHOCK WAVE COMPRESSION OF NONIDEAL PLASMAS

In Fig. 1 the possibilities of the dynamical methods are schematically

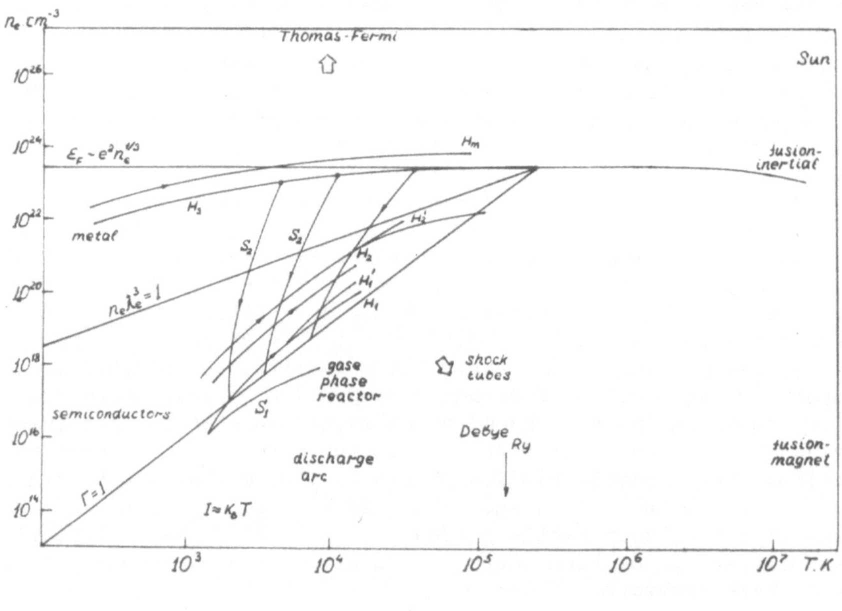

Fig. 1. Regions of nonideal plasma existence. The pointers show the directions of interaction reduction and simplification of the plasma physical properties description. Experiment: H_1, H_1 -cesium vapor compression by the incident and reflected shock waves, S_1-isentropic compression of cesium, H_2, H_2 - noble gases compression by the incident and reflected shock waves; shock compression of solid H_3 and porous H_m metals, S_2, S_3-isentropic expansion of shock compressed metals.

outlined. In addition, the boundary of strong nonideality for classical ($n_e\lambda_e^3 << 1$, $\Gamma = (e^2/K_BT)^{3/2} (8 \pi n_e)^{1/2} = 1$) and degenerate ($\varepsilon_F = (3 \pi n_e)^{2/3}/(2m_e) \sim e^2 n_e^{1/3}$) plasmas is shown. Use of various energy sources, vis. compressed[17,18] and electrically heated[19] gas, chemical[21-23] and nuclear[24,26] explosives, powerful laser[25] and electron[27] beams, pneumatic[28] and electrodynamic[29] guns has made it possible to create strongly compressed plasmas of different elements over a wide region in the phase diagram.

The Boltzmann nonideal gas-like plasma was obtained by the dynamic compression of high-pressure gases, the initial states of which were in the neighborhood of the saturation curve (cesium,[17,18] noble gases[20-22,30]) or even under supercritical conditions.[31-33] While registering the states of single (H) and double (H') compressions, one manages to obtain plasmas with supercritical parameters at pressures up to 170 kbar, temperatures up to 10^5 K and electron densities up to 10^{22} cm^{-3} and to gain access from the side of the "gas" phase to the region of a condensed state. In the case of xenon plasmas[32,33] a maximum density of 4.5 g/cc has been obtained which is 1.5 times larger than the xenon crystallographic density and near to the solid aluminum density. Adiabatic compression of saturated vapors of cesium and potassium[34,35] (adiabat S_1 in Fig. 1) leads to less intensive heating of plasmas when charge-neutral interaction prevails. It is essential that the shock-wave and the adiabatic compression product not only a nondegenerate plasma with extremely high energy concentrations over a wide range of the phase diagram, but also under these conditions to perform detailed thermodynamical[13,14,17-23] electrophysical[12,31,32] and optical[30] measurements as well as those of the laser beam reflection.[33]

The compression of metals by intense shock waves enables one to create strongly compressed plasmas with the electron's component either degenerate or partially degenerate (states H_3 on Fig. 1). For this purpose explosion and cumulation[15,36-38] methods, light-gas gun,[28] laser[25] and electron[27] beams, powerful underground explosions are applied successfully. Of special interest for plasma physics are experiments[16,43,44] in which, by using porous samples and ultraintense shock waves, the plasma states at record high temperatures and concentrations of heat energy with nondegenerate electron component at densities $n_e \sim 10^{23}$ cm^{-3} have been obtained. These experiments make it possible to investigate the thermodynamics of metal plasmas over the entire range of condensed state and to penetrate into the region of quantumstatistical description[7] up to exotic conditions, where the pressure and the energy of the equilibrium radiation become important.

The method of adiabatic expansion (curves S_2) of metals, compressed and heated by powerful shock fronts (H_3), is quite effective for the generation of plasmas with densities below those of the normal solid. This technique makes it possible to explore a wide region of the phase diagram from the strongly compressed metallic liquid to the ideal gas, including the nonideal degenerate, Boltzmann plasma region, and the critical point region.[2,13,14] These results serve as a base for wide-ranged semiempirical equations of state,[14,23] and, in combination with the data from static[46] and electroexplosive[46] experiments, enable one to draw more definite conclusions as to the form of the phase diagram of metals-which may be distorted by plasma phase transitions.[1-3,11,14]

THERMODYNAMIC MODELS OF STRONGLY COMPRESSED PLASMAS

Analysis of existing experimental thermodynamic information on high density plasmas show the absence of any noticeable anomalies,[2,12,47] that might be interpreted as phase transitions. The phase diagram of metals turns out to have a single high-temperature boiling curve and a single critical point. We point out also the strange anomaly in the shock compressibility of ion crystals in the liquid phase at $\Gamma \sim 1$.[48] Their discussion,[47] and erroneous works[49] on Wigner electronic crystals in detonation waves corrected in Ref. 50.

For the interpretation of the thermodynamic experiments the chemical picture,[1-4] based on the explicit separation of free and bound states, is used

$$F(\{N_i\},V,T) = N_e K_B T\{\mu_e - \frac{f_{3/2}(\alpha_e)}{f_{1/2}(\alpha_e)}\} + \sum_j N_j K_B T(\ln \frac{n_j \lambda_j^3}{Q_j} - 1)$$

$$+ \Delta F_{coul} + \Delta F_{ea} + \Delta F_{aa}; \; f_p(\alpha_e) = \frac{1}{\Gamma(P+1)} \int_0^\infty \frac{t^P dt}{e^{t-\alpha_e}+1} \tag{1}$$

Here $n_j = N_j/V$ the densities of the corresponding species, Q_j-partition functions of the species j, $\mu_e = \alpha_e K_B T$ -the electron chemical potential, that satisfies the condition $f_{1/2}(\alpha_e) = n_e \lambda_e^3/2$, ΔF- configurational terms. At sufficiently low plasma densities, when $n_a B_0^3 \ll 1$ ($B_0 \sim a_0$-atomic radius, n_a -density of atoms) the interatomic interaction can be neglected. In addition, far from the saturation curve, where $\xi = (n_a/K_B T) \int \phi(r) dr \ll 1$ ($\phi(r)$ -charge-neutral interaction potential), the charge-neutral interaction is also negligible and the main configurational effect is connected with the Coulomb interaction. The usual approaches based on the perturbation theory[6,51,52] are developed for determining the equation of state of reacting Coulomb gases well fitted for the weak nonideality limit $\Gamma \ll 1$. Nevertheless some of these approaches are satisfactorily extrapolated to the strongly nonideal region, where $\Gamma \sim 1$. One of these approaches is the ring Debye-Hückel approximation in the Grand canonical ensemble for which configurational term can be expressed according to[53]

$$\frac{\Delta F_{coul}}{K_B T} = - \frac{Vk^3}{24\pi} - \sum_i N_i \ln\left(1 + \frac{Z_i^2 e^2 k}{2K_B T}\right); \; k = \frac{2\sqrt{\pi} e}{\sqrt{K_B T}} \left(\sum_i Z_i^2 \exp[\frac{\mu_i}{K_B T}]\right)^{1/2} \tag{2}$$

and the partition function is given by

$$Q_j = \sum_n g_n \left[\exp(-\beta E_n) - 1 + \beta E_n\right] \; ; \; \beta = (K_B T)^{-1}$$

The comparison of the P-V-T and P-V-E data obtained in the framework of this model with experiments[18,20-22] reveals good agreement within 20%.[4] Nevertheless the difference between the experiment and the theory exceeds the experimental error[18,20-22] both for this approximation and for the other usual plasma models with different nonideality corrections (including diagrammatic terms of the higher order than ring ones) and various partition function cutoff procedures.[4] At the same time the ideal plasma model with the weight of the ground state as the atomic partition function gives the equation of state quite close to the experimental one.[18] This fact has made it necessary to assume[17] the existence of an additional repulsion in the strongly compressed plasma and (or) a deformation of bound state-effects, which are not taken into account in usual plasma models.

In fact under high pressures and temperatures a lot of atoms are excited. Their sizes can considerably exceed the size B_0 of atoms in the ground state and are comparable with the average interparticle distances. In this case the restriction of the volume available for the realization of excited bound states leads to the strong perturbation of the energy spectrum. To account for this effect in the current work the confined atom model (CA) is used, which has been recently developed[54] for determining the equation of state of dense matter.

The CA model requires each atom to be placed in a spherical cell with hard walls, hence the effective interaction energy of atomic nucleus with electrons is given by

$$U(r) = \begin{cases} -Ze^2/r & 0 \leq r \leq r_c \\ \infty & r_c \leq r \end{cases} \tag{3}$$

The calculations of energy levels for the ground and the excited states of the confined hydrogen atom[54] have shown the dependence of the excited state energies on r_c increasing sharply with the principal quantum number. To calculate the energy spectra of the confined multielectron atoms the self-consistent Hartree-Fock method (restricted variant)[55,56] is used in this work. The system of integrodifferential equations is solved for all electronic terms appearing in the framework of LS-coupling.

$$\left[\frac{d^2}{dr^2} + V_{n\ell}(r) - \varepsilon_{n\ell}\right] f_{n\ell}(r) = \int_0^{r_c} X_{n\ell}(r,r')f_{n\ell}(r')dr' + \sum_{n\neq n'} \varepsilon_{n\ell,n'\ell}\, f_{n'\ell}(r) \tag{4}$$

for the radial parts $f_{n\ell}(r)$ of the one-electron atomic wave functions. Here $V_{n\ell}(r)$ is the self-consistent potential involving interaction of electrons with the nucleus and with each other. The integral in the right-hand part of (4) is the non-local part of the potential or the exchange term. $\varepsilon_{n\ell,n'\ell}$ and $\varepsilon_{n\ell}$ are nondiagonal factors and eigen values determined from the conditions

$$f_{n\ell}(0)=0; \quad f_{n\ell}(r_c)=0$$
$$\int f_{n\ell}(r)f_{n'\ell}(r)dr = \delta_{nn'} \tag{5}$$

The system (4) is numerically solved for various values of $\varepsilon_{n\ell}$ and defines a discrete spectrum of atoms in the plasma (Fig. 2). The equilibrium value of the parameter r_c is obtained from the condition of the free energy (1) minimum

$$\frac{\partial F}{\partial r} = 0$$

The free energy depends on r_c through both the partition function $Q_a \equiv Q_a(r_c)$ and the configurational term ΔF_{aa}, which has the form[57] corresponding to the interaction of hard spheres

$$\Delta F_{aa} \equiv \Delta F_{HS} = N_a K_B T \frac{4-3y}{(1-y)^2}\, y \; ; \; y = \frac{4\pi}{3} r_c^3 \frac{N_a}{V} \tag{6}$$

approximating the results of numerical calculations by the molecular dynamics method.[58]

This thermodynamically enclosed model is in fact the synthesis of purely plasma ideas on ionization equilibrium and a model of hard spheres taken from the theory of simple fluids. In contrast to cell models of solids[59,60] the above approximation is made in the framework of a description taking explicit account of translational degrees of freedom of separate particles and which distinguishes between free and bound electrons. The comparison of this model with experiments (Figs. 3, 4) shows, that the CA model* correctly reflects the experimentally obeserved tendency towards the decrease of plasma compressibility, overestimating, in some way, the repulsion effects

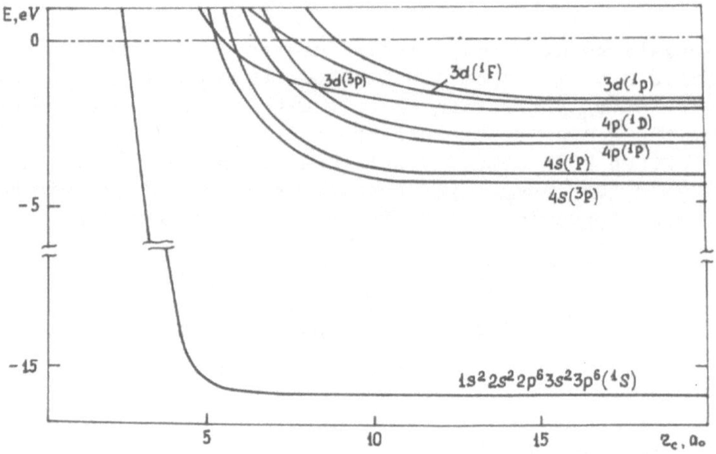

Fig. 2 Quantum-mechanical calculation of the energy spectrum of compressed argon in the framework of the confined atom model.

at high compressions. To improve the CA model both the boundary conditions for wave functions (5) and the interatomic repulsion model (6) should be replaced by more adequate ones.

One of the main problems of the quasichemical model of the strongly coupled plasma is the dividing of charges into free and bound. Separation of bound states demands the modification of the interaction between free electrons and ions at short distances. This effect in plasmas is taken into account by the introduction of the pair electron-ion pseudopotential.[61] Calculation of its parameters is very complicated problem, but in the strongly compressed plasma it can be solved semiempirically.[62] In this case the pseudopotential is given by[62]

$$\phi^*_{ei} = -\frac{e^2}{r}\left[1 - \exp\left\{-\frac{r}{\sigma}\right\}\right]; \quad \phi^*_{ee} = \phi^*_{ii} = \frac{e^2}{r}$$

Parameters of the pair correlation function

$$F \pm (r) = 1 \pm c_o \exp(-\nu r)\left[\frac{sh(\omega r)}{\omega r}\right]$$

ψ_o, ν, w are determined from the screening conditions

$$n_e \int [F_+(r) - F_-(r)]\, dV = 1$$
$$n_e \int [F_+(r) - F_-(r)]\,(r/r_D)^2\, dV = 3$$

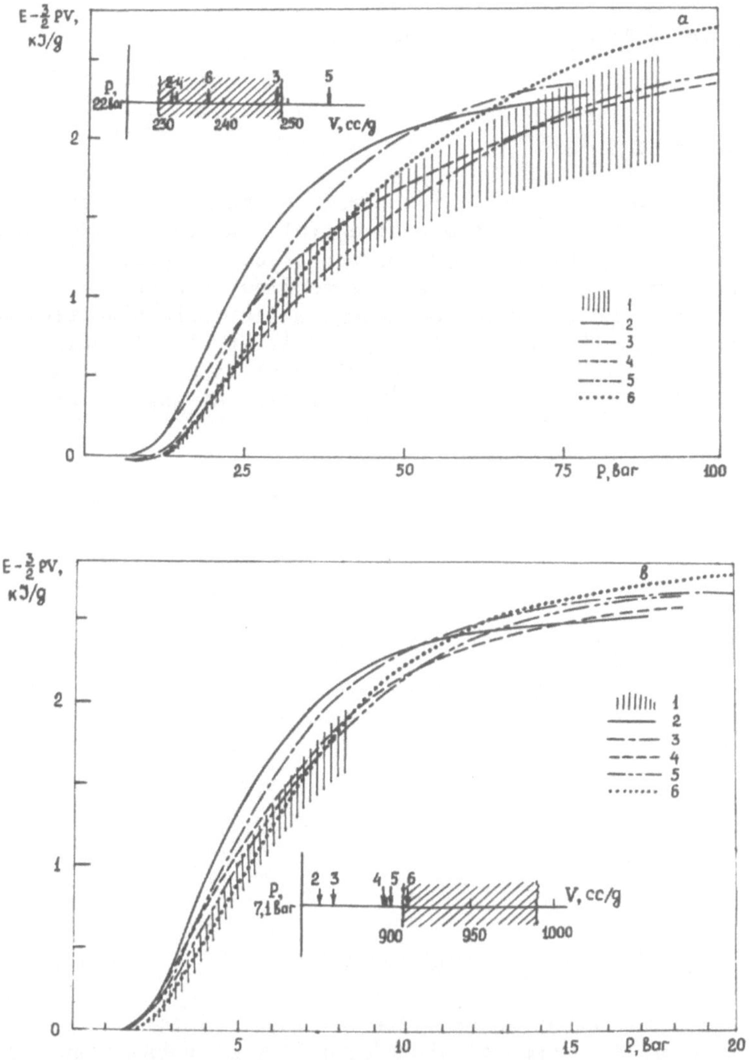

Fig. 3. Equation of state for Cesium Plasma at V=200 cc/g (a) and V=1000 cc/g (b): 1-Experimental data;[18] 2-Debye-Huckel approximation in the Grand canonical ensemble to ΔF_{coul};[53] atoms-ideal subsystem; 3-CA model; ΔF_{coul} corresponds;[53] 4-approximation[62] for ΔF_{coul}; atoms-noninteracting particles; 5-CA model with[62] the Coulomb interaction term; 6-ideal plasma model; the partition function $Q_a=g_0$ -the weight of the ground state.

and approximate relation between the amplitude of a screening cloud and the depth of the pseudopotential

$$\psi_0 \simeq \ln F_+ \simeq -\beta\,[\phi_{ei}(0) - \mu_e - \mu_i]$$

Corrections to the ideal-gas thermodynamic function are given by

$$\Pi = V n_e \int (e^2/r)\,[F_+ - F_-]\,dr$$

$$\Delta E = V n_e \int [F_+ \phi_{ei} - F_- \phi_{ii}]\,dr$$

$$\Delta PV = 1/3(2\Delta E - \Pi)$$

$$\Delta\mu_i = \Delta\mu_e \simeq (2N_e)^{-1}\Delta E$$

Where Π is the potential energy. To choose the main parameter of the model, the depth of the pseudopotential $\phi_{ei}(0)$, use is made of the experimental data on shock-wave compression of cesium plasma,[17,18] which shows that the best description of this experiment can be obtained on putting $\phi_{ei}(0)$ equal to the value of energy which divides particles into free and bound $-K_B T$. This model works properly at considerable degrees of ionization, and in combination with the CA model (4-6) shows the satisfactory description of the experiment[18] over all the experimental region (Fig. 3).

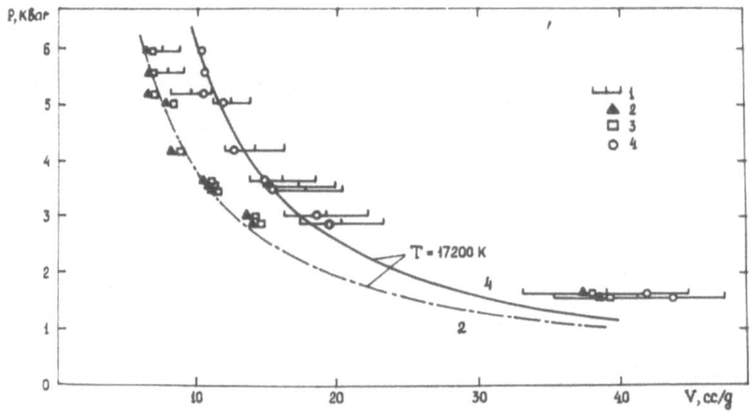

Fig. 4. Thermal Equation of State for Argon Plasma: 1-Experimental data;[20] 2-approximation[53] for charged subsystem; 3-pseudopotential model[61] with the second virial coefficient[66] for atomic interaction; 4-CA model with approximation[53] for charged subsystem.

An extremely interesting region for applying the chemical model of the plasma is the multi-megabar pressure range, realized by compression of solids by shock waves of extreme intensities. One of the examples of chemical model applications is the interpretation of the experiments on the shock compression of porous copper samples by superintensive shocks,[43,44] which make it possible to obtain extremely high concentrations of thermal energy (~ 0.75 MJ/cc) of a superdense ($n_e \sim 2 \quad 10^{23}$ cm^{-3}) plasma of high (P up to 20 Mbar) pressure. At the maximum temperatures attained, $T \sim 3\text{-}5 \cdot 10^5$ K, degeneration of electron is removed, $n_e \lambda_e^3 \sim 0.7$, the ionization degree of such plasma reaches five, while short-range and Coulomb interactions ($\Gamma \sim 2$) are strong. It should be noted that the Thomas-Fermi model with quantum and exchange corrections[63] gives values of density 20-30% exceeding the experiment[43,44] and distinction in pressure reaches up to several times. The chemical model of plasmas[9] makes it possible to analyze qualitatively some effects of the equation of state including the so-called electron-shell structure effects. In the equation of state of copper the Coulomb corrections (2), electron degeneration according to (1), and the short-range repulsion (6) of ions are taken into account. The parameter r_c of the short-range repulsion is evaluated from the Hartree-Fock calculations under boundary conditions (5) and is chosen to be 1.75 a_0. Partition functions of atoms and ions are calculated using energy levels[65] and ionization energies from.[64] Degrees of ionization up to tenth are taken into consideration.

The Hugoniots calculated by the plasma chemical model are represented in Fig. 5, for initial porousities of sample $m=\rho_{00}/\rho_0=3$(a) and 4(b). One can see the satisfactory agreement with the experiment in opposite to the Thomas-Fermi[63] model. From the chemical model effects under consideration the most essential is the short-range repulsion, the absence of which leads calculated plasma densities to be 1.5 times more than measured. The results of calculation are less depended upon the Coulomb correction and excited states, inclusion of both leads to an increase of calculated density. The least significant is the degeneration of electrons according to (1).

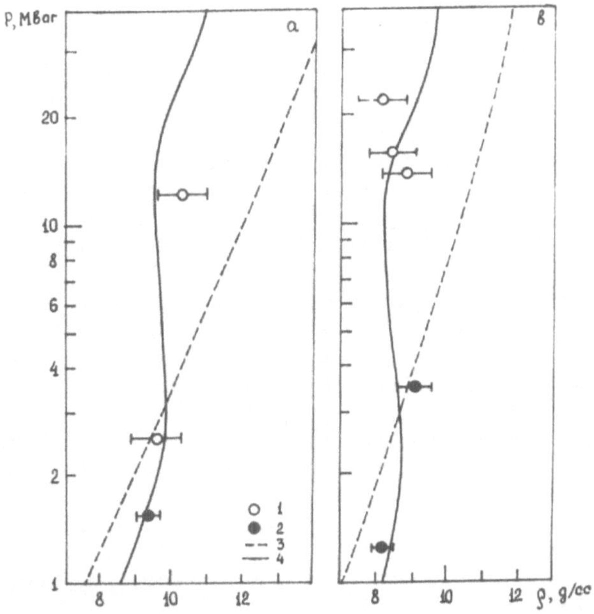

Fig. 5. Compression of Porous Copper with m=3(a) and 4(b) by Intense Shock Waves: 1-Experiment,[43] 2-experiment;[44] 3-the Thomas-Fermi model;[63] 4-current work.

REFERENCES

1. G. Kalman, P. Carini, ed. "Strongly Coupled Plasmas," Plenum, New-York-London (1978).
2. V. E. Fortov, I. T. Yakubov, "Physics of Nonideal Plasmas," Chernogolovka (1984).
3. W. Ebeling, V. E. Fortov, Yu. L. Klimontovich, et al., "Transport Properties of Dense Plasmas, Akademie-Verlag," Berlin (1983).
4. V. K. Gryaznov, I. L. Iosilevski, Yu. G. Krasnikov, et al., "Thermophysical Properties," Atomizdat, Moscow (1980).
5. V. A. Agureikin, S. I. Anisimov, A. V. Bushman, et al., Teplofiz. Vys. Temp. 22:964 (1984).
6. R. Abe, Prog. Theor. Phys. 22:213 (1959).
7. D. A. Kirzhnits, Yu. E. Lozovik, G. V. Shpatakovskaya, Usp. Fiz. Nauk 117:3 (1975).
8. J. P. Hansen, Phys. Rev. A8:3096 (1973).
9. H. E. DeWitt, G. Kalman, P. Carini, ed., in: "Strongly Coupled Plasmas," Plenum, New-York-London, p. 83 (1978).
10. V. M. Zamalin, G. E. Norman, V. S. Filinov, "Monte-Carlo Method in Statistical Thermodynamics," Nauka, Moscow (1977).
11. G. E. Norman L. S. Polak, ed. in: "Articles on Physics and Chemistry of Low-Temperature Plasmas," Nauka, Moscow (1971).
12. V. A. Alekseev, V. E. Fortov, I. T. Yakubov, Usp. Fiz. Nauk 139:193 (1983).
13. V. E. Fortov, Usp. Fiz. Nauk 138:361 (1982).
14. A. V. Bushman, V. E. Fortov, Usp. Fiz. Nauk 140:177 (1983).
15. Ya B. Zel'dovich, Yu. P. Raizer, "Physics of Shock Waves and High-Temperature Hydrodynamical Phenomena," Nauka, Moscow (1966).
16. V. A. Simonenko, N. P. Voloshin, A. S. Vladimirov, et al., Zh. Eksp. Teor. Fiz. 88:1452 (1985).
17. B. N. Lomakin, V. E. Fortov, Zh. Rksp. Teor. Fiz. 63:42 (1972).
18. A. V. Bushman, B. N. Lomakin, V. A. Sechenov, V. E. Fortov, O. E. Schekotov, I. I. Sharipdzhanov, Zh. Eksp. Teor. Fiz. 69:1624 (1975).
19. B. K. Tkachenko, et al., Fiz. Gor. Vzryva, 5:763 (1976); Teplofiz. Vys. Temp., 16:411 (1978).
20. V. E. Bespalov, V. K. Gryaznov, A. N. Dremin, V. E. Fortov, Zh. Eksp. Teor. Fiz. 69:2059 (1975).
21. V. E. Fortov, A. A. Leont'ev, V. K. Gryaznov, A. N. Dremin, Zh. Eksp. Teor. Fiz. 71:225 (1976).
22. V. K. Gryaznov, M. V. Zhernokletov, V. N. Zubarev, I. L. Iosilevski, V. E. Fortov, Zh. Eksp. Teor. Fiz. 78:573 (1980).
23. L. V. Al'tshuler, A. V. Bushman, A. A. Leont'ev, M. V. Zhernokletov, V. E. Fortov, Zh. Eksp. Teor. Fiz. 78:741 (1980).
24. C. E. Ragan III, M. G. Silbert, B. C. Diven, J. Appl. Phys. 48:2860 (1977).
25. R. J. Trainor, J. W. Shaner, J. M. Auerbach, N. C. Holmes, Phys. Rev. Lett. 42:890 (1979).
26. C. E. Ragan III, "Proc. VII Int. AIRAPT Conf.," Le Cresot (1978).
27. A. F. Akerman, A. V. Bushman, M. V. Demidov, et al., Zh. Eksp. Teor. Fiz. 89:852 (1985).
28. A. C. Mitchell, W. J. Nellis, B. Monahan in: "Shock Waves in Condensed Matter," W. J. Nellis, L. Seaman, R. A. Graham, ed., AIP, New York, 184:613 (1982).
29. R. S. Hawke, J. K. Scudder in: "High Pressure Science and Technology," B. Vodar, Marteau, Ph., Pergamon Press, v.2, p. 979, London (1980).
30. V. E. Bespalov, V. K. Gryaznov, V. E. Fortov, Zh. Eksp. Teor. Fiz. 76:141 (1979).
31. V. B. Mintsev, V. E. Fortov, V. K. Gryaznov, Zh. Eksp. Teor. Fiz. 79:116 (1980).

32. V. B. Mintsev, V. E. Fortov, V. E. Pisma, Pisma Zh. Eksp. Teor. Fiz. 30:401 (1979).
33. Yu. B. Zaporoshets, V. B. Mintsev., V. E. Fortov, O. M. Batovskii, Pisma Zh. Tehn. Fiz. 10:1139 (1984).
34. I. M. Isakov, A. A. Likal'ter, B. N. Lomakin, A. D., Lopatin, V. E. Fortov, Zh. Eksp. Teor. Fiz. 87:832 (1984).
35. A. G. Kupavin, A. V. Kirillin, Yu. S. Kopshunov, Teplofiz. vys. temp. 11:261 (1973); 13:1304 (1974).
36. L. V. Al'tshuler, Usp. fiz. nauk 85:197 (1965).
37. A. V. Bushman, I. K. Krasyuk, P. P. Pashinin, A. M. Prokhorov, V. Ya. Ternovoi, V. E. Fortov, Pisma Zh. Eksp. Teor. Fiz. 39:453 (1984).
38. O. V. Bazanov, V. E. Bespalov, A. P. Zharkov, et al., Teplofiz. 23:976 (1985).
39. I. Sh. Model, A. T. Narozhnyi, A. I., Harchenko, S. A. Holin, V. V. Hrustalev, Pisma Zh. Eksp. Teor. Fiz. 41:270 (1985).
40. C. E. Ragan III, Phys. Rev. A21:458 (1980).
41. C. E. Ragan III, Phys. Rev. A25:3360 (1982).
42. E. N. Avronin, B. K. Vodolaga, N. P. Voloshin, et al., Pisma Zh. Eksp. Teor. Fiz. 43:241 (1986).
43. S. B. Kormer, A. I. Funtikov, V. D. Urlin, A. N. Kolesnikova, Zh. Eksp. Teor. Fiz. 42:686 (1962).
44. V. N. Zubarev, M. A. Poduretz, L. V. Popova, et al., in: "Detonation, Chernogolovka," 61 (1978).
45. F. Hensel, E. U. Frank, Rev. Mod. Phys. 44:697 (1968).
46. R. S. Hixson, M. A. Winkler, J. W. Shaner, High Temperatures-High Pressures, 17:267 (1985).
47. V. E. Fortov, Teplofiz. Vys. Temp. N 1, 168 (1972).
48. S. B. Kormer, V. D., Urlin, L. T., Popova, Fiz. Tverd. Tela 3:2131 (1961).
49. M. A. Coock, J. Appl. Phys. 26, 426 (1955); 30:1881 (1959); J. Chem. Phys., 24, 60 (1955); Proc. Roy. Soc., Ser. A, 259:568 (1961).
50. V. E. Fortov, S. I. Musyankov, V. V. Yakushev, A. N. Dremin, Teplofiz. Vys. Temp. 12:957 (1974).
51. W. Ebeling, W. D. Kraeft, D. Kremp, "Theory of Bound States and Ionization Equilibrium in Plasmas and Solids," Akademie-Verlag, Berlin (1976).
52. H. E. DeWitt, J. Math. Phys. 7:616 (1966).
53. A. A. Lekal'ter, Zh. Eksp. Teor. Fiz. 56:240 (1969).
54. H. C. Graboske, D. J. Harwood, F. J. Rogers, Phys. Rev. 186:210 (1969).
55. D. R. Hartree, "The Calculation of Atomic Structures," Moscow (1960).
56. N. Glembozkis, Yu. Petkevichus, Lit. fiz. sb. 13:51 (1973).
57. N. F. Garnahan, K. E. Starling, J. Chem. Phys. 51:635 (1969).
58. B. J. Alder, T. E. Wainwright, J. Chem. Phys. 33:1439 (1960).
59. B. F. Roznyai, Phys. Rev. A5:1137 (1972).
60. A. F. Nikiforov, V. E. Novikov, Uvarov, V. B. Vopr. Atomn. nauki i Tehn., Ser: Met. i Progr. Chisl. Resh. Zadach Mat. Fiz., N 4, 16 (1979).
61. B. V. Zelener, G. E. Norman, V. S. Filinov, "Peturbation Theory and Pseudopotential in Statistical Thermodynamics," Nauka, Moscow (1981).
62. I. L. Iosilevski, Teplofiz. Vys. Temp. 18:447 (1980).
63. N. N. Kalitkin, Zh. Eksp. Teor. Fiz. 38:1534 (1960).
64. I. K. Kikoin, ed., "Tables of Physical Values," Atomizdat, Moscow (1976).
65. C. E. Moore, "Atomic Energy Levels," NBS, v. 2, Washington.
66. J. O. Hershfelder, C. F. Curtiss, R. B. Bird, "Molecular Theory of Gases and Liquids," Wiley, New York (1954).

STRONGLY-COUPLED PLASMA DIAGNOSTICS AND EXPERIMENTAL DETERMINATION OF DC ELECTRICAL CONDUCTIVITY

Marko M. Popović and
Svetozar S. Popović

Institute of Physics
Beograd, Yugoslavia

INTRODUCTION

In the last few years a remarkable effort was made to collect and analyze the observed behavior of plasmas with $0.1 < \Gamma < 10$ produced under conditions of high densities of charged particles (10^{17} cm^{-3} or more) and temperatures around 10^4-10^5 K. Dense plasmas with lower values of coupling factor are produced by stationary or pulsed electrical and optical discharges. Higher values of Γ can be obtained in shock-wave generated plasmas and there is an overlapping region where both experiments in discharges and shock tubes can be performed. This is an important fact for the reason that the results obtained by diagnostics of shock-wave produced plasmas can be compared and checked by more sophisticated methods of arc physics. This paper will try to point out some of the problems the diagnostics of dense plasmas are faced with, to show how arc physics is solving them, and to propose some experiments to be done in order to establish more reliable diagnostic methods for plasmas with higher values of Γ.

A comprehensive study of weakly nonideal plasmas from the point of view of arc physics was given by K. Gunther and R. Radtke (1). This book explained the physical background of the observed efforts and the experimental techniques used in electric arc plasma diagnostics and determination of transport properties. Although quantitative understanding of phenomena in arcs is quite satisfactory, very low Γ values do not offer many possibilities for universal conclusions.

Methods of generation of nonideal plasmas are discussed in detail by Kulik[2]. A wide variety of devices, from electric furnaces with steady or pulsed heating, electric arcs including free-burning, wall-stablized, stationary, as well as pulsed discharges were described. Dynamic compression and expansion methods are also given, but more details could be found elsewhere[3]. Plasma diagnostics could be in a sense facilitated if one of the thermodynamic properties would be kept constant during observation time. Some of the experiments do offer such possibilities and so there are isochoric gas heating in an electric discharge[1], isobaric capillary discharge,[2] isobaric expansion of an exploding wire[4], and adiabatic compression.[5]

Results of dynamic experiments that have had remarkable progress in the last decade are also reviewed in Ref. 3 and Ref. 6.

The present status of knowledge on physical phenomena in non-ideal plasmas are synthesized in the monograph by V. E. Fortov and I. T. Iakubov.[7] This systematic description and analysis of phenomena in dense plasmas revealed a lot of open questions, most of them connected with the influence of elementary processes to the macroscopic properties that are usually derived from measurements. In that sense, static electrical conductivity being the most illustrative and easiest to observe, can be used for demonstrating solutions to more general problems, provided that adequate diagnostic methods were used.

Presently there are many types of plasma devices that could under certain conditions give reliable information of plasma state and other parameters necessary to localize the measured DC electrical conductivity.

Among those that are mostly used one could point to plasma arcs ($\Gamma \leq 0.3$), shock-compression devices ($\Gamma \leq 10$), ballistic compressors ($\Gamma \leq 1$) and isobaric expansion devices ($\Gamma \leq 2$). At certain well- controlled conditions these devices could give local values of parameters with satisfying reproducibility.

Plasma parameters and macroscopic properties are evaluated from basic diagnostic data such as intensity of radiation, interferometric pattern, shock-wave velocities, electric potential, etc.

Pressure and mass density are the only parameters that are relatively free from traps consisting of certain model assumptions that are not satisfactory describing the complete plasma dynamics. Temperature particle number densities, electrical conductivity absorption coefficient and other parameters should be measured and evaluated with much more care due to their relations to the basic diagnostic data. Quantitative description of these relations is presently very doubtful in the case of plasmas with higher coupling parameter Γ.

Additional experimental difficulties arise from the fact that high particle density regions make difficult the penetration of a diagnostic tool into the plasma core. Therefore, for every particular case it is necessary to find a new way to avoid strong absorption on boundary layers and high gradients of plasma parameters along the observation path.

Insufficient knowledge of elementary processes and underdeveloped diagnostic devices complete the list of technical problems that are to be considered in order to make a properly designed experiment in this field.

TEMPERATURE MEASUREMENTS

There are two principal ways to derive temperature from basic diagnostic data in dense plasmas:

1. Quantitative spectroscopy based on elementary processes that may occur in dense plasmas. Spectra of optical radiation offer a variety of effects that could in principle be used for temperature diagnostics. Temperature-dependence of spectral line broadening and shift, spectral line intensity, relative line-to-continuum intensity and relative continuum intensity are among them. However, traditional approaches to the plasma optical properties based on impact or quasistatic microfield approximations, are no longer applicable. There are significant deviations from simple analytical expressions for density and temperature dependence of spectral line widths which make the diagnostics based on line spectrum rather doubtful.

2. Absolute measurement of radiation intensity in spectral regions where it is completely independent of atomic properties. Considering the problems concerned with the quantitative knowledge of elementary processes we would prefer using thermodynamic properties of the radiation in nonideal plasma. This method is very convenient for spatially extended plasma arrangements of constant temperature with comparatively high absorption.

At least two measurements of radiation intensity are necessary. One of these measurements should be absolute.

We will illustrate briefly the method on a few examples.

Pulsed Arc

A typical pulsed arc device is described in detail in Ref. 1. For the purpose of temperature diagnostics plasma is observed end-on through a quartz window and the observation path is following a layer of constant temperature except for the boundary region near the window. An auxiliary movable electrode is providing the possibility of two intensity measurements at two different observation lengths without disturbing isochoric nature of the discharge.

Temperature distribution along observation path is given in Fig. 1 and its typical time dependence in Fig. 2.

Intensity I (λ,x) emitted along the observation path is given by radiation transfer equation

$$\frac{d\,I\,(\lambda,x)}{d\,x} = K'B\,(\lambda,T) - I\,(\lambda,x) \tag{1}$$

where

$$K' = K(\lambda,T)(1 - e^{-\frac{hc}{KT}})$$

$B(\lambda,T)$ - blackbody intensity at the temperature T.

Measured values of radiation intensity at particular distance ℓ is introduced by boundary condition

$$I(\lambda,\ell) = I_m \tag{2}$$

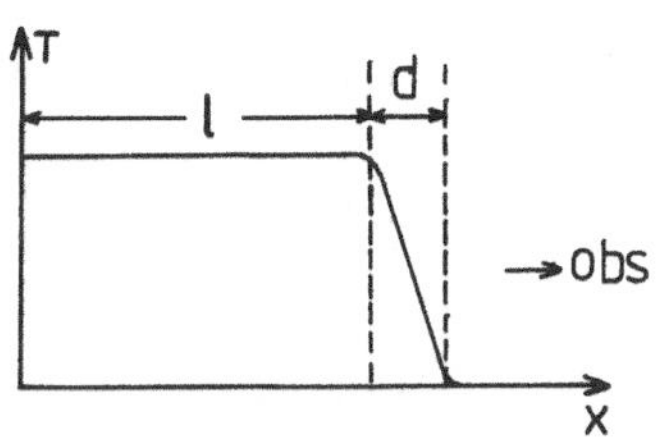

Fig. 1

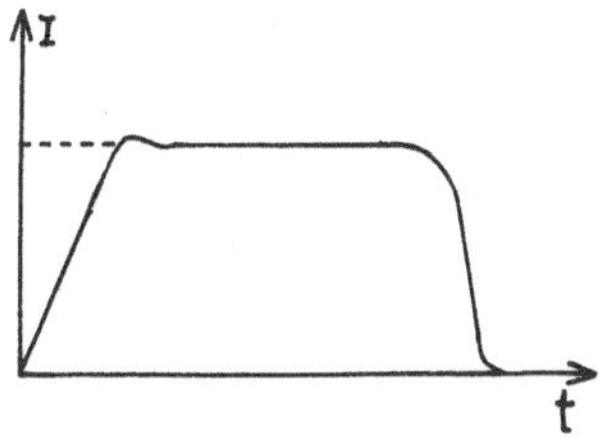

Fig. 2

By combining measured intensities with inherent boundary condition

$$I(\lambda,0) = B(\lambda,T) \tag{3}$$

one can develop a procedure for evaluation of unknown values $B(\lambda,T)$, $K(\lambda,T)$ and finally temperature T.

In certain cases absorbance in boundary inhomogeneous layer can be neglected and (1) can be reduced to

$$I(\lambda,T) = |1 - \exp(-K(\lambda,T)\ x|\ B(\lambda,T) \tag{4}$$

with unknown values $K(\lambda,T)$, $B(\lambda,T)$ and T and measured values $I_1(\lambda,T)$ at observation length ℓ_1 as an absolute intensity, and

$$\frac{I_1(\lambda,T)}{I_2(\lambda,T)} = \frac{1 - \exp(-K\ell_1)}{1 - \exp(-K\ell_2)} \tag{5}$$

Then T is obtained from blackbody function

$$B(\lambda,T) = \frac{I_1(\lambda,T)}{1 - \exp(-K\ell_1)} \tag{6}$$

Shock-Compression Tubes

There are many different constructions of explosive shock-compression devices which are reviewed in Ref. 3 and Ref. 6.

Temperature profile along the observation path is similar to the case of pulsed arc, except for the fact that both the front and rear edge of the plasma are moving during measurements with shock-wave velocity D, and free surface velocity U. Temperature profile and time dependence are given in Figs. 3 and 4.

When absorbance along shock front can be neglected, intensity of radiation emitted from a homogeneous layer along the observation path is given by

$$I(T,\lambda,t) = B(\lambda,T)\ \{1\text{-}\exp - |K'(D-U)|t\} \tag{7}$$

with K' being absorption coefficient corrected for stimulated emission, D shock-wave velocity, U free surface velocity.

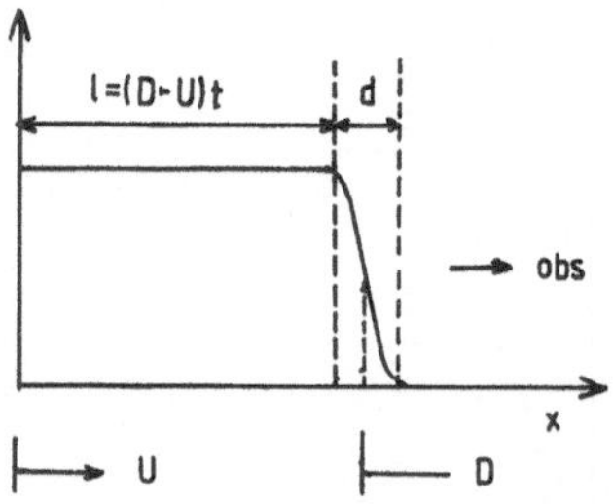

Fig. 3

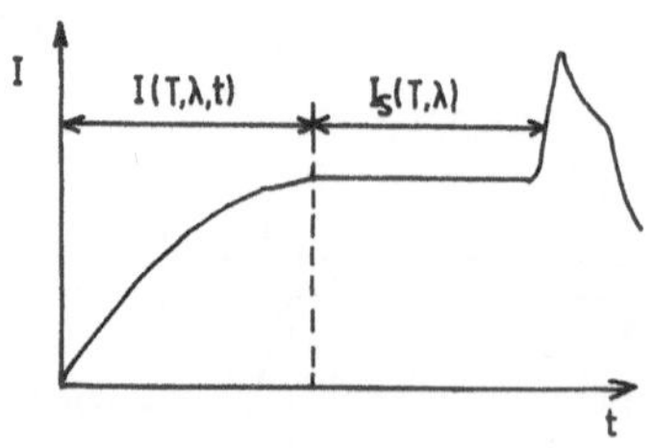

Fig. 4

Saturated values of radiation intensity in Fig. 4 $I_s(\lambda,T)$ is then interpreted as blackbody function $B(\lambda,T)$. The second measured value is the ratio of radiation intensity $I(\lambda,T,t)$ to saturated intensity $I_s(\lambda,T)$.

However, the assumption of negligible absorbance along shock-wave front should be checked carefully for every particular experimental condition. If this assumption is violated, then radiation transfer differential equation similar to (1) - (3) should be used.

Optical Discharges

Optical discharges created in the focus of laser radiation require an inversion procedure for exact profile $T(x)$ measurement. The full procedure is rather complicated and it has been performed until now only in a stationary optical discharge.[8]

In performing absolute measurements of radiation intensity one has to take care of some conditions that are to be fulfilled if simple relations as (4) or (7) are used.

First, absorbance at boundary layers should be small for the particular wavelength of observation

$$A = \int_{x}^{x+d} K(T(x))\, d\,x \ll 1 \qquad (8)$$

Although the thickness of the boundary layer is of the order of one tenth of a millimeter in cases of higher density this layer is able to reabsorb the emitted radiation. If this happens, brightness temperature that can be obtained by comparing plasma radiation to standard brightness source is different than the actual thermodynamic temperature.

Second, relative error in temperature measurement is connected to the radiation intensity by the following relation

$$\frac{dT}{T} = \frac{\lambda T}{c_2} \left| 1 - \exp\left(-\frac{c_2}{\lambda T}\right) \right| \frac{dI}{I} \qquad (9)$$

where c_2 is the constant from Planck's blackbody radiation law. This relation suggests the choice of small products of wavelength λ and plasma temperature. For a given plasma temperature observation wavelength should be as small as possible.

Also, the temperature determination method suggested here can be applied if the optical thickness of the boundary layer is negligible and the product λT so small as to diminish the experimental error.

Both conditions can be fulfilled if the observation wavelength is as short as possible.

Transverse gradient of temperature, particle number density and consequently refractive index along the observation path could cause a substantial deviation of the observation beam in the direction of increasing refraction. This deviation in radiation direction can be expressed by[9]:

$$\frac{\delta r}{\ell} = \frac{\ell}{2n} \frac{\partial n}{\partial T} \frac{\partial T}{\partial r} \tag{10}$$

where ℓ is the plasma length, n reflective index and T(r) radial temperature profile. This deviation can produce serious distortions in the measured temperature profiles.

Finally, the comparison of radiation intensity from plasma and standard radiation sources has its own rules. The ideal situation that both plasma and standard radiation source have the same temperature and directly comparable radiation intensities is never fulfilled. The problem then is consisted in the large attenuation necessary to compare the plasma intensity with a radiation standard. One or more reflections of the observation beam at a pure glass surface can yield attenuation by incidence sufficient enough for comparison of the two radiation intensities. According to Fresnel's law for one reflection one gets

$$r = \left(\frac{n-1}{n+1}\right)^2 \bullet \left(1 + \frac{n^2-1}{2-\alpha^2} \bullet \frac{\alpha^2}{n^2}\right) \tag{11}$$

where n is the refractive index of the glass and α is the angle of incidence. In this case a well calibrated glass reflector with accurately determined refractive index is necessary.

Different timing of gated detectors gives another possibility for a controlled reduction of plasma radiation intensity.

These problems connected with absolute intensity measurement should not be neglected in the design of experiments in dense plasmas. There are two main principles:

1. Absorption coefficient at the observation wavelength $K(\lambda,T)$, plasma length and the thickness of boundary layer d have to be chosen so that

   $$\ell^{-1} \ll K(\lambda,T) \ll d^{-1} \quad . \tag{12}$$

   If it is not possible, a new procedure for temperature determination should be developed, based on more treatment of radiation transfer.

   However, by choosing observation wavelength in ultraviolet region one can avoid these problems with interpretation of basic diagnostic data. For instance, for a shock-compressed argon with typical values of plasma parameters $n_e = 10^{20}$ cm^{-3}, T = 20000 K, D = 7 km/s, ℓ = 10 mm, d = 0.3 mm, Γ = 1.9, the absorption coefficient in the ultraviolet region is low enough so that condition (12) can be fulfilled at Γ < 270 mm. For higher observation wavelength a differential equation of radiative transfer has to be used.

2. The temperature of standard radiation source should be as high as possible, with the necessary precision in radiation intensity. Unfortunately, standard radiation sources with precision within 1% in intensity, tungsten ribbon lamps and carbon arcs, have much too low a temperature. Therefore, new standard sources are to be developed. A possible candidate for pulsed arc experiments could be a pulsed arc in xenon which already can reach a precision of order of 5% at 12000 K.[10] Detonation wave in some gas mixtures[11] or shock-compressed xenon could be possible secondary standards for shock tube experiments in the future.

For the reason of strong absorption in boundary layers standard brightness sources could give doubtful information on temperature in plasma core.

PRESSURE MEASUREMENT

Usual detectors for transient pressure measurements in pulsed arcs are piezoresistive or piezocapacitive probes. They have a linear response from DC to several hundred kHz, and can be calibrated in stationary regime and used for measurements with pulses longer than 10 μs. A protection against thermal stress should be provided.

In shock-compression experiments, pressure is evaluated from shock and particle velocities and mass density measurements. This procedure is very well established due to extended EOS measurements, so that pressure is not considered as a parameter which could cause any problem in dense plasma diagnostics.

MEASUREMENTS OF PARTICLE NUMBER DENSITIES

Diagnostic methods based on elementary processes inside the plasma are not to be considered as independent diagnostic tools for determination of particle number densities in high pressure plasmas

All methods that produced more or less reliable results were based on the analysis of the interaction of external radiation with the plasma. Among them, measurements of refractive index are mostly used.

In the case of large gradients over the whole observation path, as in cascade arcs, capillary discharges, and optical discharges, the Schlieren technique has a certain diagnostical capability.[12] This method is based on recording the deflection of a light beam travelling through a plasma density gradient.

The laser interferometry method is usually based on recording the temporal change of refraction index, corresponding to a simultaneous change of plasma parameters. It is still considered as a suitable method for pulsed arc diagnostics.[13]

Methods based on direct measurements of plasma frequency are suggested in literature[14, 15] but these ideas need further development.

Particle number densities are often evaluated from relations describing equilibrium plasma composition and direct measurements of pressure and temperature only. These results can be treated only as a rough estimation since the equilibrium plasma composition is itself a subject of investigation.

DETERMINATION OF ELECTRICAL CONDUCTIVITY

Local values of electrical conductivity in a plasma are usually derived from Ohm's law

$$\frac{I}{E} = \int_S \sigma(x,y)\, d\, s \qquad (13)$$

where I is the electric current, E is the electric field and $\sigma(x,y)$ local value of electrical conductivity.

In cylindrical symmetry (13) takes the form

$$I = 2 \pi E \int_0^R \sigma(r) \, r \, d r \tag{14}$$

where R is the plasma radius.

In pulsed arcs two probes for electrical measurements are enough. In shock-compression experiments two probes for current test and two potential probes are necessary.

The unknown quantity in (13) is the electrical conductivity σ which is basically a function of temperature. It can be transformed to a standard form of an inhomogeneous Volterra integral equation of the second kind[1]:

$$\sigma(T_{ax}) + \int_{T_0}^{T_{ax}} K(T,T_{ax}) \, \sigma(T) \, d T = f(T_{ax}) \tag{15}$$

where t_{ax} is the plasma axis temperature, T_0 wall temperature and we identify as the imhomogeneous part

$$f(T_{ax}) = \frac{1}{\pi R^2 g(T_{ax},T_{ax})} \frac{d}{d T_{ax}} \left(\frac{I}{E} \right) \tag{16}$$

and as the kernel

$$K(T,T_{ax}) = \frac{1}{g(T_{ax}, T_{ax})} \frac{\partial g(T, T_{ax})}{\partial T_{ax}} \tag{17}$$

where $g(T,T_{ax}) = - \frac{\partial x}{\partial T}$; $x = (r/R)^2$.

Now $\sigma(T)$ is an unknown function.

Equation (15) has a convergent numerical solution in the form:

$$\sigma(T) = \sum_n \rho_n(T) \tag{18a}$$

$$\rho_0(T) = f(T) \tag{18b}$$

$$\rho_{n+1}(T) = -\int_{T_0}^{T} K(T',T) \, \rho_n(T') \, dT' \tag{18c}$$

Measured values in this procedure are axis temperature T_{ax}, corresponding voltampere characteristics $I/E = F(T_{ax})$ and observation layer radius r.

This is a consequent procedure with well defined and controlled numerical accuracy but with high demands in accuracy of basic experimental information, such as voltampere characteristics, axis temperature and radial temperature profiles. If this cannot be fulfilled methods based on a priori assumptions on the temperature dependence of the electrical conductivity can be developed Some arbitrary constants should then be optimized so that (13) is satisfied.

In some recent papers[16-18] trial function $\sigma(T)$ was constructed in the same way as calculated $\sigma(T)$ for ideal plasmas with interaction of electrons with neutral included. A free parameter is introduced in the Coulomb logarithm through effective screening radius $r_s = xD$, where D is Debye length and x is to be changed in an iterative procedure until the relation (13) with experimental values for electric current and field strength is satisfied.

Finally, the evaluation of electrical conductivity can be trivial by applying simply $\sigma = I/\pi R^2 E$ is one assures homogeneous plasma in his experiment. It should be noted, however, that even in radiation dominated plasmas where the radiation transfer is cooling the inner parts of plasma and flatten temperature profiles, the results obtained by this simple formula proved to be incorrect.

It is the fact that experiment results for static electrical conductivity differ in plasmas with $0.1 < \Gamma < 10$ sometimes by a factor of 2 or more while authors claim accuracy in measurements within 10 to 40 percent. In that sense $\sigma(T,P)$ and $K(\lambda,T,P)$ are still not perfectly established (see Fig. 12, 14 and 15 in Ref. 6 and Fig. 8 in Ref. 5. Although pressure diagnostics has a satisfying precision, temperature diagnostics suffer from the lack of reliable absolute radiation intensity measurements and the influence of elementary processes which is still not clearly observed. Besides, unprecise radiation standards at elevated temperatures are the limiting factor for optical radiation measurements.

REFERENCES

1. K. Günther, R. Radtke, Electric Properties of Weakly Non-Ideal Plasmas, Birkhänser Verlag, Basel, Boston, Stuttgart (1984).
2. P. P. Kulik, V. A. Ryabiy, N. V. Yermokhin, Neidealynaya Plasma, Energoatomizdat, Moscow (1984) in Russian.
3. V. E. Fortov, Sov. Phys. - Uspekhi 25:781 (1982).
4. G. R. Gathers, J. W. Shaner, D. A. Young, Phys. Rev. Lett. 33:70 (1974).
5. K. Günther, H. Hess, R. Radtke, Inv. Papers, 17th ICPIG 120 (1985).
6. V. A. Alekseev, V. E. Fortov, I. T. Iakubov, Sov. Phys. - Uspekhi 26:99 (1983).
7. V. E. Fortov, I. T. Iakubov, Fizika Neidealynoy Plasmi, Cernogolovka, In Russian (1984).
8. C. Carlhoff, E. Krametz, J. H. Schäfer, J. Uhlenbusch, J. Phys. B, 19:2629 (1986).
9. S. S. Popović, N. Konjević, Zs. Naturforschung 31a:1042 (1976).
10. K. Günther, R. Radtke, J. Phys. E 8:371 (1975).
11. R. G. McQueen, J. N. Fritz, Proc 1st Conf. Shock Waves in Cond. Matter, Menlo Park 193 (1981).
12. J. Glasser, R. Villadrosa, J. Chapalle, J. Phys. D, 11:1703 (1978).
13. N. Uzelac, N. Konjević, Phys. Rev. A, 33:1349 (1986).

14. M. Skowronek, J. Rous, A. Goldstein, F. Cabannes, Phys. Fluids 13:378 (1970).
15. P. Bakshi, G. Kalman, Phys. Rev. A, 30:613 (1984).
16. K. Günther, M. M. Popović, S. S. Popović, R. Radtke, J. Phys. D, 9:1139 (1976).
17. C. Goldbach, G. Nollez, M. M. Popović, S. S. Popović, Zs. Naturf. 39a:11 (1978).
18. K. Günther, S. Lang, R. Radtke, J. Phys. D, 16:1235 (1983).

CHAPTER III

MOLECULAR DYNAMICS AND KINETIC THEORY

TWO-COMPONENT PLASMAS IN TWO AND THREE DIMENSIONS

Jean-Pierre Hansen

Laboratoire de Physique Théorique des Liquides
Université Pierre et Marie Curie
75252 Paris Cedex 05

I - INTRODUCTION

The two-component plasma (TCP) is a model system made up of N_1 positive charges q_1 and N_2 negative charges q_2 in a d-dimensional volume Ω. The corresponding number densities $n_\alpha = N_\alpha/\Omega$ satisfy the charge neutrality requirement

$$n_1 q_1 + n_2 q_2 = 0 \tag{1.1}$$

We shall be mostly concerned with point charges, and more specifically with fully stripped ions ($q_1 = + ze$) and electrons ($q_2 = - e$, where e is the elementary charge) ; the interaction is then purely Coulombic and the total Hamiltonian of the system can be cast in the form :

$$H = H_{11} + H_{22} + V_{12} \tag{1.2a}$$

where

$$H_{\alpha\alpha} = \sum_{i=1}^{N\alpha} \frac{p_i^2}{2m_\alpha} + \sum_i \sum_{<j} v_{\alpha\alpha} (|\vec{r}_i - \vec{r}_j|) \tag{1.2b}$$

$$V_{12} = \sum_{i=1}^{N_1} \sum_{j=1}^{N_2} v_{12} (|\vec{r}_i - \vec{r}_j|) \tag{1.2c}$$

$$v_{\alpha\beta}(r) = q_\alpha q_\beta \Phi(r) \tag{1.2d}$$

and $\Phi(r)$ is the solution of the d-dimensional Poisson equation, i.e. $\Phi(r) = 1/r$ in 3d and $\Phi(r) = -\ln(r/L)$ (with L an arbitrary scaling length) in 2d. Partially stripped ions (of well-defined valence Z) have a finite core, so that the ion-ion potential $v_{11}(r)$ must then include a short-range Born-Mayer repulsion, while the ion-electron interaction can be described by a pseudo-potential for distances shorter than the core radius. However, except at very high temperatures, the Coulomb repulsion between ions is sufficiently strong to prevent the cores (of diameter σ say) to come into contact (i.e. $Z^2e^2/\sigma \gg k_BT$ in 3d) so that a point ion description is generally adequate.

Now it is convenient to add opposite uniform backgrounds (of charge density $\rho_0 = \pm n_1q_1 = \pm n_2q_2$) to the ion electron system ; the corresponding ion-background, electron-background and background-background Coulomb interactions are divided among H_{11}, H_{22} and V_{12} in such a way that each term has now separately a well-defined meaning in the thermodynamic limit. The total hamiltonian takes then the form[1] :

$$H = H_{11} + H_{22} + V_{12} \tag{1.3}$$

where H_{11} is the familiar "one component plasma" (OCP) hamiltonian of the ions, H_{22} is a similar OCP (or "jellium") hamiltonian for the electrons, and V_{12} is the sum of V_{12} and of the background-background potential energy.

Dimensionality has a strong influence on the behaviour of a two-component Coulomb gas. In 1d, the Coulomb gas is a dielectric at all temperatures : oppposite charges are always bound in pairs[2]. In 3d the Coulomb gas is believed to be a conductor (or plasma) as long as it remains in a disordered (fluid phase), but Quantum Mechanics is essential to prevent collapse of opposite point charges due to the attractive r^{-1} singularity of the Coulomb potential. The 2d case is the most interesting, since the corresponding Coulomb gas is a dielectric below some density-dependent critical temperature, while it is a plasma above that temperature. This dielectric-plasma transition is the prototype of a Kosterlitz-Thouless (KT) transition in 2d systems, which is always characterized by a divergent response (susceptibility) to an external field[3,4,5].

In the next two sections, we first consider the two-component plasma (or Coulomb gas) in 3d, while the 2d Coulomb gas will be subject of the last sections.

2 - DIFFERENT REGIMES IN THREE DIMENSIONS

Two complementary approaches have been used to describe ion-electron plasmas in 3d ; they apply to different degeneracy regimes of the electronic component.

For strongly or partially degenerate electrons (i.e. for temperatures $T \lesssim T_F$, the Fermi temperature), a two-fluid description, not unlike that of liquid metals[6], appears to be the most natural. According to (1.3), the ion-electron system appears as the superposition of two independent OCP's which are coupled by V_{12}. The properties of the classical ionic OCP[7] and of the fully or partially degenerate electron "jellium"[8,9] are reasonably well known, so that it appears natural to treat the ion-electron coupling V_{12} by perturbation theory. To zeroth order in V_{12}, the plasma properties are just the sums of their ionic and electronic OCP counterparts. The first order term vanishes, due to spatial homogeneity and charge neutrality. To second order, ion-electron coupling is described by linear response theory. Such a program was carried through for the thermodynamic properties of very denses plasmas with fully degenerate (possibly relativistic) electrons[10]. The procedure has been considerably developed and improved by Ichimaru and coworkers[11], who studied static and dynamic (transport, collective modes) properties of ion-electron plasmas over an extensive range of electron degeneracy, including static local field corrections. Similar ideas have been applied to the two-temperature ion-electron plasma[12].

To go beyond the linear screening approximation in the treatment of the ion-electron coupling, one can resort to the familiar Kohn-Sham-Mermin density functional formalism, as illustrated by the work of Chihara[13] and Perrot and Dharma-Wardana[14], described elsewhere in the present procedings.

In the opposite regime of weak degeneracy ($T \gg T_F$), an alternative description of ion-electron plasmas is based on the use of effective potentials to model quantum effects at short distances, in conjunction with classical Statistical Mechanics. This approach is presented in the next section in the case of a hydrogen (electron-proton) plasma.

3 - THE SEMI-CLASSICAL HYDROGEN PLASMA

To evaluate the importance of quantum effects for electrons, we must compare the following three fundamental lengths :

$$l = \frac{e^2}{k_B T} \qquad \text{(Landau length)} \tag{3.1a}$$

$$a = (3/4\pi n_2)^{1/3} \qquad \text{(electron sphere radius)} \tag{3.1b}$$

$$\lambda_2 = \frac{\hbar}{\sqrt{2\pi m_2 k_B T}} \qquad \text{(thermal de Broglie length)} \tag{3.1c}$$

The ratio of the first two lengths defines the dimensionless Coulomb coupling parameter :

$$\Gamma = \frac{\ell}{a} = \frac{e^2}{a k_B T} \tag{3.2}$$

In the absence of interactions ($e \to 0$), only a and λ_2 are relevant length scales. The corresponding ideal Fermi gas is non-degenerate, provided $\lambda_2 < a$ (or $T > T_F$), which implies the following condition :

$$\Gamma \lesssim r_s = \frac{a}{a_{Bohr}} \tag{3.3}$$

In the presence of interactions, close electron-electron collisions can be treated classically, provided $\lambda_2 < l$, which yields the condition

$$\Gamma \gtrsim \frac{1}{r_s} \tag{3.4}$$

Hence a semi-classical description of the plasma may be expected to be reasonable as long as :

$$\frac{1}{r_s} \lesssim \Gamma \lesssim r_s \qquad (3.5)$$

which requires, in particular, that $r_s \gtrsim 1$. Notice that, for any fixed density, quantum effects eventually take over at sufficiently high temperatures, i.e. in the weak coupling limit.

Since the classical partition function for an electron-proton plasma does not exist at any temperature, due to the Coulomb collapse of oppositely charged particles, the use of an effective ion-electron potential, accounting for the "smearing" of the electron charge over a sphere of radius $\simeq \lambda_2$ (Heisenberg uncertainty principle) was suggested a long time ago by Morita[15]. A very simple form for the effective potentials, valid for sufficiently low densities, was suggested by Deutsch and collaborators[16], namely :

$$v_{\alpha\beta}(r) = \frac{q_\alpha q_\beta}{r} \quad [1 - \exp\{-r/\lambda_{\alpha\beta}\}] \qquad (3.6)$$

where :

$$\lambda^2_{\alpha\beta} = \lambda^2_\alpha + \lambda^2_\beta = \frac{\hbar^2}{2\pi\mu_{\alpha\beta}k_BT} \qquad (3.7)$$

and $\mu_{\alpha\beta}$ is the reduced mass for an α-β pair. This "primitive model"of a hydrogen plasma accounts for quantum diffraction, but not for electron symmetry (Pauli principle) effects, and is expected to be physically relevant in the thermodynamic range (3.5). Electron symmetry effects may be approximately accounted for, by adding a "Pauli repulsion" term to $v_{22}(r)$[16] ; an even finer semi-classical description distinguishes between electron pairs with parallel and antiparallel spins, with a Pauli repulsion acting only between the former[17]. This procedure may be adequate for the description of static structure, but is dubious when applied to dynamical properties (like electron transport), as shown by the recent kinetic theoretical investigation of Wallenborn et al. reported in the present procedings.

The "primitive model" (3.6) of a hydrogen plasma has been extensively studied in the strong coupling regime ($\Gamma \simeq 1$) compatible with the restrictions (3.5) by Molecular Dynamics (MD) computer simulations and by Kinetic Theory. One of the main difficulties in the simulations is that two very different time scales are involved, since the ratio of the electronic and ionic plasma frequencies ω_{p2}/ω_{p1} scales as $(m_1/m_2)^{1/2} \simeq 43$. The M.D. time step must be adapted to the faster electronic motions[18]. This means that little or no information is gained concerning the much slower ionic motions and the dynamical properties associated with them, like the shear viscosity of the plasma. This is illustrated in Fig. 1, taken from the thesis of B. Bernu, which shows the decay of the normalized autocorrelation function (ACF) $\eta(t)$ of an off-diagonal element of the stress tensor.

The initial, rapid decay may be associated with the fast electronic motion, wile the subsequent slow decay is intimately related to the ionic degrees of freedom ; this slow decay precludes any reasonable estimate of the shear viscosity $\eta(t)$.

Fortunately the fast electronic motions dominate electrical and thermal conduction and the wavenumber and frequency-dependent charge fluctuation spectrum $S_{ZZ}(k,\omega)$, which determines the dispersion and damping of the plasmon mode. The spectrum is defined by :

$$S_{ZZ}(k,\omega) = \frac{1}{2\pi} \int_{-\infty}^{+\infty} dt e^{i\omega t} \frac{1}{Ne^2} \langle \rho_{kZ}(t)\rho_{-kZ}(0)\rangle \qquad (3.8)$$

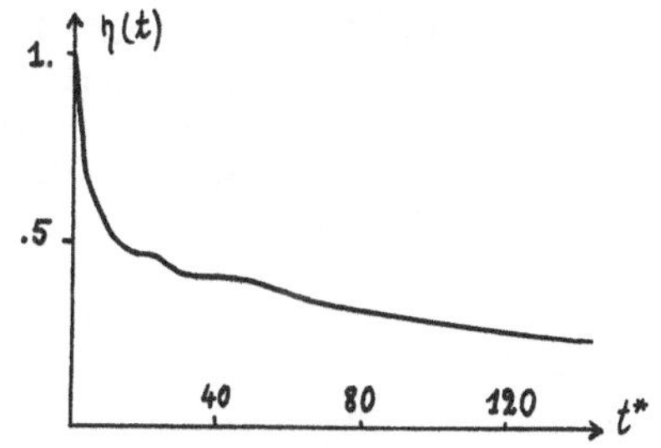

Fig.1 : Normalized ACF $\eta(t)$ versus reduced time $t^* = \omega_{p2} t$ for the semiclassical hydrogen plasma with $\Gamma = 0.5$; $r_s = 1$

where $\rho_{kZ}(t)$ is a Fourier component of the microscopic charge density at time t :

$$\rho_{kZ}(t) = \sum_{j=1}^{N} q_j \exp\{-ik.r_j(t)\} \qquad (3.9)$$

An elementary application of Ohm's law, Poisson's equation and charge conservation (continuity equation) leads to the following exact expression of the long wavelength limit of the spectrum[19] :

$$s(\omega) = \lim_{k\to 0} \frac{S_{ZZ}(k,\omega)}{S_{ZZ}(k)} = \frac{1}{\pi} \mathbb{R} \frac{1}{-i\omega + 4\pi\sigma(\omega)} \tag{3.10}$$

$$= \frac{1}{\pi} \frac{4\pi\sigma'(\omega)}{[\omega - 4\pi\sigma''(\omega)]^2 + [4\pi\sigma'(\omega)]^2}$$

where $\sigma(\omega) = \sigma'(\omega) + i\sigma''(\omega)$ is the complex a.c. conductivity which is related to the normalized electric current ACF J(t) by the standard Green-Kubo relation :

$$\sigma(\omega) = \frac{\omega_p^2}{4\pi} \int_0^\infty J(t) e^{i\omega t} \, dt \tag{3.11}$$

It is clear from (3.10) that the imaginary part $\sigma''(\omega)$ determines the position of the plasmon resonance, while the real part $\sigma'(\omega)$ determines the (collisional) damping of the mode in the long wavelength limit.

The electric current ACF J(t), as well as the electron velocity ACF $Z_2(t)$ have been calculated for several states of the hydrogen plasma around $\Gamma \sim 1$, $r_s \lesssim 1$ by MD simulations[18,20]. The ACF J(t) turns out to decay considerably more slowly than $Z_2(t)$, so that a Nernst-Einstein-like relation between the d.c. conductivity σ and the electron self-diffusion constant D_2, which follows if all cross-correlations between velocities of different electrons are neglected, namely :

$$\sigma \simeq \omega_{p2}^2 \frac{m_2}{k_B T} D_2 \tag{3.12}$$

underestimates σ by a factor of 2-3. This can be qualitatively understood, since electron-electron collisions conserve the electric current, but not individual electron velocities. For similar reasons it is found that the Kubo current associated with thermal conduction decays on roughly the same time scale as individual electron velocities, i.e. again much faster than the electric current [21].

According to (3.10) and (3.11), J(t) also determines the frequency and damping of the plasmon mode at k = 0. The most important finding is a significant shift above the plasma frequency, i.e.

$$\lim_{k\to 0} \omega(k) = \omega_{p2}[1+\Delta] \tag{3.13}$$

where Δ is typically of the order of + 0.02 for $\Gamma \sim 1$, $r_s \lesssim 1$[20]. This collisional effect persists to finite wave numbers, where the plasmon resonance in $S_{ZZ}(k,\omega)$ is considerably shifted and broadened relative to the collisionless mean-field (Vlasov) result[18]. The collisional damping is reasonably well described by a memory function analysis[18] or by generalized Fokker-Planck-like equations[22].

Kinetic theory has also been applied to the calculation of the electron collision frequency and of the transport coefficients (D_2, σ, K) of the hydrogen plasma in the framework of the "primitive model" (3.6)[23], and its extensions which include a Pauli repulsion between electrons[20,24,25]. It should be stressed once more that such calculations apply only if the conditions (3.5) are satisfied. The link between the non-degenerate and degenerate regimes has been investigated by Boercker et al.[26], while Wallenborn and his collaborators report their results on the weak coupling limit in the present proceedings.

Finally, the thermal relaxation of a two-temperature plasma (where $T_1 = T_2$ has been the object of extensive M.D. simulations[27] which are in good agreement with the predictions of the standard Landau-Spitzer result for the thermal relaxation rate, provided an adequate definition of the Coulomb logarithm is used.

4 - THE TWO-DIMENSIONAL COULOMB GAS

Due to the (weakly) binding nature of the logarithmic attraction between opposite charges, the proporties of the two-dimensional Coulomb gas (or two-component plasma) differ profoundly from those of its three-dimensional counterpart, particularly at low temperatures, as already mentionned in the introduction. The symmetric Coulomb gas is made up of oppositely charged hard disks of diameter σ ; the corresponding pair potentials are :

$$\begin{aligned} v_{\alpha\beta}(r) &= \infty \quad ; r < \sigma \\ &= - q_\alpha q_B \ln (r/L) \quad ; r > \sigma \end{aligned} \tag{4.1}$$

where $q_\alpha = \pm q$ $(1 \leq \alpha \leq 2)$.

First consider the case of point charges (σ =0) ; this plasma is thermodynamically stable as long as the Coulomb coupling

$$\Gamma = \frac{q^2}{k_B T} = \frac{1}{T^*} < 2$$

since the Boltzmann factor for a pair of opposite charges :

$$\exp \{ - \frac{q^2}{k_B T} \ln (r/L)\} = (\frac{r}{L})^{-\Gamma} \tag{4.2}$$

is clearly integrable for $\Gamma < 2$. A simple scaling argument leads to the

exact equation of state[28] :

$$\frac{\beta PS}{2N} = 1 - \frac{\Gamma}{4} \tag{4.3}$$

where S is the area containing N charges of each species. The partition function and its temperature derivatives diverge as $\Gamma \to 2^-$ due to the collapse of pairs of opposite charges. In particular the internal energy U and the specific heat at constant area, c_S, diverge as[28,29] :

$$\lim_{\Gamma\to 2^-} \frac{U}{2Nk_B T} \sim (2 - \Gamma)^{-1} \tag{4.4a}$$

$$\lim_{\Gamma\to 2^-} \frac{c_S}{2Nk_B} \sim (2 - \Gamma)^{-2} \tag{4.4b}$$

The collapse of the 2d TCP in the limit $\Gamma \to 2^-$ is intimately related to counter-ion condensation in polyelectrolytes[30]. This "recombination" of pairs of <u>opposite</u> charges has profound effects on the pair correlations between charges of the <u>same</u> sign[31]. According to Widom's conjecture[32], one would expect the pair distribution functions to have the following behaviour at short distances :

$$\lim_{r\to 0} g_{\alpha\beta}(r) \sim \exp\{ - v_{\alpha\beta}(r)/k_BT\} \sim (r/L)^{Z_\alpha Z_\beta \Gamma} \; ; \quad Z_\alpha, Z_\beta = \pm 1 \qquad (4.5)$$

This behaviour appears to be obeyed by $g_{+-}(r)$, but the formation of tight + - pairs changes the behaviour of $g_{++}(r)$, $\equiv g_{--}(r)$ relative to the prediction (4.5) for $\Gamma > 1$. In fact it is expected that[31] :

$$g_{++}(r) \sim r^{\Gamma} \quad ; \quad 0 < \Gamma < 1$$
$$\sim r^{2-\Gamma} \quad ; \quad 1 < \Gamma < 2 \qquad (4.6)$$

Thus like-particle correlations are weakened for $\Gamma > 1$ due to progressive recombination of opposite charges. One of the unexpected consequences of (4.6) is that the familiar HNC equation, which is generally quite accurate for Coulombic systems, admits no solution for $\Gamma > 1$[31].

The Coulomb gas of finite size charges ($\sigma = 0$) leads to much richer physics. The system is now thermodynamically stable and all its properties depend on two variables, which may be chosen to be Γ and the ratio $\sigma/a = 2\eta^{1/2}$, where $a = (\pi n)^{-1/2}$ $(n = n_1 \, n_2)$ is the ion-disk radius, and η is the packing fraction. This Coulomb gas is known to undergo a dielectric-plasma transition at some density-dependent critical coupling. The dielectric-plasma transition is the prototype of Kosterlitz-Thouless (KT) transitions[33,34] in 2d systems, which are characterized by a divergent response (susceptibility) to an external field ; simultaneously the decay of spatial correlation changes from exponential (screening !) in the high temperature phase to a power law below the critical temperature[35].

In the Coulomb gas the relevant static response function is obviously the dielectric function $\epsilon(k)$ which diverges as $k \to 0$ in the plasma (conducting) phase, while it is finite in the dielectric (insulator) phase. $\epsilon(k = 0)$ is related to the fluctuations of the total dipole moment $\vec{M} = \sum_i q_i \vec{r}_i$ of the sample, which have been calculated in recent Monte Carlo simultations to locate the KT transition for several densities[36]. The critical Coulomb coupling is found to be $\Gamma = 4$ in the low density $(\eta \to 0)$ limit, in agreement with the prediction of a simple mean-field calculation[33].

5 - A FIXED ION MODEL

The most natural dynamical characterization of the dielectric-plasma transition is provided by the d.c. conductivity which is non-zero only in the plasma phase. However this collective property cannot be computed with a high degree of accuracy in a M.D. simulation. By fixing the charges of one species (the positive ions) on the sites of a hexagonal lattice, while the particles of the opposite species (the negative electrons) move in the periodic field due to the former, the TCP becomes equivalent to an inhomogeneous OCP in a periodic (rather than uniform) background, and the dielectric-plasma transition is mapped onto a "delocalisation" transition[37] : at low temperatures electrons are localized, because they are individually paired with one of the fixed ions, while above the transition, the ion-electron pairs are broken, i.e. electrons become delocalized. The advantage of this transformation of the original TCP model is the emergence of a second "diagnostic" for the transition, which is more useful in practical M.D. computations, namely the electron self-diffusion coefficient D_2. If $\vec{r}_i(t)$ denotes the position of the i th electron at time t, the Einstein relation reads :

$$\lim_{t\to\infty} \langle |\vec{r}_i(t) - \vec{r}_i(0)|^2 \rangle = a + bt \qquad (5.1)$$

with

$$a = 2 \langle |\Delta \vec{r}|^2 \rangle \quad ; \quad b = 0 \quad ; \quad T < T_1 \qquad (5.2a)$$

$$b = 4 D_2 \quad ; \quad T > T_1 \qquad (5.2b)$$

where $\langle|\Delta\vec{r}|^2\rangle$ denotes the mean square displacement of an electron around its host ion in the dielectric phase and T_1 is the threshold temperature for self diffusion, which we tentatively identify with the dielectric-plasma transition temperature.

Some of the salient results of recent M.D. simulation[37,38] of the fixed ion model, for $\sigma/a = 0.1$ and 0.02, and for various couplings Γ, are the following.

a) Typical electron trajectories at a temperature slightly above T_1 are shown in Fig.2. Localized and mobile electrons are seen to coexist over time spans of several hundred plasma periods .

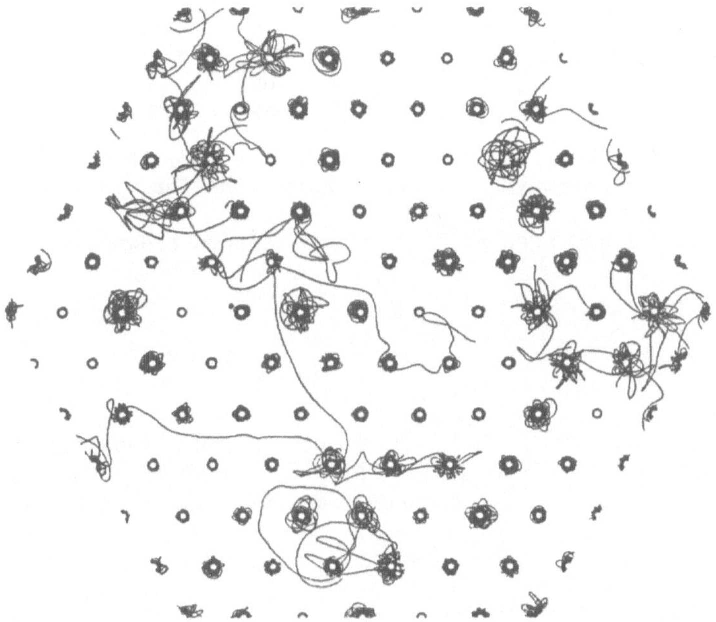

Fig.2 : Trajectories of 108 eletrons in the periodic field of as many fixed ions, at $\sigma/a = 0.1$, $\Gamma = 4.2$; exposure time is 250 ω_{p2}

b) The velocity ACF $Z(t)$ and current ACF $J(t)$ are shown in Fig.3 together with de power spectrum $\hat{Z}(\omega)$ for two states, one in the dielectric phase, and one in the plasma phase. In the former phase, $Z(t)$ and $J(t)$ are very similar and exhibit oscillations which originate in the rotational and vibrational motions of the electrons which are individually bound to ions. The two components are clearly resolved in the power spectrum $Z(\omega)$, where the low frequency peak can be associated with the rotations while the broad feature at higher frequencies is the vibrational band ; note that $\mathbb{D}_2 \sim \hat{Z}(\omega = 0)$ is strictly zero (insulator phase). In the plasma phase, on the other hand, $Z(t)$ and $J(t)$ are monotonous functions of time, with $J(t)$ decaying considerably more slowly than $Z(t)$, as already noticed for the 3d hydrogen plasma in section 3.

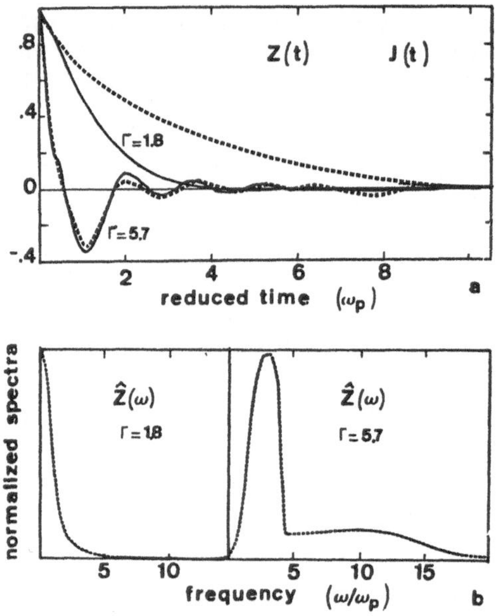

Fig.3 : a) Normalized velocity and current ACF $Z(t)$ (full curves) and $J(t)$ (dotted curves) versus reduced time for $\sigma/a = 0.1$, $\Gamma = 5.7$ and 1.8.
b) Normalized spectra $\hat{Z}(\omega)$ versus reduced frequency.

c) The coupling $\Gamma_1 = 1/T_1^*$ at which $\mathbb{D}_2$ drops to zero depends on density. For $\sigma/a = 0.1$ (i.e. $\eta = 0.01$) $\Gamma_1 \simeq 5.5$, while for $\sigma/a = 0.02$ ($\eta = 0.0004$) $\Gamma \simeq 2.8$. It is conjectured that $\lim_{\eta \to 0} \Gamma_1 = 2$, while the corresponding limit for the symmetric case, where ions are mobile, is believed to be $\Gamma_1 = 4$[33,36].

d) Ion-Electron recombination lead to a pronounced maximum in the specific heat c_S at a temperature $T_2 > T_1$.

e) J(t) can be used to calculate the a.c. conductivity $\sigma(\omega)$ according to eqn.(3.11) (with 4π replaced by 2π in 2d), and hence the long wavelength limit of the charge fluctuation spectrum according to eqn.(3.10). Some representative results are shown in Fig.4. The plasmon resonance is found to be quite sharp at high temperatures, but with increasing coupling, plasma oscillations are increasingly damped due to electron localization. The mode is overdamped long before the transition temperature is reached. The plasmon frequency at $k = 0$ drops with temperature, in qualitative agreement with the predictions of the collisionless Vlasov equation for the inhomogeneous electron gas in the periodic field of the ions[39]. Note however that this mean field kinetic equation does not lead to a KT transition, but predicts the 2d Coulomb gas to be a conductor at all temperatures.

f) M.D. computations of $S_{ZZ}(k,\omega)$ at non-zero wavenumbers in the plasma phase show that the plasmon dispersion relation $\omega(k)$ is not monotonous, but has an unusual oscillatory behaviour[38] which may be due to a coupling between the individual motion of each electron in the field of the nearest ion and the collective charge oscillation mode.

6 - CONCLUSION

Molecular Dynamics simulations have been instrumental in our present understanding of single-particle and collective dynamics in strongly coupled two-component plasmas. In 3d the usefulness of the method is however limited to the range (3.5) in the density-temperature plane, where a semi-classical modelization based on the effective potentials (3.6) may be expected to be reasonable. The 2d Coulomb gas model allows simulation of the influence of an "ionization" equilibrium on the dynamics of a two-component plasma in purely classical terms. The most striking result is the unexpectedly large importance of "recombinational" damping of plasma oscillations. More work in that direction is in progress.

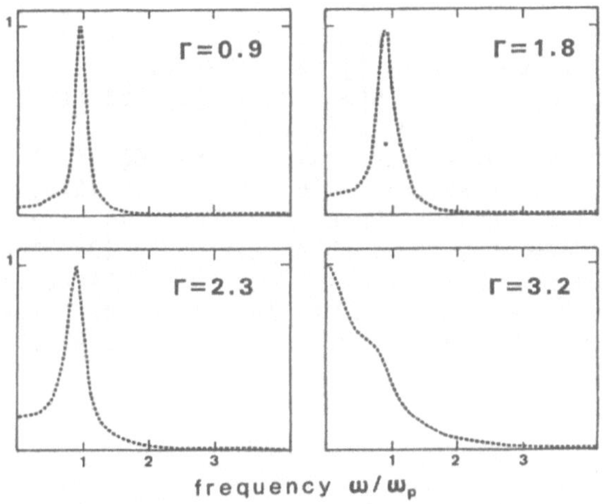

Fig.4 : $s(\omega)$ versus reduced frequency for $\sigma/a = 0.1$ and four couplings in the plasma phase.

ACKNOWLEDGEMENTS :

The author is indebted to Jean Clérouin for his efficient help during the preparation of this paper.

REFERENCES

1. N.W. Ashcroft and D. Stroud, in "Solid State Physics", vol. 33, F.Seitz and D. Turnbull, eds., Academic Press, New York (1978).
2. A. Lenard, J. Math. Phys. 2, 682 (1961).
3. J.M. Kosterlitz and D.J. Thouless, J. Phys. C6, 1181 (1973).
4. J.M. Kosterlitz, J. Phys. C7, 1046 (1974) ; 10, 3753 (1977).
5. J. Fröhlich and T. Spencer, Phys. Rev. Lett. 46, 1006 (1981).
6. M.P. Tosi, in "Electron Correlations in Solids, Molecules and Atoms", J.T. Devreese and F. Brosens, eds., Plenum Press, New York (1983).
7. M. Baus and J.P. Hansen, Phys. Rep. 59, 1 1980.
8. S. Ichimaru, Rev. Mod. Phys. 54, 1057 (1982).
9. C. Gouédard and C. Deutch, J. Math. Phys. 19, 32 (1978).
10. S. Galam and J.P. Hansen, Phys. Rev. A. 14, 816 (1976).
11. S. Ichimaru, S. Mitake, S. Tanaka and X.Z. Yan, Phys. Rev. A32, 1768, 1775, 1779, 1785, 1790 (1985).
12. D.B. Boercker and R.M. More, Phys. Rev. A33, 1859 (1986).
13. J. Chihara, J. Phys. C18, 3103 (1985).
14. F. Perrot and M.W.C. Dharma-Wardana, Phys. Rev. A29, 1378 (1984).
15. T. Morita, Progr. Theor. Phys. 22, 757 (1959).
16. Minoo, M.M. Gombert and C. Deutsch, Phys. Rev. A23, 924 (1981).
17. B. Bernu, J.P. Hansen and R. Mazighi, Phys. Lett. A100, 28 (1985).
18. J.P. Hansen and I.R. Mc Donald, Phys. Rev. A23, 2041 (1981).
19. J.P. Hansen and L. Sjögren, Phys. Fluids 25, 617 (1982).
20. L. Sjögren, J.P. Hansen and E.L. Pollock, Phys. Rev. A24, 1544 (1982).
21. B. Bernu and J.P. Hansen, Phys. Rev. Lett. 48, 1375 (1982).
22. R. Cauble and D.B. Boercker, Phys. Rev. A28, 944 (1983).
23. M. Baus, J.P. Hansen and L. Sjögren, Phys. Lett. A82, 180 (1981).
24. R. Cauble and W. Rozmus, Phys. Lett. A117, 345 (1986).
25. V. Zehnlé, B. Bernu and J. Wallenborn, Phys. Rev. A33, 2043 (1986).
26. D.B. Boercker, F.J. Rogers and H.E. DE Witt, Phys. Rev. A25, 1623, (1982).
27. J.P. Hansen and I.R. McDonald, Phys. Lett. A97, 42 (1983).
28. E.H. Hauge and P.C. Hemmer, Phys. Norv. 5, 209 (1971).
29. C. Deutsch and M. Lavaud, Phys. Rev. A9, 2598 (1974).
30. G.S. Manning, J. Chem. Phys. 51, 925 (1969).
31. J.P. Hansen and P. Viot, J. Stat. Phys. 38, 823 (1985).
32. B. Widom, J. Chem. Phys. 39, 2808 (1963).
33. J.M. Kosterlitz and D.J. Thouless, J. Phys. C6, 1181 (1973).
34. J.M. Kosterlitz, J. Phys. C7, 1046 (1974) ; 10, 3753 (1977).
35. J. Fröhlich and T. Spencer, Phys. Rev. Lett. 46, 1006 (1981).
36. J.M. Caillol and D. Levesque, Phys. Rev. B33, 499 (1986).
37. J. Clérouin and J.P. Hansen, Phys. Rev. Lett. 54, 2277 (1985).
38. J. Clérouin, J.P. Hansen and B. Piller, to be published.
39. A. Alastuey and J.P. Hansen, Europhys. Lett. 2, 97 (1986).

KINETIC THEORY OF THE INTERDIFFUSION COEFFICIENT IN DENSE PLASMAS

David B. Boercker

Lawrence Livermore National Laboratory
University of California
P. O. Box 808
Livermore, California 94550

INTRODUCTION

Ionic diffusion in dense plasma mixtures has been of interest recently for a number of reasons. In astrophysics, diffusion plays a central role in understanding the distribution of heavy elements in the atmospheres of White Dwarf stars.[1] The performance of multi-layer x-ray mirrors should be affected by diffusion, and the evaporation rate of metal "chunks" injected into the fuel of an ICF capsule by hydrodynamic instabilities is controlled by the diffusion coefficient.

In all of these applications, the plasmas can be very dense and estimates based upon the Spitzer formula[2] are often inadequate. In fact, naive applications of Spitzer's theory can lead to negative diffusion coefficients. Simple modifications, such as placing a "floor" on the value of the Coulomb logarithm[3] can eliminate such unphysical results, but they are untested under these conditions.

The interdiffusion coefficients in Binary Ionic Mixtures (two species of point ions in a uniform neutralizing background) have been calculated recently using molecular dynamics techniques by Hansen et al.[4] and by Pollock.[5] These calculations can provide useful benchmarks for theoretical evaluations of the diffusion coefficient in dense plasma mixtures. This paper gives a brief description of a kinetic theoretic approximation to the diffusion coefficient which generalizes Spitzer to high density and is in excellent agreement with the computer simulations.

DIFFUSION IN A BIM

As mentioned above, a Binary Ionic Mixture is a model of a plasma mixture with two species of classical point ions immersed in a uniform neutralizing background. The charge and mass of ion species "σ" are indicated by $Z_\sigma e$ and m_σ, respectively. Similarly, the number and mass densities are n_σ and $\rho_\sigma = m_\sigma n_\sigma$. The corresponding total densities are $n = n_1 + n_2$ and $\rho = \rho_1 + \rho_2$.

The strength of the Coulomb coupling among the ions is measured by the parameter,

$$\Gamma = \frac{e^2}{r_0 k_B T}$$

where T is the temperature and r_0 is the ion sphere radius

$$\frac{4\pi r_0^3}{3} = 1/n \quad .$$

There are two characteristic plasma frequencies in a BIM. One is the Vlasov plasma frequency defined by

$$\omega_p^2 \equiv \omega_1^2 + \omega_2^2 = \frac{4\pi n_1 Z_1^2 e^2}{m_1} + \frac{4\pi n_2 Z_2^2 e^2}{m_2} = 4\pi n e^2 \, \overline{Z^2/m} \quad . \tag{1}$$

The other is the so-called "hydrodynamic" plasma frequency defined by

$$\Omega_p^2 = 4\pi n e^2 \, \bar{Z}^2/\bar{m} \le \omega_p^2 \ . \tag{2}$$

In the above, barred quantities are number weighted averages,

$$\bar{A} = c_1 A_1 + c_2 A_2 \tag{3}$$

where $c_\sigma \equiv n_\sigma/n$ is the number concentration. The corresponding mass concentration is $X_\sigma \equiv \rho_\sigma/\rho$.

The rate at which concentration fluctuations dissipate in a mixture is governed by the interdiffusion coefficient, D, which linearly relates mass fluxes to gradients in the mass concentration. Specifically, if the center-of-mass velocity field, $\vec{u}$, is defined by

$$\vec{u}(\vec{r},t) = \sum_{\sigma=1}^{2} X_\sigma(\vec{r},t)\, \vec{u}_\sigma(\vec{r},t) \tag{4}$$

where $\vec{u}_\sigma$ is the velocity field of species "σ", then the mass flux of "σ" is

$$\vec{j}_\sigma(r,t) \equiv \rho_\sigma(\vec{r},t)(\vec{u}_\sigma(\vec{r},t) - \vec{u}(\vec{r},t)) \tag{5}$$

and the interdiffusion coefficient is defined by the relationship,[6]

$$\vec{j}_\sigma(\vec{r},t) = -\rho(\vec{r},t) D \, \vec{\nabla} X_\sigma(\vec{r},t) \quad . \tag{6}$$

As is the case with many other transport coefficients, D can be related to an equilibrium time correlation function. In particular, it can be shown that[7]

$$D = \frac{c_1 c_2}{S_{cc}(k{=}0)} \int_0^\infty dt\ V_D(t) \tag{7}$$

where

$$V_D(t) \equiv \frac{1}{3Nc_1c_2} \langle \vec{v}_D(t) \cdot \vec{v}_D(0) \rangle_0 \tag{8}$$

is the autocorrelation function for the (microscopic) diffusion velocity

$$\vec{v}_D(t) \equiv c_2 \sum_{i\epsilon 1} \vec{v}_i(t) - c_1 \sum_{i\epsilon 2} \vec{v}_i(t) \quad . \tag{9}$$

The concentration structure factor, $S_{cc}(k)$, is defined in terms of the partial structure factors, $S_{\sigma\tau}(k)$, as

$$S_{cc}(k) \equiv c_1c_2[c_2S_{11}(k) + c_1S_{22}(k) - 2\sqrt{c_1c_2}\ S_{12}(k)] \tag{10}$$

In the low-k limit, S_{cc} is related to the Gibbs free energy through

$$S_{cc}(k{=}0) = Nk_BT \Big/ \frac{\partial^2 G}{\partial c_1^2} \quad . \tag{11}$$

The appearance of the factor $c_1c_2/S_{cc}(k{=}0)$ in the expression for D, therefore, accounts for the fact that diffusion is really driven by gradients of the chemical potential, not density. For neutral gases, this factor reduces to unity for all concentrations in the low density limit. For charged particles, however, c_1c_2/S_{cc} remains concentration dependent even in the weak-coupling limit. This may be seen by using the Debye-Huckel estimates of the partial structure factors to obtain

$$c_1c_2/S_{cc}(k{=}0) \rightarrow \overline{Z^2} \;/\; \bar{Z}^2 \tag{12}$$

which is greater than unity whenever $c_1c_2 \neq 0$.

THE ENHANCEMENT FACTOR AND THE AMBIPOLAR FIELD

As seen in the previous section, the long-range nature of the Coulomb potential leads to an enhancement of ion diffusion in a binary mixture, even in the low-density limit. In this section it will be shown that an identical result follows from the usual Boltzmann theory of diffusion, provided the ambipolar field of the electron background is taken into account.

If temperature gradients are neglected, the standard Boltzmann approach gives[8]

$$\vec{j}_1 = -\frac{m_1m_2n^2}{\rho} D_0\vec{d}_1 \tag{13}$$

where D_0 is the Spitzer estimate of the diffusion coefficient and $\vec{d}_1$ is

$$\vec{d}_1 = \frac{1}{n}(\vec{\nabla}n_1 - \frac{Z_1 n_1 e}{k_B T}\vec{E}) + \frac{X_1}{p}(\sum_{\sigma=1}^{2} Z_\sigma n_\sigma e\vec{E} - \vec{\nabla}p) \quad . \tag{14}$$

The ambipolar field is $\vec{E}$ and the pressure is given by $p = nk_BT$. By assuming the system is mechanically stable,

$$\vec{\nabla}p = (Z_1 n_1 + Z_2 n_2)e\vec{E} \ , \tag{15}$$

and that it is charge-neutral over hydrodynamic scale-lengths,

$$Z_1\vec{\nabla}n_1 = - Z_2\vec{\nabla}n_2 \ , \tag{16}$$

it is straight-forward to find

$$\vec{j}_1 = -\rho \frac{\overline{Z^2}}{\bar{Z}^2} D_0\vec{\nabla}X_1 \quad . \tag{17}$$

Comparison with (12) shows that, at least in the low-density limit, the thermodynamic factor, $c_1c_2/S_{cc}(k{=}0)$, may be thought of as an enhancement of the ion diffusion due to the ambipolar electric field of the electrons.

KINETIC THEORY FOR THE TIME-CORRELATION FUNCTION

Any time correlation function may be expressed in terms of the correlations of the phase space densities

$$f_\sigma(\vec{r}\vec{p},t) = \sum_{i\epsilon\sigma} \delta(\vec{r} - \vec{r}_i(t))\delta(\vec{p} - \vec{p}_i(t)) \quad . \tag{18}$$

If δf_σ represents the deviation of f_σ from its equilibrium average value, then the phase-space correlation functions are

$$C_{\sigma\tau}(\vec{r}-\vec{r}', t|\vec{p}\vec{p}') = \langle\delta f_\sigma(\vec{r}\vec{p},t)\ \delta f_\tau(\vec{r}'\vec{p}'0)\rangle_0 \quad . \tag{19}$$

It is usually more convenient to deal with the transformed functions

$$\tilde{S}_{\sigma\tau}(kz;\vec{p}\vec{p}') = \int_0^\infty dte^{izt}\int d^3rC_{\sigma\tau}(\vec{r}-\vec{r}', t|\vec{p}\vec{p}')e^{-i\vec{k}\cdot(\vec{r}-\vec{r}')} \tag{20}$$

The diffusion coefficient may be written in terms of these latter functions as

$$D = \frac{c_1c_2}{S_{cc}} \ \mathrm{Re}\ \tilde{V}_D(0 + i\eta) \qquad (\eta\to0^+) \tag{21}$$

where

$$\tilde{V}_D(z) = \int_0^\infty dt e^{izt} V_D(t)$$

$$= \frac{1}{3}\sum_{\sigma,\tau} \gamma_\sigma \gamma_\tau \int d^3p d^3p' \tilde{S}_{\sigma\tau}(k=0,z;\vec{p}\vec{p}')\vec{p}\cdot\vec{p}' \tag{22}$$

with $\gamma_1 = \frac{1}{m_1}\sqrt{\frac{c_2}{n_1}}$ and $\gamma_2 = \frac{-1}{m_2}\sqrt{\frac{c_1}{n_2}}$.

The transformed phase-space correlation functions obey a kinetic equation of the form[9]

$$(z-\frac{\vec{k}\cdot\vec{p}}{m_\sigma}) \tilde{S}_{\sigma\tau}(kz;pp') - \sum_{\sigma'} \int d^3\vec{p}'' \, \Phi_{\sigma\sigma'}(kz;\vec{p}\vec{p}'') \, \tilde{S}_{\sigma'\tau}(kz;\vec{p}''\vec{p}')$$

$$= in_\sigma \, \phi_\sigma(p) \, [\delta_{\sigma\tau}\delta(\vec{p}-\vec{p}') + n_\tau \, \phi_\tau(p') \, \tilde{h}_{\sigma\tau}(k)] \tag{23}$$

where $\phi_\sigma(p)$ is the normalized Maxwell-Boltzmann distribution for species "σ" and $h_{\sigma\tau}(k)$ is related to the radial distribution function through Fourier transformation

$$\tilde{h}_{\sigma\tau}(k) \equiv \int d^3r e^{i\vec{k}\cdot\vec{r}}(g_{\sigma\tau}(r)-1). \tag{24}$$

The operator $\Phi_{\sigma\tau}$ is written as

$$\Phi_{\sigma\tau}(kz;\vec{p}\vec{p}') = -n_\sigma \frac{\vec{k}\cdot\vec{p}}{m_\sigma} \phi_\sigma(p) \, \tilde{c}_{\sigma\tau}(k) + \tilde{M}_{\sigma\tau}(kz;pp') \tag{25}$$

which is the sum of a mean-field term involving the direct correlation functions

$$\tilde{c}_{\sigma\tau}(k) = \tilde{h}_{\sigma\tau}(k) - \sum_{\sigma'} \tilde{c}_{\sigma\sigma'}(k) \, \tilde{h}_{\sigma'\tau}(k) \tag{26}$$

and the "memory" function, $\tilde{M}_{\sigma\tau}$, which contains the effects of collisions.

The standard procedure for solving (23) is to expand the momentum dependence of the $\tilde{S}_{\sigma\tau}$'s in terms of Hermite polynomials, which are a complete set of orthogonal polynomials with Maxwell-Boltzmann weight functions. The Hilbert space defined by these functions is then divided into two subspaces: the "hydrodynamic subspace" spanned by the ten (five for each species) functions corresponding to the hydrodynamically conserved quantities, number, three components of momentum and (kinetic) energy,[10] and its complement, the "non-hydrodynamic" subspace. Projecting the kinetic equation onto the "hydrodynamic" subspace then yields a closed set of equations for the hydrodynamic matrix elements of the $\tilde{S}_{\sigma\tau}$'s. The details of this procedure are well described in the paper by Baus[11] and will not be given here.

Applying Baus' method to the problem at hand ultimately yields

$$\tilde{V}_D(z) = \frac{1}{3}\tilde{V}_\ell(z) + \frac{2}{3}\tilde{V}_\perp(z) \tag{27}$$

where

$$\tilde{V}_\ell(z) = \frac{\rho k_B T}{n m_1 m_2} \frac{iz(z^2-\Omega_p^2)}{z^2(z^2-\omega_p^2) + iz\nu(z)(z^2-\Omega_p^2)} \tag{28}$$

is the correlation function for the longitudinal component of the diffusion velocity and

$$\tilde{V}_\perp(z) = \frac{\rho k_B T}{n m_1 m_2} \frac{iz}{z^2 + iz\nu(z)} \tag{29}$$

is the correlation function for the transverse components. If the coupling to the non-hydrodynamic subspace is completely ignored,[12] the collision frequency is

$$\nu(z) = i(\Omega^{11} + \Omega^{22}) \tag{30}$$

where

$$\Omega^{\sigma\sigma} = \frac{1}{\rho_\sigma k_B T} \int d^3p\, d^3p'\; p_\ell\; \tilde{M}_{\sigma\sigma}(k{=}0,z|\vec{p}\vec{p}')\; \phi_\sigma(p')\; p'_\ell \quad . \tag{31}$$

In terms of $\nu(z)$ we find

$$D = \frac{c_1 c_2}{S_{cc}} \frac{\rho k_B T}{n m_1 m_2 \nu(0)} \quad . \tag{32}$$

Hence, to proceed we need an expression for the memory function.

THE DISCONNECTED APPROXIMATION

The memory function may be expressed in "time space" in the form[9]

$$M_{\sigma\tau}(12;t) = -\sum_{\mu\nu} \int d1'\, d2'\, \vec{\nabla}_1 v_{\sigma\mu}(\vec{r}_1-\vec{r}_1')$$

$$\cdot \frac{\partial}{\partial\vec{p}_1} G_{\sigma\mu;\tau\nu}(11';22'|t)\vec{\nabla}_2 v_{\tau\nu}(\vec{r}_2-\vec{r}_2') \cdot \frac{\partial}{\partial\vec{p}_2}\phi_\tau^{-1}(p_2) \tag{33}$$

where the four-point function, $G_{\mu;\nu\tau}$, represents the propagation of pairs of particles between interactions. If this function is simply factored into a product representing the propagation of single particles through the plasma, then in the long time limit M reduces to the usual Lenard-Balescu collision operator.[13] In the Disconnected Approximation[14] the four-point function is factorized in such a way as to preserve its exact initial value. The principal effect of this modified factorization is to renormalize one of the potentials and replace it with a direct correlation function. An alternative form of this approximation renormalizes both potentials. This has the advantage of giving a positive definite "cross-section", but destroys the short-time behavior of the memory operator. In this paper, the first form of the Disconnected Approximation will be adopted, but some comparisons with the second form will be made.

Using this approximation the collision frequency reduces to

$$\nu(z) = -\frac{i\rho}{3m_1m_2}\int\frac{d^3k}{(2\pi)^3}k^2\tilde{v}_{12}(k)\tilde{c}_{12}(k)\int\frac{d\omega_1}{2\pi}\int\frac{d\omega_2}{2\pi}\frac{S_{11}(k,\omega_1)S_{22}(k,\omega_2)-S_{12}(k,\omega_1)S_{21}(k,\omega_2)}{z-\omega_1-\omega_2} \quad (34)$$

which for low frequencies becomes

$$\nu(0) = -\frac{Z_1Z_2e^2\rho}{3\pi m_1m_2}\int_0^\infty dk\, k^2\tilde{c}_{12}(k)\int\frac{d\omega}{2\pi}[S_{11}(k,\omega)S_{22}(k,\omega) - S_{12}^2(k,\omega)]. \quad (35)$$

To complete the calculation estimates of the dynamic structure factors are needed. These are obtained by substituting static structure factors obtained from the HNC equation into (23) with $\tilde{M}_{\sigma\tau} = 0$.

COMPARISON WITH MOLECULAR DYNAMICS

Calculations of the interdiffusion coefficient have been made using (35) in (32). The results for a 50% mixture of H^+ and He^{2+} at various Γ values are shown in Table I. The reduced diffusion coefficient, D*, is given by

$$D^* = \left[\frac{S_{cc}}{c_1c_2}\right]D/r_0^2\,\Omega_p\,. \quad (36)$$

Table I. Comparison of theoretical and numerical simulation results for D^* in 50% H^+-He^{2+} mixtures.

Γ	D^*_{MD}	D^*_T	D^*_{GM}
0.4[a]	3.00	3.18	5.25
1.0[b]	.915	.792	1.44
4.0[a]	.142	.154	.265
40.[a]	.0109	----	.0165

[a]From Ref. 4
[b]From Ref. 5

Table II. D*'s for Si^{+14}-Sr^{+36} mixtures.

%Si	D^*_{MD}	D^*_T
25	.552	.477
50	.605	.508
75	.628	.545

The subscript MD indicates the results from the Molecular Dynamics studies of Hansen et al.[4] for $\Gamma \neq 1$ and Pollock[5] for $\Gamma = 1$. D^*_T is calculated using the theory described here and D^*_{GM} is calculated from the symmetric form of the Disconnected Approximation.[15] As can be seen from the table, the agreement between D^*_T and D^*_{MD} is quite good for the lower three Γ values. At $\Gamma = 40$, the oscillations in $c_{12}(k)$ lead to a negative result for D^*_T. The symmetric theory does not run into this difficulty, but it gives results for all Γ values which are 50% too high.

Table II compares the results from the asymmetric theory to computer simulations for mixtures of Si^{+14} and Sr^{+36} at various concentrations, but all at $\Gamma = .005$. Once again the agreement is in the 10-20% range.

TIME CORRELATION FUNCTIONS

In order to study the behavior of the time-correlation function, $V_D(t)$, the high-frequency behavior of $\nu(z)$ is observed to be

$$\nu(z\to\infty) \to \frac{i}{z}\;\frac{4\pi Z_1 Z_2 e^2 \rho}{3m_1 m_2} \tag{37}$$

Using this result in (28) and (29) yields

$$\tilde{V}_\ell(z) = \frac{\rho k_B T}{nm_1 m_2}\,\frac{i}{z}\;[1 + \left(\frac{i}{z}\right)^2 \left(\Omega_p^2 - \omega_p^2 - \frac{4\pi Z_1 Z_2 e^2 \rho}{3m_1 m_2}\right) + \ldots] \tag{38}$$

and

$$\tilde{V}_\perp(z) = \frac{\rho k_B T}{nm_1 m_2}\,\frac{i}{z}\;[1 - \left(\frac{i}{z}\right)^2 \frac{4\pi Z_1 Z_2 e^2 \rho}{3m_1 m_2} + \ldots] \tag{39}$$

Hence, one notes that

$$\ddot{V}_\ell\,(t=0)/V_\ell\,(t=0) = -\;(\omega_p^2 - \Omega_p^2 + \frac{4\pi Z_1 Z_2 e^2 \rho}{3m_1 m_2}) \tag{40}$$

$$\ddot{V}_\perp(t=0)/V_\perp(t=0) = - \frac{4\pi Z_1 Z_2 e^2 \rho}{3 m_1 m_2} \tag{41}$$

Since $\omega^2_p \geq \Omega^2_p$, the correlations of the longitudinal component of the diffusion velocity have a more rapid initial decay than the transverse components. Combining (40) and (41) gives the known result[4]

$$\ddot{V}_D(t=0)/V_D(t=0) = - \frac{\Omega^2_p}{3} \frac{c_1 m_1^2 Z_2 + c_2 m_2^2 Z_1}{c_1 Z_1 + c_2 Z_2} \tag{42}$$

The more rapid initial decay of $V_\ell(t)$ illustrated by the "dashed" and "dash-dot" curves in Figure 1. The solid curve and the dots compare the theoretical estimate of $V_D(t)/V_D(t=0)$ to the corresponding simulation results for the 50% H^+-He^{2+} mixture at $\Gamma=1$. The comparison is reasonable out to about six inverse plasma frequencies, but the theoretical curve seems to miss the "shoulder" at 12 ω_p^{-1}. This may be due to V_ℓ oscillating too rapidly in this region.

DISCUSSION

The results presented here indicate that the Wallenborn and Baus form of the Disconnected Approximation[14] agrees to about 10-20%, with numerical simulation values for the interdiffusion coefficient. Such agreement is quite good, especially in view of the 10% uncertainties in the molecular dynamics results.[4,5] The only problem arises at very strong coupling where the theory apparently breaks down and gives a negative result. This is not a serious limitation, however, since most plasmas of practical interest are in the weak to moderate coupling regime. In general this calculation is another indication of the success of the Disconnected Approximation.

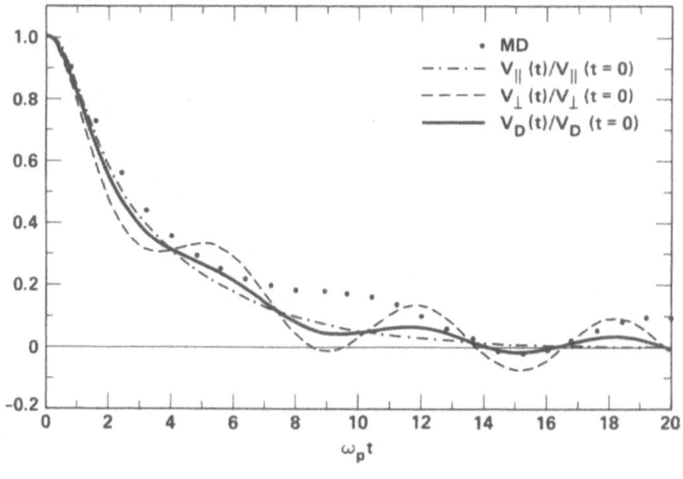

Fig. 1 Normalized velocity autocorrelation functions for 50% H^+-He^{2+} mixture with $\Gamma = 1$.

ACKNOWLEDGEMENTS

Work performed under the auspices of the U. S. Department of Energy by the Lawrence Livermore National Laboratory under contract number W-7405-ENG-48.

REFERENCES

1. C. Paquette, C. Pelletier, G. Fontaine, and G. Michaud (preprint); G. Fontaine (this proceedings).
2. L. Spitzer, Jr., Physics of Fully Ionized Gases (Interscience, New York 1956).
3. Y. T. Lee and R. M. More, Phys. Fluids 27:1273 (1984).
4. J. P. Hansen, F. Joly, and I. R. McDonald, Physica 132A:472 (1985).
5. D. B. Boercker, A. J. Ladd, and E. L. Pollock, Lawrence Livermore National Laboratory Report No. UCID-20507, Livermore, CA, 1985.
6. J. P. Hansen and I. R. McDonald, Physics of Simple Liquids, (Academic, NY, 1976)
7. C. Cohen, J. Sutherland, and J. Deutch, Phys. Chem. Liq. 2:213 (1971).
8. J. Hirschfelder, C. Curtis and R. Bird, Molecular Theory of Gases and Liquids, (John Wiley, New York, 1954).
9. See, for instance, J. I. Castresana, G. F. Mazenko and S. Yip, Ann. Phys. (N.Y.) 103:1 (1977).
10. Note that kinetic energy is conserved only in the hydrodynamic limit $k \to 0$, $z \to 0$.
11. M. Baus, Physica 38A:319 (1977).
12. This is equivalent to making a one Sonine polynomial approximation.
13. See, for instance, S. Ichimaru, Basic Principles of Plasma Physics (Academic, New York, 1973).
14. J. Wallenborn and M. Baus, Phys. Rev. A 18:1737 (1978).
15. H. Gould and G. Mazenko, Phys. Rev. A 15:1274 (1977).

TRANSPORT PROPERTIES OF A FULLY IONIZED PLASMA: SEMI-CLASSICAL OR QUANTUM MECHANICAL APPROACH

J. Wallenborn,[1] B. Bernu[2] and V. Zehnlé[1]
[1]Chimie-Physique II, C.P. 231
Université Libre de Bruxelles
B-1050 Brussels, Belgium
[2]Laboratoire de Physique Théorique des Liquides
Université Pierre et Marie Curie
F-75230 Paris Cedex 05, France

Purely classical statistical mechanics cannot describe the thermodynamics of a dense multicomponent plasma. It is necessary to take into account the quantum diffraction which avoids the collapse of electrons with ions. When computing the equilibrium properties, the quantum effects can be included in the classical partition function with the help of a temperature-dependent effective interaction potential (see e.g. Pokrant and Broyles, 1974, and references therein).

Minoo et al. (1981) have proposed an analytical expression for this effective interaction potential v_{ab} between two particles of species a and b respectively (a and b stand for e or i, electrons or ions):

$$v_{ab}(r) = v^{d}_{ab}(r) + v^{s}_{ab}(r) \tag{1}$$

where

$$v^{d}_{ab}(r) = \frac{q_a q_b}{r} (1 - e^{-r/\lambda_{ab}}) \tag{2}$$

accounts for the quantum diffraction effects as well as for the bare Coulomb interaction and where

$$v^{s}_{ab}(r) = \delta_{ae}\delta_{be}k_B T \,(\ln 2)\, e^{-r^2/\pi\lambda^2_{ee}\ln 2)} \tag{3}$$

accounts for the quantum symmetry or correlation effect. In Eqs. (2) and (3), q_a and q_b are the charge of the particles of species a and b respectively and $\lambda_{ab} = \hbar(2\pi\mu_{ab}k_B T)$ is the thermal de Broglie wavelength of the pair (a,b).

Let us point out that v_{ab} is a purely two-particle density-dependent interaction potential which does not include any collective effect. In addition, the expressions (2) and (3) are valid only under the following assumptions: s-scattering states alone are considered, there are no bound states, and the electrons are weakly degenerate. These last two assumptions imply $k_BT > 1$ Ry and $\lambda_{ee} < a$ where $a=(3/4\pi h)^{1/3}$ is the ion-sphere sphere radius, or equivalently

$$\Gamma r_s < 2 \qquad \Gamma/r_s < \pi \tag{4}$$

where Γ and r_s are the usual dimensionless parameters which define the thermodynamical state of the system:

$$\Gamma = q_e^2/(k_BTa) \qquad r_s = a/a_0 \tag{5}$$

with $a_0 = \hbar^2/mq^2{}_e$. the Bohr radius.

During the last few years, the effective potential v_{ab} (or $v^d{}_{ab}$ alone) has been used to study, in the parameters domain defined by (4), the non-equilibrium properties of a two-component plasma (Baus et al., 1981; Hansen and McDonald, 1981; Sjogren et al., 1981; Bernu, 1983; and Zehnle et al., 1986). However it is questionable that such a potential, which is constructed to evaluate static properties, will be adequate for the calculation of dynamical properties. In a recent paper (Zehnle et al., 1986), we indeed found some unexpected results. We studied the thermal conductivity of a fully ionized hydrogen plasma in the framework of the classical kinetic theory developed for strongly coupled plasmas (see e.g. Wallenborn, 1985, and references therein) using the effective potential v_{ab} of Minoo et al., [Eq. (1)]. One of our results was that the electronic part of the thermal conductivity, which classically dominates, can become of the same order of magnitude as the ionic part due to the quantum symmetry effects included in $v^s{}_{ab}$ [Eq. (3)]. This quantum reduction of the thermal conductivity is important when the product $\Gamma \times r_s$ is small, i.e. in the same domain of parameters where the quantum effects are important for equilibrium properties. However, though the thermodynamics tends to that of a perfect Fermi gas as Γ decreases, the thermal conductivity does not.

In order to test the physical validity of these surprising results, we have computed the thermal conductivity of a weakly-coupled electron one-component plasma (OCP) in two ways (Wallenborn et al., 1986): i) by a semi-classical kinetic theory, a calculation which is just the one-component analog of our previous two-component one (Zehnle et al., 1986), and ii) by a purely quantum kinetic theory starting from first principles.

In the semi-classical approach, we considered two models for the interaction potential between electrons: (cf. Eqs. (1) and (2)):

$$v_1(r) = v^d_{ee}(r) \qquad \text{and} \qquad v_2(r) = v_{ee}(r) \tag{6}$$

One may easily compute analytically the semi-classical thermal conductivity K of the OCP in the first Sonine polynomial approximation by standard methods (see e.g. Balescu, 1975). The result can be written in the following form for both model potentials $v_i(r)$ (i=1,2):

$$K_i^* = \frac{K_i}{k_B n a^2 \omega_p} = \frac{25}{24}\sqrt{3\pi}\,\Gamma^{-5/2}\,\Lambda_i^{-1} \qquad (i=1,2) \tag{7}$$

where $\omega_p = (4\pi e^2 n\ m)^{1/2}$ is the plasma frequency and where Λ_i is the so-called "Coulomb logarithm". To avoid the remaining divergence at large distance, the integrals were truncated at ℓ_c which is the Thomas-Fermi length if the system is degenerate and the Debye length if not. One then has:

$$\Lambda_1 = -\ln\epsilon + \frac{1}{2}\left[\ln\left(\epsilon^2 + \frac{\pi}{2}\right) - \frac{1}{1 + \frac{2\epsilon^2}{\pi}}\right] \tag{8}$$

$$\Lambda_2 = \Lambda_1 + \frac{\pi}{8}\frac{(\ln 2)^3}{\Gamma r_s} + \frac{\pi^{3/2}}{4}\frac{(\ln 2)^{5/2}}{(\Gamma r_s)^{1/2}} e^{\frac{\pi}{4}\ln 2} \quad E_1\left(\frac{\pi}{4}\ln 2\right) \tag{9}$$

where $\epsilon = (\beta/2m)^{1/2} h/\ell_c$ and where E_1 is the exponential integral.

The purely quantum calculation of the thermal conductivity of the OCP was made in the Wigner function formalism. The quantum Landau kinetic equation, which is based on the Coulomb potential, can be found in the literature (see e.g. diffraction, exchange and symmetry. This last effect is included in the quantum distribution function. In this formalism, the thermal conductivity K_q (in the first Sonine polynomial approximation) is written:

$$K_q^* = \frac{K_q}{k_B n a^2 \omega_p} = \frac{25}{24}\sqrt{3\pi}\ \Gamma^{-5/2}\ \Lambda_q^{-1} \tag{10}$$

$$\Lambda_q = C\ [\ -A\ \ln\epsilon + B\ ] \tag{11}$$

where ϵ is the same truncation parameter as in the semi-classical case, C is a combination of Fermi integrals, A and B are, respectively, a four- and a five-dimensional integral which are evaluated numerically. The important point is that A, B and C depend only on μ/k_BT or, equivalently, on the ratio Γ/r_s (μ is the chemical potential). Moreover, in the limit of no degeneracy ($\Gamma/r_s \ll 1$) they can be calculated analytically.

As an example of our results we show in Fig. 1 the comparison between the quantum and semi-classical evaluations of the thermal conductivity as a function of the degeneracy parameter μ/k_BT at fixed temperature. It is seen that K_2 never agrees with K_q. In particular, K_q is two orders of magnitude larger than K_2 when the electrons are degenerate ($\mu/k_BT \gtrsim 3$). This means that the reduction with Γr_s of the electronic thermal conductivity of the hydrogen plasma is a spurious effect of the symmetry potential v^s_{ee} [Eq. (3)]. On the other hand, K_1 and K_q are in agreement as far as the system is not degenerate ($\mu/k_BT \lesssim -1$) or $\Gamma/r_s < 0.25$) i.e. as far as the quantum symmetry doesn't enter into play.

These results illustrate the difficulty in including non-dynamical effects, such as the quantum correlations, in an effective potential which is used to generate the dynamics. One could be tempted to use the potential v_2 to describe equilibrium properties and the potential v_1. to describe the transport properties. This, however, implies a local

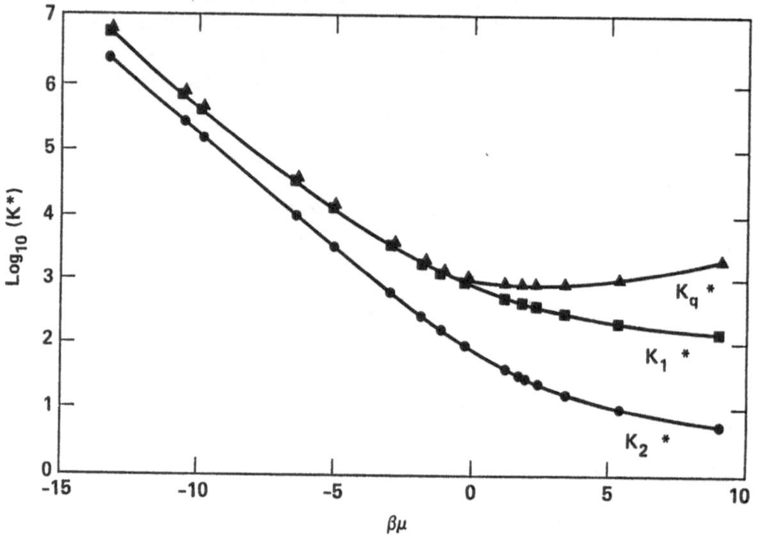

Fig. 1 Reduced thermal conductivity $K^{*}=K/(k_{B}na^{2}\omega_{p})$ of an electron gas as a function of the degeneracy parameter $\mu/k_{B}T$ for a temperature of 1360 eV; ▲ K_q, Eq. (10); ■ K_1, Eq. (7); • K_2, Eq. (7).

source term in the energy balance which certainly is not desirable. Most likely one has to choose between a good description of the statics or of the dynamics.

So far our conclusions are strictly valid only for weakly coupled plasmas. Yet, they rest on the value of the degeneracy parameter Γ/r_s and not on Γ alone. It would be very surprising if they were not correct in the case of strongly coupled plasmas.

In order to fulfill the requirement (4), the results of the molecular dynamics (Hansen and McDonald, 1981; Sjögren et al., 1981; Bernu, 1983) were obtained only for degenerate strongly coupled two-component plasmas; therefore, they must be considered at best as qualitative. A quantum calculation of the transport coefficients of these systems is thus highly desirable.

REFERENCES

Balescu, R., 1975, "Equilibrium and Non-Equilibrium Statistical Mechanics", Wiley.
Baus, M., Hansen, J. P., and Sjögren, L., 1981, Phys. Lett. 82A:180.
Bernu, B., 1983, Physica 122A:129.
Hansen, J. P. and McDonald, I. R., 1981, Phys. Rev. A 23:2041.
Minoo, H., Gombert, M. M., and Deutsch, C., 1981, Phys. Rev. A 23:924.
Pokrant, M. A., and Broyles, A. A., 1974, Phys. Rev. A 10:379.
Sjögren, L., Hansen, J. P., and Pollock, E. L., 1981, Phys. Rev. A 23:2041.
Wallenborn, J., 1985, J. Phys. C 18:4403.
Wallenborn, J., Zehnlé, V., and Bernu, B., 1986, submitted to Europhys. Lett.
Zehnlé, V., Bernu, B., and Wallenborn, J., 1986, Phys. Rev. A 33:2043.

CHAPTER IV

ASTROPHYSICS

DENSE MATTER IN ASTROPHYSICS: SELECTED TOPICS

Evry Schatzman

Observatoire de Nice
BP 139, 06003 Nice Cedex
France

INTRODUCTION

Let me first briefly report on the conditions under which a stellar object reaches the physical conditions of a highly correlated plasma. Several review papers have been devoted to this problem, but I would like to recommend especially the courses delivered by Iben, Renzini and Schramm (1977), the review papers by Iben (1974) and by Iben and Renzini (1984), which concerns isolated stars. See also Schatzman (1978, 1980). The situation of binary stars is much more complicated, due to mass exchange between the two components. This has been discussed several times, but I would like to recommend the paper by Webbink (1979) and the review paper of de Loore (1984). The effect of accretion on white dwarfs is described by Nomoto (1982,1984), at least as far as the outer layers are concerned. It should be noticed that fast accretion leads in Nomoto's models to the formation of giants.

The evolutionnary path towards the white dwarf stage determines the chemical composition of the bulk of the white dwarf and the initial radial distribution of the temperature.

The main point is that main sequence stars, after exhaustion of hydrogen in the core become giants. Further evolution leads to an important mass loss. Observational data (Weidemann and Köster, 1984) show the relation between initial and final mass. It seems that up to 8 to 10 solar masses all stars become white dwarfs, the total mass loss reaching a maximum value of 85%. For a star with $M < 2.25$ solar masses, the production of a low mass degenerate core leads to the formation of a helium white dwarf, whereas for a star of larger mass, $M > 2.25$ solar masses, the core reaches higher temperatures allowing helium burning and leading to carbon-oxygen white dwarfs.

The chemical composition of white dwarfs reflects the evolution of the parent star. A discussion by Alcock (1979) of the gravitational sorting in the outer layers of white dwarfs confirms the possible existence of two classes of white dwarfs: for $M < 0.4$ solar masses, helium white dwarfs, and carbon-oxygen white dwarfs for higher masses.

As already mentioned in other review papers (Schatzman, 1978, 1980), it is not possible to review all problems of dense matter which are raised by the study of astrophysical objects: the purpose of this paper is not to draw a list of subjects, but rather to discuss some key problems.

I would like to concentrate on the following questions:

- the thermal history of white dwarfs,
- the equation of state,
- the thermonuclear-pycnonuclear reaction rate,
- the transport process,
- some remarks about Sirius,

and try to place these questions in their astrophysical framework.

THERMAL HISTORY OF WHITE DWARFS

The thermal history of white dwarfs implies two questions: (1) the rate of heat transport in white dwarfs, compared to the classical rate of cooling, and (2) the physics of the solidification of a mixture.

The thermal conductivities of the liquid metal phase and of the solid metal phase have been calculated resp. by Itoh et al (1983) and by Itoh et al (1984), improving appreciably its value (references to former work can be found in these two papers). Even without carrying the calculation of a complete solution of the equation of heat, it is possible to estimate the rate of heat transport in white dwarfs and to compare it to the rate of cooling.

My purpose here is not to solve the problem quantitatively, but to give an idea of the problem. The argument is the following.

The rate of heat transport is given by the equation of heat :

$$\frac{1}{r^2}\frac{\partial}{\partial r} r^2 K \frac{\partial T}{\partial r} = C \frac{\partial T}{\partial t}$$

where K is the thermal conductivity and C the specific heat of the matter. An order of magnitude of the time scale of heat transport will be given by

$$t_{diff} = \left(\int \frac{dr}{(K/C)^{1/2}} \right)^2$$

This would have to be compared to the classical time of cooling :

$$t_{cool} = \frac{Q}{L}$$

where Q is the total energy available in the white dwarf and L the luminosity. This last relation can be expressed in terms of the internal temperature of the white dwarf, this temperature being defined by the relation:

$$P_{deg} = P_{gas}$$

at the bottom of the radiative, non degenerate outer zone. In this sketch of the radiative cooling of a white dwarf, it would be necessary to distinguish between the case of a pure hydrogen envelope and the case of a heavy elements rich envelope. We shall ignore this difference, as well as the effect of the surface convection zone.

It is usually assumed that the thermal diffusivity inside white dwarfs is so high that the degenerate star is isothermal during cooling. However, this is not quite the case. In order to carry properly the discussion, let us consider once more the diagram (log ρ, log T) where I have plotted (fig. 1) the boundary between degenerate and non degenerate matter, the melting boundary for ^{16}O, the Debye temperature and the lines defined by the conditions

$$\frac{q_{el}}{q_{liq}} = 1 \quad ; \quad \frac{q_{el}}{q_{solid}} = 1$$

where q_{el} is the internal heat of the electrons, given by

$$q_{el} = \frac{3\,\pi^2}{2} \; Z \left(\frac{kT}{m_e c^2}\right)^2 m_e c^2 \, \frac{1}{x} \, N_i$$

with

$$x = \frac{p_F}{m_e c}$$

valid for

$$x >> 1$$

q_{liq} is the internal heat of the ions :

$$q_{liq} = \frac{3}{2} \; \frac{\rho}{A\,H} \; k\,T$$

q solid is the internal heat of the solid,

$$q_{solid} = \frac{16\,\pi^4}{5} \left(\frac{T}{\Theta_D}\right)^3 k\,T \, \frac{\rho}{A\,H}$$

with the Debye temperature, as given by Shapiro <u>et al</u> (1984) :

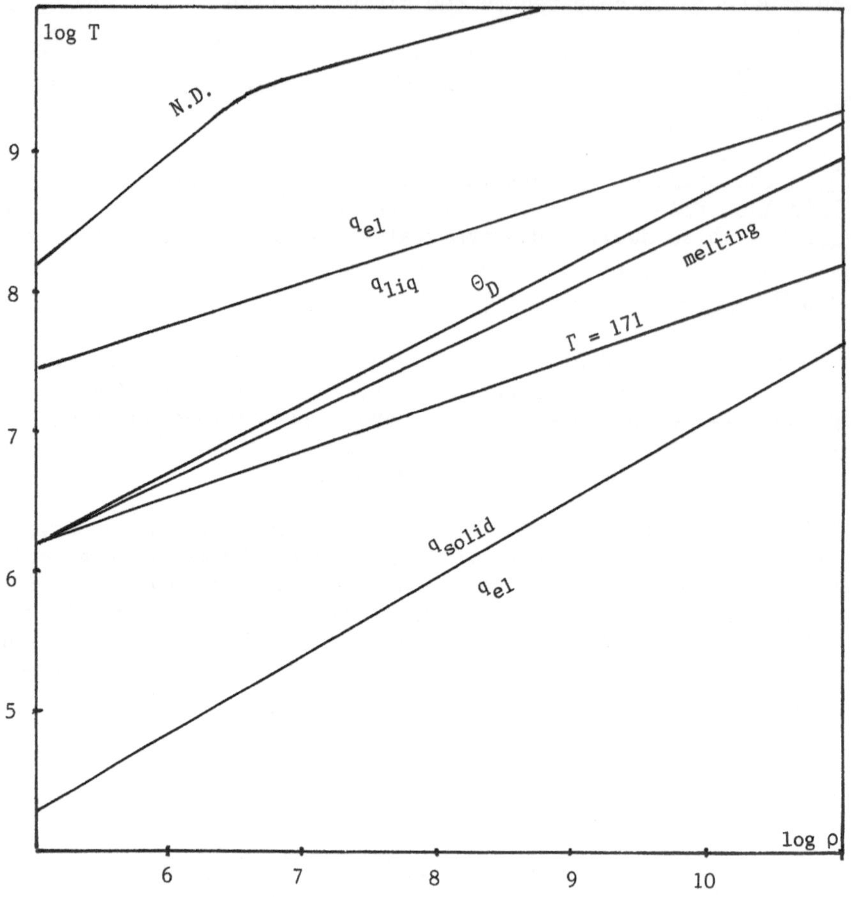

Fig.1. The different regions in the log ρ, log T plane. The curve "melting", inbetween the Debye temperature and the standard melting curve, has been obtained by applying a Debye correction to the mean kinetic energy kT (see text).

$$\Theta_D = \frac{\hbar}{k}\left[\frac{4\pi e^2 \rho}{H^2}\left(\frac{Z}{A}\right)^2\right]^{1/2}$$

where H is the atomic mass unit.

It will immediately be noticed that in the range of densities and temperatures of interest, the Debye temperature is above the fusion temperature of oxygen. As the Debye temperature is non Z dependant, the crossing point of the two curves depends on the melting temperature,

$$T_{melt} = Z^{5/3}\,\frac{e^2}{k}\left(\frac{4\pi\rho}{3H}\right)^{1/3}\frac{Z}{A}^{1/3}\,\frac{1}{\Gamma}$$

which is proportionnal to $Z^{5/3}$.

Having a Debye temperature <u>above</u> the melting temperature is obviously meaningless. This comes only from a wrong definition of the melting temperature. If we replace in the definition

$$\Gamma = \frac{Z^2 e^2}{a\, k\, T}$$

the mean kinetic energy kT by $(T/\Theta_D)^3$ k T, which is sort of taking into account the quantum nature of the solid, we obtain

$$\Gamma = \frac{Z^2 e^2}{a\, k\, T} \left(\frac{\Theta_D}{T}\right)^3$$

from which we derive, with $\Gamma = 171$, an intermediate melting curve, sitting this time <u>below</u> the curve $T = \Theta_D$,

$$T_{melt} = \left\{ \frac{1}{\Gamma} \left(\frac{4\pi}{3}\right)^{1/3} (4\pi)^{3/2} \frac{Z^5}{(A\, H)^{10/3}} \frac{e^5\, \hbar^3}{k^4} \right\}^{1/4} \rho^{11/24}$$

$$T_{melt} = 6.71 \,.\, 10^3\, (Z/8)^{5/12}\, \rho^{11/24}$$

In the range of interest, we shall consider two cases, the <u>hot</u> case ($T \simeq 10^8$ °K), where most of the star is liquid, and the <u>cold</u> case ($T \simeq 10^7$ °K), where most of the star is solid.

The internal temperature of the star will be defined by the canonical condition of degeneracy at the bottom of the outer radiative zone. This gives the two definitions :

$$\text{hot case : } T = \frac{(k/H)^{5/3}}{K} \left(\frac{3\, \sigma\, L}{4\, \pi\, a\, c\, G\, M}\right)^{2/3}$$

where σ is the diffusion coefficient, $\sigma \simeq 0.2\,(1+X)$, and

$$\text{cold case : } T = (k/\mu H)^{8/7}\, K^{-6/7} \left(\frac{12.75\, \kappa_0\, L}{16\, \pi\, a\, c\, G\, M}\right)^{2/7}$$

where K is related to the pressure of the degenerate gas,

$$P = K\, (\rho/\mu\,)^{5/3}$$

and κ_0 is related to the absorption coefficient with a Kramers law:

$$\kappa = \kappa_0\, \rho\, T^{-3.5}$$

In these two cases, we obtain the approximate expressions :

- hot case:

$$t_{diff} = 1.914 \,.\, 10^{13}\, M*^{2/3} \text{ years,}$$

$$t_{cool} = 1.332 \,.\, 10^5\, T_8^{\,1/2} \text{ years,}$$

cold case :

$$t_{diff} = 9.92 \,.\, 10^9\, T_8^{\ 3}\, \rho_{c6}^{\ -4/3}\, M*^{-2/3} \quad \text{years},$$

$$t_{cool} = 6.496 \,.\, 10^{13}\, \rho_{c6}^{\ -4/3}\, M*^{-2/3} \quad \text{years},$$

which means that in the liquid star, the time scale of heat propagation is larger than the canonical cooling time. This is due to the low conductivity and high specific heat in the liquid phase. The reverse is true in the solid star, where the conductivity is high and the specific heat is low.

This means that the hot liquid white dwarf cools first in the outer region, and that, with a thermal diffusivity

$$D_{thermal} = 3.78 \,.\, 10^{-5}\, \rho_6^{\ -2/3}$$

the central region of a dense white dwarf (close to the Kaplan-Chandrasekhar limit) may eventually never become solid. In 10^8 years, the thermal wave has reached 6% of the star, and a density of the order of 0.06 ρ_c. For $\rho_c = 10^9$ this gives $\rho = 6.10^7$.

SOLIDIFICATION AND IGNITION

Whatever is the exact solution T(r,t), the problem of the solidification of a n-component plasma has to be considered, and it has a large variety of implications on stellar evolution : novae, type I supernovae, origin of pulsars and neutron stars.

I can summarize the problem in the following way :

If we consider a C-O mixture, we find in the literature two possibilities :

(a) solidification takes place at a temperature which is intermediate between the solidification of carbon and oxygen, and the solid is an homogeneous mixture of carbon and oxygen.

Jancovici (1982) has considered an elegant approach of the problem by comparing the free energy of two carbon atoms close together in an infinite solid of oxygen to the free energy of two carbon atoms isolated in an infinite solid of oxygen (Schatzman, 1983). Assuming no deformation of the lattice we find for the difference in free energy of the two configurations

$$\Delta F = - \frac{(Z_1 - Z_2)^2\, e^2}{a}$$

If we compare to

$$\Gamma = \frac{Z_1^{\ 2}\, e^2}{a\ k\ T}$$

we can write

$$\Delta F = - \frac{(Z_1 - Z_2)^2}{Z_1^{\ 2}}\, \Gamma\, k\, T$$

For $Z_1 = 6$, $Z_2 = 8$, $\Gamma = 171$, $\Delta F = -19\ k\ T$, which means that for a low concentration of carbon in oxygen , or of oxygen in carbon, there is complete miscibility.

(b) on the other hand, Stevenson (1980) suggests that carbon and oxygen are not miscible in solid phase. An oxygen-poor eutectic is predicted, with a carbon mass fraction $X_E \simeq 0.6$ and a low freezing temperature $T_E = 0.63\ T_C$, where T_C is the freezing temperature of pure carbon at the conisdered density.

An old white dwarf, almost completely solid, accreting matter and increasing its mass will evolve differently according to the chemical composition of its central region.In case (a), the solid is an alloy of carbon and oxygen, and when the mass of the star increases, the pycno-nuclear regime of the C-C reaction can overtake the electron capture on oxygen (then explosion rather than collapse) , wheras in case (b) the fall of oxygen snow flakes towards the center (Schatzman,1982) produces a pure oxygen core,where the electron capture dominates (and then collapse). However, the final result depends on the ignition density.

For the time being, I would like to discuss briefly the problem of propagation of ignition in solid layers. Conductive velocities can be estimated from the expression (Landau and Lifshitz, 1971)

$$V = \left(\frac{K}{C\ \tau}\right)^{\frac{1}{2}}$$

where (K/C) is the thermal diffusivity and τ the characteristic time for the nucelar reactions. According to Isern et al (1986), we have, with an ignition density $\rho_c = 10^{10}$ g cm^{-3}, as given by Mochkovitch and Hernanz (1986) the following table (table 1) , with the result that the central part of the white dwarf has time enough to collapse to a neutron star, with a possible off-center ignition follown by an explosion.

PYCNO-NUCLEAR REACTIONS

Two Body Reaction Rate

It seems to me that the problem is almost entirely understood. I would like just to mention the points which are not clear to me .

In the fluid case, the rate of nuclear reactions has been studied in detail by Alastuey and Jancovici (1978) for a o.c.p., including the effect of the fluctuations of the potential.

There are two major contributions, one is the classical contribution, due to the pair correlation function, giving for the enhancement factor

$$EF(\text{classical}) = \exp C$$

with

$$C = 1.0531\ \Gamma + 2.2931\ \Gamma^{\frac{1}{4}} - 0?551\ \ln \Gamma - 2.35$$

which has been derived from Hansen (1973) simulation. The other part is the quantum part, which comes mainly from the classical screened potential

Table 1. Characteristic Times

Detonation	0.1	s
Convective Deflagration	1	s
Conductive Deflagration	3-17	s
Electron Capture	1	s

W(r). The screened potential is linear close to the potential minimum, according to De Witt et al (1973) and quadratic near the origin, according to Jancovici (1977). The departure from linearity produces a small effect. However, it seems to me that it would be useful to reach an agreement on the screening potential which is used.

If we ignore the non-linear effects, we can accept the treatment of Itoh et al (1980) for a t.c.p.. However, the values of the coefficient of the linear screening potential should be taken exactly and their weak dependance on Γ taken into account (compare the values of the coefficients of the linear approximation derived from Hansen data (equ. 2.5 of Itoh et al) and the values derived from the harmonic oscillator model (their equation 2.10).

Finally, after freezing, the potential changes and there is a discontinuity in the pycnonuclear reaction rate. In the lattice model, the screening potnetial is smaller than in the liquid model (Itoh, 1981) and I would guess that the enhancement factor is smaller after freezing (or larger after melting).

If we remember that, as underlined by Mittler (1977) the WKB approximation is not satisfactory, it is quite clear that the value of the physical parameters for ignition depend on the exact value of the thermonuclear reaction rate and not only on the efficiency factor.

Electron Polarization

Ichimaru and Utsumi (1983) first found a strong effect of the polarization of the electrons, then corrected it to a small effect (1984). Mochkovitch and Hernanz (1986) have reconsidered the polarization effect, following the perturbation method of Galam and Hansen (1976). The electronic enhancement factor is given by

$$EEF = \exp\left(\frac{\Delta F^{pol}}{k\ T}\right)$$

where ΔF^{pol} is the contribution of the electrons to the Helmoltz free energy of the plasma.

Mochkovitch and Hernanz (1986) obtain for $\Gamma = 36$, $\rho = 10^9$

$$\frac{\Delta F^{pol}}{k\ T} = 0.55$$

$$EEF = 1.7$$

Screening of Photo-desintegration Reactions

Mochkovitch and Nomoto (1986), in connection with the rate of the $^{20}Ne(\gamma,\alpha)^{16}O$ reaction have the problem of the efficiency factor resulting from the change in the potential barrier.

The major contribution to the efficiency factor for the reaction $^{16}O(\alpha,\gamma)^{20}Ne$, in ln EF , is

$$C = \mu_1 + \mu_2 - \mu_3 = \Delta\mu$$

where μ_i is the contribution to the chemical potential of nucleus i, resulting from the Coulomb interaction. On the other hand the energy threshhold of the reaction is shifted ,

$$Q = Q_0 + \Delta\mu$$

so that the reaction rate for the photodesintegration,

$$\lambda_{3\gamma} \propto < \sigma v_{12} > e^{-\beta Q}$$

is related to the low density rate $\lambda^0_{3\gamma}$ by the relation :

$$\frac{\lambda_{3\gamma}}{\lambda^0_{3\gamma}} = \frac{\lambda^0_{3\gamma} EF \exp(-\beta Q_0 - \beta \Delta\mu)}{\lambda^0_{3\gamma}}$$

The major effect is then the quantum effect. To a first approximation, for a non-resonant reaction,

$$\frac{\lambda_{3\gamma}}{\lambda^0_{3\gamma}} = \exp\left(- \frac{45}{32} \frac{\Gamma^3}{\tau^2} \right) < 1$$

For the case of neon burning,which takes place during pre-supernova evolution, the ratio is close to 1 . A similar result is obtained for the resonant case.

3-alpha reactions

In the case of the $^4He(\alpha\gamma)^8Be$ reaction, the energy threshold of the reaction is 0.094 kev. As mentionned by W.Fowler (1981) at $\rho = 6.89\ 10^9$ g cm^{-3}, ^8Be becomes stable and then the 3-α reaction proceeds on stable ^8Be. Eventually, as W.Fowler suggests, the reaction $^8Be(^8Be,\alpha)^{12}C$ will be the dominant one.

This has to be considered as an important effect when descriging the gamm-flare stars, which are presently explained by accretion on neutron stars (Woosley et al, 1982, Hameury et al 1982)(even if Woosley and Hameury do not agree on the details of the mechanism itself).

All Channels Included

Thielemann and Truran (1986) have developped a complete ser of equations for the efficiency factor, taking into account all channels of the 4-body reactions like i(j,k)n . The main question, naturally concerns the screening effect of the outgoing particle. It should be noticed first that the equilibrium concentrations are changed, due to Coulomb interactions. Thielemann and Truran notice also that the WKB approximation to Coulomb barrier penetration is not good and that the transmission coefficients calculated entirely within the WKB approximation are in error by appreciable factors. However, confirming a result of Mittler (1977), they mention that the ration of transmission coefficients (screened and unscreened) turn out to be quite accurate.

I shall limit my self to two examples, borrowed to Thielemann and Truran (1986).Introducung their notations :

μ : reduced mass of the system

$$a_{ij} = (\tfrac{1}{2})(a_i + a_j) \quad ; \quad a_i = ((3/4\pi)(Z_i / \Sigma_j Z_j n_j))^{1/3}$$

$$\Gamma_{ij} = (Z_i Z_j / a_{ij})(e^2/kT) \quad ; \quad \tau_{ij} = \left[(27\ \pi^2/4)\ 2\ \mu_{ij}\ \frac{Z_i^2\ Z_j^2\ e^4}{kT\ \hbar^2}\right]^{1/3}$$

$$V_{scr}(r) = -\ kT\ \Gamma_{ij}\ (1.25 - 0.39\ (r/a_{ij}))$$

$$V(r) = (Z_i\ Z_j\ e^2\ /\ r) - U_0 + c(r\ /a)$$

$$r_1 = (Z_i\ Z_j\ e^2\ /\ E\ )\ ;\ r_2 = (a_{ij}\ E')/(c_1\ \Gamma_{ij}\ kT)$$

$$c_1 = 0.39$$

$$\xi(E') = \exp\left[-\frac{3\ \pi}{8\ \hbar}\ \frac{r_1^{\ 2}}{r_2}\ (2\ \mu\ E'\ ^{1/2}\right]$$

We consider the existence of a final nucleus for which the Q values of individual channels are corrected by Coulomb effects:

$$Q_{scr,n} = Q_n + U_{0,j} - U_{0,n}$$

The nuclear potential in the incoming channel is lowered by the amount $U_{0,j}$ and thus a state with a higher intrinsic excitation energy is produced at r = 0. The nuclear potential in the outgoing channel is lowered by $U_{0,n}$ and the available kinetic energy at infinity is reduced by that amount.

For a capture reaction at low temperature, the dominant channel is given by the γ-transition to the ground state (n=γ,k=γ) and the whole screening factor is given by

$$\exp(H_{j,\gamma}) = \exp\left[1.25\ \Gamma_{ij} - 0.0975\ \tau_{ij}\ \left(\frac{3\ \Gamma_{ij}}{\tau_{ij}}\right)^2\right]$$

In the case of a reaction with a negative Q value, Thielemann and Truran ibtain a new result . The incoming channel is then the dominant one. The maximum value of the integrand in <σ v> occurs at Q_k(negative) + E'_{max} = $E_{Gamow,kl}$, which the Gamow energy in the outgoing channel. In thatcase, the efficiency factor becomes :

$$\exp(H_{ij,k}) = \exp\left[1.25\ \Gamma_{ij} - 0.0975\ \tau_{kn}\left(\frac{3\ \Gamma_{kn}}{\tau_{kn}}\right)^2\right],$$

Miscellaneous

Equation of state. White dwarfs with surface hydrogen (Schatzman, 1945) may have a deep hydrogen convection zone (Schatzman, 1958). However, the exact boundary is an equation of state problem. It seems, from my readings, that there remain some uncertainty in the equation of state and in the degree of ionization. A careful application of the developments due to Ebeling (1973) should be carried.

Transport problems. The deep convection zone dredges up heavy elements at the bottom of the hydrogen convection zone, allowing the presence of heavy elements in the spectrum (like Van Maanen 2). There is controversy about the rate of gravitationnal sedimentation of the elements. The exact treatment has an important effect on the M, R, T_{eff}, chemical composition of a white dwarf, withh all possible consequences on the mass function of white dwarfs, thermal evolution of white dwarfs.

Similar problems arise at the boundary of the degenerate core of giants as discussed by Iben and Tutukov (1984), and a detailed treatment to appreciably different paths of evolution.

What About Sirius?

See (1892) has carried a remarkable analysis of the informations from the ancients of the colour of Sirius. I shall notice especially the remark concerning the Almagest. Ptolemey mentions six bright red objects : Arcturus, Aldebaran, Pollux, Betelgeuse, Antares and Sirius. From Theon and Avenius, the red colour of Sirius has disappeared at the end of the 4th century, and according to Al Sûfi is definitely not red anymore at the 10 th century.

The question which has been raised several times (Brecher, 1977, 1979) is the following : was Sirius B a red supergiant which has quickly evolved to a white dwarf ? Was Sirius A temporary a bright red supergiant ? In the former case this raises an interesting problem of cooling. Anyhow, the presence of the most famous white dwarf in the binary system of Sirius, the probalby real fast evolution of one of the companions raises fascinating problems which I think deserved to be mentionned in this meeting on dense matter.

SUMMARY

A number of physical and astrophysical problems remain to be solved in order to have a better understanding of white dwarfs and of their transition to type I supernovae and neutron stars.

From the point of view of physics, I would like to list
-the Γ-dependance of the linear and quadratic part of the screening potential;
-the exact solution of the penetration factor (the efficiency factor is relatively well known, but the WKB approximation is not good enough, especially for strong screening);
-the discontinuity of the thermonuclear reaction rate at the freezing temperature;
-the properties of the eutectic for C-O mixtures (or more complicated ones)
-a better equation of state in the intermediate region and better values of the microscopic diffusion coefficients;

From the astrophysical point of view, I would like to list
-the chemical composition of the bulk of white dwarfs;
-the initial temperature distribution at the time of white dwarf formation;
-the solution of the propagation of heat, for the cooling of the star,with the description of solidification, including the oxygen snowfall mechanism.
-the propagation of ignition in the solid layers, and its application to evolution towards supernovae type I and neutron stars;
-a revised analysis of gravitationnal sorting and extension of the outer convection zone of white dwarfs.

This is certainly not a complete list,but it corresponds to the unsolved questions which come up obviously when discussing the present litterature.

REFERENCES

Alastuey A., Jancovici B., 1978, Astrophys.J. 226, 1034
Alcock C., 1979, IAU Colloquium N°53.
Brecher K., 1977, Technology Review 80,N°2,p.52
1979, Bull.Amer.Astron.Soc., 11, 660
DeWitt H.E., Graboske H.C.,Cooper M.S.,1973, Astrophys.J., 181, 439
de Loore C., 1984 a, Astrophys. Space Sci.,99,335
1984 b, Observationnal tests of stellar evolution theory, A.Maeder, A.Renzini Eds. D.Reidel Publishing Company, p.359
Ebeling W., Sandig R., 1973, Ann. Physik, 28, 289
Fowler W., 1981, private communication.
Galam S., Hansen J.P., 1976, Phys. Rev., A 14, 816
Hameury J.M., Bonazola S., Heyvaerts J., Ventura J., 1982, Astron. Astrophys.,111, 242
Hansen J.P., 1973, Phys. Rev., A 8, 3069
Iben J. Jr.,1974, Annual Rev. Astron. Astrophys.,12, 215.
Iben J. Jr., Renzini A., 1983, Annual. Rev. Astron. Astrophys.,21,
Iben J. Jr., Renzini A., Schramm D.N., 1977, Advanced stages in stellar evolution, Saas Fe 1977, P. Bouvier, A.Maeder, Eds., Observatoire de Genève.
Iben J. Jr., Tutukov A., 1984,
Ichimaru S., Utsumi K., 1983, Astrophys. J. Letter, 269, L 51
1984 Astrophys. J., 286, 363
Isern J., Labay J., Canal R., 1986 (preprint)
Itoh N., Totsuji H., Ichimaru S., De Witt H.E., 1980, Astrophys. J.,239,415
Itoh N., Mitake S., Iyetomi H., Ichimaru S., 1983, Astrophys. J., 273, 774
Itoh N., Kohyama Y., Matsumoto N., Midosi S., 1984, Astrophys. J.,285, 758
Jancovici B., 1962, Nuovo Cimento 25, 428
Jancovici B., 1977, J. Stat. Phys., 17, 357
Jancovici B., 1982, Rapport interne, RCP "Matière dense" CNRS
Landau L.L., Lifshitz E., 1959, Course of theoretical physics,
Mittler H.E., 1977, Astrophys. J., 212, 513
Mochkovitch R., Hernanz M., 1986 (preprint)
Mochkovitch R., Nomoto K., 1986 (preprint)
Nomoto K., 1982, Astrophys. J., 253, 798
Nomoto K., 1984, in Problems of collapse and numerical relativity, D.Bancel , M. Signore Eds., D.Reidel Publishing Company, p. 89
Schatzman E., 1945, Ann. d'Astrophys., 8, 143
Schatzman E. 1958, White Dwarfs, North Holland Publishing Company
Schatzman E., 1982, Proceedings of the second internatioanl colloquium on drops and bubbles, Monterey California, Nov. 19-21, 1981, Dennis H. La Croissette Ed., J.P.L. Publications 82-7, p. 222
Schatzman E., 1983, Cataclysmic variables and related objects, M.Livio, G.Shaviv Eds., D.Reidel Publishing Company, p.149
See T.J.J., 1892, Astron. Astro-physical J., XI, pp. 269, 372, 550
Shapiro S.L., Teukolsky S.A., 1983, Black Holes, white dwarfs and neutron stars, Wiley and Sons Publishing Company
Stevenson D.J., 1980, J. de Physique Supplt., N°3, 41, C2 -53
Thielemnn F.K., Truran J.W., 1986 (preprint)
Webbink R.F., in White dwarfs and variable degenerate stars, H.M. Van Horn, V. Weidemann Eds., The University of Rochester, 1979
Weidemann V., Köster D., 1984, Astron. Astrophys. 172, 195
Woosley S.E., Wallace R.K., 1982, Astrophys. J., 258, 716

TRANSPORT PROCESSES AND NEUTRINO EMISSION PROCESSES IN DENSE ASTROPHYSICAL PLASMAS

Naoki Itoh

Department of Physics
Sophia University
7-1, Kioi-cho, Chiyoda-ku Tokyo, 102, Japan

INTRODUCTION

Strongly coupled plasma physics plays an important role in the elementary processes that occur in the interior of dense stars such as white dwarfs and neutron stars. One can have a unique opportunity of observing these strongly coupled plasmas through the comparison of the X-ray observations of the neutron star surface temperature with the model calculations of the neutron star cooling (Itoh 1986). In the present review paper, I will put special emphasis on the plasma physics aspects of the elementary processes.

Transport processes and neutrino emission processes are important elementary processes which decide the evolution of dense stars. They are not only interesting from the astrophysical point of view but also from the point of view of plasma physics. Strong correlations in the dense astrophysical plasmas affect the transport processes and the neutrino emission processes in a crucial way.

Flowers and Itoh (1976, 1979) presented extensive results of the calculation of the transport properties of dense matter. Their results were widely used for the model calculations of the neutron star cooling (Nomoto and Tsuruta 1981). Later Yakovlev and Urpin (1980) and Raikh and Yakovlev (1982) made significant improvements on the results of Flowers and Itoh (1976). More recently further improvements were made by Itoh et al. (1983), Mitake, Ichimaru, and Itoh (1984), and Itoh et al. (1984c).

There are four major neutrino processes which involve electrons. They are pair neutrino, photo-neutrino, plasma neutrino, and bremsstrahlung neutrino processes. The former three processes do not involve ions. The bremsstrahlung neutrino process involves ions. Concerning pair, photo-, and plasma neutrino processes, Beaudet, Petrosian, and Salpeter (1967) calculated the neutrino energy loss rates, using the Feynman-Gell-Mann (1958) theory. Dicus (1972) calculated these neutrino energy loss rates using the Weinberg-Salam theory (Weinberg 1967; Salam 1968). However, his calculation did not cover a wide range of densities and temperatures. Thus Dicus' result could not be widely used in stellar evolution computations. To resolve this unsatisfactory situation Munakata, Kohyama, and Itoh (1985, 1986) calculated these neutrino energy loss rates using the Weinberg-Salam theory, and presented the results for a wide range of densities and temperatures.

Concerning the bremsstrahlung neutrino process Festa and Ruderman (1969) calculated the neutrino energy loss rates using the Feynman-Gell-Mann theory. Dicus et al. (1976) calculated the bremsstrahlung neutrino energy loss rate using the Weinberg-Salam theory. However, they did not take into account the ionic correlation effects accurately.

TRANSPORT PROCESSES

In the astrophysical dense plasmas relevant to white dwarfs and neutron stars electrons are generally strongly degenerate. This means that the temperature satisfies the following condition:

$$T \ll T_F = 5.930\times10^9 \{ [1 + 1.018(Z/A)^{2/3}\rho_6^{2/3}]^{1/2} - 1 [K], \qquad (1)$$

where T_F if the Fermi temperature, Z the atomic number of the nucleus, A the mass number of the nucleus, and ρ_6 the mass density in units of 10^6 gcm^{-3}. For the ionic system we consider the case that it is in the liquid state. The latest criterion corresponding to this condition is given by (Slattery, Doolen, and DeWitt 1982)

$$\Gamma \equiv \frac{Z^2e^2}{ak_BT} = 2.275\times10^{-1}\frac{Z^2}{T_8}(\frac{\rho_6}{A})^{1/3} \le 178, \qquad (2)$$

where $a=[3/(4\pi n_i)]^{1/3}$ is the ion-sphere radius, and T_8 the temperature in units of 10^8 K.

We define a parameter y such that

$$y = \frac{h^2k_F^2}{2Mk_BT}, \qquad (3)$$

where $k_F=(3\pi^2n_e)^{1/3}$ is the Fermi wave number of the electrons. For the classical ions we have

$$y \lesssim 0.01. \qquad (4)$$

For the semiclassical ions we have

$$0.01 \lesssim y \lesssim 0.1 . \qquad (5)$$

For the calculations of the transport coefficients we use the relativistic version of the Ziman formula (Flowers and Itoh 1976):

$$\sigma = 8.693\times10^{21}\frac{\rho_6}{A}\frac{1-R}{<S\sigma>} [s^{-1}] , \qquad (6)$$

$$\kappa = 2.363\times10^{17}\frac{\rho_6T_8}{A}\frac{1-R}{<S_\kappa>} [ergs\ cm^{-1}\ s^{-1}\ K^{-1}] , \qquad (7)$$

$$R = \frac{1.018(Z/A)^{2/3}\rho_6^{2/3}}{1 + 1.018(Z/A)^{2/3}\rho_6^{2/3}} \qquad (8)$$

The scattering integrals $<S_\sigma>$ and $<S_\kappa>$ are given by (Mitake, Ichimaru, and Itoh 1984)

$$<S_\sigma> = <S> - \frac{2y}{3} [<I_{+1}> - R<I_{+3}>] - \frac{2y}{3} [<S_{+1}> - R<S_{+3}>], \qquad (9)$$

$$<S_\kappa> = <S_\sigma> + \frac{2y}{\pi^2} [3<I_{-1}> - (2 + 3R) <I_{+1}> + 2R <I_{+3}>] , \qquad (10)$$

$$<S> = <S_{-1}> - R<S_{+1}> , \qquad (11)$$

$$<S_n> = \int_0^1 d \left(\frac{k}{2k_F}\right) \left(\frac{k}{2k_F}\right)^{n+4} \frac{S(k)}{[(k/2k_F)^2 \epsilon(k,0)]^2} , \qquad (12)$$

$$<I_n> = \int_0^1 d \left(\frac{k}{2k_F}\right) \left(\frac{k}{2k_F}\right)^{n+4} \frac{1}{[(k/2k_F)^2 \epsilon(k,0)]^2} , \qquad (13)$$

where S(k) is the static structure factor of the classical ions, and $\epsilon(k,0)$ is the longitudinal dielectric function due to relativistically degenerate electrons (Jancovici 1962).

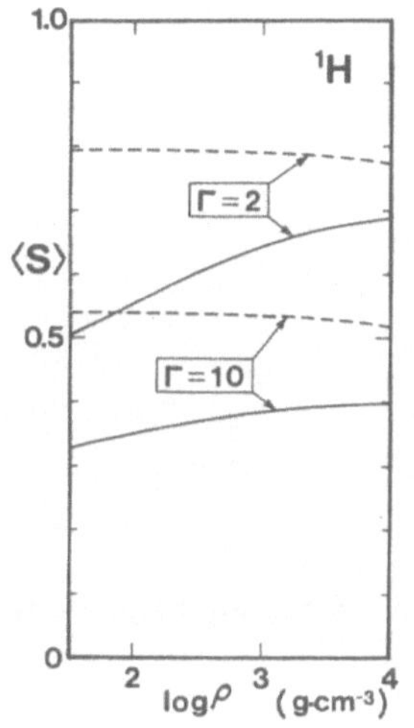

Fig. 1 Comparison of Yakovlev and Urpin's results (dashed curves) with the results of Itoh et al. (solid curves) for the ^{1}H matter.

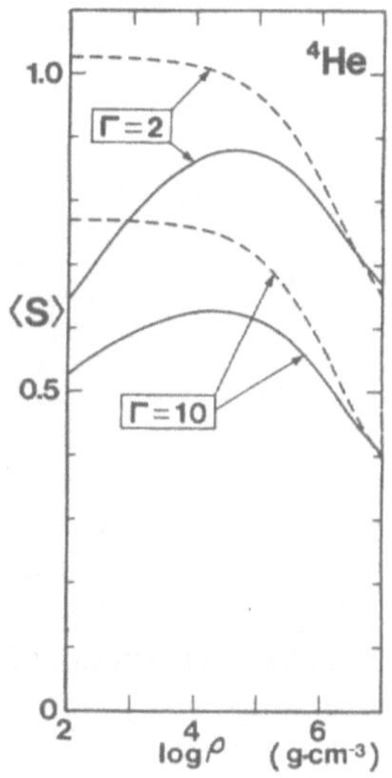

Fig. 2 Same as Fig. 1, for the ^{4}He matter.

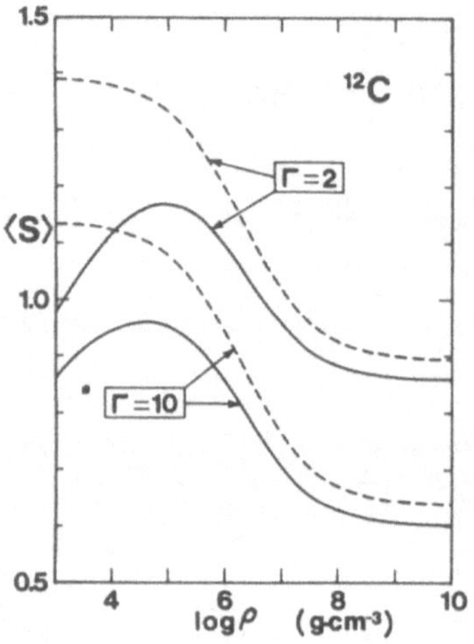

Fig. 3 Same as Figure 1, for the ^{12}C matter.

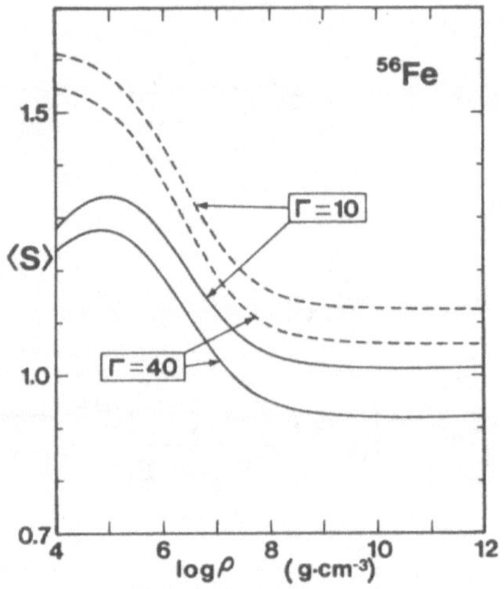

Fig. 4. Same as Figure 1, for the ^{56}Fe matter.

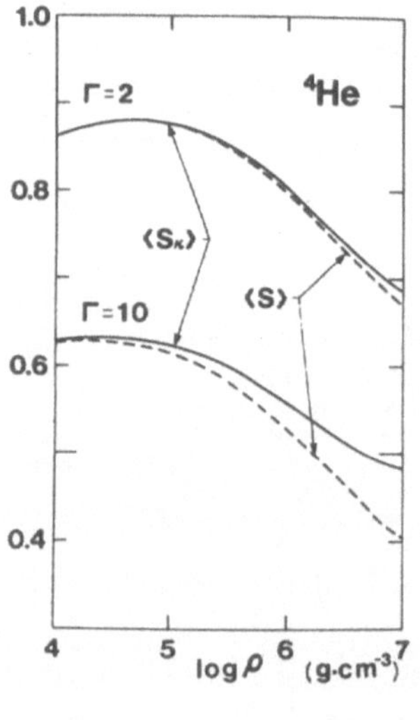

Fig. 5 Scattering integrals ⟨S⟩ and and $\langle S_k \rangle$ as functions of mass density p for the ^{4}He matter.

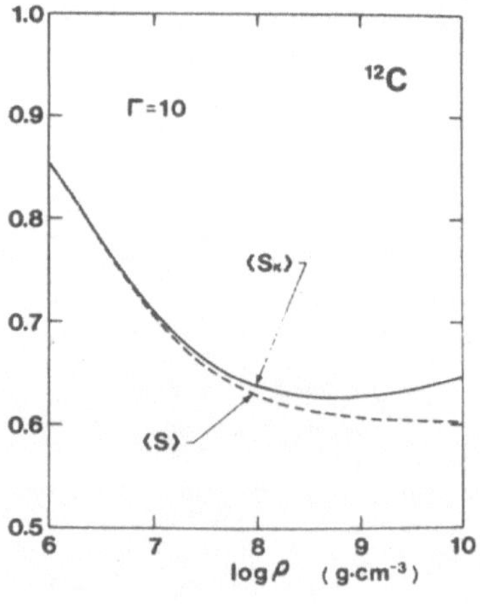

Fig. 6 Same as Figure 5, for the ^{12}C matter.

For the numerical calculation of the scattering integrals, Itoh et al. (1983) and Mitake, Ichimaru, and Itoh (1984) used the static structure factor of the classical one-component plasma calculated by the improved hypernetted chain method (Iyetomi and Ichimaru 1982 and 1983). Some examples of the numerical calculations are shown in Figures 1-4.

As is readily seen from Figures 1-4, Yakovlev and Urpin's calculation generally overestimates the resistivity (underestimates the conductivity) at low densities by 20-60%. This discrepancy is mainly caused by their neglect of electron screening.

In Figures 5-6 we show the ionic quantum correction to the resistivity. The corrections are typically 10-20%.

NEUTRINO EMISSION PROCESSES

Pair, Photo-, Plasma Neutrino Processes

The neutrino energy loss rates due to pair, photo-, and plasma neutrino processes have been calculated by Munakata, Kohyama, and Itoh (1985,1986) in the framework of the Weinberg-Salam theory. The result of the neutrino energy loss rates due to pair, photo-, and plasma neutrino processes can be written as follows:

$$Q_{pair} = \frac{1}{2}[(C_V^2 + C_A^2) + n(C_V'^2 + C_A'^2)]\, Q_{pair^+} + \frac{1}{2}[(C_V^2 - C_A^2) + n(C_V'^2 - C_A'^2)]\, Q_{pair^-} \quad (14)$$

$$Q_{photo} = \frac{1}{2}[(C_V^2 + C_A^2) + n(C_V'^2 + C_A'^2)]\, Q_{photo^+} - \frac{1}{2}[(C_V^2 - C_A^2) + n(C_V'^2 - C_A'^2)]\, Q_{photo^-} \quad (15)$$

$$Q_{plasma} = (C_V^2 + nC_V'^2)\, Q_{plasma}^{BPS}, \quad (16)$$

$$C_V = \frac{1}{2} + 2\sin^2\theta_W, \quad C_A = \frac{1}{2}, \quad (17)$$

$$C_V' = 1 - C_V, \quad C_A' = 1 - C_A, \quad (18)$$

$$\sin^2\theta_W = 0.217 \pm 0.014 \quad (19)$$

In the above n is the number of neutrino flavors whose masses are negligible compared with k_BT, and Q_{plasma}^{BPS} is the plasma neutrino energy loss rate calculated by Beaudet, Petrosian, and Salpeter (1967).

The numerical results of the calculation are shown in Figures 7-10. It has been found that the calculation of Munakata, Kohyama, and Itoh (1985, 1986) based on the Weinberg-Salam theory gives a substantially lower neutrino energy loss rate than the result of Beaudet, Petrosian, and Salpeter (1967) based on the Feynman-Gell-Mann theory. For n=0, the reduction factor α is in the range $0.35 \lesssim \alpha \lesssim 0.87$, depending on the density and temperature. For n=1, the reduction factor is in the range $0.56 \lesssim \alpha \lesssim 0.88$, and for n=2, we find $0.77 \lesssim \alpha \lesssim 0.88$.

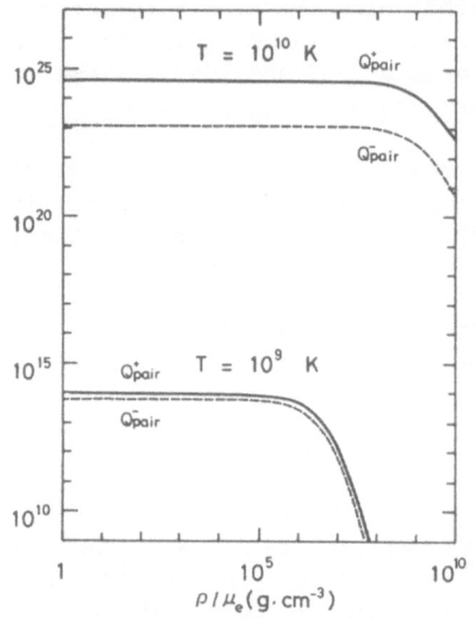

Fig. 7 Pair neutrino energy loss rate. The solid lines correspond to Q^+_{pair}, and the dashes lines correspond to Q^-_{pair}; μe is the electron mean molecular weight (Q in ergs s^{-1} cm^{-3}).

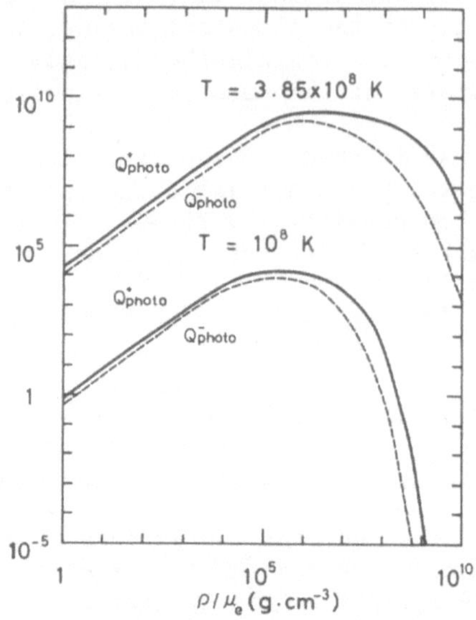

Fig. 8 Photoneutrino energy loss rate. The solid lines correspond to Q^+_{photo}, and the dashed lines correspond to Q^-_{photo} (Q in ergs s^{-1} cm^{-3}).

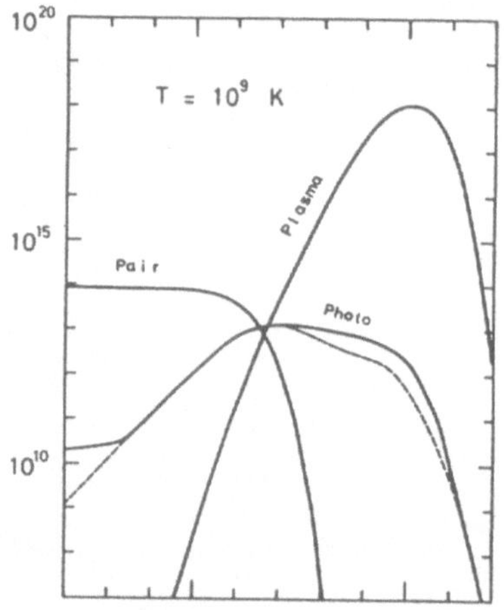

Fig. 9 Q^+_{pair}, Q^+_{photos}, and Q^{BPS}_{plasma} in ergs s^{-1} cm^{-3} for $T = 10^9$K. The solid lines correspond to the results of the numerical computations. The dashed lines correspond to the values of the fitting formulae.

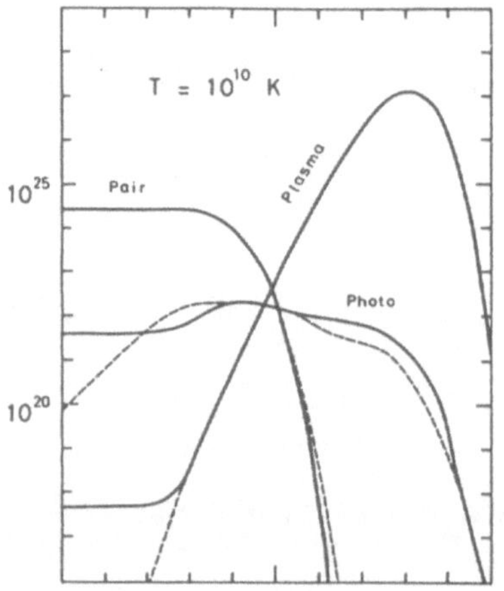

Fig. 10 Same as Figure 9, for $T = 10^{10}$k.

Bremsstrahlung Neutrino Process

The energy loss rate due to bremsstrahlung neutrino process in the liquid metal phase has been calculated by Itoh and Kohyama (1983). The result is summarized as follows:

$$\frac{Q}{\rho} = 0.5738 \text{ ergs g}^{-1}\text{s}^{-1} \frac{Z^2}{A} \left(\frac{T}{10^8\text{K}}\right)^6 \times \left\{\frac{1}{2}\left(C_V^2 + C_A^2 + nC_V'^{\,} + nC_A'^2\right)F_{liquid} - \frac{1}{2}\left(C_V^2 - C_A^2 + nC_V' - nC_A'^2\right)G_{liquid}\right\} , \quad (20)$$

$$F_{liquid} = \int_0^1 dq \frac{S(q)\ |f(q)|^2}{q^2\ |\epsilon(q,0)|^2} \frac{1}{4} \frac{1}{q} I(q) , \quad (21)$$

$$G_{liquid} = \int_0^1 dq \frac{S(q)\ |f(q)|^2}{q^2\ |\epsilon(q,0)|^2} \frac{1}{4} \frac{1}{q} J(q) , \quad (22)$$

where S(q) is the static structure factor of the ions, f(q) is the finite-nuclear-size correction factor, I(q), and J(q) are functions of q and the electron energy.

The results of the numerical calculation for ^{4}He and ^{56}Fe are shown in Figures 11-14. As is readily seen from the figures, the ionic correlation effects reduce the neutrino energy loss rate by a factor 2-20 in the liquid metal phase. The curves denoted as $\Gamma=0$ correspond to the calculation in which ionic correlation effects are totally neglected. They are essentially the same as the results obtained by Dicus et al. (1976). Therefore, strongly coupled plasma physics plays a crucial role in the bremsstrahlung neutrino process in dense stars.

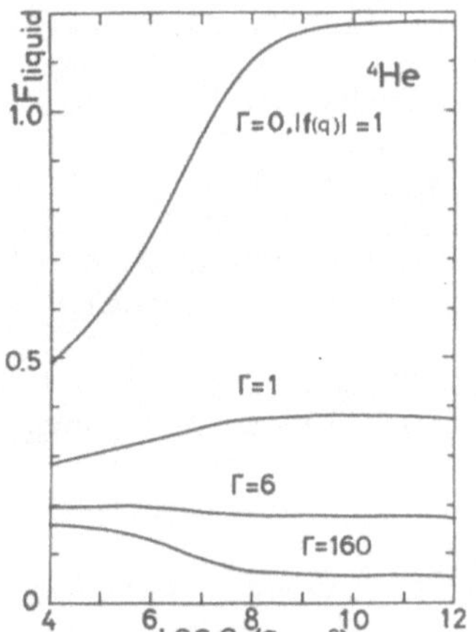

Fig. 11 F_{liquid} for the ^{4}He matter.

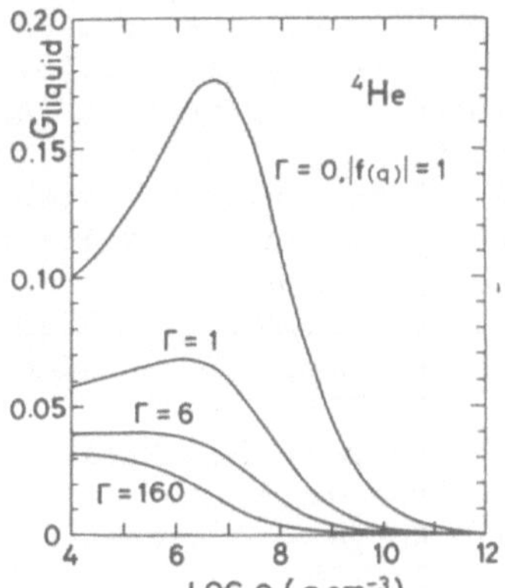

Fig. 12 G_{liquid} for the ^{4}He matter.

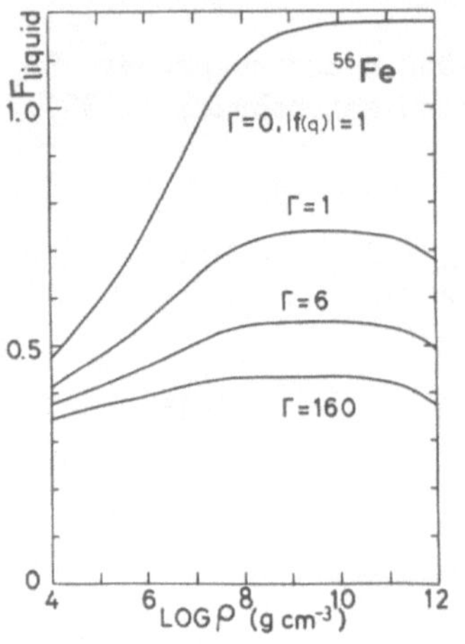

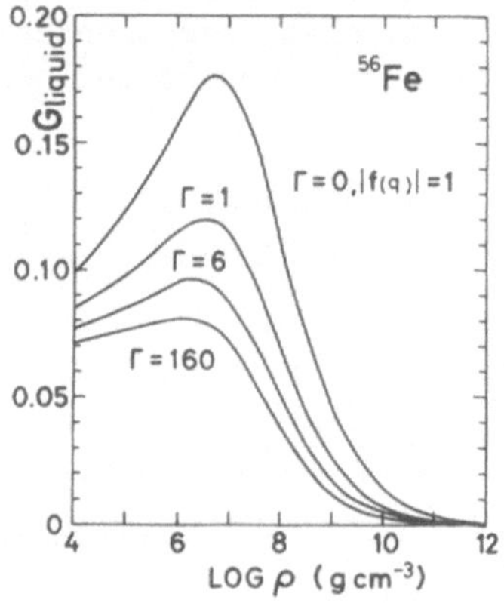

Fig. 13 F_{liquid} for the ^{56}Fe matter. Fig. 14 G_{liquid} for the ^{56}Fe matter.

CONCLUDING REMARKS

An ideal form of strongly coupled plasmas can be found in the interior of dense stars. Transport processes and neutrino emission processes in dense stars are crucially affected by the nature of strongly couple plasmas. Thus the comparison of the X-ray observations of the neutron star surface temperature with the model calculations of the neutron star cooling offers a unique opportunity of observing the behavior of the strongly coupled plasmas in dense stars. Further development of the X-ray astronomy is strongly desired.

REFERENCES

Beaudet, G., Petrosian, V., and Salpeter, E. E., Astrophys. J. 150:979

Dicus, D. A., 1972, Phys. Rev. D, 6:941.

Dicus, D., A., Kolb, E. W., Shramm, D. N., and Tubbs, D. L., 1976, Astrophys. J. 210:481.

Festa, G. G. and Rudrman, M. A., 1969, Phys. Rev. 180:1227.

Feynman, R. P. and Gell-Mann, M., 1958, Phys. Rev. 109:193.

Itoh, N., 1986, Invited review talk given at the IAU Symposium No. 125 "The Origin and Evolution of Neutron Stars" (Nanjing, China), to be published.

Itoh, N. and Kohyama, Y., 1983, Astrophys. J. 275:858.

Itoh, N., Kohyama, Y., Matsumoto, N. and Seki, M., 1984a, Astrophys. J. 280:787

Itoh, N., Kohyama, Y., Matsumoto, N. and Seki, M., 1984b, Astrophys. J. 285:304

Itoh, N., Kohyama, Y., Matsumoto, N. and Seki, M., 1984c, Astrophys. J. 285:758

Itoh, N., Kohyama, Y., Matsumoto, N. and Seki, M., 1984d, Astrophys. J. 279:413

Itoh, N., Mitake, S., Iyetomi, H. and Ichimaru, S., 1983, Astrophys. J. 273:774.

Iyetomi, H. and Ichimaru, S., 1982, Phys. Rev. A 25:2434.
Iyetomi, H. and Ichimaru, S., 1983, Phys. Rev. A 27:3241.
Jancovici, B., 1962, Nuovo Cimento, 25:428.
Mitake, S., Ichimaru, S., and Itoh, N., 1984, Astrophys. J. 277:375.
Munakata, H., Kohyama, Y. and Itoh, N., 1985, Astrophys. J. 296:197.
Munakata, H., Kohyama, Y. and Itoh, N., 1985, Astrophys. J. 304:580.
Nomoto, K. and Tsuruta, S., 1981, Astrophys. J. Lett. 250:L19.
Raikh, M. E. and Yakovlev, D. G., 1982, Astrophys. Space Sci. 87:193 .
Salam, A., 1968, "Elementary Particle Physics," edited by N. Svartholm, Almquist and Wiksells, Stockholm, p. 367.
Slattery, W. L., Doolen, G. D. and DeWitt, H. E., 1982, Phys. Rev. A 26:2255.
Weinberg, S., 1967, Phys. Rev. Lett. 19:1264.
Yakovlev, D. G. and Urpin, V. A., 1980, Soviet Astr. 24:303.

WHITE DWARF STARS: LABORATORIES FOR STRONGLY COUPLED PLASMA PHYSICS

G. Fontaine (E.W.R. Steacie Memorial Fellow)

Département de Physique
Université de Montréal
Montréal, Québec, Canada

INTRODUCTION

The white dwarf phase corresponds to the final configuration in the evolution of the vast majority of stars. The Sun, for example, will end its life as a white dwarf. When nuclear fuel is exhausted in a typical star, gravity is no longer balanced by the internal pressure generated by nuclear energy sources and the star collapses on itself. Hydrostatic equilibrium is again restored when degenerate electron pressure takes over. At that stage, further contraction is prevented, the object has reduced its radius to about 1% of the solar radius, and we speak of a white dwarf star.

It is generally believed that the immediate progenitors of white dwarfs are nuclei of planetary nebulae, themselves the products of intermediate- and low-mass main sequence evolution. Stars that begin their lifes with masses less than about 8 solar masses are expected to become white dwarfs. Those that have already ended their thermonuclear energy generation phases are objects which have burned hydrogen and helium in their interiors. Consequently, the essential of the mass of a typical white dwarf is believed to be contained in a core made of the products of helium burning: mostly carbon and oxygen. The exact proportions of C and O are unknown because of uncertainties in the rates of helium burning.

The expected structure of a typical young white dwarf is that of a stratified object with a mass of $\sim 0.6\ M_{\odot}$ consisting of a C/O core surrounded by a thin He-rich layer itself surrounded by an unprocessed H-rich layer. The respective thicknesses of these outer layers are not known a priori and must depend on the details of pre-white dwarf evolution. On theoretical grounds, however, we expect that the maximum amount of He that can survive the hot planetary nebula phase is only 10^{-2} of the total mass of the star, while the maximum amount of H is about 10^{-4} (D'Antona and Mazzitelli 1979). Although these outer layers are very thin, they are extremely opaque and play an essential role in the evolution of a white dwarf (Van Horn 1971).

The large opacity of the outer layers of a white dwarf implies that the radiation escaping from the star comes from the outermost regions -

the so-called atmosphere - containing, typically, less than 10^{-14} of the total mass of the star. Spectroscopic observations can only probe these regions which are usually dominated by hydrogen. Thus, a majority of white dwarfs are referred to as "H-rich" objects. It turns out that about 25% of the white dwarfs do not possess a H-rich layer. Those are called "He-rich" white dwarfs, with, again, the understanding that C/O cores contain essentially all of the mass even though such cores are not directly observable. Now the question why one white dwarf out of four has not retained a H-rich layer remains unanswered. It is one of the current puzzles in the theory of evolving white dwarfs.

A dying star begins the final phase of its history in the form of an extremely hot, collapsed object which can only cool off because its nuclear energy sources are depleted and gravitational energy can no longer be tapped efficiently as degenerate electron pressure prevents substantial contraction. Residual hydrogen burning may be present in the outer layers of some white dwarfs (Iben and Tutukov 1984), but this does not affect in an essential way the basic cooling picture of white dwarfs. Hence, a typical isolated white dwarf evolves with an almost constant radius, its mechanical structure being specified by the degenerate electron gas system. At the same time, the ions (largely decoupled from the electrons) provide the thermal energy which slowly leaks through to the outside, thereby producing the star's luminosity. With time, the ion system evolves from a gas to a fluid to a solid. Eventually, the whole star disappears from sight in the form of a cooled off, crystallized object known as a black dwarf.

It should be clear that dense matter physics is directly relevant to the evolution and structure of white dwarfs. For example, a detailed knowledge of the opacity and equation of state of strongly coupled plasmas is necessary to compute the rate of cooling of a white dwarf. And indeed, this rate basically depends on how much thermal energy is stored in the interior of the star and how fast this energy is transferred from the hot core to the cold interstellar medium through the opaque outer layers. Thus, a reliable description of the constitutive properties of dense plasmas is required to build a theory of evolving white dwarfs. By the same token, the observed properties of cooling white dwarfs can be used to test our theories of strongly coupled plasma physics. In particular, white dwarf stars appear to be the natural environments that best mimic the properties of the one-component plasma model.

PHYSICAL PROPERTIES OF EVOLVING WHITE DWARFS

In the context of the present meeting, it seems worthwhile to discuss some basic characteristics of evolving white dwarfs. Results are presented in Figure 1 which shows the fractional mass depth as a function of effective - or surface - temperature (lower scale) and as a function of time expressed in years (upper scale). Note that the choice of the ordinate implies a very strong bias of the figure in favor of the outermost layers. The results presented here are for the typical case of a 0.6 $M_{\odot}$ "He-rich" white dwarf model made up, initially, of a pure carbon core surrounded by a pure helium layer containing 10^{-4} of the total mass of the star.

First, the dotted line at the top of the graph corresponds to the position of the photosphere of the star, i.e. the layers from which the detectable radiation comes from. These layers are characterized by relatively low densities. The position of the photosphere changes with time because cooling changes the surface opacity thereby allowing radiation to escape from various depths.

The region defined by the two thick lines joined together by thin diagonal lines corresponds to a dense Coulomb fluid. The upper boundary is given by the condition $\Gamma = 1$, where Γ is the usual coupling parameter for dense plasmas. This condition loosely defines the transition between a gas and a liquid. The lower boundary (the curve labeled $\Gamma = \Gamma_m$) corresponds to the crystallization/melting line. Note that more than 99.99% of the mass of the white dwarf is in the form of liquid during most of its evolution. Even the surface layers show strong non-ideal behavior in the cooler phases. After some 10^{10} years of evolution, more than 99% of the mass of the star has crystallized.

The region defined by the thick curve and the thin vertical lines in the upper half of the diagram gives the location and extent of the helium convection zone that develops during the evolution. Indeed, as cooling proceeds, He begins recombining at an effective temperature of about 65,000K which leads to the formation of a superficial convection zone

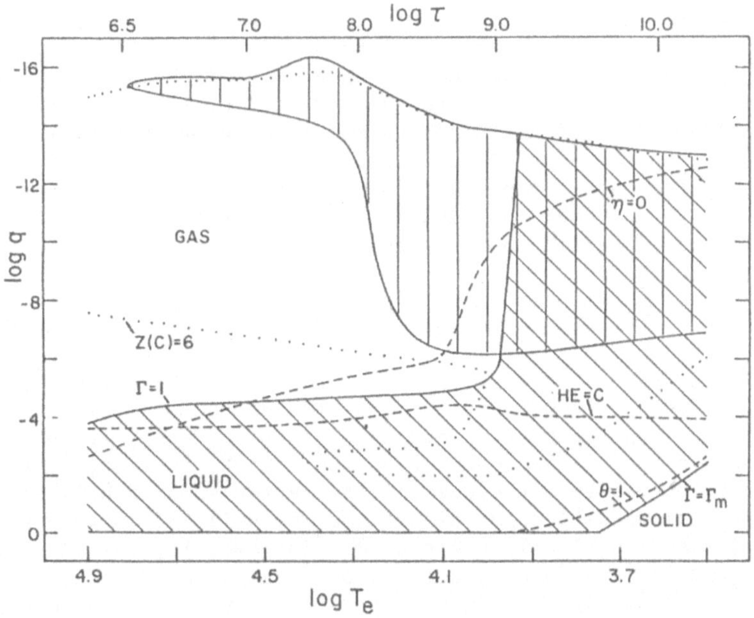

Fig. 1. Mass fraction ($q = 1- M(r)/M_*$) versus effective temperature (lower scale) or time (upper scale) for an evolutionary sequence with $M_* = 0.6\ M_\odot$ and an initial configuration consisting of a pure C core surrounded by a pure He layer containing 10^{-4} of the mass of the star. The dotted line at the top of the figure gives the location of the photosphere. The first shaded area (vertical lines) corresponds to the He superficial convection zone and the second one (diagonal lines) to regions characterized by a dense Coulomb fluid. See text for further details.

from then on. With further cooling, the base of the convection zone sinks into the star as more and more He recombines and convection becomes the dominant energy transfer mechanism in the outer layers of the white dwarf. Note that the top of the convection zone always remains at or near the photosphere. This has some importance because convection is so turbulent in white dwarfs (Böhm 1979) that any process (such as diffusion) that may change the local chemical composition at the base of the convection zone also leaves its signature on the observable surface abundances. Note also that the base of the convection zone reaches a maximum depth and then retreats toward the surface. The physical reason for this behavior is that the base of the convection zone reaches eventually the boundary of the degenerate core. Because of highly efficient electron conductivity, convection is suppressed in degenerate matter. With further cooling, the boundary of the degenerate core goes up toward the surface which forces the base of the convection zone to move up also. The dashed line labeled $\eta = 0$ (where η is the usual degeneracy parameter for non-relativistic electrons) shows precisely the behavior of the boundary of the degenerate core; below this line, electrons are degenerate. Quite clearly, it is seen that electrons can provide a uniform, neutralizing background of negative charges in the interiors of white dwarfs, a basic condition for the validity of the one-component plasma model. It is also interesting to point out that, for the coolest phases of its evolution, even the surface layers of a white dwarf are affected by electron degeneracy. In such cases, model atmosphere calculations must include conduction as well as radiation and convection as energy transport mechanisms (Böhm et al. 1977; Böhm 1979; Kapranidis 1983).

The dashed line labeled HE = C corresponds to the depth where the number of He ions is equal to the number of C ions. This line has a certain functional dependence on time because helium and carbon diffuse in a complicated way with respect to each other (see below). The other dashed line, labeled $\Theta = 1$, gives the depth where the Debye temperature of the C ions is equal to the local temperature. Quantum diffraction effects for the ionic system become important below that line. In the present case of an evolving 0.6 $M_\odot$ model, we find that these effects set in before crystallization. Hence, there is no transition to the classical solid in the core of this model.

Finally, the dotted line labeled Z(C) = 6 illustrates another feature of white dwarf physics, namely, the unusual ionization patterns that are found in such stars. It corresponds to the locus where the carbon ions are totally ionized. In the hotter phases, the degree of ionization increases monotonically with depth as in ordinary stars. (Note that He is completely ionized at the base of the convection zone which, in fact, acts as an almost perfect tracer of the partial ionization zone of He). In the cooler models, partial recombination always occurs until, at large enough depths, pressure ionization finally takes over. For example, the figure shows that, for the intermediate phases, carbon recombines at some depth before ionizing completely again. Such ionization patterns can have observational consequences (see below) and depend on the details of the envelope equation of state.

The results presented in Figure 1 are quite typical of contemporary evolutionary calculations of white dwarfs. They were taken from the detailed computations of Tassoul, Fontaine, and Winget (1986) and those of Winget, Lamb, and Van Horn (1986). These studies incorporate as good a treatment of the imput physics as any that exists in the white dwarf field. Along with the standard inclusion of the Los Alamos radiative opacities (Huebner et al. 1977) and the conductive opacities of Hubbard and Lampe (1969), they make use of the totally ionized, pure substance equation of state of Lamb and Van Horn (1975) for the liquid/solid deep interior and the non-ideal equation of state of Fontaine, Graboske, and Van Horn (1977) for the partially ionized, partially degenerate envelope. It seems appropriate here to point out that, after a decade, improvements to the equation of state and opacity data would be welcome. In particular, there is a need to further study the properties of C/O mixtures of deep white dwarf interiors after Shaviv and Kovetz (1976) and Stevenson (1977,1979). Also, it is hoped that improved models of pressure ionization such as those discussed by Hummer and Saumon at this workshop will soon be extended to the white dwarf envelope regime.

USING WHITE DWARFS AS PROBES OF DENSE MATTER PHYSICS

Under appropriate conditions, white dwarf stars can be used as proving grounds for theories of matter under extreme densities. For example, one of the very first tests of the validity of the Fermi-Dirac electron gas theory was offered when Fowler (1926) realized that stars with masses comparable to that of the Sun but with planetary dimensions could only be explained if their mechanical structure is specified by degeneracy pressure. Chandrasekhar (1931a,b, 1935) further developed this idea in a remarkable way and found the existence of a curious relationship between the mass and the radius of a white dwarf as well as the existence of a limiting mass above which a white dwarf cannot exist. Since then, the sample of known white dwarfs has increased some 100 fold and, while it is true that relatively few white dwarfs have reliable and independent estimates of their masses and radii, no example has ever been found of an isolated white dwarf that has a mechanical structure different from that predicted by Chandrasekhar.

A second method is to compare the actual observed distribution of white dwarfs (usually given in terms of the luminosity) with that predicted by detailed evolutionary calculations. For instance, at very high luminosities (i.e. in the early, hot phases of the evolution of a white dwarf), theory predicts that cooling is dominated not by thermal energy release but by neutrino processes occuring in the dense, hot core. The latest analyses (Fleming, Liebert, and Green 1986; Kawaler 1986) indicate that the observed distribution of hot white dwarfs is indeed consistent with evolutionary calculations that include neutrino emission processes and inconsistent with those that do not. Because the number of known hot white dwarfs is small, however, the statistics are not yet refined enough to distinguish between older values for the rates of neutrino emission and the newer ones discussed by Itoh at this meeting. It is hoped that future observations will provide a more stringent test.

At lower luminosities, where the bulk of the white dwarfs is found, neutrino cooling becomes negligible and the method would be used to test mostly the thermal properties of white dwarfs. In principle, one could use different sets of constitutive physics, compute several evolutionary sequences, and find the theoretical distribution that best fits the observed one. Although this method is straightforward, it cannot presently be used with a high level of confidence because the issue of white dwarf statistics is somewhat clouded by the uncertainties of the stellar birth rate function. And indeed, there are strong reasons to believe that, over the last 10^{10} years (corresponding roughly to the age of the coolest observable white dwarfs), the rate of star formation was not constant, although we do not know exactly how this rate changed with time. Therefore, white dwarfs were not fed in at a constant rate at the hot end of the temperature sequence and their distribution in terms of luminosity (or equivalently age) is affected. So in terms of testing the thermal properties of white dwarfs by comparing their observed and predicted distributions, we get into a Catch-22 situation: a perfect knowledge of the constitutive physics is required to derive the appropriate birth rate function and, at the same time, a perfect knowledge of the stellar birth rate is necessary for testing the predictions of various sets of constitutive properties. The situation will improve only if a reliable and independent means (i.e. not using white dwarf statistics) is found to derive the birth rate function.

A third method shows very promising results. It uses the fact that white dwarfs become unstable against non-radial gravity-mode oscillations during a phase of their cooling history. These instabilities manifest themselves in the form of multiperiodic luminosity variations. It appears that all "H-rich" white dwarfs become non-radial pulsators in the narrow range of effective temperature $13,000K \gtrsim T_e \gtrsim 11,000K$ (Fontaine et al. 1982), while the same may also be true for the "He-rich" white dwarfs in the range $30,000K \gtrsim T_e \gtrsim 24,000K$ (Liebert et al. 1986). As discussed by Winget and Fontaine (1982), these instability phases are related to an ionization mechanism of the main atmospheric constituent (either H or He). Comparing the observed properties of pulsating white dwarfs with those of models of such stars provides tests of both the mechanical and thermal structures of these objects. In particular, the very rich gravity-mode period spectrum of a white dwarf depends primarily on the density and pressure structures of the degenerate interior. At the same time, the composition stratification acts as a mechanical filter and only those modes that resonate with the thicknesses of the outer H- and He-rich layers can be amplified (Winget, Van Horn, and Hansen 1981). Moreover, a given mode is evanescent in the deep degenerate interior; it can only propagate in the partially degenerate outer layers. Thus, it can be driven unstable only if there exists a suitable instability mechanism in the outer layers. The hydrogen and helium ionization mechanisms discussed by Winget and Fontaine (1982) are believed to be the "driving engines" for the pulsating white dwarfs. They depend in a complicated and sensitive way on the thermal properties of the fluid envelope (adiabatic exponents, specific heats, pressure and opacity derivatives, etc.).

In practice, one could again use different sets of constitutive physics, build models of pulsating white dwarfs, and select those that best reproduce the observed properties of known pulsators. Although this method has already enjoyed some notable successes (see Winget et al. 1982a, 1982b), its full power has not yet been fully exploited because <u>actual</u> mode identification in pulsating stars remains uncertain. More detailed observations in the future should remedy to this situation. It should also be added that perhaps an even more powerful tool for testing the internal constitution of pulsating white dwarfs is provided by our ability to directly measure the cooling time scales of these objects through the observation of small but detectable changes in pulsation periods (Robinson and Kepler 1980; Winget et al. 1985). Although some years of observing will be necessary before the appropriate data are acquired, this very promising technique should allow a <u>direct</u> confrontation of theoretical and observed cooling time scales without resorting to the use of white dwarf statistics which are plagued by uncertainties related to incompleteness and small sampling.

Another method yet for probing the interior of a white dwarf is based on the realization that its atmospheric composition bears the signature of various competing mechanisms that are occuring inside the star. With cooling, the relative efficiencies of these mechanisms change, and so does the atmospheric chemical composition. While the atmospheres of white dwarfs are usually dominated by one element (either H or He), analyses of ground-based and satellite data reveal the presence of trace heavy elements that show highly unusual abundance patterns. Understanding these patterns forms the basis of the theory of the spectral evolution of white dwarfs. In the last few years, a number of us at Montreal have been working to elaborate such a theory. It turns out that it can be used as a sensitive probe of white dwarf envelopes (see, e.g., Michaud, Fontaine, and Charland 1984; Paquette et al. 1986a; Pelletier et al. 1986).

One fondamental result of these studies is that the surface abundance patterns of white dwarf stars can only be explained if diffusion processes are occuring in their envelopes. Consequently, one of the very basic ingredients in the theory of the spectral evolution of white dwarfs is a knowledge of the transport properties of the fluid envelopes. In general, in a star, the behavior of a trace element of species 2 in a background of species 1 is governed by a diffusion equation of the type:

$$\omega_{12} = D_{12}\left[-\frac{d\ln c_2}{dr} + \left(\frac{m_2}{m_1}(1+Z_1)-Z_2-1\right)\frac{d\ln P}{dr} + \alpha_t\frac{d\ln T}{dr} + \frac{m_2}{kT}g_r\right], \quad (1)$$

where D_{12} is the diffusion coefficient, r the radius, c_2 ($= n_2 / n_1$) the number fraction of element 2, m_i the mass of an atom of element i (i = 1, 2), Z_i the average charge of element i ($Z_i \leq$ atomic number), P the pressure, α_t the total thermal diffusion coefficient, T the temperature, k the Boltzmann constant, and g_r the radiative acceleration on element 2. In this convention, a negative velocity ($\omega_{12} < 0$) means that element 2 sinks into the star. With the diffusion equation written as in equation (1), the processes that induce diffusion in a star become evident. The first term, the concentration gradient term, causes ordinary diffusion. The pressure gradient term is responsible for gravitational settling, and the temperature gradient term for thermal diffusion. Finally, the last term is responsible for selective radiative acceleration on the trace element. Statistical physics must provide the values of the transport coefficients D_{12} and α_t for conditions encountered in white dwarf envelopes. In the remainder of this paper, we will concentrate on this particular aspect of the problem.

TRANSPORT PROPERTIES FOR WHITE DWARF PLASMAS

Transport coefficients appropriate for the deep interiors of white dwarfs are available from molecular dynamics and Monte Carlo studies of the one- and two-component plasma models (cf. DeWitt 1976; Hansen 1978; Hansen, Joly, and McDonald 1985). At the other extreme, the usual formulae for the diffusion coefficients of dilutes gases can be found, for example, in Chapman and Cowling (1970). White dwarf envelopes, however, are characterized by plasmas that are neither weakly or strongly coupled, a regime in which both the above theoretical models fail. Developing a theory of the spectral evolution of white dwarfs requires a knowledge of diffusion coefficients in this difficult intermediate regime. To estimate such coefficients, we have used a simple kinetic theory approach based on the numerical evaluation of collision integrals for a screened Coulomb potential of the Debye-Hückel type. The screening length is taken as the larger of the Debye length or the average interionic distance. The method can be used within the framework of Chapman-Enskog's theory (Chapman and Cowling 1970) or Burgers's (1969) method of solution of the Boltzmann equation. It becomes rigorously valid in the limit of a dilute plasma and recovers approximately the results of theories applicable at very high densities. This suggests that the region of intermediate coupling is probably reasonably bridged. The details can be found in Paquette et al. (1986b).

The transport coefficients D_{12} and α_t are expressible in terms of 8 different collision integrals which have been evaluated for a wide range of physical conditions. High-accuracy spline fits have been developed for these integrals and are presented in Paquette et al. (1986b). However, for quick estimations, the following fitting formula can be useful for the diffusion coefficient (cgs units):

$$D_{12} = 3.25 \times 10^{-15} \left[\frac{A_1 (A_1 + A_2)}{A_2} \right]^{\frac{1}{2}} \frac{1}{(Z_1 Z_2)^2} \frac{T^{2.5}}{\rho F_{12}}, \tag{2}$$

with

$$\ln F_{12} = 1.5\, \Psi_{12} - 1.56, \qquad \Psi_{12} > 0.6, \tag{3}$$

$$\ln F_{12} = 0.7\, \Psi_{12} - 1.08, \qquad \Psi_{12} \leq 0.6, \tag{4}$$

and

$$\Psi_{12} = \ln\left\{ \ln \left\{1 + 3.10 \times 10^{-10} \left(\frac{T}{Z_1 Z_2}\right)^2 \left(\frac{A_1}{\rho}\right)^{0.67} \left(1.78 \times 10^{-9} \left(\frac{T}{Z_1^2+Z_1}\right)^{1.5} \left(\frac{A_1}{\rho}\right)^{\frac{1}{2}} + 1\right)^{0.67} \right\} \right\}. \tag{5}$$

In this formula, ρ is the mass density, A_i the atomic weight of element i (i = 1,2), and the other symbols have been defined previously. The fitting formula is restricted to the case of a trace element of species 2 diffusing in a background of species 1. It recovers the results of the exact calculations of collision integrals within less than 20% over the white dwarf regime ($-7 \lesssim \Psi_{12} \lesssim 3$). At very high densities, the fitting formula assumes a functional dependence of the form $D_{12} \propto T^{1.10}/\rho^{0.53}$, which is not unlike the results of Hansen and collaborators (Hansen 1973; Pollock and Hansen 1973; Hansen, McDonald, and Pollock 1975) and those of Stevenson and Salpeter (1977). Because thermal diffusion appears negligible as compared to gravitational settling and ordinary diffusion in white dwarf plasmas (see below), no fitting formula has been derived for α_t.

It is of great interest to compare our results with the predictions of more sophisticated theories. At the time of the writing of the paper by Paquette et al. (1986b), the only results available to us were for the self-diffusion coefficient D_{11} in the limit of strong coupling. Recently, however, results for interionic diffusion (D_{12}) have become avalaible from both molecular dynamics (Hansen, Joly, and McDonald 1985 $\equiv$ HJM; Boercker, Ladd, and Pollock 1985 $\equiv$BLP) and kinetic theory (BLP). In particular, the interesting work carried out by Boercker and his collaborators at Livermore has been summarized at this meeting. In Table 1, we compare our results for D_{12} with those of BLP and HJM for the mixtures that they have considered. In the top half of the table, a Si XV - Sr XXXVII mixture of various concentrations characterized by a total ionic number density $n = 10^{22}$ and a constant coupling parameter γ is considered. Our results recover the kinetic theory results of BLP within 10% while the deviations from their molecular dynamics calculations are less than 15%. The differences are somewhat larger (10 - 30%) for the equimolar H II - He III mixture considered in the bottom half of the table. In that case, the coupling parameter was allowed to vary to span the moderate to strong coupling region. Note that we have translated the results of BLP and HJM in terms of a total ionic number density $n = 10^{26}$. As compared to BLP and HJM, we find that our results show surprisingly good agreement, even for values of the coupling parameter that are more characteristic of the high-density interior of a white dwarf than its envelope. This comparison gives us added confidence that our estimates of D_{12} for the moderately

Table 1. Interdiffusion Coefficients in Binary Ionic Mixtures

$$\gamma = \frac{e^2}{kT}\left(\frac{4\pi n}{3}\right)^{0.33}$$

Si XV - Sr XXXYVI $n = 10^{22}$	% Si	γ	D_{12}(MD) (BLP)	D_{12}(KT) (BLP)	D_{12} (This Work)
	0.1	0.005	-	1.85-1	1.90-1
	25.0	0.005	2.15-1	1.86-1	1.87-1
	50.0	0.005	2.17-1	1.82-1	1.85-1
	75.0	0.005	2.04-1	1.77-1	1.82-1
	99.9	0.005	-	1.68-1	1.80-1
H II - He III $n = 10^{26}$	**% H**	**γ**	**D_{12}(MD) (BLP,HJM)**	**D_{12}(KT) (BLP)**	**D_{12} (This Work)**
	50.0	0.4	6.67-2	7.06-2	8.53-2
	50.0	1.0	2.03-2	1.76-2	1.85-2
	50.0	4.0	3.15-3	3.41-3	2.55-3
	50.0	40.0	2.41-4	-	1.86-4

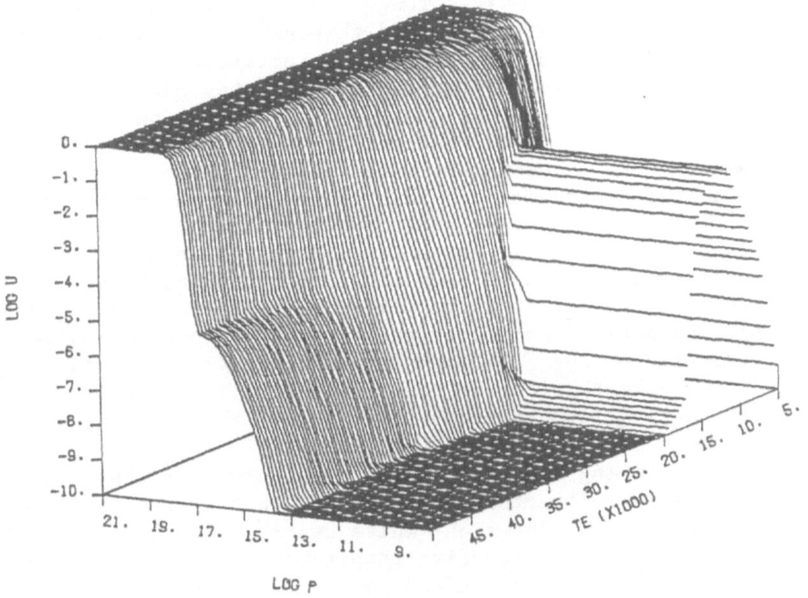

Fig. 2. Evolution of the carbon distribution for a typical white dwarf evolutionary sequence (see text). The carbon number concentration u is plotted as a function of the pressure in a given model. The third axis is the decreasing effective temperature. Note how traces of carbon can pollute the surface layers (low values of log P) in the range 20,000K $\gtrsim T_e \gtrsim$ 5,000K. Note also that pollution reaches a maximum around $T_e \simeq$ 12,000K,

coupled plasmas found in white dwarf envelopes are satisfactory. We emphasize, however, that a much more detailed comparison involving conditions encountered in actual white dwarf models remains highly desirable. In addition, it seems imperative that computations such as those presented by BLP be extended to include thermal diffusion. And indeed, our results suggest (see Fig. 3 of Paquette et al. 1986b) that the binary thermal diffusion coefficient decreases to very small absolute values with increasing density for a given isotherm. This makes thermal diffusion a negligible transport process in white dwarf plasmas. The validity of this point remains to be verified by independent calculations.

We conclude by discussing briefly an example of application in which one of the key features is a knowledge of the transport properties of white dwarf plasmas. The problem to be considered is the evolution of the carbon distribution in a "He-rich" white dwarf (details are given in Pelletier et al. 1986). From the original stratified distribution in which there is a near discontinuity in chemical composition caused by the very steep temperature dependence of the nuclear burning rate, we expect that carbon will migrate with time in a cooling white dwarf. Obviously, the evolution of the C distribution depends on both the diffusion coefficients and the local conditions. Figure 2 illustrates the results of a sample evolutionary sequence. It shows how the C distribution evolves in a 0.6 $M_{\odot}$, He/C white dwarf model with a He layer containing 10^{-4} of the total mass of the star (this is the model discussed in the second section of this paper). What is plotted is the number concentration of carbon ($u = n(C)/\left(n(C) + n\ (He)\right)$) as a function of depth - measured here in terms of the pressure - and as a function of decreasing effective temperature. In the early, high-temperature phases, the C profile "remembers" its initial condition; in the present case, a step function $\log u = 0\ /\ -5$. The kink that is observed corresponds to the former plateau at $\log u = -5$. Carbon that is located deeper than the kink diffuses upward through ordinary diffusion, while carbon located above the kink diffuses downward through gravitational settling. By the time the star has cooled down to an effective temperature of about 30,000K, the transient phase is over and the star has now "forgotten" its assumed initial configuration. At that point in time, the flow of carbon becomes unidirectional: C migrates outward. With further cooling, the base of the He convection zone sinks into the star while more C diffuses upward. Carbon that arrives at the base of the convection zone is immediately distributed uniformly across the convection zone because the latter is highly turbulent. Eventually, enough C has migrated upward that significant traces pollute the He convection zone, and, consequently, the observable atmosphere. With still further cooling, partial recombination of C (combined with the outward motion of the base of the convection zone) reverses the direction of the flow, and from an effective temperature of about 12,000K, C retreats from the convection zone. Carbon sinks back into the star as is neatly illustrated in the figure.

The most interesting aspect of these events is that they leave a definite signature which can be observed: namely, that traces of C should pollute the atmospheres of "He-rich" white dwarfs, and that these traces should show a maximum at an effective temperature of about 12,000K. The fact that these predictions are indeed confirmed by the observations (see Pelletier et al. 1986) constitutes, so far, one of the better successes of the theory of the spectral evolution of white dwarfs. It should be noted that the results are sensitive to the constitutive and transport properties of white dwarf plasmas as well as the choice of model parameters. Although it has not yet been possible to untangle these different effects, there is hope that the so-called carbon pollution phenomenon in "He-rich" white dwarfs will eventually shed additional light on the physics of white dwarf plasmas.

REFERENCES

Boercker, D.B., Ladd, A.J., and Pollock, E.L. 1985, (UCID - 20507, Lawrence Livermore National Laboratory Report).

Böhm, K.-H. 1979, in IAU Colloquium 53, White Dwarfs and Variable Degenerate Stars, ed. H.M. Van Horn and V. Weidemann (Rochester: University of Rochester), p. 223.

Böhm, K.-H., Carson, T.R., Fontaine, G., and Van Horn, H.M. 1977, Ap.J., 217, 521.

Burgers, J.M. 1969, Flow Equations for Composite Gases (New York: Academic).

Chandrasekhar, S. 1931a, M.N.R.A.S., 91, 456.

Chandrasekhar, S. 1931b, Ap.J., 74, 81.

Chandrasekhar, S. 1935, M.N.R.A.S., 95, 207.

Chapman, S., and Cowling, T.G. 1970, The Mathematical Theory of Non-Uniform Gases (3d ed.; Cambridge: Cambridge University Press).

D'Antona, F., and Mazzitelli, I. 1979, Astr. Ap., 74, 161.

DeWitt, H.E. 1976, Phys. Rev. A, 14, 1290.

Fleming, T., Liebert, J., and Green R.F. 1986, Ap.J., in press.

Fontaine, G., Graboske, H.C., Jr., and Van Horn, H.M. 1977, Ap.J. (Suppl.), 35, 293.

Fontaine, G., McGraw, J.T., Dearborn, D.S.P., Gustafson, J., and Lacombe, P. 1982, Ap.J., 258, 651.

Fowler, R.H. 1926, M.N.R.A.S., 87, 114.

Hansen, J.-P. 1973, Phys, Rev. A, 8, 3096.

Hansen, J.-P. 1978, in Strongly Coupled Plasmas, ed. G. Kalman and P. Carini (New York: Plenum), p.117.

Hansen, J.-P., Joly, F., and McDonald, I.R. 1985, Physica A, 132, 472.

Hansen, J.-P., McDonald, I.R., and Pollock, E.L. 1975, Phys. Rev. A, 11, 1025.

Huebner, W.F., Merts, A.L., Magee, N.H., and Argo, M.F. 1977, (LA-6760-M, Los Alamos National Laboratory Report).

Hubbard, W.B., and Lampe, M. 1969, Ap.J. (Suppl.), 18, 927.

Iben, I., Jr., and Tutukov, A.V. 1984, Ap.J., 282, 615.

Kapranidis, S. 1983, Ap.J., 275, 342.

Kawaler, S.D. 1986, Ph.D. Thesis, University of Texas at Austin.

Lamb, D.Q., and Van Horn, H.M. 1975, Ap.J., 200, 306.

Liebert, J., Wesemael, F., Hansen, C.J., Fontaine, G., Shipman, H.L., Sion, E.M., Winget, D.E., and Green, R.F. 1986, Ap.J., in press.

Michaud, G., Fontaine, G., and Charland, Y. 1984, Ap.J., 280, 247.

Paquette, C., Pelletier, C., Fontaine, G., and Michaud, G. 1986a, Ap.J. (Suppl.), 61, 197.

Paquette, C., Pelletier, C., Fontaine, G., and Michaud, G. 1986b, Ap.J. (Suppl.), 61, 177.

Pelletier, C., Fontaine, G., Wesemael, F., Michaud, G., and Wegner,G. 1986, Ap.J., in press.

Pollock, E.L., and Hansen, J.-P. 1973, Phys. Rev. A, 8, 3110.

Robinson, E.L., and Kepler, S.O. 1980, Sp. Sci. Rev., 27, 613.

Shaviv, G., and Kovetz, A. 1976, Astr. Ap., 51, 383.

Stevenson, D.J. 1977, Proc. Astr. Soc. Aust., 3, 167.

Stevenson, D.J. 1979, Paper presented at the C.N.R.S. International Colloquium on the Physics of Dense Matter, Paris, Sept. 17-22.

Stevenson, D.J., and Salpeter, E.E. 1977, Ap.J. (Suppl.), 35, 221.

Tassoul, M., Fontaine, G., and Winget, D.E. 1986, in preparation.

Van Horn, H.M. 1971, in IAU Symposium 42, White Dwarfs, ed. W.J. Luyten (Dordrecht: Reidel), p. 97.

Winget, D.E., and Fontaine, G. 1982, in Pulsations in Classical and Cataclysmic Variables, ed. J.P. Cox and C.J. Hansen (Boulder: University of Colorado), p. 142.

Winget, D.E., Van Horn, H.M., and Hansen, C.J. 1981, Ap.J. (Letters), 245, L33.

Winget, D.E., Van Horn, H.M., Tassoul, M., Hansen, C.J., Fontaine, G., and Carrol, B.W. 1982a, Ap.J. (Letters), 253, L29.

Winget, D.E., Robinson, E.L., Nather, R.E., and Fontaine, G. 1982b, Ap.J. (Letters), 262, L11.

Winget, D.E., Kepler, S.O., Robinson, E.L., Nather, R.E., and O'Donoghue, D. 1985, Ap.J., 292, 606.

Winget, D.E., Lamb, D.Q., and Van Horn, H.M. 1986, in preparation.

TOWARD AN IMPROVED PURE HYDROGEN EOS FOR ASTROPHYSICAL APPLICATIONS

Didier Saumon and Hugh M. Van Horn

Department of Physics and Astronomy and
C. E. K. Mees Observatory, University of Rochester
Rochester, New York 14627-0011

INTRODUCTION

The motivation for our equation of state (EOS) work derives from our interest in the structure and evolution of substellar "brown dwarf" stars. To construct an evolutionary model requires the solution of the differential equations governing stellar structure and evolution, which in turn requires knowledge of the EOS of stellar matter.

A preliminary evaluation of current versions of the EOS for pure hydrogen from the Lawrence Livermore National and Los Alamos Scientific laboratories and from the work of Fontaine *et al.* (1977) has convinced us that significant improvements (> 5%) can be made. All three of these equations of state are widely used for stellar modeling and a new, independent calculation seems appropriate at this time.

THE EQUATION OF STATE OF MOLECULAR HYDROGEN

An EOS suitable for modeling brown dwarf stars must cover the following range of densities and temperatures: $200K < T < 2x10^6K$, $\rho < 2000$ gram/cm^3. A wide variety of physical conditions is encountered under these (ρ,T) regimes. In order to develop a facility for tackling such calculations, we have first considered the relatively simple case of fluid molecular hydrogen.

To attack this problem, we have used the free energy minimization technique, employing hard sphere perturbation theory to describe the fluid properties of hydrogen. We have followed the work of Ross *et al.* (1983) closely in these computations. Our expression for the total Helmholtz free energy of H_2 is:

$$F = NkT \ln\frac{\rho}{e}\left[\frac{2\pi\hbar^2}{mkT}\right]^{3/2} - NkT \ln\left\{\frac{1}{2}\sum_{n=0}^{14}\sum_{J=0}^{J_{max}}(2J+1)\, e^{-E(n,J)/kT}\right\}$$

$$+ NkT\left[\frac{\eta(4-3\eta)}{(1-\eta)^2} - \left[\frac{\eta^4}{2} + \eta^2 + \frac{\eta}{2}\right]\right] + \frac{\rho N}{2}\int \varphi(r)g(r,\eta)d^3r$$

$$+ \frac{\hbar^2 \rho N}{24mkT} \int \nabla^2 \varphi(r) g(r,\eta) d^3r \quad . \tag{1}$$

Here, N is the total number of molecules, V is the volume of the system, $\rho=N/V$, T is the temperature, $\hbar$ is Planck's constant, k is Boltzmann's constant and m is the mass of an H_2 molecule.

The first term in (1) is the translational free energy of the ideal gas. The second is the contribution from internal vibration and rotation states of the molecules. Excited electronic states as well as dissociation are negligible for $T < 10^3$K and were not included in this calculation. The energy of each level relative to the vibrational ground state energy E(0,0) is

$$E(n,J) = \omega_e(n+\tfrac{1}{2}) - \omega_e x_e(n+\tfrac{1}{2})^2 + B_e J(J+1) - D_e J^2(J+1)^2$$
$$- \alpha_e(n+\tfrac{1}{2})J(J+1) - \omega_e/2 + \omega_e x_e/4 \tag{2}$$

where n and J are the vibrational and rotational quantum numbers, respectively. The spectroscopic constants appearing in (2) were taken from Huber and Herzberg (1979). Because of the pathological behavior of (2) for large J, the sum over rotational states is cut off at the value J_{max} which depends on n, α_e, B_e and D_e. Note that this treatment of internal structure is appropriate only for isolated molecules, i.e. for the very low densities at which the spectroscopic measurements are made.

The third term in (1) is the excess free energy of a gas of hard spheres, as given by Carnahan and Starling (1969) and modified by Ross (1979) to make the reference system closer to a $1/r^{12}$ repulsive core, which better approximates the actual potential. The hard sphere gas is characterized by one parameter, the packing fraction $\eta=\frac{\pi}{6}\rho\sigma^3$, where σ is the hard sphere diameter.

The fourth term in (1) is the first order correction to the expansion of the configuration integral, and the last term is the first order quantum correction in the Wigner-Kirkwood expansion (Landau and Lifshitz, 1958).

In the last two terms, $g(r,\eta)$ is the pair correlation function for hard spheres. We have used the analytical expression of Smith and Henderson (1970) for the Percus-Yevick approximation of $g(r,\eta)$.

The interactions between molecules are described by a spherically symmetric effective pair potential, $\varphi(r)$. The potential we used was obtained by Ross, *et al.* (1983) from shock tube experiments and static compression measurements. This effective pair potential, which implicitly includes many-body effects, agrees well with an extensive body of high compression data.

Equation (1) must be minimized with respect to the hard sphere diameter (σ), the only free parameter of the model. Any thermodynamic quantity can then be obtained by taking the appropriate derivatives of the free energy.

For reasons stated below, we have also generated an equation of state for fluid molecular deuterium, D_2. This is done simply by using the appropriate value for m, the mass of the D_2 molecule, and using a different set of spectroscopic constants in eq. (2) (Huber and Herzberg, 1979). The interaction potential is taken to be the same as for hydrogen, and thus non-ideal terms are unaffected.

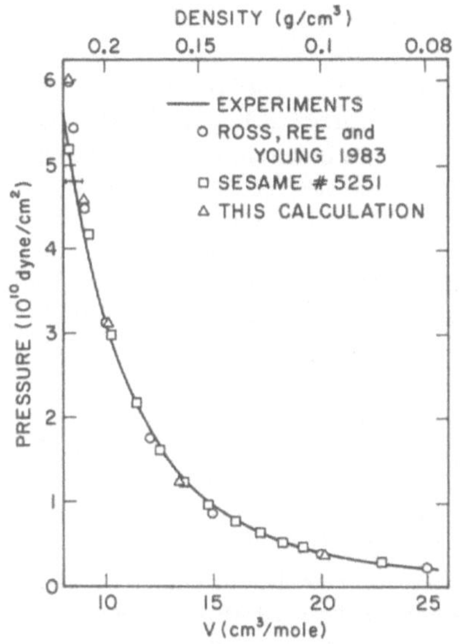

Fig. 1. T = 300K pressure isotherm for H_2.

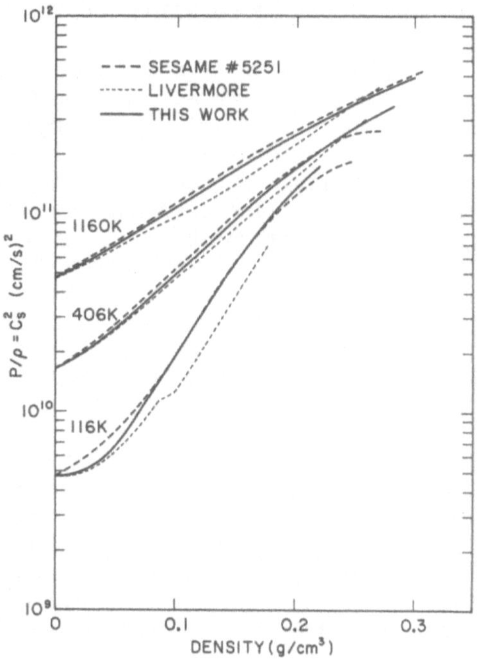

Fig. 2. Isotherms of $P/\rho = C_s^2$ for H_2.

RESULTS

To check the reliability of our computations, we have compared our results with the calculation of Ross *et al.* (1983). Figure 1 shows the 300K pressure isotherm up to the phase transition (around 5.4×10^{10} dyne/cm^2). The agreement between our calculations (triangles) and the work of Ross *et al.* (1983; open circles) is excellent. Comparison with the pressure, density and packing fraction of the fluid along the melting curve calculated by these authors again shows excellent agreement. This is strong evidence that our calculations are being done correctly.

Our computations also reproduce the existing experimental data quite well, as shown by the generally good agreement with the experimental results of Mills *et al.* (1978) and of Shimizu *et al.* (1981) (solid curve in Fig. 1). The exception to this is the region V<10 cm^3/mole, where both our calculation and that of Ross *et al.* (1983) yield too high a pressure. This is probably due to the uncertainty inherent in the theoretical calculation. Ross *et al.* (1983) report a 4% uncertainty in the predicted volume arising from the large volume error bars in the experimental data (both shock tube and diamond anvil experiments) used to fit their effective pair potential. This uncertainty is shown by the error bar on Figure 1.

We have also compared our H_2 EOS with a table provided by the Lawrence Livermore National Laboratory (Graboske and Wong, 1980) and with the Los Alamos Scientific Laboratory SESAME equation of state (Material #5251). The latter is actually obtained by Los Alamos from a density scaling of the detailed deuterium EOS in the SESAME library (Material #5263). Since the scaling method can only give an approximate EOS for hydrogen, we have also compared our D_2 EOS with the SESAME #5263 table (see below). Despite the approximate nature of the Los Alamos H_2 EOS, Figure 1 shows that the SESAME #5251 300K pressure isotherm (squares) agrees very well with the experimental curve and does not show the same high density departure as the results of our calculation and those of Ross *et al.* (1983).

Figure 2 shows $P/\rho=C_s^2$, the square of the isothermal sound speed, along isotherms for all three hydrogen EOS's. The SESAME #5251 table and our results agree very well, the largest difference being $\lesssim 6\%$. The larger deviation seen at the low density end of the 116K isotherm is probably due to the difficult interpolation between the 0K and 178K isotherms in the SESAME table. The Livermore EOS, however, shows marginal agreement with the other two calculations. There is a hint of a break in the isotherms at $\rho \sim 0.1$ gram/cm³. We compare the SESAME #5263 deuterium internal energy isotherms with ours in Figure 3. Again, the agreement is excellent if we substract what appears to be a constant, inconsequential, energy shift of $\sim 1.9 \times 10^9$ erg/gram apparent in the ideal gas regime. The difference seen as $\rho \to 0$ at 2127K is due to dissociation of the molecules, which is not included in our current model.

Figure 4 shows that the good agreement between our internal energy calculations and those tabulated in the SESAME #5263 table does not hold for the density-scaled internal energies given in SESAME table #5251. There are at least two reasons for this. First, the internal partition function of H_2 differs from that of D_2 because the rotational and vibrational constants (or energy levels) depend on the moment of inertia and the reduced mass of the molecule. Density scaling is not an exact procedure for the internal free energy, and this is the origin of the temperature dependent gap between SESAME #5251 and our isotherms as $\rho \to 0$. Second, the mass dependence of the quantum correction (eq. 1) prevents direct density scaling, which accounts for the temperature dependent deviations between the two sets of isotherms at high density. Note that the two discontinuities in the SESAME 406K isotherm occur near the fluid-solid phase transition and are probably due to interpolation on the coarse grid of a discontinuous table.

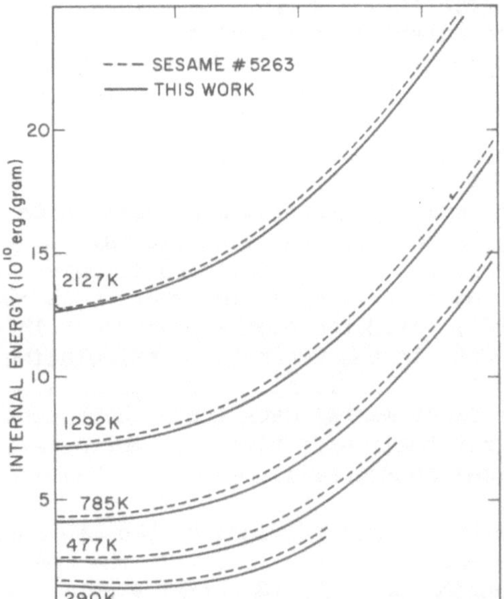

Fig. 3. Internal energy isotherms of D_2.

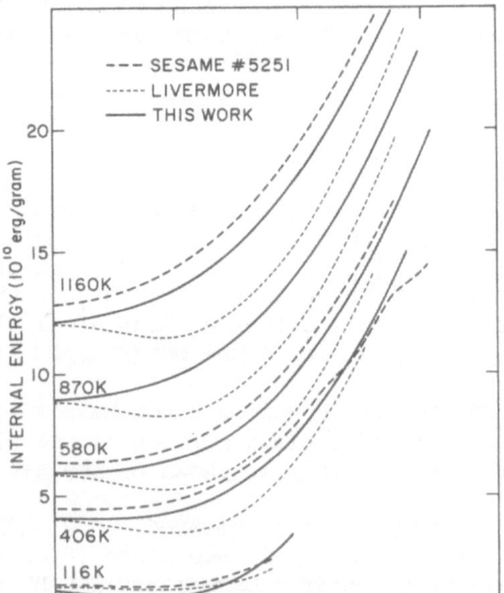

Fig. 4. Internal energy isotherms of H_2.

Internal energy isotherms show a shallow dip at low temperatures for $\rho \leq 0.1$ gram/cm^3. This is due to the long range part of the interaction potential, i.e. the van der Waals attraction. As the temperature increases, collisions between molecules occur at energies high enough to probe the hard repulsive core, and the weak attractive well becomes negligible in comparison. Therefore, the depth of the dip in the internal energy isotherms must decrease as the temperature increases. The Livermore isotherms in Figure 4 show the opposite trend quite clearly. This cannot be understood on physical grounds.

CONCLUSION

In the molecular phase, we obtain remarkably good agreement with the SESAME deuterium EOS, considering the different approaches used in each case. However, we have shown (1) that scaling the deuterium EOS in density to generate an approximate hydrogen EOS is not an accurate procedure and (2) that the Livermore EOS obviously does not include the work of Ross *et al.* (1983) (of Lawrence Livermore National Laboratory) and indeed suffers from severe difficulties at high densities.

We are now concentrating on the regime 10^3K < T < 10^6K and ρ < 1 gram/cm^3, where temperature and pressure dissociation and ionization take place. Accurate treatment of the equation of state of partial ionization zones has proven to be critical for a good understanding of the structure of white dwarfs (Fontaine, 1973) and we propose to perform a state of the art calculation of the thermodynamics of hydrogen under conditions appropriate to the envelopes of these stars and to brown dwarfs. We are currently working to adapt the hard sphere variational theory to a mixture of chemical species, including a statistical mechanically consistent treatment of internal energy levels (following Hummer and Mihalas, 1986). It is our hope that this will ensure a good treatment of the difficult phenomena of pressure dissociation and ionization.

ACKNOWLEDGMENTS

One of us (D. S.) gratefully acknowledges support from the NSERC of Canada through a 1967 Science and Engineering Scholarship. Part of this research was supported by NSF grant AST 85-11173.

REFERENCES

Carnahan, N. F., and Starling, K. E., 1969, *J. Chem. Phys.*, **51(2)**, 635.

Fontaine, G., 1973, Ph.D. thesis, University of Rochester.

Fontaine, G., Graboske, H. C., Jr., and Van Horn, H. M., 1977, *Ap.J. Supp.*, **35**, 293.

Graboske, H. C., Jr., and Wong, K. L., 1980, *UCID*-18489.

Huber, K. P., and Herzberg, G., 1979, *Molecular Spectra and Molecular Structure. IV. Constants of Diatomic Molecules.* Van Nostrand Reinhold Company, New York, New York.

Hummer, D. G., and Mihalas, D., 1986, in preparation.

Landau, L. D., and Lifshitz, E. M. 1958, *Chap. III, The Gibbs Distribution*, in: *Statistical Physics* (Reading, Mass.: Addison-Wesley).

Mills, R. L., Lienbenberg, D. H., Bronson, J. C., 1978, *J. Chem. Phys.*, **68**, 2663.

Ross, M., 1979, *J. Chem. Phys.*, **71**, 1567.

Ross, M., Ree, F. H., and Young, D. A., 1983, *J. Chem. Phys.*, **79(3)**, 1487.

Shimizu, H., Brody, E. M., Mao, H. K., and Bell, P. M. 1981, *Phys. Rev. Lett.*, **48**, 128.

Smith, W. R., and Henderson, D., 1970, *Molec. Phys.*, **19(3)**, 411.

SOLAR OSCILLATIONS AND THE EQUATION OF STATE

Werner Däppen

Observatoire de Paris-Meudon
92195 Meudon
France

INTRODUCTION

With suitable inversion techniques of solar five-minute oscillation frequencies, the local sound speed c(r) inside the Sun (r being the distance from the center) has become an observable quantity (Christensen-Dalsgaard et al., 1985). Since $c^2 = \gamma p/\rho$, where $\gamma = (\partial \ln p/\partial \ln \rho)_S$, one expects modulations of the sound speed from the lowering of γ in the zones of partial ionization of hydrogen and helium. Gough (1984) proposed to use this modulation to determine the solar helium abundance. Däppen and Gough (1984,1986) assessed the potential of this method with the help of theoretical solar models and numerically computed solar oscillation frequencies. They concluded that the quantitative influence of the He II ionization zone on the sound speed (or more precisely on the derivative of sound speed with respect to depth) is sufficiently large to serve as a calibration of the helium abundance.

By the very nature of the method, a good knowledge of the equation of state and of γ in a partially ionized hydrogen-helium plasma is crucial. It is important to know how uncertainties in the equation of state propagate into the helium abundance determination. Therefore I have repeated the analysis of Däppen and Gough (1984) with three different models of the equation of state. Since density in the solar He II ionization zone is fairly low (about $3\ 10^{-3}$ g/cm^3), the only nontrivial issue in the equation of state is the internal partition function for bound systems. The following three partition functions were used: (i) partition functions with ground-states only, (ii) partition functions truncated according to the static screened Coulomb potential (see, e.g., Rogers et al., 1970), and (iii) partition functions based on an occupation probability formalism that includes interactions by neutral and charged species (Hummer and Mihalas, 1987).

In the following I outline the essential tool of the helium abundance determination, which is the diagnostic function that relates the observed sound speed to thermodynamical quantities inside the Sun. Then I discuss the change of this function due to different models of the equation of state.

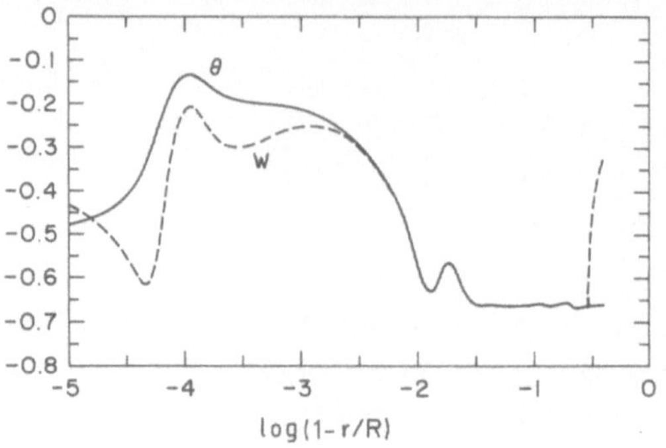

Fig. 1. The functions W and Θ defined by Equation (1) plotted against $\zeta = \log_{10}(1-r/R)$ (R being the solar radius) in a typical solar model (for details see Däppen and Gough, 1986). The hump located at $\zeta = -1.75$ is due to the He II ionization zone and its height is a measure of the helium abundance.

HELIUM ABUNDANCE DETERMINATION

The principal diagnostic equation relating the observed sound speed to thermodynamic quantities is (Gough, 1984)

$$W \equiv \frac{r^2}{Gm}\frac{dc^2}{dr} = \frac{1-\gamma_\rho-\gamma}{1-\gamma_{c^2}} \equiv \Theta \quad . \tag{1}$$

Here, $\gamma_\rho = (\partial \ln\gamma/\partial \ln\rho)_{c^2}$, $\gamma_{c^2} = (\partial\ln\gamma/\partial\ln c^2)_\rho$, m is the mass in the spherical shell of radius r and G is the gravitational constant. In deriving this equation, one uses basically only the assumptions of hydrostatic support and of adiabatic stratification. The latter assumption is well justified, because the He II ionization zone is located in a region of efficient convection, for which both standard mixing-length theory and extrapolated laboratory experiments indicate a virtually adiabatic temperature gradient.

Fig. 1. shows Θ and W for a theoretical solar model. Note that and W differ significantly for $\zeta < -2.5$, which reflects the fact that the assumption of quasi-adiabatic stratification is not satisfied in the upper parts of the convection zone. The H I and He I ionization zones produce a composite hump for $-4 < \zeta < -2$ and the He II zone produces a hump centered at about $\zeta = -1.75$. This corresponds to a depth of 12400 km. It is intuitively clear that the height of this He II humps is a measure of the Sun's helium abundance. This is indeed the case as has been shown for solar models by Däppen and Gough (1984).

The right hand side of Equation (1) indicates the importance of the equation of state in this method. The quantity γ and even more its thermodynamical derivatives depend on details of the partition functions.

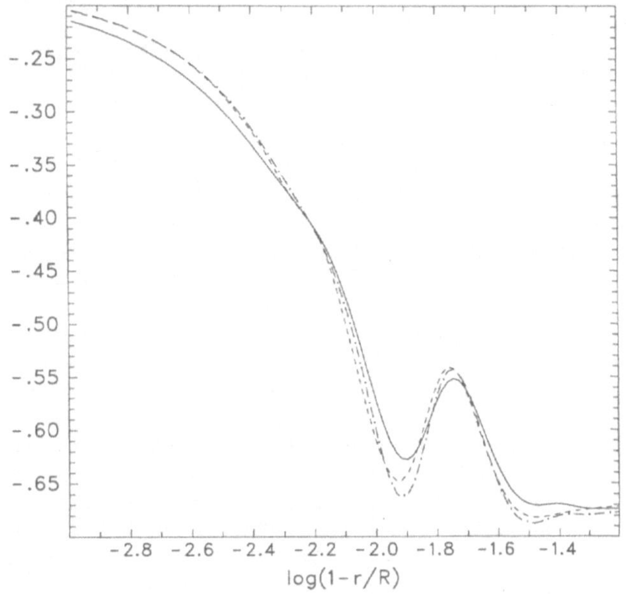

Fig. 2. The helium hump of Fig. 1. for three different equations of state, but with all other model assumptions unchanged. The solid line refers to partition functions containing ground states only, the dashed line to partition functions of the static-screened Coulomb potential, and the dotted-dashed line to the partition functions of the occupational probability formalism of Hummer and Mihalas (1987).

THE HELIUM HUMP FOR THREE DIFFERENT EQUATIONS OF STATE

I have computed three different solar envelope models, differing only in the equation of state. The other specifications, described in more detail in Däppen and Gough (1984,1986), are as in Fig. 1. The first equation of state used is a simple Saha equation with partition functions that contain only the statistical weight of the ground states of bound systems. The partition functions of the second equation of state are based on the static screened Coulomb potential. A numerical realization of this equation of state has been used for solar models by Berthomieu et al. (1980). The third equation of state is based on the occupation probability formalism introduced by Hummer and Mihalas (1987) in the framework of an ongoing opacity recomputation. The equation of state contains partition functions with probability weights assigned to the energy levels. Neutral and charged surrounding particles reduce the occupation probabilities of upper levels, the first act through an excluded-volume effect, the second through Stark ionization. The numerical realization of this equation of state has been made very smooth, the free energy and its first and second thermodynamical derivatives are analytically expressed. This gives sufficient precision to compute the derivatives of γ by numerical differentiation. First results of this equation of state will be presented elsewhere (Mihalas et al. 1987; Däppen et al. 1987).

Fig. 2. shows the resulting three different helium humps.

DISCUSSION AND CONCLUSIONS

The choice of the internal partion function does indeed influence the helium hump. Depending on how the height of the hump is defined, it can change by as much as 10-20% when different partition functions are used in the equation of state. Therefore the helium abundance determination will be critically dependent on the model of the equation of state. Nevertheless, there is hope for optimism. Since the local sound speed computed in an oscillation-frequency inversion contains more information than the helium abundance alone, it might become possible to obtain in a first step (perhaps from the form of the observed helium hump) constraints on the equation of state, and in a second step the helium abundance.

A basic issue to which solar observations might contribute is the Planck-Larkin partition function (for the recent controversy about the Planck-Larkin partition function see e.g. Ebeling et al., 1985; Rouse, 1983). While it has been recognized that the validity of the Planck-Larkin partition function cannot be tested in optical experiments (Rogers, 1986; Däppen et al., 1987), a determination of thermodynamical properties such as γ in Equation (1) could be able to distinguish between different equations of states. Fig. 2. shows the reason: although the plasma is only at a density of about $3\ 10^{-3}$ (where for most purposes the deviations from simple ideal-gas treatments and the Saha equation are negligible), W and Θ depend so sensitively on γ and especially its thermodynamical derivatives, that the partition function can leave its imprint on the helium hump.

Acknowledgments: I am grateful to Forrest Rogers and Wolf-Dieter Kraeft for stimulating discussions about the Planck-Larkin partition function.

REFERENCES

Berthomieu, G., Cooper, A.J., Gough, D.O., Osaki, Y., Provost, J., Rocca, A., 1980, in: "Nonradial and nonlinear stellar pulsation", H.A. Hill and W.A. Dziembowski, eds., Springer, Heidelberg, 307.

Christensen-Dalsgaard, J., Duvall, T.L., Jr, Gough, D.O., Harvey, J.W., Rhodes, E.J., Jr, 1985, Nature, 315:378.

Däppen, W., Gough, D.O., 1984, in: "Theoretical problems in stellar stability and oscillations", M. Gabriel and A. Noels, eds., Institut d'Astrophysique, Liège, 264.

Däppen, W., Gough, D.O., 1986, in: "Seismology of the Sun and the distant stars", D.O. Gough, ed., NATO ASI Ser., Reidel, Dordrecht, 275.

Däppen, W., Mihalas, D.M., Anderson, L.S., 1987, Astrophys. J., submitted.

Däppen, W., Mihalas, D.M., Hummer, D.G., Mihalas, B.W., 1987, Astrophys. J., submitted.

Ebeling, W., Kraeft, W.D., Kremp, D., Röpke, G., 1985, Astrophys; J., 290:24.

Gough, D.O., 1984, Mem. Soc. astr. Italiana, 55:13.

Hummer, D.G., Mihalas, D.M., 1987, Astrophys. J., submitted.

Rogers, F.J., Graboske, H.C., Jr, Harwood, D.J., 1970, Phys. Rev., A1:1577.

Rogers, F.J., 1986, Astrophys. J., in press.

Rouse, C.A., 1983, Astrophys. J., 272:377.

CHAPTER V

QUANTUM PLASMAS

THERMODYNAMIC PROPERTIES AND PHASE TRANSITIONS IN HYDROGEN AND RARE GAS PLASMAS

Helmut Hess* and Werner Ebeling**

*GDR Academy of Sciences
Central Institute of Electron Physics
Berlin 1086, GDR
**Humboldt University
Department of Physics
Berlin 1020, GDR

INTRODUCTION

Phase transitions are of great interest for the understanding of the properties of matter. The search for such transitions in hydrogen and rare gas plasmas under extreme conditions is not only important for basic research, but there are many interesting applications; for example, astrophysics. Further such studies may help to better understand the fascinating, very complex, mixed phase transitions in metal vapors which are of outstanding technological interest.

Hydrogen and noble gas strongly coupled plasmas play an unique role with respect to their thermodynamic properties. In particular, this is due to their relatively high values of ionization energy. A direct consequence of this property should be the existence of two critical points which are well separated.

$$T_c^{(1)} \approx 10^1 - 10^2 \text{ K}; \quad p_c^{(1)} \approx 10^5 - 10^7 \text{ Pa}$$

$$T_c^{(2)} \approx 10^4 - 10^5 \text{ K}; \quad p_c^{(2)} \approx 10^9 - 10^{11} \text{ Pa}$$

The second critical point which is so far hypothetical is located in the strongly coupled plasma region. The corresponding phase transition is connected with an abrupt change in density and in the degree of ionization.

The existence of a phase diagram of this type was first discussed by Landau and Zeldovich[1] in 1943. First calculations of plasma phase transitions are due to Norman and Starostin[2,3] and Ebeling and Sandig.[4,5] A more detailed discussion of this subject is given in a book[6] whereas a critical review can be found in an invited lecture on the Budapest Ionization Phenomena Conference, 1985.[7]

As a rule of thumb, a simple relation was recently given by one of us for the critical temperature of the plasma phase transition (PPT):[8]

$$kT_c^{(2)} \approx 0.1 \times E_i \quad (1)$$

where E_i is the ionization energy of the undisturbed atom. For the rare gases we get values approximately between 1×10^4 and 2×10^4 K.

A closely related estimate is based on the Debye limiting law with quantum corrections for the chemical potential of the plasma[6]

$$\mu = \mu_{id} - \frac{e^2}{r_D - \Lambda_e/8} \quad (2)$$

where μ is the ideal chemical potential, r_D is the Debye length, and Λ_e is the thermal de Broglie wavelength of the electron. The condition of thermodynamic stability

$$\frac{\partial\mu}{\partial n} \geq 0 \quad (3)$$

where n is the free electron density yields

$$\left(\frac{e^2}{kT\Lambda_e}\right)_c \leq 2 \quad (4)$$

and

$$kT_c^{(2)} \geq \frac{E_i}{4\pi} \quad (5)$$

which is close to Eq. (1).

Let us still note that in the Debye approximation without quantum corrections ($\Lambda_e = 0$) the critical temperature of the PPT would be infinite:

$$T_c^{(2)} = \propto \text{ if } h = 0 \; .$$

Coulomb interactions cause instabilities as well as van der Waals forces, the existence of repulsive forces due to quantum effects or hard cores of the charges produce a finite critical temperature. For $T < T_c$ the system splits into two fluid phases with different densities and different degrees of ionization. The gas phase is usually only weakly ionized, and the dense phase is usually strongly ionized. The transition is a typical first order thermodynamic phase transition.

Above the critical temperature there is a more or less steep increase in the degree of ionization near the critical pressure. Below the critical temperature there exists a line of coexistence connected with a sudden jump in the degree of ionization which is more pronounced the deeper the temperature is. Far below $T_c^{(2)}$ the system jumps from a dielectric atomic or molecular phase to a liquid metallic phase.

The theoretical predictions are based here on the technique of Pade approximations which integrate the existing analytical knowledge about degenerated and non-degenerated strongly coupled plasmas. An earlier result for hydrogen reads:[9]

$$T_c^{(2)} = 16500 \text{ K}, \; p_c^{(2)} = 23 \text{ GPa}, \; \rho_c^{(2)} = 0.13 \text{ gcm}^{-3}, \; \alpha = 32\%. \quad (6)$$

The predicted pressure is considerably lower than the existing estimates for the transition pressure at low temperatures which are in the region of 300 GPa.

Considerations comparably as simple as those which lead to relation (1) shows that the critical density can be related to the atomic diameter d as it is given, for example, by a Lennard-Jones or an exp-6 potential:

$$n_c^{(2)} \approx d^{-3} . \tag{7}$$

As estimation of the pressure from the critical temperature and the critical density leads to values of about some GPa for the heavy rare gases which seem to be within reach of ballistic compressors - devices which are used in our laboratory to produce high-density plasmas. Now we shall present some theoretical considerations which were restricted on an elementary model only.

ELEMENTARY MODEL

Let us assume that the free energy density of the fluid is given by

$$f = f_{ID} + f_{CI} + f_{VW} \tag{8}$$

where ID denotes the ideal part, CI the contribution of Coulomb interactions and VW that of van der Waals interactions. Assuming that we have neutral atoms in the plasma with the density n_a and electrons with the density n the free energy density is approximately

$$f_{VW} \approx n_a kT \frac{4\eta_a - 3\eta_a^2}{(1 - \eta_a)^2} - n_a A_a - 2nkT \ln (1-\eta_a) . \tag{9}$$

Here A_a is the van der Waals constant of the atoms, and

$$\eta_a = \frac{4\pi}{3} n_a R_a^3 \tag{10}$$

is the excluded volume of the atoms. The last term in Eq. (9) represents the fact that the electrons are excluded from penetrating into the interior of the atoms.[9] This term plays an essential role in the limit of high electron density.

In order to find an approximate expression for the Coulombic contribution, let us start from an expression for the chemical potential corresponding to a Debye-type contribution (as in Eq. (2)) and a Hartree-Fock contribution

$$\mu_{CI} = - \frac{e^2}{r_D + a} - \frac{e^2}{\Lambda_e} I_{-1/2}(\alpha_e). \tag{11}$$

Here I_n is the Fermi function, $\alpha_e = \mu_e^{id}/kT$ and a is a kind of minimum distance between electrons and ions. Due to Heisenberg's uncertainty relation

$$\Delta p \, \Delta r \geq \hbar/2$$

$$\Delta r \geq \frac{h}{4\pi\Delta p} = \frac{h}{4\pi} \frac{\sqrt{2\pi}}{4\sqrt{mkT}} = \frac{\Lambda e}{8} \tag{12}$$

electrons have an effective radius of about

$$R_e \approx \frac{1}{8} \Lambda_e . \tag{13}$$

This gives

$$a \approx \frac{1}{8} \Lambda_e + R_i \tag{14}$$

where R_i is an assumed ion radius.

By integration, we get for the free energy density

$$f_{CI} = \frac{kT}{12\pi r_D^3} \tau \left(\frac{a}{r_D} \right) - \frac{2e^2}{\Lambda_e^4} \int_{-\alpha}^{\alpha_e} dy\, I_{-1/2}(y) ,$$

$$\tau(X) = \frac{3}{X^3} \left[\ln (1 + x) - x + \frac{x^2}{2} \right] . \tag{15}$$

The free energy density given by Eqs. (8-15) has to be minimized at fixed total electron density (which is equal to the total ion density)

$$f(n,n_a) = \min \text{ at } n_e = n + n_a = \text{const} \tag{16}$$

The relatively simple system given by Eqs. (8-16) describes at least qualitatively the thermodynamic behavior in the whole fluid part of the phase plane. The critical temperature of the PPT is given (neglecting the Hartree-Fock term and the hard core contributions) after Eq. (3) by

$$kT_c^{(2)} = \frac{e^2}{R_i + \Lambda_e(T_c)/8} . \tag{17}$$

From Eq. (17) $T_c^{(2)}$ may be obtained by iterations. This gives for hydrogen $T_c^{(2)} \approx 13{,}000$ K. The critical density of the free electrons follows from

$$n_c^{(2)} = \frac{32}{\pi} \left(\frac{kT_c^{(2)}}{e^2} \right)^3 . \tag{18}$$

This corresponds to a relation given by Likalter[10] for the critical density of metals. In order to get the critical pressure one has to solve first the Saha equation or to carry out the minimization after Eq. (16).

Better approximations for the free energy density may be obtained by using the technique of Pade approximations.[9] In this way Ebeling and Richert obtained the critical data for hydrogen given in Eq. (6).[9].

Similar results for hydrogen were obtained by Robnik and Kundt[11] and by Kremp, Haronska, and Schlanges.[12] A survey on the quantum statistical basis of the theory of Coulomb interaction is given recently.[13] A more refined treatment of the van der Waals interaction between the neutrals was given by Nellis et al.,[14] Zisman et al.[15] and others. A survey of theoretical and experimental work was given by Fortov and Yakubov[16] and by Ross.[17]

DISCUSSION OF THEORETICAL RESULTS

The theory of Coulomb interactions which is presented here in a rudimentary form only, shows that the attractive interaction of oppositely charged particles may lead to thermodynamic instabilities as well as van der Waals forces do it. So far pure Coulombic phase transitions have been experimentally studied only in electron-hole fluids in semiconductors.[13]

In most of the phase transitions studied experimentally where charged particles are present as, for instance, in fluid metals, Coulomb and van der Waals forces both play an important role in determining the critical point. For this reason, the theoretical study of the phase transition in alkali metals and other metallic systems is extremely difficult.[13,18,19]

Another situation is met in systems which are insulators at normal conditions. Typical examples are hydrogen and the rare gases. However, to this class probably belong also water and the most organic fluids. These systems become metallic only at extremely high pressures. We expect that the liquid metallic state which should exist at high pressures and sufficiently high temperatures belongs to the same fluid phase as the plasma state of the corresponding system. Therefore, we expect for systems which show a pure van der Waals transition at low temperatures and low pressures the existence of a second, first order phase transition due to Coulomb forces at much higher temperatures and pressures: a plasma phase transition.

On the basis of the theoretical arguments discussed here these authors are convinced that the existence of a plasma phase transition is more common among natural systems (especially those which are insulators at normal pressures and temperatures) as it is believed today.

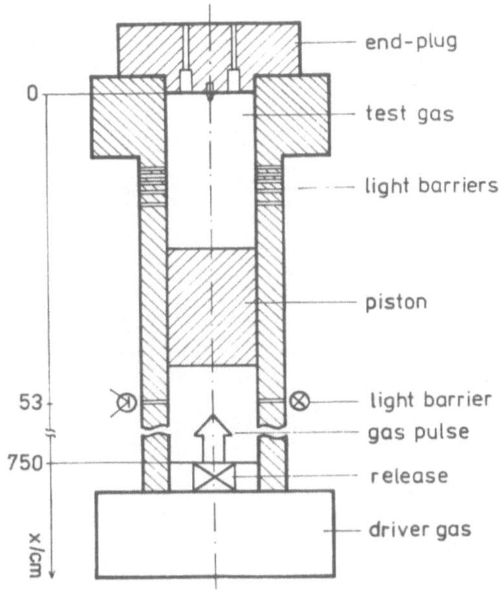

Fig. 1 Ballistic compressor, schematic view.

The difficulty of an experimental verification of this supposed transition in its pure form is connected with the enormous pressures and temperatures which are simultaneously required. Therefore, in first line dynamic methods can be successful. The parameters which should be reached in the most favorable case: pressures in the GPa range and temperatures of about 10,000 K. These parameters lie within the reach of modern shock wave technique and of ballistic piston compressors.

EXPERIMENTS

What is the difference in the possibilities between shock tubes and ballistic compressors (Fig. 1) both working in gaseous media? The shock compression in its ideal form is an adiabatic but non-isentropic process, whereas the compression of a gas by a piston (without producing shock waves, that is at a piston speed smaller than the sound speed in the compression medium and a not too long compression way) is adiabatic as well as isentropic

Therefore, the high pressure which can be reached in both devices is more due to a high temperature in shock tubes, whereas it is more due to a high density in a ballistic compressor. In other words, in shock tubes the density multiplication is limited, whereas in ballistic compressors the temperature multiplication is rather limited. Fig. 2 shows this for an ideal monatomic gas. As a result, the shock pressure is over wide ranges proportional to the temperature whereas in the piston compressor the pressure is proportional to $T^{5/2}$. This behavior, of course, is modified by the real-gas properties of the investigated species (dissociation, ionization, nonideality).

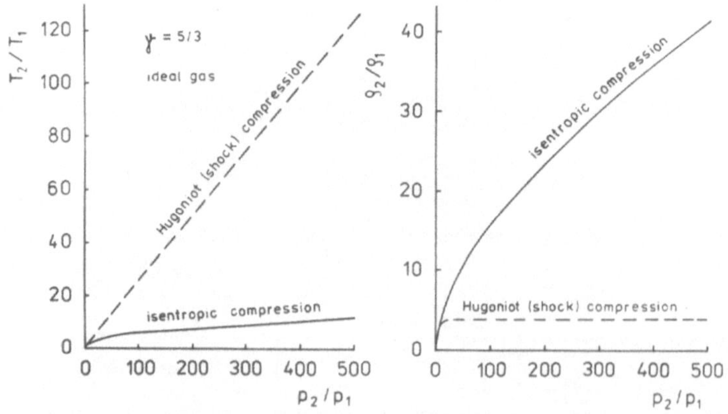

Fig. 2 Temperature and density multiplication in shock tubes and in ballistic compressors in dependence upon the obtained pressure for an ideal monatomic gas.[20]

A further difference is in the power. In shock tubes the characteristic velocity is large compared with the sound speed. In ballistic compressors it is lower than the sound speed. Therefore, the rate of energy release can differ by orders of magnitude what in the case of shock tubes often leads to destruction at each shot whereas ballistic compressors can withstand thousands of shots without severe damage.

In the Central Institute of Electron Physics in Berlin, two ballistic compressors are used for high-density plasma research especially in the field of nonideal plasmas.[20] The larger one is called AICA which means Adiabatic Impluse Compression Apparatus. It has a length of 7.5 m and an inner diameter of 150 mm (cf. Fig. 1). Due to the large diameter, a lot of diagnostics can be done simultaneously which can be used to study thermodynamic, electrical, and optical properties of nonideal plasmas.

For instance, we have found spectral line shifts in neutral xenon due to Stark effect up to 3 nm, i.e. one order of magnitude greater than earlier reported.[21-23] Deviations from the linear dependence upon the electron density are indicated (Fig. 3). Now theoretical work is under way to explain these effects.

A smaller device is called LAICA, which means little AICA. It has a length of 2 m and an inner diameter of 18 mm. Whereas the AICA piston is driven by a highly compressed gas, in LAICA cartridges are used. The

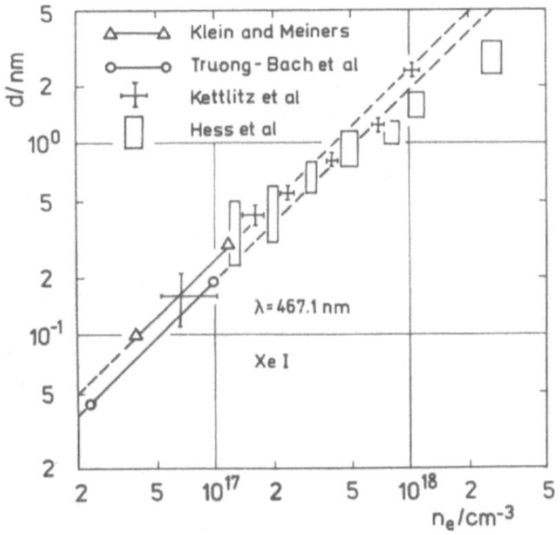

Fig. 3 Shift of the XeI line 467.1 nm in dependence on the electron density.[22]

maximum pressure obtained in LAICA was about 1.5 GPa at a temperature of about 10,000 K[20] (Fig. 4). In this device pressures of some GPa may be obtained, and therefore, the PPT line may be crossed and detected, for instance, by plasma resistance measurements.

There are, however, already experiments which seem to cross the predicted coexistence lines. In Fig. 5 a phase diagram of hydrogen in the p-T plane is shown. Beside the usual low-temperature, low-pressure part the expected behavior at higher pressure and higher temperature can be seem. The different phase-separating lines, triple and critical points are from calculations and speculations of different authors.[9,11,12] The principal behavior of the curve labelled by "Ebeling" seems to be the most realistic: if the temperature is low a higher pressure is needed to initiate the ionization catastrophe which is characteristic of the PPT. Also shown in Fig. 5 are shockwave experiments by Nellis et al.[14,24] in hydrogen and deuterium which cross the coexistence line predicted by Ebeling and Richer[9] at about 30 GPa.

We should not over stress the accuracy of the theoretical predictions as well as of the experiments. However, the question is: are there any irregularities in the measurements of Nellis et al.[14,24] in the range of some ten Gigapascals?

Unfortunately, just in the range of intersection the experiments change from single to double shock so that there is, of course, a pressure jump in the p-T plane connected, however, with the density jump behind the

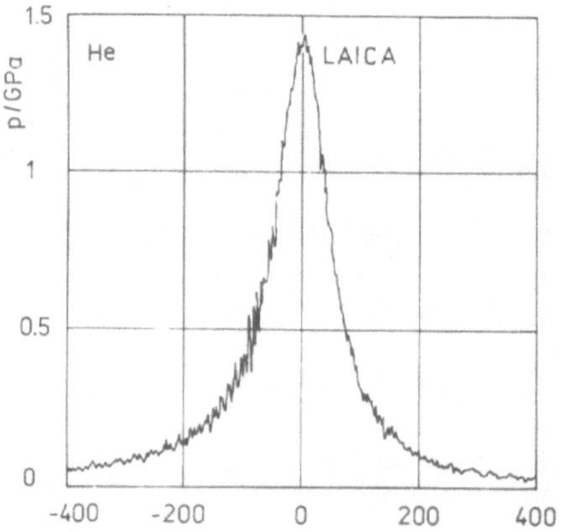

Fig. 4 Pressure pulse obtained in the ZIE ballistic compressor LAICA.

reflected shock wave.[24,31] A further irregularity can be seen in Fig. 6.[24] At the highest pressure reached the pressure decreases with increasing density. This result was obtained by two experiments done with a time difference of nearly one decade.[24,25] If this is a real effect, it could be interpreted as a run into metastable states similar as in a supercooled gas due to the fast shock process which may be faster than the corresponding relaxation time. However, it is also possible that the experimental error is so large that such speculations are irrelevant.

Further detailed studies should give the answer but these studies would be easier if the pressure were not so high. Scaling laws as mentioned in the introduction should help to find species in which a pure PPT may occur at lower pressures.

SCALING LAWS

The idea for estimating the critical temperature of the PPT is as follows: the PPT is characterized by an ionization catastrophe. The

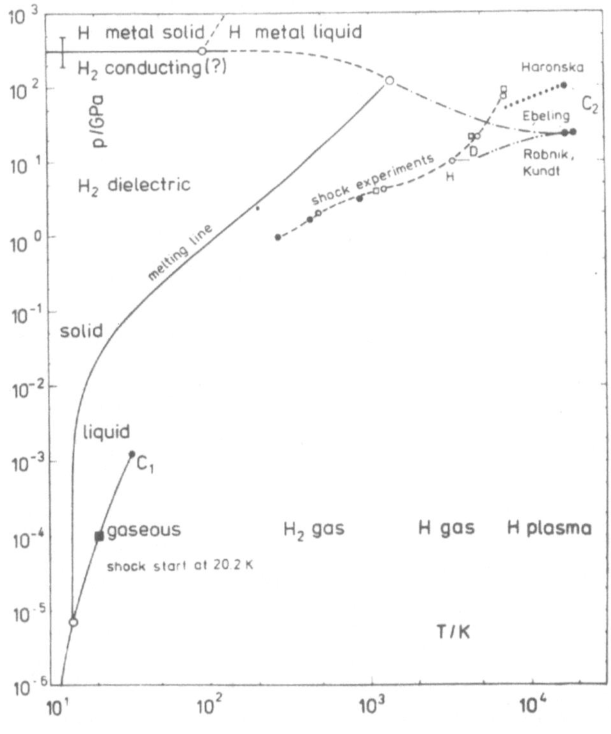

Fig. 5 Phase diagram of hydrogen in the p-T plane.

critical temperature is reached when the plasma at low pressure is already highly ionized, so that an increase in pressure or density cannot essentially increase the degree of ionization.

The temperature at which a plasma at low pressure - say at atmospheric pressure - is ionized is well-defined as is known from arc physics; it is

$$kT_c^{(2)} \approx 0.1\ E_i \tag{19}$$

where E_i is the ionization energy of the undisturbed atom. This can be seen from Fig. 7 where the degree of ionization at constant pressure is shown in dependence upon the temperature for different gases as it is given by the Saha equation. (A refinement of Eq. (19) can thus be obtained taking into account the ratios of the partition functions of the atom and the ion which are different for hydrogen and the rare gases.)

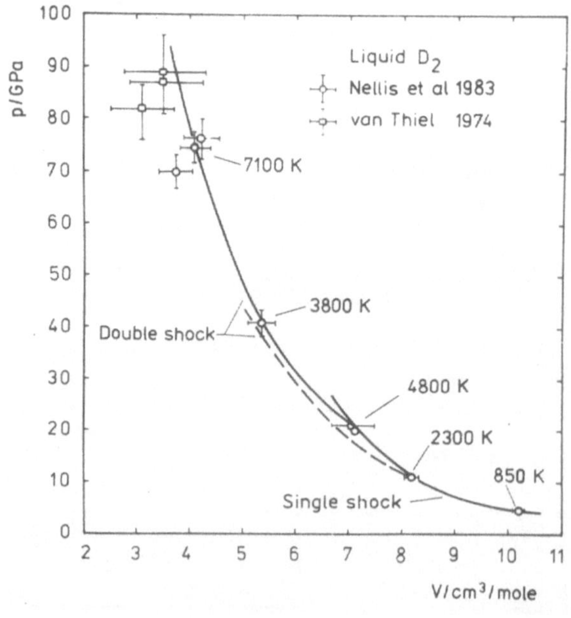

Fig. 6 p-V diagram in deuterium showing experiments (circles and squares) and calculations.

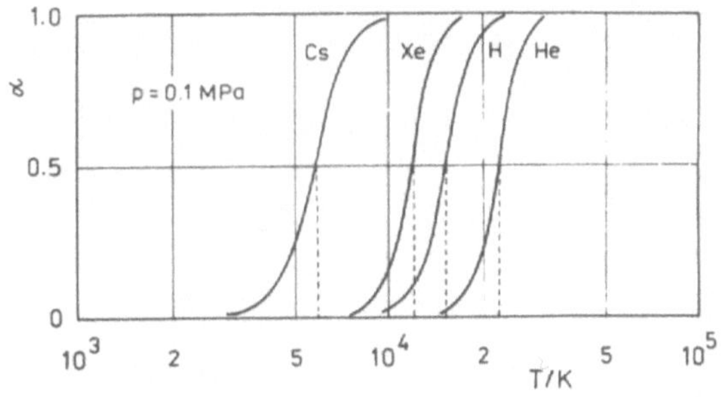

Fig. 7 Degree of ionization as a function of temperature at constant pressure for different species (SAHA-equation).

The critical temperature of different gases of the liquid-gas transition ($T_c^{(1)}$) and of the PPT according to Eq. (19) are the following:

Species	$T_c^{(1)}/K$	$T_c^{(2)}/K$
He	5	28,500
H	33	15,800
Ar	151	18,300
Xe	290	14,100
Cs	1,924	4,500

The main difference between the rare gases on the one hand and the metal vapors on the other is in the ratio of these critical temperatures or of the underlying energies. Whereas in rare gases this ratio is larger than or about 100, in metal vapors it is near unity (taking into account the lowering of the ionization potential by cluster formation) and the two kinds of phase transitions are mixed.

The density at the critical point C_2 may be nearly the same as in the liquid phase near the critical point C_1. This is an empirical result which can be seen from isochores in the p-T plane and their intersection with the calculated PPT as can be seen in Fig. 8[26-28] for xenon. (Also shown in Fig. 8 are lines of constant degree of ionization. If the coexistence line of the PPT is the locus of an ionization catastrophe, then all the lines of constant ionization degree should lie on this coexistence line in the transition region. How well it is realized can be seen here.[26])

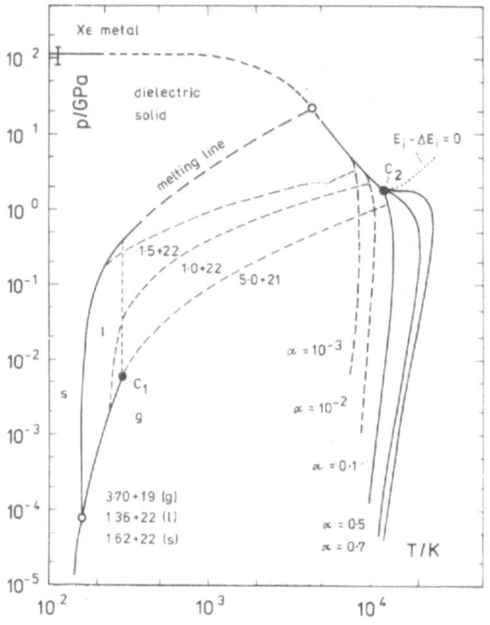

Fig. 8 p-T diagram of zenon. 1.5 + 22 read n = 1.5 x 1^{-22} cm^{-3} etc. α is the degree of ionization. ΔE_i is the lowering of the ionization energy. The dotted line E_i - ΔE_i = 0 is the continuation of the so-called Mott line beyond the coexistence line (α = 1.0!).

An estimation of the liquid state density can be obtained from an atomic diameter d:

$$n_c^{(2)} \approx d^{-3} \, . \qquad (20)$$

For different species we get (by using ideal gas (id) or hard core (hc) formulae for the pressure[35]) the following values:

Species	d/A	Reference	$n_{cr}^{(2)}/cm^{-3}$	$p_{cr,id}^{(2)}/GPa$	$p_{cr,hc}^{(2)}/GPa$
H	1.04	12	8.89 + 23	194	2910
H	1.63	9	2.31 + 23	50	750
He	2.97	30	3.82 + 22	15	225
Ar	3.85	32	1.75 + 22	4.4	66
Xe	4.47	29	1.12 + 22	2.2	33

read 8.89 + 23: 8.89 x 10^{23}; n = ρ/m,
ρ - mass density, m - atomic mass.

The critical pressures estimated here are only very rough values but the hard core pressure should be an upper limit. In general, however, this estimation explains why the critical temperature calculated by different authors are always nearly the same, whereas the critical pressure strongly depends upon the chosen atomic diameter.

The estimations for hydrogen and xenon can be compared with calculations:

HYDROGEN:				
References	11	9	12	Our Estimates
$T_c^{(2)}/K$	19,000	16,500	16,500	15,800
$n_{cr}^{(2)}/cm^{-3}$	8.38 + 22	7.78 + 22	2.57 + 23	2.31 + 23
$p_{cr}^{(2)}/GPa$	24	22.5	95	50 to 750

XENON:				
References	26	33	34	Our Estimates
$T_c^{(2)}/K$	12,000	14,000	16,000	14,100
$n_{cr}^{(2)}/cm^{-3}$	6.93 + 21	1.46 + 22	7.00 + 21	1.12 + 22
$p_{cr}^{(2)}/GPa$	1.81	46	3	2.2 to 36

Xenon due to its large diameter and its low ionization energy seems to be the favorite gas for studying the pure PPT. The predicted values of some Gigapascals and about 10^4 K seem attainable by one of our ballistic piston compressors.

The plasma phase transition should be detected by a sudden decrease of the electrical resistance of the compressed plasma corresponding to electrical conductivities which exceeds remarkably typical plasma values of $10^2\ \Omega^{-1}\ cm^{-1}$.

REFERENCES

1. L. Landau and Zeldovich, On the Relation between the Liquid and Gaseous States of Metals, Acta Physicochimica U.R.S.S. 18:194 (1943).
2. G. E. Norman and A. N. Starostin, The Invalidity of the Classical Description of a Non-Degenerated Dense Plasma, Teplofiz. Vys. Temp. 6:410 (1968).
3. G. E. Norman and A. N. Starostin, Thermodynamics of a Strongly Nonideal Plasma, Teplofiz. Vys. Temp. 8:413 (1970).
4. W. Ebeling, Quantum Statistics of Ionization and Shielding Effects in Non-Degenerate Moderately Doped Semi-conductors, Phys. Stat. Sol. (b) 46:243 (1971).

5. W. Ebeling and R. Sändig, Theory of Ionization Equilibrium in Dense Plasmas, Ann. Phys. (Leipzig) 28:289 (1973).
6. W. Ebeling, W. D. Kraeft, and D. Kremp, Theory of Bound States and Ionization Equilibrium in Plasmas and Solids, Akademie-Verlag, Berlin (1976); Revised Russ. Translation, Izd. Mir, Moscow (1979).
7. K. Guenther, H. Hess, and R. Radtke, Proc. XVII. ICPIG, Budapest (1985), Invited Papers, p. 120.
8. H. Hess, Proc. VIII. ESCAMPIG, Greifswald (1986), p. 306.
9. W. Ebeling and W. Richert, Phys. Lett. 108A:80 (1985); Contr. Plasmaphys. 25:431 (1985).
10. A. L. Likalter, Teplofiz. Vys. Temp. 23:465 (1985).
11. M. Robnik and W. Kundt, Astron. Astrophys. 120:227 (1983).
12. D. Kremp, P. Haronska, and M. Schlanges, Proc. 3rd Int. Workshop Nonideal Plasmas, Biesenthal (1984), ed. R. Radtke and H. Hess. M. Schlanges and D. Kremp, Ann. Phys. (Leipzig) 39:69 (1982). P. Haronska, Thesis, Wilhelm Pieck University, Rostock (1986).
13. W. D. Kraeft, D. Kremp, W. Ebeling, and G. Röpke, Quantum Statistics of Charged Particles, Akademie-Verlag, Berlin (1986) and Plenum Press, New York (1986).
14. W. J. Nellis et al., Phys. Rev. A 27:608 (1983).
15. A. N. Zisman et al., Phys. Rev. B 32:484 (1986).
16. V. E. Fortov and I. T. Yakubov, Physics of Nonideal Plasmas (in Russian), Chernogolovka (1984).
17. M. Ross, Rep. Progr. Phys. 48:1 (1985).
18. W. Richert, S. A. Insepov, and W. Ebeling, Ann. Phys. (Leipzig) 41:139 (1984).
19. R. Redmer and G. Röpke, Physica 130:523 (1985).
20. H. Hess, Contr. Plasma Phys. 26:209 (1986).
21. H. Hess et al., Proc. VXI. ICPIG, Dusseldorf (1983), Contr. Papers, p. 622.
22. H. Hess et al., Proc. XII. SPIG, Sibenik (1984), Contr. Papers, p. 453.
23. H. Hess et al., Proc. 3rd Int. Workshop Nonideal Plasmas, Biesenthal, (1984), p. 30.
24. W. J. Nellis et al., J. Chem. Phys. 79:1480 (1983).
25. M. van Thiel et al., Phys. Earth Planet. Inter. 9:57 (1974).
26. H. Dienemann, G. Clemens, and W. D. Kraeft, Ann. Phys. (Leipzig) 37:444 (1980).
27. C. Ronchi, J. Nucl. Mat. 96:314 (1981).
28. H. Schneidenbach, priv. comm. (1986).
29. M. Ross and A. K. McMahan, Phys. Rev. B 21:1658 (1980).
30. D. A. Young, A. K. McMahan, and M. Ross, Phys. Rev. B 24:5119 (1981).
31. M. Ross, F. Ree, and D. A. Young, J. Chem. Phys. 79:1487 (1983).
32. M. Ross, J. Chem. Phys. 73:4445 (1980).
33. W. Ebeling, H. Hess, A. Foerster, W. Richert, ICTP-Preprint, Trieste (1986).
34. T. Kahlbaum, priv. comm. (1986).
35. G. A. Mansoori, N. F. Carnahan, K. E. Starling, and T. W. Leland, J. Chem. Phys. 54:1523 (1971).

NONIDEAL PLASMAS AND BOUND STATES

W. D. Kraeft[1] and D. Kremp[2]

[1]Sektion Physik/Elecktronik der Ernst-Moritz-Arndt-Universität, 2200 Greifswald, GDR
[2]Sektion Physik der Wilhelm-Pieck-Universität 2500 Rostock, GDR

INTRODUCTION

For different applications it is necessary to know exact thermodynamic data for highly condensed plasma, e.g., highly compressed hydrogen plays an important role in present-time techniques and physics. Especially, one may mention that giant planets consist mainly of ionized dense hydrogen. Also the electron-hole plasma in optically excited semiconductors represents a dense plasma. Moreover, for many purposes it is also necessary to know transport and optical properties of dense plasmas.

For this reason it is desirable to deal with statistical mechanics which takes into account strong coupling effects, i.e., effects in which the mean value of potential energies is of the same order as the kinetic energy. Of course, strongly coupled (nonideal) plasmas were dealt with in the past; we mention here only the book edited by Kalman and Carini (1978), the Les Houches meeting in 1982, which was summarized by Baus and Hansen (1983), the article by Marvin Ross (1985) and the monograph by Kraeft et al. (1986). There the reader may find a lot of further references - besides the sources given by the present 1986 Santa Cruz plasma physics meeting.

Before going into the details we would like to list certain effects which are a consequence of taking into account nonideality effects.

While the single particle energy ϵ for a state characterized by the wavenumber k is given in the simplest case (free particle) by

$$\epsilon = \hbar^2 k^2/(2m) \qquad (1.1)$$

we must write in the strongly coupled plasma

$$\epsilon(k) = \hbar^2 k^2/(2m) + \Sigma_1(k) + i\,\Sigma_2(k) \qquad (1.2)$$

where the quantity $\Sigma_2(k)$ describes a finite lifetime of the single particle state, and Σ_1, which is negative, takes into account that the single particle interacts with its surrounding. We mention that it is impossible to write (1.2) in the shape of (1.1) using an effective mass:

$$\epsilon(k) = \hbar^2 k^2/(2\, m_{eff}). \tag{1.3}$$

Another effect simply described is the change of the two-particle energies. For an isolated pair of particles (say, electron and proton) the energy levels of bound states follow from the solution of the Schrödinger equation,

$$E_n = - \, ryd/n^2 \quad . \tag{1.4}$$

In order to destroy bound states we need the ionization energy $I = -E_o$.

If we take into account the surrounding plasma, the energy levels are charged, and the ionization energy (difference between the ground state and the continuum) decreases. At certain density, the ionization energy tends to zero and no bound state exists. This condition is referred to as Mott condition and influences a lot of physical properties, such as transport, optical and thermodynamic ones. Especially the electroconductivity changes drastically on vanishing of bound states, and spectral lines are influenced. However, in contrast to the isolated pair, two particle energies are now complex, what corresponds to a finite lifetime of states and leads, e.g., to a spectral line broadening.

Another effect of nonideality corresponds to the inclusion of higher order corrections to the electroconductivity. Here one may start from Spitzer's result which is derived from the classical Boltzmann equation for fully ionized plasmas. The quantum version leads to an inclusion of the scattering T-matix (scattering phase shifts) and gives a lowering of the conductivity as compared with Spitzer's result. The main deviation towards lower conductivity values for partially ionized plasmas is achieved by taking into account the scattering between charged and neutral particles. This latter effect is the second influence of the existing bound states (neutrals) which interferes with the first one, i.e., with the decreasing number of free charges according to the mass action law (MAL) which governs the plasma composition.

The MAL is one of the results to be derived in thermodynamic calculations, and is very sensitive to single and two particle energies. Especially, the MAL describes the Mott transition, i.e., the degree of ionization increases steeply on vanishing of bound states (see Section V).

Similarly, the pressure exhibits steep gradients, a physical quantity which is changed drastically by the formation of bound states.

The variety of effects becomes much more extended if we discuss the question of thermodynamic stability. Then a separation between two entirely different phases may occur, and especially the Mott phase transition represents a coexistence of phases with different degrees of ionization (metal-insulator); see, e.g., Ebeling, Kraeft and Kremp 1986.

As there is a rather large number of papers and monographs on strongly coupled systems which cover different areas of the field we want to outline essential features of techniques which may, in principle, be extended to describe many particle systems under arbitrary conditions.

In spite of the existence of such techniques, among which the Green function technique plays a very important role, there are fields of dense plasma physics which are not covered yet by reliable theoretical results.

To give examples, we mention that there is no quantitatively correct description of phase transition in dense plasmas. Further, if one assumes the electroconductivity σ of a plasma to be given as a function of the density n of charge carriers

$$\sigma = A \ln n + B + C\, n^{1/2} \ln n + D\, n^{1/2} + E\, n \ln n + \dots \qquad (1.5)$$

only the coefficient A is known exactly; the higher orders are known only in different approximations.

There are also no generally accepted theoretical expressions for the spectral line broadening and shift; of course there exists rather good results for special physical situations.

II. BASIC CONCEPTS

The method of Green functions is especially powerful in the thermodynamic equilibrium where the Kubo-Martin-Schwinger condition for imaginary times is fulfilled, and, consequently, a Fourier expansion with respect to Matsubara frequencies is possible (Kadanoff and Baym 1962). Using imaginary time Green functions one may solve different problems of thermodynamics including dielectric and optical properties (Kilimann 1978, Kraeft et al. 1986).

If we want to deal with nonequilibrium properties, such as kinetic equations and transport coefficients, one has to use real time Green functions as outlined by Kadanoff and Baym 1962, Keldysh 1964 and DuBois 1967; see also Kremp, Schlanges and Bornath 1985, 1986.

In this connection we have to start from the general equations of motion for Green functions according to Martin and Schwinger (1959).

The s-particle Green functions are defined in the usual way as ($h = 1$)

$$g_s(1 \dots s, 1' \dots s') = (1/i^s) < T\{\psi(1) \dots \psi(s)\psi^+(s') \dots \psi^+(1') \qquad (2.1)$$

with T being the Wick time ordering symbol and ψ, ψ^+ annihilation and creation operators, respectively.

Here, $<\dots> = \mathrm{Tr}\{\rho \dots\}$ means the average with a statistical operator which is unspecified so far.

Explicitly we have for g_1 with

$$g_1^>(1\,1') = (1/i) < \psi(1)\psi^+(1')>$$

$$g_1^<(1\,1') = \pm\, (1/i) < \psi^+(1')\psi(1)> ,$$

$$g_1(1\,1') = \theta(t_1 - t_{1'}) g_1^>(1\,1') + \theta(t_{1'} - t_1)\, g_1^<(1\,1').$$

Using for simplicity a Hamiltonian without external fields, i.e.,

$$H = \sum_i p_i^2/(2m) + \sum_{i<j} V_{ij} \quad , \tag{2.2}$$

the equation of motion for the field operators reads

$$(i \frac{\partial}{\partial t} + \frac{\nabla^2}{2m}) \psi(r\ t) = \int dr' V(r-r') \psi^+(r'\ t) \psi(r'\ t)\ \psi(r\ t)\ . \tag{2.3}$$

For the Green functions we get the hierarchy of equations

$$(i \frac{\partial}{\partial t_1} + \frac{\nabla_1^2}{2m})\ g_1(1\ 1') = \delta(1\ 1') \pm i \int d2\ V(1-2) g_2(12\ 1'2^+) \tag{2.4}$$

$$(i \frac{\partial}{\partial t_1} + \frac{\nabla_1^2}{2m})\ g_2(12\ 1'2') = \delta(1\ 1')\ g_1(2\ 2') \pm \delta(1\ 2') g_1(2\ 1')$$

$$\pm i \int d3\ V(1-3) g_3(123\ 1'2'3^+)\ . \tag{2.5}$$

and

$$(i \frac{\partial}{\partial t_1}\ \frac{\nabla_1^2}{2m})\ g_n(1\ldots n\ 1'\ldots n') = \sum_{\nu'}^{n}\ (\pm)\ \nu'-1\ \delta(1-\nu')\ .$$

$$g_{n-1}(2\ldots n\ 1'\ldots \nu'-1\ \nu'+1\ldots n')$$

$$\pm i \int d\alpha V(1-\alpha)\ g_{n+1}(1\ldots\ \alpha\ 1'\ldots\alpha^+)\ . \tag{2.6}$$

As usual, one of the main problems of the theory is the truncation of the hierarchy, i.e., to express, approximately, g_{n+1} by lower order functions, g_n, g_{n-1}, ...

As in the functional technique (Martin and Schwinger 1959), it is useful to introduce the self energy Σ by means of which a formal truncation is achieved. For example, the two particle Green function is expressed by the single particle self energy

$$\pm \int d2\ V(12) g_2(12\ 1'2^+) = \int d\bar{1}\ \Sigma\ (1\ \bar{1}) g_1(\bar{1}\ 1')\ . \tag{2.7}$$

As most of the physical properties may be expressed in terms of single and two particle operators, Σ turns out to be the most essential quantity. In this sense we get from (2.4) kinetic equations with Σ expressed approximately in terms of single particle Green functions.

Of course, this latter procedure is not always possible, especially not for partially ionized systems where we have to consider coupled sets of kinetic equations, one for each chemical species (see Klimontovich and Kremp 1981 and McLennan 1982).

On the other hand, it is possible to express thermodynamic quantities by means of the self energy Σ_a. For instance, the mean value $\langle V \rangle$ of the potential energy is given by (G means imaginary time Green function)

$$\langle V\rangle = -i(\Omega/2) \sum_a \int d\bar{1}\ \Sigma_a(1\ \bar{1})\ G_a(\bar{1}\ 1^+)\ . \qquad (2.8)$$

III. APPROXIMATIONS FOR THE SELF ENERGY

In the following section we want to give certain approximations for the self energy Σ which are of interest for two particle properties, for kinetic tasks and for thermodynamic questions. In close connection with the evaluation of any approximation of Σ, we have to consider the screening.

We do not want to go into the details (see, e.g., Kraeft et al. 1986); we mention only that classes of contributions of relevant quantities are partially summed up, and, as a consequence, instead of the bare potential V between two particles, there appears a dynamically screened potential V^s which obeys the equation

$$V^s(1\ 2) = V(1\ 2) + i \int d\bar{1}\ d\bar{2}\ V^s(1\ \bar{1})\ \Pi\ (\bar{2}\bar{1}\ \bar{2}\bar{1})V(\bar{2}\ 2)\ . \qquad (3.1)$$

In diagrams, Eq. (3.1) may be written as

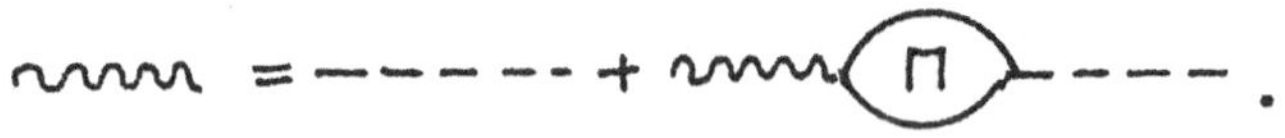

The polarization operator Π, in turn, is connected with Σ; e.g., taking into account that as a result of screening, the self energy is renormalized, $\Sigma \rightarrow \bar{\Sigma}$, we have instead of (2.7),

$$i \int d2\ V^s(1\ 2)\ \Pi\ (12\ 1'2^+) = \int d\bar{1}\ \bar{\Sigma}\ (1\ \bar{1})G_1(\bar{1}\ 1')\ . \qquad (3.2)$$

The renormalization of Σ means essentially that the self energy is now a functional of the screened potential and that the Hartree contribution does not exist.

As already mentioned above, it is necessary to consider rather restricting approximations for Σ, in dependence of the physical aim (see Kraeft et al. 1986).

If one wants to discuss macroscopic physical quantities which are essentially determined by the behavior of single (elementary) particles, and, moreover, the number of bound (composite) states is small, one has mainly to consider the "V^s-approximation" of Σ. This approximation follows from (3.2), if Π is replaced by the simplest approximation, i.e., by the product of two single particle Green functions:

$$\bar{\Sigma}\ (1\ \bar{1}) = V^s(1\ \bar{1})\ G(1\ \bar{1}) = \qquad (3.3)$$

$(G(1\ 1) \equiv$ $)$.

Approximation (3.3) is appropriate for the application in kinetic equations and leads to the level of the (quantum) Lenard-Balescu equation (see Section IV), if the number of bound states is small.

Under such conditions, Eq. (3.3) is also sufficient for the determination of the deviation of two particle bound state energies from their values at vanishing density (isolated two particle problem). This question will be discussed in Section VI.

For better approximations, however, it is necessary to take into account in Σ the formation of bound states. This is of relevance for the kinetic theory of partially ionized plasmas. In higher order approximations, it is also necessary to take into account the influence of neutrals (besides that of the elementary particles) in the vicinity of two particles on the binding energy of the latter.

The inclusion of bound states is of special importance for the determination of thermodynamic functions of partially ionized plasmas. Here especially the mere existence of bound states contributes to the pressure (see Section V).

A physically meaningful approximation is the inclusion of the bound states into the self energy Σ. In diagrams we may write (starting from (2.7) or (3.2))

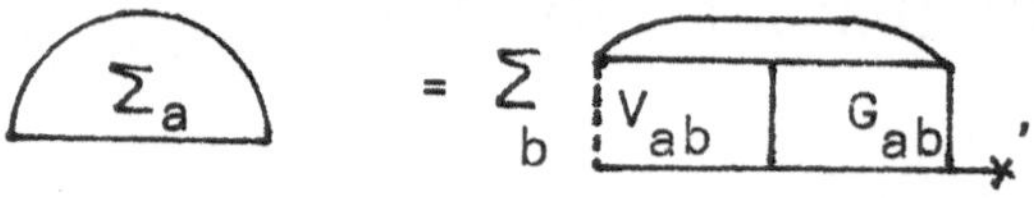

, (x means G_a^{-1}) . (3.4)

For G_{ab} we may apply a cluster expansion (Schlanges 1985, Kraeft and Luft 1984), which includes the formation of bound two and three particle states, i.e.,

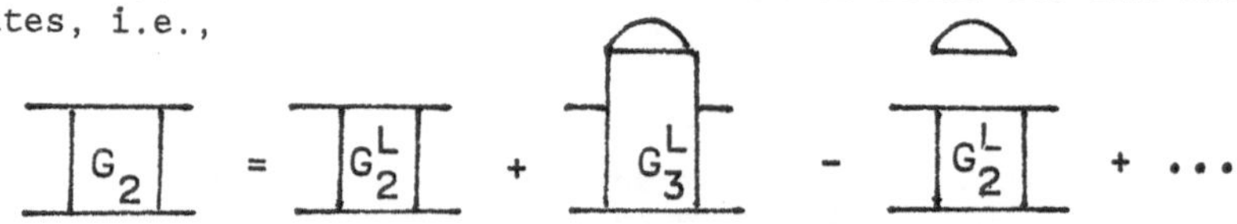

(3.5)

Here $G^L{}_2$, $G^L{}_3$ are the ladder sums of two and three particle complexes, respectively, and describe especially the formation of bound states. The relevant integral equation reads

(3.6)

The three particle Green function is given by the equation (according to (2.6))

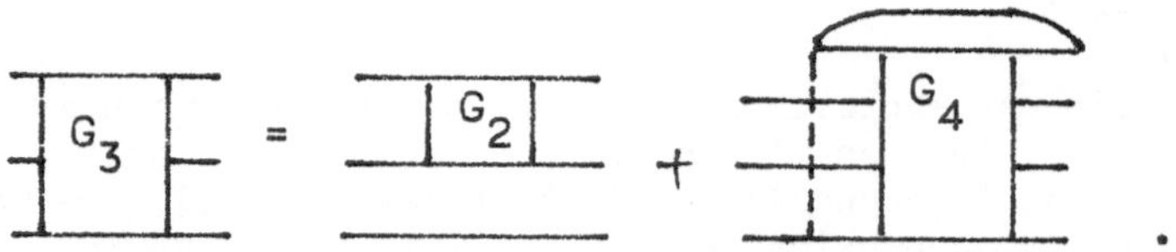

(3.7)

Here the approximation for G_4 will be used.

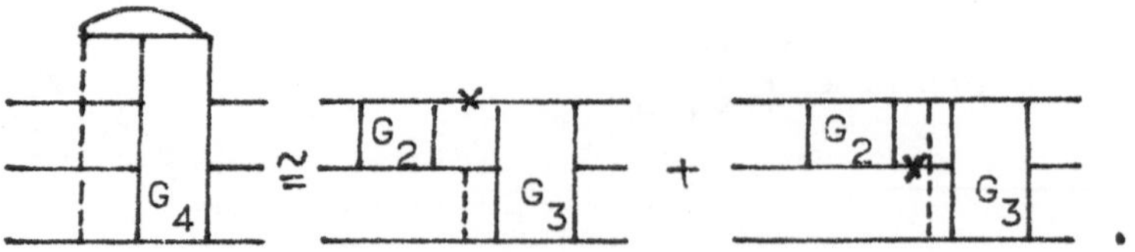

(3.8)

Again, x means the operation G^{-1}.

With the approximation (3.6) we get, from (3.7) with (3.8), the ladder approximation for G_3:

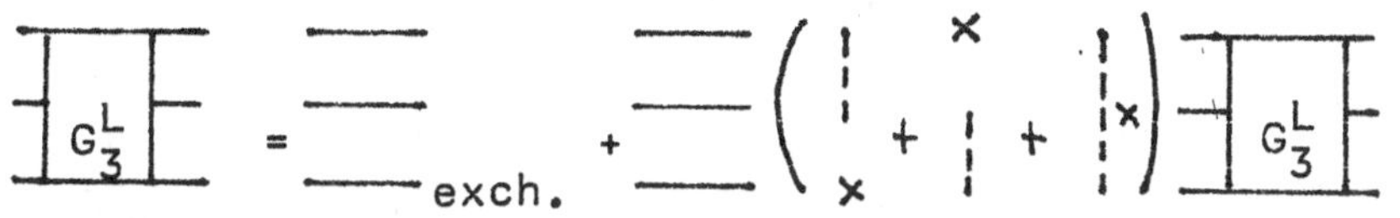

(3.9)

On the introduction of T-matrices via

(3.10)

and

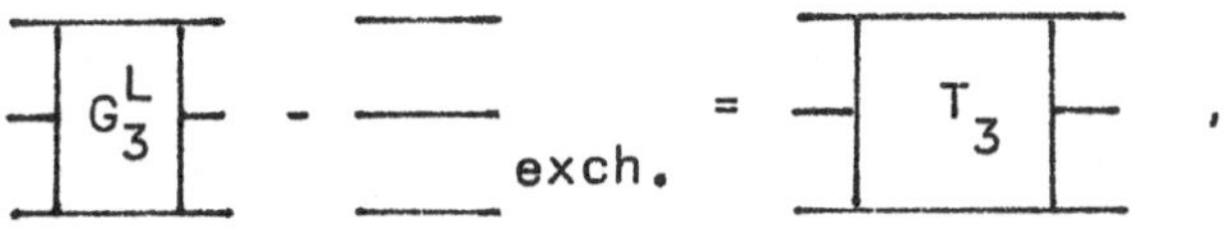

(3.11)

we get in the low density limit from (3.4) using (3.5) and (3.9)

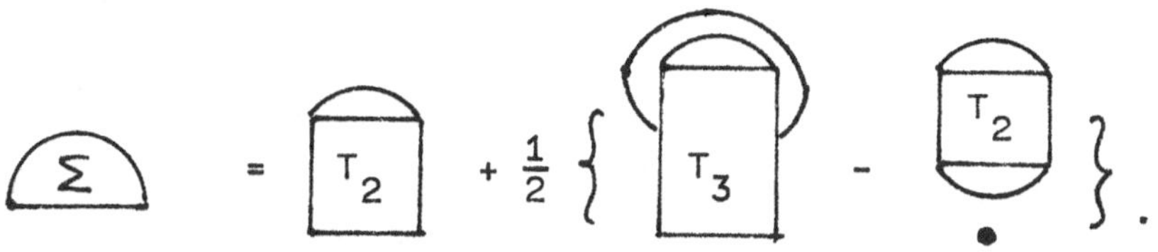

(3.12)

We used the T-matrix which is completely symmetric with respect to the channels (see Taylor 1972 and Schlanges 1985 for the discussion of multi-channel scattering).

In order to derive (3.12) one must take into account that equations of type (3.6) contain any order of the density so far. According to Section II, the single particle Green function contains, in dependence of the argument, the correlation functions g< and g>. While the lowest order of g> with respect to the particle density n is n^o, that of g< is n. For the application in (3.12), we need an approximation of $G^L{}_2$, which contains at least one pair of functions G_1< (we denote such lines by ——θ——) in order to fulfill the requirement of time integration in (3.12) (see Kraeft and Rother 1986).

From the ladder equation of $G^L{}_2$ we may derive in this approximation

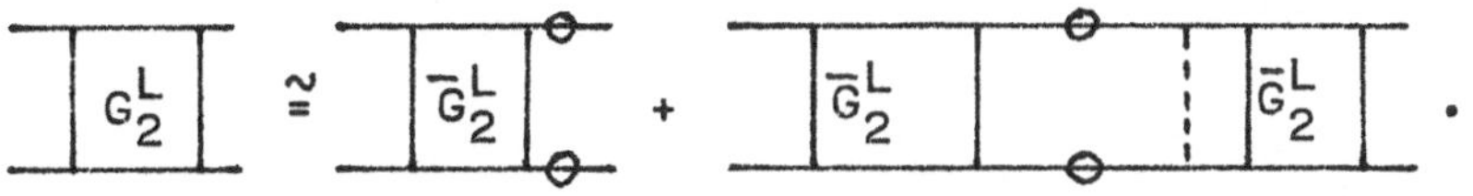

(3.13)

Here $\bar{G}^L{}_2$ represents scattering according to the order n^o in Eq. (3.10).

IV. KINETIC AND TRANSPORT PROPERTIES

The derivation of kinetic equations and the determinations of transport properties on the basis of density matrices was dealt with, e.g., by Klimontovich, Kremp and Schlanges (see Ebeling et al. 1983, 1984). Other references in this connection are McLennan 1982 and Klimontovich, Kremp and Kraeft 1986.

The linear response theory for the determination of transport quantities was developed on the basis of the Zubarev/McLennan formalism by Ropke; for references and results see Ebeling et al. 1983, 1984 and Kraeft et al. 1986.

In this connection it is very powerful to use real time Green functions (see e.g. Kremp, Schlanges and Bornath 1985, 1986) as outlined originally by Kadanoff and Baym 1962, Keldysh 1964 and DuBois 1967.

In this paper we do not want to discuss the details of kinetic equations and of the determination of transport properties of nonideal plasmas.

As a single example we give only an expression for the stopping power of charged (beam) particles entering a plasma. In principle, one has to use then a kinetic equation for the beam particles and each particle species of the plasma. As a simplification we assume the beam particles to be momoenergetic (velocity u) and the plasma to be in the equilibrium state and to obey the (quantum) Lenard-Balescu equation. In this way we consider only certain initial stage.

According to Klimontovich (1975) we write such kinetic equation

$$\frac{\partial}{\partial t} f_a(pt) = \frac{e_a^2}{\hbar(2\pi)^3} \int dp' \, d\omega dk \, \frac{1}{k^2} \, \delta(\hbar k - p + p')$$

$$\delta\Big(\hbar\omega - \frac{p^2}{2m_a} + \frac{p'^2}{2m_a}\Big) \Big\{ (\delta E \delta E)_{\omega k} \, (f_a(p't) - f_a(pt))$$

$$- 4\pi\hbar \, \frac{\mathrm{Im}\, \epsilon(\omega + i\delta, k)}{|\epsilon(\omega k)|^2} \, (f_a(p't) + f_a(pt)) \Big\} \, . \qquad (4.1)$$

Under equilibrium conditions, the field fluctuations are given by

$$(\delta E \delta E)_{\omega k} + \frac{8\pi}{\omega} \, \frac{\mathrm{Im}\, \epsilon(\omega + i\delta, k)}{|\epsilon(\omega, k)|^2} \Big(\frac{\hbar\omega}{2} + \frac{\hbar\omega}{e^{\hbar\omega/k_B T} - 1} \Big) \, . \qquad (4.2)$$

For the energy loss per unit time we get then (Kraeft and Strege 1986, see also Arista and Brandt 1981).

$$\frac{dE}{dt} = \sum_b \frac{2\, e_b^2}{\mu u} \int_0^\infty \frac{dk}{k} \int_{\omega-}^{\omega+} d\omega \, \omega \, \mathrm{Im} \, \frac{n_B(\omega)}{\epsilon(k, \omega)} \, . \qquad (4.3)$$

$$\omega_\pm = \hbar^2 k^2/(2m_a) \pm ku \, .$$

According to Lindhard (1954) an expression for the energy loss reads (see also Maynard and Deutsch 1982)

$$\frac{dE}{dt} = - \sum_b \int_0^\infty \frac{e_b^2}{\mu u} \cdot \frac{dk}{k} \int_{-ku}^{+ku} d\omega \, \omega \, \mathrm{Im} \, \frac{1}{\epsilon(k, \omega)} \, . \qquad (4.4)$$

While (4.4) is zero for zero beam particle velocity and is negatively definite, Eq. (4.3) exhibits a positive (finite) value at zero velocity and is zero for certain beam particle velocity (near the thermal velocity). At higher velocities, (4.3) and (4.4) show similar behavior.

This means the beam particles absorb energy at small velocities, and emit at higher ones. The difference between (4.4) and (4.3) is due to the neglection of field fluctuations; the latter were dealt with by Akhiezer et al. (1974).

If one wants to deal with nonequilibrium plasmas and inhomogeneities, the question becomes much more complicated. Payne and Perez (1980) used for this purpose the Gould-DeWitt (1967) kinetic equation and applied the BGK collision term (Bhatnager, Gross, Krook 1954). An evaluation of (4,3,4) is shown in Figure 1.

V. THERMODYNAMIC FUNCTIONS

In this section we will outline how to determine thermodynamic functions. We will see that single particle and two particle energies enter the calculations; the latter will be dealt with in Section VI.

A useful starting point is the connection between the density n_a and the chemical potentials of all species $\{\mu_c\}$ where c runs over the constituents. For fermions we may write

$$n_a(\{\mu_c\}) = (2\ s_a + 1) \int f_a(\omega) A_a(p,\omega) \frac{dp}{(2\pi)^3} \frac{d\omega}{2\pi} \ . \tag{5.1}$$

f_a is the Fermi function, and A_a is the spectral function of the single particle Green function, $A_a = \text{Im}\ G_a$.

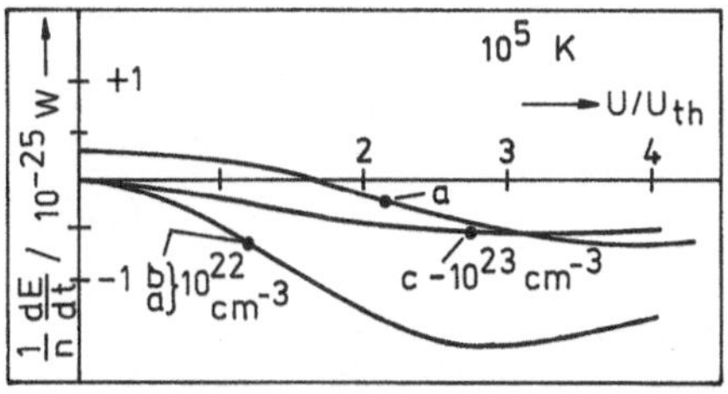

Fig. 1 Stopping power as a function of velocity u, $T = 10^5$K, a - Eq. (4.3), b - Eq. (4.4), $n_e = 10^{22}\text{cm}^{-3}$, dielectric constant in random phase approximation; c - $n_e = 10^{23}\text{cm}^{-3}$, dielectric constant after Carini and Kalman 1984. (Electrons, electron plasma).

The spectral function is connected with the real and imaginary parts of the self energy via

$$A_a(p\omega) = \frac{(2/\hbar)\ \mathrm{Im}\ \Sigma^{corr}(p,\omega + i\epsilon)}{(\omega-\epsilon_a^{HF}(p)/\hbar - \mathrm{Re}\Sigma_a^{corr}(p\omega)/\hbar)^2 + (\mathrm{Im}\Sigma_a^{corr}(p\omega)/\hbar)^2} \tag{5.2}$$

The Hartree Fock single particle energy ϵ^{HF} includes the corresponding frequency independent part of the self energy, Σ^{HF}, and Σ^{corr} is the remaining part:

$$\Sigma_a = \Sigma_a^{HF} + \Sigma_a^{corr} , \quad \epsilon_a^{HF} = \frac{\hbar^2 p^2}{2m_a} + \Sigma_a^{HF}(p) . \tag{5.3}$$

Σ^{HF} is the approximation (3.3) with $V^s \rightarrow V$.

Other thermodynamic properties, e.g., the pressure, may be determined from (5.1) by the relation

$$\frac{\partial p}{\partial \mu_a} = n_a(\{\mu_c\}) . \tag{5.4}$$

As already mentioned above, the quality in which thermodynamics is determined, depends entirely on the approximation used for Σ_a in (5.2).

If bound states shall be included, at least the approximation of (3.12)

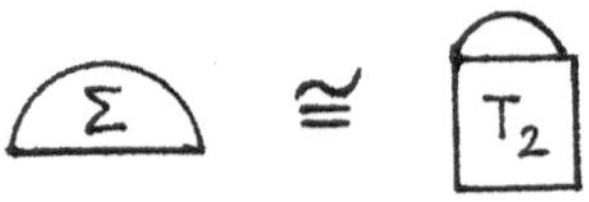

(5.5)

must be used. However, even in that case, the evaluation of (5.2) is not possible in a closed form.

According to Kremp, Kraeft and Labert 1984, Redmer 1985, we develop (5.2) to give

$$A_a(p) = \frac{2\pi\delta(\omega-\epsilon_a^{HF}/\hbar - \mathrm{Re}\Sigma_a^{corr}(p,E_a(p))/\hbar)}{1 - \frac{d}{d\omega}\mathrm{Re}\frac{1}{\hbar}\Sigma_a^{corr}(p,\omega)\Big|_{\hbar\omega = E_a(p)}}$$

$$- \frac{2}{\hbar}\ \mathrm{Im}\ \Sigma_a^{corr}(p\omega + i\epsilon)\frac{d}{d\omega}\ \frac{P}{\omega - \frac{1}{\hbar}\tilde{E}_a(p\omega)} . \tag{5.6}$$

$$E_a(p) = \epsilon_a^{HF} + \mathrm{Re}\ \Sigma_a^{corr}(p,E_a(p));\ \tilde{E}_a(p\omega) = \epsilon_a^{HF} + \mathrm{Re}\ \Sigma_a^{corr}(p\omega) .$$

We remark that $A_a(p,\omega)$ has, for vanishing Σ^{corr}, the usual δ-like shape.

In the coupled plasma, however, especially the last contribution of

(5.6) describes the formation of bound states, as we have

$$\mathrm{Im}\ \Sigma_a^{corr}(p,\omega + i\epsilon) \sim \delta(\omega - E_n/\hbar) \quad . \tag{5.7}$$

where E_n are the two particle bound state energies.

On inserting (5.6) into (5.1), we get different contributions to the density of species a; up to more refined questions which are connected with the long range divergencies of the Coulomb potential and screening and those connected with lower orders with respect to the coupling parameter we get two main contributions. The first part describes the existence of bound states, n_a^{bound}, and the second one accounts for the free (scattering) states, n_a^{scatt}:

$$n_a = n_a^{bound} + n_a^{scatt} \quad . \tag{5.8}$$

Here we have

$$n_a^{bound} = (2s_a+1) \sum_b (2s_b+1) \sum_{nl} \int_0^\infty \frac{4\pi\ dP\ P^2}{(2\pi)^3} \{g_B^{ab}(E_{nlP}^{ab}) - \Delta g(E)\} \tag{5.9}$$

and

$$n_a^{scatt} = (2s_a+1) \int \frac{dp}{(2\pi)^3}\ f_a\left(\frac{h^2p^2}{2m_a} + \Sigma_a^{HF}(p) + \Sigma_a^{MW}\left(p, \frac{\hbar^2p^2}{2m_a}\right)\right) \quad . \tag{5.10}$$

Σ^{MW} is the Montroll Ward approximation (3.3).

The quantity $\Delta g(E)$ accounts for lower orders and screening as mentioned above. At sufficiently low temperatures, $\Delta g(E)$ is of minor interest. At higher temperatures, we have in (5.9) the Planck-Larkin sum of states (Ebeling, Kraeft and Kremp 1976, 1979; Ebeling, Kraeft and Ropke 1985), what corresponds to the replacement

$$g_B^{ab}(E) - \Delta g(E) \rightarrow \exp(-\beta E) - 1 + \beta E \ . \tag{5.11}$$

The physically interesting result is that in the bound state part of the density (5.9), there occur the two particle bound state energies E^{ab}_{nlP} which must be determined under the influence of the surrounding plasma, see Section VI. On the other hand, the scattering state part of the density (5.10) contains the single particle energy $\hbar^2p^2/(2m_a) + \Sigma_a$ which accounts for the interaction with the plasma. The (arbitrary) subdivision (5.8) corresponds to the introduction of a "chemical picture" which defines certain states of the (physically) elementary particles as new species.

If we introduce the degree of ionization, i.e., the relative number of free electrons by

$$\alpha = n_e^{scatt}/(n_e^{scatt} + n_e^{bound}) = n_e^{scatt}/n_e \ . \tag{5.12}$$

we get in the non-degenerate case

$$\frac{1-\alpha}{\alpha^2} = \frac{n_e \sum_{nl} \int \frac{dp}{(2\pi)^3} \left[\exp(-\beta E^{ei}_{nl}(P)) - 1 + \beta E^{ei}_{nl}(P) \right]}{\int \frac{dp}{(2\pi)^3} \exp(-\beta (\frac{\hbar^2 p^2}{2m_e} + \Sigma^{HF}_e(p) + \Sigma^{MW}_e(p, \frac{\hbar^2 p^2}{2m_e}))) \int \frac{dp'}{(2\pi)^3} [e \rightleftharpoons i]} \quad (5.13$$

It can be seen from (5.13) that there is a competition between the bound state energies $E^{ei}{}_{nl}(P)$ which are weakly density dependent only, and the single particle energies $\hbar^2 p^2/(2m) + \Sigma$. Especially there might be a concellation; this effect is referred to as the Mott transition, i.e., in connection with the vanishing of bound states, the degree of ionization increases to unity. The details of this point will be discussed now in Section VI.

VI. TWO PARTICLE BOUND STATES

For the discussion of two particle bound states we start from the ladder equation for G_2, according to (3.6) and mention that the lines corresponding to the single particle propagation, are Green functions which are subject to the Dyson equation

$$G_1 = G^o_1 + G^o_1 \Sigma G_1 . \quad (6.1)$$

The inclusion of Σ in the approximation (3.3) into the ladder equation (3.6) gives a Bethe Salpeter equation in which the influence of free particles on the two particle states (especially also bound states) is taken into account.

The corresponding effective Schrödinger equation (homogeneous Bethe Salpeter equation) reads

$$\left(\frac{p_1^2}{2m_1} + \frac{p_2^2}{2m_2} - z \right) \psi_z(p_2 p_2) - \int \frac{dq}{(2\pi)^3} V(q) \psi_z(p_1+q, p_2-q) =$$

$$< p_1 p_2 \mid H^{pl}_{12}(z) \mid \alpha P > . \quad (6.2)$$

α are quantum numbers, P the center of mass momentum. The contribution H^{pl} to the hamiltonian represents the influence of the surrounding plasma and vanishes for the isolated pair. With the approximation (3.3), $<|H^{pl}|>$ reads

$$<|H^{pl}|> = \int \frac{dp}{(2\pi)^3} V(q) \{ [N(p_1 p_2) - 1] \psi_z(p_1+q, p_2-q) \quad (6.3a)$$

$$- [N(p_1+q, p_2-q) - 1] \psi_z(p_1 p_2) \} \quad (6.3b)$$

$$+ \int \frac{dp}{(2\pi)^3} \Delta V^{eff}(q) \{ N(p_1 p_2) \psi_z(p_1+q, p_2-q) \quad (6.3c)$$

$$- N(p_1+q, p_2-q) \psi_z(p_1 p_2) \} . \quad (6.3d)$$

Here $N(p_1 p_2) = 1 - f(p_1) - f(p_2)$.

The first line (a) corresponds to the phase space occupation (statistical correlation), (b) exchange self energy (Σ^{HF}), (c) dynamical self energy, and (d) dynamically screened effective potential.

It can be seen that there is a relative strong compensation between (a) and (b) and between (c) and (d), respectively. The compensation is exact for the q=0 contribution. From this follows that all the four contributions must be taken into account.

This compensation acts especially only for bound states, where the wave function is sharply localized in the coordinate space so that the Fourier transform does not vary strongly with q. This is not the case for scattering states, and thus there is no compensation of the type just mentioned.

In the simple version which corresponds to the approximation (3.3), the effective potential correction is given by (Zimmermann et al. 1978)

$$\Delta V^{eff}_{ep}(p_1 p_2 q E) = - \int_{-\infty}^{+\infty} \frac{d\omega}{\pi} \,\mathrm{Im}\, \frac{V(q)}{\epsilon(q,\omega+i\alpha)}$$

$$\cdot \left\{ \frac{n_B(\omega)[f_e(p_1+q) - f_e(p_1)] + f_e(p_1+q)\,[1 - f_e(p_1)]}{- {}_e(p_1+q) + {}_e(p_1)} \right.$$

$$\left. + \frac{1 + n_B(\omega)}{E - \omega - \epsilon_e(p_1) - \epsilon_p(p_2-q)} + e \leftrightarrow p,\ p_1 \leftrightarrow p_2,\ q \leftrightarrow -q \right\} . \quad (6.4)$$

It should be remarked that ΔV^{eff} does not coincide with the dynamically screened potential $V(q)/\epsilon(q,\omega)$. This fact is a consequence of the retardation of the interaction which follows from the dynamical character of the screening.

An approximation of physical relevance is the plasmon pole approximation for Im $1/\epsilon$; it reads

$$\mathrm{Im}\, \epsilon^{-1}(q,\omega+i\alpha) = - \frac{\pi}{2} \{\delta(\omega-\omega_{pl}) - \delta(\omega + \omega_{pl})\}\, \omega_{pl} , \quad (6.5)$$

ω_{pl} - plasma frequency.

The resulting two particle state energies behave, consequently, as follows.

For very low densities, the bound state energies do not vary with the density; especially they do not increase with the square root of the density. At higher densities, there is a slight decrease which is linear in the density of free charge carriers. At sufficiently high densities there appears also an energy shift linear in the number density of bound states (Redmer 1985).

In contrast, the two particle continuum states decrease for small concentrations proportional to the square root of the density.

As a result, there is a cross-over of the continuum states and the bound states, so that, at certain density, the bound states vanish. The ground state vanishes at the highest concentration; this concentration is referred to as the Mott density.

VII. CONCLUDING REMARKS

It could not be the aim, of course, to cover the full area of dense plasma physics.

Let us give some more hints to current literature of the field without achieving completeness.

First we would like to draw attention to the "International Workshop on Nonideal Plasma" held in the GDR at different mostly resort areas: 1980 Matzlow Garwitz, 1982 Wustrow (Kraeft and Rother, Eds., 1983), 1984 Biesenthal (Radtke and Hess Eds., 1985). The 1986 Workshop will be held at Greifswald.

The monographs by Günther and Radtke (1984) and by Fortov and Yakubov (1984) cover certain experimental areas. Reviews were given by Baus and Hansen 1980 and by Ichimaru 1982. Extremely dense systems and scattering problems were touched only briefly. Here we mention Rogers 1981, Rogers, Young, DeWitt and Ross 1983, Kremp et al. 1984, Perrot and Dharma wardana 1984, Ichimaru 1985 and Rother and Kraeft 1986. Certain topics of the present Santa Cruz 1986 meeting were not mentioned at all. For further references see also the invited paper by Ebeling, Kraeft, Kremp and Ropke 1986, Statphys. 16, Boston.

REFERENCES

Akhiezer, A. I., (Ed.), Akhiezer, I. A., Polovin, R. V., Sitenko, A. G., Stepanov, K. N., 1974 "Plasma Electrodynamics" (in Russian), Nauka, Moscow, 1975, Pergamon, New York.

Arista, N. R., Brandt, W., 1981, Phys. Rev. A 23:1898.

Baus, M., Hansen, J. P., 1980 Physics Reports, 59:2.

Baus, M., Hansen, J. P., 1983, J. Stat. Phys., 31:409.

Bhatnager, P. L., Gross, E. P., Krook, M., 1954, Phys. Rev., 94:511.

Carini, P., Kalman, G., 1984, Phys. Letters, 105A:229.

DuBois, D. F., 1967, Lectures in Theoretical Physics, Vol. X, p. 469, W. E. Brittain (Ed.), Gordon and Breach, New York.

Ebeling, W., Kraeft, W. D., Kremp, D., 1976, "Theory of Bound States and Ionization Equilibrium in Plasmas and Solids", Akademie-Verlag, Berlin.

Ebeling, W., Kraeft, W. D., Kremp, D., 1986, Europhys. News, 17:52.

Ebeling, W., Kraeft, W. D., Röpke, G. (Eds.), 1983, "Transport Properties of Dense Plasmas", Akademie-Verlag, Berlin; 1984, Birkhäuser, Basel, Boston, Stuttgart.

Ebeling, W., Kraeft, W. D., Kremp, D., Röpke, G., 1985, The Astrophys. J., 290:24; 1986, Invited Paper, Statphys 16, Boston.

Fortov, V. E., Yakubov, I. T., 1984, Physics of Nonideal Plasmas (in Russian), Acad. Sci., USSR, Chernogolovka.

Gunther, K., Radtke, R., 1984, Electric Properties of Weakly Nonideal Plasmas, Akademie-Verlag, Berlin; Birkhäuser, Basel, Boston, Stuttgart.

Gould, H. A., DeWitt, H. E., 1967, Phys. Rev. 155:68.

Ichimaru, S., 1982, Rev. Mod. Phys., 54:1017.

Ichimaru, S., 1985,

Kadanoff, L. P. Baym, G., 1962, Quantum Statistical Mechanics, Benjamin, New York.

Kalman, G., Carini, P., 1978, Strongly Coupled Plasmas, Plenum, New York.

Keldysh, L., 1964, Zh. Exp. Teor. Fiz. (in Russian) 47:1515.

Kilimann, K., 1978, Ph.D. Thesis II, Rostock University.

Klimontovich, Yu. L., 1975, Kinetic Theory of Nonideal Gases and Nonideal Plasmas (in Russian), Nauka, Moscow; Pergamon, 1982, New York.

Klimontovich, Yu. L., Kremp, D., 1981, Physica, 109A:517.
Klimontovich, Yu. L., Kremp, D., Kraeft, W. D., 1986, Adv. Chem. Phys., in press.
Kraeft, W. D., Luft, M., 1984, in: M. M. Popović (Ed.), The Physics of Ionized Gases, p. 681, The Institute of Physics, Beograd.
Kraeft, W. D., Rother, T. (Eds.), 1983, Contributions to the Second International Workshop on Nonideal Plasmas, Wustrow, 1982, University of Greifswald.
Kraeft, W. D., Rother, T., 1987, Ann. Physik, submitted.
Kraeft, W. D., Strege, B., 1987, Plasma Phys., submitted.
Kraeft, W. D., Kremp, D., Ebeling, W., Röpke, G., 1986, Quantum Statistics of Charged Particle Systems, Akademie-Verlag, Berlin; Plenum, New York, London.
Kremp, D., Kraeft, W. D., Lambert, A. J. D., 1984, Physica, 127A:72.
Kremp, D., Schlanges, M., Bornath, T., 1985, J. Stat. Phys., 41:661.
Kremp, D., Kilimann, K., Kraeft, W. D., Stolz, H., Zimmermann, R., 1984, Physica, 127A:646.
Lindhard, J., 1954, K. Dan. Vedensk, Selsk., Mat. Fys. Medd. 28:No. 8.
Martin, P. C., Schwinger, J., 1959, Phys. Rev., 115:1342.
Maynard, G., Deutsch, C., 1982, Phys. Rev., A26:665.
McLennan, J. A., 1982, J. Stat. Phys., 28:521.
Payne, G. L., Perez, J. D., 1980, Phys. Rev. A 21:976.
Radtke, R., Hess, H. (Eds.), 1985, Contributions to the 3rd International Workshop on Nonideal Plasmas Biesenthal 1984, Acad. Sci. of the GDR, ZIE, Berlin.
Redmer, R., 1985, Ph.D. thesis, Rostock University.
Rogers, F. J., 1981, Phys. Rev. A, 24:1531.
Rogers, F. J., 1984, Phys. Rev. A 29:868.
Rogers, F. J., Young, D. A., DeWitt, H. E., Ross, M., 1983, Phys. Rev. A 28:2990.
Ross, M., 1985, Rep. Prog. Phys. 48:1.
Rother, T., Kraeft, W. D., 1986, Phys. Stat. Sol. (b), 136.
Perrot, F., Dharma-wardana, M. W. C., 1984, Phys. Rev. A 30:2619.
Schlanges, M., 1985, Ph.D. Thesis II, Rostock University.
Taylor, J. R., 1972, Scattering Theory, Wiley, New York.
Zimmermann, R., Kilimann, K., Kraeft, W. D., Kremp, D., Ropke, G., 1978, Phys. Stat. Sol. (b), 90:175.

HIGHER-ORDER LEVINSON THEOREMS AND THE PLANCK-LARKIN PARTITION FUNCTION FOR REACTING PLASMAS

D. Bollé*

Instituut voor Theoretische Fysica - Universiteit Leuven
B-3030 Leuven, Belgium

1. INTRODUCTION

We review higher-order spectral sum rules in scattering theory for $d \leqslant 3$ dimensions. These rules generalize the well-known Levinson theorem [1] for partial-wave scattering, relating the number of bound states with the value of the scattering phase shift at the threshold energy zero, and they are valid for non-spherically symmetric short-range interactions. They involve the scattering matrix, the bound-state wave functions and some correction terms arising from the high-energy behavior of the scattering problem. These correction terms are shown to be d-dimensional generalizations of the polynomial conserved densities of the Korteweg-de Vries equation [2], [3]. The modifications of these rules necessary to allow Coulomb-type interactions are presented.

We discuss the use of these rules in the context of the S-matrix approach to statistical mechanics [4]. We show that in a high-temperature expansion of the grand-canonical partition function, the bound-state contributions explicitely cancel against scattering contributions. In this way we are able to (rigorously) derive and explain the underlying structure of the Planck-Larkin partition function for reacting plasmas.

---

*Onderzoeksleider N.F.W.O., Belgium

2. SUM RULES IN SCATTERING THEORY

For s-wave partial-wave scattering, described by the radial Schrödinger equation, Levinson proved the following theorem [1], using the theory of ordinary differential equations. If the spherically symmetric potential V(r) is short-range, i.e. if

$$\int_0 dr\, r\, |V(r)| + \int^{\infty} dr\, r^2\, |V(r)| < \infty , \tag{1}$$

then the s-wave phase shift $\delta(k)$, k the momentum, is continuous in k, the limit for $k \to \infty$, $\delta(\infty)$, exists and

$$\delta(\infty) - \delta(0) = -\pi(N + \frac{1}{2} q) , \tag{2}$$

where N is the number of s-wave negative-energy bound-states and where q = 1 if there is a zero-energy resonance, q = 0 otherwise. For higher partial waves, ℓ, (2) remains valid if we add to the r.h.s. $(-\pi N_{0,\ell})$, with $N_{0,\ell}$ the number of zero-energy bound states. Zero-energy bound states are solutions Ψ_0 of the zero-energy Schrödinger equation, $(-\Delta + V(\underline{x}))\Psi_0(\underline{x}) = 0$, with Δ the 3 d-Laplacian, satisfying

$$\int d^3x\, |\Psi_0(\underline{x})|^2 < \infty , \tag{3}$$

while zero-energy resonances are such solutions Ψ_0 obeying

$$\int d^3x\, |\Psi_0(\underline{x})|^2 = \infty \tag{4}$$

The result (2) turned out not only to provide some deep theoretical insight into the scattering problem but also to be very useful in applications. Therefore one has tried to generalize this theorem to very general scattering systems. In the following pages we briefly discuss these generalizations. For more details and a list of references, we refer to the recent review [5].

A first interesting extension is the one to non-spherically symmetric potentials. The result reads (we use natural units such that $\hbar^2/2m = 1$)

$$\int_0^{\infty} dE\, \{i\, \mathrm{Tr}[S^{\star}(E) \frac{d}{dE} S(E)] - \frac{1}{4\pi\sqrt{E}} \int d^3x\, V(\underline{x})\} = 2\pi(N_b + N_0 + \frac{1}{2} q). \tag{5}$$

Here S(E) is the on-energy-shell scattering matrix, $E = k^2$ the energy, N_b the number of negative-energy bound states, N_0 the number of zero-energy bound states and q as before. Tr is the trace on the appropriate space. (Here it is the integral with respect to the angles over the forward matrix element of the operator in the brackets [ ].) For partial-wave scattering we know that $S_\ell(E) = \exp(2i\delta_\ell(E))$, so the only difference between (5) and (2) is the presence of the potential term. This correction term arises from the high-energy Born behavior of the S-matrix : the first term in the integrand of (5) is not integrable for $E \to \infty$; by subtracting the potential term, this singular behavior is cancelled.

The method of derivation of (5) and all other rules in this section is the following. We consider the trace in configuration space of the difference of the resolvents

$$\begin{aligned} R(z) &= (H - z)^{-1} \quad , \quad H = H_0 + V \ , \\ R_0(z) &= (H_0 - z)^{-1} \quad , \quad H_0 = -\Delta \ , \end{aligned} \tag{6}$$

with z the complex energy. This quantity is a relevant one to look at since it is connected with the on-shell S-matrix, i.e.

$$2 \text{ Im Tr}[R(E + io) - R_0(E + io)] = -i \text{ Tr } S^\star(E) \frac{d}{dE} S(E) \ . \tag{7}$$

We then study the analyticity properties of the resolvent difference in the complex-energy plane, especially at low and high energies. This study is strongly dimension dependent. Finally we apply contour integration techniques and integrate the resolvent difference around the spectrum of the Hamiltonian H in the energy plane. The well-known Cauchy theorem on analytic functions then immediately leads to the result (5) : the first term in the integrand arises from the real-axis contribution together with eq (7), the second term comes from the circular contour at infinity; the r.h.s. represents the contributions from the small circle around the origin and from the circles around the (negative) discrete eigenvalues.

In fact, the generalized Levinson theorem (5) is the first rule of a whole set obtained by contour integration of higher energy-moments of the resolvent difference. We then have

$$\int_0^\infty dE \, E^N \{i \text{ Tr}[S^\star(E) \frac{d}{dE} S(E)] - \sum_{n=1}^{N+1} a_n \, E^{-n+1/2} \int d^3x \, P_n(\underline{x})\}$$

$$= 2\pi \sum_{j=1}^{N_b} (-\chi_j^2)^N + \delta_{N,0}(2\pi \, N_0 + \pi q) \ , \quad N \geqslant 0 \ , \tag{8}$$

where a_n is a numerical coefficient

$$a_n = (2n - 2)! \, [4\pi \, 2^{2n-2} \, n!(n - 1)!]^{-1} \tag{9}$$

and $P_n(\underline{x})$ is a polynomial of order n in V and its derivatives that will be discussed in Section 3. We recall that these correction terms arise from the high-energy behavior of the scattering problem. At this point it is useful to write down the partial-wave version of the rules (8) for $N = 1$, all ℓ

$$\int_0^\infty dE\ E\ \{\frac{1}{\pi}\frac{d\delta_\ell}{dE} - \frac{1}{4\pi\ E^{3/2}}\int dr\ V(r)\} = \sum_{j=1}^{N_{b,\ell}} (-\chi_{j,\ell}^2) + \frac{2\ell + 1}{4} V(0)\ . \tag{10}$$

Due to the better high-energy behavior of partial-wave scattering quantities, there is no correction term in the $N = 0$ rule (2) and only 1 correction term in the $N = 1$ rule (10) (compared with 2 in (8)).

Other extensions in $d = 3$ are possible : N arbitrary real, classical scattering, many-particle scattering .... (See [5].)

Concerning other dimensions we mention the new result on impurity scattering in $d = 1$ [6] and we briefly state the results for $d = 2$ since they are recently completed and rather surprising [7]-[10]. First of all, the 2d partial wave Levinson theorem reads

$$\begin{aligned} &\delta_\ell^{(2)}(\infty) - \delta_\ell^{(2)}(0) = -\pi\ N_{b,\ell}^{(2)} - \pi\ D_\ell\ , \\ &D_0 = 0\ ,\quad D_\ell = 0 \text{ or } 1 \quad \text{for} \quad \ell \geqslant 1\ . \end{aligned} \tag{11}$$

So contrary to $d = 3$ (and $d = 1$), possible s-wave zero-energy resonances do not contribute, while there also exist p-wave zero-energy resonances that contribute exactly like (zero-energy) bound states. The non-spherically symmetric rules read, $N \geqslant 0$,

$$\begin{aligned} &\int_0^\infty dE\ E^N\ i\ \mathrm{Tr}\ [S^\star(E)\ \frac{d}{dE}\ S(E)] \\ &= 2\pi \sum_{j=1}^{N_b} (-\chi_j^2)^N + \delta_{N,0}\ 2\pi\ D + \frac{1}{2(N + 1)}\int d^2x\ P_{N+1}(\underline{x})\ . \end{aligned} \tag{12}$$

In contrast with 3d, where a number of correction terms appeared in the integral, we now have one "surface" correction. It again involves the polynomial $P_n(\underline{x})$, which we want to discuss now.

3. HIGH-ENERGY CORRECTIONS AND KORTEWEG-DEVRIES INVARIANTS

There exist different methods to obtain the correction terms in (5), (8) and (12). (See e.g. [11], [12]). In the heat-equation (or Laplace transform) method [12], one starts from the relation

$$R(z) = \int_0^\infty d\beta\ e^{z\beta}\ e^{-\beta H}\ ,\quad \mathrm{Re}\ z < -\ |z_0|\ , \tag{13}$$

where the statistical operator $\exp(-\beta H)$ satisfies

$$(\frac{\partial}{\partial\beta} + H)e^{-\beta H}(\underline{x},\underline{x}') = 0 \quad , \quad \beta = 1/k_B T$$
$$\lim_{\beta\to 0} \; e^{-\beta H}(\underline{x},\underline{x}') = \delta(\underline{x} - \underline{x}') \quad , \tag{14}$$

with T the temperature and k_B the Boltzmann constant. One then writes the following asymptotic expansion, valid for $\beta \to 0$

$$e^{-\beta H}(\underline{x},\underline{x}') = e^{-\beta H_0}(\underline{x},\underline{x}') \sum_{n=0}^{\infty} \frac{(-\beta)^n}{n!} P_n(\underline{x},\underline{x}') \; ,$$
$$e^{-\beta H_0}(\underline{x},\underline{x}') = (4\pi\beta)^{-3/2}\exp(-(4\beta)^{-1}(\underline{x} - \underline{x}')^2) \tag{15}$$

and obtains, after insertion in (14), a recurrence relation from which the P_n can be determined. (Because of the Laplace transform (13), any expansion of $\exp(-\beta H)$ implies a related expansion for $R(z)$.) The results on the diagonal $\underline{x} = \underline{x}'$ read

$$P_1(\underline{x}) = V(\underline{x}) \; ,$$
$$P_2(\underline{x}) = V^2(\underline{x}) - \frac{1}{3}\,\Delta V(\underline{x}) \; , \tag{16}$$
$$P_3(\underline{x}) = V^3(\underline{x}) - V(\underline{x})\,\Delta V(\underline{x}) - \frac{1}{2}\,[\underline{\nabla}\,.\,V(\underline{x})]^2 + \frac{1}{10}\,\Delta^2 V(\underline{x}) .$$

We will see that these polynomials $P_n(\underline{x})$ are higher-dimensional generalizations of the invariants of the one-dimensional Korteweg-de Vries equation (cf. e.g. [2], [3]). The fact that the correction terms in the higher-order Levinson theorems(8) and (12) are nothing but thes invariants was only fully realized in the seventies ([11], [13]-[16]). The Korteweg-de Vries equation is one of the simplest non-linear dispersive wave equations and arose long ago in an approximate theory of hydrodynamic waves. It reads (subscripts denote partial differentiation)

$$u_t - uu_x + u_{xxx} = 0 \; . \tag{17}$$

This equation possesses an infinite sequence of polynomial conservation laws in the form

$$D_t + X_x = 0 \; , \tag{18}$$

where D, the conserved density, and (-X), the flux of D are polynomials (not explicitly dependent on x or t) in u and its derivatives. Two conservation laws are obtained immediately

$$u_t + (-\frac{1}{2}u^2 + u_{xx})_x = 0 ,$$
$$(\frac{1}{2}u^2)_t + (-\frac{1}{3}u^3 + uu_{xx} - \frac{1}{2}u_x^2)_x = 0 . \tag{19}$$

The first is simply the equation (17) itself rewritten and the second is obtained after multiplying it by u and rewriting. These polynomials were first derived in [3]. There is a close relationship between these invariant polynomials and constants of motion. For example, if one assumes that u and its x-derivatives vanish sufficiently rapidly at infinity, then each law (18) immediately yields a constant of motion I_n, i.e.

$$I_n[u] = \int dx\, D_n(u,u_x\ldots) , \tag{20}$$

where D_n is a polynomial in u and its x-derivatives with orders up to n - 2 which also contains u^n.

The relationship with the high-energy correction terms P_n in the 1d Schrödinger equation can be understood on the basis of the following discovery [17] (see also [13]). Consider u(x,t) to be the one-parameter family of potentials satisfying the KdV equation, then the eigenvalues of the Schrödinger equation are time-independent. Furthermore, if the evolution of u in the Schrödinger equation is governed by any equation whatsoever which leaves the eigenvalues invariant, then that equation possesses all the same polynomial conserved densities as the KdV equation. Because of (20) the D_n and P_n are equal up to total derivatives in X. The present discussion assumes smooth potentials. For a generalization to screened Coulomb lattices, see [18]. We remark that in that case P_1 and P_2 do not change.

4. MODIFICATIONS IN THE PRESENCE OF COULOMB INTERACTIONS (3d)

For the pure repulsive Coulomb scattering system, Levinson's theorem (2) is violated since $\delta_\ell(k) \to +\infty$ as $k \to 0$ and there are no bound states. For pure attractive Coulomb scattering, having an infinite number of bound states, both sides of (2) equal $+\infty$.

For a spherically symmetric short-range potential in the presence of a repulsive Coulomb interaction the form (2) stays valid when the phase shift is interpreted as the relative phase shift associated with the scattering by the total potential compared with the pure Coulomb potential; in this case q is always zero.

For a spherically symmetric short-range potential in the presence of an attractive Coulomb interaction, the equivalent of (2) reads (for a recent discussion see [19] and references therein)

$$\delta_\ell(0) = \pi \mu_\ell , \quad (21)$$

where μ_ℓ is the so-called quantum defect. It is a real number depending on the short-range potential, V_{sr}, that can be understood as follows. When the Coulomb potential is added to V_{sr}, the Balmer levels are changed according to

$$E_{n\ell} = \frac{-m e^4}{2\hbar^2(n + \ell + 1)^2} \rightarrow \tilde{E}_{n\ell} = \frac{-m e^4}{2\hbar^2(n - \mu_\ell(n,V_{sr}) + \ell + 1)^2} . \quad (22)$$

The change depends on n, ℓ and V_{sr}. In the limit $n \rightarrow \infty$, we define

$$\mu_\ell \equiv \mu_\ell(\infty,V_{sr}) . \quad (23)$$

The value of the largest integer contained in μ_ℓ represents the number of additional bound-states due to V_{sr}.

A set of higher-order sum rules analogous to (8) has also been derived in this case [20]. Because of the Coulomb potential, two changes have to be made. First, the contribution from the discrete spectrum has to be replaced by

$$\sum_{j=1}^{N_b} (-\chi_j^2)^N \rightarrow \sum_{n=1}^{\infty} |n + \ell - \mu|^{-2N} - \sum_{n=1}^{\infty} (n + \ell)^{-2N} . \quad (24)$$

Secondly, for $N \geqslant 2$, correction terms of the form $(\ell nE)^j$, j dependent on N, appear in the energy-integral of (8). For the exact expressions of the relations, which are rather involved, we refer to [20].

5. THE PLANCK-LARKIN PARTITION FUNCTION FOR REACTING PLASMAS

We now study the application of these sum rules to reacting plasmas, in the context of the S-matrix approach to statistical mechanics which formulates the statistical behavior of a system in terms of the collision processes of the constituent particles [4].

As is well-known, the grand canonical partition function Ξ can be expanded in powers of the fugacity z, using the Ursell and Mayer cluster expansion [21]

$$\Xi\,(z,V,T) = \exp\,\{V \sum_{n=1}^{\infty} b_n\, z^n\} , \quad (25)$$

with V the volume of the system and where the coefficients b_n are the cluster integrals. Assuming the latter exist in the thermodynamic limit, they read

$$b_1 = \lambda^{-3} , \tag{26}$$

$$b_2 = \frac{2^{3/2}}{2!\ \lambda^3} \mathrm{Tr}\ [e^{-\beta H_2} - e^{-\beta H_{2,0}}] , \tag{27}$$

$$b_3 = \frac{3^{3/2}}{3!\ \lambda^3} \mathrm{Tr}\ [e^{-\beta H_3} - e^{-\beta H_{3,0}} - \sum_{\alpha=1}^{3} (e^{-\beta H_{3,\alpha}} - e^{-\beta H_{3,0}})], \tag{28}$$

.....

Here λ is the thermal wavelength

$$\lambda = (2\pi\ \hbar^2 \beta/m)^{1/2} , \tag{29}$$

and H_2, $H_{2,0}$ are the total and free two-particle Hamiltonians, H_3, $H_{3,\alpha}$ and $H_{3,0}$ are the total, α-channel and free three-particle Hamiltonians. The factors in front of the trace come from the center-of-mass motion. We have assumed Boltzmann statistics. We remark that exchange effects do not introduce anything fundamentally new here but working out their details may be highly non-trivial. The $n^{\underline{th}}$ cluster coefficient only involves n- and fewer-particle effects.

Once we have an explicit form for the b_n, then the grand canonical partition function and all other thermodynamic properties of the system are determined. The equation of state e.g. can be obtained as a series in the density particles $\rho = N/V$. The result is [21]

$$\beta P = \sum_{n=1}^{\infty} a_n \rho^n , \tag{30}$$

where the coefficients a_n in this expansion, which are the viral coefficients, can be completely written down in terms of the cluster coefficients, viz.

$$a_1 = 1 , \quad a_2 = -b_2 b_1^{-2} , \quad a_3 = 4a_2^2 - 2b_3 b_1^{-3} , \ \ldots \tag{31}$$

We now want to evaluate these cluster integrals in terms of scattering quantities. The standard method is to use the Watson transform, which connects the statistical operator, $\exp(-\beta H)$, with the resolvent $R(z)$ (cfr. (6)), z the complex energy, viz.

$$e^{-\beta H} = - \frac{1}{2\pi i} \oint_C dz\ e^{-\beta z}\ R(z) , \tag{32}$$

where C is a contour around the spectrum of the Hamiltonian H (cfr. also section 2). From (27) we then get, recalling (7) and assuming for simplicity that there are no zero-energy resonances and no zero-energy bound states

$$b_2(\beta) = \sqrt{2}\,\lambda^{-3}\{\sum_{j=1}^{N_b} e^{\beta\chi_j^2} + (2\pi)^{-1}\int_0^\infty dE\, e^{-\beta E}\, \mathrm{Tr}[-i\, S^\star(E)\frac{d}{dE}S(E)]\}. \tag{33}$$

Eq. (33) is a generalization of the Beth-Uhlenbeck result [21] to non-spherically symmetric interactions. Indeed, in the case of spherical symmetry we know (see section 2) that the logarithmic derivative of the S-matrix is given by the sum of the energy derivatives of the partial-wave phase shifts. A similar treatment for b_3, ... has been discussed in the literature. There still exist some controversy about the presence of counterterms in the analogues of (33), originating e.g. from rescattering singularities in the S-matrix for three free particles going to three free particles. For more details we refer to [22]. For a recent review of various formalisms and calculation methods for the cluster integrals, containing a lot of references, see [23].

Let us look now at the continuum part of b_2 in (33) and write it as

$$b_2^S(\beta) = \sqrt{2}\,\lambda^{-3}(2\pi)^{-1}\int_0^\infty dE\, e^{-\beta E}\{\mathrm{Tr}[-i\, S^\star(E)\frac{d}{dE}S(E)] - \frac{1}{4\pi\sqrt{E}}\int d^3x\, V(\underline{x})\}$$

$$+ \sqrt{2}\,\lambda^{-3}(8\pi^2)^{-1}\int d^3x\, V(\underline{x})\int_0^\infty dE\, e^{-\beta E}\, E^{-1/2} \tag{34}$$

The second term of (34) can be worked out explicitly. The first one can be written as a total differential in the following way

$$-\sqrt{2}\,\lambda^{-3}(2\pi)^{-1}\int_0^\infty dE\, e^{-\beta E}\frac{d}{dE}\{\int_E^\infty dE_1[\mathrm{Tr}(-i\, S^\star(E_1)\frac{d}{dE_1}S(E_1)) - \frac{1}{4\pi\sqrt{E_1}}\int d^3x\, V(\underline{x})]\}.$$

Partial integration with respect to E and use of Levinson theorem (8) for N = 0 to evaluate the surface term gives then

$$b_2^S(\beta) = -\sqrt{2}\,\lambda^{-3}(2\pi)^{-1}\{2\pi N_b + (16\pi\beta)^{-1/2}\int d^3x\, V(\underline{x})$$

$$+ \beta\int_0^\infty dE\, e^{-\beta E}\int_E^\infty dE_1[\mathrm{Tr}(-i\, S^\star(E_1)\frac{d}{dE_1}S(E_1)) - \frac{1}{4\pi\sqrt{E_1}}\int d^3x\, V(\underline{x})]\}. \tag{35}$$

It is clear that this process may be repeated. For example, we now add and subtract $(16\pi)^{-1}E^{-1/2}\int d^3x\, V^2(\underline{x})$ to the last integral in (35) and we calculate this integral with the help of an integration by parts and the sum rule (8) for N = 1. In this way we arrive at

$$b_2(\beta) = \sqrt{2}\,\lambda^{-3}\,\{[\sum_{j=1}^{N_b} e^{\beta\chi_j^2} - N_b - \beta\sum_j \chi_j^2]$$

$$- (2\pi)^{-1}(16\pi\beta)^{-1/2}\int d^3x\,V(\underline{x}) - (8\pi)^{-1}\,\beta^{1/2}\int d^3x\,V^2(\underline{x})$$

$$+ \beta^2\int_0^\infty dE\,e^{-\beta E}\int_E^\infty dE_1\int_{E_1}^\infty dE_2\,[\,\mathrm{Tr}(-i\,S^\star(E_2)\,\frac{d}{dE_2}\,S(E_2)) \tag{36}$$

$$- \frac{1}{4\pi\sqrt{E_2}}\int d^3x\,V(\underline{x}) - \frac{1}{8\pi\sqrt{E_1}}\int d^3x\,V^2(\underline{x})]\}\,.$$

Continuing further in this way we finally get ($\hbar^2/2m = 1$)

$$b_2(\beta) = \sum_{n=1}^{\infty}\sqrt{2}\,(4\pi)^{-3}\,\frac{(-1)^n}{n!}\,\beta^{n-3}\int d^3x\,P_n(\underline{x})\;, \tag{37}$$

where the $P_n(\underline{x})$ are the Korteweg-de Vries invariants discussed in section 3. This result is the well-known Wigner-Kirkwood expansion that can be derived in many other ways. We remark that this type of expansions, in the presence of magnetic fields ([24], [25] and references therein) have been used e.g. to discuss magnetic properties of one component plasmas in the nearly classical case [25].

In deriving (37) we have assumed smooth, not necessarily spherically symmetric, short-range potentials. For screened Coulomb interactions, having a 1/r singularity at the origin, non-analytic terms, e.g. of order $\beta^{5/2}$ ..., appear [18], [26]. It is important to note that in this case expression (36) remains valid since P_1 and P_2 exist and the rules (8) for N = 0, 1 do not change.

What is special about our derivation is that we explicitly see a cancellation between bound-state and scattering contributions. This compensation has been studied before numerically and in WKB approximation in the framework of partial wave scattering [27]. The derivation presented here is rigorously valid on a fully quantum mechanical level. It also holds for higher cluster coefficients as can be shown e.g. for b_3 starting from (28) and the three-body sum rules derived in [28]. It stays even valid when there is an infinite number of bound states as has been verified explicitly in a model calculation for the third cluster coefficient of binary mixtures of light and heavy particles allowing for the Efimov effect [29].

For spherically symmetric interactions a similar procedure can be followed starting from (compare (33))

$$b_2(\beta) = \sqrt{2}\,\lambda^{-3} \sum_\ell (2\ell+1) \{ \sum_{j=1}^{N_\ell} e^{\beta\chi^2_{j,\ell}} + \frac{1}{\pi} \int_0^\infty dE\, e^{-\beta E} \frac{d\delta_\ell}{dE} \}. \qquad (38)$$

Using the sum rules (2) and (10) and the partial-integration technique described before we arrive at

$$b_2(\beta) = \sqrt{2}\,\lambda^{-3} \sum_\ell (2\ell+1) \sum_{j=1}^{N_\ell} (e^{\beta\chi^2_{j,\ell}} - 1 - \beta\chi^2_{j,\ell})$$

$$- (2\pi)^{-1/2}\, \lambda^{-3}\, \beta^{1/2} \sum_\ell (2\ell+1) \int dr\, V(r) + \sqrt{2}\,\lambda^{-3}\beta \sum_\ell (2\ell+1) \frac{2\ell+1}{4} V(0)$$

$$+ O(\beta^2) \,. \qquad (39)$$

Comparing with eq. (36) we see that the partial wave form has a better $\beta \to 0$ behavior, corresponding to the better high-energy behavior of the partial wave scattering quantities mentioned before.

We recognize the first three terms of (36) or (39) as the Planck-Larkin partition function

$$b_2^{PL}(\beta) = \sqrt{2}\,\lambda^{-3} \sum_\ell (2\ell+1) \sum_{j=1}^{N_\ell} (e^{\beta\chi^2_{j,\ell}} - 1 - \beta\chi^2_{j,\ell}) \,. \qquad (40)$$

See e.g. [27], [30]. We have presented a rigorous derivation of its underlying structure on the basis of higher-order scattering sum rules. We remark that for pure Coulomb scattering $b_2^{PL}(\beta)$ can be written as

$$b_{2,c}^{PL}(\beta) = \sqrt{2}\,\lambda^{-3} \sum_{n=1}^{\infty} n^2 (e^{-\beta E_n} - 1 + \beta E_n) \,, \qquad (41)$$

$$E_n = - \frac{e^2}{2a_B n^2} \,, \quad n = 1, 2, 3 \ldots \qquad (42)$$

with a_B the Bohr radius. We see that the infinite sum (41) is convergent.

For low-density hydrogen plasmas where the effective Hamiltonian can be replaced by the two-particle Coulomb Hamiltonian, we see in first instance from (41) that states with high principal quantum number having an energy below the mean thermal energy i.e. $(-E_n\beta) \ll 1$ are suppressed. In this way the Planck-Larkin form (41) provides a borderline between discrete quasi-free states near to the continuum edge, which should be treated like the scattering contributions.

For dense plasmas the Coulomb interaction is dynamically screened. This is one of the effects of the collective behavior of the charged particles. This effect can be taken into account approximately by considering static screening realized by replacing the Coulomb potential by a Debye (Yukawa) potential

$$V(r) = -e^2/r \,.\, \exp(-r/r_D) \ , \tag{43}$$

where the Debye length r_D is a function of the temperature and the density of the protons, ρ_p, viz.

$$r_D^{-2} = 8\pi\, e^2\, \rho_p \beta \ . \tag{44}$$

The non-modified bound-state sum is now finite, since the potential (43) has a finite number of levels. However, since these levels are dependent upon the screening length r_D, and hence functions of the temperature and density, the following happens. As r_D decreases the upper levels move into the continuum and the bound-state sum changes discontinously. The (-1) subtraction in (40), due to Levinson's theorem, is then necessary to resolve this unphysical difficulty. The second subtraction ensures a finite $\beta \to 0$ limit for $b_2^{PL}(\beta)$. We also remark that states below the mean thermal energy are still suppressed in (40).

So in any case the Planck-Larkin form (40) (or (41)) provides an analytical way for limiting the bound-state sum by separating it into a part associated with "composites" and a part which is treated as being delocalized. For more details and refinements of this method, and for its practical use in plasma calculations we refer to other contributions in this volume (e.g. [31]-[33]) and to the recent literature (e.g. [34]-[39]).

REFERENCES

1. R.G. Newton, Scattering Theory of Waves and Particles (Springer, N.Y. 1982) 2nd ed.
2. A.M. Perelomov, Ann. Inst. Henri Poincaré, Sect. A, 24, 161 (1976).

3. M.D. Kruskal, R.M. Miura, C.S. Gardner, and N.J. Zabusky, J. Math. Phys. 11, 952 (1970).
4. R. Dashen, S. Ma, and H. Bernstein, Phys. Rev. 187, 345 (1969).
5. D. Bollé, in Mathematics + Physics. Lectures on Recent Results, ed. L. Streit (World Scientific, Singapore 1986), Vol. 2, p. 84.
6. N.E. Firsova, Theor. Math. Phys. 62, 130 (1985).
7. M. Cheney, J. Math. Phys. 25, 1449 (1984).
8. D. Bollé, F. Gesztesy, C. Danneels and S.F.J. Wilk, Phys. Rev. Lett. 56, 900 (1986).
9. W.G. Gibson, Phys. Lett. A 117, 107 (1986).
10. R.G. Newton, Low-energy scattering for medium-range potentials, Indiana University preprint, May 1986.
11. D. Bollé, Ann. Phys. 121, 131 (1979).
12. S.F.J. Wilk, Y. Fujiwara, and T.A. Osborn, Phys. Rev. A 24, 2187 (1981).
13. V.E. Zakharov and L.D. Faddeev, Func. Anal. Appl. 5, 280 (1971).
14. M. Kac and P. Van Moerbeke, Proc. Nat. Acad. Sci. USA 71, 2350 (1974).
15. I.M. Gelfand and L.A. Dikii, Russian Math. Surveys 30, 77 (1975).
16. H.P. Kean and P. Van Moerbeke, Invent. Math. 30, 217 (1975).
17. C.S. Gardner, J.M. Greene, M.D. Kruskal, and R.M. Miura, Phys. Rev. Lett. 19, 1095 (1967).
18. J.E. Avron, Ann. Phys. 108, 448 (1977).
19. Z.Q. Ma, Phys. Rev. D 33, 1745 (1986).
20. D.R. Yafaev, Theor. Math. Phys. 11, 358 (1972).
21. K. Huang, Statistical Mechanics (Wiley, N.Y. 1963).
22. S. Servadio, "Truly-three-body" scattering and unitarity, University of Pisa, Italy preprints IFUP TH 13-14/86.
23. W.G. Gibson, in "Few-Body Methods. Principals and Applications", ed. by T.K. Lim, C.G. Bao, D.P. Hou and S. Huber, (World Scientific, Singapore 1986).
24. D. Bollé and D. Roekaerts, Phys. Rev. A 30, 2024 (1984).
25. A. Alastuey and B. Jancovici, Physica 97A, 349 (1979); 102A, 327 (1980); A. Alastuey, Physica 110A, 293 (1982).
26. H.E. DeWitt, J. Math. Phys. 3, 1003 (1962).
27. F.J. Rogers, Phys. Rev. A 19, 375 (1979) and references therein.
28. D. Bollé and T.A. Osborn, Phys. Rev. A 26, 3062 (1982).
29. W. Hoogeveen and J.A. Tjon, Physica 108A, 77 (1981).
30. W. Ebeling, Physica 73, 573 (1974).
31. W. Däppen, The Planck-Larkin partition function and solar physics applications, these proceedings.

32. W.D. Kraeft, Strongly coupled partially ionized plasmas, these proceedings.

33. F.J. Rogers, Occupation numbers for reacting plasmas, these proceedings.

34. F.J. Rogers, Phys. Rev. A 24, 1531 (1981).

35. F.J. Rogers, Phys. Rev. A 29, 868 (1984).

36. W. Ebeling, W.D. Kraeft, D. Kremp and G. Röpke, Ap. J. 290, 24 (1985).

37. C.A. Rouse, Ap. J. 272, 377 (1983).

38. R.K. Ulrich, Ap. J. 258, 404 (1982).

39. W.D. Kraeft, D. Kremp, W. Ebeling, and G. Röpke, Quantum Statistics of Charged Particle Systems (Plenum, N.Y. 1986).

METALLIC LITHIUM BY QUANTUM MONTE CARLO

G.Sugiyama, G.Zerah†, and B.J.Alder

Lawrence Livermore National Laboratory
Livermore, Ca. 94550

Lithium was chosen as the simplest known metal for the first application of quantum Monte Carlo methods in order to evaluate the accuracy of conventional one-electron band theories. Lithium has been extensively studied using such techniques. The KKR method [1], the linear muffin tin orbital method (LMTO) [2], the augmented spherical wave method (ASW) [3] and a linear combinations of gaussian type orbitals (LCGTO) method [4] agree in their predictions of the equation of state. These results are also consistent with experimental data available at low compressions [5] [6] and agree with quantum-statistical-models [7] [8] at high pressures.

Band theory calculations have certain limitations in general and specifically in their application to lithium. Results depend on such factors as charge shape approximations (muffin tins), pseudopotentials (a special problem for lithium where the lack of p core states requires a strong pseudopotential), and the form and parameters chosen for the exchange potential. The calculations are all one-electron methods in which the correlation effects are included in an ad hoc manner. This approximation may be particularly poor in the high compression regime, where the core states become delocalized. Furthermore, band theory provides only self-consistent results rather than strict limits on the energies. The quantum Monte Carlo method is a totally different technique using a many-body rather than a mean field approach which yields an upper bound on the energies.

QUANTUM MANY-BODY ALGORITHM

The Schrödinger equation was solved for a system of M fixed lithium atoms and $N=3M$ electrons using the quantum Monte Carlo algorithm previously developed for the electron gas [9] [10]. This technique does not approximate the $3N$ dimensional problem by reducing it to a set of equations of lower dimensionality, but solves it exactly within statistical error bars. The algorithm

† permanent address: Commissariat l'energie atomique, Centre d'etudes de Limeil-Valenton, France

involves several phases of progressively greater accuracy - the variational, the diffusion and the Green's function Monte Carlo methods, which yield upper bounds to the ground state energy, and the released node Green's function Monte Carlo method, which provides exact answers within statistics. In this preliminary study, various trial wavefunctions were explored in order to determine the optimum form for a fixed amount of computational time using the variational and fixed-node schemes. The choice of the trial wavefunction is a trade off between a simple analytic form which is computationally fast and a more accurate but complex form which is difficult to sample efficiently.

Variational Monte-Carlo [11]. Variationally, the total energy of a system of Hamiltonian $\mathbf{H}$ is given by the minimum with respect to the set of all possible trial functions Ψ_T of

$$E = \operatorname*{Min}_{\Psi_T} \frac{\int \Psi_T \mathbf{H} \Psi_T}{\int |\Psi_T|^2} \quad . \tag{1}$$

In practice, Ψ_T is a parameterized expression for which the integral (1) is computed using the Metropolis Monte Carlo algorithm. The result is an upper bound on the energy that is dependent upon the nature of the trial wave function. The standard form of Ψ_T consists of a Slater determinant of one-body states multiplied by a pair product Jastrow factor which incorporates two-body correlation effects. For lithium, three different forms were used for the elements of the determinant. In the simpler cases, the localized states were taken to be Gaussians with a width parameter and the delocalized states were treated as plane waves. In the third case, a more complex form was generated from band theory charge densities as discussed below.

Diffusion Monte Carlo. The diffusion Monte Carlo algorithm [12] computes a more accurate solution of the Schrödinger equation using a trial function generated by the variational Monte Carlo technique. The Schrödinger equation in imaginary time is treated as a diffusion equation with the potential acting as a branching birth and death process. The solution converges exponentially to the ground state. The wavefunction Ψ satisfies

$$\mathbf{H} \ \Psi(\vec{R},t) = [-\sum_{i=1}^{N} \frac{\hbar^2}{2m} \nabla_i^2 + V(\vec{R}) - E_T] \ \Psi(\vec{R},t) \ , \tag{2}$$

where $\vec{R}$ is the $3N$ dimensional vector of the electronic coordinates, t is the imaginary time and

$$V(\vec{R}) = \sum_{i<j}^{N} \frac{e^2}{r_{ij}} - \sum_{i,\alpha}^{N,M} \frac{Z_\alpha e^2}{r_{i\alpha}} + \sum_{\alpha<\beta}^{M} \frac{Z_\alpha Z_\beta e^2}{r_{\alpha\beta}} \ , \tag{3}$$

is the potential energy of the solid using standard Coulomb interactions. E_T is a constant trial energy which is subtracted from the potential energy for computational convenience. The sums run over the electronic coordinates i,j and the fixed atomic lattice sites α,β with $r_{ab} \equiv |r_a - r_b|$. A finite simulation cell with periodic boundary conditions is used, with the potential energies evaluated by Ewald summation.

The wavefunction Ψ can be interpreted as the density of diffusing particles as long as it is everywhere of one sign. This is not the the case for fermion statistics where the many-body wavefunction changes sign at the nodes. However, this difficulty can be overcome by using a trial wavefunction Ψ_T, whose nodes act as absorbing barriers to the diffusion process. The probablity density defined by

$$f(\vec{R},t) = \Psi_T(\vec{R})\Psi(\vec{R},t)\exp[-(E_T-E_0)t] \ , \tag{4}$$

obeys the diffusion equation

$$\hbar\frac{\partial f(\vec{R},t)}{\partial t} = -\sum_{j=1}^{N}\frac{\hbar^2}{2m}\nabla_j[\nabla_j f - f\nabla_j \ln|\Psi_T|^2] - [\frac{H\Psi_T}{\Psi_T} - E_T]f \tag{5}$$

which is derived from the Schrödinger equation. E_0 is the exact ground state eigenvalue corresponding to the eigenfunction Ψ. This equation is solved in each of the regions bounded by the trial function nodes as before, however now the trial function plays an important role in reducing the branching term - a process known as importance sampling.

The fixed-node approximation imposes the constraint that the wavefunction has the approximate nodal surface of Ψ_T leading to a upper bound criterion on the energy. In principle, the nodal surfaces could be varied to obtain the best upper bound on the energy. In general, however, it is difficult to parameterize Ψ_T in a systematic fashion. For the electron gas, the dependency upon the location of the nodes of the trial wavefunction was weak [10], however for an accurate upper bound Ψ_T should be chosen as close to the true ground state wavefunction as is feasible.

Green's Function Monte Carlo. The Schrödinger equation recast into integral form can be solved by Monte Carlo sampling of the exact Green's function G [13]. This avoids the error incurred in the diffusion Monte Carlo algorithm by the use of a short time step expansion approximation of G. Exact evaluation of G coupled with nodal release leads to the stochastically exact solution of the Schrödinger equation.

Trial wavefunction. The general form for the wavefunction that has been used successfully in previous studies consists of the product of a Slater determinant of single particle orbitals multipled by a Jastrow factor $\mathbf{J}$:

$$\Psi_T = \det|\Phi_{ij}^{+}|\det|\Phi_{ij}^{-}|\ \mathbf{J} \ , \tag{6}$$

where the Φ_{ij} are the one particle wavefunctions of the Slater determinant, with the superscripts $^+$ and $^-$ denoting the two possible spin states. The determinantal form provides the required fermion antisymmetry. The Jastrow factor

$$\mathbf{J} = \exp(-\sum_{i,j}^{N}u_{ij} - \sum_{i,\alpha}^{N,M}u_{i\alpha}) \tag{7}$$

involves a sum of the electron-nuclear and the electron-electron pair correlation factors. These exactly incorporate the cusp conditions - the singularities of the wavefunction for zero pair separation due to the coulombic divergence of the potential. The Jastrow factor is computed using the random-phase approximation [11].

Two forms of Slater determinant trial function were implemented. In the first calculation, Ψ consisted of Gaussians with parameterized widths centered about the lattices sites for localized states and plane waves for the delocalized orbitals. At extreme compressions where the 1s state was expected to be delocalized, all the Slater determinant states were taken to be plane waves - a form which should yield a lower energy. Such simple analytical forms have been successfully applied to molecules [12] the electron gas [10] and molecular and metallic hydrogen (with emphasis on the metallic transition) [14]. For highly compressed lithium, however, it was desirable to introduce a wavefunction which provides a continuous transition from a localized to a delocalized form. Conventional band theory techniques provided such a function. Specifically, the single particle orbitals in the Slater determinant part of Ψ_T were taken from an augmented plane wave (APW) calculation. The electron-nuclear Jastrow factor was suppressed as this cusp condition is correctly accounted for in the APW functions.

The APW method uses Bloch's theorem to reduce the description of a crystalline solid to a calculation in the primitive cell. This results in a set of coupled one particle Schrödinger equations which are solved by a discretization in reciprocal space. This discretization is introduced by considering only those functions which are periodic on the scale of a few unit cells.

The APW method builds solutions of the Schrödinger equation by solving the radial equation inside the muffin tin sphere and matching them to linear combinations of plane waves in the interstitial region. The APW wavefunctions are generated according to the formulae:

$$\psi_{\vec{k}}(\vec{r}+\vec{R}_j) = \sum_{\mathbf{R}} \sum_{i} c_i \, \Gamma_{11}(\mathbf{R}) \, e^{i\vec{k}\cdot\vec{X}_j} \, \phi_{\mathbf{R}\vec{k}_i}(\vec{r}) \ , \tag{8}$$

where c_i are the standard APW coefficients [15] , $\mathbf{R}$ are the set of rotations leaving the crystal invariant, $\Gamma_{11}(\mathbf{r})$ are the matrix elements of the invariant group of the vector k, $\vec{k}_i = \vec{k} + \vec{K}_i$ ($\vec{K}_i$ a reciprocal lattice vector) and $\vec{X}_j$ is a lattice site. The function

$$\phi_{\mathbf{R}\vec{k}_i}(\vec{r}) = \sum_{l,m} a_{lm}(\mathbf{R}\,\vec{k}_i)\, u_l(r)\, Y_{lm}(\hat{r}) \ , \tag{9}$$

is the muffin tin solution expressed as an expansion in spherical harmonics. Outside the muffin tin the solution is expressed as a sum of plane waves

$$\psi_{\vec{k}}(\vec{r}) = \sum_{\mathbf{R}} \sum_{i} c_i \, \Gamma_{11}(\mathbf{R}) \, e^{i\mathbf{R}\vec{k}_i\cdot\vec{r}} \ , \tag{10}$$

where the c_i are the matching constants from the APW functions of equation (8).

The APW calculations were performed using the same number of points ($\vec{k}$ values) in the Brillouin zone as there were lattice sites in the quantum Monte Carlo calculation (32 for FCC, 16 for BCC). This provided the correct number of single particle orbitals for the quantum Monte Carlo trial function. The maximum angular momentum value l was taken to be 13 in the APW calculation and restricted to 3 in the subsequent generation of the wavefunction. The l values in the wavefunction could be limited to the s,p,d, and f states since this was sufficient to represent the charge density. The APW calculations were performed for a few densities using an increased number of l and $\vec{K}$ values without significant changes in the energy and pressure. The APW results agreed with the detailed band structure at normal and 10-fold compression [4] and with the energies and pressures [3] obtained from more accurate band theory calculations, to within the uncertainty caused by number dependence and exchange potential parameters. In particular, the APW results showed that the 2s-2p band is not free-electron-like at 10-fold compression, indicating that a plane wave trial function is a poor choice. At higher compressions, both the 1s and 2s-2p bands are free-electron-like and energies and pressures approximate Thomas-Fermi model results.

Table 1. Quantum Monte Carlo variational and fixed-node diffusion energies for lithium at various densities. Energies are in Rydbergs. Densities are given in terms of the Wigner-Seitz radius r_s and the compression. The calculations are for 16 and 32 atoms in the simulation box. The trial functions are Gaussian and plane wave (g-pw), all plane wave (pw) and APW (apw). Errors in the last digit are given in parentheses.

Density		M=16 atoms			M=32 atoms		
r_s	Comp	tf	E_v	E_d	tf	E_v	E_d
3.500	0.263	g-pw	-14.78(1)	-14.97(1)			
3.000	0.418	g-pw	-14.84(1)	-15.01(1)			
2.500	0.722	g-pw	-14.90(1)	-15.03(1)			
2.260	0.978	g-pw	-14.91(1)	-15.03(1)	g-pw	-14.92(1)	-15.00(1)
					apw	-14.87(1)	-15.02(1)
2.100	1.219	g-pw	-14.90(1)	-15.03(1)			
2.000	1.412	g-pw	-14.86(1)	-15.00(1)	apw	-14.87(1)	
1.800	1.936	g-pw	-14.85(1)	-14.96(1)	apw	-14.84(1)	
1.600	2.757	g-pw	-14.74(1)	-14.88(1)	apw	-14.77(1)	
1.400	4.115				apw	-14.66(2)	
1.250	5.783	g-pw	-14.00(1)	-14.39(1)	apw	-14.49(3)	
1.000	11.289	g-pw	-13.34(1)	-13.43(1)	g-pw	-13.60(1)	-13.78(1)
		pw	-11.36(1)	-12.57(2)	apw	-13.99(1)	
0.800	22.067	g-pw	-11.22(1)	-11.32(1)	g-pw	-11.81(1)	-11.93(1)
		pw	-10.82(1)	-11.77(1)	apw	-13.99(1)	
0.750	26.766	g-pw	-10.30(1)	-10.38(1)			
		pw	-10.52(1)	-11.33(1)			
0.700	32.940				apw	-11.26(1)	
0.650	41.115	g-pw	-7.39(1)	-7.49(1)			
		pw	-9.45(1)	-10.13(1)			
0.600	52.307				apw	-9.46(1)	
0.500	90.334	pw	-5.13(2)	-5.64(1)	pw	-3.17(1)	
					apw	-5.12(1)	
0.226	978.220	pw	61.70(4)	61.50(1)			

RESULTS AND DISCUSSION

Quantum Monte Carlo calculations were performed on a supercell with periodic boundary conditions, the atoms being located at fixed crystal lattice sites. Although fixed atomic sites are not required by the quantum Monte Carlo algorithm, in contrast to band theory methods, in practice regular BCC or FCC structures were simulated. Present calculations are not sufficiently accurate to determine the type of crystal structure. Different structures were used primarily to determine the number dependence correction. Simulations were performed on cubic supercells consisting of 48 or 162 electrons (16 or 54 atoms) for the BCC structure and 96 electrons (32 atoms) for the FCC structure. Experimental data as well as band theory calculations indicate a close packing structure (HCP or FCC) for the 0° K isotherm at low pressures [4] [16] [17] [18].

Preliminary calculations of lithium using 16 atoms per simulation box and simple trial functions were performed over a compression range from 0.263 to 1000. Mixed Gaussian and plane wave trial functions were used at low to intermediate densities and planes waves alone at extreme compressions. The results are shown in Table 1 and Figure 1. The large energy gap between the variational and diffusion energies for the plane wave trial function at intermediate compressions indicates that the trial function is not of an optimal form. It is necessary to reintroduce the electron-nuclear Jastrow factor and multiply both Jastrow terms

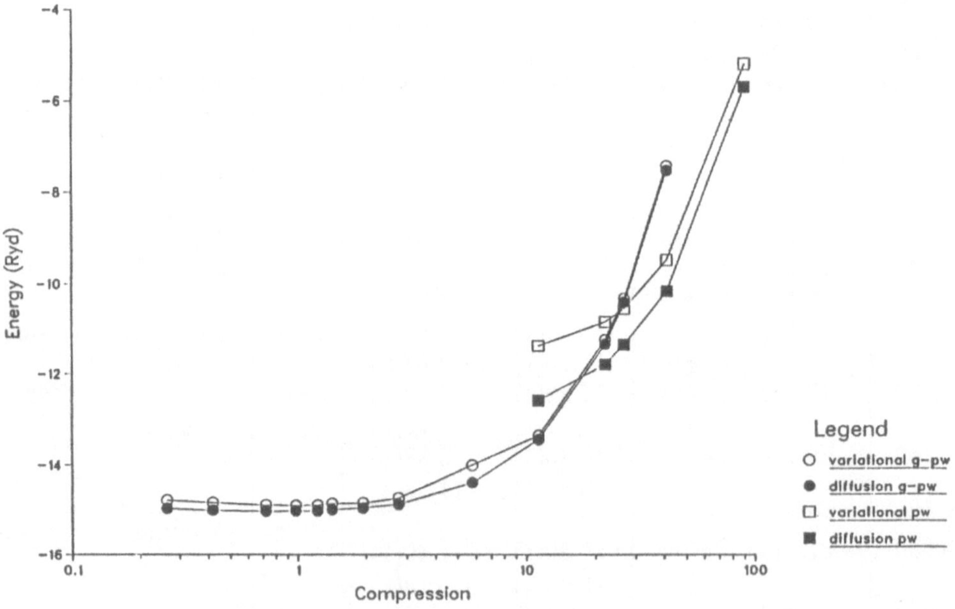

Figure 1. Variational and diffusion energies vs. Compression for quantum Monte Carlo calculations of lithium using 16 atoms in the simulation box. Trial functions are Gaussian 1s, plane wave 2s-2p (g-pw) and plane wave 1s, 2s-2p (pw) as indicated. Error bars are too small to be seen on this scale.

by a variationally optimized parameter in order to incorporate more of the pair correlation energies. This should reduce the difference in the variational and diffusion energies as well as the discontinuity in the energies of the two types of trial functions at the crossing of the energy curves. This crossing near 25-fold compression indicates the approximate location of the delocalization of the 1s core states. The result is consistent with band theory estimates of delocalization in the 30-300Mbar range.

Figure 2 shows the energies near normal density. The minimum of the curve gives an estimate for the equilibrium lattice constant which is consistent with the accepted value of 6.48 a.u. For low to intermediate densities, the difference between the variational and diffusion energies is on the order of 0.1Rydbergs (see Table 1). Previous quantum Monte Carlo simulations of the electron gas at comparable densities yielded a lowering of the energy of 0.004Rydbergs [10], indicating that the electron gas is a much simpler system than lithium. However, calculations of Li_2 showed a variational-diffusion energy difference of about 0.04Rydbergs [12]. Thus the trial functions used here for metallic lithium are of comparable quality to those used in the molecular case. The number dependence correction for metallic lithium is of the same order of magnitude as the variational-diffusion energy difference (compare the results for 16 and 32 atoms in Table 1). At normal compression, the size correction of .1Rydbergs is the same as for the electron gas of the same density.

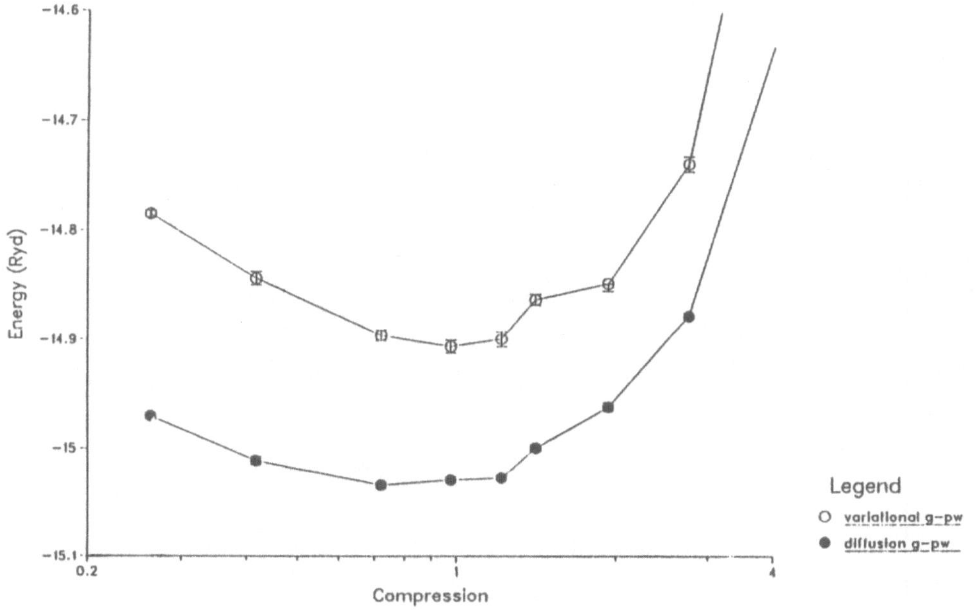

Figure 2. Variational and diffusion energies vs. Compression for quantum Monte Carlo calculations of lithium using 16 atoms in the simulation box near normal density and Gaussian 1s, plane wave 2s-2p (g-pw) trial functions.

The calculations have been carried out for crystals expanded to 4 times the normal density using trial functions with Gaussian $1s$ and plane wave $2s$ orbitals. Additional simulations are underway using localized states (Hermann-Skillman, simple Gaussian, or Wannier) for all the electrons. The density at which the energy curves for the two kinds of trial functions cross will then locate the Mott transition - the delocalization of the $2s$ states.

In order to study the core delocalization regime and for comparison with band theory, a 32 particle crystal was simulated using both simple and APW trial functions. The APW trial function was a factor of 4-5 times slower to sample than the simple Gaussian-plane wave form. This was offset by the significant lowering of the variational energies (see the preliminary results shown in Figure 3 and Table 1) and should also be reflected by a more rapid convergence in the diffusion calculations. To compare with band theory, the energies relative to that at normal density are plotted in Figure 4. The variational quantum Monte Carlo values are in reasonable agreement with both the ASW [3] and the LMTO [2] results. At extreme compressions, the variational values satisfy the $PV^{5/3}$ equation of state of Thomas-Fermi-Dirac models.

More calculations are required before the quantum Monte Carlo algorithm can be used to assess the accuracy of band theories. The Jastrow terms for both the simple and the APW trial functions must be optimized and used in diffusion and Green's function runs at selected densities. These simulations are presently underway. Dependence of the results on the number of particles must be taken into account by extrapolation to the infinite particle number limit using a fit for several supercell sizes at several densities [10]. The number dependence may not

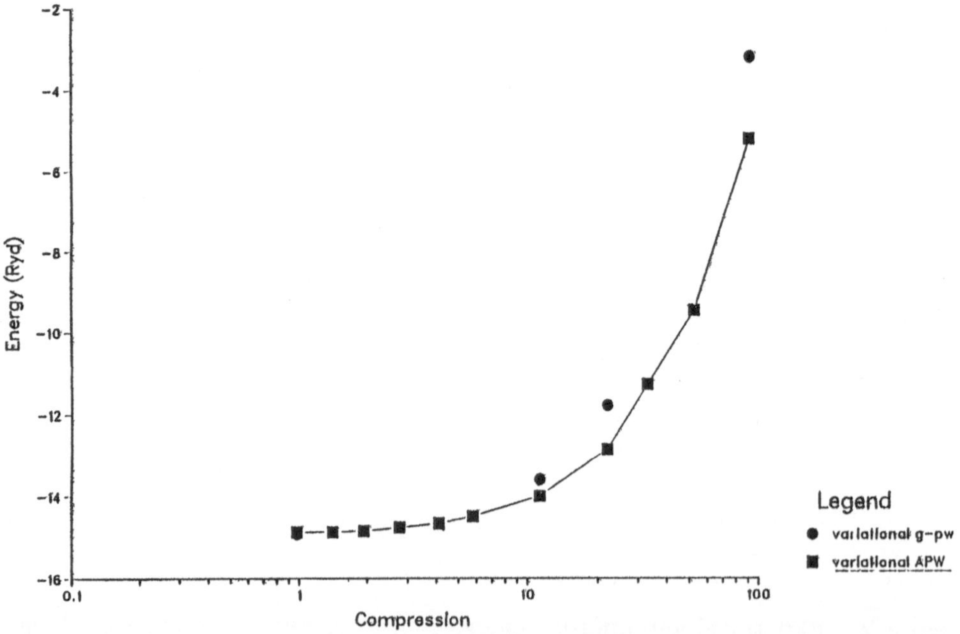

Figure 3. Preliminary variational energies vs. Compression for quantum Monte Carlo calculations of lithium using 32 atoms in the simulation box and Gaussian- plane wave (g-pw) or APW trial functions.

be the same for different trial function forms or for different numbers of delocalizated electrons. Particular emphasis will be given to the delocalization regimes of the $2s$-$2p$ electrons at low density (Mott transition) and the $1s$ electrons at high density - regions where the largest correlation effects should occur.

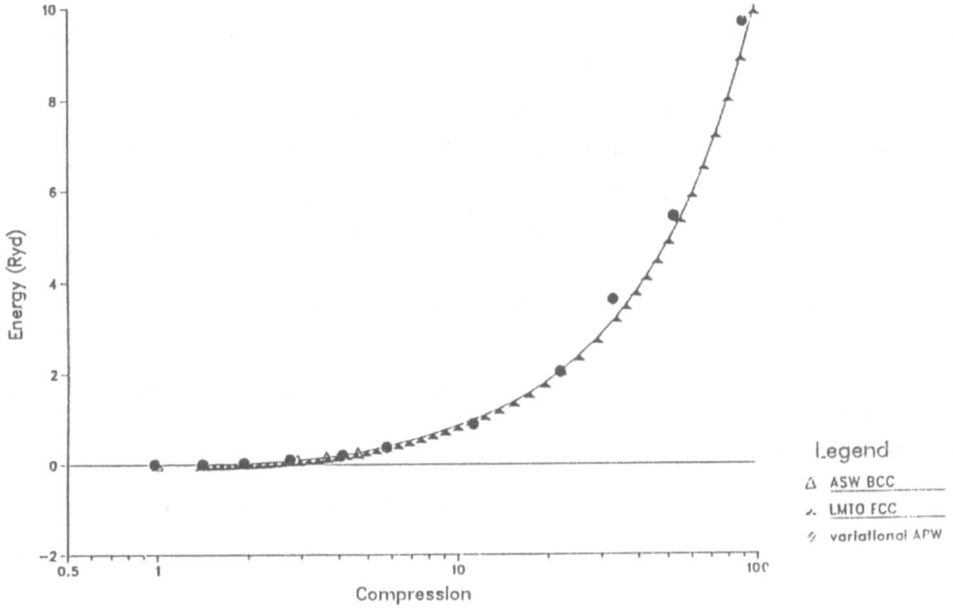

Figure 4. Relative Energy vs. Compression for band theory ASW [3] and LMTO [2] calculations and variational quantum Monte Carlo calculations using APW trial functions.

ACKNOWLEDGEMENTS

We wish to express our appreciation to M.Bukovinski for supplying the APW codes, to A. McMahan for his advice on band theory and especially to D.Ceperley for sharing his expertise on quantum Monte Carlo. We also wish to thank the San Diego Supercomputing Center for providing computer time.

This work was performed under the auspices of the Department of Energy Lawrence Livermore National Laboratory Contract No. W-7405-ENG-48.

REFERENCES

[1] D.Liberman, Colloques Internationaux de la Centre de la Recherche Scientifique No. **188**: 35 (1969)

[2] A.McMahan, private communication

[3] W.Zittel, J.Meyer-ter-Vehn, J.Boettger and S.Trickey, J. Phys. **F15**: L247 (1985)

[4] J.Boettger and S.Trickey, Phys. Rev. **B32**: 3391 (1985)

[5] A.Bakanova, I.Dudoladov, R.Trunin, Soviet Physics - Solid State **7**: 1307 (1965)

[6] R.Grover, R.Keeler, F.Roger, G.Kennnedy, J. Phys. Chem. Solids **30**: 2091 (1969)

[7] R.M.More, Phys. Rev. **A19**: 1234 (1979)

[8] F.Perrot, Physica **98A**: 555 (1979)

[9] D.Ceperley and B.J.Alder, Science **231**: 555 (1986)

[10] D.Ceperley and B.J.Alder, Phys. Rev. Lett. **45**: 566 (1980)

[11] D.Ceperley, Phys. Rev. **B18**: 3126 (1978)

[12] P.Reynolds, D.Ceperley, B.J.Alder, W.Lester, J. Chem. Phys. **77**: 5593 (1982)

[13] D.Ceperley and B.Alder, J. Chem. Phys. **81**: 5833 (1984)

[14] c.f. Reference [9] and D.Ceperley and B.Alder, publication in preparation

[15] L.Mattheis, J.Wood, A.Switendick, **Methods in Comp. Physics**, Vol. 8: 63 (1968)

[16] H.Skriver, Phys. Rev. **B31**: 1909 (1985)

[17] D.Young and M.Ross, Phys. Rev. **B29**: 682 (1984)

[18] T.Lin and K.Dunn, Phys. Rev. **B33**: 807 (1986)

THOMAS-FERMI THEORY AND ITS GENERALIZATIONS, APPLIED TO STRONGLY COUPLED PLASMAS

G. Senatore*† and N.H. March§

*Departmentof Physics and Astronomy
University of Kentucky
Lexington, KY 40506 USA

§Theoretical Chemistry Department
University of Oxford
1 South Parks Rd.
Oxford OX1 3TG U.K.

1. INTRODUCTION

Linear response theory, in liquid metals[1] and some other plasma problems, has been of major importance through the k and ω dielectric functions to which it leads. Thus, liquid metal Na and K, the former of considerable importance as a coolant in nuclear reactors, have such weak electron-ion interaction that their conduction electrons are well treated by such an approach. Indeed, the uniformity of the conduction electron distribution has led to the classical, one-component plasma model being a valuable reference liquid for treating the ion-ion pair correlation function in these metals, both in the context of liquid structure[2] and of freezing criteria.[3]

However, there are important problems, in condensed matter and in plasma physics, where such a linear response theory is not appropriate. Thus, in liquid metallic hydrogen, or in hydrogen-helium mixtures of interest in astro-physics, the electron-ion scattering is so strong that any linear theory must break down severely. In metals this was already clear from the early Hartree-Field computations on a proton embedded in an electron fluid[4], or indeed from the relative constancy of positron lifetimes in metals, in spite of a huge variation of mean conduction electron density from metallic Aℓ (mean inter-electronic spacing $r_s \sim 2a_0$ to Cs ($r_s \sim 5.5a_0$), the density $\rho_0 = 3/4\pi r_s{}^3$ varying by a factor ~20 between these two metals.

Therefore, in the present article, we shall consider the application of Thomas-Fermi theory, and its generalizations, to a number of problems involving strong electron-ion interaction. In three-dimensional problems, to which we restrict ourselves throughout, this theory has the considerable merit of non-linearity, as evidenced by the density $\rho(\vec{r})$ -potential $V(\vec{r})$ relationship in an inhomogeneous electron gas:

$$\rho(\vec{r}) = \frac{8\pi}{3h^3} (2m)^{3/2} \left[\mu - V(\vec{r})\right]^{3/2} \tag{1.1}$$

with μ as usual denoting the chemical potential of the electron cloud[4]. Of

†On leave from the University of Trieste, Italy.

course, this merit of immediate non-linearity exposed in Eq. (1.1) is purchased at the price of a theory which is useful for predicting the density only in classically allowed regions of electronic motion, and correspondingly, the energy level spectra predicted by the theory can only be fully quantitative in the regime of applicability of Bohr's correspondence principle.

The outline of our article is then as follows. In Section 2 immediately below we summarize first our own work on the energy level spectra of heavy atoms in a non-relativistic framework[5] using the self-consistent potential energy V(r) calculated by combining Eq. (1.1) with Poisson's equation of electrostatics . In the second part of Section 2, we briefly review some recent progress relating to collective oscillations in the charge cloud of heavy atoms, which can be treated by dynamical Thomas-Fermi theory and its generalizations. Then, in Section 3, an approach to liquid metals based on a generalization of Thomas-Fermi theory is presented[6], with results of an application to liquid metal Be. In Section 4, recent ideas for treating electron-electron exchange and correlation interactions will be reviewed with specific reference to the degenerate quantum fluid jellium[7], and these will be discussed in relation to the work of Cowan and Kirkwood[8] who used Thomas-Fermi theory to discuss the correlation functions in liquid metal and strongly coupled plasmas. Finally, Section 5 constitutes a summary, plus some suggestions for possible future studies in the area covered in this paper.

2. THOMAS-FERMI-THEORY OF ISOLATED ATOMS: STATICS AND COLLECTIVE MODES

2.1. WKB eigenvalues for heavy atoms in the limit of large quantum numbers

The Thomas-Fermi theory of heavy atoms was the forerunner of modern density functional theory[9]. Though its prediction must be expected to be only of qualitative character, it remains of some interest for first-principles theory. For instance, its connection with the important 1/Z expansion of Hylleeras[10] was established by March and White[11]. Also, the total energy of atoms is usefully approximated by the semiclassical Thomas-Fermi method, provided suitable corrections are applied.[12,13]
The Thomas-Fermi theory of an isolated atom (ion) is based on the combination of the density-potential relation of Eq. (1.1) with the Poisson's equation

$$\nabla^2 V(\vec{r}) = -4\pi e^2 \rho(\vec{r}) \quad , \tag{2.1}$$

supplemented by the appropriate boundary conditions. This yields a non-linear differential equation[4],

$$\frac{d^2\phi}{dx^2} = \frac{\phi^{3/2}}{x^{1/2}} \quad , \tag{2.2}$$

for the screening function ϕ, defined by

$$V(r) - \mu = -\frac{Ze^2}{r} \cdot \phi(x) \quad , \tag{2.3}$$

with x a dimensionless measure of length,

$$x = r/b \;\; ; \;\; b = \alpha Z^{-1/3} a_0 \;\; ; \;\; \alpha = (9\pi^2/2)^{1/3}/4 \tag{2.4}$$

The solution of Eq. (2.2) is all what is needed in the Thomas-Fermi theory of atoms. In fact, one can express all the atomic properties in terms of ϕ. Thus, for a neutral atom, the total binding energy E(Z) is easily related to the slope $a_2 \equiv \phi'(0)$ by

$$E(Z) = -0.48 a_2 Z^{7/3} e^2/a_o = -0.7687 Z^{7/3} e^2/a_o \; . \qquad (2.5)$$

The price paid to the great simplicity of such a description, in which all properties of the atom scale with a suitable power of Z, is represented by the absence of any discrete structure in the resulting electronic properties. In fact, Thomas-Fermi theory only provides an "average" description of the many-electron system. To overcome such a situation, Fermi[14] took the next step of regarding the potential V(r) of Eq. (2.3) as a non-selfconsistent effective one-body potential that could be used to calculate one-electron terms by solving the relative Schrodinger equation. Unfortunately, this must be done numerically[10]. However, in the limit of Z large (strictly $Z \to \infty$) some progress can be made if one resorts to the semiclassical WKB method, as we have recently shown[5]. This will be briefly discussed below.

The WKB condition determining the one-electron energies in the potential V(r) of Eq. (2.3) for a neutral atom ($\mu = 0$) can be easily written as

$$\pi(s + \tfrac{1}{2}) = Z^{1/3} \int_{x_1}^{x_2} dx \left[2\alpha \frac{\phi(x)}{x} - \frac{(\ell + 1/2)^2}{x^2} Z^{-2/3} + \varepsilon \alpha^2 Z^{-4/3}\right]^{1/2} \equiv Z^{1/3} I(Z,\ell,\varepsilon) \; . \qquad (2.6)$$

In the above equation, the scaled length x is used, x_1, x_2 are the turning points of the classical motion and the energy level ε is in Rydberg. For given Z, Eq. (2.6) implicitly gives ε as function of the integers s and ℓ. If one restricts to ε and ℓ of order Z^o, it can be shown that in the limit $Z \to \infty$

$$I(Z,\ell,\varepsilon) = \pi K + \pi i(\ell,\varepsilon) Z^{-1/3} + \ldots \; , \qquad (2.7)$$

with

$$K = \pi^{-1} \int_o^\infty dx \left(2\alpha \frac{\phi(x)}{x}\right)^{1/2} = 1.6566 \qquad (2.8)$$

and

$$i(\ell,\varepsilon) = -(\ell + \tfrac{1}{2}) \left[1 + 2f\left(\frac{\varepsilon}{(2\ell + 1)^4}\right)\right] \; . \qquad (2.9)$$

The function f appearing above is explicitly given by

$$f(y) = (1-2t)^{-1/2} \left[2E(t) - K(t)\right] / 2\pi \; , \qquad (2.10)$$

where ·

$$t = (1/2)\{1-[1-(36\pi)^2 y]^{-1/2}\} \qquad (2.11)$$

and E(t), K(t) are complete elliptic integrals of first and second kind. Substitution of Eq. (2.7) in Eq. (2.5) immediately yields, in leading order in Z, the scaled relation

$$\frac{\varepsilon}{(2\ell + 1)^4} = f^{-1}\left[(KZ^{1/3} - n)/(2\ell + 1)\right] \qquad (2.12)$$

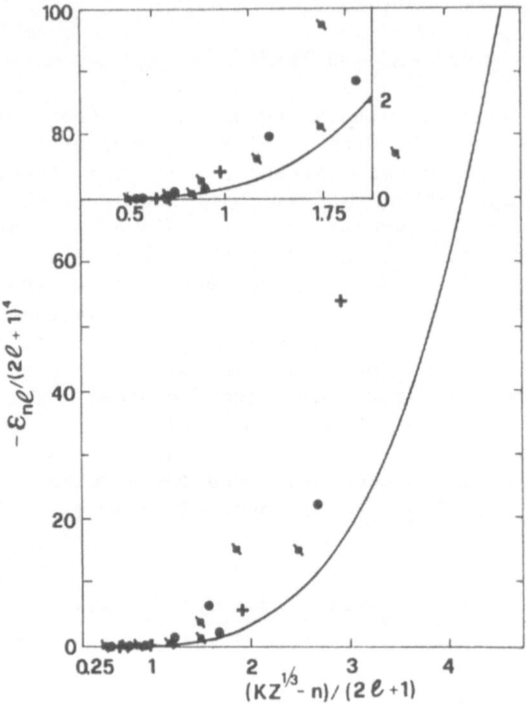

Fig. 2.1 . Comparison of the asymptotic WKB energy formula (2.10) for the Thomas-Fermi potential and numerical solution of the Schrodinger equation by Latter[10] for three values of Z in the range of the periodic Table. The full curve gives eqn. (2.12) of the text while Latter results are given by +, Z=26; ◆, Z = 65; and ●, Z = 92.

between ε and Z, ℓ, and the hydrogenic principal quantum number $n = s+\ell+1$.

We should observe that, while Eq. (2.12) gives the WKB one-electron energies in the Thomas-Fermi screened potential for Z sufficiently large and ε and ℓ finite, caution is of course needed in that the WKB levels only become precise in Bohr's correspondence limit of large quantum numbers. In Figure 2.1 the scaled relation (2.12) is shown together with results of numerical solutions of the Schrodinger's equation due to Latter[15]. Though such numerical results are for values of Z in the range of the Periodic Table and , so, not large, a reasonable agreement is found.

Another simple result, which can be simply obtained from Eq.(2.6), is a necessary condition on Z for the existence of bound states with a given ℓ. This is obtained by requiring the existence of two turning points. One finds

$$Z > 0.15641(2\ell + 1)^3 \qquad (2.13)$$

The above inequality has been previously discussed[16,17], independently from the WKB condition (2.6). Comparison with the periodic table reveals that formula (2.13) has an accuracy of few percent.

2.2. Collective oscillations of electronic cloud in atoms

Here, we shall briefly summarize recent progress on collective modes of the inhomogeneous electron gas in heavy atoms. The Thomas-Fermi method was generalized by Bloch[18] to yield a hydrodynamic theory of the collective oscillations of an inhomogeneous electron gas. This theory, and its applications, is reviewed by Lundquist[19]. The generalization of this theory in a fully quantum mechanical framework has been given by March and Tosi[20]; applications of this latter theory are now forthcoming over a wide range of systems[21,22]. Of course, it has been clear to Bloch, to Wheeler[23], and man other workers, that while the plasmon is a very well defined oscillation in extended systems like, say, Aℓ, or even Si, in finite systems the question of the width of the excitation is crucial. However, recent experiments at Los Alamos with strong laser field (Baldwin; private communication) give strong circumstantial evidence for atomic excitations in various species with large atomic number, which seem to have no natural explanation to date other than that the atomic electrons are participating collectively. The simplest viewpoint to take is that one is seeing here collective electronic density oscillations in the charge cloud of heavy atoms.

To give a full theory of such excitations, transcending the work of Bloch and other later workers is still difficult, though it has been tackled by very different, and highly complex, many-electron theories by Ambrosia[24], Fano[25], Wendin[26] and their co-workers. We want to note here, that in the spirit of liquid structure descriptions, used also, of course in Section 3 below, if we take the electron pair correlation function $g_{ee}(r)$ and its Fourier transform $S_{ee}(k)$, and define the direct correlation function for electrons by

$$c_{ee}(k) = \frac{S_{ee}(k) - 1}{S_{ee}(k)} \tag{2.14}$$

then by analogy with jellium, where

$$c_{ee}(r) \to \frac{e^2}{r(h\omega_p/2)}\ , \tag{2.15}$$

with ω_p the electronic plasma frequency, given by

$$\omega_p = \left\{\frac{4\pi e^2 n_o}{m}\right\}^{1/2}, \tag{2.16}$$

with n_o the mean electron density, we must expect in an atom that a 'local plasma frequency $\omega_p(\vec{r})$ say, defined by replacing n_o in Eq. (2.16) by the local density $n(r)$, will enter the long-range behavior of the direct correlation function. Such behavior should be accessible at least for heavy closed shell atoms, by electron scattering from such gaseous systems. This should enable progress to be made in pinning down the theory of the asymptotic form of the direct correlation between electrons in finite atomic systems.

3. INFLUENCE OF ORDER ON ELECTRON STATES IN STRONGLY COUPLED LIQUID METALS

From the problems of section 2, involving electrons coupled strongly to one atomic centre, we turn next to multicentre problems. To be specific, though the method we present below is of much wider applicability, we shall focus on the example of liquid metal Be. This is a strongly coupled liquid metal, with the 2 atomic 2s electrons forming in the condensed state a broad s-p conduction band, the 1s electrons on each atomic nucleus being in bound states, and having therefore atomic-like wave functions and discrete energy levels differing little from the 1s state of an isolated Be atom.

Below, we shall attempt to construct a description of the strongly coupled Be metal, starting again from a 'one-centre' building block. Obviously, since the 2s atomic electrons are now in a 'shared' conduction band characteristic of the liquid metal Be itself, it would be inappropriate to take the free atom, with configuration $(1s)^2(2s)^2$ as the building block. Rather, following ideas, though not methods, then fashionable in condensed matter, Rousseau, Stoddart and March (RSM) proposed[6] to plunge a single Be into an originally uniform electron gas with a mean electron density appropriate to the conduction electrons in liquid metal Be. This 'screened ion', with localized 1s electrons in atomic states, is then the desired building block. We can say that the crucial difference between this 'screened ion' building block and the free atom is that in the former the Fermi level falls in the continuum, i.e. the chemical potential μ in Eq. (1.1) is positive, if $V(r)$ is defined to tend to zero at infinity, whereas in the free atom the highest electronic state is bound, i.e. has a negative energy.

Having decided on the basic building block as a screened ion, the question, of course, is how to use it to construct some, or ideally all, of the properties of the liquid metal. The argument of RSM went as follows. They started, in essence, from the Thomas-Fermi electron density in Eq. (1.1), and enquired how to use this in a disordered system such as a liquid metal. They then emphasized that two essential steps occur in any non-perturbative calculations of electron states: (i) The electron states must be found for a given fixed configuration of ions, say $\{\vec{R}_i\}$, $\vec{R}_i$ denoting the position of the ith ion ($i=1, 2, \ldots, N$) in this particular configuration. (ii) The ensemble average with respect to ionic configurations must be taken. Following RSM, it will be assumed that the total potential $V(\vec{r})$ to be inserted in Eq. (1.1) to get the Thomas-Fermi approximation $\rho(\vec{r})$ to the electron density in configuration $\{\vec{R}_i\}$ has the form

$$V(\vec{r}) = \sum_{\vec{R}_i} U(\vec{r} - \vec{R}_i) \tag{3.1}$$

where U evidently characterizes, in a manner to be made precise below, the 'screened ion' building block discussed above. Of course, the assumption in Eq. (3.1) that U is not affected by the local environment in which the i-th ion finds itself must involve approximation, but available evidence points to it as a reasonable starting point for building a theory of the electron states in a liquid metal like Be.

3.1. Model of independent 'screened ions'

Of course, it will already be clear to the reader that the fundamentally non-linear ρ-V relation in Eq. (1.1) will not allow one to exploit the assumed superposition property (3.1) unless $|V|/\mu \ll 1$, when we can linearize Eq. (1.1). It is a basic tenet of the present work that we are dealing with just such strongly coupled plasmas that the linearization is not valid.

RSM therefore utilized as their main tool for handling the electron states in a liquid metal possessing only short-range order the so-called canonical density matrix $C(\vec{r},\vec{r}_o,\beta)$. This is defined in terms of the eigenfunctions and eigenvalues ε_i of the one-electron Hamiltonian

$$H = -\frac{1}{2}\nabla^2 + V(\vec{r}) \ , \tag{3.2}$$

with $V(\vec{r})$ as specified by the approximation (3.1), as

$$C(\vec{r},\vec{r}_o,\beta) = \sum_i \psi_i^*(\vec{r})\psi_i(\vec{r})\exp(-\beta\varepsilon_i) \quad ; \quad \beta = 1/K_BT \ . \tag{3.3}$$

We shall see below that using this tool the superposition property of $V(\vec{r})$ expressed through Eq. (3.1) can be exploited. The density matrix (3.3) is, of course, related to the partition function $Z(\beta)$ through

$$Z(\beta) = \int d\vec{r}\ C(\vec{r},\vec{r},\beta) = \sum_i \exp(-\beta\varepsilon_i) \ . \tag{3.4}$$

Next we note that, when $V(\vec{r})$ varies slowly, an approximation underlying the Thomas-Fermi theory, the wavefunctions are essentially unchanged while the eigenvalues are shifted by an energy increment V, and

$$C(\vec{r},\vec{r}_o,\beta) = C_o(\vec{r},\vec{r}_o,\beta)\ \exp[-\beta V(\vec{r})] \ , \tag{3.5}$$

C_o denoting the free-electron limit $V = 0$. Now one sees from Eq. (3.5) that the superposition form of $V(\vec{r})$ in Eq. (3.1) leads to the huge simplification

$$C(\vec{r},\vec{r},\beta) = C_o(\vec{r},\vec{r},\beta) \prod_{\vec{R}_i} \exp[-\beta U(\vec{r} - \vec{R}_i)] \ . \tag{3.6}$$

Noting the relation between C and $Z(\beta)$ from Eq. (3.4), we can conclude that, loosely, Eq. (3.6) is expressing the fact that the total partition function for configuration $\{ \vec{R}_i \}$ is a product of partition functions for the one-centre building blocks characterized by potential $U(\vec{r})$.

We now generalize (3.6), as proposed by Stoddart, Hilton and March[27], by calculating the 'one-centre' density matrix, $C_1(\vec{r},\vec{r}_o,\beta)$ say, for the single-centre scattering off $U(\vec{r})$, in the form

$$C_1(\vec{r},\vec{r}_o,\beta) = C_o(\vec{r},\vec{r}_o,\beta)\ \exp[-\beta U_1(\vec{r},\vec{r}_o,\beta)] \tag{3.7}$$

where $U_1(\vec{r},\vec{r}_o,\beta)$ is termed the effective potential matrix. The way to calculate this was discussed by Hilton, March and Curtis[28] and their results for $U_1(\vec{r},\vec{r},\beta)$ will be invoked below. It is this quantity $U_1(\vec{r},\vec{r},\beta) \equiv U_1(\vec{r},\beta)$ which constitutes our precise definition of the 'screened ion' referred to above.

Then the model of indipendent 'screened ions' motivated by the form (3.6), can be expressed quite explicitely as

$$C(\vec{r},\vec{r},\beta) = C_o(\vec{r},\vec{r},\beta) \prod_{\vec{R}_i} \exp[-\beta U_1(\vec{r} - \vec{R}_i,\beta)] \ . \tag{3.8}$$

All subsequent results for the electron states in strongly coupled plasmas to be discussed in the present paper rest on the model embodied in Eq. (3.8). Given the starting point for $V(\vec{r})$ in eqn. (3.1), the work of Ref. 27 shows the model to work well under any one of the following condition: (i) U small, (ii) β small, (iii) U_1 slowly varying, (iv) small overlap of ∇U_1 between different positions $\vec{R}_i$.

We shall discuss at the end of this Section corrections which can be applied to Eq. (3.8), should it prove necessary).

3.2. Ensemble average

Following RSM, we can make progress with the ensemble averaging of Eq. (3.8) by introducing, prompted by analogy with the Mayer function of classical statistical mechanics, the related function

$$f(r,\beta) = \exp[-\beta U_1(r,\beta)] - 1 . \tag{3.9}$$

Then one expands Eq. (3.8) in the form

$$C(\vec{r},\vec{r},\beta) = C_0(\vec{r},\vec{r},\beta)\ [1 + \sum_{\vec{R}_i} f(\vec{r} - \vec{R}_i,\beta)$$

$$+ \frac{1}{2!} \sum_{\vec{R}_i \vec{R}_j}{}' f(\vec{r} - \vec{R}_i,\beta)\ f(\vec{r} - \vec{R}_j,\beta) + \ldots] . \tag{3.10}$$

It is this expression that RSM used to perform the ensemble average.

3.2.1. Random systems

It will be useful, before tackling the short-range order in a liquid metal, to consider random averaging. One readily obtains the result[6] for the partition function $Z_r(\beta)$:

$$Z_r(\beta) = (2\pi\beta)^{-3/2}\exp[\beta\alpha_r(\beta)] \tag{3.11}$$

where $\alpha_r(\beta)$ for the random assembly is simply given by

$$\alpha_r(\beta) = \frac{\rho}{\beta} \int d\vec{r}\ \ f(\vec{r},\beta) \quad , \tag{3.12}$$

ρ being the number of ions per unit volume. Formulae (3.9), (3.11) and (3.12) relate explicitly the partition function of the random assembly to the 'screened ion building block', defined by $U_1(\vec{r},\beta)$ entering Eq. (3.9).

3.2.2. Liquid metals

As RSM point out, an attempt to perform the configuration average of Eq. (3.10) using the Kirkwood superposition approximation for the three-atom correlation function $g^{(3)}(\vec{r}_1,\vec{r}_2,\vec{r}_3)$, namely

$$g^{(3)}(\vec{r}_1,\vec{r}_2,\vec{r}_3) = g(r_{12})\ g(r_{23})\ g(r_{31}) \quad , \tag{3.13}$$

where $g(r)$ is the usual ionic pair correlation function of the liquid metal, accessible via neutron scattering, leads to a situation in which only one class of terms can be summed, which on examination does not prove dominant.

Thus, it has proved necessary, to date, to have recourse to a simplified version of the Kirkwood approximation, in which one of the pair functions in Eq. (3.13) is replaced by its asymptotic value, and the result is symmetrized to yield

$$g^{(3)}(\vec{r}_1,\vec{r}_2,\vec{r}_3) = \frac{1}{3}\ [g(r_{12})g(r_{13}) + g(r_{21})g(r_{23})$$

$$+ g(r_{31})g(r_{32})] . \tag{3.14}$$

Such a procedure is related to that of Abe[29] in passing from the

Born-Green theory of structure to the hypernetted chain theory. Furthermore, the work of Bratby, Gaskell and March[31] has also revealed serious shortcomings of the Kirkwood approximation, and again an (unsymmetrized) form related to Eq. (3.14) is more useful.

Generalizing the type of approximation (3.14) to the n-th order correlation function, RSM sum all the terms in the average of eqn. (3.10), to obtain in place of the random result in Eqs. (3.11) and (3.12):

$$\exp[\beta\alpha_{liq}(\beta)] = \int d\vec{r}_1 \; f(\vec{r}_1,\beta) \; [\exp\{\rho G(\vec{r}_1)\} - 1\,]/G(\vec{r}_1) \; + 1 \tag{3.15}$$

where

$$G(\vec{r}_1) = \int d\vec{r}_2 \; f(\vec{r}_2,\beta) \; g(r_{12}) \; . \tag{3.16}$$

Evidently, Eq. (3.15) reduces to the random 'screened ion' result in Eqs. (3.11) and (3.12) when the liquid ion-ion pair function $g(r)$ is replaced by unity.

3.2.3. Summary of the results for liquid metal Be

To our knowledge, the only calculation based on Eqs . (3.15) and (3.16) that are available so far are for liquid Be. Here, $U_1(\vec{r},\beta)$ defining the screened ion is available in approximate form from the work of Ref. 28. Figure 3.1 shows the results of RSM, for $\alpha(\beta)$ versus β for this metal. Curve 1 is the random result, 2 is for the liquid metal and 3 is for face-centred cubic Be at the same density. In the liquid case, because of the toxicity of this metal, we know of no measured pair function $g(r)$, and the hard sphere model was employed by RSM.

In principle, the partition function $Z(\beta)$ can be inverted to obtain the density $n(E)$ of electronic states. In practice, this problem has not been completely solved, as the partition is known only to finite numerical accuracy on the real axis. However, limited progress can still be made by choosing models for the density of states which can be analytically trasformed, and then compared with the calculated partition functions. In this way, RSM obtained approximate results for $n(E)$ in liquid Be which seem to be quite sensible and testify to the usefulness of the above approach in strongly coupled liquid metals.

3.2.4. Trascending independent screened ions

RMS have subsequently pointed out[6] a route for trascending the model of independent screened ions. The orthonormality of the wave functions in the definition of the canonical density matrix leads, quite generally, to the result

$$C(\vec{r},\vec{r}_o,\beta_1 + \beta_2) = \int d\vec{r}_1 \; C(\vec{r},\vec{r}_1,\beta_1) \; C(\vec{r}_1,\vec{r}_o,\beta_2) \; . \tag{3.17}$$

In the special case $\beta_1 = \beta_2 = \beta$, this equation evidently relates C at β to its value at 2β. If one substitutes into the right-hand-side of the above equation the approximate C based on independent screened ions, at a sufficiently small value of β where it is valid, then without further approximation we have the Bloch matrix at 2β.

Then RSM show that one obtains the 'liquid' partition function $Z_1(2\beta)$ as, essentially

$$\int d\vec{y} \int d\vec{x}_1 \; \Big(C_o(\vec{y},\beta)^2 \; f(\vec{x}_1,\vec{x}_1 + \vec{y},\beta) \; \{\exp[\rho G(\vec{x}_1,\vec{x}_1 + \vec{y},\beta)] - 1\}/G(\vec{x}_1,\vec{x}_1 + \vec{y},\beta) \Big) \tag{3.18}$$

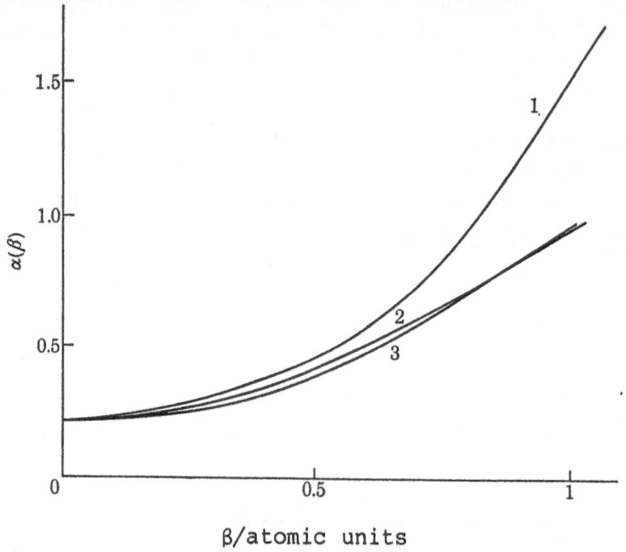

Fig. 3.1 . Partition function for method of independent pseudoatoms. Curve 1, Random assembly; 2, liquid metal; 3, crystal.

where f and G are off-diagonal generalizations of the diagonal quantities of Section 3.2.2. It should be stressed that in writing the above formula, nothing is changed concerning the higher-order correlation functions from the previous treatment of Section 3.2.2, but the quantum mechanics has been refined.

4. ELECTRON-ELECTRON PAIR FUNCTION

We have been concerned in the condensed matter studies in Section 3 with treating the electron-ion interaction in a strongly coupled liquid metal plasma.

Under normal condintions, the electronic assembly, in a liquid metal constitutes an almost totally degenerate assembly, and for such an electron liquid we want to report briefly on some recent progress which may lead to simplification of currently available methods for treating electron-electron correlations in dense liquid.

4.1. Simple interpretation of Fermi or exchange hole

We follow Dawson and March[7] in taking a starting point the calculation by Wigner and Seitz[32] of the exchange hole around an electron in noninteracting, or extremely high density jellium. If one chooses to sit on an electron at the origin in an electron liquid of mean number density n_0, then relative to this origin the density of electrons, $n_0 g(r)$, is determined by

$$g(r) = 1 - \frac{9}{2}\left\{\frac{j_1(k_F r)}{kr}\right\}^2 , \tag{4.1}$$

$j_1(x) = (\sin x - x\cos x)/x^2$ being the first-order spherical Bessel function, while the Fermi wave number k_F is determined by $n_0 = k_F^3/3\pi^2$.

Utilizing the analysis by March and Murray[33] of the free-Fermion first-order density matrix into its orbital angular momentum components,

Dawson and March focus on the p-component of the constant density n_0, namely

$$n_{\ell=1}(r) = \frac{1}{4\pi^2 r^2}\left\{k + \frac{\sin 2kr}{2r} + \frac{\cos 2kr - 1}{kr^2}\right\}, \tag{4.2}$$

where the generalization to arbitrary Fermi energy $E = h^2k^2/2m$ has been effected. It is now a straightforward matter to re-express eqn. (4.1) as

$$g(r) = 1 - \frac{9\pi^2}{k_F{}^2 r^2}\left(\frac{\partial n_1}{\partial k}\right) = 1 - \frac{3h^2}{mn_0 r^2}\,\sigma_1(r,E_F) \tag{4.3}$$

where $\sigma_1 = (\partial n_1/\partial E)_{E_F}$ is the local density of states of the component at the Fermi level[7]. This Eq. (4.3) is another example of the widely accepted view that the properties of a Fermi gas are dominated by the Fermi level behaviour.

From eqn. (4.12) of the work of March and Murray[33], one can, by differentiation with respect to energy E, obtain a differential equation from which to determine $\sigma_1(r,E)$ in the presence of a potential energy V(r), assumed to be central in isotropic jellium. This equation is explicitely

$$\frac{1}{8}\frac{\partial^3}{\partial r^3}(r^2\sigma_1) - \frac{1}{r}\frac{\partial}{\partial r}(r\sigma_1) - \frac{1}{2}\frac{\partial V}{\partial r}r^2\sigma_1$$

$$- V\frac{\partial}{\partial r}(r^2\sigma_1) + E\frac{\partial}{\partial r}(r^2\sigma_1) = 0 \tag{4.4}$$

To our knowledge, the only available integration of this equation to date is to second-order in V, as discussed in Refs. 32 and 7: however in a strongly coupled plasma we must expect that one must solve Eq. (4.4) non-perturbatively. Some discussion of ways to determine the effective potential V(r) is given in Ref. 7, but more work is required on this point.

Our final comment in this section take us back to the pioneering work of Cowan and Kirkwood[8,34]. They applied the Thomas-Fermi theory to liquid metals and plasmas. Presumably, for totally degenerate and completely ionized plasma, one improvement of their work would be to build in the Fermi hole around an electron, suitably screened as the density is lowered. The simplified re-interpretation of the exchange hole given by Dawson and March[7] for jellium may subsequently offer a way forward in that context.

5. SUMMARY AND DIRECTIONS FOR FUTURE WORK

In this article we have been predominantly concerned with Thomas-Fermi theory and its generalization as applied to (a) isolated atoms and (b) liquid metals. As to (a), we have emphasized the utility of the Thomas-Fermi neutral atom potential in generating the one-electron energy level spectrum in the regime of large quantum numbers, where Bohr's correspondence principle applies. Of course, it is just for heavy atoms that relativistic theory is needed: our own recent work[33] may offer a way forward in treating such corrections to atomic energies. Secondly, we have noted important recent progress pointing rather strongly to collective electron density oscillations in the charged cloud of heavy atoms, and the relevance of dynamical Thomas-Fermi theory and its generalizations in this context.

In (b), a model of screened ions has been summarized, for liquid metals under normal conditions. This has so far been worked out only for liquid Be, with quite encouraging results. It does have the merit of clearly separating the liquid metal problem into two parts: (i) the calculation of the electron states for a given frozen ion configuration and (ii) the configuration averaging.

As to direction for future work, we have emphasized in atoms the importance of continuing work on the relativistic theory of heavy atoms. In our opinion, the Thomas-Fermi Theory in a relativistic context is important, particularly as regard the analytic structure of the total energy $E(Z,N,\alpha)$ of heavy atomic ions with nuclear charge Ze, N electrons and fine structure constant α. In relation to collective modes, the importance of studying electron-electron pair correlations in finite atomic systems is clear: a good treatment of correlations in the ground-state should have, via the Ornstein-Zernike correlation function, a finger-print of collective (plasmon-like) oscillations impressed on it.

With regard to liquid metals, the developments of quantal hypernetted chain theory by Chihara[36] and others needs to be related, for liquid metals, to the treatment given in Section 3 of the present article. This may help to clarify the nature of the approximation underlying it. It is not clear, in particular, to the present writers that the ensemble average is performed after calculating the electron states from the Schrodinger's equation in this procedure. At least in principle, this seems an important matter to be clarified. Finally, the work of Cowan and Kirkwood using Thomas-Fermi theory may still have merit for fully ionized plasmas, with appropriate generalization. In the completely degenerate limit of their work, it would, in particular, be of interest to build in the Fermi hole, along lines surveyed in Section 4 of this article.

Encouragment to review this area, which has been invaluable in bringing this paper to fruition, is acknowledged from Drs. H. E. De Witt (Livermore) and G. Kalman (Boston College) on the theoretical aspects and Dr. G. Baldwin (Los Alamos) relating to experimental findings on collective aspects in heavy atoms. One of us (NHM) wishes to acknowledge that a visit to the International Centre for Theoretical Physics, Trieste, Italy in early 1986 afforded him the opportunity to complete his contribution to this paper. GS acknoledges partial support from the NSF grant # DMR82-16212 .

REFERENCES

1. See, for example, N. H. March and M. P. Tosi, Coulomb Liquids (Academic Press: New York, 1984).
2. D. K. Chaturvedi, M. Rovere, G. Senatore and M. P. Tosi, Physica 111B, 11 (1981).
3. A, Ferraz and N. H. March, Solid State Commun. 36, 977 (1980).
4. See, for instance, N. H. March, Selfconsistent fields in atoms (Pergamon: Oxford, 1975).
5. G. Senatore and N. H. March, Phys. Rev. A32, 1322 (1985).
6. J. S. Rosseau, J. C. Stoddart and N. H. March, Proc. Roy. Soc. A317, 211 (1970); see also the refinement proposed by the same autors: J. Phys. C4, L59 (1971).
7. K. A. Dawson and N. H. March, Phys. Chem. Liq. 14, 131 (1984).
8. R. D. Cowan and J. G. Kirkwood, J. Chem. Phys. 29, 264 (1958); N. H, March, Nuovo Cim. 15B, 308 (1973).
9. See,e.g.,Theory of Inhomogeneous electron gas (Plenum: New York,1983), Eds. S. Lundquist And N. H. March.
10. E. Hylleraas, Z. Phys. 65, 209 (1930). See also: T. Kato, Commun.

Pure Appl. Math. 10, 151 (1957); D. Layzer, Ann. of Phys. (N.Y.) 8, 271 (1959).
11. N. H. March and R. J. White, J. Phys. B5,466 (1972).
12. N. H. March and J. S. Plaskett, Proc. Roy. Soc. A235, 419 (1956).
13. J. M. C. Scott, Phil. Mag. 43, 859 (1952).
14. E. Fermi and E. Amaldi, Mem. R. Accad. Ital. Cl. Sci. Fis. Mat. Nat. Fis. 6, 117 (1934).
15. R. Latter, Phys. Rev. 99, 150, 1854 (1955).
16. L. D. Landau and E. M. Lifshitz, Quantum Mechanics (Pergamon: London, 1958), p.245-247.
17. A. A. Abrahamson, Phys. Rev A4,454 (1971).
18. F. Bloch, Z. Phys. 81,363 (1933); Helv. Phys. Acta 7,385 (1934).
19. S. Lundquist in Ref. 9.
20. N. H. March and M. P. Tosi, Proc. Roy. Soc. 330,373 (1972); see also Phil. Mag. 28, 91 (1973).
21. R. Girlanda, M. Parrinello and E. Tosatti, Phys. Rev Lett. 36, 1386 (1976).
22. G. Giuliani. E. Tosatti and M. P. Tosi, J. Phys. C12, 2769 (1979).
23. J. H. Ball, J. A Wheeler and E. L. Fireman, Rev. Mod. Phys. 333, 45 (1973).
24. M. Ya Amusia, N. A. Cherepkov and L. V. Chernysheva, Sov. Phys. Jept 33,90 (1971); M. Ya Amusia, N. A. Cherepkov, R. J. Janev, Dj Živanovic, J. Phys. B7, 1435 (1971).
25. U. Fano and J. Cooper, Rev. Mod. Phys. 40, 441 (1968).
26. G. Wendin, J. Phys. B5, 110 (1972); J. Phys. B6, 42 (1973)
27. J. C. Stoddart, D. Hilton and N. H. March, Proc. Roy. Soc. A304, 99 (1968).
28. D. Hilton, N. H. March and A. R. Curtis, Proc. Roy. Soc. A300, 391 (1967).
29. See, for example, T. L. Hill, Statistical Mechanics: principles and selected applications (McGraw-Hill: New York, 1956).
30. R. Abe, Prog. Theor. Phys. 19, 57, 407 (1958).
31. P. Bratby, T. Gaskell and N. H. March,Phys. Chem. Liq. 2, 53 (1970).
32. E. P. Wigner and F. Seitz, Phys. Rev. 43, 804 (1933); Phys. Rev. 46, 509 (1934).
33. N. H. March and A. M. Murray, Phys. Rev. 120, 830 (1960).
34. See also, N. H. March and M. P. Tosi, Ann. of Phys. (N.Y.) 81, 414 (1973); M. P. Tosi, M. Parrinello and N. H. March, Nuovo Cim. B23, 135 (1974).
35. G. Senatore and N. H. March, Phys. Rev A32, 3277 (1985).
36. J. Chihara, Phys. Rev A33, 2575 (1986).

THE THOMAS-FERMI AND RELATED MODELS

Shalom Eliezer

Plasma Physics Department
Soreq NRC
Yavne 70600, Israel

I. INTRODUCTION

The Thomas-Fermi model is essentially a statistical model for the atomic electrons put forward by Thomas[1] (1927) and Fermi[2] (1928). Originally, the model was introduced to study a many electron atom system, however, since then, it has found important applications[3-7] in molecular theory, solid state theory and in determining the contribution from the electrons to the equation of state of matter at high pressure ($P \gtrsim 10^7$ atmospheres). The last application is of considerable interest in the inertial confinement fusion problem. The advantages of the Thomas-Fermi model over other models are, its simplicity, clarity and validity over a wide range of densities and temperatures.

The Thomas-Fermi model of the atom is based mainly on the following two assumptions: (i) the electrons are considered as a degenerate gas placed in a self-consistent electrostatic field described by the electrostatic potential V(r) which varies little over a de Broglie wavelength of the electron; and (ii) the field varies slowly enough so that we can consider a volume element $d\vec{r}$ which contains a large number of particles and at the same time the field can be assumed to be approximately constant in this volume dr. For a rigorous review see Ref. 8.

The Thomas-Fermi model describes the electronic system (in an atom, in a molecule, in a perfect or defect solid, in a compressed gas or liquid, etc.) in terms of the electron density $n(\vec{r})$, $\vec{r}$ denoting the position in space. In general, this electron density is observable (e.g. n(r) can be measured by x-ray scattering). In quantum mechanics the density is obtained from solving the Schrodinger equation for the electronic wave function $\psi_i(r)$,

$$n(r) = \sum_i \psi_i^* (r)\, \psi_i(r) \tag{1.1}$$

where the wave functions are normalized and the sum is over the occupied levels. To calculate n(r) in general by using this procedure is a rather complex task. Therefore, the Thomas-Fermi model is attractive and very useful because it calculates the density n(r) directly from the knowledge of the potential V(r), by avoiding Schrodinger equation and without using Eq. (1.1).

The Thomas Fermi theory is a fluid model of the atom. The total energy associated with the electrons, at a zero temperature (T = 0) is given by

$$E = C_k \int n^{5/3} d\vec{r} - Ze^2 \int \frac{n^2}{r} d\vec{r} + \frac{e^2}{2} \int\int \frac{n(r_i)\, n(r)}{|\vec{r}_1 - \vec{r}|} \tag{1.2}$$

where the first term describes the kinetic energy E_k and $C_k = (3/8\pi)^{2/3}\, 3h^2/(10\text{ m})$, the second term is the electron-nuclear interaction while the last term is contributed by the electron-electron interaction. One requires E to be minimum with respect to variations of the electron density n, with the constraint that the total number of electrons given by

$$Z = \int n d\vec{r} \tag{1.3}$$

remains constant. Thus one has

$$\delta(E - \mu Z) = 0 \tag{1.4}$$

where μ is the Lagrange multiplier which turns out to be the chemical potential ($\mu = \partial E/\partial N$, N = Z). The variation of Eq. (1.4) with E given in (1.2) yields,

$$n(r) = \begin{cases} \dfrac{8\pi(2m)^{3/2}}{3h^3} [\mu - V(r)]^{3/2}; & \mu \geq V(r) \\ 0 & ; \mu < V(r) \end{cases} \tag{1.5}$$

which is the basic relation between the potential V(r) and the electron density n(r). Equation 1.5 together with the Poisson equation, yield the TF equation. For the spherical symmetry case the TF equation can be written in dimensionless form by

$$\frac{d^2X}{dx^2} = \frac{X^{3/2}}{x^{1/2}} \tag{1.6}$$

where $\mu - V(r) = Ze^2 X/r$, $r \simeq Z^{-1/3} a_0 x$ and a_0 is the Bohr radius.

Three types of possible solutions are given in Fig. 1 where solution I describes a neutral atom, II describes a positive ion, and III gives the solution of an atom in a material at high pressure. X satisfies the following boundary condition: (I) $X(\infty) = 0$; (II) $X(x_0) = 0$; (III) $X(x_0) = x_0 X'(x_0)$ (cell neutrality). The ionization for case II is given by Gauss law: $Z - Z^* = Z[-x_0 X'(s_0)]$ (note: prime denotes derivative).

In a similar way to TF equation for a nonzero temperature is given by[9]

$$\psi'' = a\, x\, I_{1/2}[\psi(x)/x] \tag{1.7}$$

where $\psi(x) = \{[\mu +_3 eV(r)]/kT\}\, r/r_0$, $x = r/r_0$ and $4\pi r_0^3/3 = N/V$ = number of atoms/cm^3. The constant a in (1.9) is $a = (r_0/c)^2$ with $c \simeq 1.6.10^{-9}$ cm/[kT/keV]$^{1/4}$ and I_n is the Fermi-Dirac function defined by

$$I_n(x) = \int_0^\infty \frac{y\ dy^n}{\exp(y - x) + 1} \tag{1.8}$$

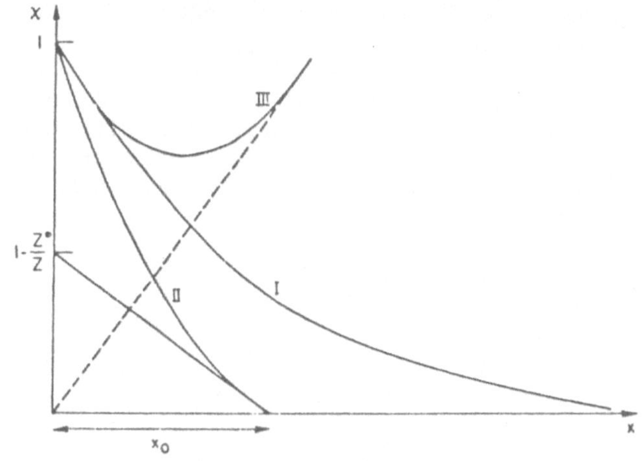

Fig. 1 Types of solutions of the dimensionless TF equation (2.22). The different solutions describe: I - a neutral atom, II - a positive ion, and III - an atom in a material at high pressure.

The boundary condition for the TF equation (1.7) is $\psi(o) = Ze^2/(kTr_0)$ and $\psi'(1) = \psi(1)$ (x = 1 implies $r = r_0$ and this condition is equivalent to $\vec{E}(r_0 = -\vec{\nabla}V(r_0) = 0$, i.e., a zero electric field at the cell boundary).

One of the most important feastures of the TF equation are the scaling laws. In particular, the atomic volume V scales as Z^{-1}: $V \sim Z^{-1}$; the temperature: $T \sim Z^{4/3}$; energy/atom: $E \sim Z^{7/3}$; pressure: $P \sim Z^{10/3}$; entropy: $S \sim Z^{-1}$; chemical potential: $\mu \sim Z^{-4/3}$. Due to the existence of these scaling laws, one has to solve the TF equation only for one atom and thus obtain an appropriate solution for any atom (any Z). Therefore, the TF model seems to be attractive and very useful. However, there are some defects with this model, and in particular it is worthwhile to point out the following deficiencies with the TF model: (a) the ion core (K shell) is not well described and the density $n(r \to 0) \sim (Z/r)^{3/2} \to \infty$. This divergence causes a systematic error in the potential V(r); (b) the outer boundary of the free atom is not well described; (c) molecular binding is not possible for molecules; and (d) no electron energy quantization.

In order to try to solve the difficulties of the TF model, various corrections have been suggested in the literature. In particular, Dirac took into account the effect of antisymmetrization of the wave functions of identical particles. This effects contributes an extra term in Eq. (1.2) given by $-C_{ex} \int n^{4/3}\, dr$ where $C_{ex} = (3/\pi)^{1/3}\, 3e^2/4$. This extra term modifies the TF equation (1.6) to the following equation (TFD)[10,11]

$$\frac{d^2\Psi}{dx^2} = x[\epsilon + (\frac{\Psi(x)}{x})^{1/2}]^3 \tag{1.9}$$

with the boundary condition $\Psi(0) = 1$, $[x\,\Psi(x)^{-1} d\Psi/dx]_{x=x0} = 1$. $\Psi(x)$ is related to the density $n(x) = \gamma[\epsilon + (\Psi(x)/x)^{1/2}]^3$ where $\gamma = 32Z^2/(9\pi^3 a_0^2)$, $\epsilon \simeq 0.2/Z^{2/3}$, etc. (see e.g. Ref. 3).

More sophisticated models were suggested in the literature in order to take into account density gradient corrections, ion corrections, etc. In particular, Kalitkin and Kirzhnitz suggested to add (to Eq. (1.2) + "exchange") the energy[12,13]

$$E_{grad} = \frac{\hbar^2}{72m} \int \left(\frac{\nabla n}{n}\right)^2 n\, d\vec{r} \tag{1.10}$$

while other authors[5] suggested the extra terms

$$-\sum_j \int \frac{Ze^2}{|\vec{r} - \vec{R}_j|} n(r) d\vec{r} + \frac{1}{2} \sum_{\substack{ij \\ i\neq j}} \frac{Z^2 e^2}{|\vec{R}_i - \vec{R}_j|}$$

where R_j is the ion coordinate and the first term describes the ion-electron energy while the second term gives the ion-ion interaction. These more sophisticated models seem to be more accurate with no divergence near $r \simeq 0$ and a better fit for pressure and energy/atom. However, the simplicity and the scaling laws are lost.

Section II describes the TF model for materials under high pressure and, in particular, the equations of state problems in inertial confinement fusion are discussed. We conclude with a short discussion in Section III.

II. EQUATION OF STATE PROBLEMS IN INERTIAL CONFINEMENT FUSION

The main idea of inertial confinement fusion is the aim of achieving very high compression using laboratory facilities, up to $\rho/\rho_0 = 10{,}000$! This concept can be easily understood by using a "realistic" equation of state. Using, for example, the Thomas-Fermi model for the hydrogen isotopes, a D-T mixture with initial (liquid) density of $\rho_0 = 0.2$ cm^3, one needs an energy of 3.0 keV per atom to increase the temperature of the fuel to 1 keV ($\simeq 10^7$ °K) without changing its density. However, for an extra energy of 0.2 keV/atom (i.e., an extra energy of about 7%!) at 1 keV temperature one gets a compression of $\rho/\rho_0 =$ 20! Using elementary knowledge of nuclear reaction cross sections it is evident that it is necessary to use the driver's energy in order to compress the material as much as possible instead of heating it by "brute force" Although the idea seems to be self-evident from the equation of state data, the way to put it into practice is still unresolved.

The most effective compression is isentropic and this might be achieved approximately if the ablation pressure could be increased according to an ideal time profile. Besides the time shaping of the driver pulse, structured targets have to be used to vary the time evolution of the pressure on the compressed material and to avoid preheating of the pellet core in order to achieve the desired isentropic compression. These considerations have shown that high compression can be achieved by (a) shaping in time the input energy of the driver, and (b) "clever" pellet design. For example, Livermore designers[14] suggested for a Nd: YAG laser driver, with maximum irradiance of 10^{14} W/cm^2, "double shell" pellet.

The properties of matter at high density and high temperature are important to explain and to calculate the compression process. In particular, the equation of state data for different materials is necessary to calculate the shock wave propagation into the pellet. For inertial confinement fusion, a knowledge of the properties of matter is needed[3] for temperatures up to 100 keV and for densities up to 10^4 times solid density.

The corresponding pressures are enormous. For example, the pressure of the degenerate electrons of hydrogen with 10^4 times liquid density (i.e. an electron density of about $n_e \approx 5 \times 10^{26}$ cm^{-3} is about 10^{12} atmospheres ($\approx 10^6$ Mbars). This estimate is obtained using the expression for the Fermi degenerate electron pressure at zero temperature. For comparison, the thermal pressure of the non-degenerate ions $P = n_e kT$ with $n_e = 5.10^{26}$ cm^{-3} and a temperature of kT -10 keV is about 5×10^{12} atmospheres. This, it can be summarized that for inertial confinement purposes, one needs the data of equations of state for many materials in the domain of

$$0 < T < 100 \text{ keV}; \; 10^{-4} < \frac{\rho}{\rho_0} < 10^4; \; 0 < P < 10^{13} \text{atm} \tag{2.1}$$

where ρ_0 is the initial liquid or solid density of the material under consideration.

The physics of inertial confinement fusion is based on the hydrodynamics of one or more fluids or equivalently on transport (e.g., Boltzmann) equations. In order to solve these equations a knowledge of the equation of state and transport coefficients (such as thermal conductivity, electrical conductivity, radiation opacities, etc.) is necessary.

For example, in a simple model,[15] the electron energy equation is

$$\left(\frac{\partial E_e}{\partial T_e}\right)_V \frac{dT_e}{dT} + \left(\frac{\partial E_e}{\partial V}\right)_T \left(-\frac{1}{\rho^2}\right) \frac{d\rho}{dt} + P_e \frac{dV}{dt} = \phi_e \quad \frac{\text{erg}}{\text{sec cm}^2 \text{g}} \tag{2.2}$$

where the quantities with subscript e refer to the electrons. E_e is the energy, e.g., $E = E(t,V)$, the specific volume V is related to the "fluid" density $\rho = 1/V$, P_e is the electron pressure and ϕ_e is the energy source term. A similar equation to (2.2) is written for the ions. The source term ϕ_e includes: (a) energy absorption from the laser in the domain of laser absorption, (b) thermal conduction of energy, (c) electron-ion energy exchange, (d) radiation losses, (e) thermonuclear energy absorption. These energy sources depend not only on equation of state data, but also on ionization average $\langle Z \rangle$ and its higher moment $\langle Z^2 \rangle$. For example, the inverse bremsstrahlung laser absorption contribution to (a), the thermal conductivity coefficient describing (b) and the electron-ion collision frequency for (c) are functions of $\langle Z \rangle$ and $\langle Z^2 \rangle$. In Thomas-Fermi model one can define the number of free electrons in the atom by counting those electrons with positive energy.[5] Those electrons are able to leave the cell according to classical physics (i.e. the kinetic energy is larger than the potential energy at the ion-sphere boundary). The number of free electrons is given by

$$\langle Z \rangle = 2 \iint_{\frac{p^2}{2m} \geq eV} \frac{d^3r \, d^3p}{h^3} \left\{1 + \exp\left[\frac{1}{kT}\left(\frac{p^2}{2m} - eV(r) - \mu\right)\right]\right\}^{-1} \tag{2.3}$$

This equation can be rewritten as

$$<Z> = C_1 \int_\Omega F_{1/2} \left(\frac{eV}{kT}, \frac{|eV - \mu|}{kT} \right) d^3r \quad (2.4)$$

where eV(r) is the electron total potential energy, C_1, is a known constant, and $F_k(x,\beta)$ is the incomplete Fermi-Dirac function of order k,

$$F_k(x,\beta) = \int_\beta^\infty \frac{y^k dy}{e^{y-x} + 1} \quad (2.5)$$

Im a similar way as <Z> is defined we suggest to define $<Z^2>$ and higher moments $<Z^k$ (k = 1, 2, ...Z) by taking the k-th power of the density of states in the phase space and integrating by counting those electrons with positive energy. Such a calculation yields

$$<Z^k> = C_k \int F_{k/2} \left(\frac{eV}{kT}, \frac{|eV - \mu|}{kT} \right) d^3r \quad (2.6)$$

with C^k a known constant.

III. DISCUSSION

Hydrodynamic codes which perform simulations in inertial confinement fusion require the use of equation of state models which describe the thermodynamic functions of both electrons and ions.[16] The main values of interest are the pressure and internal energy, as well as their respective temperature and volume derivatives as a function of amterial density and temperature. The electronic equation of state can be taken, to a good approximation, from the corrected Thomas-Fermi-Dirac model. The ion contribution to the equation of state can be described by the Debye-Gruneisen equation of state, with the appropriate density variations of the Debye temperature and the Gruneisen coefficient. At sufficiently high temperature and/or low densities the ion contribution can be described by the ideal gas equation of state. To join these two limiting cases a semi-empirical interpolation method can be used. The extent to which the inclusion of different models into the computer codes will influence the inertial confinement fusion results is a subject mostly of examination and research. In general, since the heating mechanisms, the energy transport and the effects occurring in the corona are not very dependent on the ion parameters, one would expect not much dependence of these phenomena on the ion equation of state model. However, for processes which may be more sensitive to the ion parameter, such as those taking place in the compressed solid, e.g. shock wave phenomena, one may expect changes in the calculated results due to the use of different models in computing the ion contribution to the equation of state. The equation of state also determines the velocity of shock waves and, therefore, the time scale of the whole implosion occuring in inertial confinement fusion phenomena.

For a summary and conclusion see Fig. 2 with rather negative aspects of the central ignition scheme.

A proper statistical calculation for two fluids[17,18,19] with a strong interaction between them does not separate in general into two additive contributions. Therefore, one possible approach is to define a free energy $F_e(\{R_j\}, T_e)$ for a fluid of electrons at temperature T_e, where $\{R_j\}$ is a set of coordinates describing the position of the electrons. The free energy F_e includes all the electrostatic energies (including ion-ion!) and the electronic kinetic energy. This tree energy is an effective potential for the ion motion. This approach is reasonable in the case where the ions are moving slowly on the collision time scale of electrons. The partition function in this case can be written as[5]

$$Q = \int \frac{d^3R_1 \ldots d^3R_N d^3p_1 \ldots d^3p_N}{N! h^{3N}} \exp\left\{- \frac{1}{kT_I}\left[\sum_{j=1}^{N} \frac{p_j^2}{2M_i} + F_e\left(\{R_i\}, T_e\right)\right]\right\} \tag{3.1}$$

where N is the number of ions and R_i, p_i are the canonical coordinates and momenta for the ions. The free energy of the electron ion-system is,

$$F(T_e, T_I, V) = -kT_I \log Q \tag{3.2}$$

The entropy is this approach is defined by

$$S_I = - \left(\frac{\partial F}{\partial T_I}\right)_{V,T_e} ; \quad S_e = - \left(\frac{\partial F}{\partial T_e}\right)_{V,T_I} \tag{3.3}$$

where S_I and S_e are the ion a and electron entropy. The generalized thermodynamic consistency condition is

$$\left(\frac{\partial E}{\partial V}\right)_{T_e,T_i} = T_e\left(\frac{\partial P}{\partial T_e}\right)_{T_I,V} + T_I\left(\frac{\partial P}{\partial T_I}\right)_{T_e,V} - P \tag{3.4}$$

while the thermodynamic basic relation is

$$dE = T_e dS_e + T_I dS_I - P\, dV . \tag{3.5}$$

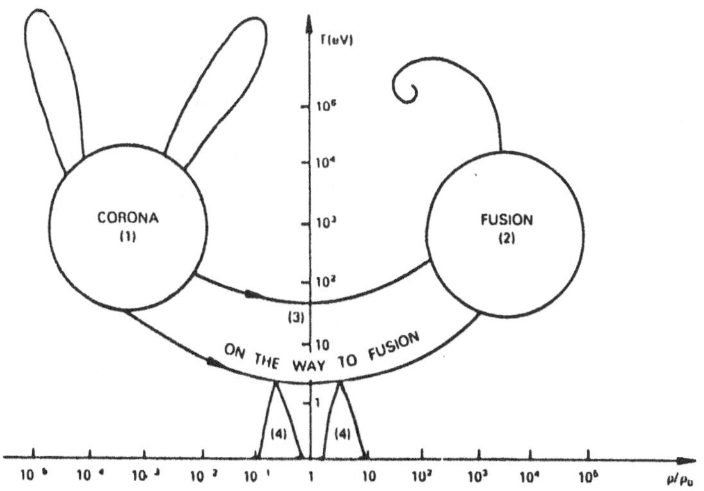

Fig. 2 The difficulties for high gain pellet fusion due to the equation of state (EOS). (1) The corona: 2 or 3 temperature EOS; Ideal gases; Weak Coulomb interation; Non-Local-Thermal Equilibrium. (2) Fusion: Strong Coulomb interation; T-F or T-F-D model. (3) On the Way to Fusion: Most difficult. (4) "Easy Experiments" regions.

Although this last approach seems to be on a more fundamental level than the phenomenological approach of adding separate electron and ion contributions, one has to solve in this case very complicated non-linear equations and to check its applicability to the hydrodynamic codes. This subject is still being researched and poses profound and conceptual difficulties for the thermodynamics of a mixture of two (or more) temperature fluids with strong interactions.

ACKNOWLEDGEMENT

I would like to thank Prof. D. Shalitin and Dr. H. Szichman for useful discussions. Prof. N. Hershkowitz' kind hospitality at the University of Wisconsin during the summer of 1986 is greatly acknowledged.

REFERENCES

1. L. H. Thomas, Proc. Camb. Phil. Soc. 23:542 (1927).
2. E. Fermi, Zeits. f. Phys. 48:73 (1928).
3. S. Eliezer, A. Ghatak and H. Hora, "An Introduction to Equations of State: Theory and Applications" (Cambridge Univ. Press, 1986).
4. S. Lundqvist and N. H. March (eds.) "Theory of the Inhomogeneous Electron Gas" (Plenum Press, 1983).
5. R. M. More, Lawrence Livermore Laboratory Report, UCRL-84991 (Parts I and II), 1981.
6. S. G. Brush, "Progress in High Temperature Physics and Chemistry" (ed. C. A. Rouse), 1:1 (1967).
7. N. H. March, Adv. Phys. 6:1 (1957).
8. E. H. Lieb, Rev. of Mod. Phys. 53:603 (1981).
9. R. Latter, Phys. Rev. 99:1854 (1955).
10. P. A. M. Dirac, Proc. Camb. Phil Soc. 26:376 (1930).
11. R. P. Feynman, N. Metropolis and E. Teller, Phys. Rev. 75:1561 (1949).
12. D. A. Kirzhnits, JETP (Soviet Physics) 8:1081 (1959).
13. S. L. McCarthy, Lawrence Livermore Laboratory Report, UCRL-14364 (1965).
14. H. G. Ahlstrom, Phys. of Laser Fusion Vol. II, Lawrence Livermore National Laboratory, p. 331 (1983).
15. A. D. Krumblein, D. Salzmann, H. Szichman and S. Eliezer, "Plasmor: A Laser-Plasma Simulation Code", IA-1396 (SOREQ, 1985).
16. S. Eliezer, A. D. Krumbein, H. Szichman and H. Hora, Laser Part. Beams 3:207 (1985).
17. H. E. DeWitt, "Equation of State of Strongly Coupled Plasma Mixtures", Lawrence Livermore National Laboratory, UCRL-90357, (1984).
18. D. B. Boercker and R. M. More, Phys. Rev. A 33:1859 (1986).
19. F. J. Rogers, Phys. Rev. A 19 (1979).

OCCUPATION NUMBERS IN PARTIALLY IONIZED PLASMAS

F. J. Rogers

University of California
Lawrence Livermore National Laboratory
Livermore, California 94550

INTRODUCTION

Discrepancies between calculation and observation of a number of astrophysical properties has led to the speculation that opacity databases may be in error. Simon,[1] for example, has noted that arbitarily increasing the heavy element opacity by a factor of 2-3 leads to Cepheid models that reproduce observed period ratios. Although other mechanisms could account for these discrepancies, an improved opacity calculation will at the least provide information on the size of the error and its impact on the corroboration with observation. In such an endeavor, one should first look at the occupation number calculations and a number or recent studies have appeared.[2-5]

For a short-ranged potential the occupation numbers are obtained directly from the statistical mechanics. In the case of the Coulomb potential there is a divergence in the internal partition function. A number of mechanisms are commonly used for removing this divergence, such as:

1) The Coulomb potential is replaced with the Debye (SSCP) potential.
2) States having an orbital radius greater than the ion sphere (or other) radius are removed due to interaction with neighbors.
3) Ion microfields broaden lines such that a certain fraction tunnel into the continuum effectively reducing the occupation probabilities.

The philosophy adopted here is that the plasma occupation numbers, just as for short-ranged potentials, should be obtained directly from statistical mechanics.

Activity Expansion for the Equation of State

The equation of state of reacting plasmas presents some difficulties not encountered for atomic and molecular gases. There are, nevertheless, similarities in the method of treatment. The equation of state and occupation numbers for reacting dense atomic and molecular gases can be obtained from a renormalized activity expansion. The renormalization is required to account for the fact that products of terms involving the fundamental particle activities multiplied by Boltzmann factors, occurring in the cluster coefficients, act like the activities for dimers, timers, etc. Everywhere a product of terms corresponding to specific composite particle occurs in $\ln\Xi$. It is relabeled and called a composite particle activity. Consequently, the expression

$$\rho_a = \frac{z_a}{V} \frac{\partial \ln \Xi}{\partial z_a} \tag{1}$$

which relate the fundamental particle activity to its number density can be factored into a set of terms that correspond to bound state occupation numbers, according to

$$\text{where } \rho_a = \rho_a^* + \sum_j \rho_j \tag{2}$$

$$\rho_a^* = \frac{z_a}{V} \frac{\partial \ln \Xi}{\partial z_a}$$

is the number density of unbound particle and

$$\rho_j = \frac{z_j}{V} \frac{\partial \ln \Xi}{\partial z_j}$$

is the number density of composite particles in state j.

The treatment for plasmas follows a similar path, but the Coulomb potential introduces some troublesome differences. For example, there is a potential infinite number of bound states which must be assigned on activity and, as a result, the corresponding renormalization transformations that Eq. (1) into Eq. (2) is complicated. In the case of the equation of state, the analytic properties of the cluster coefficients make it possible to limit the set of composite particle activities to those states lying below -kT (typically n = 3-4). This is referred to as the Planck-Larkin (PL) compensation and is discussed in detail by Bollé.[6] States lying above -kT are treated in a many-body perturbation expansion along with true scattering states. The purpose here is to indicate how the occupation numbers can be obtained from a follow-on calculation to the equation of state.

Equation of State

A multi-component activity expansion for reacting plasmas, based on PL weight factors, has been given by Rogers.[7] Only the leading terms given in that work are required for the present discussion. Truncation of the many body PL activity expansion after squared power terms gives

$$\frac{P}{kT} = \sum_i \zeta_i + S_R(\bar{\lambda}_A) + \sum_i \zeta_i \left(\frac{\partial S_R(\bar{\lambda}_A)}{\partial \zeta_i}\right)^2 + \sum_i \sum_j \zeta_i \zeta_j \, s_{ij}(\bar{\lambda}_A) + \ldots\ldots \tag{3}$$

subject to

$$\bar{\rho}_i = \zeta_i \frac{\partial P/kT}{\partial \zeta_i} \tag{4}$$

where the ρ_i are effective occupation numbers, optimized for the equation of state. In the particular case of hydrogenic plasmas i and j range over [e, α, H^{1s}, H^{2s} .....], and $H^{n\ell}$ signifies a hydrogenic ion in the state $n\ell$. The ζ_i for electrons and nuclei are real activities given by

$$\zeta_i \equiv z_i = (2s_i + 1)\, \lambda_i^{-3}\, e^{\mu_i/kT} \quad , \tag{5}$$

where μ_i is the chemical potential. For composite particles the ζ_i are effective activities defined by PL weight factors according to

$$\zeta_i = z_e\, z_\alpha\, (4\pi^{3/2}\lambda^3_{e\alpha})\, g_i\, (2s_i + 1)\, \lambda_i^{-3} \quad , \tag{6}$$

where g_i is the statistical weight and $\lambda_{e\alpha}$ is the thermal de Broglie wavelength.

Due to the non-exponential form of Eq. (6), introduced through the bound-free compensation, the composite particle activities are not the actual activities for state $n\ell$. The terms in Eq. (3) involving S_R are related to the Debye-Hückel correction that arises in a density expansion and are given by (for non-degenerate plasmas)

$$S_R = \frac{1}{12\pi\bar{\lambda}_A^3} \quad , \quad \bar{\lambda}_A = \left[\frac{kT}{4\pi e^2(\sum\limits_i Z_i^2\, \zeta_i)}\right]^{1/2} \quad ,$$

where Z_i is the charge on species i and $\bar{\lambda}_A$ is the activity dependent screening length. In the low density limit $\zeta_i \to \rho_i$, the number density, $\bar{\lambda}_A \to \lambda_D$ the effective Debye length; consequently, $S_R \to$ a Debye-Hückel-like result. The s_{ij} in Eq. (3) are closely related to the second cluster coefficients for a screened Coulomb potential having a screening length $\bar{\lambda}_A$. Explicit definitions of the s_{ij} are given in Rogers in Ref. 7. It is important to point out that, whereas the E_i that enter the fundamental particle activity expansion are screened, the E_i that go in the PL activities [Eq. (6)] are the isolated ion values. This is a consequence of the renormalization to create activities for composite particles described earlier. Due to the subtraction of the leading two terms at high temperature the ζ_i corresponding to $-\beta E_i < 1$ effectively do not contribute to P/kT. This is particularly important at high density where many terms contribute to the pressure [Eq. (3)].

It is possible to relate Eqs. (3 and 4) to the more familiar free energy minimization method. If only terms involving S_R are retained in Eq. (3) it can be shown that the free energy is given by

$$\frac{F}{kT} = - N_e \ell n\left(\frac{2e}{\bar{\rho}_e\lambda_e^3}\right) - N\, \ell n\left(\frac{2e}{\bar{\rho}_\alpha\lambda_\alpha^3}\right) - N_H \ell n\left(\frac{e\cdot \text{PLPF}}{\bar{\rho}_H\lambda_H^3}\right) - VS_R(\lambda_D) \tag{7}$$

where

$$\rho_H = \sum_k \rho_k \ , \quad k = \{H^{n\ell}\} \ , \quad \lambda_D = [kT/4\pi e^2 \sum_i Z_i^2\, \bar{\rho}_i]^{1/2} ,$$

and PLPF is the Planck-Larkin partition function. The resulting Saha-like equation that gives the effective number densities for equation of state calculations is[8]

$$\frac{N_e N_\alpha}{N_H} = \frac{2V\, \lambda_H^3\, e^{Ze^2/kT\lambda_D}}{\lambda_e^3\, \lambda_\alpha^3 \cdot \text{PLPF}} \tag{8}$$

Higher order terms in Eq. (3) can also be included in the analysis.[9]

Activity Expansion for Occupation Numbers

The previous section gave the low density Saha-like equation that should be used for equation of state calculations. This does not produce actual occupation numbers, i.e. equation of state contributions from high lying states have been canceled with the continuum contributions. This procedure has effectively redefined the continuum as starting at -kT. As a result the N_e in Eq. (8) includes electrons in weak bound states as well as electrons in continuum states. This was done for compelling procedural reasons, but in principle it is not necessary. It is nevertheless possible to carry along the compensating terms throughout the entire analysis. The equation of state calculation gives unique values of z_e z_α but, the definition of effective composite particle activities is somewhat arbitrary, i.e., slightly different definitions for the effective composite particle activities will yield identical results for P, E, z_e, and z_α provided a sufficient number of terms are used in the different versions of Eq. (3).

If instead of using PL weight factors to define composite particle activities we use the true continuum, the activities are given directly in terms of Boltzmann factors according to

$$z_k = z_e\, z_\alpha\, (4\pi^{3/2}\lambda_{e\alpha}^3)\, g_k\, e^{-\beta E_k} \quad , \quad k = \{H^{n\ell}\} \quad . \tag{9}$$

The z_k defined in Eq. (9) are real activities. Now a renormalization similar to the one leading to Eq. (3) gives

$$\frac{P}{kT} = \sum_i z_i + S_R(\lambda_A) + \sum_i \frac{z_i}{2}\left(\frac{\partial S_R(\lambda_A)}{\partial z_i}\right)^2 + \sum_i \sum_j z_i\, z_j s_{ij}(\lambda_A) + \ldots. \tag{10}$$

subject to

$$\rho_i = z_i \frac{\partial (P/kT)}{\partial z_i} \tag{11}$$

where i ranges over (e, α, $\{H^{n\ell}\}$), $i \equiv k$ for composites (see Eq. 9),

$$S_R = \frac{1}{12\,\pi\lambda_D^3} \quad , \quad \lambda_A = \left[\frac{kT}{4\pi e^2 \sum_i Z_i^2\, z_i}\right]^{1/2} \tag{12}$$

and the s_{ij} are now defined directly in terms of scattering states only, i.e., weak bound states are not included. Equation (10) is an expansion in Boltzmann activities and can be used to obtain occupation numbers. If properly carried out Eqs. (3 and 4) and Eqs. (10 and 11) will give exactly the same P, E, z_e, and z_α. Consequently, the results of a self-consistent solution of Eqs. (3 and 4) can be used directly to evaluate Eqs. (10 and 11).

To relate Eqs. (10 and 11) to the free-energy minimization method these equations are truncated after terms linear in S_R. Since most particles are in low lying states λ_A and λ_A are not very different, so that,

$$\rho_k \simeq \bar{\rho}_k\, e^{-\beta E_k}/(e^{-\beta E_k} - 1 + BE_k) \tag{13}$$

This relation does not hold for states for which higher order terms are important, nevertheless the occupation numbers can be systematically calculated from the procedure presented here. In the low density limit the occupation number version of Eq. (8) takes the form

$$\frac{N_e N_\alpha}{N_H} = \frac{2\lambda_H^3 V}{\lambda_e^3 \lambda_\alpha^3} \frac{e^{Ze^2/kT\lambda_D}}{\phi} , \tag{14}$$

where

$$\phi = \sum_k g_k e^{-\beta E_k} , \tag{15}$$

is the usual internal partition function; except Eq. (15) includes only a finite number of bound states whose energies are unshifted up to the plasma continuum starting at $-2Z/\lambda_a$. States above this continuum are rapidly subject to plasma screening and disappear at specific values of λ_a depending on $n\ell$. In the low densities limit, the occupation of individual internal states is given by

$$\frac{N_k}{N_H} = \frac{g_k e^{-\beta E_k}}{\phi} . \tag{16}$$

The sole advantage of Eq. (14) is that it estimates actual occupation numbers. However due to the slower convergence properties of Eq. (10), described earlier, percentage errors in the occupation numbers given by Eq. (14) are larger than those for the equation of state obtained through Eq. (9). This disparity can be reduced by inclusion of higher order terms.

CONCLUSION

It was shown that the equation of state and occupation numbers for dense partially ionized plasmas can be obtained from a complimentary set of activity equations, i.e., even though for the equation of state it is advantageous to treat bound electrons having energy greater than -kT as a pseudo-continuum, the occupation numbers for these states are not lost and can be obtained from a follow-on calculation. The resulting occupation numbers should be used in the calculation of dense plasma opacity. In addition, the energy levels that enter the renormalized activity expansion are unscreened up to near the plasma continuum which is another way that the current work is relevant to plasma opacity.

REFERENCES

1. N. R. Simon, Ap. J., 260:187 (1982).
2. V. Sevastyanenko, Beitr. Plasma phys., 25:15 (1985).
3. D. G. Hummer, D. Mihalas, submitted to Ap.J. (1986).
4. W. Dappen, L.Anderson, and D. Mihalas, submitted to Ap.J. (1986).
5. F. J. Rogers, Ap.J., 310:723 (1986).
6. D. Bollé, these proceedings.
7. F. J. Rogers, Phys. Rev., A24:1531 (1981).
8. F. J. Rogers, "Spectral Line Shapes" Vol 3, de Gruyter, Berlin (1985).
9. W. Ebeling, W. D. Kraeft, D. Kremp., and G. Röpke, Ap.J., 290:24 (1985).

Work performed under the auspices of the U.S. Department of Energy by Lawrence Livermore National Laboratory under contract #W-7405-Eng-48.

THOMAS FERMI CALCULATION OF THE DEGREE OF IONIZATION IN A DENSE PLASMA

Ruoxian Ying and Gabor Kalman

Department of Physics
Boston College
Chesnut Hill, MA

I. INTRODUCTION

The knowledge of the average degree of ionization of a plasma plays a significant role both in inertial confinement experiments and in astrophysical situations. The Thomas-Fermi statistical model, combining relative simplicity, clarity and excellent qualitative descriptive capacity, has been proven to be a powerful method to calculate the average properties of the atomic system, such as the equation of state and the degree of ionization. A number of different versions of Thomas-Fermi model have been adopted to calculate the degree of ionization. Kobayashi (1959) solved the Thomas-Fermi equation for positive ions and calculated the degree of ionization for zero-temperature. For finite temperature, I.J. Feng, et al. (Feng, Zakowicz and Pratt, 1981) compiled a detailed study of the degree of ionization of a dense plasma in the Thomas-Fermi (TF) and the Debye-Hückel-Thomas-Fermi (DHTF) approximation. The results of Feng's work are shown in Fig. 1. Note than an anomolous feature, viz., that the degree of ionization decreases as temperature increases, appears in both curves d and c, based on the DHTF model. In curve d, the bound electrons are defined as the electrons with a negative total energy and the number of bound electrons was calculated by integration. The dip in curve d is very obvious. In curve c, an attempt was made to remedy this defect. Here the bound electrons are restricted to be inside the sphere with the radius of mean distance between ions. The dip in the curve diminishes, but still exists.

In curve b, a condition that the total energy of bound electrons has to be less than -0.1 kT was imposed. Although a reasonable result was obtained in this way, the ad-hoc assumption itself lacks physical justification.

Curve f is based on Rozsnyai's approach (Rozsnyai and Alder, 1977). These authors defined a radius of neutrality r' in DHTF model through $4\pi\int_0^{r'} \rho(r)\, r^2 dr = Z$, where Z is the atomic number and $\rho(r)$ is the electron density in DHTF model. Then the number of free electrons $z=4/3\, \pi r'^3 \rho(r')$. A rather satisfactory result is obtained through this model, but the model itself still seems to lack sufficient physical justification.

We have established a model based on the TF method in order to reexamine the determination of the degree of ionization. We believe that this model is a more systematic approach to the problem than those used previously, and it also eliminates some of the unphysical features of the earlier models. The main features of our model are (i) separation of the bound and free electrons; (ii) a physically reasonable definition of the bound electrons; (iii) describing the system as a plasma of free electrons and TF ions; (iv) describing the source density in the Poisson equation through the electron-ion and ion-ion pair correlation function; (v) determining the degree of ionization from the minimalization of the total free energy.

II. MODEL

We write the basic Thomas-Fermi equations as:

$$\nabla^2\phi(r) = 4\pi e\, n_b\big(\psi(r)\big) \tag{1a}$$

$$\nabla^2\psi(r) = 4\pi e\, \{n_b\big(\psi(r)\big) + n_f\, \big(r,\phi(r)\big) - zn_i\big(r,\phi(r)\big)\} \tag{1b}$$

Here we distinguish between the two potentials, $\phi(r)$, created by the bound electron density n_b, and $\phi(r)$, created by the combined n_b, the free electron density n_f and the ion density n_i. The different roles the two potentials play are explained below.

In contrast to the customary TF models we confine the ion within a finite radius r_0, such that the bound electron density vanishes at r_0. Thus the bound electron density $n_b(\psi(r))$ is given by a momentum cut-off integral of the Fermi-Dirac distribution function.

$$n_b\big(\psi(r)\big) = \frac{8\pi}{h^3} \int_0^{p_m} \frac{p^2 dp}{\exp\left\{\left(\frac{p^2}{2m} - e\psi(r) - \alpha\right)/kT\right\} + 1} \tag{2}$$

where the cut-off momentum p_m is

$$p_m = 2m[\psi(r) - \psi(r_0)] \tag{3}$$

This definition is based on the following considerations. First, the bound electron density should vanish at the ion boundary. Second, the maximum energy of the bound electrons should be the same inside the ion.

$n_f(r,\phi(r))$ and $n_i(r,\phi(r))$, the number densities of the free electrons and the neighboring ions, are determined through

$$\begin{aligned} n_f(r,\phi(r)) &= \overline{n}_f(1+g_{ei}(r,\phi(r))) \\ n_i(r,\phi(r)) &= \overline{n}_f(1+g_{ei}(r,\phi(r))) \end{aligned} \tag{4}$$

where the $g_{ei}(r,\phi(r))$ and $g_{ii}(r,\phi(r))$ are the pair correlation functions for an electron-ion pair and an ion-ion pair, respectively.

The correlation functions $g_{ei}(r)$ and $g_{ii}(r)$ are to be determined self-consistently in conjunction with the interaction potential $\phi(r)$. Here we emphasize that in order to avoid double counting of the correlation effect, $g_{ei}(r)$ and $g_{ii}(r)$ are to be treated as functionals of $\phi(r)$ rather than $\psi(r)$, which plays the role of internal potential for the bound electrons only.

In (1b) z is the number of free electrons per atom. Electrical neutrality at infinity requires $\overline{n}_f = z\overline{n}_i$.

We may adopt, in a first approximation, $g_{ei}(r) = g_{ii}(r) = 0$, which means that the free electrons and neighboring ions are both assumed to be uniformly distributed and provide an electrically neutral background only.

The potential $\psi(r)$ surrounding the nucleus with the above model is now given by

$$\nabla^2\psi(r) = 4\pi e\rho(r) = 4\pi e \frac{8\pi}{h^3} \int_0^{p_m} \frac{p^2 dp}{\exp\left\{\left(\frac{p^2}{2m} - ev(r) - \alpha\right)/kT\right\} + 1} \tag{5}$$

with the boundary conditions

$$\psi_0 = \psi(r_0) = \frac{ze}{r_0}, \; \psi'(r_0) = -\frac{ze}{r_0^2} \text{ and } \psi(r) \underset{r\to 0}{=} \frac{ze}{r} \tag{6}$$

where α is a parameter to be determined (not identical to the chemical potential), Z is the atomic number, (Z-z) is the bound electron number inside r_0.

In this paper only calculations for this simple model are presented. Work with $g_{ii}(r)\neq 0$ is in progress and will be reported elsewhere. Ultimately the calculation of $g_{ei}(r)$ and $g_{ii}(r)$ has to be done through a more sophisticated self-consistent model. The STLS mean field theory is commended by its simplicity, but it is expected that it is the HNC method which can provide the more reliable approach.

III. CALCULATIONS

For the purpose of numerical calculation, it is convenient to introduce $V(x) = [\alpha+e\psi(r)]r/kTr_0$ with $x = r/r_0$.

The differential equation is now reduced to

$$V''(x) = axJ_{1/2}\left(\frac{V(x)}{x}, V(1)\right) \quad (7)$$

where

$$J_{1/2}(x,x_0) = \int_0^{x-x_0} \frac{y^{1/2}}{p^{y-x+1}} dy, \quad (8)$$

$$a = \frac{(4\pi e)^2 (2m)^{3/2}}{h^3} (kT)^{1/2} r_0^2$$

and the boundary conditions become

$$V(1) = \frac{\left(\alpha + \frac{ze}{r_0}\right)^2}{kT}, \quad V'(1) = \frac{\alpha}{kT} \text{ and } V(0) = \frac{ze^2}{kTr_0}. \quad (9)$$

It is clear that after prescribing the quantities kT, r_0, α and z, the differential equation and the boundary conditions at $x = 1$ are all specified. Therefore, an inward integration of the differential equation is feasible. The solution at $x = 0$, i.e. $\phi(o)$ would provide the value of Z.

The physical system of interest in this paper is a pure hydrogen plasma. Thus the input parameters kT, r_0, α and z must be chosen in such a way that one obtains $Z = 1$. This can be done by prescribing kT, r_0 and α, then adjusting z only. For any given set of kT, r_0 and α values, a computer program was designed to pick up the correct z with the result of $|Z-1| < 10^{-6}$ within 20 trials.

A physically reasonable and consistent model requires that the parameters α and r_0 be determined by the conditions of thermodynamic equilibrium. The equilibrium of the combined bound electron and free electron system is characterized by the fact that the total free energy F (consisting of the free energy F_1 of the bound electrons and the free energy F_2 of the free electrons) is minimal.

Among the three adjustable parameters r_0, α and z, only two are independent. Choosing α and r_0 as the two independent parameters, the conditions of minimizing the free energy F are $\partial F/\partial\alpha|_{r_0}=0$ and $\partial F/\partial r_0|_{\alpha}=0$.

The free energy of the bound electrons F_1 is expressed as

$$F_1 = (Z-z)\alpha - \frac{2}{3}\frac{a}{V(o)} ZkT\ln\left(1+eV(1)\right) \int_0^1 \left[\frac{V(x)}{x} - V(1)\right]^{3/2} x^2 dx$$

$$+ \frac{a}{3V(0)} ZkT \int_0^1 J_{1/2}\left(\frac{V(x)}{x}, V(1)\right)\left[V(x)-V'(1)x-2V(0)\right]xdx \quad (10)$$

The free energy of the free electrons F_2 is given by the ideal gas formulas:

when $\frac{(2\pi mkT)^{3/2}}{h^3}\frac{1}{n_f} \gg 1$ (n_f is the free electron density), the

Table 1. Average degree of ionization for Z = 1. The figures in the parentheses are from the Saha-equation.

kT(eV) / $n_i(cm^{-3})$	10^3	10^2	10	1
10^{23}	0.999436 (0.998932)	0.976439 (0.964683)	0.657653 (0.387420)	0.51312
10^{21}	1 (1)	0.996871 (0.999634)	0.86809 (0.962213)	0.307123 (1.933295×10^{-3})
10^{19}	1 (1)	0.999371 (1)	0.961298 (0.999634)	0.499431 (1.916533×10^{-2})

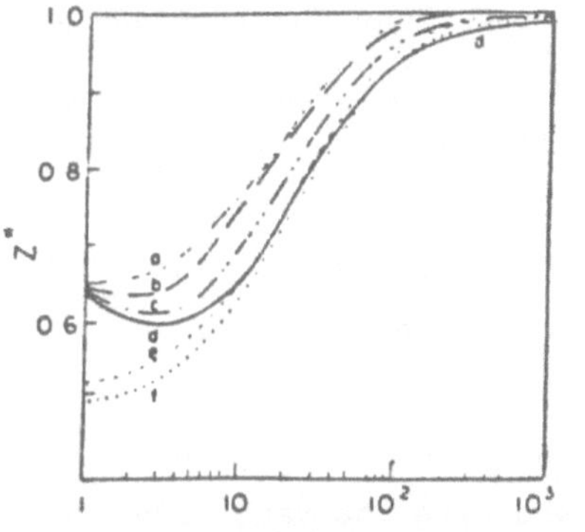

Fig. 1. Average degree of ionization Z* calculated for a pure hydrogen plasma of ion density $N_i = 10^{23} cm^{-3}$. (a) TF model with integral definition; (b) DHTF model restricting bound electrons to total energy - 0.1kT; (c) DHTF model restricting bound charge to distance less than r_0; (d) DHTF model with integral definition; (e) TF model $Z^* = 4/3\pi\ (r_0)^3 n_e(r_0)$; (f) DHTF model, Rozsnyai's approach.

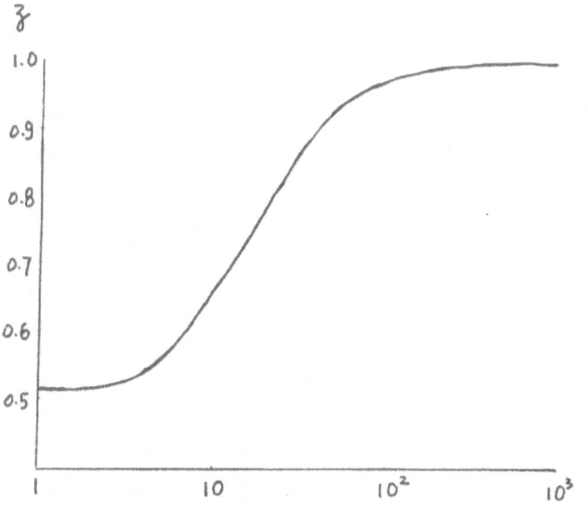

Fig. 2. Average degree of ionization z, for a pure hydrogen plasma of ion density $n_i = 10^{23} cm^{-3}$, from the present work.

Maxwell-Boltzmann distribution applies, which gives

$$F_2 = -zkT \ln \left[\frac{(2\pi mkT)^{3/2}}{h^3 nf}\right] -zkT \tag{11}$$

Otherwise the Fermi-Dirac distribution has to be used, which gives

$$F_2 = zkT \left[\frac{\mu}{kT} - \frac{2}{3}\frac{I_{3/2}\ (\frac{\mu}{kT})}{I_{1/2}\ (\frac{\mu}{kT})}\right] \tag{12}$$

The n_f is given by the overall charge neutrality condition $n_f = zn_i$ (n_i is the ion density). In the Fermi-Dirac distribution, $I_n(x) = \int_0^\infty (y^n/(e^{y-x}+1))dy$ and the chemical potential μ is determined by the normalization condition $n_f = 4\pi/h^3\ (2mkT)^{3/2}\ I_{1/2}\ (\alpha/kT)$.

The detailed numerical procedure of the calculation is as follows:

First with a fixed r_0, a run for a variety of different α-s is generated. With each α the corresponding z is determined (z is being adjusted to give Z=1). In each set of α and z values, the free energies F_1, F_2 and $F = F_1 + F_2$ are evaluated. By comparing the F-s for different α-s, the minimum value of F is selected. Thus the optimum α and the corresponding z are determined.

In the next step, r_0 is varied and the above procedure is repeated to determine the corresponding α for each specific r_0. Then the free energy F is evaluated as a function of r_0 with the condition $\partial F/\partial\alpha|_{r_0}=0$ now already being satisfied. Comparing these resulting F-s, and selecting again the minimum in principle, the optimum r_0 can be determined.

However, the total free energy F turns out to increase monotonically with the decreasing r_0; thus choosing the mean distance between ions as r_0 is appropriate.

The results of the calculated degree of ionization from our new model are given in Fig. 2 and Table 1, where comparison with the Saha-equation result is also provided.

It should be noted that in Fig. 2 the anomalous dip of I.J. Feng's results is absent.

ACKNOWLEDGEMENT

This work was partially supported by NSF Grant ECS-8315655.

REFERENCES

Kobayashi, S., 1959, J. Phys. Soc. Japan, 14:1039.

Feng, I.J., Zakowicz, W., and Pratt, R.H., 1981, Phys. Rev. A23:883.

Rozsnyai, B.F., and Alder, B.J., 1976, Phys. Rev. A14:2295.

CHAPTER VI

DENSITY FUNCTIONAL THEORY

DENSITY FUNCTIONAL METHODS IN HOT DENSE PLASMAS

Chandre Dharma-wardana

National Research Council of Canada
Ottawa, Canada K1A 0R6

1. INTRODUCTION

A plasma or a liquid metal is simply a mixture of electrons and nuclei interacting via the Coulomb potential. Here we further assume that the system is in local thermodynamic equilibrium (LTE) at a temperature T, taken for simplicity to be the same for ions and for the electrons. We further assume that the system as a whole is electrically neutral and that there is only one nuclear species of charge Z. This does not of course imply that there is only one ionic species since the nuclei will support many different bound electron configurations leading to a distribution of effective charge states $Z_c \leq Z$. The weighted mean of these charge states, $\langle Z_c \rangle$ will be called the effective charge ζ. Since the plasma is charge neutral, this somewhat artificial concept of "mean ions" of charge ζ allows us to write the electro-neutrality condition in the form

$$\bar{n} = \zeta \, \bar{\rho} \tag{1.1}$$

where $\bar{\rho}$ is the mean density of ions (or equivalently, nuclei) while $\bar{n}$ is the free electron density in the plasma.

Our objective is to set up a first principles calculation of plasma properties where, at the outset, the only things known are the temperature T, the nuclear charge Z, and either the mean matter density $\bar{\rho}$ or the mean free electron density $\bar{n}$. If the thermal energy $k_BT = 1/\beta$ is used as the energy scale, we have to consider three types of interactions, viz., $V_{ee} = \beta V_q$, $V_{ei} = -\beta Z V_q$, $V_{ii} = \beta Z^2 V_q$, where $V_q = 4\pi/q^2$ is the Coulomb potential. There are also two length scales, viz., the electron sphere radius and the ion-sphere radius defined by the relations $r_s^e = (3/4\pi\bar{n})^{1/3}$ and $r_s^i = (3/4\pi\bar{\rho})^{1/3}$ respectively. We shall sometimes use r_s for r_s^e and r_0 for r_s^i. Another parameter, Γ, useful in the classical regime, is given by $\Gamma = \zeta^2\beta/r_s^i$. In many studies of strongly coupled plasmas it is assumed that only the ions are "strongly coupled". That is, the effect of V_{ee} and V_{ei} are treated by some form of linear screening theory while the effect of V_{ii} is treated by machine simulation (MS) or by classical statistical mechanics. Such theories are unable to deal with the formation of electronic bound states and have to limp along with the tacit assumption that V_{ei} is weak.

Many body perturbation theory (MBPT) can in principle treat the case where all three interactions V_{ee}, V_{ei} and V_{ii} fall into the strong coupling regime. But in practice, as is well known, diagrammatic or, say, equations of motion methods cannot be pushed very much beyond some form of generalized random phase approximation, except in some very special model problems. Thus we need a computationally convenient but at the same time completely general, rigorous method of tackling the many-body problem posed by a system of strongly coupled electrons and nuclei at some temperature T.

Such a method, which is an exact solution to the many-body problem, is presented by density functional theory (DFT). This method has established itself during the last two decades as probably the most important theoretical method for the study of many electron systems.[1] DFT contains many of the more primitive theories like Thomas-Fermi theory or linear screening theory as special cases. The contribution to this volume by Professor Walter Kohn states the basic principles of DFT. In this paper I will confine myself to a very condensed review of the work we have done in applying DFT to a variety of plasma problems. Lack of space and time prevents me from touching upon the contributions of other workers.

In summary, density functional theory provides an energy variational principle with respect to the density distributions $n(\underset{\sim}{r})$ for the electrons and $\rho(\underset{\sim}{r})$ for the ions. This leads to two coupled equations, one of which is the Kohn-Sham equation in the external field of the ion distribution $\rho(\underset{\sim}{r})$. The other is an equation for $\rho(\underset{\sim}{r})$ in the external field of the electrons and has the form of a Gibbs-Boltzmann equation. The self consistent solution of these equations yields the equilibrium distributions $n(\underset{\sim}{r})$ and $\rho(\underset{\sim}{r})$ in the presence of the external field. These density distributions can then be used to calculate thermodynamic properties. Also, the Kohn-Sham equation provides us with a complete set of one-body eigenfunctions ϕ_ν and "energies" ε_ν where $\nu = n,\ell,m$ for bound states, and $\nu = k,\ell,m$ with $\varepsilon_\nu = k^2/2$ for "free" electron states. The "free" electron states have phase shifts $\delta_\ell(k)$. If necessary, we can now go beyond standard DFT and do many body theory using the Kohn-Sham basis rather than, say, plane waves, as is customary in ordinary linear response theory. This extended or "dressed DFT" can be used to calculate many-body objects like self energies $\Sigma(\omega)$ and polarization parts $\Pi(\omega)$ to yield lifetime effects, transport properties etc., by doing only a simple low-order MBPT calculation.

We have applied these methods to calculate a variety of interesting plasma properties. Thus (i) ion-ion and ion-electron pair distribution functions (PDF) in plasmas were calculated; the PDFs can be used to calculate thermodynamic properties (ii) Kohn-Sham energy levels of ions in plasmas were obtained and many-body perturbation theory was used to calculate level widths and shifts via the Dyson equation (iii) the ion-ion PDF, viz. $g_{ii}(r)$ has been inverted to yield the effective ion-ion potential $V_{ii}(r)$ and this has been used to generate $S(k\omega)$, the dynamic ion-ion structure factor for several hydrogen plasmas outside the classical regime (iv) the distributions $\rho(r)$ and $n(r)$ obtained from DFT have been used to calculate electric microfields in dense plasmas, thus providing a theory which can take account of internal structure of the ions and other strong coupling effects (v) the ion-ion pair distributions and electron phase shifts obtained from DFT have been used to set up a theory of the linear transport coefficients in plasmas, taking into account multiple scattering effects as well (vi) DFT in the time dependent local density approximation has been formulated to study light absorption by hot dense plasmas. Calculations have been carried out showing how channel mixing, particle correlations etc., and the consequent spectral redistribution effects, can be efficiently included in plasma light absorption calculations (vii) the contribution by François Perrot in this volume discusses how the Kohn-Sham eigenvalues ("energies") can be corrected to give reliable predictions of electronic transition energies (line positions) for bound states in plasmas, without going into MBPT.

In the following sections a more detailed discussion of the topics (i)-(vi) will be given.

2. DFT-EQUATIONS FOR A PLASMA

We consider a fixed nucleus of charge Z at the origin and study the behaviour of the plasma in the "external potential" of this nucleus. We exploit the spherical symmetry of the problem so that the electron and ion distributions around the origin can be written as $n(r)$ and $\rho(r)$ respectively, where r is the radial distance. For r sufficiently large, say $r > R$, the densities $\rho(r)$ and $n(r)$ would tend to the mean values $\bar{\rho}$ and $\bar{n}$ respectively. Hence, as far as the central ion is concerned, the infinite plasma can be replaced by a "correlation" sphere[2] of radius R such that for $r > R$, all the potentials can be taken to be zero and the particle distributions go to their mean values. R is bigger than the characteristic distances of the particle correlations in the system. Typically, for plasmas R is 5 to 10 times r_0 but rapidly becomes very large if long range order (e.g., crystallization) sets in. Then a unit cell type calculation becomes more appropriate than the correlation sphere model presented here.

The thermodynamic potential Ω is given in terms of the partition function Z, the Helmholz free-energy F, chemical potentials μ_e, μ_i and the particle numbers N_e, N_i by

$$\Omega = k_B T \log_e Z$$

$$= F - \mu_e N_e - \mu_i N_i \quad (2.1)$$

Density functional theory states that Ω is a unique functional of $n(r)$ and $\rho(r)$, viz., $\Omega[n(r),\rho(r)]$ such that it is a minimum for the true density distributions. Thus we have the functional derivative equations[3,4]

$$\frac{\delta\Omega[n,\rho]}{\delta n} = 0 \qquad \text{for electrons} \quad (2.2)$$

$$\frac{\delta\Omega[n,\rho]}{\delta \rho} = 0 \qquad \text{for ions} \quad (2.3)$$

The variational equations are subject to the condition of electro-neutrality which can be written in terms of integrals over the correlation sphere of radius R containing the nucleus of charge Z at the origin. Thus

$$\zeta\int\rho(r)d\underset{\sim}{r} = Z - \int n(r)d\underset{\sim}{r} \quad (2.4)$$

This is a generalization of Eq. (1.1). Eq. (1.1) holds outside the correlation sphere, for $r > R$. Clearly the mean charge ζ may be considered to play the role of a Lagrange multiplier chosen to satisfy Eq. (2.4) for electro-neutrality. This point of view is slightly different to that of ref. 3 but is equivalent to the computational implementation of the method. The grand potential Ω appearing in (2.1) can be written as

$$\Omega = T[n,\rho] + F_e + F_{ei} + F_i - \mu_i \int \rho(r) d\underset{\sim}{r} - \mu_e \int n(r) d\underset{\sim}{r} \tag{2.5}$$

where

$$T[n,\rho] = \int F^0[n(r),\rho(r)] d\underset{\sim}{r} \tag{2.6}$$

$$F_e = -\int \frac{Z}{r} n(r) d\underset{\sim}{r} + \frac{1}{2} \int \frac{n(r)n(r')}{|\underset{\sim}{r}-\underset{\sim}{r}'|} d\underset{\sim}{r}\, d\underset{\sim}{r}' + \int F^e_{xc}[n] d\underset{\sim}{r} \tag{2.7}$$

$$F_{ei} = -\int \zeta \frac{\rho(r)n(r')}{|\underset{\sim}{r}-\underset{\sim}{r}'|} d\underset{\sim}{r}\, d\underset{\sim}{r}' + \int F^{ei}_c[n,\rho] d\underset{\sim}{r} \tag{2.8}$$

$$F_i = \int \frac{Z\zeta}{r} \rho(r) d\underset{\sim}{r} + \int \zeta^2 \frac{\rho(r)\rho(r')}{|\underset{\sim}{r}-\underset{\sim}{r}'|} d\underset{\sim}{r}\, d\underset{\sim}{r}' + \int F^i_c[\rho(r)] d\underset{\sim}{r} \tag{2.9}$$

Eq. (2.6) is the kinetic energy functional for a non-interacting system, having the exact interacting density distributions $n(r)$, $\rho(r)$. In (2.7) the first term is the "external potential" of the central ion of charge Z acting on the electron distribution $n(r)$. The 2nd term in (2.7) is the electron-electron interaction. The last term contains $F^e_{xc}[n]$, the exchange-correlation free energy functional of density functional theory. Similarly, Eq. (2.8) and (2.9) contain F^{ei}_c and F^i_c, viz., the electron-ion and ion-ion correlation functionals respectively. These are unique but unknown functionals of the density distributions. We shall assume that most of F^{ei}_c can be incorporated into F^i_c and neglect the last term of (2.8). In view of the functional differentations indicated by Eq. (2.2) and (2.3), the exchange-correlation free energy functional gets replaced by an exchange-correlation potential defined by

$$V^e_{xc}(r) = \frac{\delta F^e_{xc}[n]}{\delta n} \tag{2.10}$$

Similarly, the ion-ion correlation potential is given by

$$V^i_c(r) = \frac{\delta F^i_c[\rho]}{\delta \rho} \tag{2.11}$$

Finally, equations (2.2) and (2.3) reduce to the form

$$\left[\frac{\delta F^0}{\delta n} + V^e(r) - \bar{\mu}_e\right]\delta n = 0 \tag{2.12}$$

$$\left[\frac{\delta F^0}{\delta \rho} + V^i(r) - \bar{\mu}_i\right]\delta \rho = 0 \tag{2.13}$$

These are equations obeyed by non-interacting particles moving in the effective one-body potentials $V^e(r)$ and $V^i(r)$ where

$$V^e(r) = -[\frac{Z}{r} + V_p(r)] + V^e_{xc}(r) - V^e_{xc}(R)$$

$$V^i(r) = \zeta[\frac{Z}{r} + V_p(r)] + V^i_c(r) - V^i_c(R) \ . \tag{2.14}$$

The electrostatic (i.e., Poisson) contribution is

$$V_p(r) = \int \frac{[\zeta\rho(r') - n(r')]}{|\underset{\sim}{r}-\underset{\sim}{r}'|} d\underset{\sim}{r}' \ .$$

The chemical potentials $\bar{\mu}_e$ and $\bar{\mu}_i$ are essentially the non-interacting electron and ion chemical potentials at the densities $\bar{n}$ and $\bar{\rho}$.

The exchange-correlation free energy F^e_{xc} and the exchange-correlation potential V^e_{xc} are modelled from the properties of the uniform electron gas at the temperature T (see Perrot and Dharma-wardana[5]) using the local density approximation (LDA). Thus the essential content of the V^e_{xc} used in our calculations is the exchange graph and the summation of ring graphs. Similarly, the ion-correlation potential V^i_c is constructed from the sum of hypernetted chain (HNC) graphs.[3] These are evaluated with the $g_{ii}(r)$ of the plasma problem itself rather than with a $g_{ii}(r)$ of a model problem (say, the one-component plasma, OCP). This has the effect of including some contributions from electron-ion correlations in V^i_c and is a better procedure than using an LDA-form based on an OCP model. Hence, if we assume trial distributions $\tilde{n}(r)$, $\tilde{\rho}(r)$, the potentials given in (2.14) can be calculated and introduced into Eqs. (2.12) and (2.13). Eq. (2.12) applies to electrons; this reduces to the Kohn-Sham equation or the Thomas-Fermi equation depending on how the kinetic energy operator is treated. The Kohn-Sham analysis leads to a Schrodinger-like equation, viz.

$$[-\frac{\nabla^2}{2} + V^e(r)]\phi_\nu(r) = \varepsilon_\nu\phi_\nu(r) \tag{2.15}$$

where $\phi_\nu(r)$ are the Kohn-Sham eigenfunctions, with "energies" ε_ν. These are NOT energy levels and $\phi_\nu(r)$ are NOT single particle orbitals. But they give, (if not for the LDA) the exact density distribution n(r), as required by DFT. For a given trial ion distribution $\tilde{\rho}(r)$, Eq. (2.15) has to be solved self-consistently, just as for the Hartree equation, to give bound states as well as scattering states, all contributing to n(r). The solution has to satisfy the Friedel sum rule and ζ has to be fixed by the requirement of electroneutrality. The Friedel sum rule ensures that the modifications in the continuum density of states produced by the external potential is correctly taken into account.

Once n(r) is obtained for a given $\tilde{\rho}(r)$, a new effective ion-potential $V^i(r)$ is obtained from (2.14). Then a new $\rho(r)$ is calculated via Eq. (2.13). For a classical system Eq. (2.13) reduces to

$$\rho(r) = \bar{\rho}\, e^{-\beta V^i(r)} \ . \tag{2.16}$$

The process is iterated until both n(r) and $\rho(r)$ become self-consistent. In actual practice, at least for $\Gamma \lesssim 10$, numerical procedures are hardly

Table 1. Energy level structure of (i) an Fe atom in vacuum (ii) an Fe-ion in an iron-plasma at a temperature of 5 keV and electron density of 60.9 electrons per atomic unit of volume. The plasma is modeled by (a) jellium, with ζ = 22.87 (b) using the most probable ion distribution of Fig. 1, here ζ = 24.85. The Wigner-Seitz radius r_0 is 0.424 a.u.

	-Energy (Ry)			Occupation		
Level	(i)	(a)	(b)	(i)	(a)	(b)
1s	515.81	556.998	478.33	2	0.968	0.860
2s	60.957	77.301	7.940	2	0.406	0.348
2p	53.084	72.656	0.428	6	1.200	1.026
3s	7.2691	9.214	-	2	0.348	-
3p	4.8912	7.070	-	6	1.044	-
3d	0.9625	2.911	-	6	1.720	-
4s	0.5451	-	-	2	-	-

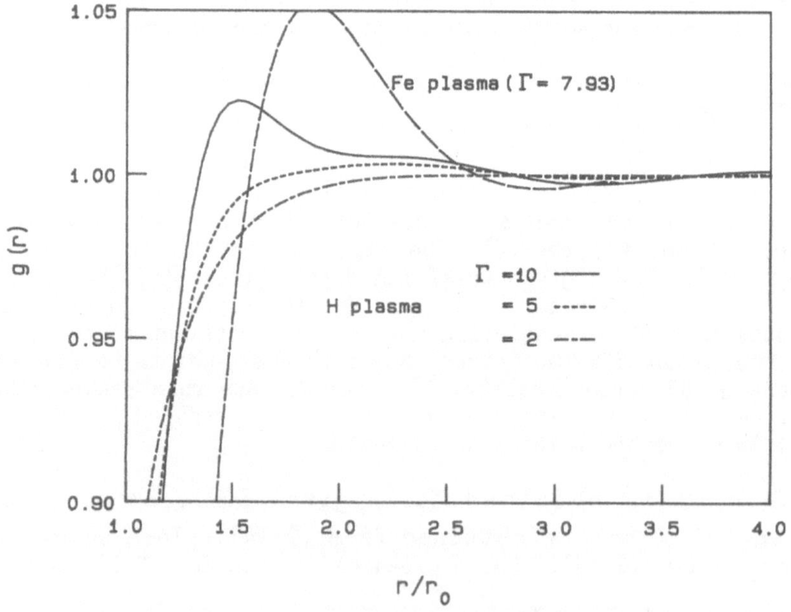

Fig. 1. The ion-ion pair distribution function g(r) for three hydrogen plasmas (r_s = 1, Γ = 2, 5, 10) and for an iron plasma (r_s = 0.42, $\Gamma \approx 8$) calculated from the coupled DFT-equations (2.15) and (2.16).

more complicated than carrying out a standard Kohn-Sham calculation or a calculation using, say, INFERNO (see Liberman[6]). For details the original literature should be consulted.

The density profiles $n(r)$ and $\rho(r)$ may be related to pair distribution functions via

$$g_{ii}(r) = \rho(r)/\bar{\rho}\ ,\ g_{ie}(r) = n(r)/\bar{n} \qquad (2.17)$$

thus providing a link with the theory of liquids. In fig. 1 we show examples of $g_{ii}(r)$ for a fully ionized H^+ plasma, with the ion sphere radius r_0 ($\equiv r_s^i$) = 1.0 (a.u.) and Γ = 10, 5 and 2. We also give the g(r) for an Fe-plasma with r_0 = 0.424, ζ = 24.85 and T = 5 keV (Γ = 7.93). In table 1 we show the Kohn-Sham "energy" eigenvalues obtained for an Fe nucleus (Z = 26) plus electrons placed in (a) a vacuum at T = 0 Kelvin; (b) a plasma modeled by a uniform ion distribution $\bar{\rho}$; (c) a plasma modeled by the most probable ion distribution $\rho(r)$, shown as the g(r) of the Fe plasma in fig. 1; this was obtained by iterating the coupled equations (2.15) and (2.16). These calculations ignore relativistic and spin-polarization effects. The "energy levels" are seen to shift as we go from the vacuum to the plasma environment. But this shift should NOT be identified as a "plasma polarization shift" since the Kohn-Sham eigenvalues do not correspond to the elementary excitation spectrum (see the paper by W. Kohn in this volume). François Perrot's paper in this volume is an attempt to correct the DFT eigenvalues so that they can be used for calculating optical transitions. Plasma polarization shifts can in principle be calculated by such a method. The rigorous alternative is to solve the Dyson equation after calculating the self-energy from many-body perturbation theory using the Kohn-Sham basis $\phi_\nu(r)$, ε_ν. Some typical results from such a calculation are given in table 2. For more details the reader should consult reference 4.

3. EFFECTIVE ION-ION POTENTIALS AND THE DYNAMIC STRUCTURE FACTOR

From the previous section it is clear that DFT can be used to calculate the ion-ion pair distribution function $g_{ii}(r)$. This enables us

Table 2. The excitation energies E_{1s}, and lifetimes γ_{1s} obtained by solving the Dyson equation for H-plasmas (at temperature T) having only one (1s) bound state. The Kohn-Sham 1s energy, ε_{1s} is also given. Atom units ($e = \hbar = m = 1$) are used throughout. E_F is the Fermi energy

r_s	T/E_F	ε_{1s}(Kohn-Sham)	E_{1s}	γ_{1s} ×100
3	1.0	-0.0528	-0.0045	3.80
	3.0	-0.1545	-0.1066	0.09
2	1.0	-0.0124	+0.0422	5.23
	3.0	-0.0938	-0.2089	0.16
1	2.0	-0.0000	-0.4571	0.61
	3.0	-0.0131	-0.6265	0.17

to construct an effective one-component classical model of the plasma as far as ions are concerned. This is the usual description of simple liquid metals where the electron coordinates have been integrated away to give a description in terms of effective pair potentials between ions. If the electron-ion interaction were weak, then the ion-ion pair potential $V_{ii}(r)$ can be written down in linear response theory (LRT). The Fourier transform of $V_{ii}(r)$ is given by

$$V_{ii}(q) = \zeta^2 V_q + |V_{ie}(q)|^2 \chi(q) \tag{3.1}$$

where $V_q = 4\pi/q^2$ and $\chi(q) = (1/\varepsilon(q)-1)V_q^{-1}$. Here $\varepsilon(q)$ is the dielectric function of the interacting uniform electron gas at a temperature T and electron density $\bar{n}$. Also, $V_{ie}(q)$ is the electron-ion pseudopotential which in the crudest approximation reduces to the bare Coulomb potential $-\zeta V_q$. If $\varepsilon(q)$ is also taken in the Debye approximation Eq. (3.1) gives $V_{ii}(r) = \zeta^2 e^{-k_D r}/r$ where k_D is the Debye screening momentum.

Unfortunately, $V_{ie}(q)$ cannot in general be replaced by the bare point-ion Coulomb potential $-\zeta^2 V_q$ if linear response theory (LRT) is to be

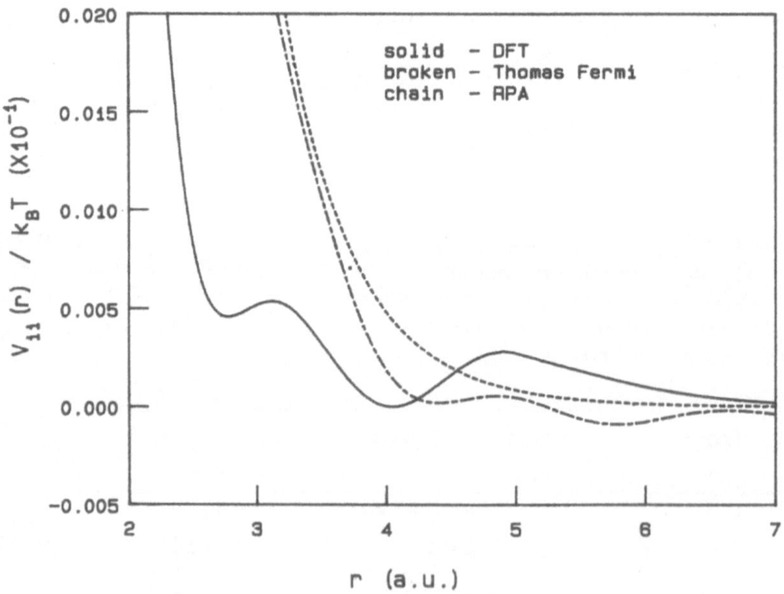

Fig. 2. The effective proton-proton interaction potential $V_{ii}(r)$ in a H-plasma at $r_s = 1$, $\Gamma = 10$ obtained by inventing the DFT-g(r) is given as a solid line. The $V_{ii}(r)$ obtained from LRT using finite temperature RPA and Thomas-Fermi screening is also given.

applied. Further, even when LRT is applicable, various formulations of the dielectric function $\varepsilon(q)$ are available and lead to different results for $V_{ii}(q)$. Hence the construction of $V_{ii}(q)$ for strongly coupled systems using (3.1) runs into difficulties.

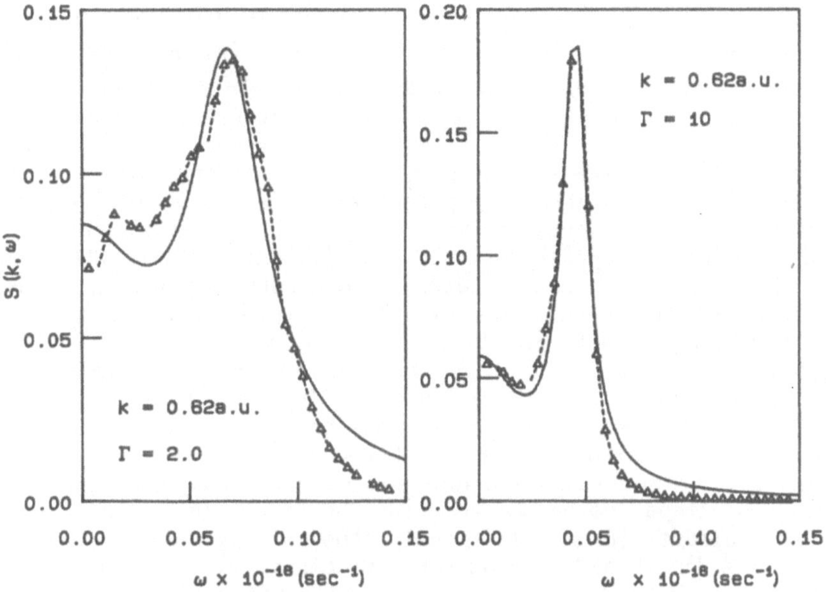

Fig. 3. The proton-proton potential $V_{ii}(r)$ obtained from the DFT-g(r) has been used in molecular dynamics to generate $S(k\omega)$ for H-plasmas (r_s = 1, Γ = 2 and Γ = 10). The MD-data are shown as triangles. The solid line is a fit to a hydrodynamic model.

A direct approach to $V_{ii}(r)$, which does not suffer from these limitations and does not assume that the response is linear is provided by the information contained in the $g_{ii}(r)$ obtained from DFT. From the statistical mechanics of liquids, $g_{ii}(r)$ is directly related to $V_{ii}(r)$ by[7]

$$g_{ii}(r) = \exp[-\beta V_{ii}(r) + h(r) - c(r) + B(r)] \tag{3.2}$$

where

$$h(r) = g_{ii}(r) - 1$$

and c(r) is such that its Fourier transform is given in terms of the Fourier transform of h(r) by

$$c(k) = h(k)/[1 + \bar{\rho}h(k)] \quad (3.3)$$

Hence, if $g_{ii}(r)$ is known $V_{ii}(r)$ is available if the "bridge function" B(r) were known. For fluids with $\Gamma \lesssim 10$ it is known that the HNC approximation of setting $B(r) = 0$ is an excellent approximation. Hence the effective ion-ion pair potential in a plasma can be determined within the HNC-approximation by inverting the $g_{ii}(r)$ obtained from DFT.[7] Such a pair potential for proton-proton interactions in a H-plasma with $r_s = 1$ and $\Gamma = 10$ is shown in fig. 2, together with $V_{ii}(r)$ obtained if $\varepsilon(q)$ is approximated by the Thomas-Fermi or RPA dielectric functions. (The use of a more sophisticated $\varepsilon(q)$ does not improve matters very much.)

A description of the plasma as a single component fluid of screened ions interacting via an effective potential $V_{ii}(r)$ is valid for processes with characteristic frequencies $\omega < \omega_p^e$ viz., the electron plasma frequency. Using the $V_{ii}(r)$ obtained from inversion of the HNC equation, machine simulation (MS) can be used to generate the dynamic structure factor $S(k\omega)$ of the ions. The static structure factor $S(k)_{MS}$ obtained from MS should agree with the S(k) of the original DFT-g(r). This was found to be true (F. Nadin et al.[8]) for all the potentials given in ref. 7. In fig. 3 we give an example of $S(k\omega)$ for hydrogen plasmas at $r_s = 1$, $\Gamma = 10$ and $\Gamma = 2$. The latter case has been directly simulated as a two component fluid by Hansen and MacDonald[9] but the ion-ion dynamic structure factor was not obtained and only $S_{ee}(k\omega)$ was reported since many more time steps are needed to relax the ions. In any case, there is reason to believe (from electrical conductivity calculations) that $r_s = 1$, $\Gamma = 2$ is already beyond the semi-classical regime for which the Hansen-MacDonald method is valid. It is clear from fig. 3 that there is considerable structure in $S(k\omega)$ but ref. 8 must be consulted for a fuller discussion.

4. ELECTRIC MICROFIELDS IN PLASMAS

Ion density fluctuations in the plasma lead to fluctuations in the electric field at a "radiating atom" placed at the origin. If some configuration of ions, given by the positions $\underset{\sim}{r}_1, \underset{\sim}{r}_2 \cdots \underset{\sim}{r}_n$ occur as an ion-density fluctuation, the corresponding electric field at the radiator is $\underset{\sim}{\varepsilon}(\underset{\sim}{r}_1,\underset{\sim}{r}_2,\cdots\underset{\sim}{r}_n)$. Ion density fluctuations are quasi-static as far as electrons are concerned and these are called low frequency microfields. The microfield distribution $W(\underset{\sim}{E})$ is the probability of occurrence of the field $\underset{\sim}{E}$, and is given by the ensemble average $\langle\delta(\underset{\sim}{E}-\underset{\sim}{\varepsilon})\rangle$. For a plasma with spherical symmetry around the radiator $W(\underset{\sim}{E}) = W(E)$ and the Fourier transform W(k) is usually the object of the theoretical analysis. The considerable theoretical sophistication that has gone into the evaluation of W(k) is clear from the paper by James Dufty in this volume. However, this sophisticated theory has been developed in the context of a very simplified model of plasmas in which the plasma is treated as a one-component classical fluid of point ions interacting via a Debye-type ion-ion pair potential. Density functional theory provides us with a more realistic and computationally practical model of the plasma necessary to deal with strong electron-ion interactions, high Z species, and internal structure due to the presence of bound states.

In the DFT approach to microfields Dharma-wardana and Perrot[10] use the Baranger-Mozer (BM) cluster expansion for W(k). Thus

$$W(k) = e^{S(k)} \tag{4.1}$$

$$S(k) = \bar{\rho}w_1(k) + \frac{1}{2}\bar{\rho}^2 w_2(k) + \frac{1}{6}\bar{\rho}^3 w_3(k) + \dots \tag{4.2}$$

The 2nd order term is of the form

$$w_2(k) = \int \phi_1(k)\phi_2(k) c_2(\underset{\sim}{r}_1, \underset{\sim}{r}_2) d\underset{\sim}{r}_1 d\underset{\sim}{r}_2 \tag{4.3}$$

$$c_2(\underset{\sim}{r}_1, \underset{\sim}{r}_2) = g(\underset{\sim}{r}_1, \underset{\sim}{r}_2) - g(\underset{\sim}{r}_1)g(\underset{\sim}{r}_2) \tag{4.4}$$

$$\phi_1(k) = \exp(i\underset{\sim}{k}\cdot\underset{\sim}{\varepsilon}_1) - 1 \tag{4.5}$$

In (4.4) we have the ion-pair distribution functions $g(\underset{\sim}{r}_1, \underset{\sim}{r}_2)$ and the one-body functions $g(\underset{\sim}{r}_1)$, $g(\underset{\sim}{r}_2)$ evaluated in the "external field" of the radiator atom held at the origin. In (4.5) ε_1 is the electric field at the origin due to the plasma ion at $\underset{\sim}{r}_1$. Thus the calculation of the microfield involves

(i) a method of calculating the relevant ion-pair distribution functions (PDFs) for plasmas of arbitrary density and electron degeneracy
(ii) a method of evaluating the electric charge at the radiator (i.e., at the origin)
(iii) a technique for approximate resummation of the higher order terms in (4.2) since explicit evaluation beyond 2nd order is impractical.

Existing methods have treated (i) and (ii) by assuming weak electron-ion coupling and using a classical model of Debye screened point ions. Electrons do not appear even in the calculation of the electric field since this is taken to be the derivative of the Debye potential.

In the DFT-model the relevant PDFs (necessary for the 2nd order B-M calculation) are obtained by solving the coupled equations (2.15) and (2.16), with the appropriate "external potentials". The electric field, say $\underset{\sim}{\varepsilon}_1$ of Eq. (4.5) is calculated using the electron charge pile up n(r) to screen the ion which generates the field. Thus the electron-ion interaction is not assumed to be weak; bound states and internal structure of perturbers and radiators are correctly taken into account automatically.

The resummation problem, viz., approximating the terms beyond 2nd order in BM is solved in the context of the BM-cluster expansion, at least for $\Gamma \lesssim 10$, by the method of Perrot and Dharma-wardana.[11] It appears that although the resummation is important for OCP and for weakly screened plasmas, it does not seem to be important for strongly screened plasmas, if the screening etc. is correctly treated as in DFT. In fig. 4(left) we show the DFT-microfields and APEX-microfields for a neutral and for a charged point in an aluminum plasma. Contributions to the DFT-microfield beyond 2nd order proved to be negligible. In fig. 4(right) we show the DFT-microfield at an H-atom (i.e., a proton carrying an electron in a 1s-bound state) in a dense hydrogen plasma (r_s = 1.0). Although such an H-atom is not stable in a dense plasma the

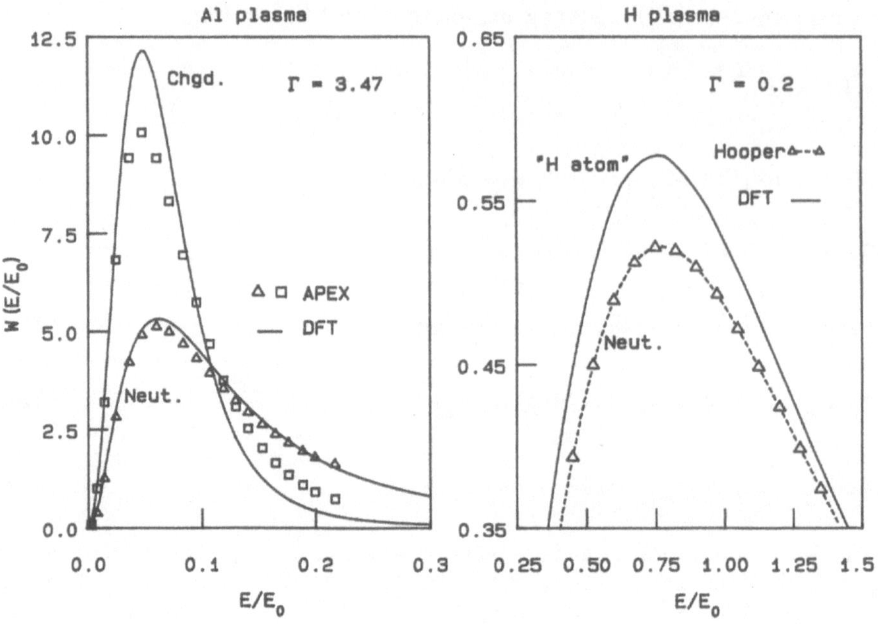

Fig. 4. The left panel shows the DFT-microfields (solid lines) at a charged test point and at a neutral test point in an Aℓ-plasma. The corresponding APEX-calculations are also given. The right panel shows the effect of structure in the test particle, taken to be an "H-atom" and a neutral point in an H-plasma. See ref. 10 for details.

high electron density is chosen to ensure validity of the linear screening approximations of standard microfield theories. However, these theories treat the H-atom as a neutral structureless point and give results different to DFT where the internal structure is taken into account in calculating the necessary distribution functions.

5. ELECTRICAL RESISTIVITY

We shall use the calculation of the electrical resistivity R of a plasma to demonstrate how linear transport coefficients (e.g., thermal conductivity, thermopower, diffusion coefficients) can be conveniently calculated using the results of DFT. Only a brief summary is given here and the reader is referred to Perrot and Dharma-wardana[12] for more details.

The main contribution to the resistivity R is the resistivity due to electron-ion scattering viz. R_{ei}. Note that we evaluate the resistivity, i.e., an inverse transport coefficient, rather than the conductivity σ which appears naturally in Kubo type theories. Although the Kubo approach is formally attractive, it leads to poor results unless extreme care is used in evaluating the vertex functions which are submerged in the integral equations for the correlation functions. On the other hand, owing to "built-in" conservation principles, even crude approximations to the Bolzmann equation and its generalizations lead to good results. Here we use rigorous generalizations to the Ziman formula which go beyond the simple Bolzmann equation and derivable using methods of multiple scattering theory.[13,14] DFT becomes relevant because the various quantities (e.g., structure factors, scattering cross sections) needed to evaluate these Ziman-type formulae are readily obtainable from density functional calculations.

Let us recall the essential approximations of the density functional calculation: (i) local density approximation for the electron exchange and correlation potential V_{xc} in the Kohn-Sham equation (ii) HNC approximation for the ion-ion correlations (iii) spherically symmetric, average-atom description of the scattering centers in the plasma; this neglects the effects of electron-configuration fluctuations and ion-distribution fluctuations in the plasma. For instance, the mean ionic charge ζ is an average over many values of the effective charge Z_c of a given short-lived electronic configuration c of the scatterer. There is some reason to believe that the neglect of fluctuations is valid to lowest order due to compensating effects.[12]

The calculation of the electrical resistivity can be classified into three types of regimes. These are (i) plasmas with weak isolated scatterers, WIS (ii) plasmas with strong isolated scatterers SIS (iii) dense plasmas where the scattering centers can no longer be considered isolated. This is the strong multiple scatterer (SMS) regime. Much of the existing calculations treat the WIS regime where the electron-ion interaction is treated by linear screening.[15] Some calculations have treated the SIS regime using a t-matrix formulation of the electron-ion interaction.[16] The DFT calculation enables us to treat the SMS regime also with no additional difficulty. Here we will not discuss the simple WIS regime but consider the more general SIS and SMS regimes.

For these two regimes we can write the generalized Ziman formula as

(i) SIS-model

$$R = \frac{\hbar}{3\pi^2\zeta^2 e^2 \bar{\rho}} \int_0^\infty d\varepsilon f'(\varepsilon) \int_0^{2k} dq\; q^3 s(q)\bar{\tau}(q) \tag{5.1}$$

with $q = 2K^2(1-\cos\theta)$, $\varepsilon = k^2/2$

(ii) SMS-model

$$R = \frac{\hbar}{3\pi^2 e^2 \bar{n}^2} \int_0^\infty d\varepsilon \tilde{f}(\varepsilon) \int_0^{2k} dq\; q^3 \tau(q) \tag{5.2}$$

In (5.1) $\bar{\tau}(q)$ is the "average-atom" scattering cross section for an isolated ion of charge ζ in a uniform neutralizing background. S(q) is the structure factor calculated from the DFT ion-ion pair distribution function $g_{ii}(r)$. Also $f'(\varepsilon)$ is the energy derivative of the Fermi-Dirac distribution at the energy ε and plasma temperature T. The scattering cross section $\bar{\tau}(q)$ is constructed from the DFT phase shifts for an "isolated ion in a uniform-background" calculation. Thus

$$\bar{t}(q) = \left|\frac{1}{2ik}\sum_{\ell}(2\ell+1)(e^{2i\delta_\ell(k)}-1)P_\ell(\cos\theta_{kk'})\right|^2 . \qquad (5.3)$$

Note that when we go to the multiple scatterer case, Eq. (5.2), the structure factor $S(q)$ has dropped out and a scattering cross section $\tau(q)$ has been introduced. In the isolated scatterer limit the structure factor $S(q)$ converts the isolated scattering cross section $\bar{t}(q)$ into the full scattering cross section $\tau(q)$ of the fluid. That is, in the isolated scatterer limit we have

$$\tau(q) = \bar{\rho}S(q)\bar{t}(q) \qquad (5.4)$$

In the SMS limit the superposition of isolated scatterers implied by (5.4) is invalid and $\tau(q)$ has to be evaluated directly. This is done in DFT, using the spherical-averaged average-configuration model, via a formula analogous to (5.3) but with the phase shifts $\delta_\ell(k)$ replaced by $\Delta_\ell(k)$. These are the phase shifts for electron scattering calculated using DFT with the scatterer ion placed in its plasma environment defined by Eq. (2.16), i.e., as given by its $g_{ii}(r)$. Also, in (5.2) we have a modified distribution $\tilde{f}'(\varepsilon)$ rather than the simple Fermi-Dirac $f'(\varepsilon)$ as in (5.1). This is because in the SMS regime the electron mean free path may become comparable to typical ion-ion correlation lengths and hence the electron momentum $\underset{\sim}{k}$ is no longer a good quantum number. For more details of such many body corrections the original paper should be consulted.[12] For most practical calculation $\tilde{f}(\varepsilon)$ may be taken to be the simple Fermi function.

In table 3 we give results for a Xe-plasma, an Fe-plasma, and for two hydrogen plasmas. For other examples see ref. 12. The Xe-plasma and the Fe-plasma involve scatterers with internal structure. The mean ionic

Table 3 DFT results for the resistivity (ohm cms) in the SIS and SMS models for some plasmas. R_{other} are results of other calculations.

	T(eV	r_s^I(a.u.)	ζ_{SIS}	ζ_{SMS}	R_{SIS}	R_{SMS}	R_{other}
Xe	2.12	7.0	0.70	0.75	(0.315	-	0.271[a]) $\times 10^{-2}$
Fe	5×10^3	0.42	22.9	24.9	(0.956	1.44	0.827[b]) $\times 10^{-6}$
H	13.6	1.0	1.0	1.0	(0.203	0.233	0.111[c]) $\times 10^{-4}$
H	25.1	2.0	0.74	0.78	(0.858	0.903	-) $\times 10^{-4}$

[a]Ref. 15; [b]Ref. 16; [c]Ref. 9

charge ζ evaluated in the SIS and SMS models differ significantly in the case of the Fe-plasma where multiple scattering effects are important. The ion-ion pair distribution function of this iron plasma was already shown in fig. 1. The hydrogen plasma at $r_s = 1$, $\Gamma = 2$ and $T = 13.6$ eV corresponds to the case studied by molecular dynamics (MD) by Hansen and MacDonald.[9] These authors used an effective electron-ion interaction with cutoffs to prevent classical collapse. The electrons are also treated as classical particles in the simulation. The use of such effective potentials for transport properties, and the use of a classical approximation for the electrons in this regime of plasma parameters are somewhat questionable and may be the source of the difference between the DFT result and the MD result.

6. LIGHT ABSORPTION BY PLASMAS

Light absorption by plasmas involves (a) bound-bound or line transitions (b) bound-free or photo-transitions and (c) free-free or brehmstrahlung processes. Such a description is possible in a simple single particle approximation,[17] in terms of a set of one-electron states $\phi_i = |i\rangle$ with energies ε_i. In practice these states are not so well defined due to the effect of ion-radiator and electron-radiator collisions. If we adopt the attitude that the ions may be treated by some quasi-static approximation, and neglect energy level widths for the moment, the light-absorption cross section in the dipole approximation can be written as

$$\sigma(\omega) = 4\pi^2(\omega/c) \sum_{i,j} (f_i - f_j)\langle i|Z|j\rangle^2 \delta(\omega + \varepsilon_i - \varepsilon_j) \qquad (6.1)$$

where the electromagnetic field of frequency ω is assumed to be polarized in the z-direction. The dipole or "length form" $\langle i|Z|j\rangle$ contained in (6.1) can be rewritten as a matrix element of the velocity operator to give the "velocity form" of $\sigma(\omega)$. Let us consider the much simpler case of an isolated Xe-atom rather than a plasma. The Kohn-Sham LDA orbitals $|i\rangle, |j\rangle$ give the same value of $\sigma(\omega)$ whether we use the length or velocity form. The Hartree-Fock orbitals do not have this property and give different values of $\sigma(\omega)$. In fig. 5 several calculations of the photo-absorption cross section of dilute atomic Xe is given.[18] The experimental points are given as triangles. It is clear that neither the Hartree-Fock nor the LDA calculation agrees with experiment.

Thus, even aside of ion and electron broadening effects which do not arise in the case of atomic Xe, the calculation of the light absorption cross section requires a many-body treatment beyond the independent particle model. A very successful method of incorporating these many-body effects for isolated atoms is due to Zangwill and Soven.[18,19] This method uses the Kohn-Sham (density functional) eigenstates calculated in the LDA but introduces the time dependent relaxation of the system whereby the external field is screened by the electrons in the system. This method, called the time dependent local density approximation, viz., TD-LDA, constructs the response of the system to the external field using the Kohn-Sham basis $|i\rangle$ rather than, say, a plane wave basis and brings in channel mixing and density fluctuation effects of the true inhomogeneous (atomic) system. In other words, the dipole operator Z appearing in (5.1) is replaced in the TD-LDA by a space and time dependent screened, complex dipole operator $Z(\mathbf{r},\omega)$.

We have adapted the TD-LDA method (see Grimaldi, Grimadi-Lecourt and Dharma-wardana[20]) for plasma calculations and applied it to the case of an iron plasma at normal compression and at a temperature of 100 eV.

Unlike in an isolated atom, bound-bound, bound-free and free-free processes all contribute and also lose their identity in the many-body picture due to channel mixing. The screened driving field is found to be significantly different to the external field in regions of strong channel mixing. Typical results are shown in fig. 6. The

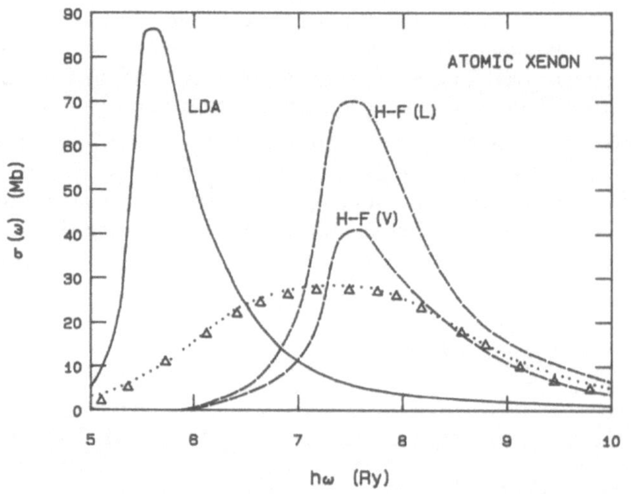

Fig. 5. The photoabsorption cross section of atomic xenon calculated using the Hartree-Fock length H-F(L) and velocity formulae H-F(V), and using Kohn-Sham LDA. Experimental points are triangles. The dotted line close to the experiment is TD-LDA.[18,19]

photoabsorption cross section as calculated using the TD-LDA (solid line) and the standard single particle picture (dotted line) are shown in fig. 7. The mixing of line processes is clearly evident here. A total cross section calculation is given in ref. 20. In these calculations we have also included electron-collision broadening effects by including self-energies in the construction of the response function. In simplified terms, this has the effect of replacing the $\delta(\omega+\varepsilon_i-\varepsilon_j)$ by a lineshape function (for details see ref. 20). Ion broadening effects will be included in a future study directly via the response function or indirectly via a microfield approach.

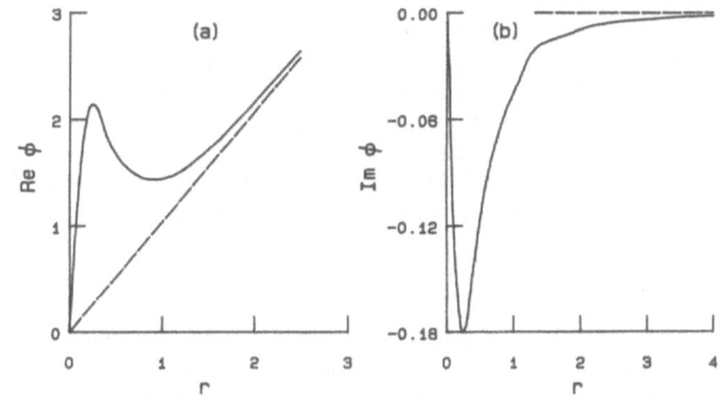

Fig. 6 The external potential (dashed line) is the dipole field given by r with zero imaginary part. The relaxation of the system converts the external potential to the driving field $\phi(r,\omega)$. Its real and imaginary parts are given[20] for ω = 27.1 a.u.

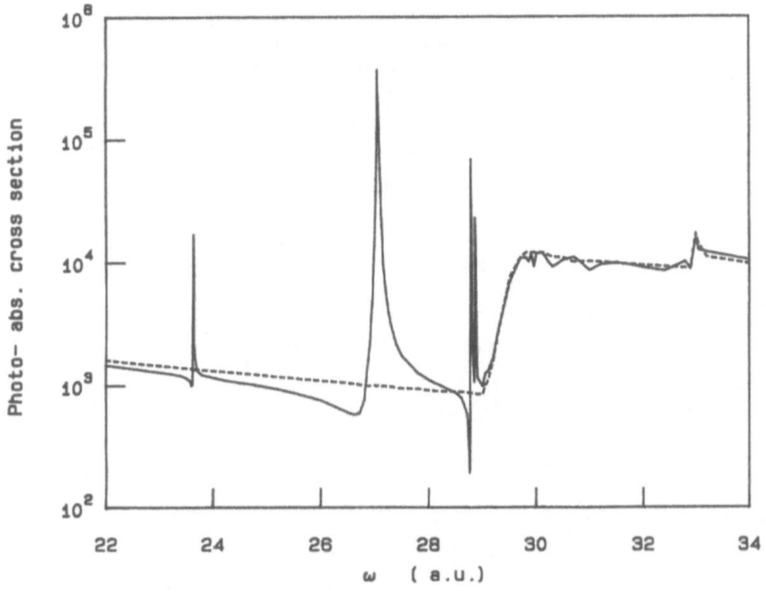

Fig. 7 The photoabsorption cross section for an Fe-plasma at 100 eV. The dotted line is the usual single particle theory. The solid line is TD-LDA.[20]

7. CONCLUSION

In this paper we have presented a brief survey of the applications of density functional theory to a variety of topics in strongly coupled plasmas of arbitrary electron degeneracy. It is evident that density functional methods, coupled with a modest amount of standard many body theory provide a powerful and computationally practical first principles method for tackling almost any property of strongly coupled plasmas. In this review we have not emphasized "straightforward" thermodynamic properties (e.g., equations of state, phase separation etc.) for which DFT is expected to be extremely good. Instead we have discussed properties which are related to fluctuations and dissipations since these are usually thought to be more difficult to formulate using DFT.

REFERENCES

1. W. Kohn and P. Vashista, in Theory of the Inhomogeneous Electron Gas, S. Lundqvist and N.H. March, ed., Plenum, New York (1983).
2. M.W.C. Dharma-wardana, J. Quant. Spectrosc. Radiat. Transfer 27:315 (1982).
3. M.W.C. Dharma-wardana and F. Perrot, Phys. Rev. A 26:2096 (1982).
4. F. Perrot and M.W.C. Dharma-wardana, Phys. Rev. A 29:1378 (1984).
5. F. Perrot and M.W.C. Dharma-wardana, Phys. Rev. A 135:231 (1985).
6. D.A. Liberman, J. Quant. Spectrosc. Radiat. Transfer 35:335 (1982).
7. M.W.C. Dharma-wardana, F. Perrot and G.C. Aers, Phys. Rev. A 28:344 (1983).
8. F. Nadin, G. Jacucci and M.W.C. Dharma-wardana (unpublished).
9. J.-P. Hansen and I.R. MacDonald, Phys. Rev. A 23:2041 (1981).
10. M.W.C. Dharma-wardana and F. Perrot, Phys. Rev. A 33:3303 (1986).
11. F. Perrot and M.W.C. Dharma-wardana, Physica A 134:231 (1985).
12. F. Perrot and M.W.C. Dharma-wardana (unpublished).
13. S.F. Edwards, Proc. Phys. Soc. 86:977 (1965).
14. A. Ferraz and N.H. March, Phys. Chem. Liq. 8:271 (1979).
15. S. Ichimaru and S. Tanaka, Phys. Rev. A32:1790 (1985).
16. G.A. Rinker, Phys. Rev. B 31:4207 (1985).
17. F. Perrot and M.W.C. Dharma-wardana, Phys. Rev. A 31:970 (1985).
18. A. Zangwill, Ph.D. Thesis, University of Pennsylvania (1981), also A. Zangwill in EXAFS and Near Edge Structure III, K.O. Hodgson, B. Hedman, and J.E. Penner-Hahn, eds., Springer-Verlag, Berlin (1984).
19. A. Zangwill and P. Soven, Phys. Rev. A 21:1561 (1980).
20. F. Grimaldi, A. Grimaldi-Lecourt and M.W.C. Dharma-wardana, Phys. Rev. A 32:1063 (1985).

DESCRIPTION OF ATOMIC SPECIES IN DENSE PLASMAS USING A DENSITY-FUNCTIONAL-THEORY APPROACH

François Perrot

Centre d'Etudes de Limeil-Valenton, B.P. 27
94190 Villeneuve St Georges, France

I. INTRODUCTION

A key quantity in plasma physics is $\langle Z^* \rangle$, the average ionization or average number of free electrons per atom. Although there is no unique definition of this quantity (what is exactly a free electron in a dense plasma ?), it is extensively used to characterize the plasma, for instance by means of the coupling parameter Γ. The average ionization may be derived from standard theories : the Saha theory for low densities, or the "Average Atom" (AA) model for dense plasmas[1].

But situations do exist, where a more detailed description of the plasma is needed. For instance, fluctuations may be important : Γ is proportional to $\langle Z^{*2} \rangle$, which may be different from $\langle Z^* \rangle^2$. The fluctuations may play a role in the study of conductivity, bremsstrahlung, and in the calculation of electric microfields also. A description of the detailed configurations of the plasma ions, which goes beyond the AA picture, may be crucial for radiative properties, such as photoeffect threshold, line opacities... Finally, non-equilibrium plasmas obviously require the knowledge of many atomic data as functions of the "true" (integer) occupation numbers of the bound electronic levels. These examples show that one needs a model for the various atomic species existing in the plasma.

In a dense plasma, it is impossible to treat one ion independently of the surrounding electron cloud. The free electrons pile up around a positive ion and tend to form a neutral atom. Also, they relax following any change in the configuration of the bound spectrum, so that there is a self-consistent modification of the whole spectrum. As the number of free states is obviously infinite, these states must be treated statistically in any practical model. In Sec. II, we shall describe such a model, combining the detailed study of the bound states and the statistical treatment of the free spectrum. The problem of bound-bound transitions will be treated in Sec. III. Average atomic species, representative of all the atomic species having a given ionization degree Z^* will be defined and calculated in Sec. IV, leading to the possibility of calculating the average of any quantity depending on Z^*.

II. THE MODEL

Definition of a species

An "atomic species", as considered in the present work, is a fictitious atom made of a given number Z_b of bound electrons distributed among a set of bound levels with integer occupation numbers $n_i = 0$ or 1, and neutralized by a cloud of delocalized electrons coming from the continuous spectrum. It is assumed that local thermal-equilibrium (LTE) is established among these quasi-free electrons. Such a species may be interpreted as the result of averaging on the LTE spectrum all the electron charges of real atoms having the fixed configuration $\{\ldots n_i \ldots\}$ in the bound spectrum.

All the models dealing with detailed ionic configurations assume that the free electrons are in LTE[2-4]. This is not a severe restriction for dense plasmas where the free spectrum is dominated by electron collisions, the rate of which increases like the square of the electron density.

Total energy and occupation numbers

The standard theories of atomic structure, at zero temperature (Slater[5], Kohn and Sham[6]) have shown that the first-order variation of the total energy of a system, when its configuration changes, is :

$$\delta E\,(\ldots\, n_j\, \ldots) = \sum_j \varepsilon_j\, \delta n_j \tag{1}$$

Eq.(1) holds only if exchange-and-correlation effects are approximated by means of a <u>local</u> functional $E_{xc}(\rho)$ of the total electron charge density $\rho(r)$. The ε_j's are the eigenvalues of a one-particle <u>effective</u> Schrödinger equation. The density $\rho(\underline{r})$ is obtained using the eigenfunctions of this equation :

$$\rho(\underline{r}) = \sum_j n_j\, \phi_j^*(\underline{r})\, \phi_j(\underline{r}) \tag{2}$$

An important property of this "Density-Functional-Theory" (DFT) is that Eq.(1) takes orbital relaxation into account (to first order in $\delta\phi_j$), as a consequence of the stationarity of E with respect to the variations of ρ. This is a significant advantage of DFT over Hartree-Fock theory where the total energy variation, as deduced from Koopman's theorem :

$$E(\ldots\, n_j = 1 \ldots) - E(\ldots\, n_j = 0\, \ldots) = \varepsilon_j$$

implies frozen orbitals. A second advantage of DFT is its much greater simplicity in practical calculations. For these reasons, we used the DFT approach for our description of atomic species.

The free-spectrum equations

We simulate an atom in the plasma in the following way. A bare nucleus of charge Z is embedded in an electron gas of uniform density $\overline{\rho}$. The plasma ions are represented by a uniform charge background which preserves electrical neutrality. In this work, we neglect the details of the ionic profile around the ion at the center. We choose the most simple shape of this profile, a spherical cavity in the uniform ionic background :

$$\rho_+(r) = \begin{cases} 0 & \text{if } r \leqslant R_c \\ -\overline{\rho} & \text{if } r \geqslant R_c \end{cases} \tag{3}$$

The radius R_c is such that the excluded charge is exactly that of the ion, as given in the LTE average atom (AA) model :

$$\frac{4}{3} \pi R_c^3 \overline{\rho} = Z_c = Z^*_{AA} \tag{4}$$

Owing to the very different time scales, this ionic profile is considered as fixed during times characteristic of the electronic relaxation.

The configuration of the immersed ion is $\{\ldots n_i \ldots\}$, with a total bound charge $Z_b = \Sigma_i n_i$. The plasma electrons polarize around it to form a neutral atom. According to the assumptions of the model, the free-electron density describes a subsystem in LTE, so that it satisfies the Kohn-Sham-Mermin[7] equations :

$$F_\ell [\rho_\ell] = F_o [\rho_\ell] + \rho_\ell \cdot v_\ell + \frac{1}{2} \rho_\ell \cdot \frac{1}{r} * \rho_\ell + F^T_{xc} [\rho_\ell] \tag{5}$$

$$\frac{\delta F_\ell [\rho_\ell]}{\delta \rho_\ell(r)} = \mu \tag{6}$$

F_ℓ is the total free-energy of the continuum electrons, a unique functional of their charge density. It includes the non-interacting free-energy $F_o [\rho_\ell]$, the interaction energy with the external field v_ℓ, the electron-electron Coulomb interaction and the exchange-and-correlation (XC) free-energy $F^T_{xc} [\rho_\ell]$. Standard notations (• for the scalar product and * for the convolution product) have been used. In this model of embedded atom, the chemical potential depends on the uniform density $\overline{\rho}$ and temperature T only. The density ρ_ℓ is determined by the stationarity condition Eq.(6), for fixed T and μ. The exact Eqs.(5,6) cannot be solved without replacing the unknown functional $F^T_{xc} [\rho_\ell]$ by an approximate one. The common approximation is the local density approximation (LDA) : $F^T_{xc} [\rho_\ell] \rightarrow F^T_{xc}(\rho_\ell)$, (the functional becomes a scalar function), frequently applied (at T = 0) in solid state physics[8]. In our calculations, we used a numerical fit of the results of Rajagopal and Gupta[9], and Perrot and Dharma-Wardana[10] for the homogeneous electron gas at finite temperature.

With this approximate treatment of XC effects, Eq.(6) is equivalent to the following system :

$$(-\frac{1}{2} \nabla^2 + U_\ell) \phi_m = \varepsilon_m \phi_m \tag{7}$$

$$U_\ell = v_\ell + \frac{1}{r} * \rho_\ell + \frac{dF^T_{xc}(\rho_\ell)}{d\rho_\ell(r)} \tag{8}$$

$$\rho_\ell = \sum_{\varepsilon_m \geqslant 0} f(\varepsilon_m) \phi^*_m \phi_m \tag{9}$$

$$f(\varepsilon_m) = [1 + \exp \{\beta(\varepsilon_m - \mu)\}]^{-1} \tag{10}$$

The interacting electron gas is equivalent to an assembly of non-interacting pseudo-particules in an effective potential U_ℓ, Eq.(8). This potential, which includes three contributions (external potential v_ℓ, Coulomb and XC potentials of the free electrons), must be determined self-consistently. The density ρ_ℓ, Eq.(9), is built using those eigenfunctions ϕ_m of the effective one-particle equation which correspond to positive eigenvalues (free spectrum). The occupation numbers of these states are average occupation numbers given by the Fermi-Dirac statistics, Eq.(10). The external field v_ℓ for the plasma electrons is created by the ions : the bare central ion and its bound electrons, the ionic background ρ_+ ; it includes also a contribution for XC effects between bound and free electrons. A detailed description of the latter is given in an extended version of this work[11].

Description of the ionic bound-electrons configuration

A particular atomic species is defined by a given configuration of the ion, i.e. a set of integer occupation numbers for the bound levels. For the bound electrons, there is no notion of temperature or of entropy since the configuration is frozen. The relevant thermodynamic quantity is thus the internal energy E. Let us note $\rho_b(r)$ the charge density of the bound electrons subsystem. This subsystem may be in the ground state of the ion, or more generally in an excited state. It feels an external potential v_b created by the bare nucleus, the background ρ_+, the free-electrons ρ_ℓ ; v_b includes also bound-free XC effects. We assume that the total energy of the subsystem is :

$$E_b[\rho_\ell] = K[\rho_b] + \rho_b \cdot v_b + \frac{1}{2} \rho_b \cdot \frac{1}{r} * \rho_b + E^0_{xc}(\rho_b) \qquad (11)$$

$K[\rho_b]$ is the non-interacting (kinetic) energy. Eq.(11) is correct for the ground state of the ion, as a result of DFT. $E^0_{xc}(\rho_b)$ is the XC energy (in the LDA approximation) ; the index 0 indicates that the functional used here is that one relevant to zero temperature. For excited states, Eq.(11) has no rigorous justification and must be considered as a convenient extrapolation. The attempts made to extend DFT to excited states have shown that E_b is in fact a functional of a linear combination of the densities of all the states lower in energy than the state of interest[12]. Recent developments also indicate that the LDA in incorrect for excited states[13]. Numerical rigorous DFT calculations for exited states are not available at the present time, so we keep the formulation of Eq.(11) for our study of atomic species.

The stationarity of E_b with respect to the variations of ρ_b leads to the set of equations :

$$\left(-\frac{1}{2}\nabla^2 + U_b\right)\phi_i = \varepsilon_i \phi_i \qquad (12)$$

$$U_b = v_b + \frac{1}{r} * \rho_b + \frac{d\,E^0_{xc}(\rho_b)}{d\,\rho_b(r)} \qquad (13)$$

$$\rho_b = \sum_{\varepsilon_i \leq 0} n_i\, \phi^*_i\, \phi_i \qquad (14)$$

Only the bound spectrum of Eq.(12) is used to build the density ρ_b. Some practical aspects of the calculation of the densities ρ_ℓ and ρ_b are mentioned in the next paragraph.

Solving the equations

The calculation of an atomic species consists in solving the two sets of equations (7-10) and (12-14) by an iterative procedure. These sets of equations are coupled by the electrostatic potential and by XC bound-free terms. The numerical technique is derived from that one used to solve the "impurity problem". It has been presented in detail elsewhere[14]. Let us recall some important points. When the self-consistent solution is reached, the central bare nucleus is totally screened by the charge $\rho_\ell + \rho_b + \rho_+$:

$$Z = \int(\rho_\ell + \rho_b + \rho_+)\, d\underset{\sim}{r} = Z_b + Z_c + \int(\rho_\ell - \overline{\rho})\, d\underset{\sim}{r} \qquad (15)$$

The Coulomb part of U_ℓ and U_b is thus short-ranged. At large distances, U_ℓ and U_b go to constants $U_{\ell o}$ and U_{bo}. The continuum eigenvalues of Eq.(7) are :

$$\varepsilon_m = \varepsilon_k = \frac{1}{2} k^2 + U_{\ell o} \qquad (16)$$

and the eigenfunctions :

$$\phi_m(\underset{\sim}{r}) = A_{kL}\; \phi_{kL}(r)\; Y_{LM}(\hat{r}) \qquad (17)$$

where the radial parts behave, at large r, like :

$$\phi_{kL}(r) \;\rightarrow\; r^{-1} \sin\left[kr - L\frac{\pi}{2} + \eta_L(k)\right] \qquad (18)$$

The phase-shifts $\eta_L(k)$ are important quantities which characterize the electronic structure of the immersed atom. They determine entirely the electronic density of states in the continuum and satisfy the "generalized Friedel sum-rule" (a consequence of charge neutrality) :

$$Z - Z_b - Z_c = \beta \cdot \frac{2}{\pi} \int_0^\infty kdk\; f_k(1 - f_k) \sum_L (2L + 1)\; \eta_L(k) \qquad (19)$$

with f_k the occupation number $f(\varepsilon_k)$ and β the inverse temperature. The free-states electron density is :

$$\rho_\ell(r) = \overline{\rho} + \frac{1}{\pi^2} \int_0^\infty f_k\, dk \sum_L (2L + 1)\; [\phi^2_{kL}(r) - k^2\, j_L^2(kr)] \qquad (20)$$

where j_L is the sperical Bessel function of order L. In numerical calculations, attention must be paid to the convergence with respect to the maximum value of the angular momentum included in the sums over L. For the very large values of L, a Thomas-Fermi like approximation can be implemented. The integrals over k are truncated at a value k_m which is such that the corresponding energy ε_{km}, Eqs.(16), is of the order of $\mu + 10\, k_B\, T$.

For the calculation of the bound states, Eqs.(12-14), the standard techniques of the self-consistent field problem can be applied with no particular difficulty.

Total energy of a species

Once the self-consistent densities ρ_ℓ and ρ_b are determined, and simultaneously the electronic spectrum and potential, the total energy of an atomic species can be straightforwardly calculated. This energy is that of embedding the atom in the electron gas :

$$\Delta E(\ldots n_i \ldots) = E(\ldots n_i \ldots) - \bar{E}\,[\bar{\rho}] \quad (21)$$

with $\bar{E}\,[\bar{\rho}]$ the energy of the homogeneous electron gas (kinetic and XC contributions) at temperature T and density $\bar{\rho}$ of the free electrons. A detailed expression of this energy can be found in Ref.11. The statistical weight of a species is related to the entropy of the delocalized electrons in LTE ; this weight is :

$$\Omega_\ell = \exp\,\{S_\ell\,[\rho_\ell]\} \quad (22)$$

The entropy $\Delta S\,[\rho_\ell] = S_\ell[\rho_\ell] - \bar{S}_\ell[\bar{\rho}]$ (where the homogeneous electron gas contribution has been substracted) includes two terms : (i) a non-interacting term $S_Q[\rho_\ell]$ which depends on the phase-shifts only, and (ii) an XC term $S^T_{xc}[\rho_\ell]$ related to the functional $F^T_{xc}(\rho_\ell)$ used in Eq.(5). Both are calculated straightforwardly.

III. BOUND-BOUND ELECTRONIC TRANSITIONS

Total energy differences

We now consider the transition of an electron from an initial bound state i to a final bound state j. The transition energy is the difference of the total energies for the two corresponding atomic species. The ionization Z* is conserved in the transition :

$$\delta E_{Z*}(i\to j) = \Delta E(..n_i{=}0,..,n_j{=}1..) - \Delta E(..n_i{=}1,..,n_j{=}0..) \quad (23)$$

The transition energy can obviously be obtained as the difference of two total energy calculations. But we shall derive a formula which requires only one self-consistent calculation to obtain $\delta E_{Z*}(i\to j)$. One of the advantages of DFT is that all the equations maintain their meaning if the n_i's are considered as continuous variables. Derivatives with respect to the occupation numbers can thus be defined. For a zero temperature system, it is well-known that the derivative of the total energy with respect to n_i is ε_i, the eigenvalue of the Kohn-Sham equation. If one assumes that the same DFT equations describe an excited state (an electron in state j and a hole in state i), the transition energy is :

$$\delta E(i\to j) = \varepsilon_j - \varepsilon_i + \sigma(\delta n^2) \quad (24)$$

where a first-order Taylor expansion of the energy difference has been made. States different from i and j do not appear in Eq.(24) (to first order) although they relax, because their occupation numbers do not change. We shall now adapt Eq.(24) to the case of atomic species. The situation is different because the free-spectrum ocupation numbers obey Fermi-Dirac statistics, so that they change with any relaxation of the system and contribute to the first-order transition energy. Let us consider a change δn_i, δn_j in occupation numbers (with $\delta n_i + \delta n_j = 0$). The corresponding total energy variation is :

$$\delta E_{Z*}(i,j) = \varepsilon_i \ \delta n_i + \varepsilon_j \ \delta n_j + \frac{\delta \ \Delta E}{\delta \rho_\ell} \cdot \delta \rho_\ell + \sigma(\delta n^2) \tag{25}$$

where the last term comes from the relaxation of the continuum sates. For these states, the internal energy is related to the free-energy, Eq.(5), by :

$$\frac{\delta \ \Delta E}{\delta \rho_\ell} \cdot \delta \rho_\ell = \frac{\delta \ F_\ell[\rho_\ell]}{\delta \rho_\ell} \cdot \delta \rho_\ell - T \ \frac{\delta \ S_\ell[\rho_\ell]}{\delta \rho_\ell} \cdot \delta \rho_\ell \tag{26}$$

Using the stationarity properties of F_ℓ, Eq.(6), and the conservation of the number of free electrons, we conclude that the first term of the right hand side vanishes. Now, the change of the small XC contribution S^T_{xc} to the entropy S_ℓ is assumed to be small : if the density deviation $\rho_\ell(r)-\bar{\rho}$ is small, the functional derivative of S^T_{xc} may be calculated for $\rho_\ell \equiv \bar{\rho}$, so it is a constant in space and the integral vanishes because $\delta\rho_\ell$ has a zero integral (total number of free electrons fixed). Thus, the important contribution in Eq.(26) comes from the non-interacting entropy S_o. Using the well-known expression of S_o in terms of the occupation numbers, the functional derivative is easily calculated and one is left with :

$$\delta E_{Z*}(i,j) = \varepsilon_i \ \delta n_i + \varepsilon_j \ \delta n_j - \beta \sum_m f_m(1-f_m) \ \varepsilon_m \ \delta\varepsilon_m \tag{27}$$

where the last term gives the contribution of the free spectrum relaxation, $\delta\varepsilon_m$ being the change in the continuum energy levels induced by the change of the density of bound electrons. This change is : $\rho_i \ \delta n_i + \rho_j \ \delta n_j$ where $\rho_i(\rho_j)$ is the charge density in state $i(j)$. To first order in occupation numbers, the change in the potential U_ℓ acting on the continuum electrons can be formally written :

$$\delta U_\ell = \delta n_i \ u_{\ell i} * \rho_i + \delta n_j \ u_{\ell j} * \rho_j \tag{28}$$

$u_{\ell i}$ is a potential including : (i) the bare Coulomb potential (ii) a potential due to the relaxation of the continuum electrons which screen the change of the bound electrons potential ; (iii) an XC contribution reflecting the change of the bound-free XC potential included in the external potential of Eq.(8). The reader will find a complete analysis of $u_{\ell i}$ in Ref.11.

Finally, the first-order estimate of $\delta\varepsilon_m$:

$$\delta\varepsilon_m = \rho_m \cdot \delta U_\ell \tag{29}$$

leads to the following expression of the total energy change :

$$\delta E_{Z*}(i,j) = \varepsilon^*_i \ \delta n_i + \varepsilon^*_j \ \delta n_j + \sigma(\delta n^2) \tag{30a}$$

$$\varepsilon^*_i = \varepsilon_i - \beta \sum_m f_m(1 - f_m) \ \varepsilon_m \ \rho_m \cdot u_{\ell i} * \rho_i \tag{30b}$$

Eqs.(30) show that, in the plasma, the change in total energy due to a change in the bound states configuration is described, <u>to first order</u>, by one-particle <u>modified effective energies ε^*</u>. These are the eigenvalues of the Kohn-Sham equation, corrected to take the continuum relaxation into account. The ε^*'s also include the bound eigenfunctions relaxation (to first order), exactly as in standard Slater theory at zero temperature. The correction $\varepsilon^* - \varepsilon$ clearly vanishes for $T \to 0$ (because $f_m(1 - f_m) \to 0$).

The accuracy of Eqs.(30) can be improved using the "transition state" technique proposed by Slater[5]. This technique uses the Taylor expansion of the total energy with respect to the occupation numbers, considered as continuous variables. If the ε^*'s are calculated in the so called "transition state", a fictitious state of the ion with configuration intermediate between the initial and final ones : $(\ldots n_i - \frac{1}{2}, \ldots, n_j + \frac{1}{2}, \ldots)$, the quadratic terms $\sigma(\delta n^2)$ vanish, so that the transition energy is :

$$\delta E_{Z^*}(i \rightarrow j) = \varepsilon^*_j - \varepsilon^*_i + \sigma(\delta n^3) \tag{31}$$

The numerical accuracy of Eq.(31) is examined in the next paragraph.

Numerical results

We first test Eq.(31) by comparison with the energy change obtained as difference of two self-consistent calculations for the final and initial species. We consider an Iron plasma with average electron density $\overline{\rho} = 7.75\ 10^{23}\ \mathrm{ecm}^{-3}$ and free electron temperature T = 100 eV. In Table 1, we show some transitions between configurations close to the full ETL configuration ; the 1s and 2p shells are complete in all cases. The estimate of the transition energy with the corrected ε^* is closer to the "exact" result difference of two self-consistent total energies everywhere. The correction is important when the initial and final one electron states have very different localizations in space. If the free-states densities ρ_m are almost uniform, the correction to $\Delta\varepsilon$ goes like the space average $\langle u_{\ell j} * \rho_j - u_{\ell i} * \rho_i \rangle$; it dominated by the bare term in $u_{\ell j}(u_{\ell i})$ and approximately equal to $\langle r^2 \rangle_j - \langle r^2 \rangle_i$. Thus, the correction is mainly governed by the difference of the average values of r^2 in the final and initial states. This conclusion is in good qualitative agreement with the results of Table 1. In the third case, where the correction is negligible, one has $\langle r^2 \rangle_{3s} = \langle r^2 \rangle_{3p} = 0.59$ a.u.

An obvious application of Eqs.(30) is the study of trends in optical transitions, when the environment changes. As an example, we show in Table 2 the energy of the transition $3p^1\ 4s^0 \rightarrow 3p^0\ 4s^1$ for Iron, in the same conditions $\overline{\rho}$, T than above. The occupation numbers of the states 3s and 3d change, leading to important variations of the transition frequency. All the configurations have a Neon-like core in common. The dispersion of $\Delta\varepsilon^*$ reaches 20 %, but the matrix element is much less sensitive and changes by 4 % only. This example shows that the model of atomic species can be useful for estimating data needed in the study of non-equilibrium situations.

IV. AVERAGE ATOMIC SPECIES

Let us consider a plasma in complete LTE and select out of it all the atomic species having a given ionization degree $Z^* = Z - \Sigma_i n_i$. Let us assume that we have calculated a particular one of these species and use it as reference species, with occupation numbers n_{io}. For any species we are interested in, we have :

$$\sum_i (n_i - n_{io}) = \sum_i \Delta n_i = 0 \tag{32}$$

Table 1. Comparison between transition energies. $\Delta\varepsilon$: difference of non-corrected one-particle eigenvalues ε in the transition state. $\Delta\varepsilon^*$: idem for corrected eigenvalues. Last column : difference of two self-consistent total energies. Rydberg units.

Configuration	$\Delta\varepsilon$	$\Delta\varepsilon^*$	Tot. en. difference
$(3s^1 3d^1 4p^0)$ $2s^2 3p^1 \rightarrow 2s^1 3p^2$	59.780	60.380	60.565
$(3s^1 3p^1 3d^1)$ $2s^2 4p^0 \rightarrow 2s^1 4p^1$	71.890	73.967	74.414
$(2s^2 3d^1 4p^0)$ $3s^1 3p^1 \rightarrow 3s^0 3p^2$	2.432	2.431	2.428
$(2s^2 3p^1 3d^1)$ $3s^1 4p^0 \rightarrow 3s^0 4p^1$	13.977	15.624	15.713
$(2s^2 3s^1 3p^1)$ $4p^0 3d^1 \rightarrow 4p^1 3d^0$	7.795	9.519	9.649

Table 2. Transition $3p^1 4s^0 \rightarrow 3p^0 4s^1$ in an Iron plasma, calculated in the transition state, non-corrected ($\Delta\varepsilon$) and corrected ($\Delta\varepsilon^*$). Rydberg units. The dipolar matrix element is also shown.

Configuration	$\Delta\varepsilon$	$\Delta\varepsilon^*$	$\|\langle 3p \| r \| 4s \rangle\|$
$3s^0$ $3d^1$	11.710	13.232	0.2031
$3s^1$ $3d^1$	10.736	12.214	0.2061
$3s^2$ $3d^1$	9.962	11.745	0.2096
$3s^0$ $3d^2$	10.784	12.399	0.2082
$3s^1$ $3d^2$	9.911	11.678	0.2105
$3s^2$ $3d^2$	9.001	10.975	0.2069

the sum running over all the bound states. A straightforward generalization of Eq.(30a) leads to :

$$\Delta E_{Z*}(..n_i..) = \Delta E_{Z*}(..n_{io}..) + \sum_i \varepsilon^*_{iZ*} \Delta n_i + \sigma(\Delta n^2) \qquad (33)$$

Eq.(33) gives the total energy (ΔE is the total energy of the atom in the plasma, Eq.(21)) of the species $(..n_i..)$ with respect to the energy of the reference species, the corrected eigenvalues of which are ε^*_{iZ*}. Eq.(33) includes a sum over the bound states only ; it gives the total energy, inclusive of the free spectrum contribution ; and any relaxation is taken into account to first order. Now, the meaning of Eq.(33) is the following : except for the constant term $\Delta E_{Z*}(..n_{i0}..)$, the energy of a particular species with ionization Z* can be interpreted as that of a system of Z-Z* <u>independent</u> particles distributed among the energy levels ε^*_{iZ*}. An average species for the ionization Z* can thus be defined. Its levels ε^*_{iZ*} are populated according to Fermi-Dirac statistics :

$$\overline{n}_i = \left[1 + \exp\left\{\beta(\varepsilon^*_{iZ*} - \mu_{Z*})\right\}\right]^{-1} \quad (34)$$

with μ_{Z*} an effective chemical potential for the bound states, fixed by the condition $\Sigma_i \overline{n}_i = Z - Z*$. The statistical weight of the average species is given by :

$$\Omega_{Z*} = \exp\left(S_{Z*} + S_\ell[\rho_\ell]\right) \quad (35)$$

where $S_\ell[\rho_\ell]$ is the entropy of the free-spectrum electrons for the reference species, and :

$$S_{Z*} = k_B \sum_i \left[\overline{n}_i \text{ Ln } \overline{n}_i + (1 - \overline{n}_i) \text{ Ln } (1 - \overline{n}_i)\right] \quad (36)$$

Finally, the probability of the ionization degree Z* is :

$$p_{Z*} = A \exp(-\beta F_{Z*}) \quad (37)$$

$$F_{Z*} = \Delta E_{Z*}(..\overline{n}_i..) - TS_{Z*} - T S_\ell[\rho_\ell] \quad (38)$$

The normalization constant A is such that $\Sigma_{Z*} p_{Z*} = 1$. For every integer value of Z*, one has to perform a self-consistent calculation of the average species. When all the p_{Z*} are known, the average value of any function of Z* can be computed straightforwardly.

Numerical example

We applied this model of average atomic species to the case of an Aluminum plasma, at T = 100 eV and $\overline{\rho} = 4.77\ 10^{23}$ ecm^{-3}. The probability of the various ionization states is shown in Table 3. The ionization states Z* = 0,1,2 do not exist in the present model because the only bound levels found in these cases are the 1s, 2s and 2p levels which can receive at most 10 electrons. The average ionization <Z*> deduced from these results is <Z*> = 7.998, in good agreement with the AA value $Z^*_{AA} = 7.720$. The mean square deviation $\sigma = \left(\langle Z^{*2}\rangle - \langle Z^*\rangle^2\right) / \langle Z^*\rangle$ is $\sigma = 0.103$.

V. CONCLUSION

We presented a model, based on DFT concepts, to describe the bound-levels structure of atoms in dense plasmas. This model goes far beyond the well-known average-atom model. Its possible applications are : (i) in the case of non-equilibrium plasmas (NLTE), the study of quantities relative to any "true" ionic configuration, in particular optical transitions. (ii) in LTE plasmas, the description of average ions for a given ionization degree and the study of fluctuations. As any DFT model, this one gives a static picture of the plasma ; future developments could concern dynamical effets.

Table 3. Probability of the various ionization degrees in Al.

Z^*	P_{Z^*}
3	$1.225\ 10^{-6}$
4	$1.892\ 10^{-4}$
5	0.0047
6	0.0427
7	0.2257
8	0.4452
9	0.2496
10	0.0313
11	$6.228\ 10^{-4}$
12	$7.413\ 10^{-11}$
13	$9.456\ 10^{-20}$

REFERENCES

1. R. M. More, "Atomic Physics in Inertial Confinement Fusion", Lawrence Livermore National Laboratory Report UCRL - 84991, unpublished, (1981).
2. H. M. Griem, "Plasma Spectroscopy", Mc Graw-Hill, New-York (1964).
3. D. Saltzman, Phys. Rev. A20 : 1713 (1979) ; Phys. Rev. A21 : 1761 (1980).
4. R. M. More, Atomic processes in high-density plasmas, in : "Atomic and Molecular Physics of Controlled Thermonuclear Fusion", C.J. Joachain and D.E. Post, ed., Plenum Publishing Corporation, New-York (1983).
5. J. C. Slater, "The Self-Consistent Field for Molecules and Solids", Mc Graw-Hill, New-York (1974).
6. W. Kohn and L. J. Sham, Phys. Rev. 140 : A1133 (1965).
7. N. D. Mermin, Phys. Rev. 137 : A1441 (1965).
8. V. L. Moruzzi, J. F. Janak and A. R. Williams, "Calculated Electronic Properties of Metals", Pergamon, New-York (1978).
9. U. Gupta and A. K. Rajagopal, Phys. Rev. A21 : 2064 (1980).
10. F. Perrot and M. W. C. Dharma-Wardana, Phys. Rev. A30 : 2619 (1984).
11. F. Perrot, to be published.
12. A. K. Theophilou, J. Phys. C 12 : 5419 (1979).
13. W. Kohn, Phys. Rev. A 34 : xxx (1986).
14. F. Perrot, Phys. Rev. A 25 : 489 (1982).

THE FREEZING OF CHARGED AND UNCHARGED HARD-SPHERE SYSTEMS

M. Baus*

Chimie-Physique II**, C.P. 231
Université Libre de Bruxelles
B-1050 Brussels, Belgium

INTRODUCTION

The phenomenon of freezing by which a fluid is transformed into a solid, when the temperature is lowered or the pressure raised, is a very general property of bulk matter. During the past decades a lot of information about this phase transition has been obtained from both laboratory experiments[1] and from computer simulations.[2] Many workers in this field now think that the freezing transition of realistic and even complex systems can be understood in terms of the freezing of a few very simple model systems. Two such systems which have been brought to the foreground as good reference systems by liquid state theories are the hard-sphere (HS) system, whose freezing is monitoring the freezing of neutral fluids, and the one-component plasma (OCP) which can be used as a reference system in the theoretical study of the freezing of ionic liquids. The theoretical study of freezing within the realm of equilibrium statistical mechanics has, after decades of stagnation,[3] made considerable progress recently. The main steps of this progress will now be summarized.

A first step forward was realized when the theory of freezing was reformulated in terms of the direct correlation function, by Ramakrishnan and Yussouff,[4] instead of the ordinary (total) correlation function as was done in the pioneering work of Kirkwood and Monroe[5]. This step, which also characterizes the modern trends in liquid state theory, is particularly important in connection with freezing because this phase transition is to a large extent insensitive to the details of the interaction potential[1,3] whereas this is in general not the case for the results based on approximations of the BGY hierarchy[6] which depend explicitly on the potential. The second step in the right direction was the description of the nonuniform solid phase in terms of the finite temperature, classical version[7] of the density functional theory of Hohenberg and Kohn[8] which has been very successful in describing nonuniform, zero temperature, quantum systems. Here the main point is that this theory gives easy access to the thermodynamic properties (free

* Chercheur Qualifié du F.N.R.S.
**Association Euratom-Etat Belge.

energy, pressure, chemical potential), which are essential ingredients of any phase transition theory, while it uses as an independent variable the least unknown of all structural properties, namely the local number density. This then may explain the recent success of the theory of freezing after many years of unsuccessful attempts.[3] In retrospect, it is clear that the modern theory of freezing is nothing but a reformulation of the old theory of Kirkwood and Monroe[5] into the modern language of the density functional theory combined with two decades of progress in the theory of liquids.[9]

THE DENSITY FUNCTIONAL THEORY

In the density functional theory[7] one starts from the Helmholtz free energy, F, which is viewed as a functional, $F=F[\rho]$, of the one particle density, $\rho(\underline{r})$, and written as the sum of three contributions: $F=F_{conf} + F_{corr} + F_{ext}$. The first term, F_{conf}, is known explicitly:

$$\beta F_{conf}[\rho] = \int_V d\underline{r}\ \rho(\underline{r})\ \{\ln(\lambda^3\rho(\underline{r})) - 1\} \tag{1}$$

and represents the configurational part of the free energy which in the appropriate limit reduces to the ideal gas free energy. It contains the configurational entropy of a system of particles of mass m and of thermal de Broglie wavelength, $\lambda = h/(2\pi m k_B T)^{1/2}$, enclosed at the inverse temperature $\beta=(k_BT)^{-1}$ in a volume V and whose one-particle configuration is described by $\rho(\underline{r})$. The second term, F_{corr}, describes the correlational contributions to the free energy and is generally unknown except for its relation[7] to the direct correlation function, $c(\underline{r},\underline{r}';[\rho])$:

$$\beta\,\frac{\delta^2 F_{corr}[\rho]}{\delta\rho(\underline{r})\ \delta\rho(\underline{r}')} = -\,c(\underline{r},\underline{r}';\ [\rho])\ . \tag{2}$$

Integrating Eq. (2) along a linear path in density space running from some reference density $\rho_R(\underline{r})$, to the actual density, $\rho(\underline{r})$, one can, however, write[7]:

$$\beta(F_{corr}[\rho] - F_{corr}[\rho_R] = \int_V d\underline{r} \int_V d\underline{r}' \int_0^1 d\lambda\ c(\underline{r},\underline{r}';[\rho_R + \lambda\Delta\rho])$$

$$\times\ \Delta\rho(\underline{r})\ \times\ \Delta\rho(\underline{r}')\ . \tag{3}$$

where $\Delta\rho(\underline{r}) = \rho(\underline{r}) - \rho_R(\underline{r})$. This last term, F_{ext}, contains the contribution:

$$\beta F_{ext}[\rho] = \int d\underline{r}\ \rho(\underline{r})\ \phi_{ext}(\underline{r}) \tag{4}$$

of the external field, $\phi_{ext}(\underline{r})$, which serves to break some of the system's symmetries[10] so as to uniquely locate the crystal phase in space and to properly take the thermodynamic limit of an infinite system with finite intensive properties, e.g. the average density:

$$\rho = \frac{1}{V}\int_V d\underline{r}\ \rho(\underline{r})\ . \tag{5}$$

The remaining thermodynamic properties are then constructed in the usual manner:[11]

$$p[\rho] = -\frac{F[\rho]}{V} + \int \frac{d\underline{r}}{V} \rho(\underline{r}) \frac{\delta F[\rho]}{\delta \rho(\underline{r})} \tag{6}$$

which is the Gibbs-Duhem equation $\Omega = -pV$, for a system of pressure p, grand potential $\Omega = F - V\rho\mu$ and chemical potential μ:

$$\rho\mu[\rho] = \int_V \frac{d\underline{r}}{V} \rho(\underline{r}) \frac{\delta F[\rho]}{\delta \rho(\underline{r})} \tag{7}$$

For the true equilibrium state we must also have,[7] $\delta\Omega[\rho]/\delta\rho(\underline{r}) = 0$ or $\mu[\rho] = \delta F[\rho]/\delta\rho(\underline{r})$, with μ a constant so that the two-phase coexistence conditions between, say, a liquid phase of density ρ_L and a solid phase of density ρ_s can be simply expressed as:

$$p[\rho_S] = p[\rho_L] \tag{8}$$

$$\mu[\rho_S] = \mu[\rho_L] \tag{9}$$

for two phase of equal temperatures, $T_L = T = T_S$.

THEORIES OF FREEZING

In order to obtain a theory of freezing (or of any other equilibrium phase transition) the preceeding density functional theory has to be implemented with a set of approximations which allow Eqs. (8-9) to be evaluated explicitly. Usually one starts from an approximate but fairly accurate theory of one of the phases and uses the density functional theory to compute the thermodynamic properties of the other phase. All density functional theories of freezing known today are liquid-phase based theories which take into account the considerable progress realized in the past decade by liquid state theory[9] in the theorteical study of thermodynamic and strucutral properties of liquids. All these theories coincide in approximating the unknown direct correlation function of the solid by the direct correlation function of some uniform reference liquid, an idea which goes back to a similar approximation of the total correlation function by Kirkwood and Monroe.[5,6] These theories differ mainly by: 1) the choice of reference state in, and the approximation of, Eq. (3); 2) the determination of the local density $\rho(\underline{r})$; and 3) the location of the phase transition point. For instance, a first group of authors[12] has considered the coexisting liquid as the reference state to be used in Eq. (3). Since for a strongly first order transition such as freezing, where the density change is considerable, this procedure usually requires the inclusion of the partly known higher order terms in the expansion of the free energy of the solid around the free energy of the coexisting liquid. The status of these higher order terms is as yet not well known. A second group of authors[13] attempts to avoid the rapid density changes characteristic of the solid phase by reformulating the theory in terms of a coarse-grained or weighted density with smoother variations. It then is a nontrivial task to find a physically reasonable weighting funtion. A third group of authors[14] has attempted to optimize the choice of the reference state by scaling the structure factor of the reference liquid to the structure of the solid or by minimizing the free energy of the solid with respect to the density of the reference liquid.

Broadly speaking all these theories have succeeded in describing the freezing transition reasonably well. Many questions have, however, been left open (e.g. the convergence of the density and the Fourier expansions used) and these should be investigated in detail before the respective merits of the different theories can be assessed.

A point which has been completely overlooked until now concerns the problem of thermodynamic inconsistency (i.e. the different results which are obtained when computing the same quantity via different routes within a given theory) which should be worse in the solid phase than in the liquid phase where it is already considerable.

THE FREEZING OF UNCHARGED HARD SPHERES

To illustrate the results we consider first the freezing of a system of (uncharged) hard spheres. This model can only mimic the repulsive potential of realistic systems but has the enormous advantage that its freezing transition has been thoroughly studied by computer simulations,[2,15] while its fluid phase can be adequately described by the Percus-Yevick equations for which an analytic solution for the direct correlation function is available.[9] The hard sphere system is also an outstanding reference system for the study of more realistic systems whose freezing transition is presumably monitored by some underlying hard sphere transition. We will describe here the results of Baus and Colot[11,14] based on Eqs. (1,3,6,7). The direct correlation function of the solid is approximated by that of some reference Percus-Yevick hard sphere fluid:

$$c(\underline{r},\underline{r}';[\rho]) = c_{PY}(|\underline{r}-\underline{r}'|;\ \rho_R(\rho)) \tag{10}$$

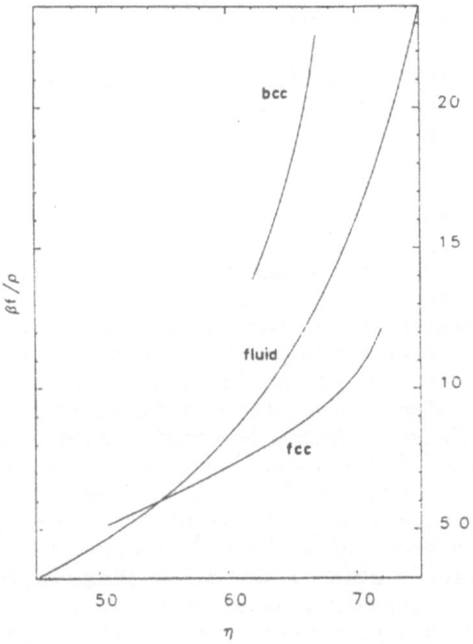

Fig. 1 The reduced free energy per particle, $\beta f/\rho$ (f is the free energy density, ρ the average density and $\beta = 1/k_BT$) versus the packing fraction, η, ($\eta = \frac{\pi}{6}\sigma^3\rho$, σ being the hard sphere diameter) for the fluid, bcc and fcc phases of the hard sphere system.[14]

whose density, ρ_R, is determined in such a manner that for each solid of average density ρ the position of the main peak of the structure factor of the reference fluid coincides with the smallest nonzero reciprocal lattice vector of the solid of density ρ. The density of the solid is parametrized by a set of Gaussians:

$$\rho(\underline{r}) = \sum_j \left(\frac{\alpha}{\pi} \right)^{3/2} \exp -\alpha|\underline{r}-\underline{r}_j|^2 \tag{11}$$

centered around the lattice sites, $\{\underline{r}_j\}$, of a given crystal structure. Combining Eqs. (10-11) with Eq. (3) the free energy of the solid can be worked out analytically as a function of the inverse width, α, of the Gaussians of Eq. (11). Minimizing the free energy with respect to α, the only variational parameter of the problem, we obtain the results[14] shown in Fig. 1. It is seen there that the bcc hard sphere solid is metastable relative to the fluid while the fcc solid is stable at high density (up to the density of close packing) and metastable at low density (up to a bifurcation point below which the solid becomes mechanically unstable). The hard sphere phase diagram in the pressure-density plane can be obtained from Eq. (6) and the tie-line can be constructed from Eqs. (8-9) using Eqs. (6-7). The result[14] for the fluid-fcc solid coexistence is compared to the simulation results[2,15] in Fig. 2. There is a systematic overestimation of the densities and of the pressures already inherent in the Percus-Yevick approximation.[14] Some of the numerical details are given in Table 1.

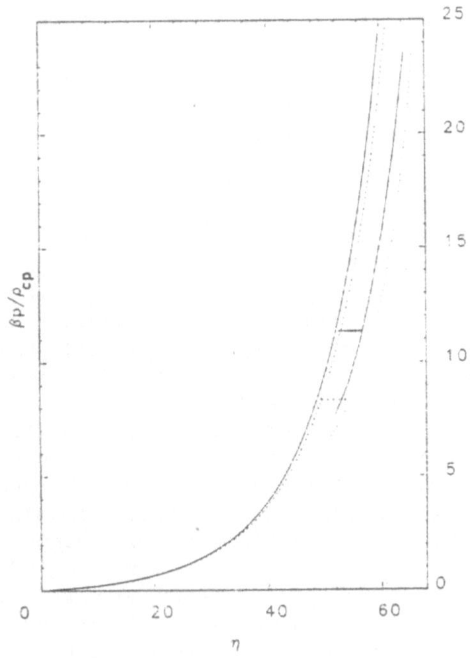

Fig. 2 The reduced pressure, $\beta p/\rho_{cp}$ (ρ_{cp} is the fcc crystal close packing density) versus the packing fraction η for the hard-sphere fluid-fcc solid coexistence. The full lines correspond to the theoretical results[14] and the dotted lines to the computer simulations.[15]

Table 1. Characteristics of the hard sphere freezing transition.

	η (fluid)	η (solid)	$\beta p/\rho$ (coex)	$(s_F - s_S)/k_B$
Theory[11]	0.567	0.520	16.2	1.36
Simulations[2]	0.545	0.494	12.4	1.16

s=entropy per particle

THE FREEZING OF CHARGED HARD SPHERES

In view of the success obtained for the hard sphere system one is tempted to consider also the extremely opposite case of the one-component plasma. Unfortunately no analytical expression for the direct correlation function of this system is as yet available and one usually considers a one-component plasma of charged hard spheres together with an additional constraint which fixes the hard sphere diameter and allows the system to mimic the one component plasma of point particles (OCP). This correspondence can only be imperfect since some of the intrinsic discontinuities of the charged hard sphere system can never be completely removed. In the fluid phase this method works well[16] as far as the computation of the thermodynamic properties (which are continuous) are concerned. We have extended[17] the previous freezing theory to the case of a system of charged hard spheres embedded in a neutralized background. The fluid phases were described within the mean spherical approximation using the analytic direct correlation function obtained by Palmer and Weeks.[18] The remaining steps are the same as for the neutral hard spheres using e.g. Eq. (11). The bcc solid was found to be always metastable leading to the fluid-fcc melting line shown in Fig. 3. This result is somewhat surprising since the system remains a hard core system crystallizing into a fcc solid even for strong Coulomb coupling parameters. It is also seen from Fig. 3 that it may be very difficult to draw conclusions about the one component plasma of point charges from the charged hard sphere plasma. Two of the routine conditions usually imposed on the charged hard sphere system in order for it to mimic the one component plasma (e.g. Gillan's condition[19]: $g(r=\alpha)=0$ or Singh's condition[20]: $((r=0)=1.33\Gamma)$ are indeed seen from Fig. 3 to lead to very different temperature-density relations which make a determination of the phase transition point of the one component plasma quite arbitrary. Some of these conclusions could, however, also result from the inadequacy of the mean spherical approximation in some of the regions of the phase plane (e.g. at low densities). Needless to say, it would be extremely useful to dispose of a genuine (continuous) one-component plasma direct correlation function. Last but not least, some of the approximations underlying the freezing theory could also be responsible for the unexpected absence of a stable bcc phase but, at present, this remains an open question.[21]

A closely related calculation has been performed by Barrat[22] who considered a mixture of oppositely charged hard spheres of equal diameters within the mean spherical approximation (i.e. the restricted primitive model of a molten salt). As shown in Fig. 4, this time the theory of

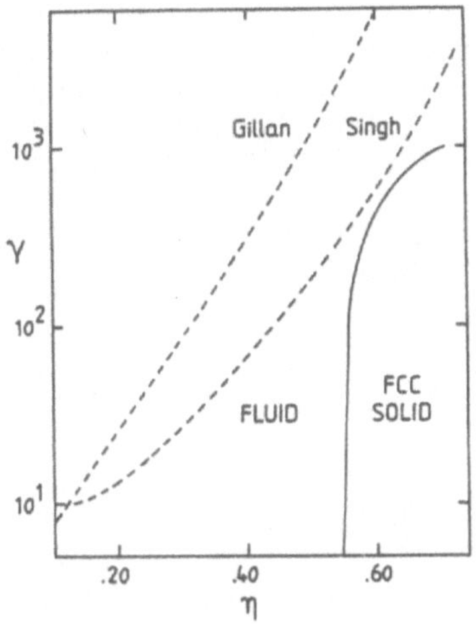

Fig. 3 The phases of the charged hard sphere one-component plasma in the inverse temperature (γ)-density (η) plane.[17] Also shown (dotted lines) are the η-γ relations following from the conditions used by Gillan[19] and Singh[20] to map the charged hard sphere system onto the point charge system.[16] As usual, $\eta = \frac{\pi}{6}\alpha^3\rho$ and $\gamma = \beta e^2/\alpha$, while $\Gamma = 2\gamma\eta^{1/3}$.

freezing predicts a structural phase transition in the solid phase from a disordered fcc structure at high temperatures where the sphere behave as uncharged to an ordered cesium chloride structure (a bcc lattice with the cations and anions on two sc sublattices) at low temperatures. The absence of the expected NaCl structure can presumably again be ascribed to the failure of the mean spherical approximation at low temperatures.

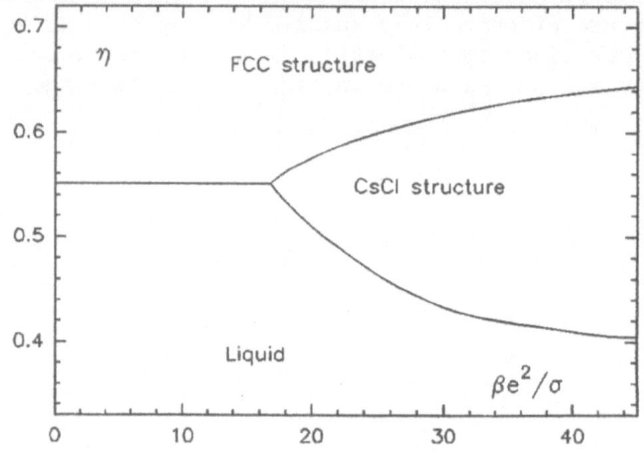

Fig. 4 The three stable phases of a system of oppositely charged hard spheres of equal diameter in the density (η) inverse temperature (γ) plane as predicted by the freezing theory.[22] (The symbols have the same meaning as Fig. 3).

CONCLUSION

From the above it is clear that a quantitative description of the liquid-solid coexistence is possible within the density functional theory of freezing. In its present form the theory seems, however, to favor the freezing into compact lattices and the major open question concerns, therefore, the correct description of the relative stability of the different crystal structures.

REFERENCES

1. S. M. Stishov, Sov. Phys. Usp. 17:625 (1975).
2. D. Frenkel and J. P. McTague, Ann. Rev. Phys. Chem. 31:491 (1980).
3. M. Baus, Molec. Phys. 50:543 (1983).
4. T. V. Ramakrishnan and M. Yussouff, Phys. Rev. B 19:2775 (1979).
5. J. G. Kirkwood and E. Monroe, J. Chem. Phys. 9:514 (1941).
6. M. Baus, Molec. Phys. 53:183 (1984).
7. N. D. Mermin, Phys. Rev. 137:A1441 (1965).
   R. Evans, Adv. Phys. 28:143 (1979).
8. P. Hohenberg and W. Kohn, Phys. Rev. 136:B864 (1964).
9. J. P. Hansen and I. R. McDonald, "Theory of Simple Liquids", Academic Press, London (1976).
10. M. Baus, Molec. Phys. 51:211 (1984).
11. M. Baus and J. L. Colot, Molec. Phys. 55:653 (1985).

12. V. N. Ryzhov and E. E. Tareyeva, Teor. Math. Phys. 48:835 (1981).
N. H. March and M. P. Tosi, Phys. Chem. Liq. 11:79 (1981).
A. D. J. Haymet and D. W. Oxtoby, J. Chem. Phys. 74:2559 (1981).
B. Bagchi, C. Cerjan and S. A. Rice, J. Chem. Phys. 79:5595 (1983).
G. L. Jones and U. Mohanty, Molec. Phys. 54:1241 (1985).
13. P. Tarazona, Molec. Phys. 52:81 (1984).
W. A. Curtin and N. W. Ashcroft, Phys. Rev. A 32:2909 (1985).
14. M. Baus and J. L. Colot, J. of Phys. C 18:L365 (1985).
J. L. Colot and M. Baus, Molec. Phys. 56:807 (1985).
J. L. Colot, M. Baus and H. Xu, Molec. Phys. 57:809 (1986).
J. L. Barrat, M. Baus, and J. P. Hansen, Phys. Rev. Lett. 56:1063 (1986).
F. Igloi and J. Hafner, J. of Phys. C (to appear).
15. W. G. Hoover and F. H. Ree, J. Chem. Phys. 49:3609 (1968).
D. A. Young and B. J. Alder, J. Chem. Phys. 70:473 (1979).
See also P. N. Posey and W. Van Megen, Nature 320:340 (1986).
16. M. Baus and J. P. Hansen, Phys. Rev. 59:1 (1980).
17. X. Wu and M. Baus (unpublished).
18. R. G. Palmer and J. D. Weeks, J. Chem. Phys. 58:4171 (1973).
19. M. J. Gillan, J. of Phys. C 7:L1 (1974).
20. H. B. Singh, J. Stat. Phys. 33:371 (1983).
21. M. Rovere and M. P. Tosi, J. of Phys. C 18:3445 (1985).
22. J. L. Barrat, J. of Phys. C (to appear).

ATOMIC STRUCTURE OF AN IMPURITY NEON IN LIQUID METALLIC HYDROGEN

Junzo Chihara

Department of Physics
Japan Atomic Energy Research Institute
Tokai-mura, Ibaraki 319-11, Japan

INTRODUCTION

The determination of the electronic structure of isolated atoms and molecules is easily performed within the framework of the density functional formalism. On the other hand, it is a difficult problem to calculate the electronic structure of an atom immersed in a liquid metal or a high-density plasma, since this contains two problems: combination of the liquid- and the atomic-structure. The "external problem" is to determine electron- and ion-density distributions around the impurity in the whole space, while the other "internal problem" is to calculate the atomic structure under this circumstance. There have been a number of studies on this problem in relevance to the inertial confinement fusion and the study of stellar interiors (for example, Skupsky 1980, Davis and Blaha 1982, Perrot and Dharma-wardana 1985, Fujima et al. 1985). A similar problem occurs in treating core-level shifts and Auger relaxation energies of an impurity in a solid metal (Williams and Lang 1978, Lang and Williams 1979). These calculations have shown that the excitation energies can be successfully obtained from the total energy difference between the initial and final states in the density-functional theory. Thus, we can expect that the electronic structure of an atom in a plasma may be calculated within the framework of the density-functional theory.

When a liquid metal or a plasma is treated as an electron-ion mixture, the ionic charge and the electron-ion interaction must be given beforehand. In laser-compressed plasmas, for example, the ionic charge and the electron-ion interaction vary over a wide range as the temperature and density change. It is a difficult problem to determine these quantities at a given temperature and density. On the other hand, if a liquid metal or a plasma is treated more fundamentally as a nucleus-electron mixture, the ionic charge and the electron-ion interaction ought to be determined without any information further than the bare inter-particle potentials (Chihara 1985). Furthermore, the electronic bound energy levels around a nucleus may be calculated from this model.

FORMULATION

Here we consider a liquid metal or a plasma to be a mixture of N_I nuclei with atomic number Z_A and N_e electrons: the nuclei behave as classical particles and the electrons constitute a quantum liquid. Let us single out one nucleus and fix it at an arbitrary position. Note that fixing a nucleus in the mixture is equivalent to imposing an external potential $v_{eN}(r)=-Z_A e^2/r$ on the mixture of (N_I-1) nuclei and $Z_A N_I$ electrons to induce density distributions, $n_I(r|N)$ and $n_e(r|N)$. In order to describe this inhomogeneous system, we take as a reference system the mixture composed of noninteracting ions, which have Z_B core-electrons around each ion with density distribution $n'_b(r)$ and $Z(N_I-1)+Z_A$ free electrons, where $Z\equiv Z_A-Z_B$ is the ionic charge of noninteracting ion. The values, Z_B and $n'_b(r)$, should be determined self-consistently with the condition mentioned below. Hereafter, this reference system will be referred to as the average ion model.

The intrinsic free energy of this reference system can be written as:

$$\mathscr{F}_0 = \frac{1}{\beta}\int n_I(r)[\ln(n_I(r)\lambda^3)-1]dr + (N_I-1)f_e^b[n'_b(r)] + \mathscr{F}_0^e[n_e(r|N)] \ , \qquad (1)$$

where $f^b{}_e[n'_b(r)]$ means the free energy of bound electrons in an ion and $\mathscr{F}^e{}_0[n_e]$ means the intrinsic free energy of free electrons with $\lambda=(h\beta/2\pi m)^{1/2}$. On the basis of this average ion model, the thermodynamic potential Ω can be expressed as

$$\Omega = \mathscr{F}_0 + \mathscr{F}_{int} - \int[\mu_e - v_{eN}(r)]n_e(r|N)dr - \int[\mu_I - v_{II}(r)]n_I(r|N)dr. \quad (2)$$

Here, $\mathscr{F}_{int}$ is the interaction part of the intrinsic free energy of this system, and v_{II} is the ion-ion potential; μ_e and μ_I denote the chemical potentials of electrons and ions, respectively. The density functional formalism gives an expression for the effective external potential which determines the electron distribution around the nucleus in terms of the interaction part of the intrinsic free energy (Chihara 1984a):

$$v_{eN}^{eff}(r) = v_{eN}(r) + \frac{\delta \mathcal{F}_{int}}{\delta n_e(r)} - \mu_e^{int} . \tag{3}$$

Therefore, the density distribution around the fixed nucleus is determined by solving the wave equation with respect to this potential in the form

$$n_e(r|N) = n_e^0(r|v_{eN}^{eff}) \equiv n_e^b(r|N) + n_e^v(r|N) \tag{4}$$

which consists of bound-electron and valence-electron parts. Here,

$$n_e^0(r|U) \equiv \sum_\ell [\exp\{\beta(\epsilon_\ell - \mu_e^0)\}+1]^{-1} |\phi_\ell(r)|^2 \tag{5}$$

$$[- \frac{\hbar^2}{2m} \nabla^2 + U(r)]\phi_\ell(r) = \epsilon_\ell \phi_\ell(r) . \tag{6}$$

Equation (4) assumes that the bound-electron distribution $n'_b(r)$ of a surrounding ion in the average ion model should be identical to the bound-electron distribution around the nucleus at the origin:

$$n_b'(r) = n_e^b(r|N) , \tag{7}$$

since we can choose any nucleus in the liquid metal as the fixed one, which should accumulate the same bound-electron distribution. As a consequence, the bound-electron number Z_B of a surrounding average ion is determined from the relation:

$$Z_B = \int n_b'(r)dr = \sum_{\ell \epsilon bound} 1/\{\exp[\beta(\epsilon_\ell - \mu_e^0)]+1\} \tag{8}$$

With this choice of Z_B and $n'_b(r)$, the average ion model becomes self-consistent with the premise at the beginning, by choosing the chemical potential μ^0_e to satisfy

$$\lim_{r\to\infty} n_e(r|N) = \frac{1}{V} \sum_{i \epsilon valence} 1/\{\exp[\beta(\epsilon_i - \mu_e^0)]+1\} = n_0^e = Zn_0^I , \tag{9}$$

which states that the electron-density distribution reduces to the uniform electron density at large distances, thus satisfying the neutrality condition. This condition leads to the determination of the chemical potential μ^0_e.

$$Z_A = \frac{1}{n_0^I} \int \frac{2}{\exp[\beta(p^2/2m - \mu_e^0)]+1} \frac{dp}{(2\pi)^3} + \sum_{\ell \epsilon bound} \frac{1}{\exp[\beta(\epsilon_\ell - \mu_e^0)]+1} \tag{10}$$

At this point, we introduce two approximations: (i) the electron-ion interaction part of the chemical potential, $\mu_{eI}(r|n_I n_e)$, is approximated by the functional expansion with respect to the density deviations up to the second order; that is the hypernetted chain (HNC) approximation:

$$\mu_{eI}(r|n_I n_e) \simeq \int \frac{C_{eI}(|r-r'|)}{-\beta} \delta n_I(r')dr' . \tag{11a}$$

(ii) the exchange-correlation potential for electrons in a liquid metal is represented by the local-density approximation to the bound electron and by the HNC approximation to the valence electrons:

$$\mu_{ee}^{XC}(r|n_I n_e) \simeq \mu_{xc}(n_e^b(r|N) + n_0^e)$$

$$- \mu_{xc}(n_0^e) + \int \frac{C_{ee}(|r-r'|)}{-\beta} \delta n_e^v(r')dr . \tag{11b}$$

Here $C_{ij}(r)$ denotes the direct correlation function (DCF) (Chihara 1984a). From Eq. (3), with these approximations, the effective potential for electrons around a fixed nucleus becomes identical to the effective electron-ion potential represented in the form (1985):

$$v_{eI}^{eff}(r) = \tilde{v}_{eI}(r) + \int \frac{C_{ee}(|r-r'|)}{-\beta} \delta n_e^v(r')dr'$$

$$+ \int \frac{C_{eI}(|r-r'|)}{-\beta} [n_I(r'|I)-n_0]dr' . \tag{12}$$

Now, the bare electron-ion interaction is given in a self-consistent manner in the form

$$\tilde{v}_{eI}(r) \equiv v_{eN}(r) + \int v_{ee}(|r-r'|)n_e^b(r'|I)dr' + \mu_{xc}(n_e^b(r) + n_0^e) - \mu_{xc}(n_0^e). \tag{13}$$

Thus, we find that the treatment of a liquid metal as a nucleus-electron mixture in the average ion model is shown to lead to the ion-electron mixture model (Chihara 1978, 84b), where the atomic structure of the ion and the electron-ion interaction are determined in a self-consistent way.

This formulation can be easily extended to treat a liquid metal composed by two kinds of nuclei and electrons. This set of integral equations turns out to be integral equations to calculate the electronic structure of an impurity in liquid metals or plasmas by taking the zero limit of impurity density (Chihara 1986b). In this way, we obtain the effective electron-impurity potential, which determines the electron density distribution and the electronic structure of the impurity immersed in a liquid metal in the form:

$$v^{eff}_{eImp}(r) = \tilde{v}_{eImp}(r) - \Gamma_{eImp}(r)/\beta \tag{14}$$

$$\Gamma_{eImp}(r) = \int C_{ee}(|r-r'|)\delta n^{v}_{e}(r'|Imp)dr'$$

$$+ \int C_{eI}(|r-r'|)\delta n_{I}(r'|Imp)dr' \ . \tag{15}$$

Here, it contains the bare electron-impurity potential as

$$\tilde{v}_{eImp}(r) = v_{eIMP}(r) + \int v_{ee}(|r-r'|)n^{b}_{e}(r'|Imp)dr'$$

$$+ \mu_{xc}(n^{b}_{e}(r|Imp) + n^{0}_{e}) - \mu_{xc}(n^{0}_{e}) \ . \tag{16}$$

The ion distribution around the impurity involved in (15) must be determined self-consistently in association with the following equations:

$$n_{I}(r|Imp)/n^{I}_{0} = \exp[-\beta v^{eff}_{IImp}(r)] \tag{17}$$

where

$$v^{eff}_{IImp}(r) = \frac{\bar{Z}_{Imp}Z_{I}e^{2}}{r} - \Gamma_{IImp}(r)/\beta \tag{18}$$

$$\Gamma_{IImp}(r) = \int C_{Ie}(|r-r'|)\delta n^{v}_{e}(r'|Imp)dr'$$

$$+ \int C_{II}(|r-r'|)\delta n_{I}(r'|Imp)dr' \ . \tag{19}$$

The ionic charge of the impurity is determined by the bound-electron distribution around the impurity

$$\bar{Z}_{Imp} = Z_{Imp} - \int n^{b}_{e}(r|Imp)dr \ . \tag{20}$$

In Eqs. (15) and (19), the direct correlation functions, C_{ee}, C_{II} and C_{eI}, are those of a liquid metal without impurity and are given beforehand by solving the QHNC equation at given temperature and density.

APPLICATION TO IMPURITY NEON IN LIQUID METALLIC HYDROGEN

Let us apply this formulation derived above to the problem of the electronic structure of an impurity Ne atom immersed in liquid metallic hydrogen (LMH). Density $n_0 = n^{e}_{0} = n^{p}_{0}$ and temperature will be specified by the plasma parameter $\Gamma = \beta e^2/a$ and the Wigner-Seitz radius $r_s = a/a_B$ with $a = (3/4\pi n_0)^{1/3}$. In this calculation, Γ and r_s are taken in such a high density region that the electrons in the system can be treated as in a fully degenerate state. In the density-functional formalism, the atomic structure of the impurity can be calculated without specifying positions of ions around it, provided that the electron- and ion-density distributions, g_{eImp} and g_{pImp}, are determined to be self-consistent with the core state of the impurity. When an impurity with atomic number Z_{Imp} in the LMH has Z_B bound electrons in the environment, the impurity ion with $\bar{Z}_{Imp}$ charge accumulates surrounding electrons, and repulses protons and yields a dip around it. Thus, the central impurity charge Z_{Imp} is neutralized by the dip in the proton distributions and the accumulated electrons, satisfying the relation

$$\bar{Z}_{Imp} = -n_0^p \int [g_{pImp}(r) - 1]dr + n_0^e \int [g_{eImp}(r)-1]dr . \tag{21}$$

In consequence, the calculation of core levels of the impurity in the LMH is reduced to the problem of how we can determine these density distributions, $g_{eImp}(r)$ and $g_{IImp}(r)$, to be consistent with its core state, in addition to the usual problem to construct a conventional self-consistent potential for the bound electrons. Our formulation provides a method to calculate the atomic structure of an impurity from this viewpoint. In the density-functional formalism, the atomic structure of an impurity can be calculated without specifying positions of ions around it as was necessary for the procedure used by Fujima et al. (1985), provided that the electron- and ion-density distributions, $n_{eImp}(r)$ and $n_{IImp}(r)$, are given.

For the sake of computational convenience, a set of Eqs. (14) - (19) are rewritten in the forms of integral equations for the DCF's, C_{eImp} and C_{pImp}

$$C_{eImp}(r) = \hat{B}^{-1}\cdot[n_e^{0v}(r|\tilde{v}_{eImp}-\Gamma_{eImp}/\beta)/n_0^e - 1] - \Gamma_{eImp}(r) \tag{22}$$

$$C_{pImp}(r) = \exp[-\beta v_{pImp}(r) + \Gamma_{pImp}(r)] - 1 - \Gamma_{pImp}(r) . \tag{23}$$

Here the Fourier transforms of $\Gamma_{eImp}(r)$ and $\Gamma_{pImp}(r)$ are obtained in terms of DCF's and the density response function of noninteracting system, χ^0_Q:

$$\Gamma_{eImp}(Q) = \{C_{eImp}(Q)-n_0[C_{eImp}(Q)C_{pp}(Q)-C_{pImp}(Q)C_{ep}(Q)]\}/ \\ D(Q) - C_{eImp}(Q) \tag{24}$$

$$\Gamma_{eImp}(Q) = \{C_{pImp}(Q)-n_0\chi_Q^0[C_{pImp}(Q)C_{ee}(Q)-C_{eImp}(Q)C_{ep}(Q)]\}/ \\ D(Q) - C_{pImp}(Q) \tag{25}$$

with

$$D(Q) = [1-n_0C_{pp}(Q)][1-n_0C_{ee}(Q)\chi_Q^0] - |n_0C_{ep}(Q)|^2\chi_Q^0 \qquad (26)$$

where $C_{pp}(Q)$, $C_{ep}(Q)$ and $C_{ee}(Q)$ are given beforehand by solving the integral equations for a LMH without impurity (Chihara 1984b, 1986a). Thus, we can calculate $C_{eImp}(r)$ and $C_{pImp}(r)$ by following the method to solve the quantal HNC equation with the use of the fast Fourier transform algorithm (Chihara 1979). The atomic structure of an impurity Ne in the LMH is now calculated in the self-consistent potential,

$$v_{eImp}^{eff}(r) = -Z_{Imp}e^2/r + \int v_{ee}(|r-r'|)n_e^b(r'|Imp)dr'$$

$$+ \mu_{xc}(n_e^b(r|Imp) + n_0) - \mu_{xc}(n_0) - \Gamma_{eImp}(r)\beta \qquad (27)$$

which takes account of the effect of surrounding ions and electrons through $\Gamma_{eImp}(r)$. With the use of this potential, the wave equation is solved by a modified Herman-Skillman program in the iterative process to obtain $C_{eImp}(r)$ and $C_{pImp}(r)$. In the present calculation, the local-density approximation for the exchange-correlation potential is taken to be of the form proposed by Gunnarsson and Lundqvist (1976),

$$\mu_{xc}(r_s) = -\frac{2}{\pi\alpha r_s}\{1 + 0.0545\, r_s \cdot \ln(1 + 11.4/r_s)\}\ \mathrm{Ry}. \qquad (28)$$

Using this approach we determined the equilibrium ionic state of an impurity Ne atom in LMH, at Γ=2 and r_s=1. Under this circumstance, an impurity Ne is shown to be ionized to Ne^{6+}. Here it is interesting to note that there are two sets of proton- and electron-density distributions, d_{eImp} and g_{pImp}, around the Ne ionized to have six valence electrons as displayed in Fig. 1. The first set of density distributions show that Z_{Imp} (the ionic charge of impurity) is screened by attracting 2.66 electrons and by pushing away 3.34 protons around it, satisfying (21) is such a way that 6=-(-2.66) + 3.34 while the other set of charge distributions satisfy the neutrality condition (21) as 6=-(-4.86) + 1.14: the former set yields a stronger screening effect than the latter, as shown by Fig. 2. Therefore, the bound energy levels of Ne ion screened by protons and electrons according to the former set are ϵ_{1s} = -58.34 Ry and ϵ_{2s} = -0.78 Ry, which are consistent with Z_B=4 and yield the stable ionic state. On the other hand, the impurity Ne screened by charge distributions given by the latter set has three bound levels: the occupied ϵ_{1s} = -64.56, ϵ_{2s} = -4.00 and the unoccupied ϵ_{2p} = -2.51 Ry. The 2p level becomes shallower as surrounding electrons fall into this level and eventually disappear. That is, the latter set of solutions for the integral equations are physically unstable and does not exist as the equilibrium state of ionized Ne.

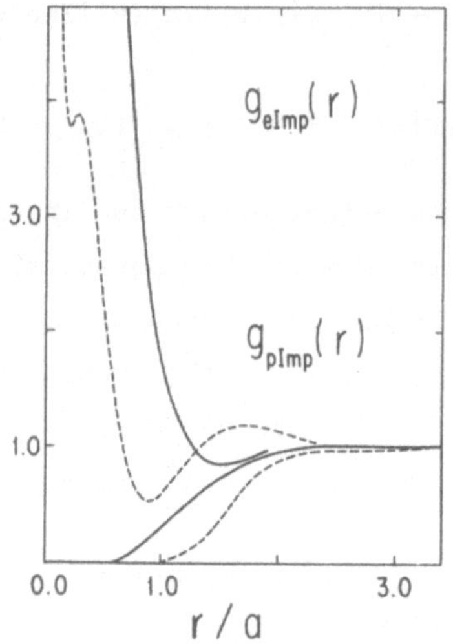

Fig. 1 Two sets of the electron- and proton-distributions around ions with Z_B=4, normalized by the density n_0: the full lines denote $g_{eImp}(r)$ and $g_{pImp}(r)$ around the stable ion (the occupied ϵ_{1s}=-58.34 Ry and ϵ_{2s}=-0.78 Ry) in the equilibrium with the environment and the dashed curves represent those around the unstable ion with three bound levels (the occupied ϵ_{1s}=-64.56, ϵ_{2s}=-4.00 and the unoccupied ϵ_{2p}=-2.51 Ry) in LMH with Γ=2 and r_s=1. The former yielding a stronger screening to the ion, produces a shallower effective potential for electrons as shown in Fig. 2.

Next, let us calculate the frequency shift of the Lyman α line arising from a hydrogen-like Ne (Z_B=1) due to the polarization of electrons and ions in LMH with Γ=2 and r_s=1:

$$\Delta E_{L\alpha} = \Delta E^{free}_{1s-2p} - \Delta E_{1s-2p} , \tag{29}$$

where $\Delta E_{1s\text{-}2p}$ denotes the frequency of the Lyman α line in LMH and $\Delta E^{free} = 100\text{-}25$ Ry is that of a free hydrogen-like Ne. The electron- and proton-density distributions around a hydrogen-like Ne are shown in Fig. 3, which reveals that $g_{pImp}(r)$ and $g_{eImp}(r)$ around it are different in the cases whether the bound-electron in the hydrogen-like Ne is in the 1s

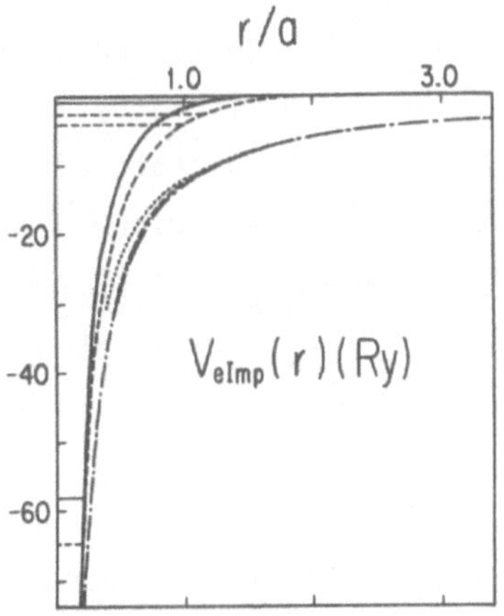

Fig. 2 Two effective potentials acting on electrons around the ions, along with the bound levels: the full curve is that around the stable ion, and the dashed line is that around the unstable ion in LMH with $\Gamma=2$ and $r_s=1$. The effective bare electron-ion potential v_{eImp} (dash-dotted line) is compared with the pure Coulombic potential (dotted line).

state ($\epsilon_{1s} = -76.19$ Ry) or in the 2p state ($\epsilon_{2p} = -7.16$ Ry), since the electron-impurity interaction given by (16) depends upon $n^b{}_e(r|Imp)$. It should be noticed that in the density-functional theory, the frequency of Lyman α line cannot be given by the difference between ϵ_{2p} and ϵ_{1s}, but it must be calculated from the difference between the total energies of these two states including the contribution of ions and electrons in LMH: the reason is that this theory can provide the exact total energy, but cannot yield real energy levels of the system.

To this end, Slater (1974) suggested that the difference in two total energies between the initial and final states of the transition can be approximately calculated from the difference between the corresponding one-electron energies in a transition state, which is defined as a state in which the occupation numbers are halfway between those of the initial and final states. In the present case to calculate the Lyman α frequency, the transition state is realized as a hydrogen-like Ne with the occupation density distributions around the hydrogen-like ion with the transition state in LMH are calculated and shown in Fig. 3, together with those around the hydrogen-like ions with $n_{1s}=1$ and with $n_{2p}=1$. In Fig. 3, the curves of these density distributions around the ion in the transition state lie between those around the other types of ions: one has an electron in the bound state of $\epsilon_{1s} = -76.19$ Ry and another has in the state of $\epsilon_{2p} = -7.16$ Ry. For this transition state of a hydrogen-like Ne in LMH with $\Gamma=2$ and $r_s=1$, a set of integral equations, (22) - (26), yield the energy levels, $\epsilon_{1s} = -76.88$ Ry and $\epsilon_{2p} = -5.63$ Ry. Therefore, we can evaluate the Lyman α frequency: $\Delta E_{1s-2p} = 76.88-5.63=71.25$ Ry, and consequently shift of the Lyman α is $\Delta e_{L\alpha} = 3.75$ Ry for Ne in LMH with $\Gamma=2$ and $r_s = 1$. Here it should be marked that this value takes account of the relaxation effects of the surrounding electrons and ions in this transition process by using Slater's transition state.

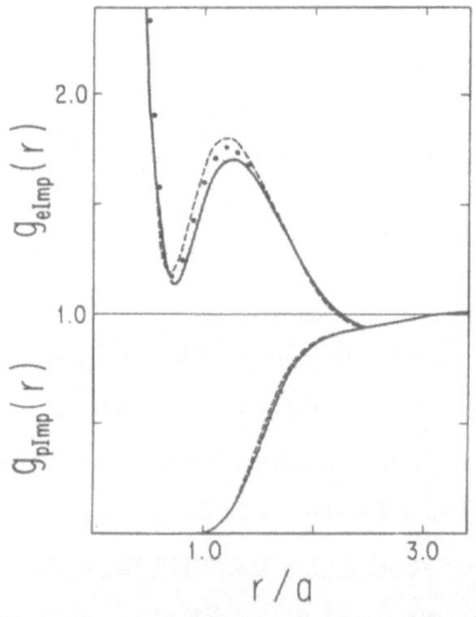

Fig. 3 The electron- and proton-density distributions around ions with $n_{1s}=1$ (full lines) and with $n_{2p}=1$ (dashed lines) and those around the ion in the transition state with $n_{1s}=1.2$ and $n_{2p}=1.2$ (full circles) in LMH with $\Gamma=2$ and $r_s=1$.

A continuum spectrum is formed when free electrons in LMH fall into bound states of an impurity Ne, and a difference between zero kinetic energy of a conduction electron and ϵ_{1s} provides the edge of the continuum spectrum. A polarization of electrons and protons in LMH around Ne causes a shift of the edge relative to the edge of continuum spectrum of a hydrogen-like Ne in the vacuum. Here we calculate the shift of the edge arising from the transition from Ne^{10+} to Ne^{9+} in LMH with the use of (22)-(27). The transition state corresponding to this problem is defined as the state with the occupation number n_{1s}=1.2, which leads to the problem of solving (22)-(27) for a Ne ion with Z_B=0.5 in LMH. As a result, the curves in Fig. 4 display the electron- and proton-density distributions around the Ne ions with Z_B=0, Z_B=1 and Z_B=0.5 in LMH. specified by Γ=2 and r_s=1. Under this circumstance, the 1s level of Ne in the transition state becomes -80.05 Ry together with two other unoccupied levels; ϵ_{2s}=-7.71 Ry and ϵ_{2p}=-7.04 Ry, while the effective potential for electrons around a bare Ne nucleus can possess three bound-energy levels; ϵ_{1s}=-81.68 Ry, ϵ_{2s}=-8.35 Ry and ϵ_{2p}=-7.95 Ry.

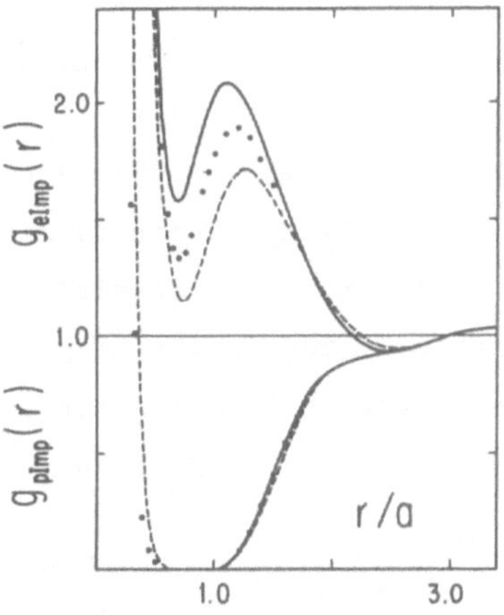

Fig. 4 The electron- and proton-density distributions around ion in the transition state with n_{1s}=1/2 (full circles) in LMH with Γ=2 and r_s=1, compared to those around a bare Ne nucleus (full lines) and around an ion with n_{1s}-1 (dashed line). The density distributions of the 1s electron (dashed lines) and the electron in the transition state (full circles) are also plotted in the normalized form $n^b{}_e(r)/n_0$.

Table 1. Edge shifts of continuum spectrum from Ne^{10} owing to the polarization effect of LMH, and the values of bound-energy level ϵ_{1s} in units of Ry. At the densities, r_s=0.4 and r_s=0.2, there is no bound state except ϵ_{1s}.

(e/cm^3) T(K)	1.58×10^5	3.16×10^5	7.90×10^5	1.58×10^6
(r_s=0.2)	(Γ=10)	(Γ=5)	(Γ=2)	(Γ=1)
ϵ_{1s}	-28.09	-28.76	-30.50	-32.61
2.01×10^{26}	71.91	71.24	69.50	67.39
(r_s=0.4)	(Γ=5)	Γ=2.5)	Γ=1)	
ϵ_{1s}	-54.77	-55.15	-56.19	
2.52×10^{25}	45.23	44.85	43.81	
(r_s=1.0)	(Γ=2)			
ϵ_{1s}	-80.05			
1.61×10^{24}	19.95			

ϵ_{2p}=-7.95 Ry. Now, the edge shift in reference to a free hydrogen-like Ne is determined as 100-80.05=19.95 Ry for LMH with Γ=2 and r_s=1. In the same procedure, the edge shifts of the continuum spectrum from Ne due to the polarization effect of the surrounding electrons and protons are calculated for LMH at the variety of densities and temperatures, and the results are tabulated in Table 1. It should be emphasized that in the higher-density region such as r_s=0.4 and 0.2, there is no bound energy level except 1s level. Therefore, it is not possible to observe the Lyman α line in the high-density region. In addition, the table suggests that the edge shift is sensitive to the density variation rather than to the temperature variation of LMH.

DISCUSSION

In some atomic structure calculations in plasmas, it is assumed that the effective potential acting on the bound electrons around the nucleus is different from that acting on the conduction electrons: in the former potential, the contribution of the bound electrons to the screening is discarded in order to avoid the self-interaction of a bound electron (Skupsky 1980, Davis and Blaha 1982, Cauble et al. 1984). In such a treatment, there appear an infinite number of bound levels for a hydrogen-like Ne in LMH, for example, at any density and temperature. Therefore, these methods might be inappropriate, especially when applied to the calculation of the Lyman α line which relates to high 2p levels.

At this point, it should be remembered that in the density-functional formalism the effective potential is determined to be common to both bound and conduction electrons, and that the self-interaction should be cancelled by the exchange-correlation potential to some extent. Furthermore, as mentioned before, the density-functional theory states that the atomic structure of an impurity in the LMH is determined as a functional of electron- and proton-densities around it. That is, only with the use of the pair correlations, $g_{ep}(r)$ and $g_{pp}(r)$, the total energy of the system can be evaluated when they are calculated self-consistently with the core-electron states. Thus, the frequency of Lyman α, for example, can be determined without specifying positions of surrounding ions, when only these pair correlations are given in the corresponding transition state, which can yield the difference of the total energies between 1s and 2p states.

In the present calculation, the HNC approximation plays the fundamental role. This approximation, when applied to the LMH, was examined and shown to give more accurate results than other approaches (Chihara 1986a). Also, it is powerful in the calculation of the atomic spectra of impurity in liquid metals as well as in solids to use Slater's transition state, which enables us to take account of the relaxation effects of the liquid structure for the excited states in core electrons. In the density-functional theory, the transition-state method is based on the relation:

$$\partial E_T/\partial n_i = \epsilon_i ,$$

where E_T is the total energy of the system and n_i is the occupation of i-th level. This relation has been proved for the system of zero temperature (Janak 1978). On the other hand, Callaway and March (1984) discussed this relation for the finite-temperature system, which includes the case of LMH under present consideration.

Also, it should be mentioned that the density-functional formalism, although being a static approach, can be taken as an appropriate method to treat the atomic structure in a liquid metal, where the electron temperature is regarded as at zero temperature, because the local order of a liquid is quite similar to that of a solid, in which the atomic structure can be calculated successfully from the density-functional theory (for example, Williams and von Barth 1983). It should be mentioned that our method can be applied to the hydrogen system in the region where the electrons are in a partially degenerate state including the classical limit, although our calculation has been restricted to the region where the electrons in LMH can be approximated as being in the fully degenerate state.

REFERENCES

Almbladh, C-O. and von Barth, U., 1978, Phys. Rev. B 18:3307-19.

Callaway, F. and March, N. M., 1984, in "Solid State Physics", Vol. 38, ed. by Ehrenreich, H., Turnbull, D. and Seitz, F., (New York: Academic Press).

Chihara, J., 1978, Prog. Theor. Phys. 59:76.

Chihara, J., 1979, Prog. Theor. Phys. 62:1533.

Chihara, J., 1984a, J. Phys. C 17:1633.

Chihara, J., 1984b, Prog. Theor. Phys. 72:940.

Chihara, J., 1985, J. Phys. C. 18:3103.

Chihara, J., 1986a, Phys. Rev. A 33:2575.

Chihara, J., 1986b, J. Phys. C (to be published).

Davis, J. and Blaha, M., 1982, J. Quant. Spectrosc. Radiat. Transfer 27:307.

Fujima, K. Watanabe, T., and Adachi, H., 1985, Phys. Rev. A 32:3585.

Gunnarssón, G., and Lundqvist, B. I., 1976, Phys. Rev. B 13:4274.

Hedin, L. and Johansson, A., 1969, J. Phys. B. 2:1336.

Janak, J. F., 1978, Phys. Rev. B 18:7165.

Lang, N. D. and Williams, A. R., 1979, Phys. Rev. B 20:1369.

Perrot, F. and Dharma-wardana, M.W.C., Phys. Rev. A 31:970.

Skupsky, S., 1980, Phys. Rev. A 21:1316.

Slater, J. C., 1974, "The Self-Consistent Field for Molecules and Solids" (New York: McGraw-Hill Book Company).

Williams, A. R. and Lang, N. D., 1978, Phys. Rev. Lett. 40:954.

Williams, A. R. and von Barth, U., 1983, "Theory of the Inhomogeneous Electron Gas", ed. S. Lundqvist and N. H. March (New York: Plenum Press) 189.

CHAPTER VII

2-D PLASMAS

SOLVABLE MODELS OF COULOMB SYSTEMS IN TWO DIMENSIONS

Angel Alastuey

Laboratoire de Physique Théorique et
Hautes Energies, Bât. 211
Université Paris-Sud
91405 Orsay Cedex, France

1. INTRODUCTION

In the last few years there has been a large amount of exact analytic results for classical two-dimensional systems with logarithmic interactions. These results concern both bulk and surface equilibrium properties. They are quite interesting for conceptual and practical reasons.

At a conceptual level, the two-dimensional (2D) models with $-\ln(r/L_s)$ potential do have some essential properties which are equivalent to those of the three-dimensional (3D) charged systems with the usual $1/r$ potential. These common properties involve the screening effects and the behavior of the correlations at large distances. They are direct consequences of the harmonic nature of the $-\ln(r/L_s)$ and $1/r$ potentials, which are the solutions of the d-dimensional Poisson equation

$$\Delta v_c(r) + S(d)\,\delta(\vec{r}) = 0 \qquad (1)$$

for d = 2 and d = 3 respectively (S(d) is the area of the unit sphere in d dimensions). Therefore, the exact solutions obtained for the 2D models, play an important role in understanding real 3D Coulomb systems.

In addition to their conceptual interest, some of these 2D models might describe physical systems. For instance, in polyelectrolytes at high densities, there are long charged molecules which are all parallel and interact via a logarithmic potential. Another example is a system of vortices in the super-fluid liquid Helium, which also interact via a logarithmic potential.

The main purpose of this short review is to illustrate the conceptual interest of the solvable models through the following few selected examples:

- The inhomogeneous one-component plasma (OCP) in one space-direction
- A line of adsorption sites in an OCP
- The two-component plasma (TCP) on a lattice.

For each example, we briefly describe the model and give the basic outlines of the exact calculations. Sum rules and characteristic features of Coulomb systems are discussed on the basis of the exact solutions. The examples mentioned above are presented in sections 3, 4, and 5. In section 2 we briefly recall exact results for the homogeneous infinite OCP. Concluding comments are given in section 6.

2. THE HOMOGENEOUS OCP

2.1 The Model

The model is nothing but the 2D version of the well-known OCP in three dimensions. One has identical point particles with charge e and number density ρ. These particles are embedded in a uniform and rigid neutralizing background with charge density $-e\rho$. The particles interact via the two-dimensional Coulomb potential $v(r) = -e^2 v_c(r) = -e^2 \ln(r/L_s)$ (instead of e^2/r in three dimensions); L_s is an irrelevant scale length which fixes the zero of $v(r)$.

The excess equilibrium properties of the homogeneous infinite system only depend on the coupling constant $\Gamma = \beta e^2$ where β is the inverse temperature $\beta = 1/k_B T$. These properties are quite similar to those of the 3D OCP. In particular, the system is in a fluid phase for $\Gamma < \Gamma_m$ and in a pseudosolid phase for $\Gamma > \Gamma_m$ (in two dimensions there is no translational long-range order); the melting value Γ_m is close to 140.

2.2 Free Energy and Correlations at $\Gamma = 2$

For the special value of the inverse temperature $\beta = \beta^* = 2/e^2$ such that the coupling constant is $\Gamma = 2$, the equilibrium quantities can be computed exactly. The starting point of the calculation (and of all the exact calculations presented in this review) is the identity

$$\exp[-\beta^* v(r_{ij})] = |z_i - z_j|^2 / L_s^2 \qquad (2)$$

where z_j is the complex number $z_j = r_j \exp(i\theta_j)$ and (r_j, θ_j) are the polar coordinates of $\vec{r}_j$ in a given frame. The simple algebraic expression (2) of the two-body Boltzmann factor allows to rewrite the N-body Boltzmann factor in terms of a Vandermonde determinant. The free energy[1] and the correlations[2] are then computed by using techniques introduced in the theory of random matrices[3,4].

The computed distribution functions are invariant under translations and rotations: this rigorously shows that the phase is fluid at $\Gamma = 2$ in agreement with the estimated value of Γ_m[5,6]. The correlations are found to decay like Gaussians at large distances[2]. Let us point out that this behavior was not expected. Indeed one usually believed that the correlations in Coulomb systems decay exponentially. These usual findings are mainly based on an analysis of the analytic properties of the structure factor in the complex plane.[7] The exact solution obtained at $\Gamma = 2$ definitely shows that this analysis is misleading and can only give some insights about the exact behaviors.[8] In fact, as indicated by a recent conditional theorem[9], the correlations in Coulomb systems should decay faster than any inverse power when they are monotonic.* The exact result at $\Gamma = 2$ does exhibit this property.

3. THE INHOMOGENEOUS OCP IN ONE DIRECTION AT $\Gamma = 2$

3.1 The Model

We now consider an inhomogeneous OCP where the background density is no longer uniform and varies in one space-direction only, i.e. $\rho_B(\vec{r}) = \rho_B(x)$. The shape of $\rho_B(x)$ is arbitrary with the only restriction $\rho_B(\pm\infty) = \rho_{1,2}$. Furthermore, the particles are subjected to an external potential $V_{ext}(x)$ which also depends on the x-direction only. The system is translationally invariant in the y-direction.

At $\Gamma = 2$, the one and two-body densities of the particles have been computed exactly by Alastuey and Lebowitz[11]. In the next subsection the main steps of the calculation are briefly described.

* There are rigorous constructive proofs[10] showing that the correlations are bounded by a decreasing exponential in the high-temperature limit (Debye-Hückel regime). These proofs do not cover the whole monotonic regime.

3.2 The Method of Calculation

One starts with the finite system described by the Canonical Ensemble. The background density $\rho_B(r)$ and the external potential $V_{ext}(r)$ only depend on the radial coordinate r. Thus the finite system has the circular symmetry.

At $\Gamma = 2$, the N-body Boltzmann factor becomes

$$\exp[-\beta^* V(\vec{r}_1,\ldots,\vec{r}_N)] = \mathrm{cst} \times \exp\left\{-\beta^* \sum_{j=1}^{N} [V_B(r_j) + V_{ext}(r_j)]\right\}$$

$$\times \left| \mathrm{Det} \begin{vmatrix} z_1^0 & \cdots & z_N^0 \\ \vdots & & \vdots \\ z_1^{N-1} & \cdots & z_N^{N-1} \end{vmatrix} \right|^2 \qquad (3)$$

where the identity (2) and the circular symmetry have been used. $V_B(r)$ is the one-body potential created by the background.

When computing the n-body density $\rho_N^{(n)}(\vec{r}_1,\ldots,\vec{r}_n)$ of the particles, one has to integrate (3) upon the angles θ_j with $n < j \leq N$. These angular integrations are performed using the same trick as for the homogeneous system[2] because the prefactor

$$\exp\left\{-\beta^* \sum_{j=1}^{N} [V_B(r_j) + V_{ext}(r_j)]\right\} \qquad (4)$$

does not depend on the angles. Expanding the Vandermonde determinant in (3) with respect to all the permutations of (1,...,N), and using the orthogonality properties of the functions $\exp(in\theta_j)$, one finds[11]

$$\rho_N^{(n)}(\vec{r}_1,\ldots,\vec{r}_n) = \exp\{-\beta^* \sum_{j=1}^{n} [V_B(r_j) + V_{ext}(r_j)]\} \,\mathrm{Det}|K_N(z_i \bar{z}_j)|_n \;. \qquad (5)$$

$K_N(z)$ is a polynomial in Z of order N and $|K_N(z_i\bar{z}_j)|_n$ is an n by n determinant ($\bar{z}_j$ is the complex conjugate of z_j).

The final step of the calculation consists of controlling the thermodynamic limit of (5). This is achieved by introducing a suitable "neutrality" radius R_0[11]. The n-body density of the inhomogeneous infinite system described in §3.1 is obtained through

$$\rho^{(n)}(\vec{r}_1,\ldots,\vec{r}_n) = \lim_{R_0\to\infty} \lim_{N\to\infty} \rho_N^{(n)}(R_0\hat{x} + \vec{r}_1,\ldots,R_0\hat{x} + \vec{r}_n) \qquad (6)$$

($\hat{x}$ is the unit vector in the x-direction). The two successive limits in (6) are taken by staying at finite distances of $R_0\hat{x}$ and by keeping fixed the shapes of the background density and of the external potential in the vicinity of R_0.

3.3 Explicit Functional Representations of the One and Two-Body Densities

Using the method exposed in §3.2, one finds the following integral representation for the one-body density[11]

$$\rho(x) = \frac{1}{\pi\sqrt{2}} \exp[-2\pi\rho_1 x^2 H(-x) - 2\pi\rho_2 x^2 H(x) - \beta^* e\phi_B(x) - \beta^* V_{ext}(x)]$$

$$\times \int_{-\infty}^{\infty} dx \exp(2xs\sqrt{2})/[\psi_<(s) + \psi_>(s)]. \qquad (7)$$

$H(x)$ is the Heavyside function, $H(x) = 0$ for $x < 0$, $H(x) = 1$ for $x > 0$. $\phi_B(x)$ is the electrostatic potential created by the charged distribution $-e[\rho_B(x) - \rho_1 H(-x) - \rho_2 H(x)]$. $\psi_<(s)$ and $\psi_>(s)$ are themselves defined by integral representations,

$$\left.\begin{aligned} \psi_<(s) &= \int_{-\infty}^{0} dv \exp[-2\pi\rho_1 v^2 + 2vs\sqrt{2} - \beta^* e\phi_B(v) - \beta^* V_{ext}(v)] \\ \psi_>(s) &= \int_{0}^{\infty} dv \exp[-2\pi\rho_2 v^2 + 2vs\sqrt{2} - \beta^* e\phi_B(v) - \beta^* V_{ext}(v)] \end{aligned}\right\} \qquad (8)$$

The truncated two-body density, $\rho^{(2,T)}(x_1,x_2,y) = \rho^{(2)}(x_1,x_2,y) - \rho(x_1)\rho(x_2)$, is given by an expression similar to (7)[11],

$$\rho^{(2,T)}(x_1,x_2,y) = -\frac{1}{2\pi^2}\exp[-2\pi\rho_1 x_1^2 H(-x_1) - 2\pi\rho_2 x_1^2 H(x_1) - \beta^* e\phi_B(x_1) - \beta^* V_{ext}(x_1)]$$

$$\times \exp[-2\pi\rho_1 x_2^2 H(-x_2) - 2\pi\rho_2 x_2^2 H(x_2) - \beta^* e\phi_B(x_2) - \beta^* V_{ext}(x_2)]$$

$$\times \left| \int_{-\infty}^{\infty} ds \exp[s(x_1+x_2-iy)\sqrt{2}]/[\psi_<(s)+\psi_>(s)] \right|^2 . \qquad (9)$$

The expressions (7, 9) are functional representations of $\rho(x)$ and $\rho^{(2,T)}(x_1,x_2,y)$ in terms of $\phi_B(x)$ and of $V_{ext}(x)$. They can be used for studying qualitative features of real charged plane interfaces. By adjusting the shapes of the background density and of the external potential, one can "simulate" a large variety of electrochemical and biological interfaces: charged, conducting or insulating walls, permeable or impermeable membranes, etc... The exact expressions obtained in these cases enable one to check sum rules, and to discuss the influence of $\rho_B(x)$ and of $V_{ext}(x)$ on the density profile and on the correlations. This exploitation of the general exact solution is illustrated in the next subsection where we consider in detail the case of a charged hard wall.

3.4 A Special Case: The Charged Hard Wall

We study the microscopic structure of the semi-infinite system near a charged wall (located at x = 0) carrying the external surface charge $e\sigma$. The density profile, which has been first computed by Jancovici[12], can also be obtained from the general expression (7) by taking[11] $\rho_B(x) = 0$ and $V_{ext}(x) = \infty$ for $x < 0$, $\rho_B(x) = \rho_B$ and $V_{ext}(x) = -\pi e^2 \sigma x$ for $x > 0$. One finds[12]

$$\rho(x) = \rho_B \frac{2}{\sqrt{\pi}} \int_{\pi\sqrt{2}\sigma a_B}^{\infty} dt \frac{\exp[-(t - x\sqrt{2}/a_B)^2]}{[1 + \mathrm{Erf}(t)]} \tag{10}$$

where a_B is the mean interparticle distance in the bulk phase (far from the wall), $a_B = 1/(\pi\rho_B)^{1/2}$, and Erf(t) is the error function

$$\mathrm{Erf}(t) = \frac{2}{\sqrt{\pi}} \int_0^t ds \exp(-s^2) . \tag{11}$$

An electrical double layer with charge density $e[\rho(x) - \rho_B]$ builds in the interface near the wall, even when $\sigma = 0$. This is due to the asymmetric nature of the model. For $\sigma = 0$ the density profile is quasimonotonic with a single oscillation and a contact density $\rho(0)$ smaller than ρ_B (repulsive effect of the wall induced by the correlations). When σ increases (respectively decreases), $\rho(0)$ decreases (respectively increases). All these qualitative characters are also typical of 3D charged double layers.

The density profile is expected to obey the overall neutrality sum rule,

$$\sigma + \int_0^{\infty} dx[\rho(x) - \rho_B] = 0 , \tag{12}$$

and the contact theorem[13]

$$P(\rho_B, \beta) = \rho(0)k_B T - e\rho_B \Delta\phi - \pi e^2 \sigma^2 ; \tag{13}$$

$P(\rho_B, \beta)$ is the thermodynamic pressure of the bulk phase and $\Delta\phi$ is the electrostatic potential drop across the double layer. The equation (12) states that the surface charge carried by the induced double layer perfectly screens the external surface charge. The equation (13) relates the thermodynamic pressure to the kinetic and radiative pressures. Both equations are well satisfied[12] by the analytic expression (10).

As well as the density profile, the correlations are given by simple integral representations in terms of the error function[14]. Using these integral representations, one explicitely checks the perfect screening sum rule,

$$\int_{x_2>0} d\vec{r}_2 \; \rho^{(2,T)}(x_1,x_2,y) = -\rho(x_1) \tag{14}$$

and a particular form of the Carnie and Chan[15] sum rule,

$$\int_0^\infty dx_1 \int_{x_2>0} d\vec{r}_2 (x_1-x_2) \; \rho^{(2,T)}(x_1,x_2,y) = \frac{k_B T}{2\pi e^2} \; . \tag{15}$$

Unlike the correlations of the homogeneous infinite system, the correlations in the direction parallel to the wall are found to have a long algebraic tail. More precisely, one obtains from the exact solution[14]

$$\rho^{(2,T)}(x_1,x_2,y) \sim \frac{c(x_1,x_2)}{y^2} \tag{16}$$

when $y \to \infty$, x_1 and x_2 fixed ($c(x_1,x_2)$ is some function of x_1 and x_2). It has been shown later[16] that this slow decay is in fact a universal feature of Coulomb systems near charged or dielectric hard walls. In the directions non-parallel to the wall, the decay of $\rho^{(2,T)}(x_1,x_2,y)$ is faster than any inverse power[14].

4. LINE OF ABSORPTION SITES IN AN OCP

4.1 The Model

We consider an infinite OCP with a uniform background density ρ, in which adsorption sites are immersed. These adsorption sites are regularly spaced on a line at points $\vec{R}_m$ with cartesian coordinates $(0, m\ell)$ in a given (x,y) frame (m is a relative integer and ℓ is the distance between two neighbor sites). Each site m creates an adsorption potential $V_m^{(a)}(\vec{r})$ which acts on the particles. The corresponding Boltzmann factor has Baxter's stick form, i.e.

$$\exp[-\beta V_m^{(a)}(\vec{r})] = 1 + \lambda\delta(\vec{r}-\vec{R}_m) \tag{17}$$

where λ is a positive constant.

At $\Gamma = 2$, the thermodynamic quantities of the system and the distribution function of the particles have been computed exactly by Rosinberg et al.[17].

4.2 Excess Surface Free Energy at $\Gamma = 2$

The excess surface free energy $\Delta f^{(s)}$ is defined as the difference (per unit of length) between the free energy of the present system and the one of the homogeneous infinite OCP described in section 2. The method of calculation is based on the resummation of the perturbative expansion of this quantity with respect to λ. The ingredients of this perturbative expansion are the Ursell functions of the homogeneous infinite OCP. The coefficients of the expansion are integrals over products of these Ursell functions by Mayer's factors $(\exp[-\beta * V_m^{(a)}(\vec{r}_j)] - 1)$. According to (17), these Mayer's factors are merely proportional to delta-functions. Thus the integrals reduce to the Ursell functions corresponding to configurations where the particles are located on the sites $\vec{R}_m$. Taking advantage of the simple Gaussian form of these Ursell functions at $\Gamma = 2$, the discrete sums upon the various configurations are then performed in terms of Jacobi's theta function,

$$\theta(\zeta,t) = 1 + 2\sum_{m=1}^{\infty} \exp(-tm^2)\cos(2\pi m\zeta) \tag{18}$$

with $t = \pi\rho\ell^2/2$. Once the coefficient of each term of order λ^n has been computed, it appears that the whole series in λ can be resumed with the result[17]

$$\Delta f^{(s)} = -\frac{e^2}{2\ell}\int_0^1 d\zeta\, \ln(1 + \lambda\theta(\zeta,t)) \;. \tag{19}$$

The other thermodynamic quantities relative to the system of the adsorbed particles can be computed from equation (19). In particular, the average number of adsorbed particles per site is given by

$$n_a = -\beta\lambda\ell\,\frac{\partial\Delta f^{(s)}}{\partial\lambda} \;, \tag{20}$$

λ playing the role of the fugacity for the system of adsorbed particles. Using (19) one immediately obtains[17]

$$n_a = \lambda\rho \int_0^1 d\zeta \frac{\theta(\zeta,t)}{[1+\lambda\rho\theta(\zeta,t)]} \; . \tag{21}$$

4.3 One and Two-Body Densities at $\Gamma = 2$

The distribution functions of the particles can be computed exactly at $\Gamma = 2$, by using a method similar to the one described in §4.2. Because of the singular form of the Boltzmann factor (17), the distribution functions become singular when one point coincides with a site. In particular, the one-body density $\rho(\vec{r}_1)$ can be rewritten as

$$\rho(\vec{r}_1) = [1 + \lambda \sum_{m=-\infty}^{\infty} \delta(\vec{r}_1 - \vec{R}_m)]\; \rho_c(\vec{r}_1) \tag{22}$$

where $\rho_c(\vec{r}_1)$ is a continuous function of $\vec{r}_1$ in the whole space. At $\Gamma = 2$, one finds[17]

$$\rho_c(\vec{r}_1) - \rho = -\lambda\rho^2 \exp(-\pi\rho|z_1|^2) \times \int_0^1 d\zeta \frac{F(\zeta,z_1,t)}{[1+\lambda\rho\theta(\zeta,t)]} \tag{23}$$

with $z_1 = x_1 + iy_1$ and

$$F(\zeta,z_1,t) = \frac{1}{2}\,[\theta(\zeta + \frac{tz_1}{\pi\ell}, t)\theta(\zeta + \frac{t\bar{z}_1}{\pi\ell}, t)$$

$$+ \theta(\zeta - \frac{tz_1}{\pi\ell}, t)\theta(\zeta - \frac{t\bar{z}_1}{\pi\ell}, t)] \; . \tag{24}$$

Similar expressions can be obtained for the continuous part of the two-body density[17]. Let us emphasize that the present system is inhomogeneous both in the x and y-directions (it is periodic in the y-direction with period ℓ). Therefore the integral representation (23) cannot be obtained from the general expression (7) which is specific to systems invariant under translations in the y-direction.

4.4 Discussion

The exact expressions described in §4.2 and §4.3 are useful for different purposes. First, one can check and illustrate thermodynamic identities and sum rules relative to the system of adsorbed particles. For instance the thermodynamic expression (21) of n_a does coincide with the microscopic expression

$$n_a = \lambda \rho_c(\vec{R}_m) \tag{25}$$

computed from (23), as it should[17]. Furthermore the compressibility derived from $\Delta f^{(s)}$ is simply proportional to the integral over the pair correlation function $(\rho_c^{(2)}(\vec{R}_m,\vec{R}_n) - \rho_c(\vec{R}_m)\, \rho_c(\vec{R}_n))$ of the adsorbed particles, as for a system with short-range forces[17].

The screening rules associated with the Coulomb nature of the system are also well satisfied by the exact expressions. In particular one has[17]

$$\int d\vec{r}_1 [\rho(\vec{r}_1) - \rho] = 0 \tag{26}$$

which states that the system is overall neutral.

Finally the influence of the various parameters ρ, ℓ, λ can be studied. In particular, it is easy to see from (21) that $n_a \to 1$ when $\lambda \to \infty$, which is the expected result. Other qualitative aspects of the exact solution might be useful in studying real adsorption problems.

5. THE TCP ON A LATTICE

5.1 The Model

Now, we consider a two-component system of positive and negative point charges (with charge e and -e respectively). The system is overall neutral. The positions of the positive (respectively negative) charges are restricted to the sites of a rectangular sublattice {X} (respectively {Y}). The sub-lattice {Y} is the image of the sublattice {X} by the translation (a/2,b/2), where a and b are the lengths of the rectangular unit cell common to both sublattices. The position of the positive charge j is defined by the complex number $x_j = (ma + inb)/2$ where m and n are even relative integers ($x_j \in \{X\}$). Similarly, the position of the negative charge j is defined by the complex number $y_j = (m'a + in'b)/2$ where m'and n' are odd relative integers ($y_j \in \{Y\}$). Two (or more) positive (respectively negative) charges cannot lie on the same site. The interaction potential of the system is the sum of the two-body logarithmic potentials between the charges.

The ratio a/b being given, the excess equilibrium properties of the model depend on the coupling constant $\Gamma = \beta e^2$ and on the packing fraction $\eta = \rho ab/2$ where ρ is the particle number density ($0 < \eta < 1$). These properties are similar to those of the continuous version of the model*. In particular the system undergoes a Kosterlitz-Thouless transition[18,19] when the temperature is lowered (at fixed η), between a conducting phase at high temperatures and a dielectric phase at low temperatures. In the conducting phase there are free charges and the system perfectly screens any static external charge. In the dielectric phase there are only dipoles and the dielectric constant of the system is finite (partial screening). When $\eta \to 0$ (limit of a very dilute system), the value of Γ at the transition[18] goes to 4; for η finite, this value is larger than 4.

The equation of state and the correlations for the isotherm $\Gamma = 2$ have been computed exactly by Gaudin[20]. The techniques involved in these calculations are very different from those used for the OCP. They are briefly described in the next subsection.

5.2 The Method of Calculation

One starts from the finite system described by the Grand-Canonical Ensemble[20]. The grand-partition function Ξ of the system is

$$\Xi = \sum_{N=0}^{M^2} \frac{z^{2N}}{(N!)^2} \sum_{\substack{x\in\{X\}\\ y\in\{Y\}}} \exp[-\beta * V_{2N}(x_1,\ldots,x_N;y_1,\ldots,y_N)] \tag{27}$$

where $V_{2N}(x_1,\ldots,x_N;y_1,\ldots,y_N)$ is the total interaction potential of N positive charges located at $x_1,\ldots,x_N$ and of N negative charges located at $y_1,\ldots,y_N$. M^2 is the number of sites of each sublattice and z is the fugacity of both species. The sum in (27) is restricted to overall neutral states; its upper limit is M^2 instead of ∞ because two particles cannot stay on the same site ($N = M^2$ corresponds to the close-packing).

---

* For the continuous model, one has to introduce a short-range repulsive potential between the positive and the negative charges in order to prevent the collapse for $\Gamma \geq 2$. In the discrete model, one overcomes this problem of the collapse between opposite charges by putting the two species on two different sublattices; this procedure corresponds to "dressing" the point charges with hard rectangles in the continuous model.

At $\Gamma = 2$, the 2N-body Boltzmann factor takes the form

$$\exp[-\beta^* V_{2N}(x_1,\ldots,x_N;y_1,\ldots,y_N)] = L_s^{2N} \frac{\prod_{i<j} |x_i-x_j|^2 \, |y_i-y_j|^2}{\prod_{i,j} |x_i-y_j|^2} . \quad (28)$$

The numerator of the rational fraction (28) comes from the interactions between charges of the same species, whereas the corresponding denominator comes from the interactions between opposite charges. Using an identity due to Cauchy, this rational fraction is rewritten as a determinant, and then the 2N-body Boltzmann factor becomes

$$\exp[-\beta^* V_{2N}(x_1,\ldots,x_N;y_1,\ldots,y_N)] = \mathrm{Det} \begin{vmatrix} 0 & \dfrac{L_s}{(y-\bar{x})} \\ \dfrac{L_s}{(x-\bar{y})} & 0 \end{vmatrix}_{2N} . \quad (29)$$

The 2N×2N matrix (29) is built with four N×N blocks: two zero-blocks and the two N×N matrices

$$\begin{vmatrix} \dfrac{L_s}{(x-\bar{y})} \end{vmatrix}_N , \qquad \begin{vmatrix} \dfrac{L_s}{(y-\bar{x})} \end{vmatrix}_N ,$$

whose generic elements are $L_s/(x_i-\bar{y}_j)$ and $L_s/(y_i-\bar{x}_j)$ respectively.

Once the 2N-body Boltzmann factor has been replaced by the determinant (29), the polynomial representation in z (27) of Ξ turns out to be identical to the polynomial obtained by expanding a Toeplitz determinant along its main diagonal[20]. The latter is the determinant of the $2M^2 \times 2M^2$ matrix $\mathbb{1} + z\,\mathbb{K}$, where $\mathbb{1}$ is the identity matrix and $\mathbb{K}$ is the Hankel matrix

$$\mathbb{K} = \begin{vmatrix} 0 & \dfrac{L_s}{(y-\bar{x})} \\ \dfrac{L_s}{(x-\bar{y})} & 0 \end{vmatrix}_{2M^2} . \quad (30)$$

Since the grand-partition function is now expressed as [20]

$$\Xi = \mathrm{Det} \left| \mathbb{1} + z\,\mathbb{K} \right|_{2M^2} , \quad (31)$$

the final step of the calculation consists of controlling the spectrum of the Hankel matrix $\mathbb{K}$ when $M \to \infty$. This analysis can be done using the invariance of the infinite system under any translation (ma, nb) (m and n relative integers). The corresponding asymptotic forms of the eigenvectors and of the eigenvalues of $\mathbb{K}$ are computed in terms of trigonometric and Jacobi's elliptic functions[20]. This allows to find explicit integral representations for the thermodynamic quantities and for the distribution functions. The results obtained for the pressure and the two-body correlations are exposed in §5.3 and §5.4.

5.3 The Equation of State at $\Gamma = 2$

The pressure P of the infinite system is given through

$$\beta^* P = \lim_{M\to\infty} \frac{\ln \Xi}{M^2 ab} = \lim_{M\to\infty} \frac{1}{M^2 ab} \sum_{\substack{\text{eigenvalues} \\ K \text{ of } \mathbb{K}}} \ln(1+zK) \; . \tag{32}$$

The first line in equation (32) is just the definition of the pressure in the Grand-Canonical Ensemble; one obtains the second line by replacing Ξ by the Toeplitz determinant (31). Using the asymptotic form (when $M \to \infty$) of the spectrum of $\mathbb{K}$, one finds[20]

$$\beta^* P = \frac{1}{\pi^2 ab} \int_0^\pi d\xi \int_0^\pi d\zeta \, \ln[1 + \frac{4z^2 L_s^2 \alpha\gamma}{ab} |ds^2(u)|] \tag{33}$$

with $u = 2(\alpha\xi + i\gamma\zeta)/\pi$. ds is Jacobi's elliptic function; α and γ are constants related to the periods of ds which is a doubly-periodic function in the complex plane. The particle number density ρ is given by[20]

$$\begin{aligned} \rho &= z \frac{\partial}{\partial z} \beta^* P(z,\beta^*) \\ &= \frac{8z^2 L_s^2 \alpha\gamma}{\pi^2 a^2 b^2} \int_0^\pi d\xi \int_0^\pi d\zeta \, \frac{|ds^2(u)|}{[1 + \frac{4z^2 L_s^2 \alpha\gamma}{ab} |ds^2(u)|]} \end{aligned} \tag{34}$$

The set of equations (33, 34) might, in principle, allow to find the equation of state, i.e. the pressure as a function of the density, by eliminating the fugacity z. In fact, this does not seem to be possible in terms of known elementary or special functions[20]. Nevertheless, one can show[20] that the pressure is an analytic function of the density in the whole range of densities corresponding to $0 < \eta < 1$. Furthermore, in the low-density and close-packing limits, the pressure behaves respectively as[20]

$$\beta^* P \sim \rho/2 \quad \text{when} \quad \rho \to 0 \tag{35}$$

and

$$\beta^* P \sim \frac{\rho_c}{2} \ln \frac{1}{(\rho_c - \rho)} \quad \text{when} \quad \rho \to \rho_c = 2/ab \ . \tag{36}$$

The asymptotic form (35) suggests that the system becomes a perfect gas of non-interacting neutral "molecules" (dipoles) in the zero-density limit. Strictly speaking, this picture is valid only for $\rho = 0$: for finite densities, there are always free charges and the system is conducting as discussed in the next subsection. The asymptotic form (36) shows that the pressure diverges in the close-packing limit, as it should.

5.4 The Correlations at $\Gamma = 2$

The distribution functions of the finite system in the Grand-Canonical Ensemble are rewritten[20] in terms of matrix elements of $z\mathbb{K}/(\mathbb{1}+z\mathbb{K})$. Using again the asymptotic form of the spectrum of $\mathbb{K}$ in the thermodynamic limit, one finds[20] integral representations for the two-body correlations $h_{++}(0,x)$ and $h_{+-}(0,y)$ ($h_{--} = h_{++}$ and $h_{-+} = h_{+-}$ for obvious symmetry reasons). These integral representations are similar to the one (33) obtained for the pressure; they also involve Jacobi's elliptic function ds.

The exact expressions obtained for the correlations at $\Gamma = 2$ do obey the compressibility sum rule

$$\rho\left[\sum_{x\in\{X\}} h_{++}(0,x) + \sum_{y\in\{Y\}} h_{+-}(0,y)\right] = z\frac{\partial\rho}{\partial z} - \rho \ , \tag{37}$$

the perfect screening sum rule

$$\sum_{x\in\{X\}} h_{++}(0,x) - \sum_{y\in\{Y\}} h_{+-}(0,y) = -1 \ , \tag{38}$$

and the second moment (Stillinger-Lovett) sum rule[21]

$$\sum_{x\in\{X\}} |x|^2 h_{++}(0,x) - \sum_{y\in\{Y\}} |y|^2 h_{+-}(0,y) = -\frac{2}{\pi\rho\beta e^2} \ . \tag{39}$$

The sum rules (37) and (38) are expected to be satisfied for any values of Γ and η. The Stillinger-Lovett sum rule (39) is violated in the dielectric phase. Since this sum rule is explicitely satisfied by the exact expressions at $\Gamma = 2$ for any density ρ, the solvable isotherm $\Gamma = 2$ is in the conducting phase in agreement with the expected phase diagram (see §5.1).

At large distances the correlations are found to decay like monotonic exponentials[20],

$$h_{++}(0,x) \sim \text{cst} \frac{\exp(-2|x|/\ell)}{|x|} \quad \text{when } |x| \to \infty, \tag{40}$$

ℓ being some correlation length[20] (h_{+-} behaves like $-h_{++}$ at large distances). This fast monotonic decay is a universal feature of Coulomb systems[9].

6. CONCLUSION

Through all this review, we have shown that the 2D solvable models give many informations about properties of Coulomb systems. Not only the exact results allow to check known sum rules or heuristic arguments, but they also suggest new sum rules and unexpected qualitative behaviors. We have given several examples which illustrate both previous aspects of the usefulness of the exact solutions. The latter have also been widely used in the literature for testing the accuracy of approximate theories.

The 2D Coulomb systems have specific properties related to the confining nature of the logarithmic potential. In particular they may exhibit Kosterlitz-Thouless transitions. This provides an additional interest to some of the exact solutions evocated here.

Finally let us mention the existence of a large number of one-dimensional solvable models with logarithmic interactions, for instance the one-component system in a uniform background[4,22] or in a periodic inhomogeneous background[23]. These models do not belong to the family of Coulomb systems as far as the properties of the correlations are concerned* (the one-dimensional Coulomb potential, solution of equation (1) for $d = 1$, is $-|x|$). Some of them[23] are quite interesting in the study of phase transitions of the Kosterlitz-Thouless type.

Unlike the review itself which is not exhaustive, the list of references is complete and includes all the 2D solvable models of Coulomb systems which have not been described in detail here[24-35].

---

* From this point of view, these models are "cousines" of 2D and 3D systems with respectively $1/r$ and $1/r^2$ potentials.

Acknowledgement

I am indebted to Joel Lebowitz for his hospitality at Rutgers University. This work is supported in part by AFOSR-82-0016D.

REFERENCES

1. A. Alastuey and B. Jancovici, On the classical two-dimensional one-component Coulomb plasma, J. Physique Paris 42:1 (1981).
2. B. Jancovici, Exact results for the two-dimensional one-component plasma, Phys. Rev. Lett. 46: 386 (1981).
3. J. Ginibre, Statistical ensembles of complex, quaternions, and real matrices, J. Math. Phys. 6: 440 (1965).
4. M.L. Mehta, Random Matrices, Academic Press, New York (1967).
5. J.M. Caillol, D. Levesque, J.J. Weis and J.P. Hansen, A Monte-Carlo study of the two-dimensional one-component plasma, J. Stat. Phys. 28: 325 (1982).
6. S.W. de Leeuw and J.W. Perram, Statistical mechanics of two-dimensional Coulomb systems II. The one-component plasma, Physica 113A: 546 (1982).
7. F. Del Rio and H.E. De Witt, Pair distribution function of charged particles, Phys. Fluids 12: 791 (1969).
8. A. Alastuey, Equilibrium properties of the classical one-component plasma in two and three dimensions, to be published in Les Annales de Physique, Paris.
9. A. Alastuey and Ph. Martin, Decay of correlations in classical fluids with long range forces, J. Stat. Phys. 39: 405 (1985).
10. D. Brydges and P. Federbush, Debye screening, Commun. Math. Phys. 73: 197 (1980).
    J. Imbrie, Debye screening for Jellium and other Coulomb systems, Commun. Math. Phys. 87: 515 (1983).
11. A. Alastuey and J.L. Lebowitz, The two-dimensional one-component plasma in an inhomogeneous background: exact results, J. Physique Paris 45: 1859 (1984).
12. B. Jancovici, Charge distribution and kinetic pressure in a plasma: a solvable model, J. Phys. Lett. Paris 42: L-223 (1981).
13. Ph. Choquard, P. Favre and Ch. Gruber, On the equation of state of classical one-component systems with long-range forces, J. Stat. Phys. 23: 405 (1980).
    H. Totsuji, Distribution of charged particles near a charged hard wall in a uniform background, J. Chem. Phys. 75: 871 (1981).
14. B. Jancovici, Classical Coulomb systems near a plane wall. I, J. Stat. Phys. 28: 43 (1982).
15. S.L. Carnie and D.Y.C. Chan, The Stillinger-Lovett condition for non-uniform systems, Chem. Phys. Lett. 77: 437 (1981).
    S.L. Carnie, On sum rules and Stillinger-Lovett conditions for inhomogeneous Coulomb systems, J. Chem. Phys. 78: 2742 (1983).
16. B. Jancovici, Classical Coulomb systems near a plasma wall. II, J. Stat. Phys. 29: 263 (1982).
17. M.L. Rosinberg, J.L. Lebowitz and L. Blum, A solvable model for localized adsorption in a Coulomb system, J. Stat. Phys. 44: 153 (1986).
18. J.M. Kosterlitz and D.J. Thouless, Ordering, metastability and phase transitions in two dimensional systems, J. Phys. C6: 1181 (1973).
19. J. Fröhlich and T. Spencer, The Kosterlitz-Thouless transition in two-dimensional abelian spin systems and the Coulomb gas, Commun. Math. Phys. 81: 527 (1981).
20. M. Gaudin, L'isotherme critique d'un plasma sur réseau, J. Physique Paris 46: 1027 (1985).

21. F.H. Stillinger and R. Lovett, Ion-pair theory of concentrated electrolytes. I. Basic concepts, J. Chem. Phys. 48: 3858 (1968); General restrictions on the distribution of ions in electrolytes, J. Chem. Phys. 49: 1991 (1968).
22. F.J. Dyson, Statistical theory of the energy levels of complex systems. III, J. Math. Phys. 3: 166 (1962).
23. P.J. Forrester, Charged rods in a periodic background: a solvable model, J. Stat. Phys. 42: 871 (1986).
24. J.M. Caillol, Exact results for a two-dimensional one-component plasma on a sphere, J. Phys. Lett. Paris 42: L-245 (1981).
25. E.R. Smith, Exact results for the electrostatic double-layer at a charged boundary of the two-dimensional one-component plasma, Phys. Rev. A24: 2851 (1981).
26. E.R. Smith, Effects of surface charge on the two-dimensional one-component plasma. I, J. Phys. A15: 1271 (1982).
27. P.J. Forrester and E.R. Smith, Effects of surface charge on the two-dimensional one-component plasma. II, J. Phys. A15: 3861 (1982).
28. P.J. Forrester, B. Jancovici and E.R. Smith, The two-dimensional one-component plasma at $\Gamma=2$: behavior of correlation functions in strip geometry, J. Stat. Phys. 31: 129 (1983).
29. Ph. Choquard, P.J. Forrester and E.R. Smith, The two-dimensional one-component plasma at $\Gamma=2$: the semiperiodic strip, J. Stat. Phys. 33: 13 (1983).
30. B. Jancovici, Surface properties of a classical two-dimensional one-component plasma: exact results, J. Stat. Phys. 34: 803 (1984).
31. L. Blum and B. Jancovici, Exactly solvable model for the interaction of two parallel charged plates in an ionic medium, J. Phys. Chem. 88: 2294 (1984).
32. L. Blum, A model for the interaction of two electric double layers in two dimensions: the metal electrolyte interface and the Donnan membrane, J. Chem. Phys. 80: 2953 (1984).
33. M.L. Rosinberg and L. Blum, The ideally polarizable interface: a solvable model and general sum rules, J. Chem. Phys. 81: 3700 (1984).
34. P.J. Forrester, The two-dimensional one-component plasma at $\Gamma=2$: Metallic boundary, J. Phys. A18: 1419 (1985).
35. A. Alastuey, B. Jancovici, L. Blum, P.J. Forrester and M.L. Rosinberg, The ideally polarizable interface: the metallic boundary limit, J. Chem. Phys. 83: 2366 (1985).

CHARGE CORRELATIONS AND SUM RULES IN COULOMB SYSTEMS I

Bernard Jancovici

Laboratoire de Physique Théorique et Hautes Energies*
Université de Paris Sud
91405 Orsay, France

INTRODUCTION

We discuss Coulomb fluids (systems of particles interacting through Coulomb's law, plus perhaps some short-range interaction) in the framework of classical equilibrium statistical mechanics. Coulomb fluids exhibit the fundamental property of screening. Screening means that a charge q introduced into the fluid induces a polarization cloud of charge -q, and therefore the excess charge distribution (defined as the sum of the external charge distribution and the induced one) has zero total charge. More generally, all the electrical moments of the excess charge distribution vanish[1,2] *provided* the charge correlations in the fluid have good decay properties at long distances. We do not attempt here to derive screening from first principles, but we take it for granted.

A consequence of screening is that the charge correlation function obeys a variety of sum rules. The purpose of this lecture is to give a review of these sum rules, from an as unified as possible point of view, in the general case of inhomogeneous fluids, i.e. fluids not invariant by translations (for instance because of the presence of walls or of external charges). We only discuss statics (dynamics is considered in the lecture by Ph. A. Martin with the same title).

The object of our study, the static charge correlation function (or structure function) S is defined, in terms of the microscopic charge density $\rho(r)$ at point r, as

$$S(r,r') = <\rho(r)\rho(r')> - <\rho(r)> <\rho(r')> ,$$

where < > denotes an equilibrium average (this definition includes a self part proportional to $\delta(r-r')$).

HOMOGENEOUS FLUIDS : STILLINGER-LOVETT RULES

In the special case of homogeneous fluids, S depends only on $|r-r'|$ and obeys the well-known Stillinger-Lovett sum rules[3]

*Laboratoire Associé au CNRS.

$$\int dr' \, S(|r-r'|) = 0 \,, \tag{1a}$$

$$\frac{2\pi\beta}{3} \int dr' |r-r'|^2 S(|r-r'|) = 1 \tag{1b}$$

(where β is the inverse temperature). Equivalently, in terms of the Fourier transform

$$S(k) = \int dr' e^{ik\cdot(r'-r)} S(|r-r'|) \,,$$

one has

$$\lim_{k\to 0} 4\pi\beta \frac{S(k)}{k^2} = 1 \,. \tag{2}$$

The sum rules to be reviewed here can be regarded as generalizations of the above to inhomogeneous systems.

CARNIE AND CHAN'S GENERALIZATION

The Carnie-Chan sum rule[4,5] reads

$$\beta \int dr \int dr' \frac{1}{|r'|} S(r,r') = 1. \tag{3}$$

It should be noted that the integral is not absolutely convergent and therefore the order of the integrations cannot be changed ; actually

$$\int dr \, S(r,r') = 0$$

(this is just screening for the charge at r').

A simple proof of (3) can be based on screening plus a linear response argument. It goes as follows. Let us perturb the fluid by introducing a small test charge q at the origin. The corresponding perturbing Hamiltonian is

$$H' = \int dr' \frac{q}{|r'|} \rho(r').$$

Static linear response theory states that the response of the average charge density at r is a change

$$\delta\rho(r) = -\beta \, [< \rho(r)H' > - < \rho(r) > < H' >] = -\beta \int dr' \frac{q}{|r'|} S(r,r').$$

Screening states that

$$\int dr \, \delta\rho(r) = -q \,,$$

and this leads immediately to (3).

FURTHER GENERALIZATION : MULTIPOLE SUM RULE

By considering the response to a point multipole rather than to a point charge, one obtains the very general sum rule[6]

$$\frac{4\pi\beta}{2\ell+1} \int dr |r|^{\ell} Y_{\ell m}(\hat{r}) \int dr' \frac{1}{|r'|^{\ell'+1}} Y^*_{\ell' m'}(\hat{r}') S(r,r') = \delta_{\ell\ell'}\delta_{mm'} , \quad (4)$$

where $\hat{r} = r/|r|$ and $Y_{\ell m}$ is a spherical harmonic.

It should be emphasized that this general sum rule holds only when the underlying assumptions are satisfied. In particular, near an insulating wall, the rule holds only for the case $\ell=\ell'=0$, because the correlations do not decay fast enough for a multipole of higher order to be screened.

SPECIAL CASES ($\ell=\ell'=0$)

The Carnie-Chan sum rule (3) can be applied to a variety of special geometries. We shall show that along insulating boundaries the charge correlations have a slow decay, in contrast with the decay in the bulk or near a conducting wall which is believed to be faster than any power law.

Homogeneous Fluid.

In this case, the Carnie-Chan sum rule (3) and its generalization (4) are simply equivalent to the Stillinger-Lovett sum rule (2), as it can be seen[7] by rewriting (3) or (4) in terms of the Fourier transforms of $1/|r|$ and S.

Slab.

We consider a Coulomb fluid confined in a slab of thickness a. A three-dimensional position vector r will be written as $r = (x,y)$ where x is the component in the direction normal to the slab faces (located at $x = 0$ and $x = a$) and y stands for the other two components (along the slab). S is now a function $S(x,x',|y-y'|)$ for which we define a Fourier transform with respect to $y-y'$:

$$S(x,x',k) = \int dy' \, e^{ik\cdot(y'-y)} S(x,x',|y-y'|),$$

where k is a two-dimensional vector along the slab. In terms of such Fourier transforms, the Carnie-Chan sum rule (3) becomes

$$\beta \lim_{k\to 0} \int_0^a dx \int_0^a dx' \, \frac{2\pi}{|k|} e^{-|k|x'} \, S(x,x',k) = 1,$$

which means that $2\pi\beta\int_0^a dx\int_0^a dx' S(x,x',k)$ behaves like $|k|$ for small $|k|$. This singularity (assumed to be the only one) governs the asymptotic behavior along the slab, and by an inverse Fourier transformation one finds[8]

$$\int_0^a dx \int_0^a dx' \, S(x,x',|y-y'|) \underset{|y-y'|\to\infty}{\sim} -\frac{1}{4\pi^2\beta} \frac{1}{|y-y'|^3} \quad (5)$$

Thus, S has a slow (algebraic) decay along the slab.

Plane Wall

A limiting case of the previous one is a Coulomb fluid occupying the half-space $x > 0$; the plane $x = 0$ is assumed to be an insulating wall. In that case, after having substracted[9] a bulk contribution, and assuming there are no subtle cancellations, one finds for S an asymptotic behavior along the slab of the form

$$S(x,x',|y-y'|) \underset{|y-y'|\to\infty}{\sim} \frac{f(x,x')}{|y-y'|^3}, \tag{6a}$$

where $f(x,x')$ is a function which is important only for small x and x' and which obeys the sum rule

$$\int_0^\infty dx \int_0^\infty dx' \, f(x,x') = -\frac{1}{8\pi^2\beta}. \tag{6b}$$

If the wall has a (finite) dielectric constant ε_W, the right-hand side of (6b) must be multiplied by ε_W.

Using another scheme of substraction of a bulk contribution, one can obtain[8] another sum rule

$$4\pi\beta \int_0^\infty dx \int_{x'>0} dr' \, x' \, S(r,r') = -1. \tag{7}$$

If the wall has a (finite) dielectric constant ε_W, it does *not* affect (7). The sum rule (7) can also be derived by other methods[5,10].

Interface

Similarly, if the plane $x = 0$ is the interface between two different Coulomb fluids, one finds[8]

$$-4\pi\beta \int_{-\infty}^{\infty} dx \int_{x'>0} dr' \, x' \, S(r,r') = 4\pi\beta \int_{-\infty}^{\infty} dx \int_{x'<0} dr' \, x' \, S(r,r') = 1. \tag{8}$$

Cylinder

We now consider a Coulomb fluid confined in a cylinder with a cross-section of arbitrary shape. It is now convenient to write a position vector r as $r = (R,z)$, where the z axis is along the cylinder and R stands for the two other components of r (thus the two-dimensional vector R is normal to the z axis). One now finds

$$\int dR \int dR' \, S(R,R',|z-z'|) \underset{|z-z'|\to\infty}{\sim} -\frac{1}{4\beta} \frac{1}{(z-z')[\ln|z-z'|]^2}, \tag{9}$$

a very slow decay.

Wedge

Similarly, one can consider a Coulomb fluid inside a wedge (two half-planes intersecting along the z axis); let α be the angle between these half-planes. After substractions[8] of bulk and plane wall contributions, one finds for S an asymptotic behavior along z of the form

$$S(R,R',|z-z'|) \underset{|z-z'|\to\infty}{\sim} \frac{f(R,R')}{|z-z'|[\ln|z-z'|]^2}, \tag{10a}$$

where f(R,R') is a function which is important only near the edge of the wedge and obeys the sum rule

$$\int dR \int dR' f(R,R') = \frac{\alpha-\pi}{8\pi\beta}. \tag{10b}$$

SPECIAL CASES ($\ell=\ell'=1$)

We now use the sum rule (4) for $\ell=\ell'=1$.

Interface

Considering again the interface case (the plane x = 0 separates two different Coulomb fluids), we now find from (4)

$$4\pi\beta \int_{-\infty}^{\infty} dx\ x \int_{x'>0} dr'\ S(r,r') = -\,4\pi\beta \int_{-\infty}^{\infty} dx\ x \int_{x'<0} dr'\ S(r,r') = 1. \tag{11}$$

It should be noted that, at an interface, both (8) and (11) hold. They are different, not equivalent, sum rules.

Ideal Conductor Wall

In the limiting case of a Coulomb fluid occupying the half-space x>0 and bounded at x=0 by an ideal conductor wall (i.e. a Coulomb fluid of vanishing Debye length), we find

$$4\pi\beta \int_{0}^{\infty} dx\ x \int_{x'>0} dr'\ S(r,r') = 1. \tag{12}$$

This result can also be derived directly[6] by taking into account the image forces in the Hamiltonian.

It should be noted that the sum rule (7) which holds for an insulating wall and the sum rule (12) which holds for an ideal conductor vall are different.

DIELECTRIC SUSCEPTIBILITY AND CORRELATIONS[12]

It is amusing to note that the long-range correlations along an insulating wall are necessary for the consistency between macroscopic electrostatics and statistical mechanics.

Let us consider a ν-dimensional material in a box Λ of some shape, for instance spherical. The electrical susceptibility χ (defined as such as the polarization per unit volume induced by a weak *external* electric field E be χE) and the dielectric constant ε are related by the Clausius-

Mossotti equation

$$\varepsilon = \frac{1 - \frac{\nu-1}{\nu}\, 2\pi\chi}{1 + \frac{(\nu-1)^2}{\nu}\, 2\pi\chi}, \qquad \nu = 2,3.$$

For a Coulomb fluid, $\varepsilon = \infty$, and therefore

$$\chi = \frac{\nu}{(\nu-1)2\pi}.$$

A simple linear response argument relates χ to the correlation function S_Λ defined in the finite system

$$\chi = -\frac{\beta}{2\nu}\frac{1}{|\Lambda|}\int_\Lambda dr \int_\Lambda dr' |r-r'|^2\, S_\Lambda(r,r').$$

When the system is large, one is tempted to forget the surface contributions, to write

$$\chi = -\frac{\beta}{2\nu}\int dr' |r-r'|^2\, S(r,r'),$$

and to take for S the infinite system correlation function which obeys the Stillinger-Lovett rule (1b). This approach gives $\chi = 1/(\nu-1)2\pi$ instead of the correct result $\nu/(\nu-1)2\pi$.

The explanation of this apparent paradox is that it is not permissible to replace the finite system correlation function S_Λ by the infinite system one, before performing the integrals ; whatever large the system may be, near the boundaries S_Λ has a long-ranged part which gives to χ a surface contribution of the same order of magnitude as the bulk one. When this effect is properly taken into account, one does find, for a solvable two-dimensional model, the correct result $1/\pi$, as predicted by electrostatics.

Similar considerations apply to other shapes of the box.

CONCLUSION

We are making progress towards a better understanding of the interplay between electrostatics and statistical mechanics.

REFERENCES

1. Ch. Gruber, J.L. Lebowitz, and Ph. A. Martin, Sum rules for inhomogeneous Coulomb systems, J. Chem. Phys. 75 : 944 (1981).
2. L. Blum, C. Gruber, J.L. Lebowitz, and P. Martin, Perfect screening for charged systems, Phys. Rev. Lett. 48 : 1769 (1982).
3. F.H. Stillinger and R. Lovett, General restriction on the distribution of ions in electrolytes, J. Chem. Phys. 49 : 1991 (1968).
4. S.L. Carnie and D.Y.C. Chan, The Stillinger-Lovett condition for non-uniform electrolytes, Chem. Phys. Lett. 77 : 437 (1981).
5. S.L. Carnie, On sum rules and Stillinger-Lovett conditions for inhomogeneous Coulomb systems, J. Chem. Phys. 78 : 2742 (1983).
6. B. Jancovici, Sum rules for inhomogenous Coulomb fluids, and ideal conductor boundary conditions, J. Physique 47 : 389 (1986).

7. C.W. Outhwaite, Comment on the second moment condition of Stillinger and Lovett, Chem. Phys. Lett. 24 : 73 (1974).
8. B. Jancovici, J.L. Lebowitz, and Ph. A. Martin, Time-dependent correlations in an inhomogeneous one-component plasma, J. Stat. Phys. 41 : 941 (1985).
9. B. Jancovici, Classical Coulomb systems near a plane wall. II, J. Stat. Phys. 29 : 263 (1982).
10. L. Blum, D. Henderson, J.L. Lebowitz, Ch. Gruber, and Ph. A. Martin, A sum rule for an inhomogeneous electrolyte, J. Chem. Phys. 75 : 5974 (1981).
11. B. Jancovici and X. Artru, Weakly coupled classical Coulomb systems near a cylindrical wall, Molec. Phys. 49 : 487 (1983).
12. Ph. Choquard, B. Piller, and R. Rentsch, On the dielectric susceptibility of classical Coulomb systems, J. Stat. Phys. 43 : 197 (1986), and to be published.

CHARGE CORRELATIONS AND SUM RULES IN COULOMB SYSTEMS II: DYNAMICS

Philippe A. Martin

Institut de Physique Théorique
Ecole Polytechnique Fédérale de Lausanne
CH-1015 Lausanna, Switzerland

I. A SUM RULE FOR THE TIME DEPENDENT STRUCTURE FUNCTION OF THE OCP

We present in this paper some exact results on the dynamical structure function of a class of inhomogeneous one component plasma (OCP). This generalizes in a natural way some of the static sum rules discussed in[1] and summarizes essentially the papers[2,3,4] where more detailed proofs and additional comments can be found.

The time dependent structure function of an OCP of particles of charge e, mass m and background density $\rho_b(r)$ is defined by

$$S(r,r',t) = e^2 (\langle N(r,t)\, N(r',0)\rangle - \langle N(r,t)\rangle \langle N(r',0)\rangle) \tag{1}$$

where N(r,t) is the microscopic particle number density at time t and <...> is the equilibrium average at temperature $k_B T=\beta^{-1}$. It is known that the static structure function S(r,r') = S(r,r',t=0) obeys the perfect screening condition written in the Carnie and Chan form[1,5]

$$\beta \int dr \int dr' \frac{1}{|r'|} S(r,r') = 1 \tag{2}$$

We consider here a classical OCP having a background density which is asymptotically constant in all directions, i.e.

$$\lim_{|r| \to \infty} \rho_b(|r|,\Omega) = \rho_b(\Omega) \tag{3}$$

with Ω the angles of r (precisely we require $\int_0^\infty d|r| \left| \frac{\partial}{\partial |r|} \rho_b(|r|,\Omega)\right| < \infty$ insuring that the limit is obtained without fast oscillations).

Particular examples treated in Section III are the semi-infinite OCP bounded by a plane wall:

$$\rho_b(r) = \begin{cases} 0, x < 0 \\ \rho_b, x > 0 \end{cases} \tag{4}$$

$r = (x,y)$ y the component of r parallel to the wall, and the two densities OCP,

$$\rho_b(r) = \begin{cases} \rho_-, & x < 0 \\ \rho_+, & x > 0 \end{cases}, \qquad (5)$$

Then, the static perfect screening condition has the following simple time dependent generalization[2,3]

$$\beta \int dr \int dr' \frac{1}{|r'|} S(r,r',t) = \cos \bar{\omega}_p t \qquad (6)$$

$\bar{\omega}_p^2$ being the angular average of the asymptotic plasmon frequencies

$$\bar{\omega}_p^2 = \frac{e^2}{m} \int d\Omega \, \rho_b(\Omega) \qquad (7)$$

The formula (6) deserves the following comments.

(i) The pure $\cos\bar{\omega}_p t$ oscillation in (6) is due to the fact that in an OCP satisfying the property (3), the long wave length mode oscillates undamped with a single frequency $\bar{\omega}_p$. This feature does not extend to the electron gas in the periodic field of an (infinite) ionic lattice. In this case, the plasmon mode is shifted and damped as a consequence of the coupling of the electrons to the ionic lattic.[6]

(ii) There is no generalization of (6) to multicomponent systems, which show dissipation also in the long wave limit because of interparticle collisions (for a review on the properties of the structure function, see Ref. 7).

(iii) For a quantum mechanical OCP satisfying (3), the equivalent of formula (6) becomes[3]

$$\int dr \int dr' \frac{1}{|r'|} S(r,r',t) = \frac{\hbar\bar{\omega}_p}{2} \left(\frac{\exp(-i\bar{\omega}_p t)}{1-\exp(-\beta\hbar\bar{\omega}_p)} - \frac{\exp(i\bar{\omega}_p t)}{1-\exp(\beta\hbar\bar{\omega}_p)} \right) \qquad (8)$$

which reduces to (6) as $\hbar \to 0$.

(iv) Contrary to the static case, there does not exist up to now solvable models for the time dependent structure function where dynamical sum rules could be checked. The assertion that (6) is exact is justified by the fact that it can be derived from the microscopic equations of motion (the BBGKY hierarchy) without approximation, as sketched in Section IV.

Finally, there are sum rules analogous to (6) when the plasma is submitted to an external uniform magnetic field.[8]

II. MACROSCOPIC ELECTRODYNAMICS AND LINEAR RESPONSE

A simple derivation of (6) can be given if one assumes that macroscopic electrodynamics is valid in the long wave length limit. We consider electric field amplitude $E(\omega,r)$ due to an oscillating external charge in the plasma $e_o \exp(-i\omega t)$ located at the origin:

$$\nabla \cdot (\epsilon(\omega,r)\, E(\omega,r)) = 4\pi\, e_o \delta(r) \tag{9}$$

where

$$\epsilon(\omega,r) = 1 - \frac{\omega_p^2(r)}{\omega^2}, \quad \omega_p^2(r) = \frac{4\pi e^2}{m} \rho_b(r),$$

is the local dielectric functions.

Assuming that the induced charge density decays sufficiently fast at large distances, the electric field is asymptotically radial

$$E(\omega,r) \simeq \frac{e_o + Q^{ind}(\omega)}{|r|^2} \hat{r}, \quad |\hat{r}| = 1, \quad |r| \to \infty \tag{10}$$

where $Q^{ind}(\omega)$ is the total net charge induced in the plasma by the external charge. Then, integrating (9) on a large sphere and using Gauss theorem yields

$$\int_{|r| = R} R^2\, d\Omega\, \epsilon(\omega,R,\Omega) \left(\frac{e_o + Q^{ind}(\omega)}{R^2} \right) = 4\pi e_o \tag{11}$$

Letting $R \to \infty$, we obtain from (11) with (3)

$$Q^{ind}(\omega) = \frac{\bar{\omega}_p^2}{\omega^2 - \bar{\omega}_p^2} e_o \tag{12}$$

Finally, we obtain from the fluctuation-dissipation theorem and (12) (ω being understood as having an infinitesimal positive imaginary part corresponding to switching on the perturbation adiabatically)

$$\begin{aligned} \text{Im}\, Q^{ind}(\omega) &= -\pi\beta\omega \int dr \int dr' \frac{e_o}{|r'|} S(r,r',\omega) \\ &= -\pi\beta \frac{\omega}{2} e_o \left(\delta(\omega - \bar{\omega}_p) + \delta(\omega + \bar{\omega}_p)\right) \end{aligned} \tag{13}$$

which gives (6) by Fourier transform. The formula (8) follows from the quantum mechanical version of the fluctuation dissipation theorem.

III. PARTICULAR CASES

(1) Uniform OCP

$\bar{\omega}_p$ is identical to the usual plasmon frequency $\omega_p = \left(\frac{4\pi e^2\rho}{m}\right)^{1/2}$. Since $S(r,r',t) = S(r-r',t)$ is translation invariant, one can use the convolution theorem to transform (6) to Fourier space (with $\frac{4\pi}{|k|^2}$ the Fourier transform of $\frac{1}{|r|}$) giving

$$4\pi\beta S(k,t) = |k|^2 \cos \omega_p t, \quad |k| \to 0 \tag{14}$$

This is the well known long wave length behavior of the structure factor (for instance see Ref. 7).

(2) Impenetrable Wall

Here, from (4) and (7) one has $\bar{\omega}_p^2 = \frac{2\pi e^2\rho}{m}$ and $\bar{\omega}_p = \frac{\omega_p}{\sqrt{2}}$ is the surface plasmon mode. The plasma being still homogeneous in the y directions (parallel to the wall), one can introduce the two-dimensional y-Fourier transform

$$\int dy \exp\left(i\, k\cdot(y-y')\right) S\,(xy,x'y',t) = S(x,x',k,t)$$

Since $\int dy \exp(-ik\cdot y)\,(y^2 + x^2)^{-1/2} = \frac{2\pi}{|k|} \exp(-|k|\,|x|)$, Eq. (6) is equivalent with[4]

$$2\pi\beta\int dx \int dx' \exp(-|k||x'|)\, S(x,x',k,t) = |k| \cos\bar{\omega}_p t, \quad |k| \to 0 \tag{15}$$

The non analytic behavior as $|k| \to 0$ is linked to a slow decay of $S(xy,x'y',t)$ parallel to the wall. This is best exhibited by considering its difference with the pure bulk structure function (the difference being assumed to be jointly integrable in x and x'). Coming back to the space variables and noting that the two-dimensional Fourier transform of $|k|$ is $\frac{1}{2|y|^3}$, 15 implies after some algebra[4]

$$2\pi\beta\int_0^\infty dx \int_0^\infty dx' \left(S\,(xy,x'y',t) - S^{bulk}(xy,x'y',t)\right)$$

$$= -\frac{1}{2\pi|y-y'|^3}\left(\cos \frac{\omega_p t}{\sqrt{2}} - 1/2 \cos \omega_p t\right), \quad |y-y'| \to \infty \tag{16}$$

At $t = 0$ one recovers the spatial decay already obtained in the static case.[1] It is interesting to compare (15) and (16) which involve the surface plasmon to another kind of sum rule, of dipolar type, which also holds for the semi-infinite plasma[3]:

$$\int_0^\infty dx \int dy \int_0^\infty dx'\; x'\, S\,(xy,\, x'o,t) = \frac{1}{4\pi\beta} \cos \omega_p t \tag{17}$$

Here only the bulk plasma frequency occurs: this generalizes the classical static dipole sum rule at a wall.[1]

(3) Two Densities OCP

With (5) and (7), $\bar{\omega}_p^2 = \frac{1}{2}(\omega_+^2 + \omega_-^2)$, $\omega_\pm^2 = \frac{4\pi e^2 \rho_\pm}{m}$, and (15) implies the asymptotic behavior[4]

$$2\pi\beta\int_{-\infty}^{\infty} dx \int_{-\infty}^{\infty} dx' \, (S(xy,x'y',t) - \vartheta(x')\, S_+(xy,x'y',t) - \vartheta(-x')S_-(xy,x'y',t))$$

$$= -\frac{1}{2\pi}\frac{1}{|y-y'|^3}(\cos\bar{\omega}_p t - \frac{1}{2}(\cos\omega_+ t + \cos\omega_- t)), \quad |y-y'| \to \infty \qquad (18)$$

S_+ and S_- are the bulk functions corresponding to ρ_+ and ρ_- and $\vartheta(x) = \begin{matrix} 0, & x<0 \\ 1, & x>0 \end{matrix}$. One notices here that the long tail disappears at $t = 0$, in accordance with the fact that the screening cloud of a test particle in a metallic interface should remain localized. However, this is no more true when $t \neq 0$, or quantum mechanically even at $t = 0$ where the dynamics cannot be disentangled from the statics. The two densities OCP satisfies also dipole sum rules similar to (18) involving the individual bulk frequencies ω_+ (resp. ω_-) for $x > 0$ (resp. $x<0$).[3]

(4) Slab of Thickness a

Since $\rho_b(r)$ vanishes in all directions not parallel to the walls, $\bar{\omega}_p = 0$. One finds that for all times[3]

$$2\pi\beta\int_0^a dx \int_0^a dx' \, S(xy, x'y',t) = -\frac{1}{2\pi|y-y'|^3}, \quad |y-y'| \to \infty \qquad (19)$$

The same result holds quantum mechanically. The slab behaves essentially as a two-dimensional electron gas where the plasmon frequency vanishes as $|k|^{1/2}$ as $|k| \to 0$. This explains why the right hand side of (19) is time independent.

(5) Cylinder of Base Σ

Here again $\bar{\omega}_p = 0$. Calling z (resp. u) the variable parallel (resp. perpendicular) to the axis of the cylinder and performing the Fourier transfrom on z, one finds from (6)

$$\beta\int_\Sigma du \int_\Sigma du' \, S(u,u',k,t) = \frac{1}{2\ln|k|}, \quad |k| \to 0 \qquad (20)$$

This leads to a spatial decay of the form[3]

$$\beta\int du \int du' \, S(uz,u'z',t) = -\frac{1}{4|z-z'|\,(\ln|z-z'|)^2},$$

$$|z-z'| \to \infty \qquad (21)$$

In this geometry also, the plasmon frequency vanishes as $|k| \to 0$.

Other cases (cone, wedge, dielectric walls) are described in Ref. 3.

IV. AN EXACT CLOSURE OF THE BBGKY HIERARCHY

We show that the dynamical sum rule (6) is an exact consequence of the microscopic equations of motion.[3] For simplicity we present here the case of the uniform OCP with density ρ. We introduce higher order correction functions of the type

$$\rho(rvt|r_1v_1, \ldots, r_nv_n) = \langle N(r,v,t)\, N(r_1,v_1,o) \ldots N(r_n,v_n,o)\rangle$$

where v is the velocity and N(r,v,t) is the phase space microscopic density. The quantity

$$\hat{\rho}(rt\,|r_1v_1, \ldots, r_nv_n) = e \frac{\int dv\, \rho(rvt|r_1v_1,\ldots,r_1v_1}{\rho(r_1v_1,\ldots,r_nv_n)} - \rho \tag{22}$$

represents the excess charge density in the plasma at r and time t, given that n particles were located in $r_1,\ldots,r_n$ at time t = 0 with initial velocities $v_1,\ldots v_n$. It is known[9,10] that in the static case (t = 0), the excess charge density carries no multipoles of any order, i.e. the multipoles induced by specifying any arrangement of the system's charge are shielded. In the homogeneous OCP, the charge sum rule

$$\int dr\, \hat{\rho}\, (rt \mid r_1v_1,\ldots,r_nv_n) = 0 \tag{23}$$

at the dipole sum rule

$$\int dr\, r\, \hat{\rho}\, (rt|r_1,\ldots,r_n) = 0 \tag{24}$$

remain valid for all times (suppression of the velocity arguments means that they have been integrated out). The interpretation of (24) is that the screening cloud evolves without developing a dipole moment (this is obvious for n = 1 because of isotropy, but non trivial for $n \geq 2$). One should stress that this property holds only for the OCP where the electric current is proportional to the mass current, which is a conserved quantity.

To establish (23) and (24), we consider the first BBGKY equation

$$\frac{\partial}{\partial t} \rho\, (rvt|\upsilon) = -\, v \cdot \nabla_r \rho(rvt|\upsilon)$$

$$-\, \frac{e^2}{m} \int dr'\, F(r-r') \cdot \nabla_v\, (\rho(rv,r',t|\upsilon) - \rho(rvt|\upsilon)\rho) \tag{25}$$

where $\upsilon = (r_1v_1,\ldots r_nv_n)$. The evolution of a general velocity independent function f(r) is obtained by integrating out the velocity in (25), the ∇_v-terms giving no contribution

$$\frac{\partial}{\partial t} \int drf(r)\, \rho_T(r\, t|\upsilon) = -\int drf(r)\nabla_r \cdot \int dvv\, \rho_T(rv\, t|\upsilon) \tag{26}$$

We have introduced the truncated functions $\rho_T(rvt|\upsilon) = \rho(rvt|\upsilon) - \rho(rv)\rho(\upsilon)\ldots$ defined in the usual way.

Setting now f(r) = 1, the r.h.s. of (26) vanishes, assuming that the truncated functions vanish at large spatial separation. Taking into account the static charge sum rule $e\int dr\rho_T(rt|\upsilon)|_{t=0} = \rho(\upsilon)\int dr\hat{\rho}(rt|\upsilon)|_{t=0}=0$ gives (23). To derive the dipole sum rule, it is useful to consider the second time derivative obtained from (25), (26) and partial integrations on velocities

$$\frac{\partial^2}{\partial t^2}\int drf(r)\ \rho_T(rt|\upsilon) = -\ \omega_p^2\int drf(r)\ \rho_T(rt|\upsilon) \tag{27}$$

$$-\int drdv\ (v\cdot\nabla_r f(r))\ (v\cdot\nabla_r\ \rho_T(rv,t|\upsilon)) \tag{28}$$

$$+\frac{e^2}{m}\int dr\int dr'\ (\nabla_r f(r))\cdot F\ (r-r')\ \rho_T(r,r',t|\upsilon) \tag{29}$$

The first term of the r.h.s. results of Poisson equation $\nabla_r\cdot F(r) = 4\pi\delta(r)$ and $\omega_p = (\frac{4\pi e^2\rho}{m})^{1/2}$ is the plasmon frequency. Choosing now f(r) = r, the terms (28) and (29) vanish, (28) being the integral of a gradient and (29) because of the antisymmetry of the force. So (27) becomes the ordinary differential equation with f(r) = r

$$\frac{\partial^2}{\partial t^2}\int dr\ f(r)\ \rho_T(r\ t|\upsilon) = -\ \omega_p^2\int dr\ f(r)\ \rho_T(rt|\upsilon) \tag{30}$$

From the static dipole sum rules and (26) the initial conditions are found to be

$$\int dr\ r\ \rho_T(rt|\upsilon)\Big|_{t=0} = 0$$

$$\frac{\partial}{\partial t}\int dr\ r\ \rho_T(rt|\upsilon)\Big|_{t=0} = (\sum_{j=1}^{n} V_j)\ \rho(\upsilon)$$

and the solution is thus

$$\int dr\ r\ \rho_T(rt|\upsilon) = \frac{1}{\omega_p}(\sum_{j=1}^{n} V_j)\rho(\upsilon)\sin\omega_p t \tag{31}$$

Averaging (31) on the initial velocities givès the dipole sum rule (24).

We can now establish the sum rule (6) for the structure function by noting that in the homogeneous situation we have the identity

$$\int dr\int dr'\ \frac{1}{|r'|}\ S\ (r-r',t) = -\ \frac{2\pi}{3}\int dr|r|^2\ S\ (r,t)$$

Since $S(r,r',t) = e^2 \rho_T(rt|r')$, we set $f(r) = |r|^2$ and $\upsilon = (ov)$ in (27), and integrate over v. The key point is that the terms (28) and (29) vanish again as a consequence of the relations (23) and (24). This provides an exact closure of the BBGKY equation (27), which reduces to the simple differential equation (30) with now $f(r) = |r|^2$. Supplemented with the initial condition (2) and $\frac{\partial}{\partial t} \int dr\, |r|^2\, S\,(r,t)\Big|_{t=0}$, we find (6).

To show that the contribution of the three point function (29) vanishes, we make the change of variables $r = r_1 - r$, $r' = -r$ and use the translation invariance (exchanging the r_1, r integrals)

$$(29) = \frac{2e^2}{m} \int dr_1\, F(r_1) \cdot \int d\bar{r}\, (r_1 - \bar{r})\, \rho_T(r_1, 0, t|\bar{r}) =$$

$$\frac{2e^2}{m} \int dr_1\, F(r_1) \cdot \int d\bar{r}\, (r_1 - \bar{r})\, \rho_T(\bar{r}, -t|r_1, 0) \qquad (32)$$

We also used $\rho(r_1, r_2|r,t) = \rho(r_1 - t|r_1, r_2)$ which follows from the stationnarity of the equilibrium state. It is easily checked that (23) and (24) imply the same relations for $\rho_T(r,t|r_1,r_2)$ instead of $\rho(r,t|r_1,r_2)$, thus (32) vanishes. The term (28) is treated in the same way.

In the case of an inhomogeneous OCP with the property (3), we make the choice $f(r) = \frac{1}{|r|}$ in (27) and show[3] again that the BBGKY equation reduces to (30). The assumption used here is that at any fixed time the correlations of the inhomogeneous OCP with all arguments going to infinity in a fixed direction Ω converge sufficiently fast to those of an uniform OCP with density $\rho_b(\Omega)$.

REFERENCES

1. B. Jancovici, "Charge Correlations and Sum Rules in Coulomb Systems I: Statics" in this volume.
2. J. L. Lebowitz and Ph. A. Martin, Phys. Rev. Lett. 54:1506 (1985).
3. B. Jancovici, J. L. Lebowitz and Ph. A. Martin, J. Stat. Phys. 41:941 (1985).
4. B. Jancovici, J. Stat. Phys. 39:427 (1985).
5. S. L. Carnie and D. Y. C. Chan, Chem. Phys. Lett. 77:438 (1981).
6. A. Alastuey and J. P. Hansen, Europhys. Lett. 2:97 (1986).
7. M. Baus and J. P. Hansen, Phys. Rep. 59:72 (1980).
8. B. Jancovici, N. Macris and Ph. A. Martin, in preparation.
9. Ch. Gruber, J. L. Lebowitz and Ph. A. Martin, J. Chem. Phys. 75:944 (1981).
10. L. Blum, Ch. Gruber, J. L. Lebowitz and Ph. A. Martin, Phys. Rev. Lett. 48:1769 (1982).

THEORY OF THE STRONGLY COUPLED 2-D PLASMA WITH THE 1/r POTENTIAL

Kenneth I. Golden

Department of Computer Science and Electrical Engineering
The University of Vermont
Burlington, VT 05405

INTRODUCTION

The classical two-dimensional (2-d) Coulomb fluid of electrons trapped in surface states above a liquid-helium surface is a fascinating one-component-plasma (ocp) configuration which has been of experimental and theoretical interest since the early 1970's. In the actual laboratory setup[1], extra electrons are deposited in a monolayer just above the free surface of liquid helium and are confined there by means of image binding. The electrons are accordingly constrained to execute only horizontal (parallel-to-the-surface) motions, and they interact via the $\phi(r) = e^2/r$ potential (r is the horizontal range). A compensating uniform positive background is provided by an electrode placed below the liquid surface. At a temperature $T \sim 0.5K$ and over the range $10^5 < n < 10^9$ cm^{-2} of areal densities realized in the Grimes-Adams experiments[1c,1d], $\beta E_F < .06$ and $1.5 < \Gamma = \beta e^2\sqrt{\pi n} < 150$; we may therefore consider the electrons to be classical strongly correlated particles.

Numerous computer experimental and theoretical efforts have provided a great deal of information about the static and dynamical properties of the 2-d ocp over a wide range of γ values ($\gamma = 2\pi n e^4\beta^2 = 2\Gamma^2$ is the 2-d plasma parameter). On the static level, liquid-solid boundaries have been established from molecular dynamics (MD)[2] and Monte Carlo (MC)[3] simulations and theories have been proposed about the mechanism which causes melting of the 2-d Wigner lattice.[4-6] As to the liquid state theories, Fetter[7], Chalupa[8], and Totsuji[9a,b] have formulated thermodynamic functions and equations of state from perturbation expansion (in γ) schemes. The expansion has been further improved and refined to the point where the calculated pair correlation function, static structure function and correlation energy density accurately reproduce the experimental data for values of γ up to 0.5.[9c] For $\gamma>1$, reliable values of the pair correlation function have been calculated from the hypernetted chain (HNC) equation.[10]

On the dynamical level, structure function data and curves for the 2-d ocp collective modes have been generated from MD computer simulations.[11] Formulas for the dispersion and damping of the collective modes have been derived (i) by calculating the frequency spectrum of the lattice vibrations of a finite 2-d electron crystal[12], (ii) in the random-phase approximation (RPA)[7,13], (iii) by following the Singwi-Tosi-Land-Sjolander[14] mean-field-theory approach[15], (iv) from the Boltzmann equation[9b], (v) from modern hydrodynamic calculations[16,17], and most recently, (vi) by following the Golden-Kalman[18] velocity-average-approximation (VAA) nonlinear-response-function approach.[19]

In this lecture, I shall review the status of the long-wavelength plasma mode structure in the 2-d ocp over a range of Γ values spanning the entire fluid regime. The following are of particular interest:

(a) The plasmon structure at weak coupling ($\Gamma \ll 1$) : (iv) and (vi) predict a structure which differs dramatically from (v).

(b) The critical Γ value marking the crossover from plasmon (Vlasov)-like to longitudinal-phonon (crystal)-like dispersion.

(c) A definitive fluid-state formulation of the longitudinal phonon dispersion at very strong coupling ($\Gamma \gg 1$).

This lecture is based on the Ref. 19 work by Professor G. Kalman, Professor De-xin Lu and myself.

The long-wavelength plasmon structure in the 2-d ocp with a 1/r potential exhibits pecularities which make it quite different from the corresponding problem in three dimensions. These pecularities can be traced to two easily identifiable differences between the two- and three-dimensional systems:

(1) In three dimensions, the plasma frequency ω_p is a constant; in two dimensions, it is k-dependent, viz., $\omega_p(k) \propto \sqrt{k}$.

(2) Short-range binary collisions profoundly affect the damping of the plasma mode in the 2-d ocp -- even in the weak coupling ($\gamma \ll 1$) limit; in three dimensions the corresponding effect is negligible (i.e., $O(\gamma)$) in this limit.

The importance of the binary collisions manifests itself in the γ dependence of the dielectric response function $\varepsilon(\underset{\sim}{k}\omega)$. Various independent calculations[9b,16,20] indicate the appearance of γ-independent terms of non-RPA origin. This fact, together with the one that the coefficient of the ω^{-6} term in the high-frequency expansion of $\varepsilon(\underset{\sim}{k}\omega)$ exhibits marked $1/\gamma$

and log γ divergences[21], indicate that the Vlasov approximation is unphysical for the two-dimensional electron fluid in the $\gamma = 0$ limit.

There have been three different approaches to the calculation of $\varepsilon(\underset{\sim}{k}\omega)$ beyond the Vlasov approximation. The first approach is based on a linearized Vlasov-Boltzmann equation.[9b] The second is a systematic formal expansion (in γ)[20] of the first two equations of the BBGKY hierarchy, similar to the method[22] used some time ago in three-dimensional calculations. Finally, the third is a hydrodynamic nonlinear-response approach[17] formulated recently by De-xin Lu and myself.

If we define

$$\delta\varepsilon(\underset{\sim}{k}\omega) = \varepsilon(\underset{\sim}{k}\to 0,\omega;\ \gamma = 0) - \varepsilon_{RPA}(\underset{\sim}{k}\to 0,\omega), \tag{1}$$

the results of the three calculations can be listed as follows:

$$\text{Ref. 9b:}\quad \delta\varepsilon(\underset{\sim}{k}\omega) = i\,\frac{3}{8}\,\frac{\omega_p^5(k)}{\omega^5}\sqrt{\frac{\pi k}{k_D}}\ , \tag{2}$$

$$\text{Ref. 20:}\quad \delta\varepsilon(\underset{\sim}{k}\omega) = i\,\frac{1}{4}\,\frac{\omega_p^5(k)}{\omega^5}\sqrt{\frac{\pi k}{k_D}}\ , \tag{3}$$

$$\text{Ref. 17:}\quad \delta\varepsilon(\underset{\sim}{k}\omega) = i\,\frac{\zeta k^2}{mn}\,\frac{\omega_p^2(k)}{\omega^3} + \frac{\omega_p^4(k)}{\omega^4}\,\frac{k}{k_D}\ , \tag{4}$$

where $\omega_p(k) = (2\pi n e^2 k/m)^{1/2}$ is the plasma frequency, $k_D = 2\pi n e^2\beta$ is the Debye wavenumber, and ζ is the coefficient of viscosity. Each of the above expressions leads to the corresponding γ-independent frequency shifts

$$\text{Ref. 9b:}\quad \delta\omega(k\to 0) = \omega_p(k)\left\{\frac{21\pi}{256}\frac{k}{k_D} - i\,\frac{3}{16}\sqrt{\frac{\pi k}{k_D}}\right\}\ , \tag{5}$$

$$\text{Ref. 20:}\quad \delta\omega(k\to 0) = \omega_p(k)\left\{\frac{7\pi}{128}\frac{k}{k_D} - i\,\frac{1}{8}\sqrt{\frac{\pi k}{k_D}}\right\}\ , \tag{6}$$

$$\text{Ref. 17:}\quad \delta\omega(k\to 0) = \omega_p(k)\left\{-\frac{1}{2}\frac{k}{k_D} - i\,\frac{\zeta k^2}{2mn}\right\}\ . \tag{7}$$

It is especially interesting to observe the behavior of the k-dependent correction to the RPA plasmon dispersion in the various approximation schemes. In the model of Ref. 17, the importance of collisions is taken into account and local thermodynamic equilibrium is assumed, producing a two-dimensional adiabatic compression with a negative "adiabatic" shift of $-(k/2k_D)\omega_p(k)$ from the RPA value. A similar result obtains in the hydrodynamic theory of Baus.[16] The approximation schemes of Refs. 9b and 20 in contrast, generate a positive "super-adiabatic" shift from the RPA plasmon frequency (this shift is calculated in Ref. 20 but is not explicitly noted in Ref. 9b). The physical mechanism responsible for this shift is not immediately recognizable, but its formal origin is traced to the appearance of the γ -independent damping comprising Eqs. (2) and (3).* Thus, there is a fundamental and certainly not easily reconciliable difference in the modification brought about in the plasmon dispersion by the hydrodynamic model[17] and by its kinetic equation-based counterpart[20], even though both of them can point at the increased importance of binary collisions as the origin of the modification.

Now the phenomenological Navier-Stokes closure condition which allows the formulation of the hydrodynamical approximation of Ref. 17 is predicated on the observation that in the 2-d ocp with a $1/r$ potential the coupling parameter is independent of the k-independent quantity $\bar{\nu} \sim \omega_o = \sqrt{2\pi ne^2 k_D/m}$ quoted in Refs. 16 and 17 as the relevant collision frequency for comparison with ω. If $\bar{\nu}$ were the relevant quantity, then at frequencies $\omega = \omega_p(k\to 0) \ll \bar{\nu}$, local thermodynamic equilibrium would indeed prevail all the way down to the weak coupling ($\gamma \ll 1$) limit. My colleagues (G. Kalman and De-xin Lu) and I now contend, however, that it is the k-dependent quantity**

$$\nu(k\to 0,\omega) = \frac{1}{4}\frac{\omega_p^3(k)}{\omega^2}\sqrt{\frac{\pi k}{k_D}} \ll \omega_p(k) \lesssim \omega \tag{8}$$

calculated from Eq. (3), which comes closer to being the relevant collision frequency. Consequently, the 2-d ocp cannot be in a state of local

---

*There is some question as to whether the numerical coefficient in Eq. (3) is correct to any order in the perturbation expansion of the BBGKY hierarchy. In fact, we cannot really know the answer without continuing the calculation of Ref. 20 through order γ^2. It is not inconceivable that such an arduous calculation -- which is beyond the scope of the present work -- might possibly introduce additional positive or negative $O(k^3/\omega^5)$ corrections into (3).

**The collision frequency ν is defined through the 2-d phenomenological a.c. conductivity formula $(\omega/2\pi k)\,\mathrm{Im}\,\delta\varepsilon = \mathrm{Re}\,\delta\sigma = ne^2\nu/(m\omega^2)$.

thermodynamic equilibrium for $\omega \simeq \omega_p(k \to 0) >> \nu(k \to 0, \omega_p(k))$ and $\gamma << 1$. It then follows that the small - k plasma mode behavior is not correctly described at weak coupling ($\gamma << 1$) by the current hydrodynamic theories.[16,17] Indeed, we would reach this same conclusion even if Eq. (8) were not quite correct: momentum conservation in an ocp requires that the collision frequency be k-dependent; the very existence of a long-lived[1c] plasma mode then ensures that $\nu(k \to 0) << \omega_p(k \to 0)$.

In the work of Ref. 19, my colleagues and I formulated anew a 2-d ocp dynamical theory without invoking the hypothesis of local thermodynamic equilibrium. We did this by adapting our earlier 3-d ocp VAA-nonlinear response formalism[18] to the 2-d 1/r situtation. The VAA formalism avoids the previously discussed flaws -- especially the γ-independent negative dispersive shift -- which characterize the current hydrodynamic theories[16,17] at weak coupling, while at the same time retaining the excellent collective mode features of the Ref. 17 calculation at very strong coupling.

The VAA approximation scheme is developed in three stages. The first-stage calculations lead to a formula for the linear polarizability in terms of dynamical and static three-point structure functions. The second-stage development converts the latter into more accessible quadratic polarizabilities by means of the nonlinear fluctuation-dissipation theorem.[23] Self consistency is then guaranteed in the third stage by approximating the quadratic polarizabilities in terms of linear ones. The resulting VAA formula for $\varepsilon(\underset{\sim}{k}\omega)$, when analyzed at the low frequencies $\omega \simeq \omega_p(k \to 0)$ characteristic of the long-wavelength 2-d plasma mode, leads to a dispersion formula which expresses Re $\omega(k \to 0; \gamma)$ entirely in terms of the (γ-dependent) correlation energy density E_c and static structure function $S(\underset{\sim}{q}, t=0)$. Considering the low-frequency character of the excitations, it is hardly surprising that the dispersive corrections turn out to be wholly thermodynamic, a feature that is not shared by the high-frequency plasma mode in three dimensions.

STRUCTURE FUNCTIONS AND POLARIZABILITIES

Structure functions and polarizability response functions are the relevant quantities in the VAA approximation scheme. The former are customarily defined as follows:

$$\frac{1}{N} < n_{\underset{\sim}{k}}(t) n_{\underset{\sim}{p}}^*(0) >^{(0)} = \delta_{\underset{\sim}{k}-\underset{\sim}{p}} \{S(\underset{\sim}{k}, t) + N\delta_{\underset{\sim}{p}}\} , \tag{9}$$

$$\frac{1}{N} < n_{\underset{\sim}{k}}(t) n_{\underset{\sim}{q}}^{*}(0) n_{\underset{\sim}{p}}^{*}(0) >^{(0)} = \delta_{\underset{\sim}{k}-\underset{\sim}{q}-\underset{\sim}{p}} \{S(\underset{\sim}{q}\ t; \underset{\sim}{p}\ t)$$

$$+ N[\delta_{\underset{\sim}{k}} S(\underset{\sim}{q}\ t = 0) + (\delta_{\underset{\sim}{q}} + \delta_{\underset{\sim}{p}})\, S(\underset{\sim}{k}\ t)] + N^2 \delta_{\underset{\sim}{q}} \delta_{\underset{\sim}{p}}\} \quad ; \tag{10}$$

$n_{\underset{\sim}{k}}(t) = \Sigma_i \exp\{-i\, \underset{\sim}{k} \cdot \underset{\sim}{x}_i(t)\}$ is a microscopic density; the angular brackets denote ensemble averaging, and the zero superscript indicates that the average is to be taken over the unperturbed (equilibrium) system.

"External" polarizability response functions are next defined through the 2-d ocp constitutive relations

$$< n_{\underset{\sim}{k}} >^{(1)}(\omega) = -\hat{\alpha}(k\omega)\hat{n}(\underset{\sim}{k}\omega) \quad , \tag{11}$$

$$< n_{\underset{\sim}{k}} >^{(2)}(\omega) = \frac{in}{2A} \sum_{\underset{\sim}{q}} \beta^2 \phi_{\underset{\sim}{q}} \phi_{\underset{\sim}{k}-\underset{\sim}{q}} \int_{-\infty}^{\infty} \frac{d\mu}{2\pi}\, \hat{\alpha}(\underset{\sim}{q}\ \mu;\ \underset{\sim}{k}-\underset{\sim}{q}\ \omega-\mu)$$

$$\times\ \hat{n}(\underset{\sim}{q}\ \mu)\ \hat{n}\ (\underset{\sim}{k}-\underset{\sim}{q}\ \omega-\mu) \tag{12}$$

connecting the average density response $< n_{\underset{\sim}{k}} > (\omega)$* to the weak external density perturbation $\hat{n}$; $\phi_q = 2\pi e^2/q$ is the Fourier-transformed Coulomb energy. "Total" polarizabilities connect $<n_{\underset{\sim}{k}}> (\omega)$ to the total density perturbation $\eta = \hat{n} + <n>$; they are defined as follows:

$$< n_{\underset{\sim}{k}} >^{(1)}(\omega) = -\alpha(\underset{\sim}{k}\omega)\eta^{(1)}(\underset{\sim}{k}\,\omega) \quad , \tag{13}$$

$$< n_{\underset{\sim}{k}} >^{(2)}(\omega) = \frac{in}{2A} \sum_{\underset{\sim}{q}} \beta^2 \phi_{\underset{\sim}{q}} \phi_{\underset{\sim}{k}-\underset{\sim}{q}} \int_{-\infty}^{\infty} \frac{d\mu}{2\pi} \frac{a(\underset{\sim}{q}\mu;\ \underset{\sim}{k}-\underset{\sim}{q}\ \omega-\mu)}{\varepsilon(\underset{\sim}{k}\omega)}$$

$$\times\ \eta^{(1)}(\underset{\sim}{q}\mu)\,\eta^{(1)}(\underset{\sim}{k}-\underset{\sim}{q}\ \omega-\mu). \tag{14}$$

Since $\eta^{(1)} = \hat{n}/\varepsilon$, we have from (11) to (14) that

$$\hat{\alpha}(\underset{\sim}{k}\omega) = \frac{\alpha(\underset{\sim}{k}\ \omega)}{\varepsilon(\underset{\sim}{k}\ \omega)} = 1 - \frac{1}{\varepsilon(\underset{\sim}{k}\ \omega)} \quad , \tag{15}$$

$$\hat{a}(\underset{\sim}{q}\mu;\underset{\sim}{p}\nu) = \frac{a(\underset{\sim}{q}\mu;\underset{\sim}{p}\nu)}{\varepsilon(\underset{\sim}{q}\mu)\varepsilon(\underset{\sim}{p}\nu)\varepsilon(\underset{\sim}{q}+\underset{\sim}{p},\mu+\nu)} \quad . \tag{16}$$

---

* The ensemble averaging notation $< \ >(\omega)$ refers to the frequency-domain evolution carried by the phase-space distribution function.

APPROXIMATION SCHEME

The three-stage development of the approximation scheme starts from the linearized VAA-kinetic equation

$$(\omega-\underset{\sim}{k}.\underset{\sim}{v})F^{(1)}(\underset{\sim}{k}\omega;\underset{\sim}{v}) + \frac{1}{m}\ \underset{\sim}{k}\phi_k\ \eta^{(1)}(\underset{\sim}{k}\omega)\cdot\frac{\partial}{\partial\underset{\sim}{v}}F^{(0)}(v)$$

$$= -\frac{1}{m}\frac{\partial}{\partial\underset{\sim}{v}}F^{(0)}(v)\cdot\frac{1}{N}\sum_{\underset{\sim}{q}\neq\underset{\sim}{k}}\underset{\sim}{q}\phi_q < n_{\underset{\sim}{k}-\underset{\sim}{q}}\ n_{\underset{\sim}{q}} >^{(1)}(\omega) \tag{17}$$

for the first-order incremental response $F^{(1)}(\underset{\sim}{k}\omega;\underset{\sim}{v})$ [over and above the Maxwellian distribution $F^{(0)}(v) = (\beta mn/2\pi)\exp(-\beta mv^2/2)$] to the weak external density perturbation $\hat{n}(k\omega)$.*

The right-hand-side nonequilibrium two-point function is next converted into equilibrium three-point functions by means of the linear fluctuation-dissipation relation

$$< n_{\underset{\sim}{k}-\underset{\sim}{q}}\ n_{\underset{\sim}{q}} >^{(1)}(\omega)\Big|_{\underset{\sim}{q}\neq\underset{\sim}{k}}$$

$$= -\frac{k_D}{k}\hat{n}(\underset{\sim}{k}\omega)\{i\omega\int_0^\infty dt e^{i\omega t} S(\underset{\sim}{q}t;\underset{\sim}{k}-\underset{\sim}{q}t) + S(\underset{\sim}{q}t=0;\underset{\sim}{k}-\underset{\sim}{q}t=0)\} \tag{18}$$

mentioned in Professor Kalman's lecture.

The routine calculation of the first-order average density response is then carried out by substituting (18) into (17), solving for $F^{(1)}$, and integrating over velocity space. Comparison with (11) results in the first-stage polarizability formula

$$\hat{\alpha}(\underset{\sim}{k}\omega) = \hat{\alpha}_{RPA}(\underset{\sim}{k}\omega)\ \{1 + \hat{v}(\underset{\sim}{k}\omega)\}\quad , \tag{19}$$

where

$$\hat{v}(\underset{\sim}{k}\omega) = -\frac{k_D}{k}\frac{1}{N}\sum_{\underset{\sim}{q}\neq\underset{\sim}{k}}\chi\{i\omega\int_0^\infty dt e^{i\omega t}S(\underset{\sim}{q}t;\underset{\sim}{k}-\underset{\sim}{q}t) + S(\underset{\sim}{q}t=0;\underset{\sim}{k}-\underset{\sim}{q}t=0)\} \tag{20}$$

and $\chi = (\underset{\sim}{k}\cdot\underset{\sim}{q})/(kq)$. It is especially illuminating to think of Eq. (19) as the first equation in the VAA hierarchy linking dynamical two- and three-point structure functions. The VAA expression (19) is exact in the

---

*Eq. (17) is exact when $\hat{n}$ is a static perturbation, i.e., $\hat{n}(\underset{\sim}{k}\omega) = 2\pi\delta(\omega)\ \hat{n}(\underset{\sim}{k}\ t=0)$. This follows from the fact that the perturbed phase-space distribution function is still a canonical distribution in terms of the perturbed Hamiltonian, and that it factorizes into velocity- and coordinate-dependent contributions.[24]

static ($\omega = 0$) limit[24] and at high frequencies ($\omega \to \infty$) it reproduces the exact sum rule expansion[17,21]

$$\mathrm{Re}\ \hat{\alpha}(\underset{\sim}{k}\omega \to \infty) = -\frac{\omega_p^2(k)}{\omega^2} - \frac{\Omega_4(\underset{\sim}{k})}{\omega^4} - \ldots \tag{21a}$$

$$\Omega_4(\underset{\sim}{k}) = \omega_p^4(k)\ \{1 + 3\frac{k}{k_D} + \frac{1}{N}\sum_{\underset{\sim}{q} \neq \underset{\sim}{k}} \chi^2(q/k)\ [\ S(\underset{\sim}{k}-\underset{\sim}{q}t=0) - S(\underset{\sim}{q}t=0)]\ \}$$

$$(k,\ \gamma \text{ arbitrary}) \tag{21b}$$

through order $1/\omega^4$. At long wavelengths, (21b) further simplifies to

$$\Omega_4(k \to 0) = \omega_p^4(k)\ \{\ 1 + 3\ \frac{k}{k_D} + \frac{5}{8}\ \frac{k}{k_D}\ \frac{\beta E_c(\gamma)}{n}\}\quad ; \tag{22}$$

$E_c(\gamma)$ is the correlation energy density. The collective mode analysis of Ref. 17 reveals that it is the correlational part of (22) which primarily controls the $k \to 0$ longitudinal phonon dispersion characteristic of the 2-d hexagonal lattice[25]. This important feature is also highlighted by the present theory.*

The second-stage calculation consists in converting the three-point structure function in (20) into quadratic polarizabilities by application of the nonlinear fluctuation-dissipation theorem[17,23]. The resulting polarizability formula

$$\alpha(\underset{\sim}{k}\omega) = \alpha_{RPA}(\underset{\sim}{k}\omega)\ \{1 + v(\underset{\sim}{k}\omega)\}\quad , \tag{23}$$

$$v(\underset{\sim}{k}\omega) = \varepsilon(\underset{\sim}{k}\omega)\hat{v}(\underset{\sim}{k}\omega)$$

$$= \frac{i}{N}\frac{k_D}{k}\sum_{\underset{\sim}{q}}\chi\int_{-\infty}^{\infty} d\mu\ \delta_-(\mu)\{\frac{a(\underset{\sim}{q}\mu;\underset{\sim}{k}-\underset{\sim}{q}\omega-\mu)}{\varepsilon(\underset{\sim}{q}\mu)\varepsilon(\underset{\sim}{k}-\underset{\sim}{q}\omega-\mu)} + \frac{a(\underset{\sim}{q}\omega-\mu;\underset{\sim}{k}-\underset{\sim}{q}\mu)}{\varepsilon(\underset{\sim}{q}\omega-\mu)\varepsilon(\underset{\sim}{k}-\underset{\sim}{q}\mu)}\} \tag{24}$$

(see Ref. 18 for the detailed mathematical steps) is the previously mentioned first response function equation of the VAA chain. The quadratic polarizability $a(\underset{\sim}{q}\mu;\underset{\sim}{k}-\underset{\sim}{q}\omega-\mu)$ is defined through the constitutive equation (14). The formal operations which transform (19) and (20) into (23) and

---

*Generally speaking, dynamical theories which fail to reproduce the correct sum-rule coefficient $\Omega_4(\underset{\sim}{k})$ also fail to reproduce the correct numerical coefficient for the long-wavelength dispersion of the longitudinal phonon mode at very strong coupling ($\Gamma >> 1$); the Refs. 14 3-d and Ref. 15 2-d mean field theories fall into this category. The Ref. 16 modern hydrodynamic approach also fails to reproduce the correct dispersive coefficient at large Γ values[17], but the origin of the defect in that work may not be sum-rule related.

(24) do not entail any approximations or restrictions whatsoever. Consequently, Eqs. (23) and (24) are also valid for arbitrary k, ω and γ values. They are exact at $\omega = 0$ and they reproduce the exact sum-rule expansion for $\alpha(k, \omega\to\infty)$ through order $1/\omega^4$.

The final stage in the formulation of the approximation scheme consists in making Eqs. (23) and (24) self consistent by approximating the quadratic polarizability in terms of linear ones. The steps leading to the resulting long-wavelength dynamical superposition formulas

$$v(k\to 0, \omega) = \frac{\omega_{p(k)}^2}{\omega^2} \frac{k}{k_D} \left\{ \frac{5}{8} \frac{\beta E_c(\gamma)}{n} - \frac{1}{8N} \sum_{\underset{\sim}{q}} \int_{-\infty}^{\infty} d\mu \, \underset{-}{\delta}(\mu) \frac{\alpha(\underset{\sim}{q}\mu)\,\alpha(\underset{\sim}{q}\omega-\mu)}{\varepsilon(\underset{\sim}{q}\mu)\,\varepsilon(\underset{\sim}{q}\omega-\mu)} \right\} , \quad (25)$$

$$v(k\to 0, \omega_p(k)) \approx \frac{k}{k_D} \left\{ \frac{5}{8} \frac{\beta E_c(\gamma)}{n} - \frac{1}{16} \frac{1}{N} \sum_{\underset{\sim}{q}} \frac{k_D^2}{q^2} S^2(\underset{\sim}{q}t=0) - \frac{i\omega_0}{32N} \left(\frac{k}{k_D}\right)^{\frac{3}{2}} \sum_{\underset{\sim}{q}} \frac{k_D^2}{q^2} \int_{-\infty}^{\infty} \frac{d\mu}{2\pi} S^2(\underset{\sim}{q}\mu) \right\} \quad (26)$$

are detailed in Refs. 17 and 19. The explicit re-emergence of the correlational part of $\Omega_4(k\to 0)$ guarantees internal consistency between the second- and third-stage constructions of the approximation scheme. The correlation energy contribution is, of course, exact; it is only the $(\alpha\alpha/\varepsilon\varepsilon)$ - cluster contribution which is approximate. At weak coupling $(\gamma \ll 1)$, the occurrence of a divergence in the q - summations is due to the fact that the RPA cannot be used at short range $r \lesssim \beta e^2$ where the electrons are strongly correlated. We have accordingly avoided this unphysical divergence by imposing the customary $q_{max} = 1/(\beta e^2)$ cutoff. At strong coupling $(\Gamma = \sqrt{\gamma/2} > 1)$, one takes $q_{max} = \sqrt{\pi n} > 1/(\beta e^2)$.

Eqs. (23) and (25) comprise the self-consistent approximation scheme in which $E_c(\gamma)$ is considered to be a given known input. Eq. (26) is needed for the collective mode analysis which follows.

PLASMON STRUCTURE

The calculation of the long-wavelength plasma mode frequency is carried out according to the procedure of Ref. 17. One readily obtains

$$\omega(k\to 0) = \left\{ 1 + \frac{3}{2} \frac{k}{k_D} + \frac{1}{2} v(k\to 0, \omega_p(k)) \right\} \omega_p(k) , \quad (27)$$

leading, in turn, to the following mode structure (see Eq. (26):

$$\mathrm{Re}\ \omega(k\to 0) = (1 + \lambda(\gamma)\frac{k}{k_D})\ \omega_p(k) \quad , \tag{28}$$

$$\mathrm{Im}\ \omega(k\to 0) = -\delta(\gamma)\ (\frac{k}{k_D})^{3/2}\omega_p(k) \quad , \tag{29}$$

$$\lambda(\gamma) = \frac{3}{2} + \frac{5}{16}\ \frac{\beta E_c(\gamma)}{n} - \frac{1}{32}\ \frac{1}{N} \sum_{\underset{\sim}{q}} (\frac{k_D}{q})^2\ S^2(\underset{\sim}{q}t=0) \quad , \tag{30}$$

$$\delta(\gamma) = \frac{\omega_o}{32}\ \frac{1}{N} \sum_{\underset{\sim}{q}} (\frac{k_D}{q})^2 \int_0^\infty \frac{d\mu}{2\pi}\ S^2(\underset{\sim}{q}\mu)\ . \tag{31}$$

Since the VAA is exact at $\omega = 0$, it is entirely justifiable to input the dispersive coefficient (30) with structure function and correlation energy formulas which are constructed from accurate computer experiments or from independent theoretical approaches. Inserting Totsuji's cluster expansion[9b,26a] and Monte Carlo[26b] correlation energy formulas (quoted in Ref. 17) accordingly leads to

$$\lambda(\gamma) = \frac{3}{2} - 0.188\gamma\ (\ln\gamma^{-1} - 1) \quad \text{for } \gamma \ll 1 \quad , \tag{32}$$

and

$$\lambda(\Gamma) = 1.381 - 0.35\Gamma + 0.222\Gamma^{1/4}$$

$$- \frac{1}{64A}\ (\frac{1}{1-A} + \frac{1}{A}\ln(1-A)) \quad \text{for } (1/\sqrt{2}) < \Gamma < 50 \quad ;$$

$$A = 0.42 - \frac{0.405}{\Gamma} - \frac{0.355}{2\Gamma^{3/4}} - \frac{0.044}{2\Gamma^{7/4}}\ . \tag{33}$$

As the plasma parameter increases from zero, λ decreases from its maximum value $\lambda(\gamma = 0) = 3/2$ to zero and then becomes negative. Eq. (33) predicts that this transition from plasmon-like ($\lambda > 0$) to longitudinal phonon-like ($\lambda < 0$) dispersion occurs at $\Gamma^{(2d)}_{crit} = 4.85$, somewhat higher than the previous Baus[16] and Golden-Lu[17] values of 3.55 and 3.76, respectively, but still well below the 3-d ocp $\Gamma^{(3d)}_{crit} = 8.8$[27]; the Totsuji-Kakeya MD experiments[11] predict that $\Gamma^{(2d)}_{crit} > 2.29$. This transition from positive to negative dispersion is preceded by the onset of a liquid state short-range order signalled by the development of oscillations in the equilibrium pair correlation function g(r)[26b,10] somewhere in the range $2.2 < \Gamma < 2.9$. The substantially lower critical value of Γ in the two-dimensional, compared to three-dimensional case[27], is interesting to observe: here it comes closer to the Γ-range where g(r) develops oscillatory behavior, while this is not the case in the 3-d ocp where $\omega(k)$ remains finite as $k\to 0$.

In the $\Gamma \to \infty$ limit, both $\lambda(\Gamma)$ and the sum rule coefficient Ω_4 $(k \to 0)$ are dominated--similarly to the three-dimensional case--by the correlation energy term and go as Γ. Strictly speaking, our theory can reach this limit only by employing the fluid HNC formula[10]

$$\frac{\beta E_c(\Gamma)}{n} = -1.0952\Gamma + 0.9851 \text{ for } \Gamma > 30 \quad , \tag{34}$$

since both the VAA and HNC approaches are translation-invariant. The resulting dispersion then saturates to the value

$$\mathrm{Re}\ \omega(k \to 0)\Big|_{\substack{\Gamma\to\infty \\ \mathrm{HNC}}} = (1 - 0.1711 \frac{k}{\sqrt{\pi n}})\ \omega_p(k) \ . \tag{35}$$

We note that the value of the dispersive coefficient changes only slightly (to -0.1728) if (30) is instead used with the Gann-Chakravarty-Chester[3] solid phase Monte Carlo correlation energy formula quoted in Ref. 17. Eq. (35) is identical to its Ref. 17 counterpart; both very nearly reproduce the Bonsall-Maradudin longitudinal-phonon-dispersion formula[25]

$$\omega(k \to 0)\Big| = (1 - 0.173 \frac{k}{\sqrt{\pi n}})\ \omega_p(k) \tag{36}$$

for the 2-d hexagonal lattice[5] at long wavelengths.

The present VAA and previous Ref. 17 hydrodynamic approaches, while they are conceptually quite different at weak coupling $\gamma \ll 1$, nevertheless, lead to structurally similar expressions for the dynamical polarizability and dispersive coefficient $\lambda(\gamma)$. The two treatments share one crucially important feature: both satisfy the third-frequency-moment sum rule whose correlational part dominates the long-wavelength dispersion of the 2-d longitudinal phonon excitations at extreme coupling.

DISCUSSION

Starting from the VAA kinetic equation, we constructed a self-consistent approximation scheme for the calculation of the dynamical polarizability and long-wavelength plasma mode structure in the 2-d ocp with a 1/r potential. The nonlinear fluctuation-dissipation theorem and dynamical superposition formula were central elements in the approximation scheme.

The principal results in the three-stage development of the theory are Eqs. (23) - (26). Eqs. (23) and (24) constitute the first of a chain of VAA dynamical polarizability relations; the first relation links linear and quadratic polarizabilities. It is valid for arbitrary k, ω and γ values. It is exact at $\omega = 0$ and it reproduces the exact sum rule expansion for $\alpha(k, \omega \to \infty)$ through order $1/\omega^4$. Eq. (23) and the $k \to 0$ dynamical superposition formula (25) comprise the self-consistent approximation scheme in which the correlation energy density is considered to be a given known input. Eq. (23) is structurally quite similar to its hydrodynamic counterpart

$$\alpha(\underset{\sim}{k}\omega) = \alpha_H(\underset{\sim}{k}\omega) \quad (1 + v(\underset{\sim}{k}\omega)) \quad ,$$

$$\alpha_H(\underset{\sim}{k}\omega) = \frac{\omega_p{}^2(k)}{\omega^2} - 2\frac{\omega_p{}^4(k)}{\omega^4}\frac{k}{k_D} + \frac{i\zeta k^2\omega_p{}^2(k)}{mn\omega^3}$$

in Ref. 17. The conceptual difference between the Ref. 19 and Ref. 17 approaches, while it is especially pronounced at weak coupling ($\gamma < 1$), makes itself manifest at Γ values greater than unity--even up to $\Gamma^{(2d)}_{crit}$ marking the crossover from plasmon-like to longitudinal-phonon-like dispersion; the Ref. 19 predicts $\Gamma^{(2d)}_{crit} = 4.85$ which is to be compared with the Ref. 17 value of 3.76.

The one major question which remains unanswered is whether the damping coefficient and super-adiabatic shift in Eq. (6) actually describe the $\gamma=0$, $k\to 0$ 2-d plasma mode structure. It is, after all, not inconceivable that future higher-order-in-γ calculations might further increase or decrease or leave unchanged the expression (3) for $\delta\varepsilon(k\to 0,\omega;\ \gamma=0)$; or perhaps they might even act to decrease it all the way down to zero (but not below zero since the system is stable). This last scenario would lead us right back to the RPA which is, at the same time, the $\gamma = 0$ limit of the VAA.

Should future calculations re-affirm the dielectric correction (3) as a structural feature of 2-d electron plasmas, then the present VAA formalism might be improved by phenomenologically replacing α_{RPA} in Eq. (23) with the "super-RPA" expression

$$\alpha_{SRPA}(\underset{\sim}{k}\omega) = \alpha_{RPA}(\underset{\sim}{k}\omega) + \frac{i}{4}\frac{\omega_p{}^5(k)}{\omega^5}\sqrt{\frac{\pi k}{k_D}}$$

The higher critical Γ value, 5.37, which results then leads us to conclude that $\Gamma^{(2d}_{crit} > 4.85$.

ACKNOWLEDGEMENTS

This work was partially supported by U.S. National Science Foundation Grants No. ECS-8315801 and No. ECS-8315655. This manuscript was prepared under the sponsorship of NSF Grant No. ECS-8645108.

REFERENCES

1. (a) T. R. Brown and C. C. Grimes, Phys. Rev. Lett. 29:1233 (1972);
   (b) C. C. Grimes and T. R. Brown, Phys. Rev. Lett. 32:280 (1974);
   (c) C. C. Grimes and G. Adams, Phys. Rev. Lett. 36:145 (1976);
   (d) 42:797 (1979).
2. R. W. Hockney and T. R. Brown, J. Phys. C8:1813 (1975).
3. R. C. Gann, S. Chakravarty and G. V. Chester, Phys. Rev. B20:324 (1979).
4. M. W. Cole, Rev. Mod. Phys. 46:451 (1974).
5. P. M. Platzman and H. Fukuyama, Phys. Rev. B10:3150 (1974).
6. D. J. Thouless, J. Phys. C11:L189 (1978);
   R. H. Morf, Phys. Rev. Lett. 13:931 (1979).
7. A. L. Fetter, Phys. Rev. B10:3739 (1974).
8. J. Chalupa, Phys. Rev. B12:4 (1975).
9. (a) H. Totsuji, J. Phys. Soc. Japan 39:253 (1975);
   (b) J. Phys. Soc. Japan 40:857 (1976);
   (c) Phys. Rev. A19:889 (1979).
10. F. Lado, Phys. Rev. B17:2827 (1978).
11. H. Totsuji and N. Kakeya, Phys. Rev. A22:1220 (1980).
12. R. S. Crandall, Phys. Rev. A8:2136 (1973).
13. P. M. Platzman and N. Tzoar, Phys. Rev. B13:3197 (1976).
14. K. S. Singwi, M. P. Tosi, R. H. Land and A. Sjolander, Phys. Rev. 176:589 (1968); K. S. Singwi, A. Sjolander, M. P. Tosi and R. H. Land, Sol. State Commun. 7:1503 (1969); K. S. Singwi, A. Sjolander, M. P. Tosi and R. H. Land, Phys. Rev. B1:1044 (1970).
15. N. Studart and O. Hipolito, Phys. Rev. A22:2860 (1980).
16. M. Baus, J. Stat. Phys. 19:163 (1978).
17. K. I. Golden and De-xin Lu, Phys. Rev. A31:1763 (1985).
18. K. I. Golden and G. Kalman, Phys. Rev. A19:2112 (1979).
19. K. I.Golden, G. Kalman and De-xin Lu, "Velocity-Average-Approximation Approach to Two-Dimensional Classical Electron Plasmas", submitted to Phys. Rev. A.
20. De-xin Lu and K. I. Golden, Commun. in Theor. Phys. (Beijing, China) 4:21 (1985).
21. De-xin Lu and K. I. Golden, Phys. Rev. A28:976 (1983).
22. J. Coste, Nucl. Fusion 5:284 (1965); 5:293 (1965).
23. De-xin Lu and K. I. Golden, Phys. Lett. 97A:391 (1983).
24. K. I. Golden and G. Kalman, Ann. Phys. (N.Y.) 143:160 (1982).
25. L. Bonsall and A. A. Maradudin, Phys. Rev. B15:1959 (1977).
26. (a) H. Totsuji, Phys. Rev. A19:889 (1979);
    (b) A17:399 (1978).
27. P. Carini and G. Kalman, Phys. Lett. 105A:229 (1984); G. Kalman, "Plasmon Dispersion in Strongly Coupled Plasmas", invited paper, XIIth Symposium on the Physics of Ionized Gases, Sibenik, Yugoslavia, 1984.

CHAPTER VIII

METALLIC FLUIDS

METALLIC FLUIDS IN THE CRITICAL REGION

F. Hensel

Institute of Physical Chemistry
Philipps-University of Marburg
D-3550 Marburg, FRG

INTRODUCTION

During the last 15 years, considerable effort has been put into investigating the properties of dense plasmas of metal vapors at relatively low temperatures ($T<2500$ K), but considerably high pressures ($10\ \text{bar} \lesssim p \lesssim 10^4$ bar). Much of this effort has been motivated by the fact that dense fluid metals, because of their high latent heats of vaporization and heat-transfer coefficients, are prominent candidates for high temperature working fluids and for heat-transfer media. In addition to this technological interest, dense metal vapors offer the valuable opportunity to investigate the continuous evolution of electrical, magnetic, and optical properties with continuously varying density, from the low density weakly ionized domain, through the pressure induced ionization transition (the metal-insulator transition), which becomes significant as the average distance between atoms approaches the order of the atomic diameter, and into the degenerated fully ionized plasma (metallic) regime. Such continuous variation of the density is only possible, however, if the vapor is compressed above the critical temperature T_c which terminates the liquid-vapor phase separation. As table 1 shows, the critical point is at low enough temperature and pressure to be studied with conventional static high temperature-high pressure techniques for only a few metals (Hg, Cs, Rb, and K). The location of the critical point of Mo is well above that of the above metals; the values have been determined in a transient experiment with exploding wires[4]. One of the oldest question, that is still unresolved is, what is the relationship between the ionization transition (the metal-nonmetal transition) and the liquid-vapor phase transition? Is it, for instance, necessary connected with crossing the vapor pressure curve or its prolongation into the critical isochore? Or does it occur wholly in the liquid or wholly in the vapor phase. Questions of this kind were raised as long ago as 1943 by Landau and Zeldovitch[5] in the context of the liquid-vapor equilibrium in mercury. They suggested the possibility of separate first-order electronic and liquid-vapor transitions in fluid metals.

Table 1. Critical data of fluid metals

metal	T_c (K)	p_c (bar)	d_c (g/cm^3)	reference
Hg	1750	1673	5.8	1
Cs	1924	92.5	0.38	2
Rb	2017	124.5	0.29	2
K	2280	161	0.19	3
Mo	14300	5700	2.9	4

THE CRITICAL POINT PHASE TRANSITION OF ALKALI METALS

From the experimental standpoint it is clear that there is no discontinuous conductivity change except across the liquid-vapor phase change. This is convincingly demonstrated by a comparison of fig.1 and fig.2 which present a selection of the most recent and most accurate density ρ [6] and electrical conductivity σ [6] results for fluid cesium in form of isotherms as a function of pressure at sub- and supercritical conditions. The conductivities of the coexisting liquid phase are given by the dashed line, the arrows indicate the abrupt transition to the vapor phase!

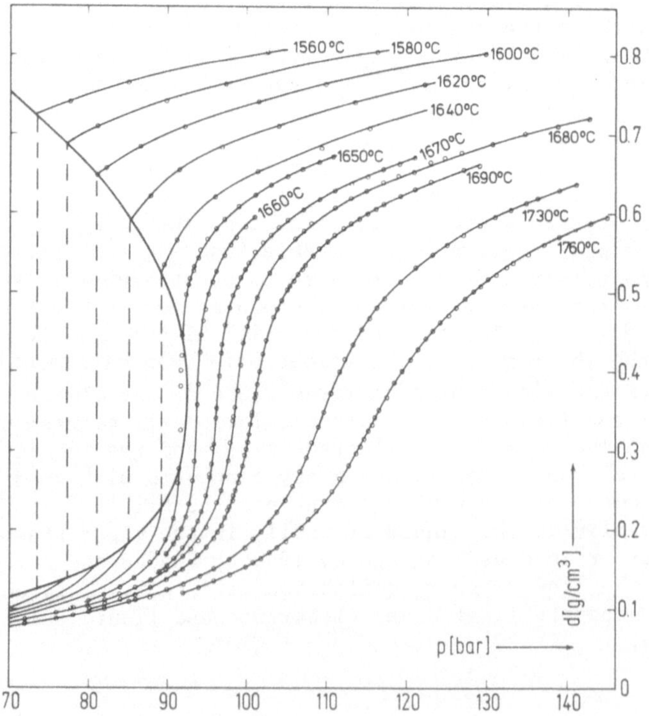

Figure 1. Equation of state data of fluid cesium near the critical point.

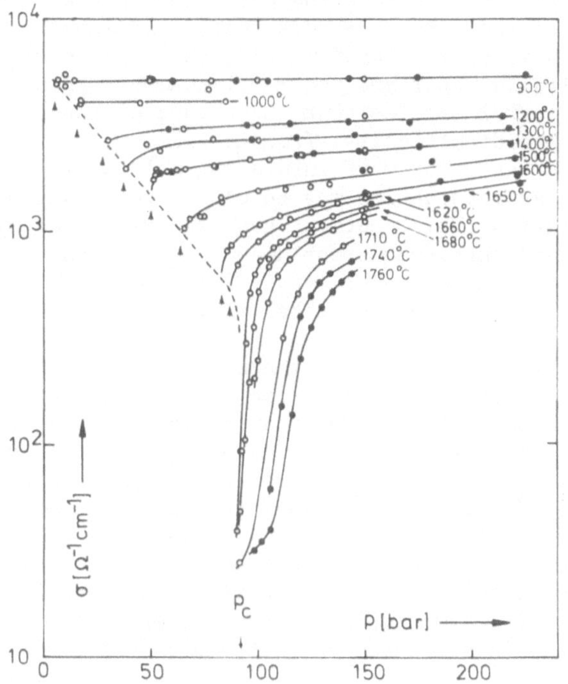

Figure 2. Conductivity isotherms of fluid cesium near the critical point.

It is tempting to speculate that this behavior of σ, which is also observed for the alkali metals Cs and K, indicates that the phase separation tends to separate the nonmetallic vapor and the metallic liquid. One criterion often invoked for the onset of metallic behavior is the minimum metallic conductivity σ_{min} suggested by Mott[7] to characterize the boundary between delocalized states and states localized by disorder (the Anderson transition[8]). For liquids, σ_{min} typically has values between 200 and 300 $ohm^{-1}\ cm^{-1}$ which correlates roughly with the value of σ at the critical density ρ_c of Cs, Rb and K.

Another important problem for the understanding of the metal-nonmetal transition in expanded monovalent alkali metals is the role of electron-electron interactions which favor localization to reduce double occupancy of individual sites[9], i.e. the Mott-Hubbard transition. The role of the short-range intraatomic Coulomb interaction in a metal has been discussed by Brinkman and Rice[10]. They showed that the metallic state near the metal-nonmetal transition should be highly correlated, having a low instantaneous fraction of doubly occupied sites. The correlated metal has an enhanced density of states and, consequently, enhanced values for the paramagnetic susceptibility. The presence of large correlation effects in the alkali metals was first convincingly demonstrated by the observation of a strong enhancement of the total mass susceptibility for expanded liquid cesium[11]. Similar susceptibility enhancements have been observed in subsequent works for expanded rubidium[12] and sodium[13]. The low-density enhancement in Cs has been observed also in nuclear-magnetic-resonance experiments[14].

The experimental observations indicate that both effects, electron-electron interaction and structural disorder, play important roles for the metal-nonmetal transition in alkali metals, and that there exists a clear link between the liquid-vapor and the metal-nonmetal transition. The occurrence of the latter implies that the nature of the interparticle interaction must change dramatically, from metallic to a van der Waals-type interaction. By contrast, for most insulating molecular substances like Xe the intermolecular interaction may be considered independent of density to a good approximation. This contrast has been discussed by Goldstein and Ashcroft[15] who argued that the strong dependence on density, near ρ_c, of the electronic structure may considerably influence the thermodynamic features of fluid metals in the critical region.

Since the available equation of state data for fluid Cs and Rb[2] approach the critical point close enough we are able to test experimentally the validity of this hypothesis. For that purpose we compare in fig.3 [2,16] the reduced densities of gaseous and liquid Xenon (inner curves) and rubidium (outer curves) as a function of $\tau = |T_c-T|/T_c$. Pure reduced correlation between Rb and Xe is observed. Thus the experimental evidence shows that metals and nonmetals cannot be included together in a group obeying a principle of corresponding states. The coexistence curve of Rb is remarkably asymmetric compared to that of the simple nonmetallic fluid Xe. The asymmetry, however, is very similar to that observed for other metallic systems in which the interactions vary strongly with state. In particular, the metal-ammonia[17] and the electron-hole[18] liquid phase diagrams have asymmetries[19] which have been attributed to the change in the range of the screened interactions near the critical points, due to the proximity of metal-nonmetal transitions. It is obvious from fig.3 that Rb violates the law of rectilinear diameter over a surprisingly large temperature range. By contrast, this law is experimentally valid for the coexistence curves of nearly all simple nonmetallic one-component fluids to within the capacity of present-day experimentation. Thus far the only other one-component

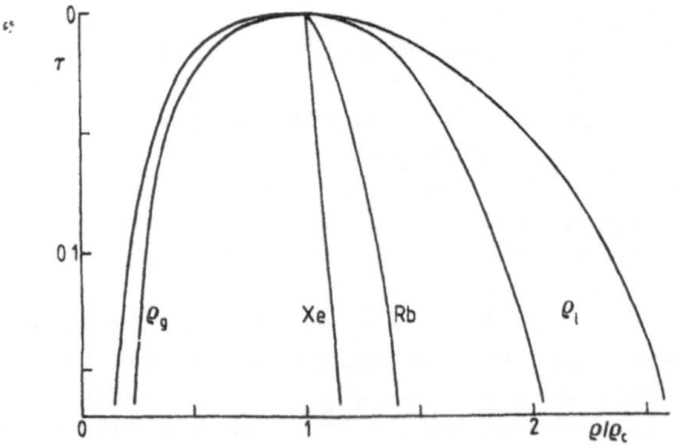

Figure 3. The reduced densities of gaseous and liquid Xenon (inner curves) and rubidium (outer curves).

system exhibit an appreciable deviation from rectilinear diameter behavior is SF_6, as studied by Weiner et al.[20].

As mentioned above, it has been suggested[15] that the contrast between the diameter data of metals, and the apparent experimental linearity of the diameters of essentially all nonmetallic one-component fluids arises from many-body effects whose magnitudes distinguish the particle interactions in metallic fluids from those in nonmetallic fluids. In particular, it is argued that the strong thermodynamic state dependence of the effective interactions in a metal, especially as the metal-nonmetal transition is traversed, corresponds to the mixing of thermodynamic fields present in certain solvable lattice models[21,22,23,24,25]. These models, thermodynamic arguments[26,27,28], and renormalization-group studies[29,30,31] predict that the average value of the density, i.e. the diameter, will have the asymptotic form

$$\rho_d = (\rho_L + \rho_V)/2 = \rho_c + D(\tau)^{1-\alpha} \qquad (1)$$

where $\tau = |T_c - T|/T_c$ and the exponent α is the same as that of the divergence of the constant volume specific heat for a pure fluid. Thus the densities $\rho_{L,V}$ in the two branches of the coexistence curve are expected to behave like

$$\rho_{L,V} = \rho_c \pm B(\tau)^{\beta} + A(\tau)^{1-\alpha} + \ldots \qquad (2)$$

This implies that a single power law analysis results in effective exponents

$$\beta^{eff}_{L,V} = \partial \ln(|\rho_{L,V} - \rho_d|)/\partial \ln(\tau) \qquad (3)$$

in the two branches of the coexistence curve which may differ[6] because of the different relative importance of the $A\cdot(\tau)^{1-\alpha}$ term in eq.(2) and which may strongly deviate from the values calculated with the renormalization-group approach[6]. However, the asymptotic exponents

$$\beta^{asym}_{L,V} = \lim_{\tau \to 0} \beta^{eff}_{L,V} \qquad (4)$$

must be the same.

It is an empirical fact that the higher order terms in eq.(2) cancel to a large extent when the difference $\rho_L - \rho_V$ is formed. For this difference a power law with the same exponent $\beta = \beta_L = \beta_V$ is found to hold, and the asymptotic range of validity is normally quite large in τ. The latter fact is of great help in the analysis of coexistence curves. In fig.4 we have plotted $\log(\rho_L - \rho_V/2\rho_c)$ versus $\log|\tau|$ 2 for Rb. Fitting to the equation

$$(\rho_L - \rho_V)/2\rho_c = B|\tau|^{\beta} \qquad (5)$$

we find $B = 2.45$ and $\beta = 0.36$. A single power law applies over a range $10^{-3} < |\tau| < 10^{-1}$ and the apparent experimentally determined β-value is very close to those observed for normal nonmetallic fluids which belong to the same static universality class as an uniaxial ferromagnet represented by the three-dimensional Ising model or the Landau-Ginzburg-Wilson model with an one-component order parameter. The main difference between the coexistence curves of nonmetallic and metallic fluids seems to be the different magnitude of the $A|\tau|^{1-\alpha}$ term in eq.(2) which manifests itself in large amplitudes of the singularities in the coexistence curve diameters[15].

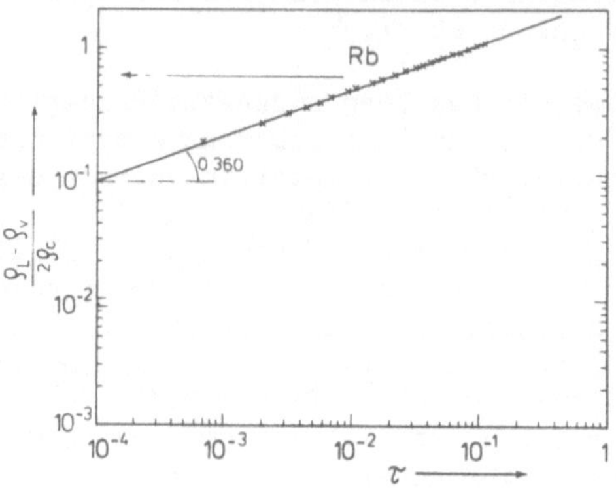

Figure 4. Single power law analysis of the coexisting liquid and vapor densities of Rb.

The predicted singularity in the diameters (see eq.(1)) is difficult to verify experimentally for several reasons. Firstly, it has been shown[32] that if one particular function, e.g. ρ, has a $|\tau|^{1-\alpha}$ singularity, then any less symmetric function ρ', where ρ' is an analytic function of ρ (e.g. $\rho = V^{-1}$), behaves as $|\tau|^{2\beta}$. Thus the sought-for effect will be missed unless the correct function is chosen. The mass density has long been known empirically to give a more symmetric coexistence curve than volume[33]. However, this is only a strong, but not a conclusive argument for supposing that ρ is the appropriate function for the order parameter. Secondly, the size of the asymptotic range (or, equivalently, the amplitudes of the correction terms to eq.(2)) depends on the choice of the order parameter.

Renormalization-group theory has been used to produce series expansions for representing data over a wide range of thermodynamic space. The expansions provide the following correction terms to eq.(2) for the diameter

$$\rho_d = \rho_L + \rho_V/2\rho_c = 1 + D_0|\tau|^{1-\alpha} + D_1|\tau| + \ldots \qquad (6)$$

Since $(1-\alpha)$ is not very different from unity, the true singularity is difficult to separate out from the analytic temperature term of eq.(6). The coefficient D_1 does not even have to be much larger than D_0 for the analytic term to dominate over the entire range accessible to experimentation. The latter certainly causes the invisibility of the $|\tau|^{1-\alpha}$ anomaly in most nonmetallic fluids.

Up to now the only convincing experimental evidence for the existence of a $(1-\alpha)$ term for one-component systems is the analysis of the diameters of Cs and Rb by Jüngst et al.[2]. A plot for Rb is shown in fig.5. A single power law applies over a range $10^{-3}<|\tau|<10^{-1}$ and the apparent experimentally determined $(1-\alpha)$ values are very close to the value 0.89 predicted by the renormalization-group theory. This finding strongly supports the suggestion[15] that the strong state-dependence of the effective interparticle interactions, and especially the changes in such forces in course

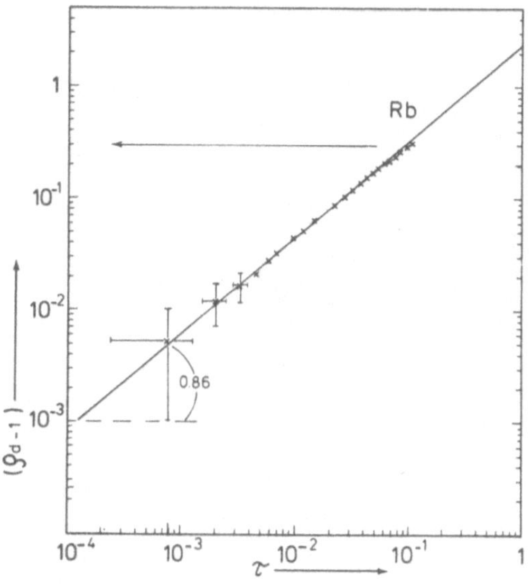

Figure 5. Single power law analysis of the diameter of Rb.

of the metal-nonmetal transition, lead to very large amplitudes of the $(1-\alpha)$ anomaly in the diameters of liquid-vapor coexistence curves of metals.

ELECTRONIC PROPERTIES NEAR THE LIQUID-VAPOR CRITICAL POINT OF MERCURY

During the last 15 years, considerable effort has been put into investigating the properties of mercury near its liquid-vapor critical point. Much of this effort has focussed on the metal-nonmetal transition which occurs at reduced densities of the fluid metal. Experimental measurements such as those of the Knight shift[34,35] and optical reflectivity[36,37] show that as the density is lowered, a gap in the density of states appears to develop at the relatively high density of 9 g/cm^3. The opening of this gap means that mercury changes macroscopically to a nonmetallic, effectively "semiconducting" state at the same density, i.e. before the critical point density $\rho_c = 5.8\ g/cm^3$ is approached. On the other hand, most theoretical calculations[38,39,40] have yielded values for the gap closing density that are very close to ρ_c. It should be noted, however, that these calculations did not take into account the situation close to the critical point where critical density fluctuations may strongly affect the electronic properties. The importance of density fluctuations for fluid mercury densities smaller than 9 g/cm^3 and larger than 4 g/cm^3 in the temperature region around T_c is clearly demonstrated by figure 6 which shows the isothermal compressibility $\chi_T = 1/\rho\cdot(\partial\rho/\partial p)_T$ at constant temperatures as a function of density. χ_T begins to rise quite rapidly as the density falls below 9 g/cm^3.

Knowledge of the interplay between these critical point fluctuations and the electronic properties is especially important for the understanding of the ionization catastrophe and its relationship to the liquid-vapor phase transition in fluid mercury. The strong interplay between the critical point and the electrical transport properties of mercury becomes immediately evident when σ is plotted in the vicinity of the critical point as a function of density ρ at constant temperature T, as shown in

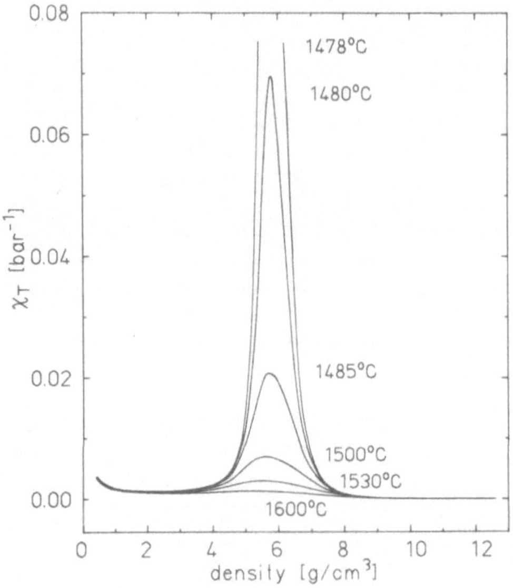

Figure 6. Isothermal compressibility $\chi_T = 1/\rho\cdot(\partial\rho/\partial p)_T$ of fluid mercury at constant temperatures T (T_c = 1478°C) as a function of density (ρ_c = 5.8 g/cm^3).

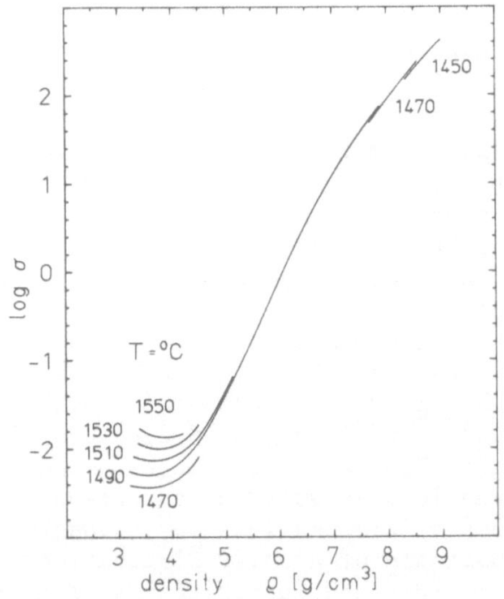

Figure 7. Electrical conductivity σ of fluid mercury at constant temperatures close to T_c = 1478°C as a function of density ρ.

fig.7 for a number of selected isotherms. At a density of 9 g/cm^3 the conductivity is about the "minimum metallic conductivity" (i.e. about 200 ohm^{-1} cm^{-1}). For densities smaller than 9 g/cm^3, σ falls more rapidly and approaches a value of about 1 ohm^{-1} cm^{-1} at the critical density ρ_c = 5.8 g/cm^3. The more rapid fall of σ for densities smaller than 9 g/cm^3 has been considered as a strong indication for the onset of the transition to a nonmetallic state. It is clear from the results of figures 7 and 8 that there is a close correlation between the slope of the σ-ρ-curves and the

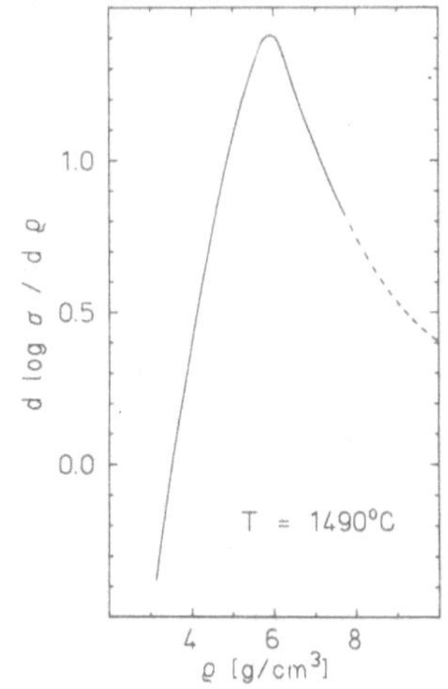

Figure 8. The density coefficient $(\partial \ln\sigma/\partial\rho)_T$ of fluid mercury at constant temperatures as a function of density (T_c = 1478°C, ρ_c = 5.8 g/cm^3).

critical point. There is no doubt that the steepest fall in the conductivity of Hg is observed at the critical point.

The pattern of the σ-ρ-curve is especially interesting for densities below the critical density where fluid mercury forms a dense partially ionized plasma consisting of neutral species, ionic species and electrons. In this case, in addition to screened Coulomb interaction among the charges, electron-neutral interaction plays an important role for the transport[41,42]. If the density of neutrals is high enough, the electron can interact with many atoms at the same time; i.e. it can be captured temporarily by density fluctuations or clusters[43,44]. This effect may be expected to be large in the region of small degree of ionization, i.e. for densities well below the ionization catastrophe, and for low temperatures. Both the minimum in the σ-ρ-curves at around 4 g/cm^3 (ρ_c = 5.8 g/cm^3) and the strong positive temperature dependence of σ in this range are completely consistent with this model.

As is well known, thermoelectric measurements are especially suited to study changes in the nature of the electrical transport process[42,45]. Several experiments have suggested that in Hg at pressures and temperatures near and above critical values the thermoelectric power vanishes[46,47,48,49,50]. From simultaneous measurements of S and density σ Götzlaff[50] was able to evaluate the density dependence of the thermoelectric power at a constant temperature near T_c (fig.9). A remarkably strong increase of S up to large positive values is observed in the density region where electron-neutral interaction becomes important (cp. also fig.7). The energy transport by neutrals induced by the electron current seems to give a large contribution (a drag effect) to the thermoelectric power in the region of small degree of ionization. For densities above the ionization catastrophe density the thermopower tends to small negative values

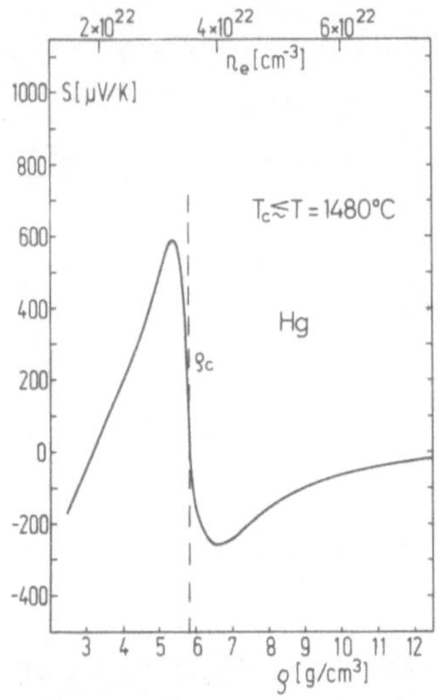

Figure 9. Thermopower S of fluid mercury at a constant temperature as a function of density ρ (T_c = 1478°C, ρ_c = 5.8 g/cm^3).

(fully ionized degenerate plasma). From the results in fig.9 it is clear that the change in the electrical transport behavior of expanded Hg is intimately related to the vapor-liquid critical point.

The interplay between the critical point phase transition and the rapid changes in the electronic structure, when the ionization catastrophe is traversed, becomes especially evident when the critical point is approached from the insulating vapor side. This approach has recently been extensively studied by measurements of the density- and temperature dependence of the optical properties of the vapor. At very low densities a line-spectrum is observed with the main absorption lines at 4.89 eV and 6.7 eV corresponding to transitions between the 6s ground state and the 6p triplet and singlet state of the Hg-atom. As the density is increased the sharp lines broaden due to interactions with neighboring atoms resulting in a relatively steep absorption edge which moves rapidly to lower energies with increasing density. Fig.10 gives a few selected data[51] for the spectral dependence of the extinction coefficient K of mercury at a constant temperature close to T_c = 1478°C and densities ρ between 2.6 g/cm^3 and 4.5 g/cm^3 (ρ = 5.8 g/cm^3). The most striking feature of the data at low densities is a very abrupt edge in absorption which moves rapidly to lower energies with increasing density. At densities larger than 4 g/cm^3, the optical absorption qualitatively changes, aquiring a low frequency foot, which extends at least to 0.5 eV.

The closing of the effective gap in the excitation spectrum (fig.10) can also be viewed in terms of a nonlinear enhancement of the real part of the dielectric constant ε_1 [52] with increasing density as demonstrated in fig.2 which shows results for ε_1 at the constant photon energy 1.27 eV in the form of isotherms plotted versus pressure. As the pressure is increased at a constant temperature T larger than about 0.96 T_c (i.e.

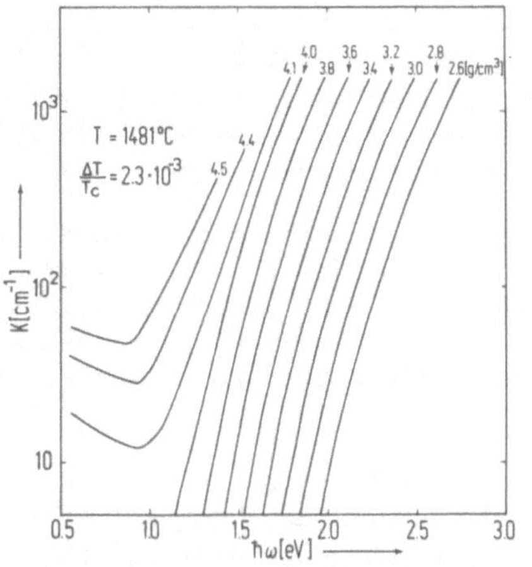

Figure 10. Spectral dependence of the extinction coefficient K of fluid mercury at a constant temperature close to the critical temperature of 1478°C as a function of density.

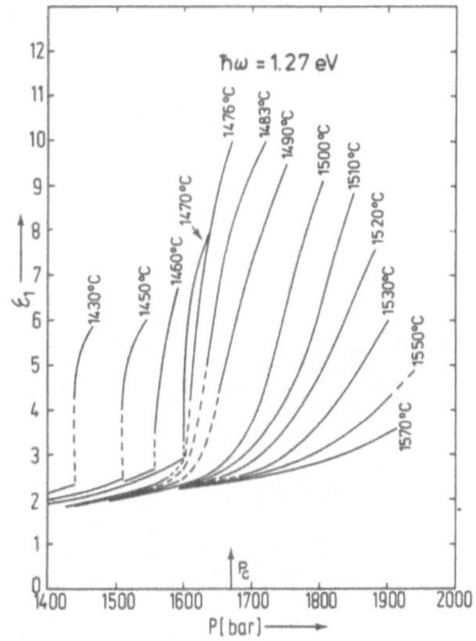

Figure 11. The real part of the dielectric constant ε_1 at the constant photon energy 1.27 eV as a function of pressure p for different sub- and supercritical temperatures p = 1673 bar.

1400°C) ε_1 initially follows the Clausius Mosotti model of the polarisibility of induced dipoles before it shows a strong upward deviation from Clausius-Mosotti behavior. It is obvious from the pattern of the ε_1-curves in fig.11 that the strong dielectric anomaly is inextricably related to the critical point phase transition of mercury.

This becomes especially obvious when ε_1 is studied in the critical region as a function of temperature at constant density, as shown in fig. 12 for a number of selected isochores. The data show close to the critical isochore clearly the presence of a large anomalous contribution in the dielectric constant reaching a magnitude of about 70% of the background at $\Delta T/T_c=10^{-3}$. This is in contrast to the behavior of nonmetallic fluids for which a comparatively weak dielectric anomaly (smaller than 0.1% for CO at $\Delta T/T_c<10^{-4}$ [53]) occurs[54]. We consider the finding of a large amplitude of the dielectric constant anomaly for mercury as evidence for a strong interplay between the vapor-liquid critical point phase transition and the large changes in the electronic structure as the metal-nonmetal transition is approached.

As mentioned before the enhancement of ε_1 with density ρ or temperature T can also be viewed in terms of changes in the shape of the absorption spectrum. In fact, ε_1 may be obtained from the standard Kramers-Kronig integral over the optical absorption coefficient. Therefore, we plotted in fig.13 the extinction coefficient K of fluid mercury at the constant photon energy 1.27 eV as a function of temperature. A comparison of figures 12 and 13 shows that the temperature anomalies of ε_1 and K close to the critical point of Hg are remarkably similar. This similarity is strong indication that both features have the same physical origin, i.e. they are consequences of the strong interplay between the critical point phase transition and the changes in electronic structure as the metal-nonmetal transition is approached.

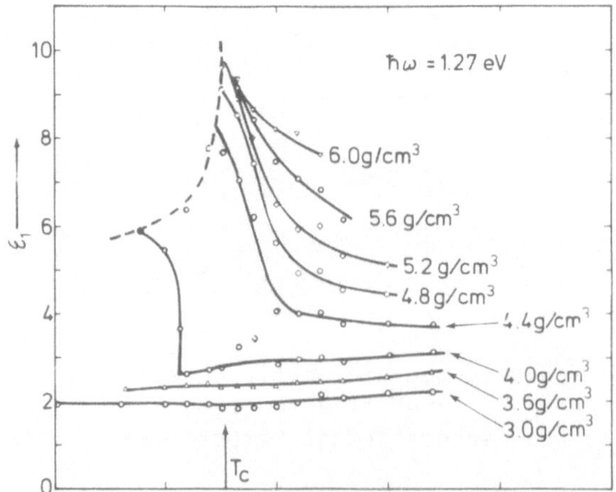

Figure 12. The real part of dielectric constant ε_1 at constant density versus temperature (T_c = 1478°C, ρ_c = 5.8 g/cm³).

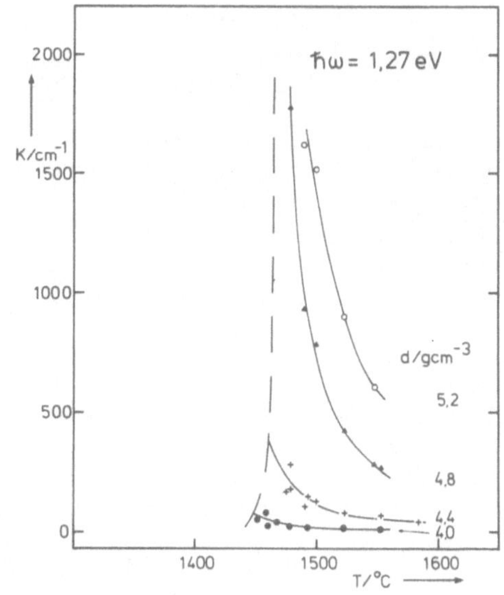

Figure 13. The extinction coefficient K at constant photon energy 1.27 eV and at constant density ρ as a function of temperature (T_c = 1478°C, ρ_c = 5.8 g/cm^3).

ACKNOWLEDGEMENT

Financial support by the Deutsche Forschungsgemeinschaft and the Bundesministerium für Forschung und Technologie is gratefully acknowledged.

REFERENCES

1. W.Götzlaff, G.Schönherr, and F.Hensel, Proceedings of the 6th International Conference on Liquid and Amorphous Metals, Garmisch-Partenkirchen (1986).
2. S.Jüngst, B.Knuth, and F.Hensel, Phys.Rev.Letters 55:2160 (1985).
3. W.Freyland and F.Hensel, Ber.Bunsenges.Phys.Chem. 76:347 (1972).
4. U.Seydel and W.Fucke, J.Phys.F: Metal Phys. 8:L157 (1978).
5. L.Landau and G.Zeldovitch, Acta Phys.Chem.USSR 18:194 (19743).
6. F.Hensel, S.Jüngst, F.Noll, and R.Winter, in: "Localization and Metal-Insulator Transition", D.Adler and H.Fritzsche, eds., Plenum Press, New York (1985), p.109.
7. N.F.Mott, Phil.Mag. 19:835 (1969).
8. P.W.Anderson, Phys.Rev. 109:1492 (1958).
9. N.F.Mott, "Metal-Insulator Transition", Taylor and Francis, London (1974).
10. W.F.Brinkman and T.M.Rice, Phys.Rev.B 2:4302 (1970).
11. W.Freyland, Phys.Rev.B 20:5104 (1979).
12. W.Freyland, J.Phys.(Paris)Colloq. 41:C8-74 (1980).
13. L.Bottyan, R.Dupree, and W.Freyland, J.Phys.F: Metal Phys. 13:L173 (1983).
14. U.El-Hanany, G.F.Brennert, and W.W.Warren, Phys.Rev.Letters 50:540 (1983).
15. R.E.Goldstein and N.W.Ashcroft, Phys.Rev.Letters 55:2164 (1985).
16. J.R.Rowlinson, Nature 319:362 (1986).
17. P.Chieux, P.Damay, J.Dupuy, and J.P..Jal, J.Phys.Chem. 84:1211 (1980).

18. G.A.Thomas, Nuovo Cimento 39:561 (1977).
19. G.A.Thomas, J.Phys.Chem. 88:3749 (1984).
20. J.Weiner, K.H.Langley, and N.C.Ford,Jr., Phys.Rev.Letters 32:879 (1974).
21. J.S.Rowlinson, Adv.Chem.Phys. 41:1 (1980).
22. B.Widom and J.S.Rowlinson, J.Chem.Phys. 52:2670 (1970).
23. N.D.Mermin, Phys.Rev.Letters 26:957 (1971).
24. N.D.Mermin, Phys.Rev.Letters 26:169 (1971).
25. P.C.Hemmer and G.Stell, Phys.Rev.Letters 24:1284 (1970).
26. M.S.Green, M.J.Cooper, and J.M.H.Levelt-Sengers, Phys.Rev.Letters 26:492 (1971).
27. N.D.Mermin and J.J.Rehr, Phys.Rev.Letters 26:1155 (1971).
28. J.J.Rehr and N.D.Mermin, Phys.Rev.A 8:472 (1973).
29. M.Ley-Koo and M.S.Green, Phys.Rev.A 16:2483 (1977).
30. J.F.Nicoll, Phys.Rev.A 24:2203 (1981).
31. J.F.Nicoll and P.C.Albright, Proceedings of the 8th Symposium on Thermophysical Properties, J.V.Sengers, ed., ASME, New York (1982), p.377.
32. M.J.Buckingham, in: "Phase Transitions and Critical Phenomena", C.Domb and M.S.Green, eds., Academic Press, London (1972),Vol.2.
33. J.M.H.Levelt-Sengers, Physica 73:73 (1974).
34. U.El-Hanany and W.W.Warren, Phys.Rev.Letters 34:1276 (1975).
35. W.W.Warren and F.Hensel, Phys.Rev.B 26:5980 (1982).
36. H.Ikezi, K.Schwarzenegger, A.L.Passner, and S.L.McCall, Phys.Rev. 18:2494 (1978).
37. W.Hefner, R.W.Schmutzler, and F.Hensel, J.Phys.(Paris)Colloq. 41:C8-62 (1980).
38. L.R.Matheiss and W.W.Warren, Phys.Rev.B 16:624 (1977).
39. F.Yonezawa and T.Ogawa, Supp.Prog.Theor.Phys. 72:1 (1982).
40. H.Overhof, H.Uchtmann, and F.Hensel, J.Phys.F: Metal Phys. 6:523 (1976).
41. J.P.Hernandez, Phys.Rev. 53:2320 (1984).
42. F.E.Höhne, R.Redner, G.Röpke, and H.Wegner, Physica 128A:643 (1984).
43. W.Hefner and F.Hensel, Phys.Rev.Letters 48:1026 (1982).
44. J.P.Hernandez, Phys.Rev.Letters 48:1682 (1982).
45. T.C.Harman and J.M.Honig, "Thermoelectric and Thermomagnetic Effects and Applications", McGraw-Hill Book Company, New York (1967).
46. L.J.Duckers and R.G.Ross, Phys.Letters A 38:291 (1972).
47. V.A.A.Alexees, A.A.Vedenov, V.G.Orcharenkov, and Yu F.Ryzhkov, Sov.Phys.JETP Lett. 16:49 (1972).
48. M.Yao and H.Endo, J.Phys.Soc.Japan 51:1504 (1982).
49. F.E.Neale and N.E.Cusack, J.Phys.F: Metal Phys. 9:85 (1978).
50. W.Götzlaff, Doctoral Thesis, University Marburg (1987).
51. U.Brusius, Doctoral Thesis, University Marburg (1986).
52. F.Hensel, S.Jüngst, B.Knuth, H.Uchtmann, and M.Yao, Physica 139B:90 (1986).
53. M.W.Pestak and M.H.W.Chang, Phys.Rev.Letters 46:939 (1981).
54. J.V.Sengers, D.Bedeaux, P.Mazur, and S.C.Greer, Physica 104A:573 (1980).

ACOUSTIC VELOCITY MEASUREMENTS ON FLUID METALS FROM TWO-FOLD COMPRESSIONS TO TWO-FOLD EXPANSIONS

J. W. Shaner, R. S. Hixson, and M. A. Winkler

Los Alamos National Laboratory
Los Alamos, NM 87545

J. M. Brown

University of Washington
Seattle, WA

I. INTRODUCTION

Fluid metals around normal density can be thought of as low temperature non-ideal plasmas. They are plasmas in that the Coulomb interactions among the constituent particles are important in determining thermodynamic and transport properties. They are low temperature in that even up to 1 eV the temperature is less than or comparable to the average interionic electrostatic energy. The resulting high Γ's ($\Gamma = Z^2e^2/\langle r\rangle kT$) cause the pair distribution function to look more like that of a normal liquid than that of an ideal gas. For the present studies $\Gamma \sim 10\text{-}100\ Z^2$, where Z is the effective charge of the ions. The real metallic systems have several important differences from idealizations such as the one component plasma model. For example, since the mean distance between electrons, r_s, varies between 2 and 3 in atomic units with temperatures between 0.1 and 1 eV, the electron gas is degenerate, but polarizable. Therefore, unless the electron screening is well known, the effective Coulomb interaction between ions cannot be specified. Furthermore, according to the model of Ashcroft and Lekner,[1] the packing fraction along the liquidus remains roughly constant at 0.46. As a result the excluded volume of the ions probably cannot be ignored anywhere over the density range from two-fold compressed to four-fold expanded. These complexities make realistic modeling of dense fluid metals very difficult in practice.

Over the past several decades, we have developed techniques for accurate thermodynamic and transport measurements over a wide range of densities and temperatures. The two regions we shall concentrate on here are shock compression and heating to 1 eV and two-fold density increases, and isobaric expansion to 1 eV and four-fold density decreases. These experimental capabilities can now produce metal samples in stable, equilibrium states, defined to 1-2% in thermodynamic parameters, for times long enough to perform other experiments.

There are several reasons for choosing the acoustic velocity as a parameter to measure for dense fluid metals. First, the acoustic velocity is directly related to the adiabatic bulk modulus, a derivative of the equation of state surface in P,V,E space. Since the equation of state surface may be subtly affected by changes in interparticle potential, measurements of the derivative will be much more sensitive to these changes. For example, there is growing evidence that at least some of the conduction electrons are localized in fluid mercury as the density decreases below 9 gm/cc.[2] Since screening by the degenerate electron gas is accomplished by electrons near the Fermi surface, changes due to localization in the density of states at the Fermi Surface will certainly change the effective interparticle potential. There is indication that this effect can be seen in the acoustic velocity.[3]

Another reason for interest in acoustic velocities is that, when combined with accurate data along shock compression or isobaric expansion paths on the equation of state surface, these measurements allow a purely experimental determination of most of the important thermodynamic quantities. These include the heat capacity at constant volume, C_V, the isothermal bulk modulus, B_T, and Grüneisen's gamma, γ_G. The latter two quantities can be thought of roughly as the orthogonal derivatives of the P,V,T equation of state surface in that

$$B_T = -V \left.\frac{\partial P}{\partial V}\right)_T \quad ,$$

and (1)

$$\gamma_G = \frac{V}{C_V} \left.\frac{\partial P}{\partial T}\right)_V \quad .$$

In the following sections we will outline the methods of making acoustic velocity measurements on samples which are destroyed in time scales of milliseconds or less, the analytical techniques for calculating the thermodynamic quantities of interest, and new results indicating a linear relationship of acoustic velocity with density over a very large density range.

II. SHOCK COMPRESSION

Shock compression is one way of obtaining fluid metals at high temperature and density and in well defined thermodynamic states. The irreversible nature of shock compression results in large entropy or temperature increases, so that for sufficiently strong shocks the compressed material will be molten. The thermodynamic state is determined by the Hugoniot relations, which are nothing more than statements of conservation of mass, momentum, and energy across the shock front.[4] One complexity often encountered for solids is avoided when the shocked material is molten. Since the momentum conservation condition gives a Hugoniot equation referring to the longitudinal stress (i.e., in the direction of shock propagation), the stress tensor in the shocked solid may be undetermined. In the solid, longitudinal and transverse stresses may be different. However, if the shock melts the sample, the liquid cannot support significant deviatoric stresses, so the longitudinal and tangential stresses are identical. For this case the pressure is well defined.

The question of equilibrium often arises in shock wave physics. How does one know that the P,V,E point determined by dynamical measurements in a shock wave experiment determines a point on the equilibrium equation of state surface? In answer to that, we have a wealth of empirical evidence that dynamic and static high pressure data normally agree to within experimental uncertainties.[5] Furthermore, molecular dynamics calculations for monatomic systems, such as metals, indicate a very rapid equilibration of the translational degrees of freedom following an abrupt departure from equilibrium.[6] For these reasons we consider that the thermodynamic state behind a shock is both well defined and in equilibrium. We have developed techniques for measuring acoustic velocity in shock compressed materials primarily to determine high pressure melting points. If the compressed material is solid, this velocity is the longitudinal elastic wave velocity, while, if the material is partially molten, the highest velocity acoustic disturbance travels with the bulk wave velocity.[7] Since these velocities typically differ by 20-30%, it is easy to determine where the Hugoniot curve crosses the solidus.[8]

The basis for acoustic velocity measurements in shock melted metals is the use of a short shock, as illustrated in Fig. 1. When a thin plate hits a target, shocks move forward (to the right on Fig. 1) from the impact surface into the target and backward in the plate. When the shock reaches the free rear surface of the plate, the zero pressure boundary condition requires that a rarefaction wave propagates forward in the direction of the shock in the target. This situation corresponds to t = 0 in Fig. 1. The rarefaction wave is dispersive since the leading edge is moving into hot compressed material, in which the acoustic velocity normally exceeds the original shock velocity, while the trailing edge of the rarefaction is moving into decompressed material. When the rarefaction overtakes the shock wave the peak shock pressure decreases, as does the shock velocity. Previous attempts to measure the overtake by observing the decrease in shock velocity have often been ambiguous because the measured wave velocity scales roughly as $P^{1/2}$. In this case small changes in peak pressure result in even smaller changes in wave velocity.

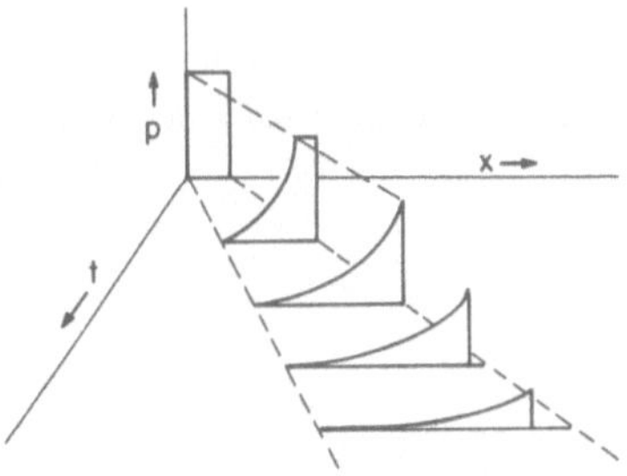

Figure 1. Evolution of a short shock. At t = 0 the release from a free surface begins to overtake the shock. When overtake occurs the peak pressure and shock velocity decrease.

The improvement we have introduced, which allows us to measure the acoustic velocity in hot materials at very high pressure, involves measuring the thermal radiation from the shock front in a transparent medium.[9] Since the thermal radiation intensity varies as a high power of the temperature at the shock front, or equivalently the shock pressure, small decreases in peak pressure result in much larger fractional decreases in detected light intensity. By choosing a transparent medium, the optical analyzer, in which the shock front is opaque, we can assure that we are measuring the leading edge of the shock wave structure. By varying the metal target thickness through which the shock wave structure must pass before entering the optical analyzer we can make measurements which allow us to extrapolate to a target thickness for which the rarefaction overtakes the shock at the target-optical-analyzer interface. This experimental technique eliminates the need of complex hydrodynamic calculations to account for perturbations due to waves reflecting from the metal-optical-analyzer interface.

The calculations required to obtain the acoustic velocity from the measured time for a release to overtake the shock have been presented in detail elsewhere.[10] Since the acoustic velocity in a fluid determines the slope of an isentrope centered on the shock state, and the P-V relation determined by the Hugoniot equations (referred to as the Hugoniot) determines the slope of a different, stiffer curve on the equation of state surface, by differencing them one can obtain an expression for the Grüneisen parameter:

$$\gamma_G = V \left.\frac{\partial P}{\partial E}\right)_V = V \frac{\partial P/\partial V)_S - \partial P/\partial V)_H}{\partial E/\partial V)_S - \partial P/\partial V)_H} \quad , \tag{2}$$

where the subscript S refers to the isentropic derivatives, while the subscript H refers to derivatives along the Hugoniot. If one also measures or calculates the temperature along the Hugoniot, the constant volume heat capacity can be calculated from[11]

$$C_V = \frac{(\partial E/\partial V)_H + P}{(\partial T/\partial V)_H + (\rho\gamma)T} \quad . \tag{3}$$

The isothermal bulk modulus is then derived from

$$B_T = B_S - \alpha B_T \gamma T = B_S - (\rho\gamma)^2 (C_V/\rho)T \quad . \tag{4}$$

These equations have been used to obtain all the useful thermodynamic data for fluid CsI, for example, up to 1.5 Mbar and 10,000 K.[12]

Some of the results we have obtained for acoustic velocity in shock compressed metals are shown in Fig. 2. In the case of iron we found evidence of a solid-solid transition at a shock pressure of 2 Mbar and melting at 2.5 Mbar. The data along the extension of the bulk sound velocity curve (C_B) are for the liquid phase. In the case of tantalum we observed only the melting transition. In both of these cases the acoustic velocity is a linear function of density along the Hugoniot in the liquid phase.

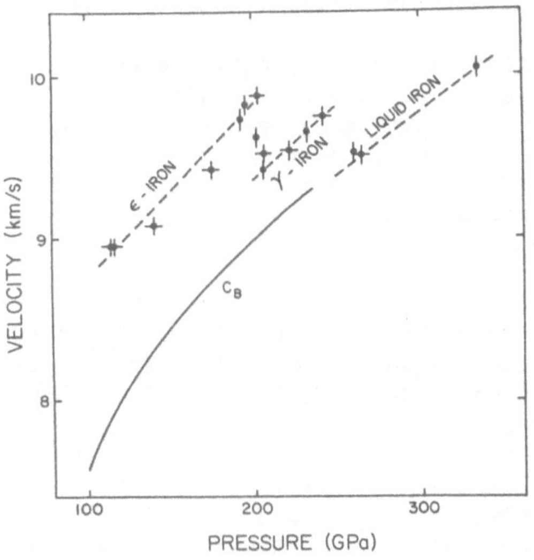

(a)

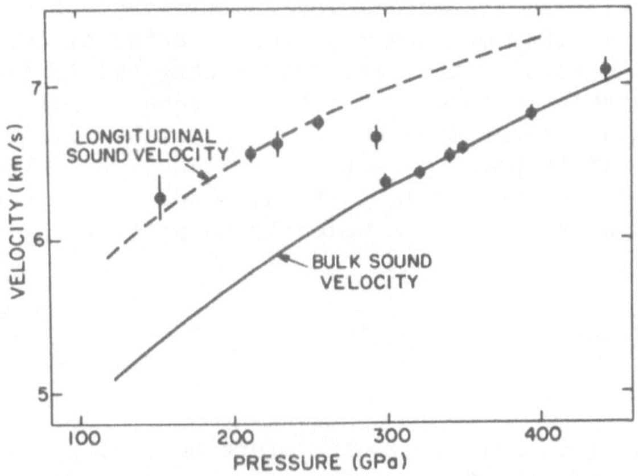

(b)

Figure 2. Rarefaction wave velocity as a function of shock pressure. (a) Iron (Ref. 13); (b) tantalum (Ref. 14). C_B refers to the bulk sound velocity.

III. ISOBARIC EXPANSION

The isobaric expansion experiment has been developed over the past decade to provide stable, equilibrium liquid metal samples up to four-fold expansion and 10,000 K.[15] The experiment consists of a wire electrically heated in an inert gas filled pressure vessel. The roughly square electrical pulse, with 5 μsec rise and fall times and 30-100 μs duration is chosen so that an approximately 1-mm-diameter wire is heated quickly enough to avoid hydrodynamic instabilities, such as the capillary instability. On the other hand the heating pulse is slow enough so extra heating of the surface through the skin effect is avoided, and the pressure throughout the sample is nearly constant during the expansion. Enthalpy is calculated as the time integral of the current times the voltage, with the current and the voltage measured during the heating pulse by a four probe method. Temperature is determined by multicolor optical pyrometry of the wire surface, so the inert gas prevents chemical reaction giving non-thermal radiation as well as providing the pressure medium. By increasing the pressure to supercritical, stable expansions can be obtained to roughly the critical density (~4-fold expansion).

With this capability we have demonstrated that one can obtain good pressure, density, enthalpy, and resistivity data along an isobaric expansion curve. With the addition of pyrometric temperature determinations, the constant pressure heat capacity is also available.[16] For dense metals, such as tantalum and lead, the liquid column remains stable for tens of microseconds after the current is turned off, even at temperatures above 8000 K.

The basis for acoustic velocity measurements in this thermodynamic regime is a laser induced stress wave. After the current pulse has been stopped, but before the liquid column falls apart, we irradiate one side of the sample with a 0.1-0.5 J, 25 ns, pulsed ruby laser propagating radially. With a focal spot of 100 μ diameter we can generate a 10 kbar stress wave, the velocity of which rapidly decays to the sonic velocity. This wave propagates across a diameter of the wire and emerges at the opposite side, causing a compression wave in the gas. With Schlieren photography, we can photograph this wave in the gas and determine when it breaks out of the sample. The sound velocity is then calculated from the time interval and the sample diameter.[17]

Data from many experiments on lead and tantalum are shown in Fig. 3. We have chosen density as the independent variable to show the linear dependence of sound velocity on density, over a factor of two expansion in the lead data. Several pressures are represented in Fig. 3a, and, at least in the low density data we can obtain a rough upper limit on the intrinsic temperature dependence of the sound velocity. Since at a density of 5-6 gm/cc the temperature spread between 1 and 3 kbar isobars is greater than 500 K and the accuracy of the sound velocity data is better than 4%, the independence of sound velocity on pressure gives

$$\frac{\partial c}{\partial \ln T} < 0.3 \text{ km/sec} \quad .$$

On the other hand, from Fig. 3a we can determine that

$$\frac{\partial c}{\partial \ln \rho} \sim 1 \text{ km/sec} \quad .$$

Therefore the intrinsic temperature dependence at large expansions appears to be much weaker than the density dependence.

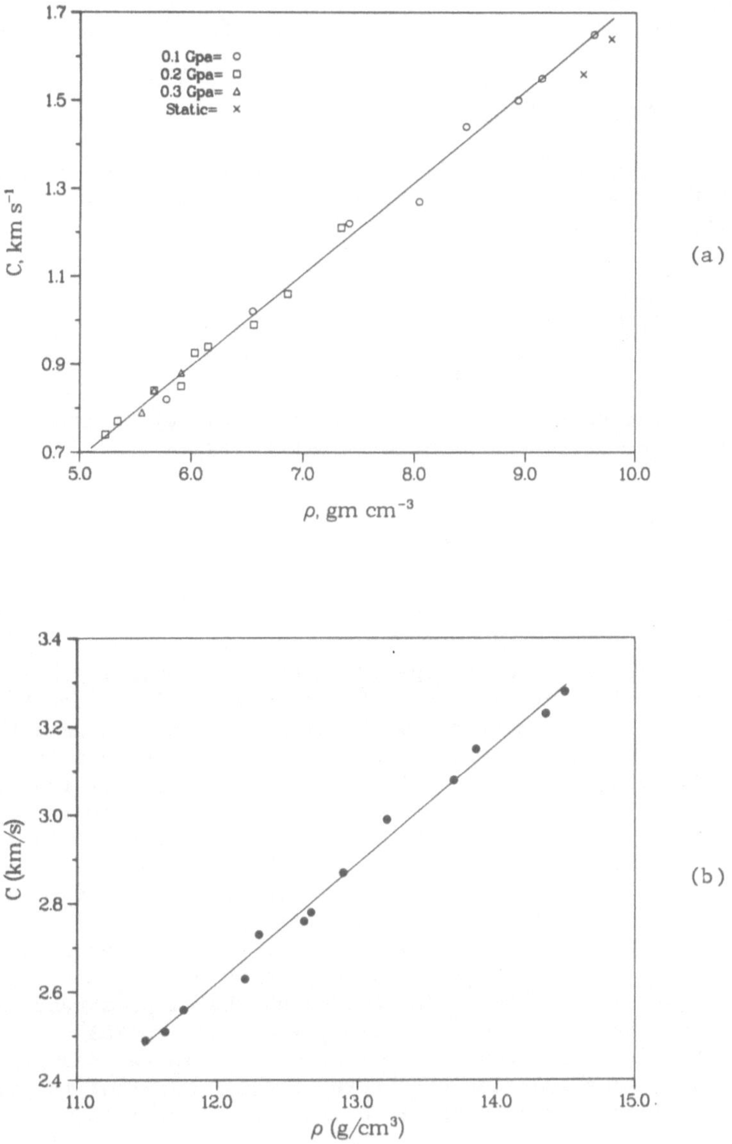

Figure 3. Acoustic velocity in heated, expanded liquid metals: (a) lead (Ref. 18); (b) tantalum (Ref.19).

The linearity of the present data with density is consistent with other measurements of sound velocity in fluid metals over much narrower density ranges.[20] Typically C is reported as linear in T at constant pressure.[21] However, the apparent temperature dependence is probably implicit through thermal expansion.

Given the previously obtained thermodynamic properties and the sound velocity, we can use the relations

$$C_P = C_V + \alpha^2 VTB_T \quad , \tag{5}$$

with α the thermal expansion coefficient, and

$$c^2 = VB_T\, C_P/C_V \quad , \tag{6}$$

to determine experimentally both C_V and B_T. Also, we can express Grüneisen's gamma as

$$\gamma_G = \left(\frac{\partial V}{\partial H}\right)_P c^2 \quad , \tag{7}$$

so the isobaric expansion data and sound velocity completely determine the derivatives of the equation of state surface.

IV. BIRCH'S LAW

With a linear density dependence of sound velocity both in compression and expansion for fluid metals, it is natural to plot both sets of data together. The only fluid metal for which we have data both in shock compression and isobaric expansion is tantalum, and this is shown along with other available data in Fig. 4. These data show that for tantalum the same linear relation fits both sets of data from 10 to 30 gm/cc. The lead data from isobaric expansion is extrapolated to high density by using the common assumptions for shocks in condensed media: shock velocity linear in material velocity and $\rho\gamma_G$ constant. The reasonable linearity of sound velocity vs density is again to be seen. The iron and aluminum shock data are augmented by one-atmosphere data on molten liquids.[22,23] The one material which appears to fall outside the uniform collection of data is lanthanum, for which the slope appears steep. However, we know that the acoustic velocity of this metal is affected by changes in the electronic band structure in the density range shown here.[10] The other systematic feature of Fig. 4 is that the slope of the linear relations is monotonically decreasing in atomic mass.

This kind of plot has been presented first by Birch in an attempt to derive the average atomic number of materials deep in the earth from seismic velocities. Birch started by measuring elastic wave velocities in rocks to 10 kbar, representing a density change of less than 10%.[24] He later included shock compression data, from which isentropic moduli were calculated or the shock wave velocity was used directly, to establish Birch's Law: for isostructural materials the wave velocity decreases as the square root of the mean atomic mass, while for a given material the wave velocity increases linearly with density.[25] Since the original work, several people have tried to explain this linear relation as an approximation to more general solid state models over a limited density range.[26] The present data show an apparent linearity over a much wider density range than has been previously considered.

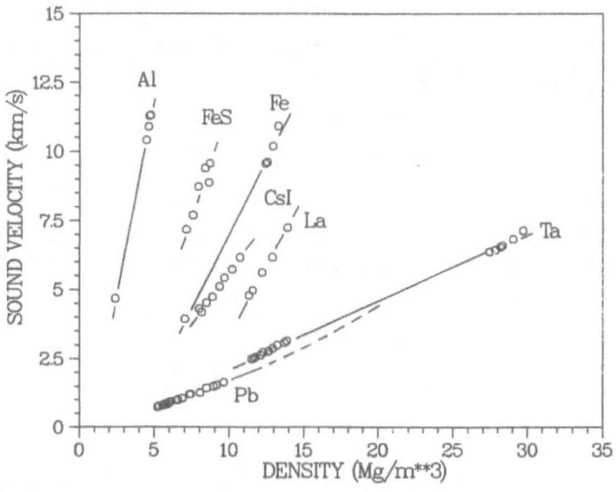

Figure 4. Density dependence of sound velocity for liquid metals.

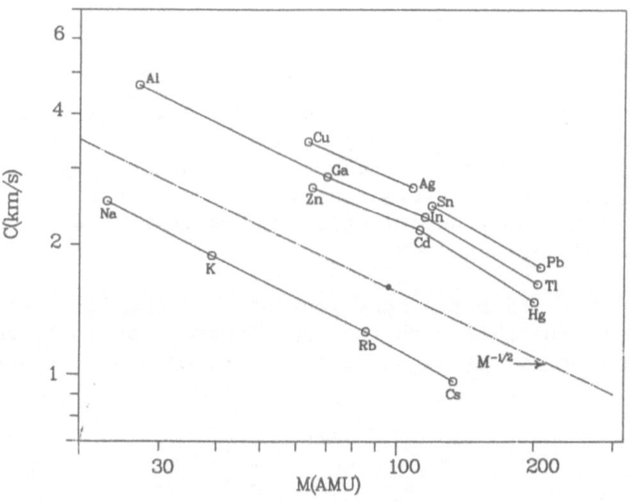

Figure 5. Acoustic velocity of liquid metals as a function of atomic mass. The thermodynamic state is the 1 atmosphere melting point in all cases (Ref. 23).

The first half of Birch's law is that isostructural materials should have a wave velocity which scales inversely with the square root of the mean atomic weight. In Fig. 5 we show that this scaling works very well for liquid metals at the 1 atmosphere liquidus. According to the work of Ashcroft and Lekner,[1] all metals at this point have roughly the same packing fraction, so the structures are closely similar. The metals group naturally according to column in the periodic table so the offsets must be determined by a combination of the valence, or core charge, and the core electronic structure. These figures show that we have rediscovered Birch's law for fluid metals.

V. DISCUSSION

A starting point for understanding acoustic velocity in fluid metals is to consider the disturbance as an ion plasma wave. The unscreened ion plasma wave frequency is given by

$$\Omega_p^2 = \frac{4\pi n(Ze)^2}{M} \quad , \tag{8}$$

where n is the ion density, Z is the ionic charge and M is the ionic mass. Screening by a gas of free electrons introduces a q dependence through the dielectric function which results in a dispersion relation

$$\omega_p = cq \quad , \tag{9}$$

where

$$c = \left(\frac{nZ^2}{MN(E_F)}\right)^{1/2} \quad , \tag{10}$$

and $N(E_F)$ is the density of states at the Fermi surface. $N(E_F)$ scales as $n^{1/3}$, so c, although it has the proper dependence on ionic mass and temperature, should vary as the cube root of the density in this model, and not linear with density. This model, developed by Bohm and Staver, also fails to give proper quantitative values to within a factor of two except for the alkali metals.[23]

By putting terms which are functions of Γ in the free energy of the system, one gets no closer to the experimental data. If the Helmholz free energy is a function of Γ, as in the one-component plasma model,[27] then

$$\begin{aligned} B_T &= \rho \left.\frac{\partial P}{\partial \rho}\right)_T \\ &= \rho \frac{\partial}{\partial \rho}\left[\rho^2\left(\frac{\partial F}{\partial \rho}\right)_T\right]_T \\ &= \frac{1}{9}\frac{\rho^{5/3}}{(kT)^2}\left(\frac{\partial^2 F}{\partial \Gamma^2}\right)_T \quad . \end{aligned} \tag{11}$$

Ignoring the distinction between adiabatic and isothermal moduli, this expression suggests a stronger temperature than density dependence for the sound velocity.

Two models which do reproduce the experimental results are variations on the van der Waals equation of state. They are the hard sphere van der Waals theory of Young and Alder[28] and a more realistic soft sphere version.[29] Both of these models are of the form

$$P = \frac{NkT}{V}\left[1 + A(\rho,T)\right] - B(\rho) \quad ,$$

and (12)

$$E = E_0 + NkT\left[\frac{3}{2} + C(\rho,T)\right] - D(\rho) \quad ,$$

where the functions A, B, C, and D are determined by fits to Monte Carlo simulations of systems with hard sphere or soft sphere potentials. The various parameters, such as hard sphere radius, or the power of the soft sphere potential, can be fit consistently to the data for fluid metals, including the linear density dependence of the sound velocity. Two differences between these models and the Bohm-Staver or one component plasma model are that the potentials are stiffer and that the non-zero core size is explicitly included. It is not clear which of these differences is most important in correcting the point charge models. The van der Waals models are semi-empirical, however. The more fundamental theories, such as those developed by Ashcroft and Langreth[30] have not yet been applied to the density dependence of the bulk modulus.

Recent measurements by Shaw and Caldwell on fluid alkali metals up to 7 kbar and 450 K have given similar results to ours, although over a much more limited thermodynamic range.[21] They find a weak temperature dependence and a roughly linear density dependence of acoustic velocity. However, their data is presented in terms of "experimentalist variables" - pressure and temperature - instead of density and temperature. Until the inversion is done the accuracy of Birch's law cannot be confirmed.

We have shown that with current experimental techniques thermodynamic quantities can be measured accurately for fluid metals over a four or five-fold density range up to temperatures of 10,000 K. The result we have presented here is that the acoustic velocity in these systems is apparently relatively insensitive to temperature and linear in density over a very wide thermodynamic range. We believe that the ionic core properties are important in determining this result both through their finite size and in the stiffness of the potential relative to a point change system. However, the question still remains: Why is Birch's law so good?

REFERENCES

1. N. W. Ashcroft and J. Lekner, Phys. Rev. 145, 83 (1966).
2. V. El-Hanany and W. W. Warren, Phys. Rev. Lett. 34, 1276 (1975).

3. K. Suzuki, et al., J. Phys. Colloq. 40 #C-7, 685 (1979).
4. See, for example, M. H. Rice, et al., in Solid State Physics, v. 6, F. Seitz and D. Turnbull, eds., Academic Press, NY, 1957.
5. H. K. Mao, et al., J. Appl. Phys. 49, 3276 (1978).
6. B. L. Holian, et al., Phys. Rev. A22, 2789 (1980).
7. J. Asay and D. B. Hayes, J. Appl. Phys. 46, 4789 (1975).
8. J. M. Brown and R. G. McQueen, Geophys. Res. Lett. 7, 533 (1980).
9. R. G. McQueen, J. W. Hopson, and J. N. Fritz, Rev. Sci. Inst. 53, 245 (1982).
10. J. M. Brown, J. W. Shaner, and C. A. Swenson, Phys. Rev. B32, 4507 (1985).
11. R. G. McQueen in "Laboratory Techniques for Very High Pressures and the Behavior of Metals," edited by K. A. Geschneidner, M. T. Hepworth, and N.A.D. Parlee, Metal. Soc. Conf. 22, 44 (1964).
12. C. A. Swenson, J. W. Shaner, and J. M. Brown, Phys. Rev. (in press).
13. J. W. Shaner, J. M. Brown, and R. G. McQueen, Mat. Res. Soc. Symp. Proc. 22, 137 (1984).
14. J. M. Brown and J. W. Shaner in "Shock Waves in Condensed Matter - 1983," J. Asay, R. A. Graham, and G. K. Straub, eds., Elsevier Science Publishers, 1984.
15. G. R. Gathers, J. W. Shaner, and R. L. Brier, Rev. Sci. Inst. 47, 471 (1976).
16. J. W. Shaner, G. R. Gathers, and C. Minichino, High-Temp./High-Press. 9, 331 (1977), and G. R. Gathers, J. W. Shaner, R. S. Hixson, and D. A. Young, High-Temp./High-Press. 11, 653 (1979).
17. R. S. Hixson, M. A. Winkler, and J. W. Shaner, High-Temp./High-Press. 17, 267 (1985).
18. R. S. Hixson, M. A. Winkler, and J. W. Shaner, Int. J. of Thermophys. 7, 161 (1986).
19. R. S. Hixson, M. A. Winkler, and J. W. Shaner, High-Temp./High-Press. (in press).
20. H. N. Spetzler, M. D. Meyer, and T. Chin, High-Temp./High-Press. 7, 481 (1975).
21. G. H. Shaw and D. A. Caldwell, Phys. Rev. B32, 7937 (1985).
22. W. Kurz and B. Lux, High-Temp./High-Press. 1, 387 (1969).
23. G.M.B. Webber and R.W.B. Stephens in "Physical Acoustics," vol. IVB, W. Mason, ed., Academic Press, NY, 1968.
24. F. Birch, J. Geophys. Res. 66, 2199 (1961).
25. F. Birch, Phys. Earth and Planet. Interiors 1, 141 (1968).
26. T. J. Shankland, J. Geophys. Res. 77, 3750 (1972), D. H. Chung, Science 177, 261 (1972), and O. L. Anderson, J. Geophys. Res. 78, 4901 (1973).
27. W. L. Slattery, G. D. Doolen, and H. E. DeWill, Phys. Rev. A21, 2087 (1980).
28. D. A. Young and B. J. Alder, Phys. Rev. A3, 364 (1971).
29. G. R. Gathers, J. W. Shaner, and D. A. Young, Phys. Rev. Lett. 33, 70 (1974).
30. N. W. Ashcroft and D. Langreth, Phys. Rev. 155, 682 (1967).

STRUCTURE OF THE JOVIAN ENVELOPE AND THE

EQUATION OF STATE OF DENSE HYDROGEN

W. B. Hubbard and M. S. Marley

Department of Planetary Sciences
Lunar and Planetary Laboratory
University of Arizona
Tucson, Arizona 85721

INTRODUCTION

The interior composition of the planet Jupiter is deduced by comparing models generated from the equation of hydrostatic equilibrium and high-pressure equations of state with the known mass, equatorial radius, and gravitational multipole moments of the planet. The planet is primarily composed of liquid metallic hydrogen, but there appears to be a substantial admixture of denser elements present as well. Jupiter's hydrogen-rich envelope is substantially enriched in material other than hydrogen and helium, containing approximately 20 to 60 earth masses of such material, in addition to 6 to 4 earth masses of such material in a distinct core. Thus Jupiter's bulk composition differs from that of the sun. These conclusions are heavily dependent upon accurate pressure-density relations for pure metallic hydrogen in the pressure range from about 3 to 40 megabars, and at temperatures ranging from about 10000 to 20000 K. Experimental results for compression of hydrogen in the nonmetallic pressure range are helpful in constraining models, but accurate theoretical calculations of the thermodynamics of the liquid metallic phase provide the most help in constraining models. We discuss the state of the strongly-coupled plasma in the Jovian interior, and propose a phase diagram for dense liquid hydrogen.

Physical conditions in the deep interior of Jupiter provide a laboratory of sorts for the study of metallic hydrogen. In this planet there exists the largest natural reservoir of this substance in the solar system, comprising approximately 230 earth masses, where

$$1 \text{ earth mass} = M_E = 5.976 \times 10^{27} \text{ g}$$

out of a total planetary mass of 317.7 M_E.

These numbers are of course model-dependent, and are not actually available independently of thermodynamic models of metallic hydrogen. On the other hand, they are fairly reasonable in the light of all available evidence, and show that currently-available theories for the thermodynamics of metallic hydrogen are not grossly in error. The purpose of this discussion is to show how theoretical calculations of the thermodynamics of metallic calculations are incorporated in Jovian interior models, and to exhibit the most crucial (and uncertain) aspects of these calculations from the point of view of such models.

The temperature-pressure regime of interest in the deep Jovian interior is likely to be moderately close to an adiabat of pure metallic hydrogen, with a specific entropy close to that of the deep convective atmosphere, a region of the planet accessible to direct observation. The entropy of the latter region is specified by a starting temperature $T_1 = 165$ K at a pressure of 1 bar (10^5 Pa; Lindal et al., 1981). Taking the thermal conductivity of liquid metallic hydrogen to be about 10^8 ergs/cm/s/K at a mass density of 1 g/cm^3 (Hubbard and Lampe, 1969), the temperature gradient required to transport the observed interior heat flow $H_i = 5400 \pm 400$ erg/cm^2/sec (Hanel et al., 1981) would be substantially superadiabatic and therefore lead to convection. Efficient convection results in an interior temperature profile which is close to isentropic. Melting temperatures of metallic hydrogen in the pressure range of interest (3-40 Mbar; or 3-40 x 10^{11} Pa) are uncertain because of their great sensitivity to the effective interproton potential, but there is little doubt that with likely values in the range of about 10^3 K, they fall well below the temperatures estimated for isentropic Jovian interior models (about 10^4 K). The calculation of interior properties of Jovian models therefore requires an accurate liquid-state theory for metallic hydrogen.

THERMODYNAMICS OF DENSE LIQUID HYDROGEN

Current Jovian interior models are based upon the assumption of a "conventional" phase diagram for dense hydrogen (for examples of other phase diagrams, see related papers in these proceedings). Fig. 1 shows a phase diagram for pure hydrogen (Marley and Hubbard 1986) which is calculated in the following manner. The Helmholtz free energy of liquid metallic hydrogen (H+) is evaluated using the formula of Hubbard and DeWitt (1986), which is in turn adjusted to give the best fit to a large number of Monte Carlo evaluations of the pressure P, internal energy E, heat capacity C_V, and generalized Gruneisen parameter, for an effective potential between protons which is described by the zero-temperature Lindhard dielectric function. As a result of the fitting procedure, the coefficients are obtained for the expression

$$f = -a\Gamma_e + b\Gamma_e^{1/4} - c \ln \Gamma_e + d,$$

where

$$f = F_I/NkT,$$

$$\Gamma_e = e^2/(sa_0kT),$$

$$s = (3V/4\pi N\langle Z\rangle)^{1/3} / a_0.$$

Here F_I is the interaction part of the Helmholtz free energy, e is the electron charge, V is the volume, N is the number of nuclei, a_0 is the Bohr radius, and <Z> is the average charge of the nuclei (=1 for H+). A complete free energy expression is obtained by adding to F_I the ideal-gas free energies of a fully-degenerate electron gas and a Maxwell-Boltzmann gas of nuclei, together with the electron exchange energy and a nonlocal electron correlation energy E_c which is represented by the expression

$$\frac{E_c}{N\langle Z\rangle e^2/a_0} = -0.0311 \ln \frac{(1+4.69/s)(1+0.676\sqrt{s})}{(1+0.236\sqrt{s})}$$

(Salpeter and Zapolsky, 1967).

The interaction component of the free energy of liquid molecular hydrogen (H_2) is similarly fitted to results of Monte Carlo calculations which utilize an H_2-H_2 pair potential derived from shock wave experiments to 0.76 Mbar (Nellis et al., 1983). We then compute chemical potentials and solve for the H_2 - H+ boundary between the two liquid phases. We find $(dT/dP) > 0$ along this boundary, hence

$$\Delta S(H_2 \rightarrow H+) < 0$$

at the presumed first-order transition between liquid phases. For a Jovian adiabat (T_1 = 165 K), the H_2 model predicts a transition temperature of 7500 K at a pressure of 5.5 Mbar (5.5×10^{11} Pa). This result is quite sensitive to the assumed form of the electron correlation energy, as discussed by Ross and Shishkevish (1977). An uncertainty in the predicted transition pressure of about a factor of two is not at all unlikely. Moreover, the value of the latent heat of transition between the two phases (and even its sign) is fairly sensitive to the assumed internal partition functions of the H_2 molecules in the vicinity of the transition. However, it is significant that this model predicts substantial temperature-dependence in the transition pressure, such that the transition in the deep Jovian and Saturnian interiors may take place at substantially higher pressures than the corresponding transition at zero (or at least $T \ll 10^3$ K) temperature.

As discussed by Stevenson and Salpeter (1977), the transition from molecular to metallic hydrogen in Jupiter and Saturn is likely to occur under conditions of constant temperature across the phase boundary, as opposed to the case where the temperature follows the phase boundary within a finite pressure range until the entropy is equalized and all of the H_2 has been converted to H+. In this case the specific entropy must decrease from the H_2 envelope to the H+ interior, with the result that the interior temperature values predicted for a fully isentropic model are overestimated by a factor of about 1.3 in the case of Jupiter. Thus the maximum temperature (in the deepest metallic-hydrogen layer in Jupiter) would be about 13000 K at 40 Mbar, rather than 17000 as predicted by the isentropic model. The mass density of hydrogen at this level is about 3.6 g/cm^3, corresponding to an electron spacing parameter s = 0.9, and a plasma coupling parameter

$$\Gamma_e \simeq 21 \text{ to } 27,$$

depending on the precise temperature distribution. At constant specific entropy,

$$T \propto \rho^{\gamma},$$

where

$$\rho = \text{mass density},$$

and in this pressure and density range,

$$\gamma \simeq 0.6.$$

Since

$$\Gamma_e \propto \rho^{1/3}/T,$$

it follows that Γ_e will be slightly larger in the layers close to the transition region than in the deeper metallic-hydrogen envelope.

Corrections for the presence of helium and other dense elements will change the preceding numbers to some extent, but not most of the qualitative results. Calculation of the transition of H_2 to H+ in the presence of helium introduces an additional thermodynamic degree of freedom, and reliable models for the phase transition in this case are so far not available.

Thus, models of hydrogen-rich planets which assume complete adiabaticity throughout the interior may overestimate the temperature in the H+ region. The temperature and initial pressure of transition for two pure hydrogen adiabats are given in Table 1. Also given are the change in entropy and density across the transition. The ratio T(ad)/T gives the factor by which the assumption of constant specific entropy all the way from a pressure P = 1 bar to pressures characteristic of the deep Jovian interior, overestimates the actual temperature in the H+ region.

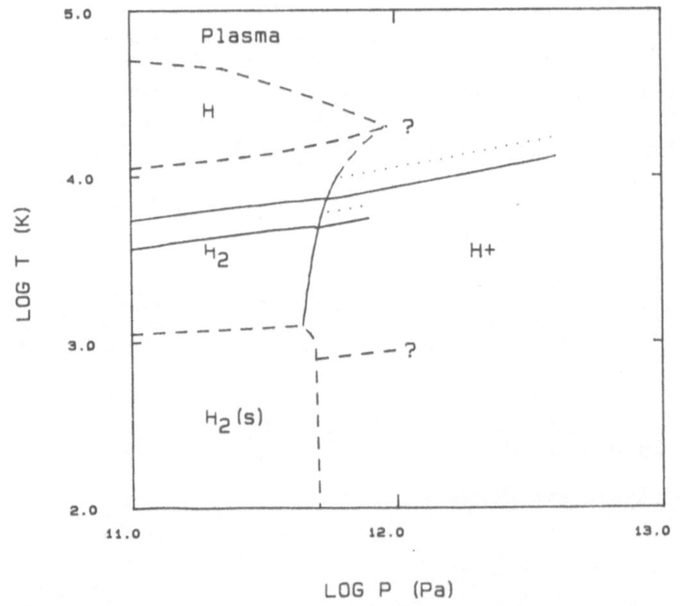

Fig. 1. Hydrogen phase diagram and interior temperature profiles in Jupiter (upper solid curve) and Saturn (lower solid curve). The transition between liquid H_2 and liquid H+, presumed first-order, is shown as a solid, near-vertical line sloping to the right. Other possible phase boundaries, whose location is not important for Jovian interior models, are shown dashed. The dotted curves are extensions of the H_2 adiabats, assuming constant entropy across the phase boundary.

Table 1. Parameters for molecular-metallic phase transitions for two jovian-planet adiabats (pure hydrogen assumed).

T(@1 bar)	T(K)	P(Mbar)	ΔS(k/N)	$\Delta\rho$(g/cm^3)	T(ad)/T
90	5000	5.2	-0.42	0.29	1.21
165	7500	5.5	-0.59	0.27	1.31

As experimentally-attained pressures are still well below predicted transition pressures, the possibility of a gradual H_2 - H+ transformation at Jovian temperatures in the pressure range of 1-6 Mbar cannot be excluded. In this case the absence of a sharp first-order transition between H_2 and H+ would eliminate the necessity for a discontinuity in specific entropy in the deep Jovian or Saturnian interior, and temperatures in this region would be correspondingly higher by the factors given in Table 1.

INTERIOR MODELS OF JUPITER

The primary constraints on Jovian interior models are the gravitational mass of the planet GM, the equatorial radius at a pressure level of 1 bar a_1, and the dimensionless gravitational multipole moments of the planet J_n. The latter represent the response of the planet to the rotational perturbation parameter

$$q = \omega^2 a_1^3/GM,$$

where ω is the planet's angular rotation rate.

The modeling procedure is approximately as follows. A proposed pressure-density profile (or profiles in the case of a stepwise-continuous planet) is substituted into the equation of hydrostatic equilibrium to compute the radius a_1 of a nonrotating planet of mass GM. The two-dimensional structure of rotating planet with a rotation parameter q is then calculated, yielding the J_n (in practice, only J_2 and J_4 are known with sufficient precision to constrain models; J_6 has been determined to about $\pm$65%; Campbell and Synnott, 1985). An iterative procedure is then imposed on the adjustable model parameters until satisfactory agreement with all observational parameters is obtained (Hubbard and Horedt, 1983).

Because of uncertainties in the hydrogen phase diagram discussed above, the temperature distribution in the deep Jovian interior is uncertain to within approximately $\pm$25%. The temperature-dependent part of the pressure in the corresponding pressure and temperature domain comprises about 20%; thus the uncertainty in the total pressure due to inadequacies in the thermal model is about 5%. This error may seem acceptably small, but one of the major goals of constructing models of the Jovian interior is to obtain the relative abundance of materials other than hydrogen.

One may estimate the impact of a given error in the pure-hydrogen pressure on the inferred interior abundances, as follows. As shown by Hubbard and DeWitt (1985), the equation of state of a mixture of metallic hydrogen and helium under jovian interior conditions is given to excellent approximation by the so-called additive volume law for a mixture containing mass fraction Y of helium:

$$\rho^{-1} = (1-Y)\rho_H^{-1} + Y\rho_Y^{-1},$$

where ρ_H, ρ_Y are the mass densities of pure H and He at given P.

The additive-volume is valid to within < 1% for pressures greater than 10 Mbar. It can also be used for mixtures of hydrogen with other elements in addition to helium, in which case Y would represent the mass fraction of all non-hydrogen components. No systematic study of the validity of the additive-volume law has been made for this more general case, but it is expected to be at least as good an approximation as for the hydrogen-helium system alone.

Letting δ be the fractional error in the pure-hydrogen density, and letting δY be the error in the inferred Y from the additive volume law, we find

$$\frac{\delta Y}{Y} \simeq -\delta \frac{\rho_H}{\rho} / Y \simeq -\delta/Y$$

since $\rho_H \simeq \rho$ because of the predominance of hydrogen.

We conclude that, since Y is on the order of 0.2 to 0.3 in Jupiter's hydrogen-rich envelope, uncertainties on the order of 5-10% in the pure-hydrogen equation of state can lead to uncertainties on the order of 25 to 50% in the inferred value of Y.

Studies of the abundances in Jupiter's hydrogen-rich envelope have been carried out using the theory of Hubbard and DeWitt (1985), and also using an alternative equation of state based on Thomas-Fermi-Dirac (TFD) theory (Hubbard and MacFarlane, 1985). The TFD equation of state gives a higher pressure for given composition, temperature, and density, than does the equation of state based on the Lindhard dielectric function. For pure hydrogen, the TFD pressure is about 10% higher at P = 22 Mbar. Of the two equations of state, that of Hubbard and DeWitt (Lindhard) requires a smaller admixture of non-hydrogen material in Jupiter, and gives a total hydrogen abundance in Jupiter corresponding to the result given in the introduction. In contrast, the TFD equation of state gives only approximately 180 M_E of hydrogen in Jupiter, an implausible result.

CONCLUSIONS

Neither of the two theories for liquid metallic hydrogen yields a Jupiter model of precisely solar composition, and so it seems a safe conclusion that Jupiter is enriched in heavy elements with respect to the sun. However, there are many remaining uncertainties which must be addressed before this result can be made more precise. On the basis of the work carried out so far, the liquid-metallic hydrogen equation of state based on Monte Carlo runs using an effective potential derived from the Lindhard dielectric function, seems to give more reasonable Jupiter models than does a TFD-based equation of state. Thus, in this limited sense, one can use Jupiter interior models to discriminate between hydrogen equations of state. Further work on this topic, incorporating more extensive experimental results on hydrogen and helium compression, is in progress.

ACKNOWLEDGMENT

This work was supported in part by NASA Grants NSG-7045 and NGT-50049.

REFERENCES

Campbell, J. K., and Synnott, S. P., 1985, Gravity field of the Jovian system from Pioneer and Voyager tracking data, Astron. J., 90:364.

Hanel, R. A., Conrath, B. J., Herath, L. W., Kunde, V. G., and Pirraglia, J.

A., 1981, Albedo, internal heat, and energy balance of Jupiter: preliminary results from the Voyager infrared investigation, J. Geophys. Res. 86:8705.

Hubbard, W. B., and DeWitt, H. E., 1985, Statistical mechanics of light elements at high pressure. VII. A perturbative free energy for arbitrary mixtures of H and He, Astrophys. J., 290:388.

Hubbard, W. B., and Horedt, G. P., 1983, Computation of Jupiter interior models from gravitational inversion theory, Icarus, 54:456.

Hubbard, W. B., and Lampe, M., 1969, Thermal conduction by electrons in stellar matter, Astrophys. J. Suppl., 18:297.

Hubbard, W. B., and MacFarlane, J. J., 1985, Statistical mechanics of light elements at high pressure. VIII. Thomas-Fermi-Dirac theory for binary mixtures of H with He, C, and O, Astrophys. J., 297:133.

Lindal, G. F., Wood, G. E., Levy, G. S., Anderson, J. D., Sweetnam, D. N., Hotz, H. B., Buckles, B. J., Holmes, D. P., Doms, P. E., Eshleman, V. R., Tyler, G. L., and Croft, T. A., 1981, The atmosphere of Jupiter: an analysis of the Voyager radio occultation measurements, J. Geophys. Res., 86:8721.

Marley, M. S., and Hubbard, W. B., 1986, Calculation of the molecular-metallic hydrogen transition, Bull. Amer. Astron. Soc., 18:780.

Nellis, W. J., Ross., M., Mitchell, A. C., van Thiel, M., Young, D. A., Ree, F. H., and Trainor, R. J., 1983, Equation of state of molecular hydrogen and deuterium from shock-wave experiments to 760 kbar, Phys. Rev. A, 27:608.

Ross, M., and Shishkevish, C., 1977, Molecular and metallic hydrogen, DARPA Report R-2056-ARPA.

Salpeter, E. E., and Zapolsky, H. S., 1967, Theoretical high pressure equations of state, including correlation energy, Phys. Rev., 158:876.

Stevenson, D. J., and Salpeter, E. E., 1977, The dynamics and helium distribution in hydrogen-helium fluid planets, Astrophys. J. Suppl., 35:221.

CHAPTER IX

RESPONSE FUNCTIONS AND ION STOPPING POWER

NONLINEAR FLUCTUATION-DISSIPATION THEOREMS AND THEIR APPLICATIONS TO DYNAMICAL PROBLEMS IN STRONGLY COUPLED PLASMAS

G. Kalman

Department of Physics
Boston College
Chestnut Hill, MA

I. INTRODUCTION

This paper discusses the methodology and some applications of a dynamical mean field theory type approach developed by K.I. Golden and the author to the problem of dynamical response functions and collective mode dispersion in a strongly coupled plasma. Additional applications are discussed by Golden (Golden, this Volume).

Central to the approximation scheme is the Quadratic Fluctuation-Dissipation Theorem, which relates quadratic response functions to the three-point equilibrium correlations and also "response functions of the second kind" (see below) to three-point equilibrium correlations.

We discuss the Quadratic Fluctuation-Dissipation Theorem in a generalized quantum language, recently developed by us (Kalman and Gu, 1986) and its classical limit, which has been known for some time (Golden, Kalman, Silevitch,.1972) in Section II. In Section III, we outline the general philosophy of the Dynamical Mean Field Theory, due to Golden and Kalman (GK). In Sections IV and V, we review application of the scheme to the problem of plasmon dispersion in the classical one-component plasma (OCP) and binary ion mixture (BIM), respectively.

II. QUADRATIC FLUCTUATION-DISSIPATION THEOREM

The relationships commonly known as "Fluctuation-Dissipation Theorems" (FDT), establishing a link between the linear response (a non-equilibrium property) of the system and equilibrium correlations of fluctuating quantities, have become a powerful tool in modern statistical physics and many-body theory. The primitive idea was due to Nyquist (1928) who studied the relationship between the resistivity and noise of electrical networks. The establishment of the FDT in its modern form is, however, due to Kubo (1957, 1959, 1966). (See also Callen and Welton, 1951; Martin, 1968; Golden and Kalman, 1969.) While Kubo's formalism focuses on the linear response of the system, it is clear that, in general, the system's response is not restricted to be linear. Thus, in addition to the well-explored linear response functions, one can examine the properties of higher order (quadratic, cubic, etc.) response functions, which relate the system's response to higher powers of the perturbing field. Moreover, once one relaxes the restriction of concentrating on the simplest response characteristics of the system, the very concept of "response" can be generalized. The conventional response functions, to which we will refer as "response functions of the first kind", relate the perturbed averages of physical quantities (density, current, etc.) at a given space-time point to the perturbing field. The effect of the perturbation on the system is, however, further characterized by the perturbation of averages of correlated physical quantities taken at two, three, etc. space-time points. The relationships between these perturbed two-, three-, etc. point functions, which also can exhibit both linear and higher order behavior, and the perturbing field define (Golden and Kalman, 1982) "response functions of the second kind", "response functions of the third kind", etc. That all these higher order response functions, and also the response functions of higher kind, would satisfy some kind of fluctuation-dissipation-like theorem, i.e., should be related to averages of equilibrium correlations, is a rather obvious expectation. Even a cursory reflection over the derivation of the linear FDT should suggest a correlation, e.g., between the quadratic response function or the linear response function of the second kind, on the one hand, and the equilibrium three-point function, on the other.

Over the last fifteen years, a series of quadratic FDT-s have been established along these lines. The basic relationship between the quadratic conductivity and the three-point current-current correlations for a one-component classical plasma was derived by Golden, Kalman and Silevitch (1972) and independently by Sitenko (1978). Relationships for the current-current response function of the second kind were given by Golden and Kalman (1982).

The generalized quantum mechanical quadratic FDT has recently been derived by Kalman and Gu (1986); there have been earlier works by Soviet authors on related topics (Efremov, 1968; Bochkov and Kuzovlev, 1977; Stratanovich, 1970). The derivation is based on standard perturbation expansion of the quantum von Neumann-Liouville equation for the statistical operator Ω

$$\frac{\partial\Omega}{\partial t} + iL\Omega = 0 \tag{1}$$

The Liouville super-operator is related to the Hamiltonian

$$L = \frac{1}{\hbar}[H,\ldots]$$
$$L = L^{(0)} + L^{(1)} \tag{2}$$

where $L^{(1)}$ is generated through $H^{(1)}$ by the external perturbation $\hat{\phi}_{\underset{\sim}{k}}(t)$

$$H^{(1)} = \frac{1}{V}\sum_{\underset{\sim}{k}} \hat{\phi}_{\underset{\sim}{k}}(t) n^{+}_{\underset{\sim}{k}} \tag{3}$$

One proceeds to calculate $\Omega^{(2)}$, from that the second order average current $j^{(2)}$, which then determines the quadratic (external) conductivity $\hat{\sigma}$, in terms of the three-point dynamical structure functions, defined through

$$\frac{1}{2\pi}\langle n_{\underset{\sim}{k}_1}(\omega_1)\, n_{\underset{\sim}{k}_2}(\omega_2)\, n_{-\underset{\sim}{k}}(-\omega)\rangle = N\delta_{\underset{\sim}{k}-\underset{\sim}{k}_1-\underset{\sim}{k}_2}\delta(\omega-\omega_1-\omega_2)\,\{S(120)$$
$$+ N[\delta_{\underset{\sim}{k}_1}\delta(\omega_1)\, S(\underset{\sim}{k}_2\omega_2) + \delta_{\underset{\sim}{k}_2}\delta(\omega_2)\, S(\underset{\sim}{k}_1\omega_1)$$
$$+ \delta_{\underset{\sim}{k}}\delta(\omega)S(\underset{\sim}{k}_1\omega_1)] + N^2\delta_{\underset{\sim}{k}_1}\,\delta_{\underset{\sim}{k}_2}\delta(\omega_1)\delta(\omega_2)\} \tag{4}$$

with

$$\frac{1}{2\pi}\langle n_{\underset{\sim}{k}_1}(\omega_1) n_{-\underset{\sim}{k}_2}(-\omega_2)\rangle = N\delta_{\underset{\sim}{k}_1-\underset{\sim}{k}_2}\delta(\omega_1-\omega_2)\{S(\underset{\sim}{k}_1\omega_1) + N\delta_{\underset{\sim}{k}_1}\delta(\omega_1)\} \tag{5}$$

and

$$\frac{1}{2\pi}\langle j_{\underset{\sim}{k}_1}(\omega_1) j_{\underset{\sim}{k}_2}(\omega_2) j_{-\underset{\sim}{k}}(-\omega)\rangle = N\delta_{\underset{\sim}{k}-\underset{\sim}{k}_1-\underset{\sim}{k}_2}\delta(\omega-\omega_1-\omega_2)\; Q(120) \tag{6}$$

Here $n_{\underset{\sim}{k}}(\omega)$ and $j_{\underset{\sim}{k}}(\omega)$ are the Fourier-transforms of the local density and current-density operators. Note that the ordering of these operators is of crucial importance and constitutes the major difference between the quantum and classical derivations.

The expression for $\hat{\sigma}$ is

$$\hat{\sigma}(\underset{\sim}{k}_1\omega_1;\, \underset{\sim}{k}_2\omega_2) = \frac{n}{2\hbar^2}\iint d\omega'_1 d\omega'_2\; \delta_+(\omega_2-\omega'_2)\; \delta_+(\omega_1+\omega_2-\omega'_1-\omega'_2)\; \frac{1}{\omega'_1\omega'_2}$$
$$\times\, \{[Q(012)+Q(210)]-[Q(102)+Q(201)]\}$$
$$+ \frac{n}{2\hbar^2}\int d\omega'_1 d\omega'_2\; \delta_+(\omega_1-\omega'_1)\; \delta_+(\omega_1+\omega_2-\omega'_1-\omega'_2)\frac{1}{\omega'_1\omega'_2}$$
$$\times\, \{[Q(021)+Q(120)]-[Q(102)+Q(201)]\} \tag{7}$$

Eq. (7) is the primitive form of the QFDT. It provides a link between the two principal objects, the quadratic conductivity $\hat{\underset{2}{\sigma}}$ and the three-point current correlations. It is still not of the form that could constitute the desired formulation of the QFDT. The main reason for this is that the r.h.s. of Eq. (7) is an integral relationship which, in fact, generates an integral equation for the combination of the Q-functions that appear under the integral.

We proceed to solve the integral equation by first introducing an auxiliary quantity

$$\Psi(\underset{\sim}{k}_1\omega_1;\ \underset{\sim}{k}_2\omega_2) \equiv \hat{\sigma}(\underset{\sim}{k}_1\omega_1;\underset{\sim}{k}_2\omega_2)$$

$$+ \int_{-\infty}^{+\infty} d\mu\ \delta_+(\omega_1-\mu)\ \hat{\sigma}^*\ (\underset{\sim}{k}_1\mu;\ -\underset{\sim}{k}-\omega_2-\mu)$$

$$+ \int_{-\infty}^{+\infty} d\mu\ \delta_+(\omega_2-\mu)\ \hat{\sigma}^*\ (\underset{\sim}{k}_2\mu;\ -\underset{\sim}{k}-\omega_1-\mu)$$

$$= -\frac{n}{2\hbar^2}\int_{-\infty}^{+\infty}d\mu_1\int_{-\infty}^{+\infty}d\mu_2\ \ \delta_+(\omega_1-\mu_1)\ \delta_+\ (\omega_2-\mu_2)\ S\left\{\frac{Q(\underset{\sim}{k}_1\mu_1;\ \underset{\sim}{k}_2\mu_2)}{\mu_1\mu_2}\right\} \tag{8}$$

Here S { } represents full symmetrization with respect to the permutation of the arguments:

$$S\left\{\frac{Q(\underset{\sim}{k}_1\omega_1;\ \underset{\sim}{k}_2\omega_2)}{\omega_1\omega_2}\right\} = \frac{Q(102)+Q(201)}{\omega_1\omega_2} - \frac{Q(012)+Q(210)}{\omega\omega_2} - \frac{Q(021)+Q(120)}{\omega\omega_1} \tag{9}$$

The Q-functions are real: thus taking the real part of (5) leads to a more explicit form of the integral equation:

$$\Psi'(\underset{\sim}{k}_1\omega_1;\ \underset{\sim}{k}_2\omega_2) = -\frac{n}{8\hbar^2}\,S\left\{\frac{Q(\underset{\sim}{k}_1\omega_1;\ \underset{\sim}{k}_2\omega_2)}{\omega_1\omega_2}\right\} + \frac{n}{8\hbar^2\pi^2}\ PP\int_{-\infty}^{+\infty}d\mu_1\int_{-\infty}^{+\infty}d\mu_2\ \frac{S\left\{\frac{Q(\underset{\sim}{k}_1\mu_1;\underset{\sim}{k}_2\mu_2)}{\mu_1\mu_2}\right\}}{(\omega_1-\mu_1)(\omega_2-\mu_2)} \tag{10}$$

Prime (') and double prime (") represent real and imaginary parts, respectively.

Noting the fact that the real part of $\hat{\underset{2}{\sigma}}$ has odd parity and the imaginary part has even parity with respect to the simultaneous sign reversals of its frequency arguments, we obtain from (8)

$$\Psi'(\underset{\sim}{k}_1\omega_1;\ \underset{\sim}{k}_2\omega_2) = \hat{\sigma}'(\underset{\sim}{k}_1\omega_1;\ \underset{\sim}{k}_2\omega_2) - \frac{1}{2}\,\hat{\sigma}'(\underset{\sim}{k}_1-\omega_1;\ -\underset{\sim}{k}\omega) - \frac{1}{2}\,\hat{\sigma}'(-\underset{\sim}{k}\omega;\ \underset{\sim}{k}_2-\omega_2) \tag{11}$$

$$+ \frac{1}{2\pi}\int_{-\infty}^{+\infty}\frac{d\mu}{\omega_1+\mu}\ \hat{\sigma}''(\underset{\sim}{k}_1\mu;\ -\underset{\sim}{k}\omega_2-\mu)$$

$$+ \frac{1}{2\pi} P \int_{-\infty}^{+\infty} \frac{d\mu}{\omega_2+\mu} \hat{\sigma}''(-\underset{\sim}{k}\omega_1-\mu;\ \underset{\sim}{k}_2\mu)$$

Next we rotate the arguments of Ψ and form the rotationally symmetrized combination

$$R[\Psi(\underset{\sim}{k}_1\omega_1;\ \underset{\sim}{k}_2\omega_2)] \equiv \Psi(\underset{\sim}{k}_1\omega_1;\ \underset{\sim}{k}_2\omega_2) + \Psi(\underset{\sim}{k}_2\omega_2;\ -\underset{\sim}{k}-\omega) + \Psi(-\underset{\sim}{k}-\omega;\ \underset{\sim}{k}_1\omega_1) \tag{12}$$

which now possesses the same "triangle-symmetry" as S {Q(012)}.

Application of the R operation to the r.h.s. of (12) eliminates the Hilbert-transform terms. It has the similar effect on the r.h.s. of (8); after considerable algebra it leaves one with the result

$$R[\hat{\sigma}'(\underset{\sim}{k}_1\omega_1;\ \underset{\sim}{k}_2\omega_2)] = -\frac{n}{4\hbar^2} S\left\{\frac{Q(\underset{\sim}{k}_1\omega_1;\ \underset{\sim}{k}_2\omega_2)}{\omega_1\omega_2}\right\} \tag{13}$$

Alternative forms of the QFDT can be obtained by trading, on the one hand, the conductivity for the polarizability α or for the density response function χ, (Kalman, 1978) and, on the other hand, the current three-point function for the density three-point functions S(120) etc. defined in (5), and by working in terms of the "internal" response functions (Kalman, 1978; Golden and Kalman, 1979). Then

$$S\{\omega\ S(\underset{\sim}{k}_1\omega_1;\ \underset{\sim}{k}_2\omega_2)\} = -\frac{\hbar^2}{n\pi} \frac{k_1k_2k}{e^3} \operatorname{Im} \frac{1}{\varepsilon(\underset{\sim}{k}_1\omega_1)\ \varepsilon(\underset{\sim}{k}_2\omega_2)\ \varepsilon^*(\underset{\sim}{k}\omega)} \tag{14}$$

$$\times\ R[\omega\alpha\ (\underset{\sim}{k}_1\omega_1;\ \underset{\sim}{k}_2\omega_2)]$$

$$= \frac{4\hbar^2}{n} \operatorname{Re} \frac{1}{\varepsilon(\underset{\sim}{k}_1\omega_1)\ \varepsilon(\underset{\sim}{k}_2\omega_2)\ \varepsilon^*(\underset{\sim}{k}\omega)} R[\omega\chi\ (\underset{\sim}{k}_1\omega_1;\ \underset{\sim}{k}_2\omega_2)]$$

We now demonstrate that the classical $h\to0$ limit (which is manifestly equivalent to the high temperature $\beta\to0$ limit) of Eq. (4) reproduces the classical result, which has been known for some time (Golden, Kalman and Silevitch, 1972; Sitenko, 1978). In order to accomplish this, we need an expansion of Eq. (7) to order $\hbar^2$, which then yields

$$S\{\omega S(\underset{\sim}{k}_1\omega_1;\underset{\sim}{k}_2\omega_2)\} = -\ \omega\omega_1\omega_2 \quad \frac{e^{\beta\hbar\omega_2}S(012) + e^{\beta\hbar\omega_1}S(021)}{\omega_1\omega_2}$$

$$- \frac{e^{\beta\hbar\omega}S(021) + S(012)}{\omega\omega_2}$$

$$- \frac{e^{\beta\hbar\omega}S(012) + S(021)}{\omega\omega_1}$$

$$= \frac{\beta^2\hbar^2}{2} \omega\omega_1\omega_2 \left[S(012) + S(021)\right] \tag{15}$$

and thus

$$R\left[\frac{\hat{\alpha}''(\underset{\sim}{k}_1\omega_1;\underset{\sim}{k}_2\omega_2)}{\omega_1\omega_2}\right] = \frac{\hat{\alpha}''(\underset{\sim}{k}_1\omega_1;\underset{\sim}{k}_2\omega_2)}{\omega_1\omega_2} - \frac{\hat{\alpha}''(\underset{\sim}{k}_1\omega_1;-\underset{\sim}{k}-\omega)}{\omega\omega_1} - \frac{\hat{\alpha}''(\underset{\sim}{k}_2\omega_2;-\underset{\sim}{k}-\omega)}{\omega\omega_2}$$

$$= \frac{\pi\beta^2 e^3 n}{kk_1k_2} S(\underset{\sim}{k}_1\omega_1;\underset{\sim}{k}_2\omega_2) \tag{16}$$

This is exactly the classical result given by Golden, Kalman and Silevitch (1972). In terms of the density response function χ

$$S(\underset{\sim}{k}_1\omega_1;\underset{\sim}{k}_2\omega_2) = -\frac{4}{\beta^2 n} \mathrm{Re} \frac{1}{\varepsilon(\underset{\sim}{k}_1\omega_1)\, \varepsilon(\underset{\sim}{k}_2\omega_2)\, \varepsilon^*(\underset{\sim}{k}\omega)}$$

$$\times \frac{\chi'(\underset{\sim}{k}_1\omega_1;\underset{\sim}{k}_2\omega_2)}{\omega_1\omega_2} - \frac{\chi'(\underset{\sim}{k}_2\omega_2;-\underset{\sim}{k}-\omega)}{\omega\omega_2} - \frac{\chi'(-\underset{\sim}{k}-\omega;\underset{\sim}{k}_1\omega_1)}{\omega\omega_1} \tag{17}$$

Eq. (17) above reproduces the result given by (Kalman, 1978) with the exception of the coefficient $4/\beta^2 n$ which is incorrectly given as $2/\beta^2$ in that reference.

We now list the second relationship ("FDT of the second kind") relating a "response function of the second kind", defined below, to the equilibrium three-point functions. The response function of the second kind Ξ $(\underset{\sim}{k}_1\omega_1;\ \underset{\sim}{k}_2\omega_2)$ is defined as the response of the system in terms of its perturbed correlations to the external perturbation:

$$\frac{1}{2}\{\langle n_{\underset{\sim}{k}_1}(\omega_1)\, n_{\underset{\sim}{k}_2}(\omega_2)\rangle^{(1)} + \langle n_{\underset{\sim}{k}_2}(\omega_2) n_{\underset{\sim}{k}_1}(\omega_1)\rangle^{(1)}\}$$

$$- n_{\underset{\sim}{k}}(\omega)\{\delta \underset{\sim}{k}_1 \delta(\omega_1) + \delta \underset{\sim}{k}_2 \delta(\omega_2) \tag{18}$$

$$= \frac{1}{V}\, \Xi\ (\underset{\sim}{k}_1\omega_1;\ \underset{\sim}{k}_2\omega_2)\, \hat{\Phi}(\underset{\sim}{k}\omega)$$

$$\underset{\sim}{k} = \underset{\sim}{k}_1 + \underset{\sim}{k}_2 \quad \omega = \omega_1 + \omega_2$$

The FDT of the second kind now states

$$\Xi''(\underset{\sim}{k}_1\omega_1;\ \underset{\sim}{k}_2\omega_2) = -\frac{1}{8\pi\hbar}\{S(120)-S(012)+S(210)-S(012)\} \tag{19}$$

Of particular interest is the classical limit for equal times:

$$\Xi(\underset{\sim}{k}_1,\underset{\sim}{k}_2;\omega) = i\beta\omega \int \frac{d\nu_1}{2\pi} \int \frac{d\nu_2}{2\pi}\, \delta_+(\omega+\nu_1+\nu_2)\ S(012) \tag{20}$$

III. FORMALISM

The philosophy of the approach developed (Golden, Kalman and Silevitch, 1974; Kalman, 1978; Golden, 1978; Golden and Kalman, 1979) can be best understood by comparing it with the way the most prominent mean field theory (MFT), due to Singwi, Tosi, Land and Sjolander (STLS) (Singwi, Tosi, Land and Sjolander, 1968; Singwi, Sjolander, Tosi and Land,

1969, 1970) can be derived. In this latter case, one starts with the linearized first BBGKY equation in the presence of external perturbation

$$\left(\frac{\partial}{\partial t} + \underset{\sim}{v}_1 \cdot \frac{\partial}{\partial \underset{\sim}{x}_1}\right) F^{(1)}(1) - \frac{e}{m} \hat{\underset{\sim}{E}}(1) \cdot \frac{\partial}{\partial \underset{\sim}{v}_1} F^{(0)}(1)$$

$$= \frac{1}{m} \int \underset{\sim}{K}(12) \cdot \frac{\partial}{\partial \underset{\sim}{v}_1} G^{(1)}(12)\, d2 \tag{21}$$

where $\hat{\underset{\sim}{E}}$ is the external electric field, $\underset{\sim}{K}(12)$ the microscopic Coulomb-field acting between particles 1 and 2, F(1) and G(12) are the one-particle and two-particle distribution functions and the superscripts $^{(0)}$ and $^{(1)}$ refer to unperturbed and to first order perturbed quantities. $G^{(1)}(12)$ can be decomposed as

$$G^{(1)}(12) = \left\{ F^{(0)}(1)\, F^{(1)}(2) + F^{(1)}(1)\, F^{(0)}(2) \right\}$$
$$\times \left\{ 1 + g^{o}(12) \right\} \tag{22}$$
$$+ \Gamma^{(1)}(12)$$

where $g^{o}(12)$ is the pair equilibrium correlation function and $\Gamma^{(1)}(12)$ is the "irreducible" part of the correlation function. The STLS MFT is obtained by neglecting this latter contribution entirely. The truncated equation then can be used to calculate $\varepsilon(\underset{\sim}{k}\omega)$ which obviously becomes a functional of $g^{(0)}{}_{\underset{\sim}{k}}$: $\varepsilon = \varepsilon \left\{g^{(0)}\right\}$. The scheme can be made self-consistent through the application of the linear Fluctuation-Dissipation Theorem. The result is a dielectric function, with an effective static potential, which leads to the deficiencies already noted.

Our theory, deviates from the MFT primarily by retaining $\Gamma^{(1)}(12)$ in (22). Nevertheless, in order to arrive at a self-consistent scheme, an approximation is needed. This is provided by the "velocity average approximation" (VAA), which amounts to the replacement

$$G\,(\underset{\sim}{x}_1\underset{\sim}{v}_1\,;\,\underset{\sim}{x}_2\underset{\sim}{v}_2) \rightarrow \tag{23}$$

$$\frac{F(\underset{\sim}{x}_1\underset{\sim}{v}_1)\, F(\underset{\sim}{x}_2\underset{\sim}{v}_2)}{n(\underset{\sim}{x}_1)\, n(\underset{\sim}{x}_2)} \int d\underset{\sim}{v}_1{}'\, d\underset{\sim}{v}_2{}'\, G\,(\underset{\sim}{x}_1\underset{\sim}{v}_1{}';\,\underset{\sim}{x}_2\underset{\sim}{v}_2{}')$$

$$= \frac{F(\underset{\sim}{x}_1\underset{\sim}{v}_1)\, F\,(\underset{\sim}{x}_2\underset{\sim}{v}_2)}{n(\underset{\sim}{x}_1)\, n(\underset{\sim}{x}_2)} \left\{ \langle n(\underset{\sim}{x}_1)\, n(\underset{\sim}{x}_2)\rangle - \delta(\underset{\sim}{x}_1-\underset{\sim}{x}_2)\, n(\underset{\sim}{x}_1) \right\}$$

The correlation of density fluctuation $n(\underset{\sim}{x})$ in this expression has to be calculated, in general, to arbitrary order, i.e. over the perturbed distribution function: thus to first order one has to deal with

$\langle n(\underset{\sim}{x}_1)\ n(\underset{\sim}{x}_2)\rangle^{(1)}$

The further reduction of $\langle n(\underset{\sim}{x}_1)\ n(\underset{\sim}{x}_2)\rangle^{(1)}$ takes place with the aid of the quadratic or nonlinear Fluctuation Dissipation Theorem (NLFDT) discussed in the previous Section.

Expressing $\langle nn\rangle^{(1)}$ in Fourier transform language with the aid of (20) we find that the density response function can be expressed in the form

$$\hat{\chi}(\underset{\sim}{k}\omega) \equiv \frac{\chi(\underset{\sim}{k}\omega)}{\varepsilon(\underset{\sim}{k}\omega)} = \hat{\chi}^{0}(\underset{\sim}{k}\omega)\left[1+\hat{v}(\underset{\sim}{k}\omega)\right] \tag{24}$$

Here $\hat{\chi}^{0}(k\omega)$ designates the RPA (Vlasov) value and

$$\hat{v}(\underset{\sim}{k}\omega) = \frac{\kappa^2}{k^2}\frac{1}{N\beta}\sum_{\underset{\sim}{q}}\frac{\underset{\sim}{k}\cdot\underset{\sim}{q}}{q^2}\int d\mu\ \Xi\ (\underset{\sim}{k}-\underset{\sim}{q},\mu;\ \underset{\sim}{q},\mu) \tag{25}$$

$$\kappa^2 = 4\pi e^2 n\beta$$

Further reduction takes place by a repeated application of the NLFDT expresses the three-point function in terms of quadratic polarizabilities. Considerable algebra leads to

$$\hat{v}\ (\underset{\sim}{k}\omega) = i\ \frac{\kappa^2}{k^2}\frac{1}{N}\sum_{\underset{\sim}{p}}\frac{\underset{\sim}{k}\cdot\underset{\sim}{p}}{p^2}\int_{-\infty}^{+\infty} d\mu\ \delta_-\ (\mu) \tag{26}$$

$$\times\ \{\ \hat{\alpha}\ (\underset{\sim}{p}\ \mu,\ \underset{\sim}{k}-\underset{\sim}{p}\ \omega-\mu) + \hat{\alpha}(\underset{\sim}{p}\ \omega-\mu,\ \underset{\sim}{k}-\underset{\sim}{p}\ \mu)\}$$

Combination of (26) and (24) now provides a link between the linear and quadratic polarizabilities. From this a self-consistent relationship for the polarizability can be generated, if a closure is provided by independently reducing the quadratic polarizability $\underset{2}{\alpha}$ to a combination of linear polarizabilities. This is accomplished by an additional approximation, the dynamical superposition approximation (DSA). This approximation is based upon the $k/\omega \to 0$, $\gamma \to 0$ behavior of $\underset{2}{\alpha}(\underset{\sim}{k}-\underset{\sim}{q},\ \mu;\ \underset{\sim}{q},\ \nu)$ which allows one to decompose $\underset{2}{\alpha}$ into clusters of α. This structure then, which is exact in the above limit, is adopted as an approximation in the $k/\omega \to 0$, but arbitrary γ situation: this is the DSA, whose resemblance to the customary superposition approximation (for the triplet correlation function in terms of doublet correlations) in equilibrium statistical mechanics can be noted. The resulting expression for $\hat{v}(\underset{\sim}{k}\omega)$ is

$$\hat{v}(\underset{\sim}{k}\omega) = \frac{1}{\varepsilon(\underset{\sim}{k}\ \omega)}\frac{\omega_0^2}{\omega^2}\frac{k^2}{\kappa^2}\frac{1}{N}\frac{4}{15}\beta E_{corr} \tag{27}$$

$$- \frac{6}{5} \frac{\omega_0^2}{\omega^2} \frac{k^2}{\kappa^2} \sum_{\underset{\sim}{p}} \int_{-\infty}^{+\infty} d\mu \delta_-(\mu) \hat{\alpha}(\underset{\sim}{p}, \omega-\mu) \hat{\alpha}(\underset{\sim}{p}\mu)$$

(where E_{corr} is the correlation energy of the system). (24) and (27) together now constitute a self-consistent approximation scheme for $\alpha(\underset{\sim}{k}\omega)$, valid for arbitrary coupling in the long wavelength (high frequency) limit. It should be noted that the first term in (27) is simply the (exact) coefficient of the ω^{-4} term in the high-frequency sum rule expansion of $\varepsilon(\underset{\sim}{k}\omega)$, guaranteeing the satisfaction of this sum rule.

IV. ONE-COMPONENT PLASMA

Eq. (24) and (27) constitute a highly complicated integral equation for $\alpha(\underset{\sim}{k}\omega)$. Since our main interest is in the behavior of the plasmon dispersion, we look for an approximation valid in the vicinity of ω_0. In this domain $\hat{\alpha}(\underset{\sim}{p}\mu)$ can be represented by a two-pole structure

$$\hat{\alpha}(\underset{\sim}{p}\mu) = \frac{\omega_0^2}{(\mu+\mu_{\underset{\sim}{p}}+i\nu_{\underset{\sim}{p}})(\mu-\mu_{\underset{\sim}{p}}+i\nu_{\underset{\sim}{p}})} \qquad (28)$$

$$\mu_{\underset{\sim}{p}} = \omega_0 \{1 + A(\gamma) (\tfrac{p}{\kappa})^2\}$$

$$\nu_{\underset{\sim}{p}} = \omega_0 B(\gamma) (\tfrac{p}{\kappa})^2$$

Thus the dispersion and damping of plasmons is represented by the coefficients $A(\gamma)$ and $B(\gamma)$, respectively.

By adopting the above structure for $\hat{\alpha}$ in the integral equation, it reduces to an algebraic equation for the unkonwn functions $A(\gamma)$ and $B(\gamma)$. We impose the "boundary conditions"

$$A(0) = 3/2 \qquad B(0) = 0$$
$$A(\gamma) = B(\gamma) = 0 \text{ for } p > p_c \qquad (29)$$

The maximum wavenumber cutoff p_c is required on physical grounds. Such an assumed behavior for B is closer to the expected, probably exponentially, vanishing damping at high frequencies than an unbounded increase. The more direct method of introducing a frequency, rather than wavenumber dependent cutoff leads to almost insurmountable algebraic difficulties in the resulting integral equation. The vanishing of A, for $p > p_c$ assures that even for negative dispersion no unphysical negative frequencies are generated.

The resulting coupled algebraic equations for $A(\gamma;p_c)$ and $B(\gamma;p_c)$ have been solved numerically (Carini, Golden and Kalman, 1979; Carini and Kalman, 1984). Physical considerations dictate p_c to be of the order of the inverse Landau length $(e^2\beta)^{-1}$, i.e. it is coupling dependent; best agreement with computer data was obtained by choosing $p_c = 0.77(e^2\beta)^{-1}$. The solution obtained exhibits some noteworthy features: (1) The dispersion coefficient $A(\gamma)$ monotonically decreases from its Bohm-Gross value 3/2, as γ increases from zero, in agreement with the perturbation calculations in the weak coupling limit. (2) $A(\gamma)$ is in very good agreement with MD data up to $\gamma = 52$ and in reasonably good agreement for larger γ values up to $\gamma = 3258$ ($\Gamma = 152$). (3) The critical γ value, at which the dispersion becomes negative is found to be $\gamma_c = 45$ ($\Gamma_c = 8.8$). (4) For $\gamma \to \infty$ the sum rule term dominates, and $A(\gamma) \sim \gamma^{2/3}$; this also indicates that in the high frequency sum rule expansion for $\gamma \to \infty$ the coefficient of the ω^{-4} term becomes dominant. (5) The comparison with MD data for the damping coefficient $B(\gamma)$ demonstrates less satisfactory agreement, which is partly due to the inherent scatter and inaccuracy of the MD data for the half width of the plasmon peak. (6) $B(\gamma)$ exhibits a peak around $\gamma \approx 52$, in agreement with a recent suggestion of Hansen (1981).

All these features can be observed in Figs. 1, 2 and 3. Finally, it

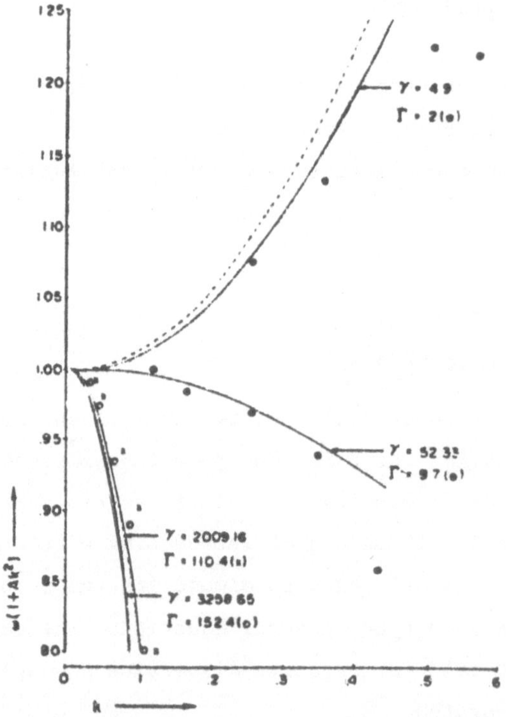

Fig. 1. Comparison of our results for the dispersion coefficient $A(\gamma)$ with the MD data of HPM for $p_c = 0.77$. The full lines are plots of the real part of the dispersion equation $\omega = 1 + A(\gamma)k^2$ for the values of γ indicated. The MD data are represented by (•xo) for the same γ values. The dashed line is the Bohm-Gross dispersion curve for $\gamma = 0(A(0)) = 3/2$

should be noted that the introduction of the cutoff wavenumber p_c is critical to obtain good agreement with MD data; in an earlier version of the work with $p_c \to \infty$ the agreement was markedly poorer.

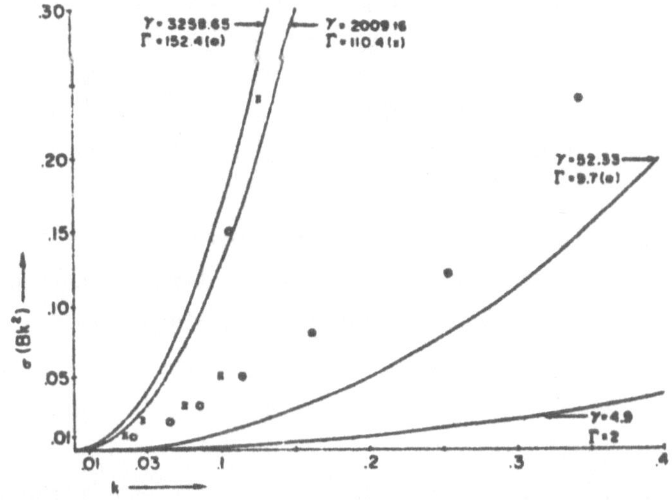

Fig. 2. Comparison of our results for the damping coefficient $A(\gamma)$ with the MD data of HPM for $p_c = 0.77$. The full lines are plots of the imaginary part of the dispersion equation $B = B(\gamma)k^2$ for the values of γ indicated. The MD data are represented by (●xo) for the same γ values.

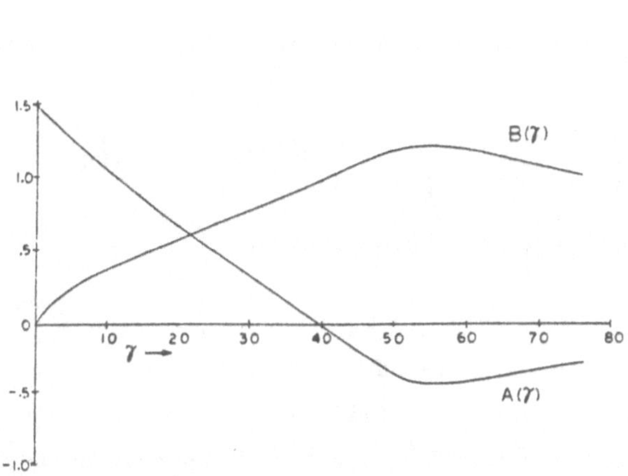

Fig. 3. Plot of the dispersion coefficient, $A(\gamma)$, and damping coefficient, $B(\gamma)$, as a function of γ when the maximum wavenumber, p_c, is scaled proportional to the inverse Landau length, $(e^2\beta)^{-1}$.

V. BINARY IONIC MIXTURES

In a binary ionic mixture two distinct species of positively charged particles are immersed in an inert neutralizing background. The system is obviously more complex than the simple OCP and requires an appropriate generalization of the formalism. As to the NLFDT, this has been done recently (Golden and Lu, 1982). Furthermore, the formalism has to be generated in terms of partial response functions (χ_{AB}), rather than in terms of the physical response functions. These latter represent the response of a species (A) to the perturbation exerted by the other species (B) (Kalman and Golden, 1984). Now the equivalent of (24) is the matrix relationship

$$\chi_{AB}(\underset{\sim}{k}\omega) = \sum_{\bar{C}} \chi^{o}{}_{A\bar{C}}(k\omega)\ \{\delta_{\bar{C}B} + \hat{v}_{\bar{C}B}(k\omega)\} \tag{30}$$

The rest of the formal program can be carried out along this line, but it results in an expression of much greater complexity than the equivalent OCP relation (27) (Golden, Green and Neilson, 1985a; 1985b). Instead of the general Γ-dependence which has not been evaluated yet, the plasmon dispersion has been studied in the $\gamma<<1$ and $\Gamma>>1$ limits. We now quote the principal results of this analysis.

(1) In general, there is a shift in the plasmon frequency at $k = 0$ from its $\gamma = 0$ value $\Omega = (\Sigma\ 4\pi Z_A{}^2e^2n_A{}^2/m_A)^{1/2}$ to $\Omega + \Delta\Omega$. (2) For $\gamma<<1$ $\Delta\Omega$ has been evaluated for 50%-50% H^+-He^{++} mixture with the result

$$\Delta\Omega = 0.008\gamma\Omega = 0.053\Gamma^{3/2}\Omega. \tag{31}$$

In contrast to the OCP case, where a comnparison with the exact perturbative result to $O(\gamma)$ is feasible (Carini, Golden and Kalman, 1982), no such standard is available at the present time. However, the result can be compared with the value given by Baus (1978): $\Delta\Omega = 0.08\Gamma^{3/2}\Omega$. (3) For $\Gamma>>1$, the frequency shift is given solely by the sum-rule contribution (cf. Eq. (27).

$$\Delta\Omega = \left\{\frac{1}{2}\ \left[1+\left(1+\frac{4}{3}\ \frac{\Omega_1{}^2\ \Omega_2{}^2}{\Omega^4}\ \left[\frac{Z_1}{m_1} - \frac{Z_2}{m_2}\right]^2\right)^{1/2}\right]^{1/2} - 1\right\}\Omega \tag{32}$$

For the same H^+-He^{++} mixture, this provides

$$\Delta\Omega = 0.0198\Omega \tag{33}$$

in good agreement with molecular dynamics data. Note that in the $\Gamma\to\infty$ limit $\Delta\Omega$ is Γ-independent. (4) Similarly to the OCP, there is a trend for the coefficient A in $\omega = (\Omega+\Delta\Omega)(1 + A\ k^2/\kappa^2)$ to change from positive (for $\gamma<<1$) to a negative value (for $\Gamma>>1$). However, whether A indeed becomes negative, depends also on the Z_1/Z_2, m_1/m_2 and n_1/n_2 ratios.

VI. CONCLUSIONS

The dynamical mean field theory scheme, based on the application of NLFDT, on the velocity average approximation, and on the dynamical superposition approximation, when applied to a variety of dynamical systems has provided (see also Golden, this Volume) reasonably good results for the plasmon dispersion over a wide range of γ-values. The establishment of the quantum NLFDT now opens the way towards the application of the scheme to the analysis of the collective motions in degenerate systems (3-d and 2-d electrongas, electron-hole liquids).

ACKNOWLEDGEMENTS

Special thanks are due to Xiao-yue Gu, who has greatly contributed to the preparation of Section II of this paper. This work was partially supported by NSF Grant ECS-8315655.

REFERENCES

Baus, M., 1978, Phys. Rev. Lett., 40:793.

Bochkov, A. N. and Kuzovlev, Yu. E., 1977, Zh. Eksp. Teor. Fiz., Z2:238 [Sov. Phys. JETP, 1977, 45:125].

Callen, H. B. and Welton, T. A., 1951, Phys. Rev., 83:34.

Carini, P., Golden, K. I. and Kalman, G., 1980, Phys. Lett., 78A:450.

Carini, P., Kalman, G. and Golden, K. I., 1982, Phys. Rev., A26:1686.

Carini, P., and Kalman, G., 1984, Phys. Lett., 105A:229.

Efremov, A. F., 1968, Zh. Eksp. Teor. Fiz., 55:2322 [Sov. Phys. JETP, 1969, 28:1232].

Golden, K. I., This Volume.

Golden, K. I. 1978, in "Strongly Coupled Plasmas", edited by G. Kalman, Plenum Press, New York.

Golden, K. I. and Lu, Dexin, 1982, J. Stat. Phys., 29:281.

Golden, K. I., Green, F. and Neilson, D., 1985a, Phys. Rev., A31:3529; 1985b, Phys. Rev., A32:1669.

Golden, K. I. and Kalman, G., 1969, J. Stat. Phys., 1:415.

Golden, K. I., Kalman, G. and Silevitch, M. B., 1972, J. Stat. Phys., 6:87.

Golden, K. I. and Kalman, G., 1974, Phys. Rev. Lett., 33:1544.

Golden, K. I. and Kalman, G., 1979, Phys. Rev., A19:2112.

Golden, K. I. and Kalman, G., 1982, Am. Phys., New York, 143:160.

Hansen, J.-P., 1981, J. Phys. Lett., (Paris), 42:397.

Kalman, G., 1978, in "Strongly Coupled Plasmas", edited by G. Kalman, Plenum Press, New York.

Kalman, G. and Golden, K. I., 1984, Phys. Rev. A29:844.

Kalman, G. and Gu, Xiaoyue, 1986, submitted to Phys. Rev. A.

Kubo, R., 1957, J. Phys. Soc., Japan 12:570.

Kubo, R., 1959, in "Lectures on Theoretical Physics", edited by W. E. Brittin and L. G. Dunham, Vol. I, Interscience, New York.

Kubo, R., 1966, Rep. Progr. Phys., Inst. of Phys. and the Phys. Soc., London, Part I, 29:263.

Martin, P. C., 1968, in "Many-Body Physics - Les Houches, 1967, edited by C. DeWitt, Gordon and Breach, New York.

Nyquist, H., 1928, Phys. Rev., 32:110.

Singwi, K. S., Sjolander, A., Tosi, M. P. and Land, R. H., 1969, Solid State Commun. 7:1503.

Singwi, K. S., Sjolander, A., Tosi, M. P. and Land, R. H., 1970, Phys. Rev., 131:1044.

Singwi, K. S., Tosi, M. P., Land, R. H. and Sjolander, A., 1968, Phys. Rev., 176:589.

Sitenko, A. G., 1978, Zh. Eksp. Teor. Fiz., 75:104 [Sov. Phys. JETP, 1978, 48:51].

Stratonovich, R. L., 1970, Zh. Eksp. Teor. Fiz., 58:1612 [Sov. Phys. JETP, 1970, 31:864]

COLLECTIVE MODES AND MODE COUPLING FOR A DENSE PLASMA IN A MAGNETIC FIELD

L.G. Suttorp

Institute for Theoretical Physics
University of Amsterdam
Valckenierstraat 65
1018 XE Amsterdam, The Netherlands

1. INTRODUCTION

Collective modes play an important role in the dynamical response of macroscopic systems to external disturbances. In particular, the modes with small wavenumber determine the large-scale behaviour of the dynamical structure factor.

To derive collective modes various methods are available. From the point of view of kinetic theory the modes follow from the memory kernel of the formal kinetic equation for the one-particle time correlation function in momentum space. In a recent paper [Suttorp and Cohen, 1985] the complete set of modes for a dense plasma in a magnetic field has been derived along these lines. An alternative approach to the evaluation of the mode spectrum starts from the microscopic balance equations of particle number, momentum and energy. After establishing fluctuation formulae for the densities and the flows of these quantities the mode frequencies and the associated amplitudes may be derived by using projection operator techniques. This method has been employed before to determine the collective modes for a neutral fluid [Kadanoff and Swift, 1968; Résibois, 1972] and for an unmagnetized plasma [Marchetti and Kirkpatrick, 1985]. Recently, the collective modes of a magnetized plasma have been derived in this way [Suttorp and Schoolderman, 1986].

The collective modes of a plasma in a magnetic field depend on the angle between the wave vector and the field. Transverse and longitudinal modes can no longer be distinguished. As a consequence both the mode amplitudes and the mode frequencies are given by expressions that are rather more complicated than those for an unmagnetized plasma. In particular, the viscous modes and the plasmon modes of an unmagnetized plasma merge in a set of four mixed 'gyro-plasmon' modes, if a magnetic field is turned on.

The amplitudes and the frequencies of the collective modes are essential ingredients in the theory of mode-coupling, which may be used to analyse the long-time behaviour of time correlation functions like the velocity autocorrelation function of a tagged particle. For systems of neutral particles this method is well-established [Kawasaki, 1970; Ernst e.a., 1971,1976]. Furthermore it has been employed to determine the long-time tails of the

Green-Kubo integrands of an unmagnetized plasma [Marchetti and Kirkpatrick, 1985]. In a magnetized plasma the velocity autocorrelation function depends on the direction of the velocities owing to the anisotropy of the system; one should distinguish therefore a longitudinal and a transverse velocity autocorrelation function. Recently [Suttorp and Schoolderman, 1986] mode-coupling theory has been used to derive expressions for the long-time tails of both autocorrelation functions. It has been found that these tails are dominated by the coupling with the gyro-plasmon modes.

In the following a review will be given of the derivation of the collective modes and of the long-time behaviour of the velocity autocorrelation functions for a magnetized plasma. As a model we shall adopt the classical one-component plasma, consisting of charged particles which are immersed in a neutralizing inert background and which interact through a Coulomb potential. The external magnetic field is assumed to be static and uniform in space.

2. COLLECTIVE MODES

To obtain the modes for a one-component plasma in a magnetic field one may start from the microscopic balance equations of particle number, momentum and energy. These give the time derivative of the particle density $n(\vec{k})$, the momentum density $\vec{g}(\vec{k})$ and the energy density $\varepsilon(\vec{k})$ in Fourier space. These time derivatives are conveniently written in terms of the Liouville operator L in phase space, which determines for an arbitrary function F its time derivative as $\dot{F} = iLF$.

The microscopic momentum balance equation contains, apart from a pressure term, with a pressure tensor $\overleftrightarrow{\tau}(\vec{k})$, and a consistent field term depending on the electric field generated by the charge density fluctuations, a Lorentz force term depending on the direction and the strength of the magnetic field, as given by the unit vector $\hat{B}$ and the Larmor frequency $\omega_B = eB/mc$, with e the charge and m the mass of the particles. The energy balance equation contains an energy flow $\vec{j}_\varepsilon(\vec{k})$, which is the sum of a kinetic and a potential contribution, as is the case for the pressure tensor.

The collective modes are particular linear combinations of the particle density, the momentum density and the energy density. Let $a_i(\vec{k})$ denote a set of five independent linear combinations of these quantities, with adjoints $\bar{a}_i(\vec{k})$ such that

$$\frac{1}{V} < \bar{a}_i^*(\vec{k})\ a_j(\vec{k}) > = \delta_{ij} \quad . \tag{2.1}$$

Here the brackets denote a canonical ensemble average; V is the volume of the system. In the course of time $a_i(\vec{k})$ evolves into $a_i(\vec{k},t)$ of which the Laplace transform

$$a_i(\vec{k},z) = -\,i \int_0^\infty dt\ e^{izt}\ a_i(\vec{k},t) \tag{2.2}$$

satisfies the equation

$$(z+L)a_i(\vec{k},z) = a_i(\vec{k}) \quad . \tag{2.3}$$

Introducing a projection operator P by writing

$$Pf(\vec{k}) = \sum_i \frac{1}{V} < \bar{a}_i^*(\vec{k}) f(\vec{k}) > a_i(\vec{k}) \tag{2.4}$$

for an arbitrary function $f(\vec{k})$ in phase space, one derives an equation for the hydrodynamic propagators:

$$G_{ij}(\vec{k},z) = \frac{1}{V} < \bar{a}_i^*(\vec{k}) \frac{1}{z+L} a_j(\vec{k}) > \tag{2.5}$$

in the form

$$\sum_\ell [z\delta_{i\ell} - \Omega_{i\ell}(\vec{k},z)] G_{\ell j}(\vec{k},z) = \delta_{ij} \quad . \tag{2.6}$$

The frequency matrix is given by

$$\Omega_{ij}(\vec{k},z) = \Omega_{ij}^{(1)}(\vec{k},z) + \Omega_{ij}^{(2)}(\vec{k},z) \quad , \tag{2.7}$$

where the direct and the indirect parts are

$$\Omega_{ij}^{(1)}(\vec{k},z) = -\frac{1}{V} < \bar{a}_i^*(\vec{k}) L\, a_j(\vec{k}) > \quad , \tag{2.8}$$

$$\Omega_{ij}^{(2)}(\vec{k},z) = \frac{1}{V} < \bar{a}_i^*(\vec{k}) LQ \frac{1}{z+QLQ} QL\, a_j(\vec{k}) > \quad , \tag{2.9}$$

with $Q = 1 - P$.

The collective mode frequencies follow as the eigenfrequencies of the frequency matrix for small values of the wave number $\vec{k}$. The modes themselves are the corresponding eigenvectors.

For vanishing wavenumber the five mode frequencies are:

$$z_T^{(0)} = 0 \quad , \tag{2.10}$$

$$z_{\lambda\rho}^{(0)} = \rho\, w_\lambda \quad , \tag{2.11}$$

with $\lambda = \pm 1$, $\rho = \pm 1$. Here w_λ is given by:

$$w_\lambda = \tfrac{1}{2}(\omega_p^2 + \omega_B^2 + 2\omega_p\omega_B \hat{k}_{//})^{\frac{1}{2}} + \tfrac{1}{2}\lambda(\omega_p^2 + \omega_B^2 - 2\omega_p\omega_B \hat{k}_{//})^{\frac{1}{2}} \quad , \tag{2.12}$$

with $\hat{k}_{//} = \vec{k}\cdot\hat{\vec{B}}/k$ and ω_p the plasma frequency. Choosing as a basis set $k^{-1}n(\vec{k})$, $\vec{g}(\vec{k})$ and $\varepsilon(\vec{k})$, one finds for the modes up to first order in $\vec{k}$:

$$a_T(\vec{k}) = C_T(\hat{\vec{k}})[\varepsilon(\vec{k}) - hn(\vec{k})] \quad , \tag{2.13}$$

$$a_{\lambda\rho}(\vec{k}) = C_\lambda(\hat{\vec{k}})[\frac{k_D}{k} n(\vec{k}) + \frac{1}{k_B T c_V}(\frac{1}{3}c_V + \tfrac{1}{2}k_B)\frac{k}{k_D}\varepsilon(\vec{k})$$
$$+ \frac{1}{(mk_B T)^{\frac{1}{2}}} \vec{v}_{\lambda\rho}(\vec{k})\cdot\vec{g}(\vec{k})] \quad . \tag{2.14}$$

The heat mode (2.13) contains the enthalphy h per particle, which is related to the specific heat c_V and the isothermal compressibility κ_T of the plasma:

$$h = - k_B T + \frac{1}{3} c_V T + 3/(n\kappa_T) \quad , \tag{2.15}$$

with n the particle density and T the temperature. The vectors $\vec{v}_{\lambda\rho}$ occurring in the gyro-plasmon modes (2.14) are defined as:

$$\vec{v}_{\lambda\rho}(\vec{k}) = \frac{\rho w_\lambda \omega_p}{w_\lambda^2 - \omega_B^2} \hat{k}_\perp + \frac{\rho \omega_p}{w_\lambda} \hat{k}_{/\!/} - \frac{i\omega_p \omega_B}{w_\lambda^2 - \omega_B^2} \hat{k} \wedge \hat{B} \quad . \tag{2.16}$$

Here $\hat{k}$ is a unit vector in the direction of the wave vector with components parallel and perpendicular to $\hat{B}$ denoted by $\hat{k}_{/\!/}$ and $\hat{k}_\perp$, respectively. Furthermore k_D is the Debye wave vector, while $C_T(\vec{k})$ and $C_\lambda(\vec{k})$ are normalization constants that should be chosen such that (2.1) is satisfied.

The corrections to the mode frequencies that are of order k^2 follow by applying perturbation theory. Up to order k^2 one gets:

$$z_T = \frac{1}{V} < a_T^*(\vec{k}) LQ \frac{1}{z+QLQ} QLa_T(\vec{k}) > \quad , \tag{2.17}$$

$$z_{\lambda\rho} = \rho w_\lambda \left[1 + \tfrac{1}{2}k^2 c_s^2 \frac{w_\lambda^2 - \omega_B^2 \hat{k}_{/\!/}^2}{w_\lambda^2(\omega_p^2 + \omega_B^2) - 2\omega_p^2 \omega_B^2 \hat{k}_{/\!/}^2}\right]$$

$$+ \frac{1}{V} < a_{\lambda\rho}^*(\vec{k}) LQ \frac{1}{z+QLQ} QLa_{\lambda\rho}(\vec{k}) > \quad . \tag{2.18}$$

with c_s the sound velocity. The Laplace variable z stands for the zeroth-order frequency $z^{(0)}$, as given by (2.10) and (2.11).

The dependence of the mode frequencies (2.17) and (2.18) on the wave vector $\vec{k}$ can be analyzed by using the balance equations and the symmetry properties of the system. The frequency of the thermal mode is found to contain the static thermal conductivities in the longitudinal and transverse directions:

$$z_T = \frac{-ik^2}{nc_V} (\hat{k}_\perp^2 \lambda_\perp + \hat{k}_{/\!/}^2 \lambda_{/\!/}) \quad . \tag{2.19}$$

These are defined by writing

$$\hat{k}_\perp^2 \lambda_\perp + \hat{k}_{/\!/}^2 \lambda_{/\!/} = \frac{i}{k_B T^2} \lim_{z \to i0} \lim_{k \to 0} \frac{1}{Vk^2}$$

$$\times < \vec{k}\cdot\vec{j}_\varepsilon^*(\vec{k}) \, Q \frac{1}{z+QLQ} Q \, \vec{k}\cdot\vec{j}_\varepsilon(\vec{k}) > \quad . \tag{2.20}$$

The mode frequencies $z_{\lambda\rho}$ can be analyzed in a similar way. From (2.14) one obtains up to second order in $\vec{k}$:

$$\frac{1}{V} < a_{\lambda\rho}^{*}(\vec{k}) LQ \frac{1}{z+QLQ} QLa_{\lambda\rho}(\vec{k}) > =$$

$$= \frac{C_\lambda^2}{mk_BT} \frac{1}{V} < [\vec{k}\cdot\vec{\tau}(\vec{k})\cdot \vec{v}_{\lambda\rho}(\vec{k})]^{*} Q \frac{1}{z+QLQ} Q\vec{k}\cdot\vec{\tau}(\vec{k})\cdot \vec{v}_{\lambda\rho}(\vec{k}) > \quad . \qquad (2.21)$$

Employing again the symmetry properties one may write up to second order in $\vec{k}$:

$$\frac{1}{V} < \vec{k}\cdot\vec{\tau}^{*}(\vec{k}) Q \frac{1}{z+QLQ} Q\vec{k}\cdot\vec{\tau}(\vec{k}) > = - ik_BTk^2 \vec{T}(\hat{k},z) \quad , \qquad (2.22)$$

with

$$T_{ij}(\hat{k},z) = f_1(z)\delta_{ij} + f_2(z)\hat{k}_i\hat{k}_j$$
$$+ f_3(z)(\hat{k}_i\hat{B}_j + \hat{k}_j\hat{B}_i)\hat{k}\cdot\hat{B} + f_4(z)[\hat{B}_i\hat{B}_j + \delta_{ij}(\hat{k}\cdot\hat{B})^2]$$
$$+ f_5(z)\hat{B}_i\hat{B}_j(\hat{k}\cdot\hat{B})^2 + f_6(z)[\hat{k}_i(\hat{k}\wedge\hat{B})_j - (\hat{k}\wedge\hat{B})_i\hat{k}_j - \varepsilon_{ijm}\hat{B}_m]$$
$$+ f_7(z)[\hat{B}_i(\hat{k}\wedge\hat{B})_j - (\hat{k}\wedge\hat{B})_i\hat{B}_j - \varepsilon_{ijm}\hat{B}_m\hat{k}\cdot\hat{B}]\hat{k}\cdot\hat{B} \quad . \qquad (2.23)$$

The coefficients f_i depend on z and on the Larmor frequency ω_B . Instead of f_i one may introduce dynamical viscosity coefficients by writing

$$f_1 = -\eta_1 + 2\eta_2 \quad , \quad f_2 = \frac{1}{3}\eta_1 + \eta_V - 2\zeta \quad , \quad f_3 = -\eta_1 + \eta_3 + 3\zeta \quad ,$$
$$f_4 = \eta_1 - 2\eta_2 + \eta_3 \quad , \quad f_5 = 2\eta_1 + 2\eta_2 - 4\eta_3 \quad ,$$
$$f_6 = \tfrac{1}{2}\eta_4 \quad , \quad f_7 = -\tfrac{1}{2}\eta_4 - \eta_5 \quad . \qquad (2.24)$$

The coefficients $\eta_1,\ldots,\eta_5$ are the shear viscosities, η_V is the volume viscosity, while ζ describes a cross effect between shear stresses and volume strains and vice versa [de Groot and Mazur, 1962].

Substituting (2.16) and (2.22), with (2.23) and (2.24), into (2.18) with (2.21) we obtain the mode frequencies:

$$z_{\lambda\rho} = \rho\, w_\lambda[1 + \tfrac{1}{2} k^2 c_s^2 \frac{w_\lambda^2 - \omega_B^2 \hat{k}_{//}^2}{w_\lambda^2(\omega_p^2+\omega_B^2)-2\omega_p^2\omega_B^2\hat{k}_{//}^2}]$$
$$+ \frac{k^2}{2nm[w_\lambda^2(\omega_p^2+\omega_B^2)-2\omega_p^2\omega_B^2\hat{k}_{//}^2]} \{ \rho w_\lambda^3\omega_B[2(\eta_4+\eta_5)\hat{k}_{//}^2-2\eta_4]$$
$$+ iw_\lambda^2\{\omega_p^2[-2(\eta_1+\eta_2-2\eta_3)\hat{k}_{//}^4 + 2(2\eta_2-2\eta_3-3\zeta)\hat{k}_{//}^2 + \frac{2}{3}\eta_1-2\eta_2-\eta_V+2\zeta]$$
$$+ \omega_B^2[(-\frac{5}{3}\eta_1+4\eta_2-2\eta_3+\eta_V-2\zeta)\hat{k}_{//}^2 + \frac{5}{3}\eta_1-4\eta_2-\eta_V+2\zeta]\}$$
$$+ \rho w_\lambda\,\omega_p^2\omega_B\,\hat{k}_{//}^2[-2(\eta_4+2\eta_5)\hat{k}_{//}^2 + 2(\eta_4+\eta_5)]$$
$$+ i\omega_p^2\omega_B^2\,\hat{k}_{//}^2[(3\eta_1-4\eta_2+\eta_3+6\zeta)\hat{k}_{//}^2 - \frac{5}{3}\eta_1+4\eta_2+\eta_3+\eta_V-2\zeta]\} \quad , \qquad (2.25)$$

where the viscosities η_i, η_V, ζ are to be evaluated at $z = \rho\, w_\lambda$.

The expressions (2.19) and (2.25) for the mode frequencies may be compared to those obtained by alternative methods. A macroscopic magnetohydrodynamical treatment leads to expressions for the mode frequencies of a similar form. However, they contain phenomenological transport coefficients that are static real quantities, defined at frequency zero. On the other hand, in (2.25) dynamical complex-valued viscosities at the finite frequency $\rho\, w_\lambda$ show up.

Another method to derive the mode spectrum is furnished by the kinetic theory for time correlation functions. The results obtained in this way contain frequency-dependent transport coefficients that are given in terms of matrix elements of a kinetic kernel. A detailed comparison [Suttorp and Schoolderman, 1986] shows that these frequency-dependent transport coefficients do not coincide with those introduced in the present treatment. The reason is that the projection operator used in kinetic theory has no simple relation to that defined in (2.4).

In a recent paper [Marchetti, Kirkpatrick and Dorfman, 1984] expressions for the frequencies of the oscillating modes of a strongly magnetized plasma have been presented. However, the terms of order k^2 are not given explicitly in terms of the seven anisotropic viscosity coefficients, so that a comparison is difficult. Expressions for the modes that might be compared to (2.13) and (2.14) are not given either.

3. MODE COUPLING AND LONG-TIME TAILS OF THE VELOCITY AUTOCORRELATION FUNCTION

The velocity autocorrelation function of a tagged particle is defined as

$$F(\vec{k},t) = \lim_{\vec{k} \to 0} \frac{1}{k^2} < \vec{k} \cdot \frac{\vec{g}_s^{\,*}(\vec{k})}{m}\, e^{iLt}\, \vec{k} \cdot \frac{\vec{g}_s(\vec{k})}{m} > \quad , \tag{3.1}$$

where the tagged-particle momentum density is given by

$$\vec{g}_s(\vec{k}) = \vec{p}_s\, e^{-i\vec{k}\cdot\vec{r}_s} \quad , \tag{3.2}$$

with $\vec{r}_s$ and $\vec{p}_s$ the position and momentum of the particle. The autocorrelation function (3.1) is anisotropic, as it depends on the angle between the wave vector $\vec{k}$ and the magnetic field $\vec{B}$. In fact, one may write:

$$F(\vec{k},t) = \hat{k}_{/\!/}^{2}\, F_{/\!/}(t) + \hat{k}_{\perp}^{2}\, F_{\perp}(t) \quad , \tag{3.3}$$

which defines the longitudinal and the transverse velocity autocorrelation functions $F_i(t)$, with $i = /\!/\, , \perp$.

The long-time behaviour of the velocity autocorrelation functions $F_i(t)$ can be determined if one assumes it to be adequately described by mode-coupling theory. According to mode-coupling theory the long-time behaviour of the velocity autocorrelation function is dominated by contributions originating from the coupling of the tagged-particle momentum density to the product of a collective mode and of the tagged-particle density:

$$F(\vec{k},t) \simeq \lim_{\vec{k}\to 0} \frac{1}{k^2 V} \sum_i \sum_{\vec{q}} |A_i(\vec{k},\vec{q})|^2 \, e^{-i[z_i(\vec{q}) + z_s(\vec{k}-\vec{q})]t} \quad . \qquad (3.4)$$

The sums are extended over the five collective modes (with label $i = T, \lambda\rho$) and over all values of the wave vector $\vec{q}$ of these modes. The amplitudes A_i are given as:

$$A_i(\vec{k},\vec{q}) = \langle a_i^*(\vec{q}) \, a_s^*(\vec{k}-\vec{q}) \, \vec{k} \cdot \frac{\vec{g}_s(\vec{k})}{m} \rangle \quad . \qquad (3.5)$$

The collective modes $a_i(\vec{q})$ and the corresponding frequencies $z_i(\vec{q})$ have been discussed in the first part of this paper. The tagged-particle density mode:

$$a_s(\vec{q}) = e^{-i\vec{q}\cdot\vec{r}_s} \qquad (3.6)$$

is a dissipative mode, with a frequency

$$z_s(\vec{q}) = - iq_{/\!/}^2 D_{/\!/} - iq_\perp^2 D_\perp \qquad (3.7)$$

that is determined by the longitudinal and the transverse self-diffusion constants, $D_{/\!/}$ and $D_\perp$.

As a consequence of the symmetry of the momentum integration, which is implied in the average (3.5), only the gyro-plasmon modes, with $i = \lambda\rho$, contribute to the mode-coupling expression (3.4). The corresponding amplitudes are easily evaluated with the help of (2.14):

$$A_{\lambda\rho}(\vec{k},\vec{q}) = C_\lambda(\hat{q})(k_B T/m)^{\frac{1}{2}} \, \vec{v}_{\lambda\rho}^{\,*}(\vec{q}) \cdot \vec{k} \quad . \qquad (3.8)$$

The mode frequencies $z_{\lambda\rho}(\vec{q})$ have the form:

$$z_{\lambda,1}(\vec{q}) = w_\lambda(\hat{q}) - iq^2 D_\lambda(\hat{q}) \quad , \quad z_{\lambda,-1}(\vec{q}) = - [z_{\lambda,1}(\vec{q})]^* \quad , \qquad (3.9)$$

with (complex) damping coefficients D_λ . Inserting (3.8) and (3.9) in (3.4), taking the limit $\vec{k} \to 0$, averaging over the azimuthal angle of $\vec{q}$ (in a spherical coordinate system with a polar axis in the direction of the magnetic field) and integrating over $|\vec{q}|$ we obtain an expression for $F_i(\vec{k},t)$, or, with the help of (3.3), for $F_i(t)$:

$$F_i(t) \simeq \frac{k_B T}{m(4\pi t)^{3/2}} \sum_{\lambda = \pm 1} \mathrm{Re} \int_{-1}^{1} d\hat{q}_{/\!/} \, C_\lambda^2(\hat{q}) \Phi_i(\hat{q})$$

$$\times \frac{e^{-iw_\lambda(\hat{q})t}}{[\hat{q}_{/\!/}^2 D_{/\!/} + \hat{q}_\perp^2 D_\perp + D_\lambda(\hat{q})]^{3/2}} \quad . \qquad (3.10)$$

Here we introduced the abbreviations:

$$\Phi_{/\!/}(\hat{q}) = \frac{\bar{\omega}_p^2}{w^2} \hat{q}_{/\!/}^2 \quad , \qquad (3.11)$$

$$\Phi_\perp(\hat{\vec{q}}) = \frac{1}{2} \frac{(w^2 + \omega_B^2)\omega_p^2}{(w^2 - \omega_B^2)^2} \hat{q}_\perp^2 \quad , \tag{3.12}$$

with $w = w_\lambda(\hat{\vec{q}})$.

Let us consider first the contribution $F_i^{(+)}$ of the modes with $\lambda = +1$. Choosing the new integration variable $w = w_\lambda(\hat{\vec{q}})$ we obtain:

$$F_i^{(+)}(t) \simeq \frac{k_B T}{nm(4\pi t)^{3/2}} \operatorname{Re} \int_{w_1}^{w_2} dw \frac{w^2 - \omega_B^2}{\omega_p \omega_B (\omega_p^2 + \omega_B^2 - w^2)^{\frac{1}{2}}} \Phi_i \frac{e^{-iwt}}{(\hat{q}_{/\!/}^2 D_{/\!/} + \hat{q}_\perp^2 D_\perp + D_1)^{3/2}} \quad , \tag{3.13}$$

with $w_1 = \omega_M \equiv \mathrm{Max}(\omega_p, \omega_B)$ and $w_2 = \omega_0 \equiv (\omega_p^2 + \omega_B^2)^{\frac{1}{2}}$. In the integrand one should substitute:

$$\hat{q}_{/\!/} = \frac{w}{\omega_p \omega_B} (\omega_p^2 + \omega_B^2 - w^2)^{\frac{1}{2}} \tag{3.14}$$

and $\hat{q}_\perp = (1 - \hat{q}_{/\!/}^2)^{\frac{1}{2}}$.

The integrand in (3.13) is a regular nonvanishing function for all w in the open interval (w_1, w_2). At the upper boundary of the integration domain the integrand is proportional to $(w_2 - w)^{\frac{1}{2}}$ for $i = /\!/$ and to $(w_2 - w)^{-\frac{1}{2}}$ for $i = \perp$. At the lower boundary it is proportional to $(w - \omega_B)$ for $i = /\!/$ and to $(w - \omega_p)$ for $i = \perp$.

For large values of t the contribution of the interior of the integration domain in (3.13) may be disregarded. In fact, as a consequence of the phase factor exp(-iwt) destructive interference damps all contributions from the interior region. The main contributions to the asymptotic expression for the integral originate from the boundaries of the integration domain, since there the interference is not completely destructive. One may derive the following asymptotic expression for $F_i^{(+)}(t)$:

$$F_i^{(+)}(t) \simeq A_{2,i}\, t^{-\nu_{2,i}} \cos(\omega_0 t + \theta_{2,i}) - A_{1,i}\, t^{-\nu_{1,i}} \cos(\omega_M t + \theta_{1,i}) \quad , \tag{3.15}$$

valid for large t. The indices 1,2 indicate contributions from the boundaries at w_1, w_2, respectively. The exponent $\nu_{2,i}$ equals 3 for $i = /\!/$ and 2 for $i = \perp$. The other exponent $\nu_{1,i}$ depends on the relative magnitude of ω_p and ω_B ; for $\omega_p > \omega_B$ one has $\nu_{1,/\!/} = \frac{5}{2}$, $\nu_{1,\perp} = \frac{7}{2}$, while for $\omega_p < \omega_B$ these values are interchanged.

Likewise one may derive the asymptotic expression for the contribution $F_i^{(-)}(t)$ that results by taking $\lambda = -1$ in (3.10). As before the dominant contribution to the integral for large t stems from the boundaries of the integration domain. From the behaviour of the integrand at these boundaries one finds:

$$F_i^{(-)}(t) \propto t^{-\nu_i} \cos(\omega_m + \theta_i) \quad , \tag{3.16}$$

with a frequency $\omega_m = \mathrm{Min}(\omega_p, \omega_B)$ and exponents ν_i that are related to $\nu_{1,i}$ in (3.15) as $\nu_{/\!/} = \nu_{1,\perp}$ and $\nu_\perp = \nu_{1,/\!/}$.

Comparing the exponents we conclude that the dominant terms in $F_i(t)$ have the form:

$$F_{/\!/}(t) \propto t^{-5/2} \cos(\omega_p t + \theta_{/\!/}) \quad , \tag{3.17}$$

$$F_{\perp}(t) \propto t^{-2} \cos(\omega_0 t + \theta_{\perp}) \quad . \tag{3.18}$$

A detailed calculation leads to explicit expressions for the proportionality factors and for the phase angles. In fact, the result for the longitudinal velocity autocorrelation function is:

$$F_{/\!/}(t) \simeq \frac{k_B T(\omega_B^2 - \omega_p^2)}{8\pi^{3/2} nm\omega_p \omega_B^2 t^{5/2}} \operatorname{Re}\left\{ \frac{i\, e^{-i\omega_p t}}{\left[D_{/\!/} + \frac{i}{2}\frac{c_s^2}{\omega_p} + \frac{1}{2nm}\left(\frac{4}{3}\eta_1 + \eta_V + 4\zeta\right)\right]^{3/2}} \right\} , \tag{3.19}$$

with the (complex) dynamical viscosities η_i , ζ at the frequency $z = \omega_p$ and the (real) static self-diffusion coefficient $D_{/\!/}$. For the transverse velocity autocorrelation function one obtains:

$$F_{\perp}(t) \simeq \frac{k_B T(\omega_p^2 + 2\omega_B^2)}{16\sqrt{2}\, \pi\, nm\omega_p \omega_B \omega_0^{1/2} t^2}$$
$$\times \operatorname{Re}\left[\frac{\exp(i\frac{\pi}{4} - i\omega_0 t)}{\left\{D_{\perp} + \frac{i}{2}\frac{c_s^2}{\omega_0} - \frac{i\eta_4 \omega_B}{nm\omega_0} - \frac{1}{2nm}\left[\frac{2}{3}\eta_1 - 2\eta_2 - \eta_V + 2\zeta + \frac{\omega_B^2}{\omega_0^2}(\eta_1 - 2\eta_2)\right]\right\}^{3/2}} \right] \tag{3.20}$$

with the dynamical viscosities η_i, ζ at the frequency $z = \omega_0$ and the static self-diffusion coefficient $D_{\perp}$.

In the case of resonance, with $\omega_p = \omega_B$, the asymptotic expression for $F_{/\!/}(t)$ is no longer given by (3.17) or (3.19). Instead, the dominant term stems then from the upper boundary of the integral for $F^{(+)}_{/\!/}(t)$. The asymptotic expression for $F_{/\!/}(t)$ reads in this case:

$$F_{/\!/}(t) \simeq \frac{k_B T}{8\pi\, 2^{1/4}\, nm\omega_p^{3/2}\, t^3}$$
$$\times \operatorname{Re}\left[\frac{i \exp(i\frac{\pi}{4} - i\sqrt{2}\,\omega_p t)}{\left\{D_{\perp} + \frac{i}{2}\frac{c_s^2}{\omega_0} - \frac{i\eta_4}{\sqrt{2}\, nm} - \frac{1}{2nm}\left(\frac{7}{6}\eta_1 - 3\eta_2 - \eta_V + 2\zeta\right)\right\}^{3/2}} \right] \quad . \tag{3.21}$$

As is well-known [Alder and Wainwright, 1970; Kawasaki, 1970; Ernst e.a., 1971, 1976] the velocity autocorrelation function for a fluid of neutral particles has a tail proportional to $t^{-d/2}$ with $d = 3$. For an unmagnetized one-component plasma the tail of the velocity autocorrelation function is the sum of a term proportional to $t^{-3/2}$ and a term proportional to $t^{-3/2} \cos(\omega_p t + \theta)$ [Gould and Mazenko, 1975; Giaquinta e.a., 1976; Varley, 1977; Gaskell, 1982; Marchetti and Kirkpatrick, 1985]. For a magnetized one-component plasma we have found that the tails behave qualitatively differently. The anisotropy of the mode spectrum leads to interference effects in the coupling of the modes. As a consequence the tails drop off more rapidly than those of the correlation functions for an unmagnetized plasma. Moreover, a second frequency, viz. $\omega_0 = (\omega_p^2+\omega_B^2)^{\frac{1}{2}}$ shows up on a par with the plasma frequency. In the general off-resonant case this frequency determines the oscillations of the tail of the transverse velocity autocorrelation function, whereas the tail of the longitudinal function still oscillates at the plasma frequency, as in the unmagnetized case. As a consequence one expects a peak at ω_0 in the power spectrum of the transverse autocorrelation function and similarly a peak at ω_p for the longitudinal function. The latter will be less pronounced, however, since the tail of the longitudinal function drops off more rapidly, so that its contribution to the power spectrum is less important.

In the particular case of resonance the frequency $\omega_0 = \sqrt{2}\,\omega_p$ determines the oscillations of the tails of both the transverse and the longitudinal autocorrelation functions. As before it is expected that in the power spectrum the peak at ω_0 is more pronounced for the transverse case; in the longitudinal case the tail is damped by an extra factor t^{-1} so that its influence is less important.

In a paper that appeared several years ago [Bernu, 1981] molecular dynamics computations for the velocity autocorrelation functions of a one-component plasma in a magnetic field have been reported on. It was found that the power spectrum of the transverse velocity autocorrelation function for strongly coupled plasmas (Γ = 10 or 100) in a magnetic field with a resonant Larmor frequency ($\omega_B = \omega_p$) indeed shows a peak structure at a frequency $\omega \simeq 1.3\,\omega_p$, which is quite near to $\sqrt{2}\,\omega_p$. The plasmon peak, which is present for vanishing magnetic fields, turned out to be suppressed completely in the resonant case. This result is corroborated by the mode-coupling calculation of the tails, as presented here. As to the power spectrum of the longitudinal velocity autocorrelation function, it turned out to be rather flat. Apparently, the influence of the tail, which would have led to a peak at $\sqrt{2}\,\omega_p$ as well, is rather weak, as was anticipated above, in view of the strong damping ($\propto t^{-3}$) of the tail in this case.

REFERENCES

Alder, B.J. and T.E. Wainwright, 1970, Phys.Rev., A1:18.

Bernu, B., 1981, J. de Phys.Lett., 42:L253.

Ernst, M.H., E.H. Hauge and J.M.J. van Leeuwen, 1971, Phys.Lett., 34A:419.

Ernst, M.H., E.H. Hauge and J.M.J. van Leeuwen, 1976, J.Stat.Phys., 15:7.

Gaskell, T., 1982, J.Phys., C15:1601.

Giaquinta, P.V., M. Parrinello, M.P. Tosi and N.H. March, 1976, Phys.Chem. Liquids, 5:197.

Gould, H. and G.F. Mazenko, 1975, Phys.Rev.Lett., 35:1455.

de Groot, S.R. and P. Mazur, 1962, "Non-equilibrium Thermodynamics", North-Holland, Amsterdam.

Kadanoff, L.P. and J. Swift, 1968, Phys.Rev., 166:89.

Kawasaki, K., 1970, Phys.Lett., 32A:379.

Marchetti, M.C., T.R. Kirkpatrick and J.R. Dorfman, 1984, Phys.Rev., A29:2960.

Marchetti, M.C. and T.R. Kirkpatrick, 1985, Phys.Rev., A32:2981.

Résibois, P., 1972, in: "Irreversibility in the Many-Body Problem", J. Biel and J. Rae, eds., Plenum, New York, 273.

Suttorp, L.G. and J.S. Cohen, 1985, Physica, 133A:357,370.

Suttorp, L.G. and A.J. Schoolderman, 1986, preprint.

Varley, R.L., 1977, Phys.Lett., 62A:340.

ION STOPPING POWER IN DENSE PARTIALLY DEGENERATE PLASMAS

Claude Deutsch

Laboratoire de Physique des Gaz et des Plasmas,*
Bat. 212
Universite Paris XI
91405 Orsay Cedex, France

I. INTRODUCTION

In close connection with beam-target interaction problems encountered in inertial confinement fusion (ICF) driven by particle beams [Deutsch, 1986], we intend to solve exactly the model for the stopping of nonrelativistic pointlike and positive ions in a homogeneous, and dense electron fluid taken at any temperature. Such a model is usually considered as the simplest in providing a coherent theoretical framework with reliable estimates for the beam-target interaction parameters. The rational underlying this view is based on the observation that many, if not most, of the compressed pellet states encountered during a full compression lie in the parameter space close to weakly coupled systems indexed by a dimensionless quantity

$$\chi^2 = \frac{1}{\pi q_F a_o} = \frac{V_o}{mV_F} = \frac{1}{\pi}\sqrt{\frac{I_H}{k_B T_F}} = \frac{\alpha r_s}{\pi} \quad , \tag{1.1}$$

with q_F, V_F, T_F denoting Fermi wave number, velocity and temperature respectively. a_o, V_o, I_H refer to Bohr wavelength, velocity and energy $r_s = (4/3\,\pi n)^{-1/3}\,a_o^{-1}$ in terms of the free electron number density n, while $\alpha = (9\pi/4)^{-1/3}$. At high temperature ($T \gg T_F$), eq. (1.1) becomes ($T_e = T/T_F$)

$$\frac{3\chi^2}{2T_e} = \frac{e^2}{\pi k_B T R_{ee}} = \frac{\Gamma_e}{\pi} \quad , \tag{1.2}$$

in terms of $R_{ee} = (4/3\,\pi n)^{-1/3}$ and of the classical plasma parameter Γ_e. At any degeneracy (or temperature), the Random Phase Approximation (R.P.A.) is valid in a (T,n) domain defined by [Lindhard, 1954; Dar et al., 1974]

$$\frac{\chi^2}{1+T_e} \ll 1 \quad , \tag{1.3}$$

*Associé au CNRS

so that the potential energy content of an electron pair located at the screening distance always remains much smaller than the kinetic energy per particle. As restricted as it looks at first sight, inequality (1.3) allows us to encompass a huge number of different systems ranging from high-temperature Tokomaks to dense and moderately hot plasmas envisioned in particle beam driven ICF.

Another fundamental point, stressing the basic importance of a simple but efficient modelling for the free electron component of an otherwise strongly coupled ionic mixture in the target, lies on the observation that although the bound electrons always provide a non-negligible amount to stopping, the free electrons are expected to give the largest part within the usual temperature range of interest i.e. [Deutsch, 1986; Deutsch, Maynard, Minoo, 1983] .

$$50 \text{ eV} < k_B T < 200 \text{ eV} .$$

Therefore, energetic ions impinging on the target are supposed to yield most of their energy to free electrons, which display more flexibility in exchanging momentum and energy during elastic collisions with projectiles. In this respect, the Born approximation is fundamental to treat the electron-ion encounter. The projectile is then considered, at variance, as pointlike, or as a quantum plane wave-packet. At last, it should be mentioned that we are fully entitled to reduce the complex beam-target interaction to a single ion-target interaction, in agreement with the fact that whatever its intensity (kiloamps up to megaamps/cm^2), any beam will appear as dilute in dense matter. The inbeam ion-ion average distance is likely to remain at least two orders of magnitude larger than the Thomas-Fermi screening length in cold matter. Previous analyses of this problem were based on the quantum-mechanical dielectric theory or the classical binary-collision approximation. The results were applied to the medium at zero temperature. Thermal as well as quantum-mechanical effects were taken into account by **Skupsky [1977], who made use of the dielectric formalism to derive the energy loss of charged particles with velocities lower than those of the electrons in the plasma.** These results were applied to the **slowing down of the 3.5 MeV α particles produced in the dominant deuterium-tritium (DT) reaction in ICF.** Current feasibility studies of different inertial confinement fusion programs require a complete and accurate description of the energy-loss process for a variety of ionic species, over a wide range of nonrelativistic ion velocities in very dense and hot plasmas with partially ionized species included.

II. R.P.A. DIELECTRIC FUNCTION

We start with the usual assumption that the Coulomb interaction between a projectile and the stopping free electron is essentially elastic, so there are no such things as electron pair creation or other inelastic processes. So, we are entitled to consider the given interaction within the standard framework of linear response theory satisfying the usual relation

$$\vec{J}_{ind} = - \frac{i\omega}{4\pi} (\varepsilon(q,\omega) - 1) \vec{E}(q,\omega) ,$$

and it remains to compute the fully dynamical dielectric function $\varepsilon(q,\omega)$. For this goal, we shall follow the exact R.P.A. treatment previously worked out by Gouedard and Deutsch [1978].

A. General Results

They pertain to an homogeneous electron fluid which remains weakly coupled for any degeneracy

$k_B T/\varepsilon_F$

It is the obvious finite-temperature extension of the standard Lindhard quantity valid at T = 0, for $r_s < 1$. It smoothly joins the $T \to \infty$ and classical Fried-Conte expressions. Within the framework of linear response theory, it is also introduced as

$$\varepsilon(q,\omega) = 1 - V(q)\ \chi^0(q,\omega), \ldots \tag{2.1}$$

with

$$V(q) = \frac{4\pi e^2}{q^2}$$

and a free electron response

$$X^o(q,\omega) = -2\int \frac{d^3k}{(2\omega)^3}\ \frac{[n^0(k+q) - n^0(k)]}{(h\omega+i\eta) - (\varepsilon^0_{k+q}-\varepsilon^0_k)} \tag{2.2}$$

where η is a small positive quantity,

$$\varepsilon^0_k = \frac{\hbar^2 k^2}{2m_e}$$

$$n^0_k(k) = \{e^{\beta(\varepsilon^0_k-\mu)} + 1\}^{-1} \quad ,$$

$\beta = 1/k_B T$, and μ is the chemical potential.

To simplify the discussion, we make use of the dimensionless variables

$$z = \frac{q}{2q_F} \qquad \text{and} \qquad u = \frac{\omega}{qv_F}$$

so that

$$X^o(z,u) = -\frac{\alpha r_s}{\pi^2} G(z,u)$$

$$G(z,u) = f_1(z,u) + if_2(z,u) \quad , \tag{2.3}$$

$$f_2(z,u) = -\frac{\pi T_e}{8} \operatorname{Log} \frac{1 + \exp\left(\frac{\nu^e - p_+^2}{T_e}\right)}{1 + \exp\left(\frac{\nu^e - p_-^2}{T_e}\right)} \tag{2.4}$$

The other dimensionless parameters are

$$T_e = \frac{T}{T_F} , \quad \nu^e = \frac{\mu}{\varepsilon_F^o} = \alpha^e T_e , \quad P_{\pm} = u \pm z$$

$f_1(z,u)$ is computed through the Kramers-Kronig relation

$$f_1(z,u) = - \frac{1}{\pi} \text{P.P} \int_{-\infty}^{+\infty} \frac{f_2(z,u')}{(u - u')} du' \qquad (2.5)$$

which can be transformed through

$$f_1(z,u) = - \frac{T_e}{8z} [F(p_+) - F(p_-)] \qquad (2.6)$$

into

$$F(p) = - \frac{1}{\pi} \text{P.P.} \int_{-\infty}^{+\infty} \frac{h(p')dp'}{p - p'}$$

$$h(z) = \text{Log} \left(1 + \exp \left[\frac{\nu^e - z^2}{T_e}\right]\right)$$

With equations (2.4), (2.5), (2.6) one recovers the two well-known temperature limits:

- $T_e << 1$ [Lindhard, 1954]

$$F(p) = 2p \left[\frac{1}{2} + \frac{1 - p^2}{4p} \text{Log} \frac{p + 1}{p - 1}\right] \qquad (2.7)$$

- $T_e >> 1$

$$F(p) = \frac{4}{3\pi T_e^{3/2}} Z(p/\sqrt{T_e}) \qquad (2.8)$$

Z(p) being the usual Fried and Conte function [Jackson, 1975]

$$Z(p) = - \frac{1}{\sqrt{\pi}} \int_{-\infty}^{+\infty} dt \frac{\exp(- t^2)}{(p - t)} \qquad (2.9)$$

which can also be easily computed through Padé approximants [Nemet et al., 1981].

At arbitrary temperatures, the following technical remarks are useful:

- f_1 and f_2 are essentially significant on a range in u(or z) measured by $a_o(T_e)$, with

$$a_o(T_e) = \frac{1}{\sqrt{2}} [\nu^e + (\nu^{e2} + \pi^2 T_o^2)^{1/2}]^{1/2} \tag{2.10}$$

The thermal velocity reads $V_{th} \simeq V_F a_o(T_e)$.

- f_1 and f_2 have their respective maxima, in u and z, located between 0 and $1/(1+T_e)$, so

$$\chi^2 f_1(u,z) \ll a_o(T_e) \; .$$

- $f_2(u,z) \simeq 0$ as soon as $|z - u| > 2\, a_o(T_e)$

- $f_1(u,z) < 0$ for $u > a_o(T_e)$. (2.11)

Another important parameter is the location of the resonance ($\varepsilon(z,u) = 0$) given by ($\chi^2 = \alpha r_s/\pi$)

$$z_r^2 = - \chi^2 f_1(z_r,u_r) \quad \text{and} \quad f_2(z_r,u_r) = 0$$

$$z_r^2 = \frac{\chi^2}{3u_r^2} [1 + \frac{T_e F_{3/2}(\alpha^e)}{u_r^2 F_{1/2}(\alpha^e)} + \frac{T_e^2 F_{5/2}(\alpha^e)}{u_r^4 F_{1/2}(\alpha^e)}] \quad , \tag{2.12}$$

III. ENERGY LOSS AT FINITE TEMPERATURES

A comprehensive treatment of the energy-loss problem, in terms of the equilibrium dielectric function $\varepsilon(q,\omega)$, can be formulated by starting from the scattering rate [Arista-Brandt,1981]

$$R(\vec{q},\omega) = \left(\frac{4\pi Z e^2}{q^2}\right)^2 \frac{2\pi}{\hbar^2} S(\vec{q},\omega) \quad , \tag{3.1}$$

for energy transfer $h\omega = E(\vec{p}\,')-E(\vec{p})$ and momentum transfert $h\vec{q}=\vec{p}\,'-\vec{p}$, which applies to the scattering of a particle of charge Ze, with initial momentum p and energy E(p), to the final state given by p', E(p'). The dynamical structure factor $S(q,\omega)$ is related to the dielectric function $\varepsilon(q,\omega)$ through

$$S(\vec{q},\omega) = \frac{\hbar^2 q^2}{4\pi^2 e^2} N(\omega) \; \mathrm{Im} \left(\frac{-1}{\varepsilon(\vec{q},\omega)}\right) \quad , \tag{3.2}$$

where $N(\omega) \equiv [\exp(\beta h\omega) - 1]^{-1}$ and $\beta = 1/kT$.

The temperature dependence is contained in the dielectric function $\varepsilon(q,\omega)$ and in the Planck function $N(\omega)$. The energy-loss rate is given by

$$\frac{dE}{dt} = \int \frac{d^3p'}{(2\pi h)^3} \hbar\omega \; R(\vec{q},\omega) \tag{3.3}$$

$$= \left(\frac{Ze}{\pi}\right)^2 \int d^3q \, \frac{\omega N(\omega)}{q^2} \, \mathrm{Im} \left(\frac{-1}{\varepsilon(\vec{q},\omega)}\right) ,$$

where $\omega \equiv \omega(p,q)$ is determined from

$$\hbar\omega(\vec{p},\vec{q}) \equiv E(\vec{p}\,') - E(\vec{p}) = h\vec{q}.\vec{v} + \frac{\hbar^2 q^2}{2M} \tag{3.4}$$

in terms of the incident velocity $v = p/M$ and the mass M of the projectile. For heavy particles $M >> m$, recoil effects are small and we can expand Eq. (5) in terms of $\Delta\omega \equiv \hbar^2 q^2/2M$ to obtain

$$\frac{dE}{dt} = \left(\frac{dE}{dt}\right)_0 + \left(\frac{dE}{dt}\right)_1 + \ldots , \tag{3.5}$$

where the first two terms are

$$\left(\frac{dE}{dt}\right)_0 = \left(\frac{Ze}{\pi}\right)^2 \int d^3q \frac{\omega\ N(\omega)}{q^2} \operatorname{Im} \left.\frac{-1}{\varepsilon(\vec{q},\omega)}\right|_{\omega = \vec{q}.\vec{v}} , \tag{3.6}$$

$$\left(\frac{dE}{dt}\right)_1 = \left(\frac{Ze}{\pi}\right)^2 \frac{\hbar}{2M} \cdot \int d^3q \frac{\partial}{\partial\omega} \left.\left[\omega N(\omega) \operatorname{Im} \frac{-1}{\varepsilon(\vec{q},\omega)}\right]\right|_{\omega = \vec{q}.\vec{v}} . \tag{3.7}$$

The integrals range over both negative frequencies (loss processes) and positive frequencies (gain processes), but it is here more instructive to transform them into integrals over positive frequencies only.

We can simplify the expression for the main term $(dE/dt)_0$, by splitting the integral into the $\omega > 0$ and $\omega < 0$ parts, and then making use of the relations $N(\omega) + N(-\omega) = -1$ and $\varepsilon(q,-\omega) = \varepsilon^*(q,\omega)$; this leads to an expression of the form

$$\left(\frac{dE}{dt}\right)_0 = \int_{\omega>0} d^3q N(\omega) f(\vec{q},\omega) - \int_{\omega>0} d^3q [N(\omega) + 1]\ f(\vec{q},\omega) . \tag{3.8}$$

The two terms in $N(\omega)$ cancel exactly, with the result for the stopping power S,

$$S \equiv -\frac{dE}{dx} \simeq \frac{-1}{v} \left(\frac{dE}{dt}\right)_0 = \frac{2}{\pi} \left(\frac{Ze}{v}\right)^2 \int_0^\infty \frac{dq}{q} \int_\infty^\infty d\omega\omega \operatorname{Im} \left(\frac{-1}{\varepsilon(\vec{q},\omega)}\right) . \tag{3.9}$$

The only temperature dependence is now contained in the energy-loss function $\operatorname{Im} -1/\varepsilon(q,\omega)$, and arises from a thermal redistribution of the oscillator strengths in the medium. One can interpret this result as a cancellation between the processes of stimulated absorption and stimulated emission of energy $h\omega$ by the projectile, since both processes are proportional to the Planck distribution $N(\omega)$ that characterizes the thermal equilibrium of excitation quanta in the medium. Thus, the energy-loss rate is only determined by spontaneous emission processes, which are independent of $N(\omega)$.

A similar analysis can be made for the energy loss straggling Ω,

$$\Omega^2 = \int \frac{d^3\vec{p}'}{(2\pi h)^3} (\hbar\omega)^2 R(\vec{q},\omega) = \frac{<(dE)^2> - <dE>^2}{\Delta x} \quad , \tag{3.10}$$

which can be expanded as

$$\Omega^2 = \Omega_o^2 + \Omega_1^2 + \ldots ,$$

with

$$\Omega_o^2 = \frac{Z^2 e^2 \hbar}{\pi^2 v} \int d^3q \frac{\omega^2}{q^2} N(\omega) \operatorname{Im} \left(\frac{-1}{\varepsilon(\vec{q},\omega)}\right)\Bigg|_{\omega=\vec{q}.\vec{v}} \quad , \tag{3.11}$$

$$\Omega_1^2 = \frac{Z^2 e^2 \hbar^2}{2\pi^2 Mv} \int d^3q \frac{\partial}{\partial\omega} \left[\omega^2 N(\omega) \operatorname{Im} \left(\frac{-1}{\varepsilon(\vec{q},\omega)}\right)\right]\Bigg|_{\omega = \vec{q}.\vec{v}} \quad . \tag{3.12}$$

For the balance between positive and negative frequencies in the Ω_o^2 term, all the contributions from stimulated absorption ($\omega > 0$), proportional to $N(\omega)$, and those from stimulated and spontaneous emission ($\omega < 0$), proportional to $[N(\omega) + 1]$, are collected, and one obtains

$$\Omega_o^2 = \frac{2Z^2 e^2 \hbar}{\pi v^2} \int_o^\infty \frac{dq}{q} \int_o^\infty d\omega\, \omega^2 [2N(\omega)+1] \operatorname{Im} \left(\frac{-1}{\varepsilon(\vec{q},\omega)}\right) \quad , \tag{3.13}$$

The temperature dependence of Ω_o^2 is contained in $N(\omega)$ and $\varepsilon(q,\omega)$.

We discuss now our results for low and high temperatures. When $k_B T \ll h\omega$, $N(\omega) \to 0$, and we retrieve the expression for the energy straggling in a degenerate electron gas.

In the opposite limit $k_B T \gg \hbar\omega$, we can approximate $[2N(\omega) + 1] \simeq 2k_B T/\hbar\omega$. The straggling integral Eq. (3.3) then becomes identical to the stopping integral Eq. (3.9) multiplied by $2k_B T$, i.e., straggling Ω and stopping power S are related as

$$\Omega^2(v,n,T) \simeq 2k_B T S(v,n,T) \quad , \tag{3.14}$$

for all values of v, n, and T such that the condition $\hbar\omega \ll k_B T$ is fulfilled. Since the frequencies of interest fall in the integration range from zero to $\omega_{max} = 2mv\,(v+v_e)/\hbar$, Eq. (3.14) will apply when

$$2mv(v+v_e) < k_B T \quad . \tag{3.15}$$

In the limit $T_e \gg 1$ one approaches $1/2\, mv_o^2 \simeq 3/2\, k_B T$, and Eq. (3.15) defines the domain $v < 0.15\, v_e$, corresponding to projectiles much slower than the thermal electrons in the plasma. The velocity dependence of Ω^2 is the same as that of S, viz., $\Omega^2 \sim v$. By contrast, in a degenerate electron gas at low velocities, Ω^2 is a quadratic function of v.

The applicability of Eq. (3.14) to a hot plasma $k_BT >> E_F$ accords with a classical description, in terms of the Fokker-Planck equation, for the fluctuations in the energy of a slow particle in a thermalized medium. It pertains, moreover, to a general quantum-mechanical relation between the generalized resistance and voltage fluctuations in linear dissipative systems.

IV. BORN R.P.A. (B.R.P.A.) STOPPING POWER

In dimensionless units (z and u) the B.P.R.A. stopping power

$$S \equiv -\frac{dE}{dx} = \frac{2}{\pi}\left(\frac{ze}{V}\right)^2 \int_0^\infty \frac{dq}{q} \int_0^\infty d\omega\, \omega\, \mathrm{Im}\left(\frac{1}{\varepsilon(q,\omega)}\right) , \tag{4.1}$$

is written in the form

$$\frac{dE}{dx} = -\frac{z^2 e^4}{4\pi\varepsilon_0^2 m_e V^2} n_e L_e \tag{4.2}$$

$$L_e = \frac{6}{\pi\chi^2} \int_0^{V+V_F} u\, du \int_0^\infty z\, dz\, \mathrm{Im}\frac{1}{\varepsilon(z,u)}$$

$$= \frac{6}{\pi\chi^2} \int_0^{V/V_F} u\, du \int_0^\infty z^3\, dz \frac{\chi^2 f_2(z,u)}{\{z^2+\chi^2 f_1(z,u)\}^2 + \{\chi^2 f_2(z,u)\}^2} , \tag{4.3}$$

with L_e also dimensionless L_e depends on T_e through $\varepsilon(z,u)$ only.

On the other hand, the energy loss straggling (3.13) writes

$$\Omega_0^2 = \frac{Z^2 e^4}{4\pi\varepsilon_0^2} n_e . L_\Omega$$

$$L_\Omega = \frac{12}{\pi}\left(\frac{V_F}{V}\right)^2 \int_0^{v/v_F} u^2 du \int_0^\infty \frac{dz\, z^4 f_2(z,u)}{[z^2+\chi^2 f_1(z,u)]^2+[\chi^2 f_2(z,u)]^2} \times \tag{4.4}$$

$$\times \left[\frac{2}{e^{\frac{4zu}{T_e}}-1}\right] ,$$

At this point, we have to make clear a few obvious assumptions.

On most part of their range the projectiles are more energetic than target particles. So, their trajectory may be taken as linear, in view of the very small energy exchange at each encounter. The projectile ions are supposed to be pointlike with a given charge.

Eqs. (4.3), (4.4) are free from divergences at $z \gg a_0(T_e)$, diffraction effects yield $f_2 = 0$, while shielding through $|\varepsilon(z,u)|^2$ secures the opposite limit $z \ll a_0(T_e)$. With R.P.A., one has

$$|\chi^2 f_1| \quad , \quad |\chi^2 f_2| \ll a_0 (T_e)$$

and one divides the z-domain into two regions:

. $z \ll a_0(T_e)$, where the test charge yields its energy to the collective modes with a resonance at $z = z_r$ when $u > a_0(T_e)$, and energy exchange close to $\hbar\omega_r \simeq \hbar\omega_p$.
. $|z-u| < a_0(T_e)$, which pertains to binary collisions. For $u > a_0(T_e)$, shielding vanishes. The corresponding energy exchange is now $\hbar\omega = \hbar^2 q^2/2m$.

These two domains remain distinct when $u > a_0(T_e)$; i.e. for an energy exchange larger than the kinetic energy $\simeq k_B T_F(1 + T_e)$. This basic property accounts for the weak coupling character of the R.P.A. Collective modes retain less energy than the particle kinetic energy.

Moreover the usual Z^2-dependence of the stopping formula yields the well-known scaling relation

$$\frac{dE'}{dx} (Z',M',E') = \frac{Z'^2}{Z^2} \frac{dE}{dx} (Z,M, \frac{M}{M'} E')$$

so we restrict to protons in the sequel.

It should be appreciated that one of the main outputs of the present work is the possibility to compute S for any velocity ratio V/V_{th}, because partial degeneracy is treated exactly.

For instance, in the large V limit

$$\frac{V}{a_0(T_e)\, V_F} \gg 1$$

one may check that, for $T_e \neq 0$, there are, as in the $T_e = 0$ case two equal contributions of S
. exchange of energy with a plasmon mode around $z \simeq z_r$
. exchange of energy through binary encounters around $z = u$.

V. NUMERICAL RESULTS AND APPROXIMATIONS [Maynard-Deutsch, 1985]

In order to get orders of magnitude for the most relevant parameters, we put them into numerical correspondence in Table I.

Table I- Relations between α'_e and T_e, and n, x^2 and T_F, α'_e is given the normalization condition for the Boltzmann statistics.

α^e	-5	-1.5	0	1.5	5.	
T_e	23.22	2.361	0.9887	0.4973	0.1934	
α'_e	-5.002	-1.573	-0.268	0.763	2.18	
$n\,(\mathrm{cm}^{-3})$	10^{22}	10^{23}	10^{24}	10^{25}	10^{26}	10^{27}
x^2	0.902	0.419	0.194	0.902×10^{-1}	0.419×10^{-1}	0.194×10^{-1}
T_F (K)	0.196×10^5	0.912×10^5	0.423×10^6	0.196×10^7	0.912×10^7	0.423×10^8

The numerical analysis of Eq. (4.3) is mostly performed through

$$F(z,u) = -z \operatorname{Im} \frac{1}{\varepsilon(z,u)}$$

and

$$\int_0^\infty dz\, F(u,z) = \int_0^{z_1} dz\, F(z,u) + \int_{z_1}^{z_2} dz\, F(z,u) + \int_0^\infty dz\, F(z,u) \equiv I_1 + I_2 + I_3 \tag{5.1}$$

with $z_1 = z_2 = 0$, $u \le a_0(T_e)$
$z_1 = \mathrm{Max}\,(0,\ z_r(u) - \varepsilon)$
$z_2 = z_r(u) + \varepsilon$ $u > a_0(T_e)$
$\varepsilon \simeq 0.01\ a_0(T_e)$

I_1 and I_3 are evaluated numerically, while I_2 can be given an analytic expression.

$S = -dE/n\,dx$ is displayed on Fig. 1 for various densities n in the target as a function of the projectile (proton) energy, for a given degeneracy parameter α^e. Basic trends are as follows:

- Maximum stopping efficiency is achieved for $V \simeq V_{th}$.
- $dS/dT \simeq 0$ for $V \gg V_{th}$
- $S \simeq n^{-1}$ for $V \ll V_{th}$

A. <u>Low Projectile Velocity</u> ($x = V/V_{th} \ll 1$)

Eq. (4.3) then becomes:

$$L_e = \frac{V^3}{V_F^3} \int_0^\infty \frac{dz\, z^3}{(z^2 + Z_c(z))^2 \left(1 + \exp\left[\frac{z^2}{T_e} - \alpha^e\right]\right)} \equiv \left(\frac{V}{V_F}\right)^3 C(\chi^2, \alpha^e) \quad , \tag{5.2}$$

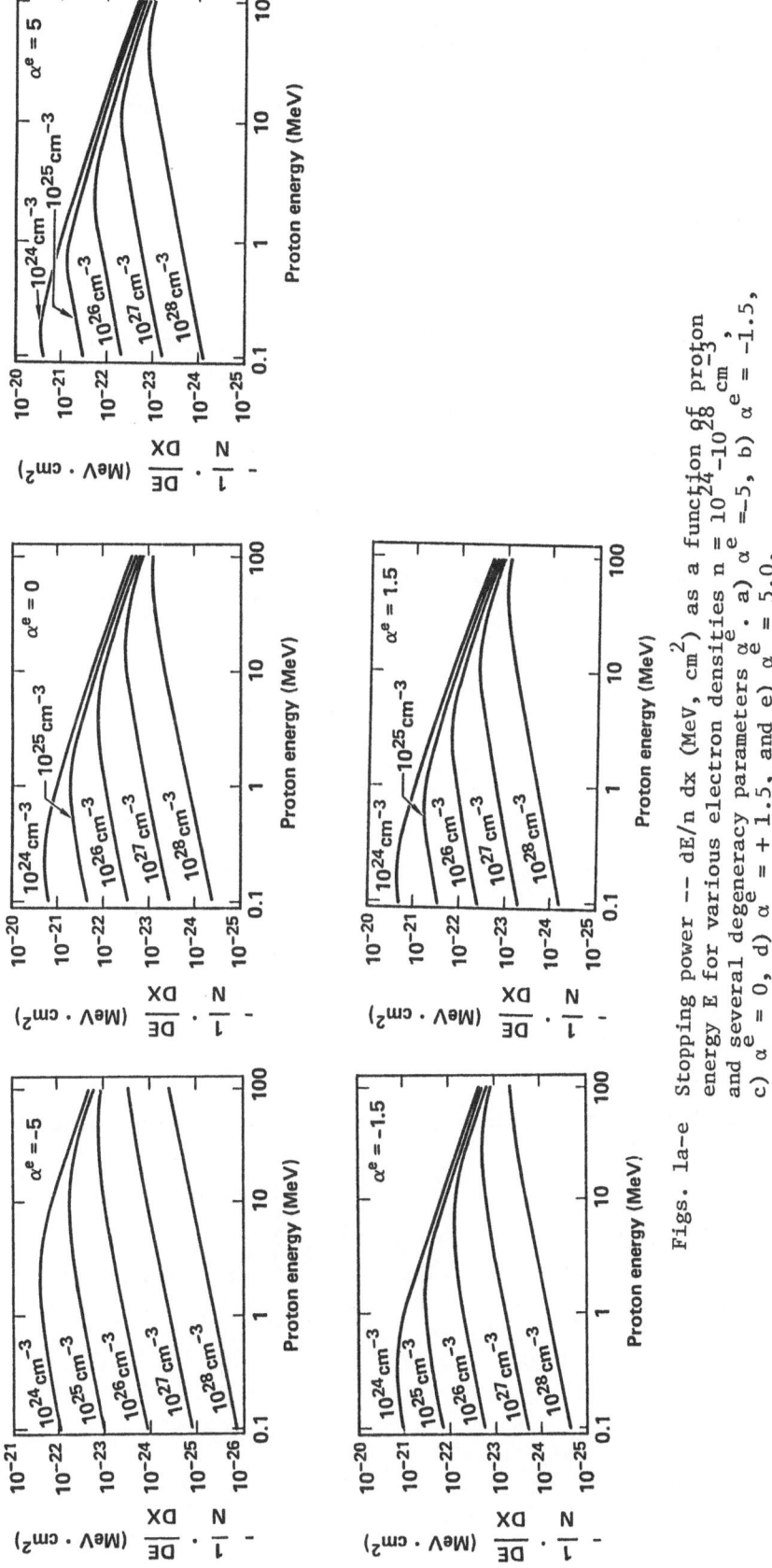

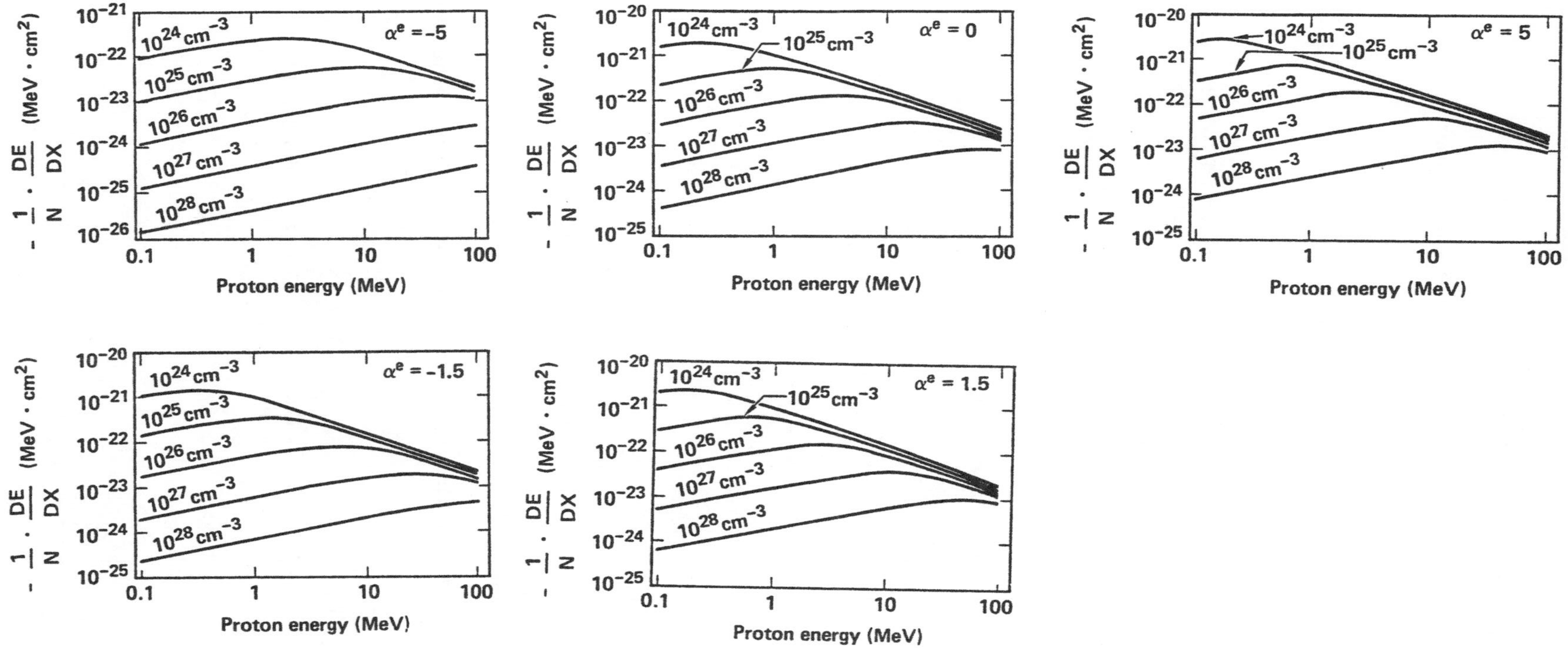

Figs. 1a-e Stopping power -- dE/n dx (MeV, cm^2) as a function of proton energy E for various electron densities $n_e = 10^{24}$-10^{28} cm^{-3}, and several degeneracy parameters α^e. a) $\alpha^e = -5$, b) $\alpha^e = -1.5$, c) $\alpha^e = 0$, d) $\alpha^e = +1.5$, and e) $\alpha^e = 5.0$.

where

$$Z_c(z) = \chi^2 f_1(z,0)$$

when $\chi^2/(1 + T_e) \ll 1$ we can use the additional assumption:

$$Z_c(z) \simeq \chi^2 f_1(0,0) = \frac{T_e^{1/2}}{2} F_{1/2}(\alpha^e) = Z_c$$

$$= \frac{1}{4\, q_F^2\, \lambda_{TF}^2}, \quad \lambda_{TF}^2 = \frac{V_F^2}{3\omega_p^2}, \quad \text{Thomas-Fermi} \quad T_e \ll 1 \tag{5.3}$$

$$= \frac{1}{4\, q_F^2\, \lambda_D^2}, \quad \lambda_D^2 = \frac{k_B T}{4\,\pi n e^2}, \quad \text{Debye} \quad T_e \gg 1 \ .$$

Eq. (5.3) for Z_c^2 allows to rewrite L_e as

$$L_e = \left(\frac{V}{V_F}\right)^3 \frac{1}{2}\left(\ln \frac{1 + Z_c^2}{Z_c^2} - \frac{1}{1 + Z_c^2}\right), \quad T_e \ll 1$$

$$= \frac{2}{\sqrt{\pi}} \frac{x^3}{3} \ln \frac{1}{4\,\delta e \nu}, \quad T_e \gg 1 \tag{5.4}$$

where $\nu = e^{0.517}$, $\delta = \frac{\lambda\!\!\!^{-2}}{16\,\lambda_D^2}$, $\lambda\!\!\!^- = \frac{\hbar}{m\, V_{th}}$

Approximation (5.4) is excellent for $\chi^2/(1 + T_e) \leq 1$ and lags within 15% in a cold solid.

B. <u>High Projectile Velocity</u> ($x \gg 1$)

Extending the T = 0 Lindhard-Winther procedure to any temperature we make use of

$$\lim_{V\to\infty} \left\{ \int_0^\infty dz \int_0^{V/V_F} du F(u,z) \right\} = \lim_{V\to\infty} \left\{ \int_{z_r(V/V_F)}^{V/V_F} dz \int_0^\infty du F(u,z) \right\}, \tag{5.5}$$

and

$$\int_0^\infty d\omega\, \omega\, \mathrm{Im} \frac{1}{\varepsilon(q,\omega)} = \frac{\omega_p^2}{2}, \tag{5.6}$$

to derive

$$\lim_{V\to\infty} L_e = \int_{Z_r(V/V_F)}^{V/V_F} \frac{dz}{z} = \ln \left(\frac{V}{V_F\, Z_r\left(\frac{V}{V_F}\right)} \right) \tag{5.7}$$

which, when combined to (4.2) through

$$\left\langle \frac{V_e^{2n}}{V_F^2} \right\rangle = \frac{T_e^n\, F_{n+1/2}(\alpha^e)}{F_{1/2}(\alpha^e)}$$

yields (m = electron mass)

$$\lim_{V\to\infty} L_e = \ln \frac{2mV^2}{h\omega_p} - \frac{\langle V_e^2\rangle}{V^2} - \frac{\langle V_e^4\rangle - 0.5\, \langle V_e^2\rangle^2}{V^4} + \ldots \tag{5.8}$$

The full V^{-2}-expansions are thus recovered from $L_e(V_1)-L_e(V_2)$ with $V_{1,2} \gg V_{th}$ so that

$$\lim_{V\to\infty} L_e = \ln \frac{2mV^2}{h\omega_p} - \frac{\langle V_e^2\rangle}{V^2} - \frac{\langle V_e^4\rangle}{2\, V^4} + \ldots \quad , \tag{5.9}$$

(5.9) already gives a one percent accuracy for $V > V_{th}$. The sum rule result (5.8) lies remarkably close to this full asymptotic one.

C. <u>Interpolation Formula</u> (any V)

To a large extent, the numerical gap between (5.9) and (5.2) (i.e. between low V and high V) can be bridged through

$$\begin{aligned} L_e(V) &= \left(\frac{V}{V_F}\right)^3 C(\chi^2,\alpha^e) \times \frac{1}{(1+GV^2)} = L_e^1(V), \qquad V \leq V_{int} \\ &= \ln\left(\frac{2mV^2}{h\omega_p}\right) - \frac{\langle V_e^2\rangle}{V^2} - \frac{\langle V_e^4\rangle}{2\, V^4} = L_e^2(V), \qquad V \geq V_{int} \end{aligned} \tag{5.10}$$

where G is fixed by $L_e^1(V_{int}) = L_e^2(V_{int})$.

and $V_{int} = \sqrt{1.5\, \langle V_e^2\rangle + \frac{3\hbar\omega_p}{2m}}$,

with a relative error (any T) smaller than five percent for $\chi^2/(1 + T_e) < 0.3$.

D. Statistical Effects

At this point, we think it worthwhile to investigate quantitative modifications of stopping when one replaces Fermi statistics by Boltzmann at an arbitrary temperature. For instance, on Fig. 2 we compare dE/dx respectively computed with a Fermi distribution $f(E) = (1 + e^{E/T-\alpha^e})^{-1}$ and a Boltzmann distribution $f_B(E)=e^{\alpha'}e^{-E/T}$ for the same (n,T) data through $F(p) = 4/3\ T_e^{3/2}\ Z\ (P/T_e^{1/2})$ (Z(x) = Fried-Conte expression) and

$$f_2(z,u) = \frac{\pi T_e}{8\,z}\left(e^{\alpha'_e - p_-^2/T_e} - e^{\alpha'_e - p_+^2/T_e}\right) \quad , \tag{5.11}$$

with α'_e plotted on the last line in table I. $\alpha^e = \alpha'_e$ at $T_e >> 1$.

Moreover, we recover $dE/dx = dE^B/dx$ at high velocity.

Statistical effects are thus mostly significant in the low velocity regime.

. As expected, discrepancies increase with increasing α^e. For instance maxima exhibit a 6% discrepancy for $T_e = 1$ and a 12% one for $T_e = 0.5$ respectively. Also, the maximum and slope of dE^B/dx tend to shift gradually away from their Fermi homologues.

. Fermi statistics gradually freezes out the free electron degrees of freedom, altogether with the corresponding stopping. all in all, a Fermi plasma tends to be more transparent than a Maxwellian one.

However, it should be noted that for low T and V, the classical Debye screening is more efficient thant the Thomas-Fermi screening, which reduces dE^B/dx.

VI. LOCAL FIELD CORRECTIONS (LFC)

Up to now, we retain the weakly coupled jellium with $r_s \leq 1$. To consider actual cases with $r_s > 1$, we have to include local field corrections G(q),, which correct for exchange and correlation effects at short distances. We first restrict to $T_e = 0$ and static corrections [cf. Hubbard, see Kugler, 1975].

Amongst the various proposals, a recent one due to Ichimaru and Utsumi [1981] appears particularly interesting. It also [Nagy et al., 1985, Tanaka-Ichimaru (1985)] allows for a more accurate treatment of the short-ranged (large q) interactions in

$$\varepsilon(q,\omega) = 1 - \frac{v(q)\chi_o(q,\omega)}{1+v(q)G(q)\chi_o(q,\omega)} \tag{6.1}$$

We begin by noting the long wavelength behavior [Kugler, 1975] ($q << q_F$, the Fermi wave number),

$$G(q) \rightarrow \gamma_o Q^2 \quad , \quad (Q \equiv q/q_F) \quad , \tag{6.2}$$

where the coefficient γ_o is connected to the correlation energy $E_c(r_s)$ in rydbergs per electron via the compressibility relation

$$\gamma_o = \frac{1}{4} - \frac{\pi\alpha}{24} r_s^5 \frac{d}{dr_s}\left[r_s^{-2} \frac{d}{dr_s} E_c(r_s)\right] \quad , \tag{6.3}$$

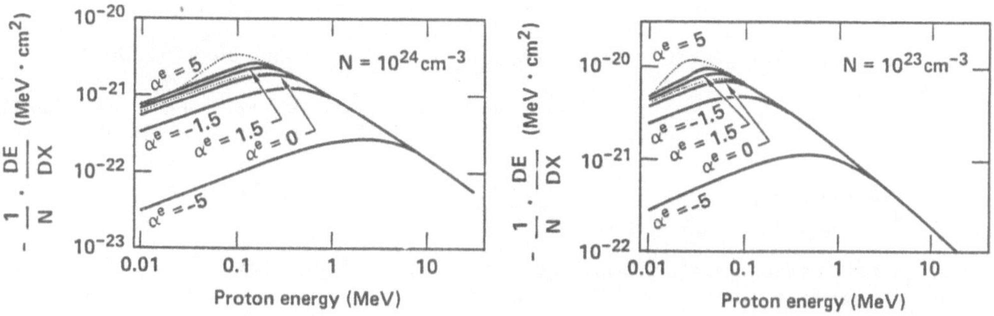

Figs. 2a-b Comparison of Fermi (full line) and Boltzmann (dotted line) stopping power for same target density and projectile velocity, at various degeneracies.

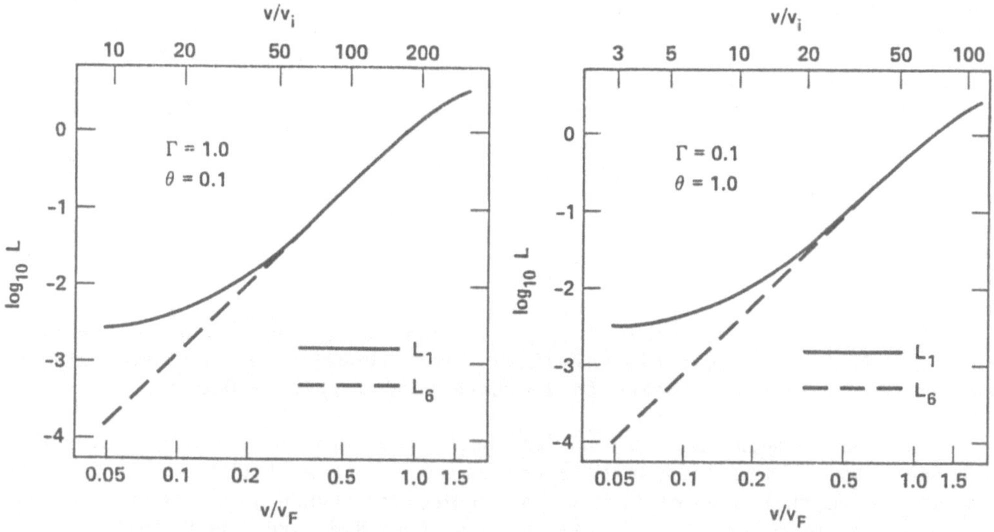

(a) $\Gamma = 1.0$ and $\theta = 0.1$ (b) $\Gamma = 0.1$ and $\theta = 1.0$

Fig. 3 Stopping number L versus v/v_F or v/v_i. L_1 is a TCP calculation with DLFC; L_6 is an electron OCP calculation in the RPA [Yan, Tanaka, Mitake and Ichimaru, 1985].

with $\alpha \equiv (4/9\pi)^{1/3}$. For an electron liquid in the paramagnetic state, which we are here concerned with, one has

$$r_s \frac{dE_c(r_s)}{dr_s} = b_o \frac{1 + b_1 x}{1+b_1 x+b_2 x^2+b_3 x^3} \quad , \quad (x \equiv \sqrt{r_s}) \quad , \tag{6.4}$$

where $b_o = 0.0621814$, $b_1 = 9.81379$, $b_2 = 2.82224$, and $b_3 = 9.736411$.

The short wavelength behavior of G(q) is related to the radial distribution function g(r) as

$$\lim_{q \to \infty} G(q) = 1 - g(0) \ . \tag{6.5}$$

The short range correlation can be described by the electron-electron ladder interactions

$$g(0) = \frac{1}{8} [\frac{z}{I_1(z)}]^2 \quad , \quad z \equiv 4(\alpha r_s/\pi)^{1/2} \quad , \tag{6.6}$$

where $I_1(z)$ is a modified Bessel function of the first order. Use of (6.6) in (6.5) thus determines the short wavelength behavior of G(q). To simulate the numerical results of the microscopic theory as well as to accommodate the boundary conditions, (6.2) and (6.5), it is appropriate to express

$$G(q) = Aq^4 + Bq^2 + [Aq^4 + (B + \frac{8}{3}A)q^2 - C] \frac{4-q^2}{4q} \ell n \left|\frac{2+q}{2-q}\right| \quad , \tag{6.7}$$

where

$$A = 0.029 \ , \ (0 \leq r_s \leq 15) \ , \tag{6.8}$$

$$B = \frac{9}{16} \gamma_o - \frac{3}{64} [1 - g(0)] - \frac{16}{15} A \quad , \tag{6.9}$$

$$C = - \frac{3}{4} \gamma_o + \frac{9}{16} [1 - g(0)] - \frac{16}{5} A \quad . \tag{6.10}$$

Eqs. (6.9) and (6.10) derive from (6.2) and (6.5). Eq. (6.8) is adapted so that Eq. (6.7) closely simulates the results of the microscopic theory. For $r_s > 15$, A begins to decrease gradually from 0.029.

The one-component and static LFC G(q) may be straightforwardly extended to a two-component electron-ion system (TCP) with indexed $G_{\mu\nu}(q)$. A further generalization is dynamical (DLFC), so that all the microscopic correlaction effects beyond the RPA are thus lumped into $G_{\mu\nu}(q,\omega)$. Once those LFC's are determined, the strong-coupling theory of a dense plasma is completed.

The $G_{\mu\nu}$'s may be viewed as relating the effective potentiel $\phi_{\mu\nu}(q,\omega)$ on a μ-species particle produced by the density fluctuation $\delta\rho_\nu(q,\omega)$ in the ν-species particles, which may be written as

$$\phi_{\mu\nu}(q,\omega) = Z_\mu Z_\nu v(q) \ [1 - G_{\mu\nu}(q,\omega)] \ \delta\rho_\nu(q,\omega) \tag{6.11}$$

This potential generally differs from the bare Coulomb potential $Z_\mu Z_\nu v(q)\ \delta\rho_\nu(q,\omega)$ because of the microscopic correlation effects involved; the difference is here measured by the dynamic LFC, $G_{\mu\nu}(q,\omega)$.

To elucidate the effects of LFC's and the contribution of the ion component to the stopping power, we consider with Yan et al. [1985] the stopping number L for several different cases of physical settings. First, we may regard the plasma either as a TCP or as a one-component plasma (OCP) of the electrons. Second, we may take account of the LFC's by the dynamic expressions [Yan et al., 1985], by the static values $G_{\mu\nu}(k)$, or by assuming $G_{\mu\nu}(q,\omega) = 0$; those will be designated, respectively, by DLFC, SLFC, or RPA. Combinations of those two sets of the prescriptions yield the following six cases:

L_1 : TCP and DLFC,
L_2 : TCP and SLFC,
L_3 : TCP and RPA,
L_4 : OCP and DLFC,
L_5 : OCP and SLFC,
L_6 : OCP and RPA.

One is inclined to regard L_1 as providing the most accurate evaluation of L for a given plasma.

The strength of Coulomb coupling in the classical ion system may be measured by a dimensionless parameter [Deutsch, 1982]

$$\Gamma \equiv \langle Z^{5/3}\rangle e^2/a_1 k_B T \quad , \quad a_1 = (\tfrac{4}{3}\pi n_1)^{-1/3}$$

where

$$\langle Z^{5/3}\rangle = \sum_{\mu\geq 2} n_\mu Z_\mu^{5/3} \Big/ \sum_{\mu\geq 2} n_\mu \ .$$

Following the schemes shown above we display the values of the stopping number for the two parametric combinations of hydrogen plasmas: (i) $\Gamma = 1.0$, $T_e = 0.1$; and (ii) $\Gamma = 0.1$, $T_e = 10$. The numerical results for L_1 and L_6 are plotted in Figs. 3 as functions of v/v_F and v/v_i, where

$$v_F = (3\pi^2 n)^{1/3}\ (h/m) \ ,$$

$$v_i = (k_B T/m_i)^{1/2} \ ,$$

and m_i refers to the mass of an ion (i.e., a proton).

In the low-velocity regime $v/v_F < 1$, the effects of both the LFC's and the ion component become significant so as to increase the magnitude of L, as Figs. 3 illustrate.

VII. STOPPING BY BOUND ELECTRONS

We are now concerned with an accurate determination of the mean excitation energy I which figures in the well-known Bethe formula for the stopping power

$$-\frac{dE}{dx} = 4\pi n \frac{(Z_1 e^2)^2}{m_e v_1^2} \ln \frac{2 m_e v_1^2}{I} \tag{7.1}$$

where Z_1 and v_1 refer to the atomic number of an incident particle and its non-relativistic velocity, respectively. n is the ion density in the target. Up to now, most of measurements of the ion stopping have been restricted to cold solid or gaseous targets. In order to optimize the efficiency of a pellet compression, however, an accurate estimation of the stopping parameters in dense and hot plasmas is required. The target material is likely to be heated up to temperatures of 50 - 200 eV, before fuel ignition starts up. Under these conditions, such tamper elements as Pb will be only partially ionized. Consequently, the stopping by bound electrons is no longer negligible and thus, we have to anticipate an energy range of a given ion beam in a cold matter and in an equivalent plasma for the same line density $n\ell$.

We are thus obviously required to extend to nonhydrogenic cold targets, the usual knowledge already used for neutral atoms and hydrogenic ions. Furthermore, we shall compute the corresponding I in hot and dense plasmas of atomic number Z_T.

A. I in Cold Targets

It has been shown recently that the standard Thomas-Fermi estimate $I \simeq 10.32\ Z_T$ eV may be extended to any ionicity through the radial electron density distribution [Green, Sellin and Zachor, 1969]

$$4\pi r^2\ \rho(r) = \frac{Nx}{d}\left(\frac{He^x}{(HT+1)^2}\right)\left(-1+\frac{2\ He^x}{1+HT}\right) \tag{7.2}$$

where $T = e^x - 1$ and $x = r/d$. N is the number of bound electrons is a given target atom (ion). The two parameters (a_o = Bohr radius).

$$0.5 \le d/a_o \le 1.3$$

$$H = 1.05 \times d \times N^{0.4}$$

specify Eq. (7.2) for a given target atom.

An average <I> may be given under the form

$$\langle I\rangle^2 = \frac{Z_T N}{6d}\ \frac{\left(1-\frac{N}{6Z_T}\left(1+\frac{1}{H}\right)\right)}{F\left(1-\frac{1}{H}\right)}\ ,\quad F(\alpha) = \sum_{n=0}^{\infty}\frac{\alpha^n}{(n+1)^2}\ \text{with } 0 \le \alpha \le 1, \tag{7.3}$$

with the atomic limit (Z_T = N)

$$\langle I\rangle^2 = \frac{Z^2}{6d}\cdot\frac{9+\alpha}{12F(\alpha)}\ , \tag{7.4}$$

In the TF limit ($Z_T \to \infty$) one recovers

$$\alpha = 1 \ , \qquad F(\alpha) = \frac{\pi^2}{6}$$

$$\frac{\langle I \rangle}{Z_T} = \sqrt{\frac{5}{6\,\pi^2 d}} \ , \tag{7.5}$$

which when combined to the most likely estimate $d/a_o \simeq 0.7$ provides

$$\langle I \rangle / Z_T \simeq 9.4462 \text{ eV} \ , \tag{7.6}$$

in fair agreement with cold TF estimates. The main interest of Eq. (7.3) lies in its flexibility. It applies to any ionicity and can be extended to partially stripped ions in plasmas.

Estimates (7.3) are compared in Table II with previous and more thorough (variational) calculations. Last column refers to the interpolation formula

$$\langle I \rangle (Z_T, q) \simeq \left(\frac{Z_T}{Z_T - q}\right)^2 \ I\ (Z_T - q, 0) \quad , \tag{7.7}$$

where q is the ionicity. This expression is accurate in both limits $q = 0$ (neutral) and $q = Z_T - 1$ (hydrogenic). Results (7.3) fall in pretty good agreement with the accurate ones (columns 4 and 2). The variational method is detailed elsewhere [Garbet-Deutsch, 1986].

Table II- Log I (au) for isolated ion Al^{n+}. Maximum discrepancy for the variational results are given within parenthese.

Ionicity	McGuire et al. (1982)	GSZ Eq. (7.3)	Variational method (1986)	Eq. (7.7)
0	1.51	1.61	1.62 (7.0%)	1.52
1	1.77	1.91	1.87 (4.5%)	1.65
2	2.11	2.13	2.16 (4.2%)	1.85
3	2.44	2.39	2.47 (3.5%)	2.10
4	2.48	2.53	2.59 (3.0%)	2.15
5	2.66	2.58	2.72 (2.7%)	2.20
6	2.82	2.75	2.87 (3.2%)	2.28
7	2.97	2.88	3.04 (3.2%)	2.37
8	3.16	3.04	3.24 (3.0%)	2.50
9	3.40	3.26	3.51 (3.0%)	2.71
10	3.82	3.68		3.15
11	4.47			4.10

B. I in Hot Targets

With the intention of extending the above results to the plasma case, we implement the average atom model.

The basic assumption in both cases is that during an elementary stopping process, the target ion is nearly instantaneously neutralized by plasma fluctuations. This hypothesis is particularly relevant to compressed matter by intense ion beams, in view of the relatively long pulse time (20-30 ns) allowing to consider the target species in local thermodynamic equilibrium (L.T.E.) with $T_e \neq T_i$.

The numerical procedure is initialized by taking each target ion embedded in a spherical neutral box of radius $R_o = (3/4\ \pi n)^{1/3}$.

Bound electrons are considered independent. They are supposed to move in a spherically symmetric are considered independent $V_{eff}(r)$, the same for all electrons. However, exchange and correlation are still retained in $V_{eff}(r)$. The total electron density

$$\rho(r) = \rho_b(r) + \rho_f(r) \qquad \text{fulfils}$$
$$4\pi \int_0^{R_0} dr\, r^2\, \rho(r) = Z_T \tag{7.8}$$

with $\rho_b(r)$ in terms of single-electron eigenquantities (ε_i, ψ_i).

The average atom assumption allows us to replace the various excitation states in target by those of a fictitious atom with noninteger occupation numbers for excited orbitals.
$V_{eff}(r)$ is taken constant within each subshell.

The number $\overline{Z}(n,T)$ of free electrons per nuclei in target may be initialized with a Thomas-Fermi approximation.

In Fig. 4, I_{av} computed at different temperatures ($T_e = T_i$) for high pressure aluminium are compared with several cold (standard atomic physics) matter methods: variational (the most accurate), local plasma approximation (LPA)

$$\log I_{av} = \int \log(\nu\hbar\omega_p(r))\, \rho_b(r)\, dr \quad ,$$

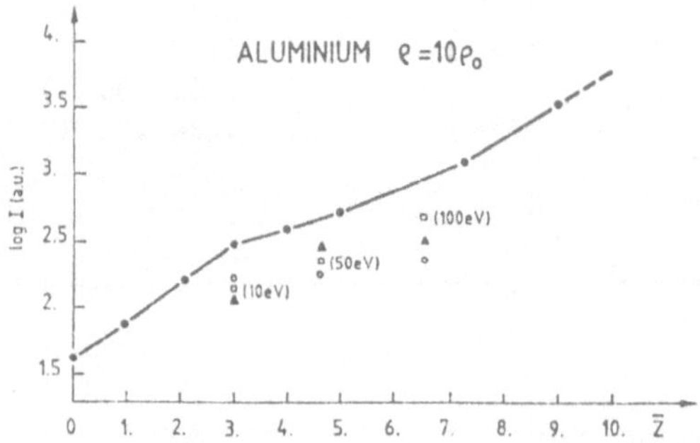

Figure 4- Mean excitation energy as a function of ionization with several approximations: (•) Isolated ions; (o) Local plasma approximation (LPA (ν = 2)); (□) Variational method; (▲) Spline interpolation with variational method; in Al at 10 times the solid density and high temperature.

$(\nu = \sqrt{2})$

with

$$\omega_p^2 = \frac{4\rho_b(r)e^2}{m_e} \tag{7.9}$$

and a Spline interpolation (cubic) based on variational results with (for subshell α)

$$\log I^{\alpha} \equiv \frac{d}{d\mu} \ln S_{\alpha}^{p}(\mu) \Big|_{\mu=0} \quad , \tag{7.10}$$

in terms of

$$S_{\alpha}^{p}(\mu) = \sum_{\alpha'} (1 - h_{\alpha'}) f_{\alpha\alpha'} E_{\alpha\alpha'}^{\mu} \quad ,$$

where $h_{\alpha'} = g_{\alpha'}/d_{\alpha'}$. $g_{\alpha'}$ is the number of electrons $\{\alpha'\}$, while $d_{\alpha'}$ refers to level degeneracy.

$f_{\alpha\alpha'}$ and $E_{\alpha\alpha'}$ denote the usual oscillator strength and energy difference for transition $\alpha \to \alpha'$, respectively.

All the results agree in showing [Garbet-Deutsch, 1986]

$$I\ (Z_T, n, T > 10 \text{ eV}) < I_{av}\ (Z_T,\ Z(n,T)) \ . \tag{7.11}$$

Up to a temperature of 10 eV, it is acceptable to deduce I_{av} from a Saha determination of ionization Z in target. However, at a higher temperature, occupation of higher subshells has to be taken into account, and their contribution to stopping is larger than that arising from more tightly bound electrons on lower subshells. The Spline interpolation adequately improved with variational inputs seems to provide fair data. As stated above the LPA produces the least accurate results.

At this juncture, it has to be appreciated that the accuracy of the variational results is not restricted by the numerical procedure. For instance, the mesh used for computing the basis wave functions may be taken as dense as required. Convergence criteria are mostly governed by convexity inequalities. The variational accuracy is motsly limited by the neglect of any electron exchange between neighbouring ions.

ACKNOWLEDGEMENTS

I express my gratitude to my collaborators and friends P. Fromy, X. Garbet and G. Maynard.

REFERENCES

Arista, N.R. and Brandt, W., 1981, Phys. Rev., **A23**, 1898.
Arista, N.R. and Brandt, W., 1985, J. Phys., **C18**, L1169.
Dar A., Grunzwei-Genossar, J., Peres, A., Revzen, M. and Ron, A., 1974, Phys. Rev. Lett., **32**, 1299.
Deutsch, C., 1982, Phys. script., **T2**, 192.
Deutsch, C., Maynard, G. and Minoo, H., 1983, J. Physique, **C8**, 67.
Deutsch, C., Laser and part. beams, 1984, **2**, 449.
Deutsch, C., 1986, Ann. Phys. Fr., **11**, 1.
Garbet, X. and Deutsch, C., 1986, Phys. Rev. A, submitted.
Green, A.E.S., Sellin, D.L. and Zachor, A.S., 1969, Phys. Rev., **184**, 1.
Gouedard, C. and Deutsch, C., 1978, J. Math. Phys., **19**, 32.
Ichimaru, S. and Utsumi, D., 1981, Phys. Rev., **B24**, 32.
Jackson, J.D., 1975, Classical Electrodynamics, (Wiley, New YOrk).
Kugler, A.A., 1975, J. Stat. Phys., **12**, 35.
Lindhard, J., 1954, Mat. Fis. Medd. Dan. Vid., Sel., **28**, 1.
Maynard, G. and Deutsch, C., 1982, J. Physique, **46**, 1113.
McGuire, E.J., Peek, J.M. and Pitchford, L., 1982, Phys. Rev., **A26**, 1318.
Nagy, I., Laszlo, J. and Giber, J., 1985, Z. Phys. **A321**, 221.
Nemet, G., AG, A. and Paris, G.Y., 1981, J. Math. Phys., **22**, 1192.
Skupsky, S., 1977, Phys. Rev., **A16**, 727.
Tanaka, S. and Ichimaru, S., 1985, J. Phys. soc. Jpn., **54**, 2537.
Yan, X.Z., Tanaka, S., Mitake, S. and Ichimaru, S., 1985, Phys. Rev., **A32**, 1785.

EXACT ASYMPTOTIC EXPRESSION FOR THE STATIC DIELECTRIC FUNCTION OF A UNIFORM ELECTRON LIQUID AT LARGE WAVE VECTORS

A. Holas

Institute of Physical Chemistry of the
Polish Academy of Sciences
Kasprzaka 44/52
01-224 Warsaw, Poland

I. INTRODUCTION

Although dielectric properties of a uniform electron liquid have been studied for more than 30 years, they are still known only approximately. Therefore, exact relations connecting various characteristic functions of an electron system are very important and useful for checking the accuracy and consistency of various approximations. As an example, let us recall the frequency-moment sum rules, which express the response function and other screening properties of an electron liquid at high frequencies in terms of such general characteristics as the plasma frequency, the mean kinetic energy and the static structure factor. Section II reviews some of these properties.

In this paper we will show that similar expressions, giving the response and other functions at large wave vectors and low frequencies, can be derived from the already known relations. In Section III we obtain such expressions for the response and dielectric functions, while in Section IV - for the local-field correction function. In Section V the Niklasson's relations are discussed and shown to be in agreement with expansions discussed in Sections II-IV. In Section VI numerical values of various coefficients, occurring in the mentioned expansions, are estimated. Some results obtain in the Appendix are helpful for these estimates. Conclusions are given in Section VII.

II. GENERAL PROPERTIES OF THE RESPONSE FUNCTION

A. Relation to the Dielectric Function. Units

A basic function, from which other characteristics may be derived, is the density-density response function $\chi(k,\omega)$, see e.g. [Pines and Nozieres, 1966]. The dielectric function $\epsilon(k,\omega)$ is related to is according to

$$1/\epsilon(k,\omega) = 1 + v_k \, \chi(k,\omega) \tag{2.1}$$

where

$$v_k = 4\pi e^2/k^2 \tag{2.2}$$

is the Fourier transform of the Coulomb potential.

Throughout the paper $k_F=(\alpha a_B r_s)^{-1}$ is used as the unit for wave vectors, $2E_F=\hbar^2 k_F{}^2 m_e{}^{-1}$ for energy and $2E_F\hbar^{-1}$ for frequency, where $\alpha = [4/9\pi)]^{1/3}$ $=0.521$, a_B is the Bohr radius and r_s characterizes the density of electrons $\rho = [r_s a_B)^3\ 4\pi/3]^{-1}$.

B. Analyticity and High Frequency Expansion

Analytical properties of the response function, based on the principle of causality, allow to write it in the form of the spectral representation, see e.g. [Kugler, 1975].

$$\chi(k,z) = \frac{1}{\pi}\int_{-\infty}^{\infty} \frac{d\omega\ \mathrm{Im}\ \chi(k,\omega)}{\omega - z} = \frac{2}{\pi}\int_0^{\infty} \frac{d\omega\ \omega\ \mathrm{Im}\ \chi(k,\omega)}{\omega^2 - z^2} \qquad \mathrm{Im}\ z > 0 \quad . \tag{2.3}$$

In writing the second form of Eq. (2.3) the odd symmetry of Im $\chi(k,\omega)$, was used. The physical response function $\chi(k,\omega)$, for real frequency ω, is obtained as a limit

$$\chi(k,\omega) = \lim_{0<\eta\to 0} \chi(k,\omega+i\eta) \tag{2.4}$$

From the spectral representation, Eq. (2.3), it is easy to obtain the asymptotic expansion for high frequencies ($|z| \to \infty$)

$$\chi(k,z) = -\frac{1}{z^2}\sum_{n=0}^{\infty} \frac{\chi_{M2n+1}(k)}{z^{2n}} \tag{2.5}$$

Here $\chi_{M\ell}(k)$ is the frequency moment, defined for any real ℓ as

$$\chi_{M\ell}(k) = \frac{2}{\pi}\int_0^{\infty} d\omega\ \mathrm{Im}\ \chi(k,\omega)\ \omega^{\ell} \quad . \tag{2.6}$$

Note that only integer-odd moments occur in the expansion (2.5).

For a uniform electron liquid the above integral is convergent, however, for limited ℓ only

$$-2 < \ell < 9/2 \tag{2.7}$$

because, for small ω [Pines and Nosieres, 1966]

$$\mathrm{Im}\ \chi(k,\omega) \propto \omega^1 \tag{2.8}$$

and for large ω [Glick and Long, 1971]

$$\mathrm{Im}\ \chi(k,\omega) \propto \omega^{-11/2} \tag{2.9}$$

Therefore, the series (2.5) for the system under consideration is actually limited to two regular terms and a small remainder:

$$\chi(k,z) = -\frac{\chi_{M1}(k)}{z^2} - \frac{\chi_{M3}(k)}{z^4} + o(\frac{1}{z^4}) \tag{2.10}$$

C. <u>The 1-st and 3-rd Moment Sum Rules</u>

The values of the 1-st and 3-rd moment may be calculated directly from the Hamiltonian as equal-time commutators, see e.g. [Kugler, 1975], [Pathak and Vashishta, 1973].

$$\chi_{M1}(k) = -\frac{\omega_p^2}{v_k} , \tag{2.11}$$

$$\chi_{M3}(k) = -\frac{\omega_p^2 k^4}{v_k^4} \{ 1 + 8\langle E_K\rangle \frac{1}{k^2} + 4\omega_p^2[1-G^{PV}(k)] \frac{1}{k^4} \} , \tag{2.12}$$

where

$$\omega_p^2 = 4\alpha r_s/(3\pi) \tag{2.13}$$

is the squared plasma frequency, $\langle E_K\rangle$ is the exact mean kinetic energy per electron, and the function $G^{PV}(k)$ (PV denotes Pathak and Vashishta [1973]) is defined in terms of the exact static structure factor S(k) according to:

$$G^{PV}(k) = \frac{3}{4}\int_0^\infty dq\, q^2[1-S(q)] [\frac{5}{6} - \frac{q^2}{2k^2} + \frac{(k^2-q^2)^2}{4\, k^3\, q} \ln \left| \frac{k+q}{k-q} \right|] \tag{2.14}$$

A finite limit of the function $G^{PV}(k)$ exists

$$\lim_{k\to\infty} G^{PV}(k) = G^{PV}(\infty) = \frac{2}{3} [1 - g(0)] \tag{2.15}$$

where g(r) is the pair correlation function. Note that $\langle E_K\rangle$, $G^{PV}(k)$, S(k), g(r) are implicit functions of the electron gas density (or r_s).

Equations (2.11) and (2.12) are called the 1-st and 3-rd frequency-moment sum rules. Equation (2.10) together with (2.11) and (2.12) gives an expression for the response function in the high frequency range, in terms of $\langle E_K\rangle$ and $G^{PV}(k)$ - two characteristics of the electron system.

D. <u>The 0-th Moment Sum Rule</u>

The dynamic structure factor is expressed in terms of the response function, for T=0, [Pines and Nozieres, 1966] as

$$S(k,\omega) = \begin{cases} 0 \text{ for } \omega<0 \\ -[v_k k^2/(\omega_p^2\pi)] \,\mathrm{Im}\, \chi(k,\omega) \text{ for } \omega \geq 0 \end{cases} \tag{2.16}$$

Therefore the static structure factor, defined as

$$S(k) = \int_{-\infty}^{\infty} d\omega\ S(k,\omega) \tag{2.17}$$

is simply related to the 0-th moment:

$$\chi_{M0}(k) = -\frac{2\omega_p^2}{v_k k^2} S(k) \quad . \tag{2.18}$$

The asymptotic form of the static structure factor for large k is known due to Kimball [1973, 1975]

$$S(k) = 1 + \frac{C}{k^4} + o(\frac{1}{k^4}) \tag{2.19}$$

where

$$C = -\frac{8}{3\pi}\frac{dg(r)}{dr}\bigg|_{r=0} = -2\ \omega_p^2\ g(0) \tag{2.20}$$

Note that the dynamic structure factor $S(K,\omega)$ (loss function) must be non-negative, therefore, according to Eq. (2.16) Im $\chi(k,\omega) \leq 0$ within the limits of integration in Eq. (2.6). So that

$$\chi_{M\ell}(k) < 0 \tag{2.21}$$

for any k and any ℓ.

III. THE RESPONSE FUNCTION AT LOW FREQUENCIES

A. Asymptotic Expansion at Low Frequencies

From the spectral representation of $\chi(k,z)$, Eq. (2.3), the following asymptotic expansion at low frequencies ($|z| \to 0$) may be obtained

$$\chi(k,z) = \sum_{n=0}^{\infty} \chi_{M(-1-2n)}(k)\ z^{2n} \tag{3.1}$$

in terms of the odd-negative frequency moments.

For a uniform electron liquid, because of the relation (2.7), only one such moment exists, χ_{M-1}, so in this case expansion (3.1) is reduced to

$$\chi(k-z) = \chi_{M-1}(k) + o(z^0) \tag{3.2}$$

or, equivalently, to

$$\chi(k,0) = \chi_{M-1}(k) \quad . \tag{3.3}$$

As a result, that static response function may be calculated as the (-1)-st frequency moment.

It must be noted, however, that for different systems or in some approximate theories, a few odd-negative moments may exist. In this case the expansion (3.1) is more useful, because a larger number of terms may be retained in it. In particular, in RPA or in the first order perturbation theory (FOPT) of a uniform electron gas [Holas, Aravind, and Singwi, 1979], for $k>2$, Im $\chi(k,\omega)$ is exactly zero for $0 \leq \omega < (\frac{k^2}{2} - k)$ and for $\omega > (\frac{k^2}{2} + k)$ so all negative and all positive moments do exist. In such cases other odd-negative moments may be calculated (for large k) using the same method as developed in Section III.D. for $\chi_{M-1}(k)$.

B. Reduced-Frequency Moments

According to Eq. (2.6) the (-1)-st moment is given by

$$\chi_{M-1}(k) = \frac{2}{\pi} \int_0^\infty \frac{d\omega \; \text{Im} \; \chi(k,\omega)}{\omega} \tag{3.4}$$

Evaluating this integral for large k, we see that the main contribution comes from the frequency range close to the peak position of Im $\chi(k,\omega)$, namely around $\omega = k^2/2$. The width of this peak is of the order of k. In this way characteristic frequencies $k^2/2 \pm k$ appear in the denominator of the integrand in Eq. (3.4). Owing to the fact that the ratio of the peak width to the peak position, ~1/k, is small at large k, asymptotic expansion of $\chi_{M-1}(k)$ will be possible.

As we already mentioned, within the RPA or in the FOPT, Im $\chi_{(k,\omega)}=0$ outside the frequency range $[\frac{k^2}{2} - k, \frac{k^2}{2} + k]$. In the exact theory, weak wings are present on both sides of the main peak, due to multi-pair scattering processes (contributions higher than the 1-st order). Therefore,
expressing Im $\chi(k,\omega)$ in terms of the reduced frequency ν, instead of ω,

$$\nu = \nu(k,\omega) = (\omega - \frac{k^2}{2})/k = -\frac{k}{2}(1 - \frac{2}{k^2}\omega) \tag{3.5}$$

we make the shape of this function (vs ν), weakly dependent on k in the region of large k.

In order to take advantage of this fact, let us introduce frequency moments, connected with this reduced frequency

$$\chi_{\mu\ell}(k) = \frac{2}{\pi} \int_0^\infty d\omega \; \text{Im} \; \chi(k,\omega) \; [\nu(k,\omega)]^\ell \tag{3.6}$$

The above definition of the reduced-frequency moment is similar to the previous definition (2.6) of $\chi_{M\ell}(k)$.

For non-negative integer ℓ, a moment of one type is a linear combination of the finite number of moments of other type, and vice versa. Namely, from Eq. (3.6) and (3.5) it follows that

$$\chi_{\mu\ell}(k) = (-\frac{k}{2})^{\ell} \sum_{n=0}^{\ell} \binom{\ell}{n} (-\frac{2}{k^2})^n \chi_{Mn}(k) \quad . \tag{3.7}$$

From Eq. (3.7) and the limitations (2.7), pertaining $\chi_{M\ell}(k)$, we conclude that for a uniform electron gas $\chi_{\mu\ell}(k)$ exists for

$$\ell \leq 4 \tag{3.8}$$

Inverting the relation (3.5)

$$\omega = \omega(k,\nu) = \frac{k^2}{2} + k\,\nu = \frac{k^2}{2}\,(1 + \frac{2}{k}\,\nu) \tag{3.9}$$

and substituting it into Eq. (2.6), we obtain a relation opposite to (3.7)

$$\chi_{M\ell}(k) = (\frac{k^2}{2})^{\ell} \sum_{n=0}^{\ell} \binom{\ell}{n} (\frac{2}{k})^n \chi_{\mu n}(k) \tag{3.10}$$

C. <u>Properties of the Reduced-Frequency Moments at Large K</u>

In order to investigate $\chi_{\mu\ell}(k)$ vs k, let us rewrite Eq. (3.6) using ν as a variable of integration, Eq. (3.9),

$$\chi_{\mu\ell}(k) = \frac{2}{\pi}\int_{-\infty}^{\infty} d\nu\ \tilde{\chi}(k,\nu)\ \nu^{\ell} \tag{3.11}$$

Here

$$\tilde{\chi}(k,\nu) = k \,\mathrm{Im}\,\chi(k,\frac{k^2}{2} + k\nu)\ \theta(\nu + \frac{k}{2}) \quad , \tag{3.12}$$

and $\theta(x)$ is a unit step function.

Comparing definition (3.6) with (2.6) at ℓ=0 we see that

$$\chi_{\mu 0}(k) = \chi_{M0}(k) \quad . \tag{3.13}$$

At large k, using Eq. (2.18) and (2.19), we find that a finite limit exists

$$\lim_{k\to\infty} \chi_{\mu 0}(k) = \chi_{\mu 0}(\infty) = -\frac{2\omega_p^2 k_F^2}{4\pi e^2} \quad . \tag{3.14}$$

On the other hand, from Eq. (3.11) we have for the same limit the expression

$$\lim_{k\to\infty} \chi_{\mu 0}(k) = \frac{2}{\pi}\int_{-\infty}^{\infty} d\nu\ \tilde{\chi}(\infty,\nu) \quad . \tag{3.15}$$

This means that $\tilde{\chi}(\infty,\nu)$ must exist as an integrable function. So all other moments, Eq. (3.11), must exist in this limit

$$\lim_{k \to \infty} \chi_{\mu\ell}(k) = \frac{2}{\pi} \int_{-\infty}^{\infty} d\nu \, \tilde{\chi}(\infty,\nu) \, \nu^{\ell} = \chi_{\mu\ell}(\infty) \quad . \tag{3.16}$$

From this property we conclude immediately that the leading term in the sum (3.10) corresponds to n=0. So

$$\chi_{M\ell}(k) = (\frac{k^2}{2})^{\ell} \, \chi_{\mu 0}(\infty) \, [\, 1 + 0(\frac{1}{k}) \,] \quad , \tag{3.17}$$

i.e. for any ℓ the leading term of $\chi_{M\ell}(k)$ for large k is expressed in terms of one common constant - $\chi_{\mu 0}(\infty)$, Eq. (3.14), while its magnitude is of the order of $k^{2\ell}$. So it will be convenient to introduce the "normalized" moments

$$h_{\ell}(k) = \chi_{M\ell}(k) \; / \; [\; (\frac{k^2}{2})^{\ell} \, \chi_{M0}(\infty) \;] \quad , \tag{3.18}$$

asymptotic expansion of which must be in the form

$$h_{\ell}(k) = \sum_{n=0}^{\infty} \frac{a_n(\ell)}{k^n} \tag{3.19}$$

with

$$a_0(\ell) = 1 \quad . \tag{3.20}$$

Frequency moments with ℓ=0,1,3 are known exactly, Eqs. (2.18), (2.11), and (2.12). We may expand them in the series (3.19) and find their coefficients $a_n(\ell)$. They are listed in Table I in the columns headed ℓ=0,1,3.

It is interesting and important that some coefficients $a_n(\ell)$ of the second and 4-th moments (which, as it will be shown later, are necessary for evaluation of the static response) can be determined from the already listed coefficients. This possibility follows from the properties of Eq. (3.7). Its l.h.s. is $0(k^0)$ (because of Eq. (3.16)), while the r.h.s. is $0(k^{\ell})$ (because of Eq. (3.17)). In order to remove this contradiction, we must require that the coefficients at k^{ℓ}, $k^{\ell-1}$, ..., k^1, be zero. For example, let us take ℓ=3, the coefficient at k^2:

$$a_2(0) - 3\, a_2(1) + 3\, a_2(2) - a_2(3) = 0 \quad . \tag{3.21}$$

So the coefficient of the second moment $a_2(2)$ can be determined from the coefficients on the 0-th, 1-st, and 3-rd one.

In general, at fixed n, among coefficients $a_n(\ell)$, only $a_n(0)$, $a_n(1)$,...,$a_n(n)$ are independent. For ℓ>n, $a_n(\ell)$ is a linear combination of $a_n(m)$, $0 \leq m \leq n$. In this way some other coefficients for ℓ=2 and ℓ=4 are obtained (see corresponding columns in Table I). Unfortunately, three of them - $a_3(2)$, $a_4(2)$ and $a_4(4)$ - are unknown.

D. Asymptotic Expansion of $\chi(k,0)$ at Large K

In order to obtain an asymptotic expansion of the static response function, Eq. (3.3), we calculate $\chi_{M-1}(k)$, Eq. (3.4), in which ω is substituted according to Eq. (3.9)

$$\chi(k,0) = \frac{2}{\pi}\int_0^\infty \frac{d\omega \, \mathrm{Im}\,\chi(k,\omega)}{\frac{k^2}{2}\,(1+\frac{2}{k}\nu)} = \frac{2}{k^2}\sum_{\ell=0}^{\infty}\left(-\frac{2}{k}\right)^\ell \chi_{\mu\ell}(k). \tag{3.22}$$

Since for the system being considered the power index ℓ is bounded by Eq. (3.8), the series is limited in this case to

$$\chi(k,0) = \frac{2}{k^2}\left[\sum_{\ell=0}^{4}\left(-\frac{2}{k}\right)^\ell \chi_{\mu\ell}(k) = o\left(\frac{1}{k^4}\right)\right]. \tag{3.23}$$

Note that because of the property (3.16), the order of the magnitude of each term of the expansion (3.22) is (1/k)-times smaller than the order of the preceeding term.

By means of simple algebra, expansion of $\chi_{\mu\ell}(k)$ is obtained from expansions of $\chi_{Mn}(k)$, $0 \le n \le \ell$,. Eq. (3.7), and finally, expansion of $\chi(k,0)$ from expansions of $\chi_{\mu\ell}(k)$, Eq. (3.23). The result is

$$\chi(k,0) = -\frac{4\,\omega_p^2}{v_k k^4}\left[1 + \frac{8}{3}\langle E_K\rangle\frac{1}{k^2} + a_3(-1)\frac{1}{k^3} + a_4(-1)\frac{1}{k^4} + o\left(\frac{1}{k^4}\right)\right] \tag{3.24}$$

where

$$a_3(-1) = 4\,a_3(2)\ , \tag{3.25}$$

$$a_4(-1) = -\frac{10}{3}\,\omega_p^2\,[\,2 - 7\,g(0)\,] + 10\,a_4(2) + a_4(4)\ . \tag{3.26}$$

These coefficients are obtained from $a_n(\ell)$, $0<\ell<4$, listed in Table I. For systems with different coefficients the general formula is

$$a_n(-1) = \sum_{\ell=0}^{n}(-1)^\ell\binom{n+1}{\ell+1}a_n(\ell)\ . \tag{3.27}$$

Table I. Expansion coefficients $a_n(\ell)$ for exact theory.

$n \backslash \ell$	-1	0	1	2	3	4
0	1	1	1	1	1	1
1	0	0	0	0	0	0
2	$\frac{8}{3}\langle E_K\rangle$	0	0	$\frac{8}{3}\langle E_K\rangle$	$8\langle E_K\rangle$	$16\langle E_K\rangle$
3	$4\,a_3(2)$	0	0	$a_3(2)$	0	$-6a_3(2)$
4	$a_4(-1)$, Eq. (3.26)	$-2\omega_p^2 g(0)$	0	$a_4(2)$	$\frac{4}{3}\,\omega_p^2[1+2g(0)]$	$a_4(4)$

Expansion (3.24) is the main result of the present paper. It gives explicitly two leading terms of the exact asymptotic expansion of the static response function.

E. Inferring the Parameters $a_3(-1)$ and $a_4(-1)$

Expansion coefficients of the 2-nd and 4-th moment, $a_3(2)$, $a_4(2)$, $a_4(4)$, necessary to find the exact $a_3(-1)$ and $a_4(-1)$, Eqs. (3.25) and (3.26), are not available. Therefore, we are going to infer these parameters employing arguments of continuity.

We notice from Table I that for known moments (ℓ=0,1,3) expansion coefficients $a_n(\ell)$ with odd n, are zero. But, according to the definition (2.6), $\chi_{M\ell}(k)$ at fixed k is a continuous, smooth function of ℓ. Therefore, $a_n(\ell)$ must be also such a function. If we disregard such unlikely possibility that $a_3(\ell)$ vs ℓ oscillates, going through zero exactly at ℓ=0,1,3, we should conclude that it is identically zero for all ℓ, so

$$a_3(-1) = 0 \quad . \tag{3.28}$$

In order to infer $a_4(-1)$ let us employ arguments based on continuity with respect to coupling parameter r_s. For a free (non-interacting, r_s=0) electron gas Im $\chi(k,\omega)$ is very simply and all its moments may be easily calculated. Performing the integrations of Eq. (2.6) we obtain the following expansion coefficients $a^0{}_n(\ell)$ (there is no restriction on ℓ):

$$a_2^o(\ell) = \ell(\ell-1)\,\frac{4}{3}\,\langle E_K\rangle_o \quad , \tag{3.29}$$

$$a_4^o(\ell) = \ell(\ell-1)(\ell-2)(\ell-3)\,\frac{8}{15}\,\langle E_K^2\rangle_o \quad , \tag{3.30}$$

$$a^0_{2n+1}(\ell) = 0 \quad . \tag{3.31}$$

The quantities $\langle E_K\rangle_o$ and $\langle E^2{}_K\rangle_o$ are mean values per one electron of the kinetic energy and its square, respectively, calculated for noninteracting electron system. To facilitate further comparison, we put some values of the parameters $a^0{}_n(\ell)$ into Table II.

Table II. Expansion coefficients $a^0{}_n(\ell)$ for free electrons.

n \ ℓ	-1	0	1	2	3	4
0	1	1	1	1	1	1
1	0	0	0	0	0	0
2	$\frac{8}{3}\langle E_K\rangle_o$	0	0	$\frac{8}{3}\langle E_K\rangle_o$	$8\langle E_K\rangle_o$	$16\langle E_K\rangle_o$
3	0	0	0	0	0	0
4	$\frac{64}{5}\langle E_K^2\rangle_o$	0	0	0	0	$\frac{64}{5}\langle E_K^2\rangle_o$

In Table III we list coefficients derived from the results of Holas, Aravind and Singwi [1979] in the FOPT, for ℓ=-1,0,1,3. Values for ℓ=2 and 4 are determined with the help of Eq. (3.21) and (3.27). Here $g_o(r)$ is the pair correlation function for non-interacting electron gas.

We see that both in the free-electron and FOPT case, all $a_3(\ell)$ are zero, supporting our previous conjecture (3.28). Next, whenever $\langle E_K\rangle_o$ appears in Table II, it is also present at the same position in Table III, and, replaced by $\langle E_K\rangle$ (i.e. the mean value for <u>interacting</u> electrons), in the same position in Table I. The same is true about $\langle E^2_K\rangle_o$ in Tables II and III. Finally, whenever the combination of ω^2_p and $\omega^2_p\, g_o(0)$ appears in Table III, it enters also Table I with $g_o(0)$ replaced by $g(0)$ - its counterpart for interacting electrons. We see also that expressions for $a_n(\ell)$ in Table I contain no other terms than ones present in Table III. If we apply the established rule that $\langle\ldots\rangle_o$ must be replaced by $\langle\ldots\rangle$ and $g_o(0)$ by $g(0)$, when going from Table III to Table I, we obtain for $a_4(-1)$:

$$a_4(-1) = \frac{64}{5} \langle E_K^2\rangle - \frac{4}{3}\omega_p^2 [1 + 2g(0)] \tag{3.32}$$

Note that for any other coefficient of Table I no assumption of weak interaction was made. In this sense the expression (3.32) for $a_4(-1)$ should hold also for any r_s, while the use of the FOPT allowed us to infer its form only. Because the established rules of extrapolation from the FOPT to the exact theory are quite natural and obvious, we consider the results (3.32) and (3.28) as "almost exact".

F. <u>Asymptotic Expansion of $\epsilon(k,0)$</u>

Using Eq. (2.1) we may calculate from Eq. (3.24) with (3.32) and (3.28) the asymptotic expansion of the static dielectric function $\epsilon(k,0)$ and static proper polarizability $Q(k,0)$:

$$\epsilon(k,0) - 1 = Q(k,0) = \frac{4\,\omega_p^2}{k^4} \{1 + \frac{8}{3}\langle E_K\rangle \frac{1}{k^2} + \left[\frac{64}{5}\langle E_K^2\rangle + \frac{8}{3}\omega_p^2 (1-g(o)) \right] \frac{1}{k^4} + o(\frac{1}{k^4}) \} \tag{3.33}$$

Table III. Expansion coefficients $a^{FO}_n(\ell)$ obtained from the FOPT.

n \ ℓ	-1	0	1	2	3	4
0	1	1	1	1	1	1
1	0	0	0	0	0	0
2	$\frac{8}{3}\langle E_K\rangle_o$	0	0	$\frac{8}{3}\langle E_K\rangle_o$	$8\langle E_K\rangle_o$	$16\langle E_K\rangle_o$
3	0	0	0	0	0	0
4	$\frac{64}{5}\langle E_K^2\rangle_o-\frac{4}{3}\omega_p^2[1+2g_o(0)]$	$-2\omega_p^2 g_o(0)$	0	$a_4^{FO}(2)$	$\frac{4}{3}\omega_p^2[1+2g_o(0)]$	$\frac{64}{5}\langle E_K^2\rangle_o + \frac{\omega_p^2}{3}[16+62g_o(0)] - 10a_4^{FO}(2)$

IV. THE LOCAL-FIELD CORRECTION FUNCTION

A. Definition

All alternations, which must be introduced to the RPA response function, in order to include the exchange and correlation effects in the exact response function, are usually incorporated in the so called local-field correction function $G(k,\omega)$, in terms of which the response function is written as

$$\chi(k,\omega) = \chi_0(k,\omega) \;/\; \{ 1 - v_k [1 - G(k,\omega)] \chi_0(k,\omega) \} \tag{4.1}$$

where $\chi_0(k,\omega)$ is the Lindhard's function (the response function of free electrons)

$$\chi_0(k,\omega) = \frac{\omega_p^2}{v_k} < (\omega+i0^+ - \vec{k}\cdot\vec{q})^2 - \frac{k^4}{4}]^{-1} >_0 \quad . \tag{4.2}$$

The average $<...>_0$ is defined as

$$< f(\vec{q}) >_0 = \frac{2}{N} \sum_{\vec{q}} n_0(|\vec{q}|) f(\vec{q}) = \frac{3}{4\pi} \int_\infty d^3q \, n_0(|\vec{q}|) f(\vec{q})$$

where $n_0(q)$ is an occupation number fraction for free (non-interacting) electrons

$$n_0(q) = \theta(1-q) \tag{4.4}$$

Equation (4.1) servers as a definition of $G(k,\omega)$. We may solve it with respect to $G(k,\omega)$

$$G(k,\omega) = 1 + [v_k\chi(k,\omega)]^{-1} - [v_k\chi_0(k,\omega)]^{-1} \tag{4.5}$$

B. Asymptotics of G(k,0) for Large k

Now we are going to obtain an asymptotic expansion of $G(k,0)$ for large k, using Eq. (4.5). Such expansion of $\chi(k,0)$ is already found, Eq. (3.24) with (3.28) and (3.32), while expansion of $\chi_0(k,0)$ may be easily obtained from Eq. (4.2)

$$\chi_0(k,0) = - \frac{\omega_p^2}{v_k} \frac{4}{k^4} < [1 - \frac{4(\vec{k}\cdot\vec{q})^2}{k^4}]^{-1} >_0$$

$$= - \frac{4\omega_p^2}{k^4 v_k} [1 + \frac{4}{3}<q^2>_0 \frac{1}{k^2} + \frac{16}{5}<q^4>_0 \frac{1}{k^4} + 0(\frac{1}{k^6})] \tag{4.6}$$

(it agrees, of course, with Table II, Column ℓ = -1). So, from Eq. (4.5) at $\omega = 0$:

$$G(k,0) = \frac{2}{3\omega_p^2} [\langle E_K \rangle - \langle \frac{q^2}{2} \rangle_o] k^2 + \frac{1}{4\omega_p^2} \{ \frac{8}{3} \omega_p^2 [1-g(0)] + \frac{64}{5}$$

$$[\langle E_K^2 \rangle - \langle \frac{q^4}{4} \rangle_o] - \frac{64}{9} [\langle E_K \rangle^2 - \langle \frac{q^2}{2} \rangle_o^2] \} + o(k^0) \quad . \qquad (4.7)$$

The leading terms, $\sim k^4$, cancelled out exactly. Note that $E(q)=q^2/2$ is the leading kinetic energy of an electron characterized by the wave vector q. Therefore, $\langle \frac{q^2}{2} \rangle_o$ is the mean kinetic energy of free electrons, while $\langle \frac{q^4}{4} \rangle_o$ is $\langle E_K^2 \rangle_o$. It is easy to calculate from Eq. (4.3) using (4.4) that

$$\langle E_K \rangle_o = \frac{3}{10} \quad , \quad \langle E_K^2 \rangle_o = \frac{3}{28} \quad . \qquad (4.8)$$

In terms of the relative change, due to interaction, of the mean kinetic energy and the mean squared kinetic energy

$$\delta_K = [\langle E_K \rangle - \langle E_K \rangle_o] / \langle E_K \rangle_o \quad , \qquad (4.9)$$

$$\delta_{KK} = [\langle E_K^2 \rangle - \langle E_K^2 \rangle_o] / \langle E_K^2 \rangle_o \quad , \qquad (4.10)$$

we obtain finally the following exact result for the static local-field correction in the large-k limit (which is equivalent to the result (3.24) or (3.33))

$$G(k,0) = \frac{\delta_K}{5\omega_p^2} k^2 + \frac{2}{3} [1-g(0)] + \frac{12\ \delta_{KK}}{35\ \omega_p^2} - \frac{4(2\delta_K + \delta_K^2)}{25\ \omega_p^2} + o(k^0) \quad . \qquad (4.11)$$

C. Asymptotic Behavior of $G(k,\omega)$ for Large ω

Although the asymptotic behavior is well known (see e.g. [Kugler, 1975]), we are going to derive it again in order to demonstrate the similarities between the two (large-k and large-ω) expansions. We need the asymptotic expansion of the Lindhard's function. It may be immediately obtained from the definition (4.2) by expansion of the denominator in powers of $1/\omega$. The results may be written as

$$\chi_o(k,\omega) = - \chi_{oM1}(k) \frac{1}{\omega^2} - \chi_{oM3}(k) \frac{1}{\omega^4} + O(\frac{1}{\omega^6}) \qquad (4.12)$$

where

$$\chi_{oM1}(k) = - \frac{\omega_p^2}{v_k} \qquad (4.13)$$

$$\chi_{oM3}(k) = - \frac{\omega_p^2 k^4}{v_k 4} [1 + 8 \langle \frac{q^2}{2} \rangle_0 \frac{1}{k^2} \qquad (4.14)$$

(this agrees, of course, with Table II, columns $\ell=1,3$). Substituting (4.12) and (2.10) into Eq. (4.5) we get

$$G(k,\omega) = 1 + \frac{1}{v_k} \{ [- \frac{1}{\chi_{M1}(k)} + \frac{1}{\chi_{oM1}(k)}] \omega^2 + [\frac{\chi_{M3}(k)}{(\chi_{M1}(k))^2} - \frac{\chi_{oM3}(k)}{(\chi_{oM1}(k))^2}] \omega^0 + o(\omega^0) \} \quad (4.15)$$

The terms $\sim\omega^2$ cancel out because of 1-st moments are equal, Eq. (2.11) and (4.13). Using (2.12) and (4.14) we get

$$\lim_{\omega\to\infty} G(k,\omega) = - \frac{2}{\omega_p^2} [\langle E_K \rangle - \langle \frac{q^2}{2} \rangle_0] k^2 + G^{PV}(k) \quad (4.16)$$

Note that terms $\sim k^4$ cancelled out in $\chi_{M3} - \chi_{oM3}$. Using (2.15), (4.9) and (4.8) we may rewrite (4.16) on the form analogous to (4.11) for large k

$$G(k,\infty) = - \frac{3\delta_K}{5\omega_p^2} k^2 + \frac{2}{3} [1-g(0)] + o(k^0) \quad (4.17)$$

D. The Leading Term of G(k,0) and G(k,∞) Expansions.

By comparing Eqs. (4.11) and (4.17) we see that the leading term in each case has essentially the same structure. The latter is obtained from the former when multiplied by a numerical factor (-3). In both cases the presence of δ_K reflects the fact that the mean kinetic energy $\langle E_K \rangle$, occurring in the exact response function $\chi(k,\omega)$ (see Eq. (3.24) or (2.10) with (2.13)) differs from the mean kinetic energy $\langle E_K \rangle_0$ in the reference system - free electron gas, represented in (3.5) by its response function $\chi_0(k,\omega)$.

It may be proven that the quantity δ_K is always positive (see Appendix, Eq. (A.14)). Then from (4.11) and (4.17) we see that the local field correction function for large k is of the order of k^2, positive in the low-frequency range and negative in the high-frequency range.

V. NIKLASSON'S RELATIONS

A. Apparent contradiction with our results and its solution.

Niklasson [1974] has derived two relations concerning particular limits of the local-field correction, namely at finite k,

$$\lim_{\omega\to\infty} G_I(k,\omega) = G^{PV}(k) \quad (5.1)$$

where the function $G^{PV}(k)$ was already introduced in Eq. (2.14), and finite ω,

$$\lim_{k\to\infty} G_I(k,\omega) = \frac{2}{3} [1-g(0)] = G^{PV}(\infty) \quad (5.2)$$

The relation (5.1) differs drastically from the result (4.16), and the same may be said about relation (5.2) (at ω=0) and our result (4.11). In both cases the Niklasson expressions lack the leading, $\sim k^2$, term.

But apparent contradiction may be solved easily, because the definition of the local-field correction, used by Niklasson

$$\chi(k,\omega) = \chi_{Io}(k,\omega) / \{ 1 - v_k[1-G_I(k,\omega) \chi_{Io}(k,\omega) \} \qquad (5.3)$$

differs from the traditional definition (4.1). Namely, the "reference" function $\chi_o(k,\omega)$ in Eq. (4.1), which is the Lindhard function, Eq. (4.2), is replaced by Niklasson by another function - $\chi_{Io}(k,\omega)$, defined analogically as in Eq. (4.2), but with the average $<...>_o$ replaced by $<...>_I$. The meaning of this average is the same as in definition (4.3), accept that $n_o(q)$ is to be replaced by $n_I(q)$ - the exact occupation number function for interacting (I) electrons. Because Eqs. (4.1) and (5.3) define the same response $\chi(k,\omega)$, the two local-field corrections G and G_I must be related:

$$G(k,\omega) = G_I(k,\omega) + [v_k \chi_{Io}(k,\omega)]^{-1} - [v_k \chi_o(k,\omega)]^{-1} \qquad (5.4)$$

The average <...>, used in all previous chapters, has exactly the same meaning as $<...>_I$, introduced now.

B. High-Frequency Case

Now we are able to calculate $G(k,\omega)$ in the same limiting cases, as given in the Niklasson's relations (5.1) and (5.2).

The expansion of $\chi_o(k,\omega)$ for large ω is already found, Eq. (4.12). The expansion for $\chi_{Io}(k,\omega)$ will be the same, except the average $<...>_o$ will be replaced by $<...>_I$ in (4.14). After substituting these expansions and the relation (5.1) into (5.4) we get

$$\lim_{\omega\to\infty} G(k,\omega) = - \frac{3\delta_K}{5\omega_p^2} k^2 + G^{PV}(k) \qquad (5.5)$$

which is exactly the same as Eq. (4.16) and (4.17). The term $\sim\delta_K k^2$ arose in (5.5) due to $[\chi_{IoM3} - \chi_{oM3}] \propto [<\frac{q^2}{2}>_I - <\frac{q^2}{2}>_o] k^2$.

C. Large-k, Low-ω Case

Let us investigate the large-k limit of G(km0). The expansion of $\chi_o(k,\omega)$ was found in Eq. (4.6). The expansion for $\chi_{Io}(k,\omega)$ needs the average $<...>_o$ to be replaced by $<...>_I$. Substituting these expansions and Niklasson's relation (5.2) into Eq. (5.4), we obtain after some algebra

$$G(k,0) = \frac{\delta_K}{5\omega_p^2} k^2 + \frac{2}{3} [1-g(0)] + \frac{12\delta_{KK}}{35\omega_p^2} - \frac{4(2+\delta_K)\delta_K}{25\omega_p^2} + o(k^0) \qquad (5.6)$$

Again the term $\sim\delta_K k^2$ arises due to the difference between $<q^2>_I$ and $<q^2>_o$, present in χ_{Io}^{-1} and χ_o^{-1} in Eq. (5.4).

We see that the expansion (5.6), based on Niklasson's relation (5.2), is exactly the same as our result (4.11). Therefore, the arguments used by Niklasson to derive his relations (5.1) and (5.2) serve also as arguments in favor of validity of the expressions (3.28) and (3.32) for $a_3(-1)$ and $a_4(-1)$ - the "almost exact" part of our results.

VI. NUMERICAL ESTIMATES

As it was shown in previous chapters, asymptotic expansions of χ, ϵ, G involve the following characteristics of an electron liquid at given density: $G^{PV}(k)$ (or, equivalently, S(k)), g(0), $<E_K>$ (or δ_K) and $<E^2_{PVK}>$ (or δ_{KK}). Iwamoto, Krotcheck and Pines [1984] calculated δ_K, g(0) and G^{PV} (= I(k) in their notation) quite accurately, "combining the best currently available microscopic data on the ground-state properties from the Green's-function Monte Carlo calculations". We used their values below in Table IV. For evaluation $<E^2_K>$ the occupation-number function is necessary, but it is not known at present with sufficient accuracy. Therefore, we are going to use the lower bound of the estimate (A.21) as the representative value of $\delta_{KK}=\delta_4$.

It will be convenient to have rewritten here the earlier obtained expressions for asymptotic expansions of G(k,0), $\chi(k,0)$, and $\epsilon(k,0)$ and for corresponding coefficients:

$$G(k,0) = b_{-2}k^2 + b_0 + o(k^0) , \qquad (6.1)$$

$$b_{-2} = \frac{\delta_K}{5\omega_p^2} \qquad (6.2)$$

$$b_0 = b_0^A + b_0^B + b_0^C \qquad (6.3)$$

$$b_0^A = \frac{2}{3}\,[1-g(0)] , \qquad (6.4)$$

Table IV. Characteristics of an electron liquid and expansion coefficients of G(k,0) Eqs. (6.1) - (6.6).

r_s	ω_p^2	g(0)	δ_K	δ_{KK}	b_{-2}	b_0^A	b_0^B	b_0^C	b_0
0	0	.50	0	0	0	.33	0	0	.33
1	.22	.28	.036	.10	.033	.48	.16	-.05	.59
2	.44	.17	.091	.26	.041	.55	.20	-.07	.68
5	1.12	.045	.292	.82	.052	.64	.25	-.10	.79
10	2.21	~.00	.619	1.73	.056	.67	.27	-.12	.82

$$b_0^B = \frac{12\delta_{KK}}{35\,\omega_p^2} \qquad (6.5)$$

$$b_0^C = -\frac{8\,\delta_K\,(1 + \frac{1}{2}\delta_K)}{25\,\omega_p^2} \qquad (6.6)$$

$$\chi(k,0) = -\frac{4\,\omega_p^2}{k^4 v_k}\left[\,1 + \frac{a_2}{k^2} + \frac{a_4}{k^4} + o(\frac{1}{k^4})\,\right] \qquad (6.7)$$

$$a_2 = \frac{8}{3}\langle E_K\rangle = \frac{4}{5}(1 + \delta_K) \qquad (6.8)$$

$$a_4 = a_4^A + a_4^B \qquad (6.9)$$

$$a_4^A = \frac{64}{5}\langle E_K^2\rangle = \frac{48}{35}(1 + \delta_{KK}) \qquad (6.10)$$

$$a_4^B = -\frac{4}{3}\omega_p^2\,[1 + 2g(0)] \qquad (6.11)$$

$$\epsilon(k,0)-1 = Q(k,0) = \frac{4\omega_p^2}{k^4}\left[\,1 + \frac{a_2}{k^2} + \frac{a_4 + 4\omega_p^2}{k^4} + o(\frac{1}{k^4})\,\right] \qquad (6.12)$$

Table IV contains expansion coefficients of G(k,0) and their constituents, while Table V the same concerning $\chi(k,0)$ and $\epsilon(k,0)$. We see that, in general, all expansion coefficients change significantly with increasing r_s, although quite smoothly. The full coefficients: b_{-2}, b_0, a_1, a_4, $(a_4 - 4\omega_p^2)$, all are positive, increasing functions of r_s, except a_4, which is decreasing.

Table V. Expansion coefficients of (k,0) and (k,0), Eqs. (6.7) - (6.12).

r_s	a_2	a_4^A	a_4^B	a_4	$4\omega_p^2$	$a_4+4\omega_p^2$
0	.80	1.37	0	1.37	0	1.37
1	.83	1.51	- .46	1.05	.88	1.93
2	.87	1.73	- .79	.94	1.76	2.70
5	1.03	2.50	-1.63	.87	4.48	5.35
10	1.30	3.74	-2.95	.79	8.84	9.63

VII. CONCLUSIONS

It was noticed that for a general system in the large k region there exists a direct relation between high-frequency and low-frequency properties. This is due to analgesic is the response function.

This idea was applied to the system of a uniform electron gas. The large-k asymptotic expansions of the static response, dielectric and local-field correlation functions was obtained. These results are exact, valid for arbitrary r_s. For evaluation of the expansion coefficients the information about the 1-st and 3-rd frequency moments and the static structure factor, all in the large-k region, was used.

The leading term of the expansion of the local-field function was found to be proportional to k^2, with the coefficient being the difference between the mean kinetic energy of the interacting and the free electron gas.

But the Niklasson's relation shows that the leading term of the mentioned expansion should be a constant. The apparent contradiction is immediately removed if the difference in the definition of the local field function, adopted by Niklasson and that traditionally used, is taken into account. Then his and our results become equivalent.

ACKNOWLEDGEMENTS

The author would like to thank Prof. Abdus Salam, the International Atomic Energy Agency and UNESCO for hospitality at the International Centre for Theoretical Physics, Trieste, where this work was initiated and a significant part of the results were obtained.

REFERENCES

Glick, A. J., Long, W. F., 1971, Phys. Rev. B 4:3455.
Holas, A., Aravind, P. K., and Singwi, K. S., 1979, Phys. Rev. B 20:4912.
Iwamoto, N., Krotscheck, E., and Pines, D., 1984, Phys. Rev. B 29:3936.
Kimball, J. C., 1973, Phys. Rev. A 7:1648.
Kimball, J. C., 1975, Phys. Rev. A 8:1513.
Kugler, A. A., 1975, J. Stat. Phys. 12:35.
Niklasson, G., 1974, Phys. Rev. B 10:3052.
Pathak, K. N., and Vashishta, P., 1973, Phys. Rev. B 7:3649.
Pines, D. and Nozieres, P., 1966, "The Theory of Quantum Liquids", Benjamin, New York.
Yasuhara, H. and Kawazoe, Y., 1976, Physica 85A:416.

APPENDIX A

WAVE-NUMBER MOMENTS

Let us investigate the properties of the wave-number moments of the occupation function, defined as (Comp. Eq. (4.3))

$$<q^\ell> = \frac{2}{N} \sum_{\vec{q}} n(|\vec{q}|)\ |\vec{q}|^\ell = 3 \int_0^\infty dq\ q^2 n(q)\ q^\ell \tag{A.1}$$

for non-interacting, $<q^\ell>_o$, and interacting $<q^\ell>_I$, electron gas, with $n(q)=n_o(q)$, Eq. (4.4), or $n(q)=n_I(q)$, respectively.

We introduce the difference of the moments due to the interaction

$$\Delta_\ell = <q^\ell>_I - <q^\ell>_o , \tag{A.2}$$

and the relative difference

$$\delta_\ell = \Delta_\ell / <q^\ell>_o \tag{A.3}$$

Introduced earlier δ_K and δ_{KK} are particular cases of δ_ℓ, namely $\delta_K=\delta_2$ and $\delta_{KK}=\delta_4$.

The moment $<q^\ell>_I$ is finite for ℓ belonging to the range

$$-3 < \ell < 5 \tag{A.4}$$

The lower bound in (A.4) is determined by the behavior of the integrand in Eq. (A.1) at small q, since

$$\lim_{q\to\infty} n_I(q) = n_I(0) > 0 \tag{A.5}$$

The upper bound is due to the following behavior of the occupation function at large q

$$n_I(q) = \frac{\omega_p^4\ g(0)}{2} \frac{1}{q^8} + o(\frac{1}{q^8}) \tag{A.6}$$

shown by Yasuhara and Kawazoe [1976]. For the moments $<q^\ell>_o$, the parameter ℓ is limited from below only, $\ell > -3$.

Because the occupation functions are normalized, we have

$$<q^0>_I = <q^0>_o = 1 \tag{A.7}$$

$$\Delta_0 = \delta_0 = 0 \tag{A.8}$$

Using definition (A.2), (A.1) and the expression (4.4), we may split Δ_ℓ into two parts:

$$\Delta_\ell = -3 \int_0^1 dq\ [1-n_I(q)]\ q^{\ell+2} + 3 \int_1^\infty dq\ n_I(q)\ q^{\ell+2} \tag{A.9}$$

The integrands in both integrals of Eq. (A.9) are obviously positive. Let us consider the power index ℓ to be a continuous variable from the range (A.4). We may calculate the derivatives of Δ_ℓ with respect to ℓ

$$(\frac{d}{d\ell})^n \Delta_\ell = -(-1)^n\ 3 \int_0^1 dq\ [1-n_I(q)]\ q^{\ell+2}\ [\ell n(\frac{1}{q})]^n$$

$$+ 3 \int_1^\infty dq\ n_I(q)\ q^{\ell+2}\ [\ell n(q)]^n. \tag{A.10}$$

We see that for the odd derivative

$$(\frac{d}{d\ell})^{2m+1} \Delta_\ell > 0 \tag{A.11}$$

because it is a sum of two positive integrals. From (A.10) it is obvious that

$$\lim_{\ell \to 5-0^+} (\frac{d}{d\ell})^n \Delta_\ell = +\infty \tag{A.12}$$

$$\lim_{\ell \to -3+0^+} (\frac{d}{d\ell})^n \Delta_\ell = \begin{cases} +\infty \ , & \text{for } n=2m+1 \ , \\ -\infty \ , & \text{for } n=2m \ . \end{cases} \tag{A.13}$$

From the fact that $\Delta_0 = 0$ and $\frac{d\Delta_\ell}{d\ell} > 0$, we conclude that

$$\text{sgn}(\Delta_\ell) = \text{sgn}(\ell) \ . \tag{A.14}$$

Now we estimate the 4-th moment, assuming that the 0-th and 2-nd ones are given. Because Δ_ℓ is increasing function of ℓ, therefore

$$\Delta_4 > \Delta_2 \tag{A.15}$$

Taking into account that

$$\langle q^\ell \rangle_o = \frac{3}{3+\ell} \tag{A.16}$$

we may rewrite (A.15) in terms of relative changes, (A.3), in the form

$$\delta_4 > \frac{7}{5}\ \delta_2 \tag{A.17}$$

We believe that an estimate stronger than (a.15) holds, namely

$$\Delta_4 \gtrsim 2\ \Delta_2 \tag{A.18}$$

It can be obtained from the relation

$$\Delta_4 = \Delta_2 + \int_2^4 d\ell \frac{d\Delta_\ell}{d\ell} \qquad \text{(A.19)}$$

together with the estimate, for $2 < \ell < 4$, that

$$\frac{d\Delta_\ell}{d\ell} \gtrsim \left.\frac{d\Delta_\ell}{d\ell}\right|_{\ell=2} \simeq \frac{\Delta_2 - \Delta_0}{2} = \frac{1}{2}\Delta_2 \qquad \text{(A.20)}$$

where we take into account the proximity of the considered range of to the bound ℓ-5, for which the derivative becomes infinite, Eq. (A.12). From (A.18) it follows that

$$\delta_4 \gtrsim 2.8\ \delta_2 \qquad \text{(A.21)}$$

Another estimate of δ_4 is possibly due to the Cauchy-Schwarz inequality

$$\langle q^a\rangle \langle q^b\rangle \geq [\ \langle q^{(a+b)/2}\rangle\ ]^2 \qquad \text{(A.22)}$$

taken for a=0, b=4, from which, using (A.7) and (A.16), we get

$$\delta_4 \geq \frac{21}{25}(1 + \delta_2)^2 - 1 \qquad \text{(A.23)}$$

The estimate (A.23) is stronger than the estimate (A.17) for $\delta_2 >$ 0.30, and stronger than (A.21) for $\delta_2 > 1.46$.

THE STRONGLY COUPLED OCP PLASMON DISPERSION FOR FINITE WAVENUMBERS

M. Minella and G. Kalman

Department of Physics
Boston College
Chestnut Hill, MA

INTRODUCTION

In a strongly coupled OCP the plasmon dispersion becomes negative $(d\omega/dk)_{k=0} < 0$ for $\Gamma > \Gamma_{crit}$, $2 < \Gamma_{crit} < 10$. This has been established both by MD simulations and analytic studies (Hansen, Pollock and MacDonald, 1974; Hansen, Pollock and MacDonald 1975; Abramo and Parrinello, 1975).

The analytic work, however, has been restricted to the long wavelength domain, more precisely to the $O(k^2)$ expansion of the frequency-wavenumber relationship, $\omega(k)$. However, the analysis of the dispersion relation for finite wavenumbers is certainly of interest: one would expect that while for $k \to o$ the short range order, induced by the strong coupling, determines the character of the dispersion, for finite wavenumbers it is the thermal dispersion, being of $O(kv)$, that has the decisive influence. It is the former that is responsible for the negative dispersion, and the latter causes $d\omega/dk > 0$. Thus one expects that $\omega(k)$ develops a minimum say $\omega_* = \omega(k_*)$ at some intermediate $k = k_*$ value. On the other hand, the MD calculations of Hansen, Pollock and McDonald, (1975) show a maximum for higher k values before the disappearance of the solution of the dispersion relation due to the strong Landau damping.

In this paper we study the finite k dispersion features of the strongly coupled OCP, by adopting a relatively simple approach justified on the basis of the physical observations made above.

We use two models for the calculation of the plasmon dispersion. The first model is based on the dynamical mean field theory of Golden and Kalman (1979), which provides that the polarizability α ($\varepsilon = 1 + \alpha$) is given by

$$\alpha(\underset{\sim}{k}\omega) = \alpha_0(\underset{\sim}{k}\omega)\{1 + v(\underset{\sim}{k}\omega)\} \tag{1}$$

where $\alpha_0(\underset{\sim}{k}\omega)$ is the Vlasov polarizability valid for γ=o and arbitrary k, and $v(\underset{\sim}{k}\omega)$ is the dynamical screening function. This latter has been calculated by Carini, Golden and Kalman in the "two pole approximation" to $O(k^2)$ in the vicinity of ω_0 (the plasmon frequency) for arbitrary γ (Carini, Golden and Kalman, 1980; Kalman, this volume). The combination of the two different approximations is justified in view of the observations already noted.

In the second model we write $\alpha(\underset{\sim}{k}\omega)$ as

$$\alpha(\underset{\sim}{k}\omega) = \alpha o(\underset{\sim}{k}\omega) - \frac{\omega_0{}^4}{\omega^4} \quad \Delta_4(k) \tag{2}$$

where $\Delta_4(k)$ is the exact correlational contribution to the coefficient of the ω^{-4} term in the high frequency sum-rule expansion of $\alpha(\underset{\sim}{k}\omega)$, which can be calculated for arbitrary k-values from the pair correlation function. This model is adopted, because there is strong indication that to $O(k^2)$ the sum rule expansion of $\alpha(\underset{\sim}{k}\omega)$ is exhausted as $\gamma\to\infty$ by the ω^{-2} and ω^{-4} terms.

This feature is exhibited by CGK calculation (Carini and Kalman, 1980) and also by the $\alpha(\underset{\sim}{k}\omega)$ for a 2-d electron liquid as calculated by Golden and Lu 1982), (Golden, this Volume) on the basis of the GK scheme. Moreover, the dispersion of optical phonons in a 2-d hexagonal lattice, as calculated by Bonsall and Maradudin (1977) shows this same feature as well.

1. SUM RULE APPROXIMATION

The high frequency sum rules provide the coefficients of the inverse powers of ω in the asymptotic expansion of the real part of the dielectric function $\varepsilon(\underset{\sim}{k},\omega)$ through equations of motions (i.e., conservation laws), the Fluctuation-Dissipation Theorem and the Kramers-Kronig relations. The result to order ω^{-4} is

$$\varepsilon(\underset{\sim}{k},\omega) = 1 - \frac{\omega_0{}^2}{\omega^2} - \frac{\omega_0{}^4}{\omega^4}\left[3\frac{k^2}{k_0{}^2} + \Delta_4(k,\gamma)\right]$$

where

$$\Delta_4(k,\gamma) = \frac{1}{V} \sum_{\underset{\sim}{p}} \frac{(\underset{\sim}{k}\cdot\underset{\sim}{p})^2}{k^2p^2} \quad (g_{\underset{\sim}{k}-\underset{\sim}{p}} - g_{\underset{\sim}{p}}) \tag{3}$$

with $g_{\underset{\sim}{p}}$ being the Fourier-transformed pair distribution function, which is reasonably well-known up to very high γ-values (from MD simulations, e.g.,

Hansen 1973, Ng 1974). In this preliminary study, however, we report on the results of a study where g_p is chosen as the Debye pair distribution function for the calculation of Δ_4.

Recalling that $\gamma = k_o^3/4\pi n$. Where K_0^{-1} is the Debye length and adopting normalized frequencies and wavevectors $\omega = \omega/\omega_0$, $k = k/k_0$, the result for (3) is

$$\Delta_4(k;\gamma) = -\frac{\gamma}{4}\left[\frac{(1+k^2)^2}{k^3}\, tg^{-1}_{k} - \frac{1}{k^2} - \frac{5}{3}\right] \qquad (4)$$

The function within parentheses is a monotonically increasing function of k whose asymptotic behaviors are k^2 for small k and k for large k. Even though the Debye approximation obviously represents a small γ expansion, for the purpose of the present calculation we adopt (3) for arbitrary γ values.

The complete real part of ε is given by

$$\acute{\varepsilon}(k\omega) = \acute{\varepsilon}_0(\underset{\sim}{k}\omega) + \frac{1}{\omega^4}\frac{\gamma}{4}\left[\frac{(1+k^2)^2}{k^3}\, tg^{-1}_{k} - \frac{1}{k^2} - \frac{5}{3}\right]$$

We will ignore the imaginary part of $\varepsilon(\underset{\sim}{k}\omega)$ and seek solution of the dispersion relation $\acute{\varepsilon}(\underset{\sim}{k}\omega) = 0$ only. It is apparent, for a given γ there is maximum value of k beyond which the curve no longer intersects the ω=0 axis and there is no more solution of the dispersion relation.

The solution of $\acute{\varepsilon}(k,\omega) = 0$ is then plotted in the usual ω versus k fashion, with γ as a parameter this time, in Figure 1.

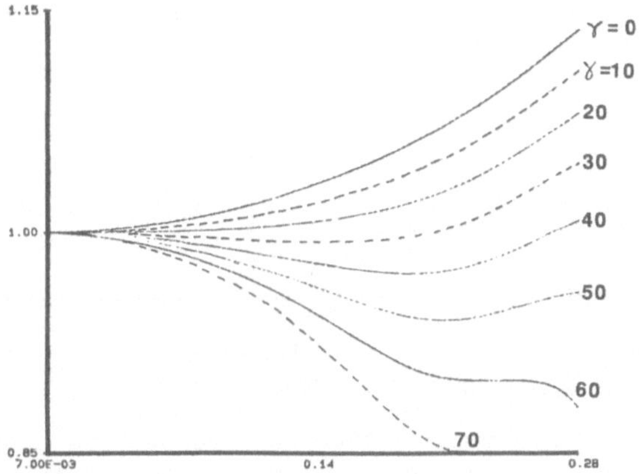

Fig. 1. Sum rule, Dispersion relation ω versus k with γ varying from 0 to 70 in steps of 10.

All the graphs are obtained numerically. The solution of $\acute{\varepsilon}(k,\omega) = 0$ on which we concentrate is only the one closest to $\omega = \omega_0$. Figure 3 deserves some comment:

1. The higher the coupling, the shorter the range of k's for which in our model we can find a solution of the dispersion relation; such a range is finite also for the Vlasov $\gamma = 0$ case in our approximations. This is consistent with the replacement of the full ε by its real part. (Fried and Conte, 1961).

2. Starting from a critical value of $\gamma = 25$ and up to $\gamma = 60$ the dispersion relation shows a minimum lower than the plasma frequency ($\omega = 1$ in the figure).

As γ approaches 60 there are no more solutions of the dispersion relation at the point where we expect the minimum. This is an expectable failure of our crude model.

The value of the normalized minimum frequency $\Omega(n) = \omega_*/\omega_0$ is plotted in Figure 2 versus the coupling parameter γ.

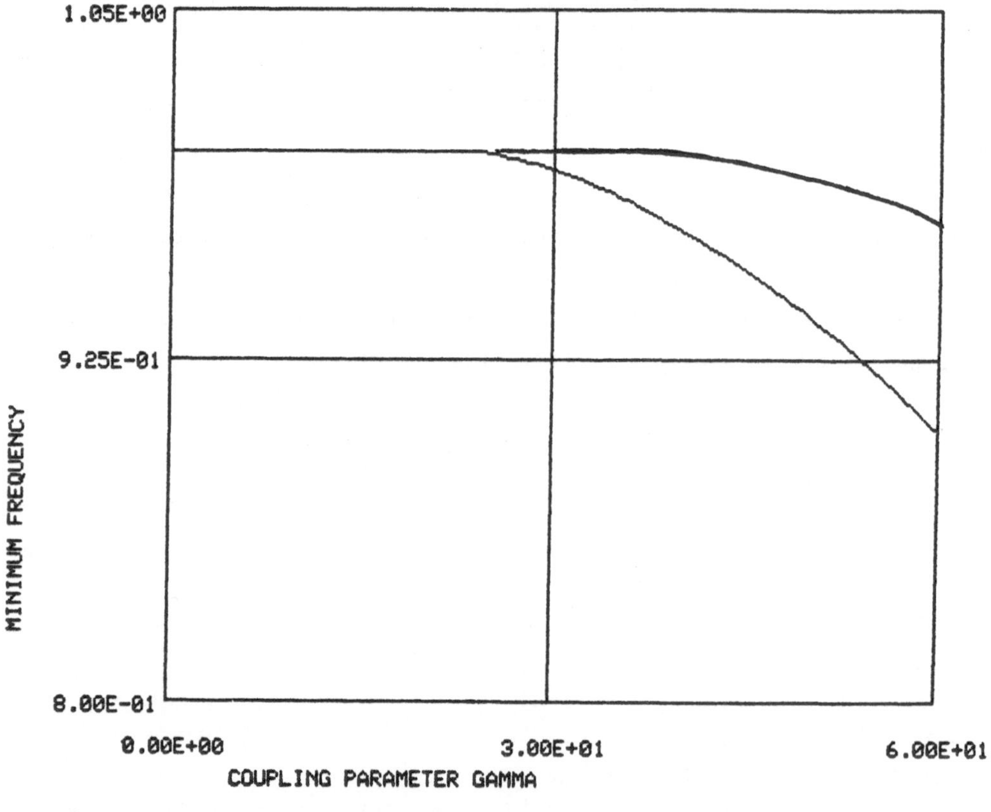

Fig. 2. Position of the minimum frequency versus the coupling parameter. Thick line, Carini-Kalman: shaded line sum rule model.

We also plot the minimum frequency $\omega_*(n) = \Omega(n)\ \omega_0(n)$ versus the density, in Figure 3.

Since γ depends on $n^{1/2}\ T^{-3/2}$ the density at which the two lines of Fig. 5 start splitting (determined by the condition $\gamma \cong 27$) will be higher at higher temperatures, by a factor of 3 in the logarithmic scale.

III. GOLDEN-KALMAN MODEL: APPROXIMATION SCHEME AND RESULTS

The Golden-Kalman formulation for the ocp dynamical polarizability is given in (1). We can assume a simple hydrodynamical structure for the external polarizability (Kalman and Carini 1980).

The dispersion relation derived numerically from the solution of $\acute{\varepsilon}(\underset{\sim}{k}\omega) = 0$ is plotted in Fig. 4. Similarly to the previous case the dispersion relation exhibits a minimum at some finite k. All observations

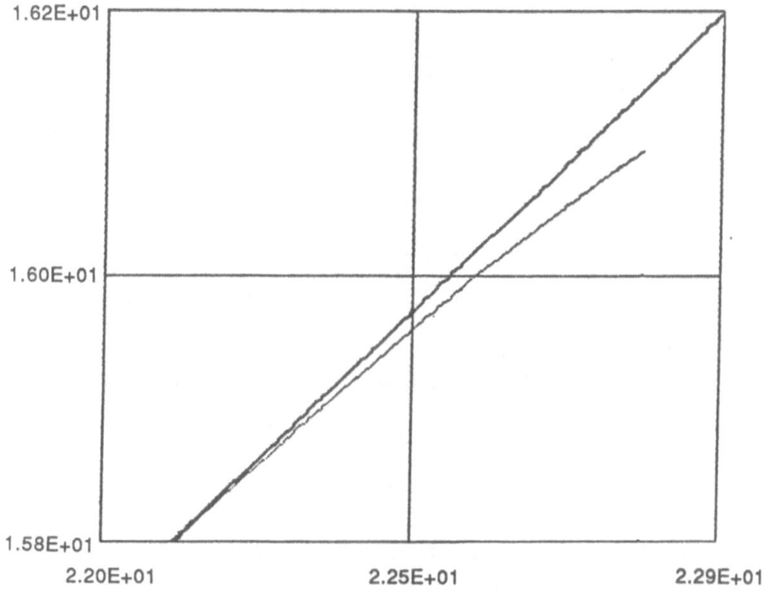

Fig. 3. Minimum frequency of propagation versus density (logarithmic plot). The straight line is the plasma frequency versus the density, $\omega_0 = 5.6.\ 10^4.n\ 1/2$.

made for the plot of Fig. 1 are valid for Fig. 4 as well. The minimum of these dispersion relations is also plotted in Fig. 2 to allow comparison with the previous model.

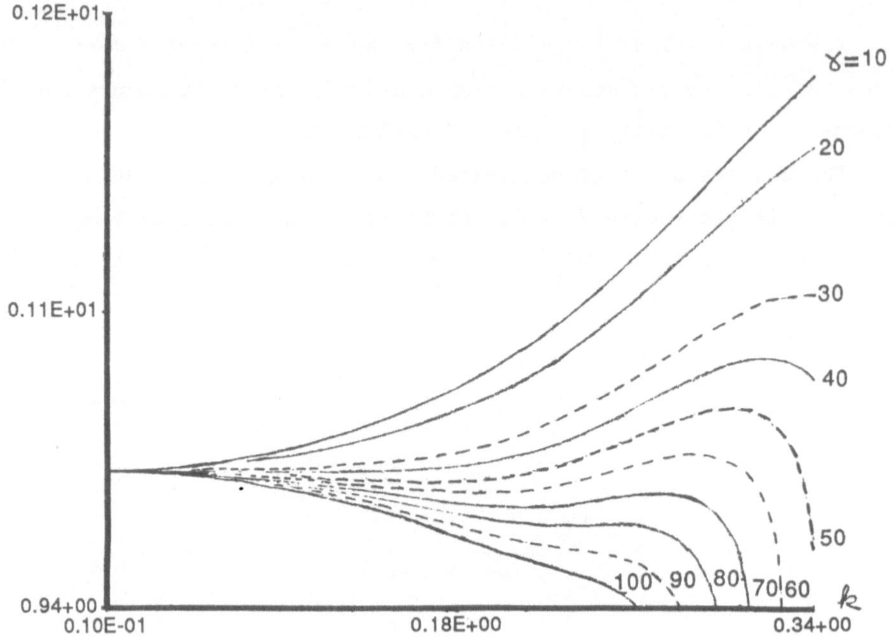

Fig. 4. CARINI-KALMAN. Dispersion relation ω versus k with γ ranging from 0 to 100 in steps of 10.

V. CONCLUSIONS

We have found that two independent models for the polarizability of the strongly coupled ocp give remarkably similar results for the plasmon dispersion relation dependence on the coupling parameter:

1) As it is well known there is a critical γ after which the slope of the dispersion curve becomes negative; the critical value of γ_{crit} is ≃ 27 for the sum rule model and ≃ 36 for the GK approximation.

2) For $\gamma_{crit} < \gamma < \gamma^x{}_{crit}$ the dispersion curve develops a minimum. $\gamma^x{}_{crit}$ is ≃ 60 for the sum rule model and ≃ 80 for the GK approximation. The sum rule model predicts a stronger dependence of the position of the minimum versus the plasma parameter.

Figure 5 suggests the possible misinterpretation of results of experiments such as the ones reported by Fortov et al. (Fortov, this volume).

In their coherent scattering experiments, the frequency of the laser beam reflected from the plasma at a certain point is assumed to be the plasma frequency at that point. This is true only when the plasma frequency is the minimum frequency; but if there is a local minimum, lower than the plasma frequency, the identification of the laser frequency with

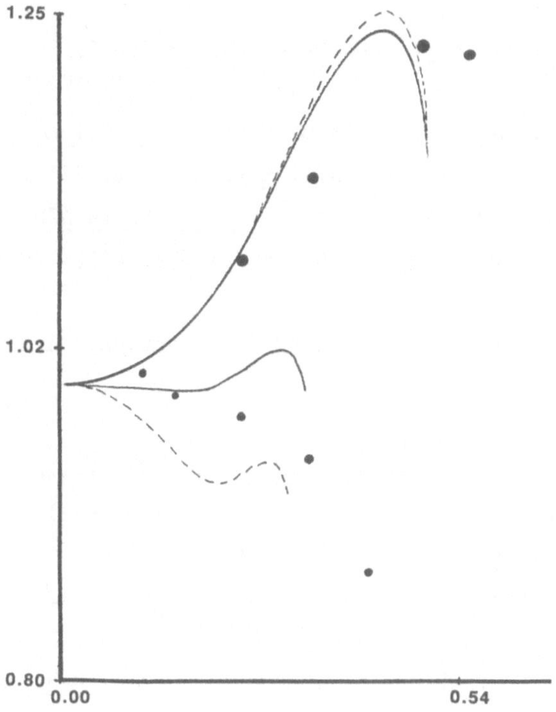

Fig. 5. Comparison between the maxima in the dispersion relation of our models (thick, Carini-Kalman; shaded, sum rule) and the MD calculations of HPM (1974).

the plasma frequency is unjustified: the laser frequency is lower than the local plasma frequency and therefore, the local density is actually higher than the value inferred without accounting for the effect described on this paper.

Finally, we note that the maxima of the dispersion relation in both models agree qualitatively with the molecular dynamics results of Hansen, Pollock and McDonald (1975). In particular, all approaches have a maximum in the dispersion curve at $.46 \lesssim k \lesssim .54$. (Fig. 5)

ACKNOWLEDGEMENT

This work was partially supported by NSF Grant ECS-8315655.

REFERENCES

Abramo, M.C. and M. Parrinello, 1975, Lett. Nuovo Cimento, 12, 667

Bonsall, L. and A.A. Maradudin, Phys. Rev. B 13, (1977) 1959

Carini, P., G. Kalman and K. Golden, Physics Letters A78, (1980), 450

Carini, P. & G. Kalman, Physics Letters, A105 (1984), 229.

Fortov, this volume

Fried, B., S. Conte "The Plasma Dispersion Function," Academic Press, New York, (1961)

Golden, This Volume

Golden, K.I. and G. Kalman, Phys. Rev. A19 (1979), 2112.

Golden, K.I. and Dexin Lu, J. of Stat. Phys. 29, (1982) 281

Hansen, J.P., Phys. Rev. A8, (1973), 3096

Hansen, J.P., E.L. Pollock and G. R. McDonald, Phys. Rev. Lett. 32 (1974) 277

Hansen, J.P., E.L. Pollock and G.R. McDonald, Phys. Rev. A11, (1975) 1025

Kalman, G, in: "Strongly Coupled Plasmas", edit. G. Kalman, Plenum Press, New York, 1978

Kalman, G. & K. Golden, Phys. Rev. A20, (1979), p. 2638

Kalman, this volume

Ng, K.C., J. Chem. Phys. 61, (1974), 2680.

CHAPTER X

ELECTRIC MICROFIELD AND OPTICAL PROPERTIES

ELECTRIC MICROFIELD DISTRIBUTIONS

James W. Dufty

Department of Physics
University of Florida
Gainesville, FL 32611

I. INTRODUCTION

Recent research on controlled fusion by inertial confinement has stimulated renewed consideration of atomic phenomena in hot, dense plasmas. In many cases the dominant coupling of the atom to its plasma environment is through the atomic dipole interaction with the local electric microfield. The electric microfield distribution (probability density for a given field value) is therefore an important property of the plasma for description of such atomic processes as emission and absorption of radiation. Prior to laser-produced plasmas, typical laboratory experiments involved only weakly coupled plasmas and quite accurate theories were available to calculate the microfield distributions under such conditions[1,2]. These theories fail for strongly-coupled plasmas and present research has focused on calculations for more extreme plasma conditions. Progress in this direction was initiated by Iglesias, et. al.[3,4] who proposed a method for ion fields at charged points, which gives excellent agreement with computer simulations of strongly coupled classical one component plasmas (OCP) in two and three dimensions. Extension of this method to multi-component plasmas also proved fruitful[5]. Subsequently, others have addressed related aspects of microfields in dense plasmas such as ion fields at neutral points[6], electron fields[7], effects of atomic structure[8], and quantum degeneracy[9,10]. The objective here is to review these developments in a general context, to clarify some of the approximations used, and to identify some of the remaining problems.

There are two distinct parts to the microfield distribution - its definition and its calculation. With regard to the definition, some specific microscopic field must be chosen, whose distribution of values under given state conditions is desired. For example, in a two component plasma it could be the Coulomb field of the ions, or that of the electrons, the total field, or some given screened field. In general the choice depends on the particular application for which the microfield distribution is to be used. This ambiguity in the definition of microfield has not received much attention in the past since applications were focused on the specific problem of static ion spectral line broadening by weakly coupled plasmas. In that case, simple Debye shielded ion fields are adequate[11]. The current interest in strong coupling and ion dynamical effects requires a more careful selection of the appropriate microfield distribution to be calculated.

Another part of the definition is the model used for the plasma and "radiator" (atom or ion at which the field distribution is to be calculated). Most results in the literature have been obtained under very restricted conditions (e.g., classical OCP) but the formulations actually apply more widely. In typical inertial confinement experiments there are many highly charged ionic species, possibly having bound electronic structure. The plasma state may be nonequilibrium, or at best local equilibrium and, while the ions are typically nondegenerate, electron degeneracy may be important. In the third section, the Baranger-Mozer formulation[1] is extended to apply to this general situation. In this formulation the problem is transformed into a series whose terms characterize the plasma-radiator interaction through their correlation functions. For weakly coupled plasmas higher order correlation functions are negligible, leading to a computationally simple and quite accurate expression for the microfield distributions. However, as the thermal de Broglie wavelength increases or Debye screening length decreases (relative to the average interparticle spacing) higher order correlation functions become significant and the plasma is strongly coupled. Some partial resummation, or other rearrangement, of the Baranger-Mozer series must then be carried out in general for an adequate approximation.

The relevant parameter space is indicated qualitatively in the density-temperature plane of Figure 1. The importance of Coulomb

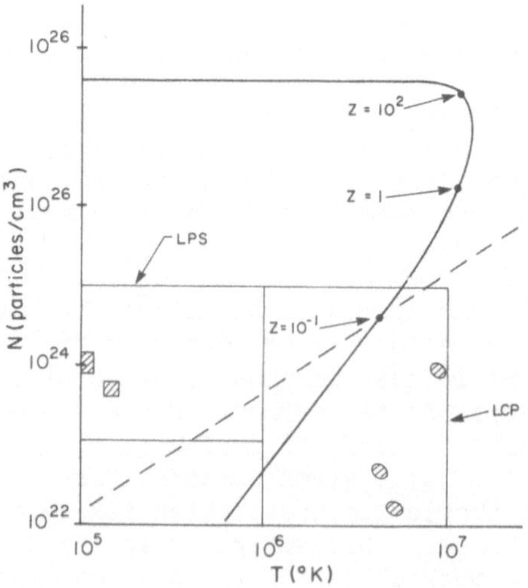

Figure 1: Density-temperature plane with contours of Γ=0.1 (——) and z=0.1 (---).

coupling for electrons is estimated by the plasma parameter Γ = average Coulomb energy/average kinetic energy[12], and the importance of quantum degeneracy is determined by the activity, z. The solid line denotes a contour for $\Gamma = 0.1$ and the dashed line refers to $z = 0.1$. The common region to the right of both curves is therefore a weakly coupled electron system. Otherwise the electrons are strongly coupled in one or more respects. For example the point marked $z = 10^2$ on the $\Gamma = 0.1$ curve has weak Coulomb coupling but strongly degeneracy coupling. The circles represent conditions achieved in typical laser compression experiments[13] (LCP) and the associated box estimates the region accessible in the near future. Also shown are corresponding quantities for laser produced shock experiments[14] (LPS). In LCP experiments strong coupling of electrons is not yet as important as in those of LPS. For ions the most important strong coupling effects are due to Coulomb interactions, rather than degeneracy. The thermal de Broglie wavelength is reduced almost two orders of magnitude by the large mass of the ions relative to electrons, so strong coupling from exchange effects on the ions can be neglected. In contrast, for Coulomb coupling the plasma parameter for ion-ion correlations is $Z^2\Gamma$ so the ions can be strongly coupled even when the electrons are not. For example hydrogenic neon and argon plasmas have charges Ze with Z = 9 and 17 respectively. The ion-electron coupling is $Z\Gamma$, and may not be classified clearly in either category.

The basic ideas of Iglesias, et. al. are described in section V to show how the Baranger-Mozer series can be "renormalized" for strong Coulomb coupling, while still retaining the computational simplicity of the weak coupling approximations. Variations of this procedure and alternative treatments with comparable accuracy are also indicated. In all cases the primary feature of these strong coupling approximations is that all plasma properties occur only through the pair correlation functions. In this context the calculation of microfield distributions is essentially a special application in the theory of plasma correlation functions. Correlation functions have been studied extensively for the classical OCP, where integral equations provide accurate pair correlation functions up to quite large plasma parameters.[15] Accurate results also can be obtained for multi-ion systems in a neutralizing background.[16] Difficulties arise when a proper treatment of bound and free electrons is attempted. The case of degenerate electrons and small plasma parameter can be treated adequately by a modified random phase approximation. For large plasma parameters and/or bound electrons, more sophisticated methods are required[17] and some of those applied recently to the microfield calculation are described in section VII[18,19]. Finally, some aspects of real plasmas encountered in current experiments are noted, and the current accomplishments and remaining problems for microfield calculations are summarized.

II HIGH FREQUENCY AND LOW FREQUENCY MICROFIELDS

The microfield distribution gives the probability density for values, $\vec{\varepsilon}$, of a given microscopic field $\vec{E}^*$. The asterisk on $\vec{E}^*$ has been included to emphasize that the form of this field need not be the Coulomb field. The microfield distribution function is then

$$Q(\vec{\varepsilon}) = \mathrm{Tr}\, \rho\, \delta(\vec{\varepsilon}-\vec{E}^*) \equiv \langle\, \delta(\vec{\varepsilon}-\vec{E}^*)\, \rangle \qquad (2.1)$$

where ρ is the statistical density operator characterizing the state of the plasma (equilibrium or nonequilibrium), and the trace extends over all degrees of freedom of the system. In subsequent sections attention will be restricted to evaluation of $Q(\vec{\varepsilon})$ for a given plasma state and given form of $\vec{E}^*$. Before doing so, some comments on the choice of $\vec{E}^*$ are offered to put the calculation of microfield distributions in the context of how they are to be used.

As a specific important illustration, consider the evaluation of the spectral line profile for an atomic or ionic radiator, in an equilibrium two component plasma of electrons and ions. The line shape function is[20]

$$I(w) = \pi^{-1} \mathrm{Re} \int_0^\infty dt\ e^{iwt} < \vec{d}(t)\cdot\vec{d}(o) > \tag{2.2}$$

where $\vec{d}$ is the radiator dipole operator[21]. Usually, the dipole autocorrelation function in Eq.(2.2) is evaluated in a two step process,

$$< \vec{d}(t)\cdot\vec{d}(o) > = \int d\vec{\varepsilon}\ Q(\vec{\varepsilon}) < \vec{d}(t)\cdot\vec{d}(o) >_{\varepsilon} \tag{2.3}$$

where $< >_{\varepsilon}$ represents an average constrained to have $\vec{E}^* = \vec{\varepsilon}$. A final average over all values of $\vec{\varepsilon}$ is then performed. There is no approximation implied by the representation (2.3) in general (as shown in reference 11), but rather only an expectation that both $Q(\vec{\varepsilon})$ and the constrained dipole correlation function can be calculated more accurately than $< \vec{d}(t)\cdot\vec{d}(o) >$, directly. The basis for this expectation is the widely separated time scales for the electrons and ions, due to their mass difference. Over much of the line profile the ions essentially form a static background during the relevant radiation time, while the electrons make nearly completed collisons with the radiator. Typically, stronger approximations are also invoked, i.e., that these two processes are statistically independent[22]. These latter approximations imply that $< >_{\varepsilon}$ is replaced by a charge neutral electron gas average, and that $Q(\vec{\varepsilon})$ is the ion microfield distribution for an OCP. To mitigate this drastic decoupling of ions and electrons, it is recognized that the effective ion field at the radiator is in fact screened by the fast moving electrons. To include these ion-electron correlations $\vec{E}^*$ is chosen to be a Debye-screened ion field, and a corresponding screened ion-ion potential is used to define the OCP equilibrium density matrix. Such a screened microfield distribution is sometimes referred to as a "low frequency" distribution, while the $Q(\vec{\varepsilon})$ for a Coulomb field and Coulomb OCP is called the "high frequency" distribution.

The use of Debye screened interactions and field $\vec{E}^*$ in the OCP calculation of $Q(\vec{\varepsilon})$ is plausible only for weakly coupled plasmas; its justification and generalization to complex plasma conditions requires a more systemmatic analysis. An outline of how such an analysis can be carried out is given in reference 11, where the right side of (2.3) is studied without any decoupling of the ion-electron-radiator subsystems. The essential conclusions of that study are threefold: 1) The microfield formulation in Eq.(2.3) can be accomplished for a wide class of choices of $\vec{E}^*$; however, different choices of $\vec{E}^*$ imply physically different interpretations and properties for $< \vec{d}(t)\cdot\vec{d}(o) >_{\varepsilon}$; 2) If, as in the standard theories, $< \vec{d}(t)\cdot\vec{d}(o) >_{\varepsilon}$ is to represent electron broadening in an ion Stark field, $\vec{\varepsilon}$, then the choice of $\vec{E}^*$ should be

$$\vec{E}^* = \mathrm{Tr}_e\ \rho\vec{E} / \mathrm{Tr}_e\ \rho \tag{2.4}$$

where the partial trace is taken over the electron states and $\vec{E}$ is the total Coulomb field; 3) Generally, $Q(\vec{\varepsilon})$ is defined with an average over the two component electron-ion system; however, an equivalent $Q(\vec{\varepsilon})$ for an effective OCP can be introduced with shielded ion-ion interactions given by

$$V^s_{ii} = V_{ii} + \int_0^1 d\lambda\ \mathrm{Tr}_{re}\ \rho_\lambda V_{ie} \tag{2.5}$$

where V_{ii} and V_{ie} are the ion-ion and ion-electron Coulomb potentials, and ρ_λ denotes the equilibrium density operator with the replacement

$V_{ie} \to \lambda V_{ie}$. The trace in (2.5) extends over both radiator and electron states. In the limit of a weakly coupled plasma the results (2.4) and (2.5) reduce to the Debye shielded field and potential, respectively.[11] Consequently, the standard model for spectral line shapes with low frequency microfield distribution is recovered in this limit. More generally, however, the expressions (2.4) and (2.5) show how the microfield choice should be generalized for other plasma conditions. Some observations can be made regarding the general case without further calculation: The field $\vec{E}^*$ is not simply related to the gradient of V^s_{ii}, it is not a sum of single particle fields, and it can depend on the coupling of the plasma to the radiator. Such complications may not always be important, but it should be clear that the definition of the low frequency microfield distribution is a difficult problem in itself.

The above discussion refers to the method of calculating spectral line shapes over the portion of the profile for which ions are static. When ion dynamic effects are important alternatives to the formulation (2.3) are more relevant. A novel approach to this problem, the model microfield method[23], has been applied with considerable success[24]. The idea is based on the fact that the plasma environment of the radiator is entirely characterized by the dynamics of the total field there. Microfield distributions also play an important role in the method, but now both electron and ion fields are required. The appropriate choice for $\vec{E}$ might then seem to be the Coulomb fields for all charges. In practice a low frequency OCP ion microfield distribution and a high frequency electron gas microfield distribution are used. The statistical mechanical basis for the model microfield method has been outlined[25], but a detailed determination of the fields to be used has not yet been given.

In summary, there is no "correct" definition of the field $\vec{E}^*$ whose distribution of values is to be computed. Rather, the particular choice (shielded, unshielded, ion, electron, total, . . .) depends on the theory in which the microfield distribution is to be used. Some caution should be used in discussing or comparing results outside of this context. In the following sections it is assumed that an appropriate field has been given, and the technical problem of calculating its distribution of values is considered.

III. BARANGER-MOZER FORMULATION

The original formulation of electric microfield distributions by Baranger and Mozer leads to a series representation ordered according to the correlation functions for the plasma. The purpose here is to introduce this formulation in its most general context. The plasma consists of an identifiable "dominent" set of ions, electrons, and neutrals (in general there is the difficult problem of determining relative frequencies of various bound states in the ensemble). It is not necessary to assume the plasma is charge neutral or even in equilibrium, so the formulation applies, for example, to physically important states such as local equilibrium, two temperature, or even turbulent plasmas. Ultimately, attention is focused on the equilibrium state. A specific particle is identified as the radiator and its degrees of freedom are denoted by subscript o. In general the radiator will have atomic structure so there may be many degrees of freedom associated with it. The remaining charged particles, electrons and ions, will be referred to as the perturbers. An electric field at the radiator due to the perturbers is assumed given, (the asterisk on the field will be deleted in the following for simplicity),

$$\vec{E} = \sum_{\sigma} \sum_{\alpha_\sigma}^{N_\sigma} \vec{E}(\alpha_\sigma) \tag{3.1}$$

where σ denotes a species label and α_σ is the αth perturber of type σ Also $\vec{E}(\alpha_\sigma)$ is the contribution from the αth perturber and depends only on the center of mass coordinate of the αth perturber relative to that of the radiator. Here N_σ is the number of particles of type σ, $N = \sum_\sigma N_\sigma$, and the trace is defined over suitably symmetrized N-particle states. To determine the microfield distribution it is convenient to evaluate the associated generating function, G,

$$Q(\vec{\varepsilon}) = \int \frac{d\vec{\lambda}}{(2\pi)^3} e^{-i\vec{\lambda}\cdot\vec{\varepsilon}} e^{G(\vec{\lambda})} \quad , \quad G(\vec{\lambda}) = \ell n < \pi \, e^{i\vec{\lambda}\cdot\vec{E}(\alpha_\sigma)} > \tag{3.2}$$

where the product extends over all particles of the system. For equilibrium states, $Q(\vec{\varepsilon})$ depends only on the magnitude of $\vec{\varepsilon}$, and the probability density for the magnitude of the field is defined by

$$P(\varepsilon) \equiv 4\pi\varepsilon^2 \, Q(\vec{\varepsilon}) \tag{3.3}$$

The equilibrium data is reported in terms of $P(\varepsilon)$ rather than $Q(\vec{\varepsilon})$.

In coordinate representation the operators exp $i\vec{\lambda}\cdot\vec{E}(\alpha_\sigma)$ represent long-ranged functions approaching a value of unity far from the radiator. For practical purposes it is better to express $G(\vec{\lambda})$ in terms of the shorter-ranged functions

$$\phi(\vec{\lambda};\alpha_\sigma) \equiv e^{i\vec{\lambda}\cdot\vec{E}(\alpha_\sigma)} - 1 \tag{3.4}$$

that essentially measure the volume around the radiator that contributes to the microfield distribution. Then since all of the λ-dependence occurs through $\phi(\vec{\lambda})$, the problem is to calculate $G(\vec{\lambda}) = G[\phi]$. The Baranger-Mozer representation follows directly from a functional Taylor series expansion in products of ϕ,

$$G\,[\phi] = \sum_{n_1} \cdots \sum_{n_s} \int G^{(n+1)}(\{n_1\},\cdots\{n_s\}) \pi_\sigma \frac{1}{n_\sigma!} \pi_{\alpha_\sigma}^{n_\sigma} \; \phi(\alpha_\sigma) d\{n_\sigma\} \tag{3.5}$$

Here $n = \sum_\sigma n_\sigma$ and $\{n_\sigma\}$ denotes the coordinates of a set of n_σ particles for each species σ. The functional derivatives $G^{(n+1)}(\{n_1\},\cdots,\{n_s\})$ are the cluster functions associated with the reduced density matrix, or correlation function, for the n perturbers plus radiator (see references 9 for further details). The important property of these cluster functions is that they vanish unless there is a correlation among <u>all</u> n+1 particles. Consequently, for weak correlations this property of the cluster functions leads to rapid convergence of the series, and low order truncation can be made for practical calculations. For example, the first order terms in ϕ are,

$$G[\phi] = \sum_\sigma \int d\vec{r}_\sigma \; g^{(2)}(\vec{r}_\sigma)\phi(\vec{\lambda};\vec{r}_\sigma) + \ldots \tag{3.6}$$

Here $g^{(2)}$ are the 2-particle correlation functions, defined as the diagonal matrix elements in coordinate representation of the 2-particle reduced density matrix. For Debye-Huckle plasmas, the term linear in ϕ gives an excellent approximation to the distribution of ion fields at charged radiators, while both linear and quadratic terms are required for similar accuracy at neutral radiators.

The result here differs from the original Baranger-Mozer formulation only in its generality: no restrictions on the plasma state have been made (e.g., equilibrium or not), all quantum effects are accounted for, and the possibility of bound states is implicitly included. Actually, the details of such complications are only supressed, and they appear more explicitely when the correlation functions must be determined.

IV STRONG COUPLING VIA DEGENERACY

The Baranger-Mozer formulation is obviously useful if the series can be truncated at first or second order. In the classical (high temperature) limit correlations are due to the Coulomb interactions among the particles, and the terms of the series are ordered according to powers of the plasma parameter, Γ. However, for values of the degeneracy parameter $z \gtrsim 1$ the effects of quantum statistics are important and a new correlation length, the de Broglie wavelength, appears. As Figure 1 shows, at high densities and low temperatures there can be large contributions to the correlations arising from quantum statistics, even when $\Gamma \ll 1$. To calculate these effects of degeneracy and illustrate a rearrangement of the Baranger-Mozer series for strong coupling, the electron microfield distribution at a neutral point for an electron gas with $\Gamma \ll 1$ is considered.[9]

The correlations due to the Fermi statistics for electrons are present even in the absence of Coulomb forces. This suggests extracting from each term in the series a part whose correlations are due entirely to statistics. It is then possible to sum all such contributions to a closed form[9]. The remaining terms of the series then have correlations that vanish as $\Gamma \to 0$. This part of the series can be truncated for small Γ. The result obtained in this way is valid for all degrees of degeneracy (strong quantum coupling) and $\Gamma<1$,

$$G[\phi] = G^{(o)}[\phi] + G_c[\phi]. \tag{4.1}$$

$G^{(o)}[\phi]$ is the contribution from quantum statistics alone,

$$G^{(o)}[\phi] = \mathrm{Tr}_1 \{\ln [1+\hat{n}(p)\hat{\phi}(1)]\} \tag{4.2}$$

where $\hat{n}$ and $\hat{\phi}$ are the single particle operators corresponding to the Fermi function and (3.4), respectively. The trace in (4.2) extends over single particle states, so that $G^{(o)}[\phi]$ expresses a reduction of the N-body statistics to an effective one body form. The second term of (4.1), $G_c[\phi]$, represents the effects of quantum Coulomb correlations,

$$G_c[\phi] = 1/2 \int d\vec{r}_1 d\vec{r}_2 [g^{(2)}(\vec{r}_1,\vec{r}_2) - g_o^{(2)}(\vec{r}_1,\vec{r}_2)]\, \phi(\vec{r}_1)\phi(\vec{r}_2) \tag{4.3}$$

Here $g^{(2)}$ and $g_o^{(2)}$ are the two particle equilibrium correlation functions for the interacting and noninteracting electron gas. For small plasma parameter these correlation functions can be evaluated (including finite temperature quantum effects) in the generalized random phase or chain approximation[26]. The detailed results in a form suitable for numerical evaluation are given in reference 9. It is found that as the degeneracy increases at fixed Γ, the peak of the microfield distribution shifts monotonically to smaller fields. Similar qualitative features have been observed from small degeneracy expansions as well[10].

V STRONG COUPLING VIA COULOMB INTERACTION

In the last section, the Baranger-Mozer series required a partial resummation because the deBroglie wavelength could be large compared to the interparticle distance, although the plasma parameter, Γ, was chosen to be small. Here, the opposite case is considered, large plasma parameter but small deBroglie wavelength. These conditions are more appropriate for ion microfield distributions since the ions are typically in their classical limit but can be strongly coupled by Coulomb interactions. When such long range correlations are important the Baranger-Mozer series again must be rearranged to extract the dominant contribution. Several methods are now available to describe strong coupling in this sense, and are briefly described below.

a) Effective field renormalization

The first successful calculation of ion microfields at large plasma parameters was given by Iglesias, et. al.[4]. Their basic idea was to retain the structural simplicity of the low order approximation to the Baranger-Mozer series, while incorporating higher order correlations (approximately) through an appropriately screened microfield. To motivate this, note that the size of the integrals in the Baranger-Mozer series are controlled by both the functions $G^{(n)}$ and the functions $\phi(\alpha_\sigma)$, so truncation is possible when either of these is small. At large plasma parameters the correlation functions do not provide a small parameter and their contribution can be large. However, it is possible that correlations in higher order terms shield $\vec{E}$ in such a way that ϕ becomes small outside a sphere of radius r_0. Then the series would be ordered according to powers of (r_0/ℓ_0), were ℓ_0 is the interparticle spacing, and it might be truncated by virtue of the range of ϕ rather than that of the correlation functions. To illustrate how higher order terms in the Baranger-Mozer series can lead to such a screened field, a direct expansion of $G(\vec{\lambda})$ to order λ^2 can be written in the suggestive form

$$G(\lambda) = \sum_\sigma \int d\vec{r}_\sigma g^{(2)}(\vec{r}_\sigma)\, R(r_\sigma) \quad i\vec{\lambda}\cdot\vec{E}'(r_\sigma)[1 + \tfrac{1}{2}i\vec{\lambda}\cdot\vec{E}'(\vec{r}_\sigma) + \text{order } \lambda^2] \tag{5.1}$$

where R is the ratio of field magnitudes, E/E', and $\vec{E}'(\vec{r})$ is the screened field,

$$\vec{E}'(\vec{r}_\sigma) \equiv \vec{E}(\vec{r}_\sigma) + \sum_{\sigma'}\int d\vec{r}_{\sigma'}\vec{E}(\vec{r}_{\sigma'})[g^{(3)}(\vec{r}_\sigma,\vec{r}'_{\sigma'}) - g^{(2)}(\vec{r}_\sigma)\, g^{(2)}(\vec{r}'_{\sigma'})]/g^{(2)}(\vec{r}_\sigma) \tag{5.2}$$

But, to the same order in λ, Eq. (5.1) can be written

$$G(\vec{\lambda}) = \sum_\sigma \int d\vec{r}_\sigma\, g^{(2)}(\vec{r}_\sigma)\, R(r_\sigma)\, [e^{i\vec{\lambda}\cdot\vec{E}'(r_\sigma)}-1] + . . \tag{5.3}$$

This is similar to the first term of the Baranger-Mozer series, with $\vec{E}$ replaced by the screened field $\vec{E}'$.

To formulate these ideas more precisely[6], a new function, $\tilde{\phi}\,(\vec{\lambda};\alpha_\sigma)$, is introduced by the definition,

$$\tilde{\phi}\,(\vec{\lambda};\alpha_\sigma) \equiv e^{i\vec{\lambda}\cdot\tilde{\vec{E}}(\alpha_\sigma)} -1 \tag{5.4}$$

where $\tilde{\vec{E}}\,(\alpha_\sigma)$ is a new "renormalized" field. The functional relationship of $\tilde{\phi}$ to ϕ is easily found to be

$$1+\phi = (1+\tilde{\phi})^R, \qquad R \equiv |\vec{E}|/|\tilde{\vec{E}}| \tag{5.5}$$

Consequently, the generating functional for the microfield distribution can be considered as a functional of $\tilde{\phi}$ instead of ϕ, $G(\vec{\lambda}) \equiv \tilde{G}[\tilde{\phi}]$. A renormalized Baranger-Mozer series is obtained analogous to Eq. (3.5) by a functional Taylor series expansion in $\tilde{\phi}$ instead of ϕ, where $\tilde{G}^{(n+1)}$ are a new set of cluster functions, obtained from the functional derivatives of $\tilde{G}$ with respect to $\tilde{\phi}$. The idea now is to choose $\tilde{E}$ such that this new series is rapidly convergent, even when the original Baranger-Mozer series is not. Since $\tilde{E}$ is essentially arbitrary at this point there are many possible ways to proceed. In principle, it is possible to require that all corrections to the leading term of the series vanish. In practice, it is convenient to choose $\tilde{E}$ real and independent of λ such that the leading term of the series describes $G(\lambda)$ exactly to order λ^2. This leads to the condition,

$$\sum_\sigma \int d\vec{r}_\sigma g^{(2)}(\vec{r}_\sigma) E_i(\vec{r}_\sigma) \tilde{E}_j(\vec{r}_\sigma) = \langle E_i (E_j - \langle E_j \rangle) \rangle \tag{5.6}$$

The right side of Eq. (5.6) is the second moment of the total electric microfield. Assuming that this condition provides the desired rapid convergence, the first approximation to the renormalized Baranger-Mozer series is given by

$$G(\vec{\lambda}) \cong \sum_\sigma \int d\vec{r}_\sigma g^{(2)}(\vec{r}_\sigma) R(\vec{r}_\sigma) \tilde{\phi}(\vec{\lambda};\vec{r}_\sigma) \tag{5.7}$$

which justifies the form (5.3). This is essentially the result of Iglesias, Lebowitz, and MacGowan[4].

There are two differences between this expression and the leading term of (3.6). First, $\tilde{\phi}(\vec{r}_\sigma)$ involves the field $\tilde{E}$ rather than E, and consequently $\tilde{\phi}$ is expected to be shorter-ranged than ϕ. Also, the ratio of fields, R, appears. Iglesias et. al. suggest interpreting (5.7) as the generating functional for a system of independent perturbers coupled to the radiator. For such a system (in the classical limit) all terms in the Baranger-Mozer series vanish except the first. These independent "quasiparticles" are chosen to have an effective density around the radiator given by $\tilde{g}^{(2)}(\vec{r}_\sigma) \equiv R(\vec{r}_\sigma) g(\vec{r}_\sigma)$ and an effective field $\tilde{E}(\vec{r}_\sigma)$ constrained by the second moment condition, (5.6). The effective charge density has the property that the average field produced by the quasiparticles is the same as the true field, i.e., $\tilde{E}\,\tilde{g}^{(2)} = E\, g^{(2)}$. Also, the second moment condition is reasonable in the following sense. The large field behavior of the distribution function is dominated by configurations with a single perturber close to the radiator, so an independent particle model should be a good approximation. Also, small fields are associated with many particles at large distances. Asymptotically, these large distance many-particle fields are Gaussian distributed, with covariance determined by the second moment of the electric microfield. The second moment condition on the effective field, $\tilde{E}$, assures that this collective effect is preserved by the independent particle model. In this way both large and small field limits are adequately represented, and the primary uncertainty is how well the approximation interpolates between these two limits. For ion perturbers or highly charged positive radiators the strong Coulomb repulsion minimizes the importance of configurations with several particles relatively close to the radiator. The model should work well in this case. The same reasoning suggests that the least favorable conditions would be for calculation of microfields at a neutral point. In the following Eq. (5.7) will be referred to as the <u>renormalized independent particle model</u>.

To illustrate the effectiveness of this approximation, consider the special case of the high frequency ion microfield distribution for the equilibrium classical OCP. The form of the field $\tilde{E}$ is still arbitrary beyond the constraint (5.6). The effect of correlations is expected to make $\tilde{E}$ more short ranged so Iglesias et. al. introduce a screened field of the form,

$$\vec{\tilde{E}}(\vec{r}) \equiv \vec{E}(\vec{r})(1+\alpha r)e^{-\alpha r} \tag{5.8}$$

where α is a parameter adjusted to fit the condition, (5.6). For reasons to be made clear below, this is called the adjustable parameter exponential approximation, or APEX. This model now has the same simplicity as the weak coupling approximations: only the pair correlation function is required. For the OCP, accurate correlation functions are obtained from the hypernetted chain integral equation even for quite large plasma parameters. The agreement of APEX with computer simulation results is excellent over a wide range of plasma parameters, Γ, and charge ratios, Z/Z_0, even for the extreme case of $\Gamma = 100$. Next, consider the case of a neutral radiator[6]. Figure 2 shows the results for $\Gamma = 10$. There is now a significant difference between the APEX calculation and the computer simulation results, although APEX is a considerable improvement over the first two terms of the usual Baranger-Mozer series. Also shown are the first two terms of the renormalized series (corrected APEX), which is again in reasonable agreement with the computer simulation data. These results support the above suggestion that renormalized independent particle models are expected to be least accurate for the neutral case, but still the leading term in a rapidly convergent series.

b) Direct evaluation of $G(\vec{\lambda})$

An alternative approach is the direct evaluation of $G(\vec{\lambda})$ from a closed form rather than a series representation. Such a form is easily obtained as follows,

$$G(\vec{\lambda}) = \int_0^{\lambda} d\ell \; \partial G(\vec{\ell})/\partial \ell$$

$$= \sum_{\sigma} \int_0^{\lambda} d\ell \int d\vec{r}_{\sigma} \; i\vec{\ell}\cdot\vec{E}(\vec{r}_{\sigma}) g^{(2)}(\vec{\ell};\vec{r}_{\sigma}) \tag{5.9}$$

where,

$$g^{(2)}(\vec{\ell};\vec{r}_{\sigma}) \equiv \frac{\delta G[\phi]}{\delta \phi(\vec{\ell};\vec{r}_{\sigma})} e^{i\vec{\ell}\cdot\vec{E}(\vec{r}_{\sigma})} \tag{5.10}$$

This form of the generating functional was first noted by Igelsias[3] who suggested that integral equation methods from the theory of liquids[15]

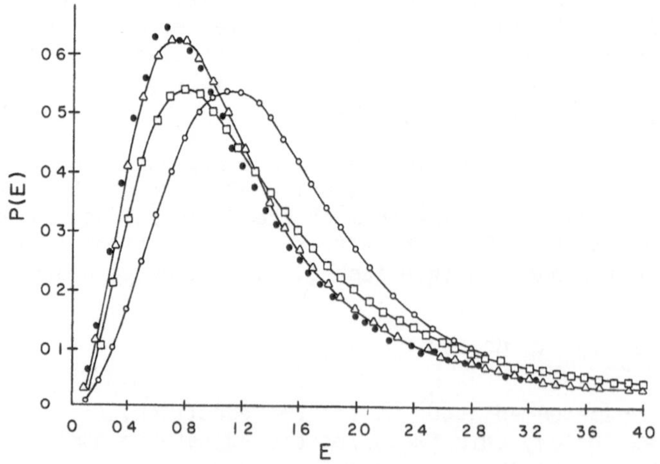

Figure 2: $P(\varepsilon)$ for neutral point at Γ=10, Baranger-Mozer (-o-o-), APEX (- - -), and renomalized Baranger-Mozer (-Δ-Δ-).

might be applicable to the evaluation of $g^{(2)}(\vec{\ell};\vec{r}_\sigma)$. The reason for this suggestion is that in the classical limit $g^{(2)}(\vec{\ell};\vec{r})$ is indeed a pair correlation function for a system of particles interacting via the complex potential, $V(\vec{\ell}) = V - i\vec{\ell}\cdot\vec{E}/\beta$. An application of integral equation methods has been given recently by Lado.[30] He first expands $g^{(2)}(\vec{\ell};\vec{r})$ in Legendre polynomials and then determines the coefficients by an extension of the mean spherical approximation.[31] The resulting microfield distribution is a Gaussian characterized by the correct second moment for the OCP, a relatively poor approximation. Subsequently, however, Lado has employed improved integral equation approximations based on the generalized bridge function for the mean spherical approximation leading to excellent results of comparable accuracy to APEX.

An approach similar in spirit to the effective renormalized field expansion is obtained in terms of a new "field" $\tilde{E}(\vec{\ell};\vec{r}_\sigma)$ defined by

$$g^{(2)}(\vec{\ell};\vec{r}_\sigma) \equiv e^{i\vec{\ell}\cdot\vec{\tilde{E}}}\, g^{(2)}(\vec{r}_\sigma) \tag{5.11}$$

Direct expansion to order ℓ gives the result (5.2), $\vec{\tilde{E}} = \vec{E}'$, as might have been anticipated. Use of Eqs. (5.11) in (5.9) again leads to the renormalized independent particle model. In the limit of classical mechanics and for the special case where $\vec{E}'(\vec{r}_\sigma)$ is derivable from the interparticle potential, $Z_0 e\vec{E}'$ is the average or mean force field,[32]

$$Z_0 e\vec{E}'(\vec{r}_\sigma) = k_B T\, \vec{\nabla}_0 \ln g^{(2)}(\vec{r}_\sigma) \tag{5.12}$$

More generally, the form (5.2) must be used in order to satisfy the second moment condition when quantum effects are important and/or when the microfield is not determined from the interparticle potential (e.g., Eqs. (2.4) and (2.5)). As expected, with the second moment condition satisfied the mean force field gives results similar to APEX. Corrections obtained by expanding to order ℓ^2 have also been determined[28].

VI A MODEL FOR MICROFIELD DISTRIBUTIONS

In the last two sections, two limiting cases of strong coupling have been treated successfully, but by different methods. Large plasma parameter effects in classical OCP's are well-described by the models of section V, while exchange correlations associated with large deBroglie wavelengths can be described by a partial resummation of the ideal Fermi gas contributions. It appears that a single model for microfield distributions that can describe both degenerate electrons and strongly coupled ions would have to combine both of the above methods. Instead, it is interesting to inquire if the renormalized independent particle models, designed for Coulomb correlations, also adequately account for quantum correlations. If so, a model for microfield distributions is obtained with unusual generality and simplicity.

To investigate this possibility, the renormalized independent particle model can be applied to calculate the ideal Fermi gas microfield distribution. All correlations are then due to quantum statistics, including those responsible for screening the field $\tilde{E}$. At high temperatures these correlations vanish and (5.7) approaches the correct Holtzmark limit. The relevant test is therefore at low temperatures, or extreme degeneracy. This calculation has been performed recently by Boercker[32] using the APEX choice for $\tilde{E}$, and by Pollock[33] using a quantum Monte Carlo simulation. Figure 3 shows the results for a neutral point with degeneracy parameter, $z \sim 10^5$ (representative of $T \sim 0$). Also shown are results using the first two terms of the quantum Baranger-Mozer series and the high temperature Holtzmark limit. Although this exploration is still incomplete, the relative agreement of APEX and the Monte Carlo data

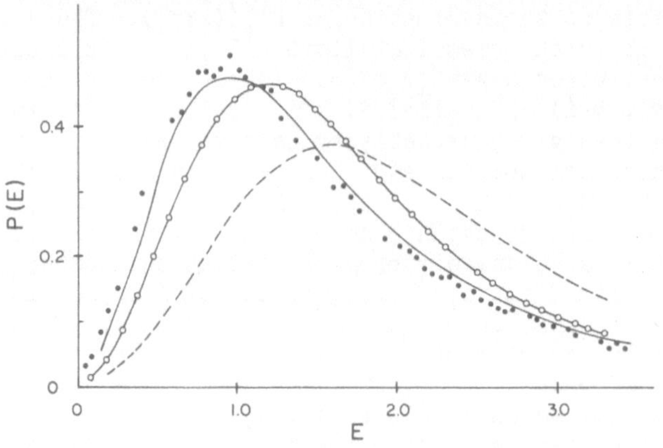

Figure 3: P(ε) for Fermi gas, Monte Carlo (•••), APEX (——), Baranger-Mozer (-o-o-), and Holtzmark (---).

in this "worst case" situation is striking and indicates that the model accounts for degeneracy as well as Coulomb correlations.

It appears reasonable to adopt the renormalized independent particle model for calculating microfield distributions under current experimental conditions and those of the near future, for both classical and quantum effects. The field, $\tilde{E}$, in these definitions is constrained by the second moment condition, but is otherwise free to be chosen on additional mathematical, physical, or practical grounds. The only input required is the pair correlation functions and second moments of the given microfields. The domain of applicability is expected to include electron and ion microfields, weakly and strongly coupled plasmas, classical and quantum effects, and (as discussed below) even some nonequilibrium states.

VII QUANTUM ELECTRON EFFECTS

The primary difficulties in application are associated with the quantum effects of electrons. The simplest conditions are those of high temperature and low density, since both electrons and ions are essentially classical according to Figure 1. The relavant pair correlation functions can then be computed from a multicomponent hypernetted chain set of integral equations. Even here, however, there is a classical short range divergence from electron-ion attraction. The resolution of this problem is through a proper quantum mechanical analysis, but then the advantage of the hypernetted chain equations for strong coupling of the ions is lost. A compromise is possible if suitable effective pair potentials, representing the short range quantum effects, are used in the classical integral equations. These effective pair potentials are obtained from the quantum pair correlation function for two isolated particles, $g_o^{(2)}$,

$$V(\vec{r}) \equiv -k_B T \ln g_o^{(2)}(\vec{r}) \qquad (7.1)$$

In the classical limit this reduces to the usual Coulomb potential, but more generally $V(\vec{r})$ includes quantum effects necessary for finite potential energy of oppositely charged particles at short distances. The effects of spin, bound and free states, and two particle exchange symmetry are conveniently included in this way. Detailed calculations of such effective potentials have been given recently by Gombert, Minoo, and Deutsch[35]. The Coulomb potential is modified only at distances less than the de Broglie wavelength so that Coulomb correlations are still essentially classical in the nondegenerate domain.

For weakly coupled plasmas quantum effects of electrons on interparticle correlations are adequately described by the random phase approximation (RPA) extended to finite temperatures[26]. For the electron gas considered in section IV, the RPA approximation to the density response function, $\chi(\vec{k},w)$, is

$$\chi_{RPA}(\vec{k},w) = \chi^{(o)}(\vec{k},w)/[1- V(k)\ \chi^{(o)}(\vec{k},w)], \tag{7.2}$$

where $\chi^{(o)}(\vec{k},w)$ is the ideal gas response function and $V(k)$ is the Fourier transformed Coulomb potential. The pair correlation function, $g(\vec{r})$, can be obtained from $\chi(\vec{k},w)$ by suitable integration over $\vec{k}$ and w. It is straight forward to extend (7.2) to a multicomponent plasma so that a proper quantum mechanical description of all correlation functions can be given. However, this description breaks down for the ions since they can be strongly coupled even when the RPA applies for the electrons. The quantum aspects of (7.2) are entirely contained in the ideal gas response functions, $\chi^{(o)}(\vec{k},w)$. Consequently, it is desireable to wed this structure with the classical strong coupling hypernetted chain approximation in a form that accounts for both strong ion coupling and quantum electron effects. One such synthesis has been given recently by Ichimaru and coworkers[18]. They modify the random phase approximation with local field corrections to account for strong Coulomb coupling. For the electron gas, Eq. (7.2) is replaced by[36]

$$\chi(\vec{k},w) \to \chi^{(o)}(\vec{k},w)/[1-V(\vec{k})\chi^{(o)}(\vec{k},w)(1-G(\vec{k})] \tag{7.3}$$

In the classical limit, $G(k)$ is simply related to the correlation function, $g^{(2)}(\vec{r})$. An approximation that includes both quantum and strong coupling effects is now obtained as follows: Calculate the classical $g^{(2)}(r)$ using the hypernetted chain approximation to determine $G(k)$. Use this classical $G(k)$ in (7.3) to get an approximate $\chi(\vec{k},w)$. Finally determine a new $g^{(2)}(\vec{r})$ from this $\chi(\vec{k},w)$, with quantum and classical strong coupling thereby included.

The procedure is considerably more complex in the multicomponent case of interest. The simplest realistic plasma would require three components: electrons, ions, and radiator. Although the spirit of imbedding the classical strong coupling results in the quantum density response formualism is the same as above, additional assumptions regarding the electron-ion coupling are imposed. These assumptions require that the electron-ion subsystems be weakly coupled so that a linear response approximation (similar to that used for liquid metals) can be employed. The local field corrections for the interference of these two subsystems then vanishes, and the dielectric response function for the plasma has the RPA form with additive polarizabilities for the electrons and the ions. As noted in the introduction, the ion-electron plasma parameter is proportional to the charge number for the ions and this weak coupling approximation can fail for high Z plasmas.

The limitations of weak electron-ion correlations are most severe when bound states must be accounted for in detail. This could be accomplished by using the effective potentials of Gombert et. al. in conjunction with the method of Ichimaru et. al. A more sophisticated approach capable of describing effects of such atomic structure is density functional theory (DFT), and has been explored in the present context by Dharma-wardana and Perrot[19,8]. Schrodinger equations for the local electron and ion charge densities around the radiator are obtained formally in terms of effective one-particle potentials. To describe strong ion-ion coupling, the effective ion potential is determined from the classical hypernetted chain approximation with a screened potential defined in terms of the electron charge density. The latter is determined from the Schrodinger equation with an approximate effective potential. Local density approximations for the latter typically parallel the RPA for a uniform electron gas and neglect direct electron-ion interactions. Self-consistent solution to the Schrodinger equations provides both free and bound charge densities near the ion. For example, in a hydrogen plasma at $r_s = 2$, and T/Fermi temperature = 5.08 they find a reduction of the effective ionic charge by almost 10%, due to formation of weak bound states. Dharma-wardana and Perrot have also compared microfield distributions calculated from a second order Baranger-Mozer model using DFT correlation functions, with APEX using the hypernetted chain approximation and Debye screened potentials. However, for reasons stressed in section II it is difficult to interpret their comparison since the microfields chosen are different in each case, and the method of calculation (APEX or Baranger-Mozer) is different. Discrepancies of 10-20% for the charged point case may or may not be attributable to improvements of DFT over the hypernetted chain approximation for the correlation functions. A more meaningful comparison would be the application of both sets of correlation functions to a single model with the same definition for the microfield.

In summary, it appears that reasonable methods are now available to include qualitatively (and often quantitatively) the most important effects in the parameter space of Figure 1, for calculation of pair correlation functions, and hence, microfield distributions.

VIII REAL PLASMAS

Although the formalism developed here is quite general, all specific applications have been for equilibrium states. The experiments for which strong coupling effects would be most important entail a variety of non-equilibrium conditions as well. These include different electron and ion temperatures, indefinite or changing compositions (e.g., populations of bound states), radiative transfer, and related failures of local thermodynamic equilibrium. In some cases, numerical hydrodynamic codes can be used to track the plasma under implosion or shock formation for better estimates of the state conditions. Generally, however, the theoretical objects of interest (like microfield distributions) are used in part as diagnostics to deduce state conditions such as temperature and density. In this context the theory and experiment rely on mutual feedback, rather than a comparison of two determinations under precisely specified conditions. It is important to have a theory that includes the effects of interest, but which does not depend too sensitively on the details of their calculation. After all, the microfield distribution is a simple structureless curve characterized largely by its peak field value, and by the high and low field asymptotes. For complex plasmas, a means to calculate how these characteristic features change with state conditions is sufficient.

To illustrate this spirit, consider a radiator in a multicomponent plasma of indefinite charges $\{Z_\sigma e\}$ and composition $\{n_\sigma/n\}$. Without more precise information, a detailed theoretical treatment of this complex plasma would be fruitless. However, it is possible that the radiator is sensitive only to some average interaction from all different species that could be characterized by an overall effective charge, $\bar{Z}e$. The microfield distribution could then be characterized by this single parameter (perhaps determined experimentally) rather than a large number of theoretical parameters. This possibility has been tested in some detail by Iglesias and Lebowitz for a classical plasma consisting of a radiator, two species of ions, and a uniform neutralizing background[5]. The multicomponent version of APEX was compared to a corresponding OCP calculation with $\bar{Z} = \Sigma Z_g n_g/n$. The results are shown in Figure 4, where the open circles represent the effective OCP results. Clearly, the detailed form of the microfield is well-described by the OCP. This example illustrates the need to understand exactly what minimum set of parameters (e.g. $\Gamma, z, Z_o, \bar{Z}$ . . .) is required to "tune" the microfield distribution.

The "quasi-equilibrium" condition of different electron and ion temperatures has been considered for microfield distributions by Tighe and Hooper[37]. They calculate low frequency microfield distributions for weakly coupled ions, using a Debye shielded OCP. Since the shielding is due to electrons, the temperature difference of ions and electrons can be introduced by $\lambda_e = \lambda_i/\sqrt{T_R}$ where λ is the Debye length and T_R is the ratio of ion to electron temperature. As T_R increases (at constant density) the effective screening length decreases, representing a more strongly coupled plasma. For strongly coupled plasmas or two component plasmas this approach breaks down, and a more complete description of the non-equilibrium state is required. Progress in this direction has been made recently by Boercker and More[38], who propose a model for the complete density matrix of a two temperature plasma. Explicit expressions for the pair correlation functions are given for the limit of weak ion-electron coupling, but including strong ion-ion coupling. These results allow Tighe and Hooper's calculations to be extended to the conditions of interest here.

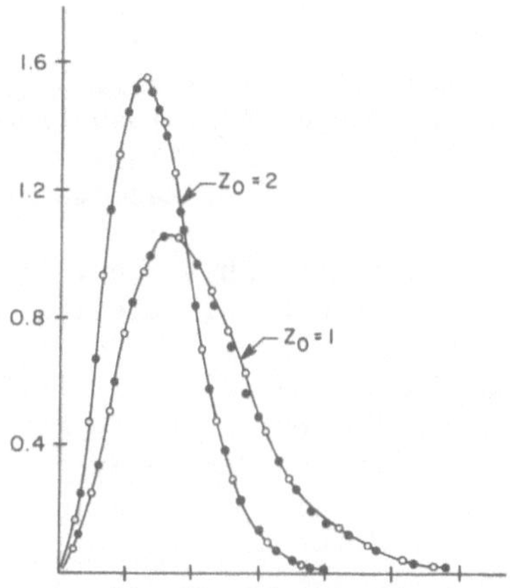

Figure 4: $P(\varepsilon)$ for binary mixture ($Z_1=2, Z_2=1$) at $\Gamma=4.88$, Monte Carlo (•••), APEX binary (——), APEX effective OCP (ooo).

Other complications such as micro-instabilities, turbulence, and coupling to the radiation field may be important in a given experiment. The ability to handle such factors emphasizes the desirability of a relatively general and simple formulation for the microfield distribution.

IX SUMMARY

The renormalized independent particle models discussed here provide a practical and accurate method to calculate microfield distribution functions in both the weak and strong coupling limits. Futhermore, this accuracy is maintained for a variety of physical effects that may be of interest such as quantum degeneracy, multicomponent plasmas, and bound electronic states. The primary input for the model is the pair correlation functions for perturbers and the radiation, and an appropriate choice for the microfield. This relative simplicity provides the potential to analyze microfield distribution functions for quite complex conditions. Clearly, significant progress has been made in recent years. Remaining problems, beyond improved accuracy of the correlation functions, include calculation of the second moment condition, Eq.(5.6), for quantum systems (since three particle correlations are then required) and a better determination of the low frequency microfield. As indicated in Section II, the latter problem requires a more extensive theoretical investigation of the specific application of microfield distributions intended.

ACKNOWLEDGEMENTS

The author is indebted to D. Boercker, H. DeWitt, C. Hooper, C. Iglesias, E. Pollock, and F. Rogers for helpful discussions and for allowing some of their results to be presented here. He is also indebted to M.W.C. Dharma-wardana, S. Ichimaru, and F. Lado for providing results prior to publication. This work was supported by National Science Foundation grant CHE-8411932.

REFERENCES

1. M. Baranger and B. Mozer, Phys. Rev. 115, 521 (1959); B. Mozer and M. Baranger, Phys. Rev. 118, 626 (1960); H. Margenau and M. Lewis, Rev. Mod. Phys. 31, 569 (1959); H. Pfenning and E. Trefftz, Z. Naturf 219, 697 (1966).
2. C. F. Hooper, Jr. Phys. Rev. 149, 77 (1966); 165, 215 (1968); C. A. Iglesias and C. F. Hooper A25, 1049 (1982).
3. C. A. Iglesias, Phys. Rev. A27, 2705 (1983).
4. C. A. Iglesias, J. L. Lebowitz, and D. MacGowan, Phys. Rev. A28, 1667 (1983).
5. C. A. Iglesias and J. L. Lebowitz, Phys. Rev. A30, 2001 (1984).
6. J. W. Dufty, D. B. Boercker, and C. A. Iglesias, Phys. Rev. A31, 1681 (1985).
7. X-Z Yan and S. Ichimaru, Phys. Rev. A34, 2167 (1986).
8. M. W. C. Dharma-wardana and F. Perrot, Phys. Rev. A33, 3303 (1986).
9. D. B. Boercker and J. W. Dufty, in Radiactive Properties of Hot, Dense Matter, J. Davis, et. al., editors (World Scientific Pub. Co., Singapore, 1985); and in Spectral Line Shapes, Vol. 2, K. Burnett, editor (W. de Gruyter, NY, 1983).
10. B. Held, M. Gombert, and C. Deutsch, Phys. Rev. A31, 921 (1985).
11. J. W. Dufty in Proceedings of CECAM Workshops on U.V. and X-Ray Spectra of Hot and Dense Plasmas (Orsay, France, 1976); C. A. Iglesias and J. W. Dufty in Spectral Line Shapes, Vol. 2, K. Burnett, editor (de Gruyter, NY, 1983).

12. The plasma parameter is defined here by $\Gamma = (e^2/\ell_o)/(2K/3)$, where ℓ_o is the average interparticle distance and K is the average kinetic energy per particle. In the high temperature limit $\Gamma \to (e^2/\ell_o)/(k_B T)$ which is the usual classical plasma parameter. In the low temperature limit $\Gamma \to 1.35\, r_s$ where r_s is the usual zero temperature coupling constant.
13. See, for example, A. Haur in Spectral Line Shapes, B. Wende, editor (de Gruyter, Berlin, 1981).
14. R. J. Trainor, J. W. Shaner, J. M. Auerbach, and N. C. Holmes, Phys. Rev. Lett 42, 1154 (1979).
15. J. P. Hansen and I. McDonald, Theory of Simple Liquids, (Academic Press, NY, 1976).
16. F. J. Rogers, J. Chem. Phys. 73, 6272 (1980).
17. S. Ichimaru, Rev. Mod. Phys. 54, 1017 (1982); M. Baus and J. P. Hansen, Phys. Rev. 59, 1 (1980).
18. S. Ichimaru, S. Mitake, S. Tanaka, and X-Z Yan, Phys. Rev. A32, 1768 (1985); ibid A32, 1775 (1985).
19. M. W. C. Dharma-wardana and F. Perrot, Phys. Rev. A26, 2096 (1982); ibid A29, 1378 (1984).
20. H. R. Griem, Spectral Line Broadening by Plasmas, (Academic Press, NY, 1974).
21. Doppler effects have been deleted from this expression for notational simplicity; their inclusion is straightforward and does not change any of the following discussion.
22. E. Smith, J. Cooper, and C. Vidal, Phys. Rev. 185, 140 (1969); D. Voslamber, Z. Naturf 249, 1458 (1969); T. Hussey, J. W. Dufty, and C. F. Hooper, Phys. Rev. A12, 1084 (1975); J. W. Dufty and D. B. Boercker, J. Quant. Spectrosc. Radiat. Trans. 16, 1065 (1976).
23. A. Brissaud and V. Frisch, J. Quant. Spectrosc. Radiat. Transf. 11, 1761 (1971).
24. J. Seidel, Z. Naturforsch 32a, 1207 (1977).
25. J. W. Dufty in Spectral Line Shapes, Vol. 1, B. Wende, editor (W. de Gruyter, Berlin, 1981).
26. A. Isihara, Statistical Physics (Academic Press, N.Y., 1971).
27. The discussion here is limited to methods applicable even for very strongly coupled plasmas (e.g., $\Gamma > 10$). Some other methods are available for the intermediate coupling range of $\Gamma \sim 1$. See, for example, F. Perrot and M. W. C. Dharma-wardana, Physica 134A, 231 (1985).
28. A. Alastuey, C. Iglesias, J. Lebowitz, and D. Levesque, Phys. Rev. A30, 2537 (1984).
29. H. C. Anderson and D. Chandler, J. Chem. Phys. 57, 1918 (1972).
30. F. Lado (preprint, and elsewhere in this volume).
31. J. L. Lebowitz and J. K. Percus, Phys. Rev. 144, 251 (1966).
32. T. L. Hill, Statistical Mechanics, (McGraw-Hill, NY, 1956).
33. D. B. Boercker (private communication).
34. E. Pollock (private communication).
35. M. Gombert, H. Minoo, and C. Deutsch, Phys. Rev. A29, 940 (1984); A23, 924 (1981).
36. More generally the local field correction is a function of frequency as well as wavevector; see Ref. 17 for a discussion of approximations that neglect the frequency.
37. R. J. Tighe and C. F. Hooper, Phys. Rev. A15, 1773 (1977).
38. D. B. Boercker and R. M. More, Phys. Rev. A33, 1859 (1986).

OPTICAL PROPERTIES OF NON-IDEAL PLASMAS

G. A. Kobzev

Institute for High Temperatures
U.S.S.R. Academy of Sciences
Izhorskaya str., 17/19, Moscow, 127412, U.S.S.R.

INTRODUCTION

Optical properties are of considerable interest for the physics of dense plasmas. As the density grows, the optical consequences of non-ideality begin to be displayed prior to corresponding changes in thermodynamic and transport properties and manifest themselves in the shift and broadening of spectral lines, as well as in the shift of the thresholds of photoionization continuums. With the further growth of non-ideality, the energy state (primarily weakly bound states) is markedly restructured leading to experimentally observed "relative brightening" of dense plasma, the emergence of "transparency windows" near the photoionization thresholds, effective reduction of the photodetachment continuums for weakly bound negative ions.

Today there is no consistent theoretical explanation of the effects observed in the experiment. The description is based on qualitative physical models. As a rule, the aforementioned effects are not taken into account in mass opacity calculations.

This paper deals with some experimental manifestations of the effects of non-ideality on the optical properties of dense plasma and discusses the technique of their theoretical description.

WEAK NON-IDEALITY APPROXIMATION (WNA) IN PLASMA OPACITY CALCULATIONS

Absorption Coefficient. Plasma Composition Calculation.

The spectral absorption coefficient (with the frequency exceeding that of plasma) in a plasma of moderate density (collisional, but not dense, i.e., weakly non-ideal) is determined as the sum of the products of cross-sections $\delta_{ij}(\nu)$ for the photoabsorption of light by the atomic species of the ith sort in the jth state by the population N_{ij} of these states:

$$K_\nu(T) = \sum_{ij} \delta_{ij}(\nu)\, N_{ij}(T) \qquad (2.1)$$

The total cross-section $\delta_i(\nu,T)$ is also introduced for the photo-absorption in the field of the ion

$$\delta_i(\nu,T) = \sum_j \delta_{ij}(\nu)\frac{N_{ij}}{N_i} = \frac{1}{\sum_i(T)}\sum_j \delta_{ij}(\nu)g_{ij}e^{-E_{ij}/T} \quad , \qquad (2.2)$$

where $\sum_i(T)$ is the internal partition function of the ith ion; g_{ij} and E_{ij} are the statistical weight and the excitation energy respectively, of, the jth state of the ith ion. The summation in Eq. (2.2) is performed with the Boltzmann distribution for the given ion states. With due account of Eq. (2.2) one has

$$k_\nu(T) = \sum_j \delta_i(\nu,T)\ N_i(T) \qquad (2.3)$$

where N_i (T) is the total concentration of the ith ions.

Thus, the calculation of the absorption coefficient k_ν in the case of dilute gas reduces to calculating the atomic cross-section and the composition of the gas mixture in appropriate approximation. Consideration of the effect of non-ideality, (i.e., interparticle interaction) on the optical properties amounts, in this case, to corrections to be made when calculating the composition.

Extensive literature is devoted to studying the thermodynamics of strongly coupled plasma. In recent years the research into these problems has been discussed in detail, e.g.. in Ref. 1–4. Therefore, we will not consider the problem of thermodynamic properties and the composition of plasma in detail in this paper, rather we will discuss very briefly the case of weak non-ideal plasma. The degree of plasma non-ideality is characterized by plasma coupling parameters $\gamma = (e^2/kT)/\langle\tau\rangle$ or $\Gamma = (e^2/kT)/\tau_D$, where $\langle\tau\rangle = (2N_e)^{-1/3}$ is the mean interparticle distance for charged particles, τ_D is Debye shielding length $\tau_D = (kT/8\pi N_e e^2)^{1/2}$.

The low temperature plasma under consideration is non-degenerate and is described by Boltzmann statistics. In the language of criteria this means that the relation between DeBroglie thermal electron wavelength $\lambda_e = \hbar/(2\pi mkT)^{1/2}$ and mean interparticle distance is less than one, i.e., - $\lambda_e/\langle\tau\rangle < 1$.

In addition, quantum effects do not play any noticeable role in the interaction of particles in the plasma under study.

To obtain data on the ionization composition of the plasma a quasi-chemical method of description is usually used in case of weak non-ideal "gas systems". According to this method the plasma is regarded as a mixture of particles of different sorts. The relation between concentrations of particles is described by the equations of ionization equilibrium. Interparticle interaction is taken into account as corrections to ideal-gas formulas. In this case the free energy F of the system can be written as

$$F(\nu,T) = kT\ \{N_e \ln\left(\frac{N_e\lambda_e^3}{2e}\right) + \sum_k N_k \ln\left(\frac{N_k\lambda_k^3}{eQ_k}\right)\} + \Delta F \quad , \qquad (2.4)$$

where Q_k is partition function of the kth atomic or ionic component, F is a correction for non-ideality.

For example, Debye-like corrections were taken into account in Ref. 5 in the formula of ΔF to calculate the composition of an air plasma at T up to 3 10^6 K. The first twelve energy levels were taken into account to calculate the internal atomic partition functions Q_k. The contribution of a large number of levels at low temperatures (compared with the ionization energy of the ion under study) is not essential, and at higher temperature it is necessary to cut off the partition function according to the Planck-Larkin method which corresponds to rejecting the excited levels with ionization energies lower than temperature.

Experiments in dense cesium plasma with $\Gamma < 1$ corroborated a sufficient accuracy of these approximations.[4] At higher Coulombic non-idealities one has to use the idea of the deformation of atomic bound states to describe the experimental data on thermal and caloric equations of state of heavy inert gases.

Elementary Radiative Processes. Free-Free Transitions in Fields of Ions and Neutrals

Included as co-factors in the sum total in Eqs. (2.1 or 2.3) describing the absorption coefficient are the cross-sections of photoabsorption which correspond to different elementary radiative processes. In atomic and ionic plasmas, several processes are responsible for the absorption of radiation: bound-bound transitions in atoms and ions which spectral lines correspond to in spectrograms: bound-free transitions with the involvement of atoms or ions, i.e., the photodecomposition of negative ions and photoionization of atoms and ions, which absorption continuums having long-wave thresholds for each discrete state correspond to in the observed spectra; and free-free transitions in the fields of atoms and positive ions - inverse bremsstrahlung. Extensive literature is devoted to methods of calculating the cross-sections of elementary radiative processes, therefore, we have restricted ourselves only to the remarks concerning bremsstrahlung.

Traditionally, bremsstrahlung is calculated on the basis of quasi-classical Kramers' formula with the Gaunt correction factor taking into account quantum-mechanical effects.[6] In the case of collisions with non-hydrogenic ions, as the radiation frequency grows, the radiation cross-section is described by the Kramers' formula with an effective charge varying from the ion charge to nucleus charge. For x-ray radiation there are discussed the effects of screening with bound electrons.[7] At the same time, unfairly forgotten is the method proposed by Biberman and Norman taking into account the non-hydrogenic character of atomic energy spectra in calculating the recombination and bremsstrahlung radiation with the aid of so-called ζ-function. The latter markedly differs from one and has non-monotonic dependence on frequency not only for atoms, but also for many-charged ions. For ions with charge $Z \geq 10$ "soft" photons already belong to the x-ray spectrum.

The reliability of series expansions with parameter ω/Z^2 (ω is the frequency of radiation in rydbergs) and extrapolations of quantum defects to energy continuum which are used in Ref. 8 proves to be sufficiently high. The method allows using analytic formulas and calculated phases of elastic scattering on the ionic core to calculate the ζ-functions.[9]

The second remark on elementary processes is not associated with the effects of non-ideality either. In low temperature plasma ($T \simeq 10^4$ K), as pressure goes up, the processes involving negative ions play an increasingly greater role in forming continuous spectra.[10] In the past, the radiation of unstable negative ions manifesting themselves in electron-scattering as resonances was not studied properly. In Refs. 11 and 12 it was shown that the contribution of low-energy resonance to the radiation continuum is described in terms of the resonance bremsstrahlung theory. When the resonance's effect on bremsstrahlung is substantial, the principal term of the radiation cross-section can be presented as:

$$\frac{d\delta_{EaEb}}{d\omega} = \frac{\Gamma/2\pi}{(E_b-E_o)^2+(\Gamma/2)^2} \left(\frac{\hbar\omega}{E_a-E_o}\right)^3 \delta_{E_a, n\ell_o} , \qquad (2.5)$$

where δ_{Ea}, $n\ell_o$ has the form of the cross-section of photoattachment to a normal discrete level

$$\delta_{E,n\ell_o} = \frac{4\pi^2 d(E-E_o)^3}{3mc^2E} \left[\ell_o \left(R^{n\ell o}_{E,\ell_o^1}\right)^2 + (\ell_o + 1)\left(R^{n\ell_o}_{E,\ell_o+1}\right)^2\right] \qquad (2.6)$$

Here, the bremsstrahlung cross-section $d\delta_{EaEo}/d\omega$ is expressed through the square of matrix element

$$R^{E_a\ell_o}_{E_a\ell_a} = \int_0^\infty P_{E_a\ell_a} P_{E_b\ell_b} rdr$$

Where $P_{E\ell}$ is the radial wave function of the electron with the energy E and the orbital moment ℓ, normalized by the condition of $\int \rho_{E\ell}P_{E'\ell'}dr = \delta\ (e-E')$ where $\omega = (E_a - E_b)/\hbar$ is the radiation frequency. To isolate the radiation fraction associated with resonance, and to follow the Fano method, $P_E\ell_o$ is presented as a superposition of the localized state wave function $P_{n\ell o}$ and functions $P_{E\ell o}$ corresponding to the potential scattering disregarding resonance:

$$P_{E\ell o} = u_{nE} \tilde{P}_{n\ell_o} + \int dE' U_{E'E} \tilde{P}_{E'\ell_o} \qquad (2.7)$$

Therefore, the resonance in electron-atom scattering plays almost the same part in forming plasma radiation spectra as the weakly bound state of a negative ion. This statement is important for the interpretation of experiments in the radiation of nitrogen and air plasma. These experiments were conducted in a wide range of temperature and pressures. Measurements performed under a pressure of from one to a thousand atmospheres and a temperature of 9,000 to 15,000 K (cf., e.g., Refs. 13 and 14) give the intensity of continuous radiation of nitrogen or air plasma in the visible and ultraviolet spectrum which exceed considerably the theoretically calculated total intensity of radiation. The excess of experimentally measured radiation over theory was associated with the process of electron photoattachment to nitrogen atom. An analogous situation is observed in the case of oxygen plasma radiation where the photoattachment involving the formation of negative ion O^- makes a marked contribution to the spectral radiation density. Unlike oxygen, the energy spectrum of a negative nitrogen ion has no bound state, but according to the theory of resonance bremsstrahlung from the standpoint of calculation of the total intensity of electron radiation in the field of the nitrogen atom it does not matter whether the bound state of the electron (negative ion) occurs or not.

Line Merging. Destruction of Weakly Bound Atomic States by Plasma Microfields. Spectroscopic Stability Principle (SSP)

In a plasma of moderate density (even Debye-like) non-ideality effects, besides their influence through thermodynamics, manifest themselves directly in optical spectra. At low charge densities the effect on total cross-sections of photoabsorption is insignificant and it affects separate small spectral intervals and practically it does not exert any influence on an integral radiative energy flux. It is weakly bound atomic states that are most affected by perturbing plasma microfields. In principle, in a plasma of any density there are such high atomic states for which the perturbation effects of the surroundings is more intense than the binding with the parent core (ion of charge Z_r), and it is not correct to regard these states as one-particle. The size of the perturbed spectral region of the radiating atom or ion with the charge $Z_r - 1$ in a plasma of the electron density N_e which is connected with ion densities by the quasineutrality relation

$$N_e = \sum_p Z_p N_p = \bar{Z}_p \bar{N}_p \quad , \tag{2.8}$$

can be estimated by one of the formulas:

$$\Delta E = (Z_r \bar{Z}_p)^{1/2} \bar{N}_p^{1/3} \, 7.64 \cdot 10^{-7} \mathrm{eV} \tag{2.9}$$

obtained from the estimation of the potential barrier lowering in a uniform electric microfield with the Holtsmark distribution function, or

$$\Delta \nu \equiv \Delta E = (Z_r^4 \bar{Z}_p^2)^{1/5} \bar{N}_p^{4/15} \, 1.03 \cdot 10^{-5} \mathrm{eV} \quad , \tag{2.10}$$

corresponding to the consideration of the splitting of Stark sublevels of a hydrogenic ion in the Holtsmark microfield, i.e., to the Inglis-Teller effect,[15] or a similar relation obtained also on the basis of the high levels merging criterion with due account of electron broadening.

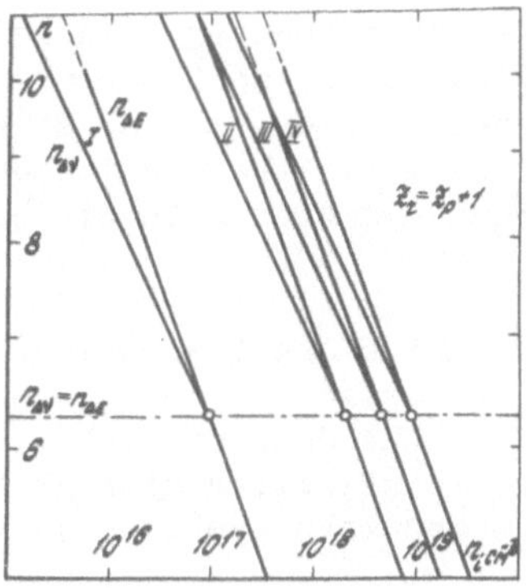

Fig. 1

Figure 1 shows the dependencies of $n_{\delta\nu}$ - principal quantum level number corresponding to Eq. (2.10 and $n_{\Delta E}$, corresponding to Eq. (2.9) for the atom (signed by Fig. 1) and ions of different charge (II, ...) according to Ref. 16. As charge density increases at $N_i = N_1 = 9.14 \cdot 10^{16}$ $(Z_r\ Zp)^{3/2}$ the values of $n_{\Delta\nu}$ and $n_{\Delta E}$ are equal and at $N_i\ N_1$ the effect of line nonrealization[17] prevaileds in a sufficiently dense plasma. The considered (Eq. 2.10) effect of merging the highest levels broadened by the interaction with the surrounding particles is a predominant one, as shown in Fig. 1, at $N_i\ N_1^*$, and is the only considered manifestation of the effect of non-ideality on optical properties (ENOP-referred to herein below) in the weak non-ideality approximation (WNA). This approximation is used traditionally in opacity calculations. The calculation of thermodynamic functions and plasma composition is performed in Debye or in slightly more sophisticated approximations, and the Inglis-Teller shift of photoionization threshold is taken into account in the photoionization cross-section. Bremsstrahlung cross-section, as a rule, is assumed to be invariable. The arising problem of extending the photoionization cross sections for individual energy levels is solved by using the spectroscopic stability principle (SSP)[18] for photoionization cross-sections.

According to the SSP the total oscillator strength is conserved for the transition between two states under the perturbation satisfying the unitarity condition. In the theory of line broadening, this thesis is known as a conservation of the integral over the spectrum for an individual spectral line. Successful application of generalized SSP for determining the photoionization cross-sections on the basis of measured oscillator strengths of respective spectral series and the solution of the inverse problem[19] for gas and Debye plasma have facilitated the further application of WNA in opacity calculations. In the conditions of weakly nonideal plasma there occurs the displacement of the apparent photoionization threshold towards a long-wavelength side, which is stipulated by an interparticle interaction in the plasma. The higher members of spectral series are assumed to be broadened to such an extent that they are superposed and merged to form a continuous spectrum. This explanation is not quite correct, therefore, what in in question is the transformation of the lines into a continuous spectrum rather than the merging of these lines.

For an approximate consideration of the photoionization threshold displacement, it is assumed that the influence of interparticle interaction on the continuous spectra is reduced only to the fact that the higher members of spectral series are transformed into continuum in keeping with an unperturbed oscillator strength density. Thus, the photoionization cross-section from an individual mth level turns out to be continued to a long wavelength side beyond the ideal threshold frequency ν_m^o to a certain quantity of $\Delta\nu$. The photoionization cross-section is considered to be invariable for frequencies greater than ν_m^o. In Ref. 19 the available experimental data on individual cross-sections are analyzed. It is necessary to compare the density of oscillator strengths and the photoionization cross-section measured in the conditions of a small shift, and photoionization cross-sections measured with different values of the shift $\Delta\nu$ (Fig. 2).

The sum total of data enables one to conclude that the theoretical curve serves as the real photoionization and photoexcitation cross-sections of an isolated atom. Besides, the data of Fig. 2 support the idea that the assumption of the spectroscopic stability of high members of the spectral series, as these are transformed into continuum, is reasonable.

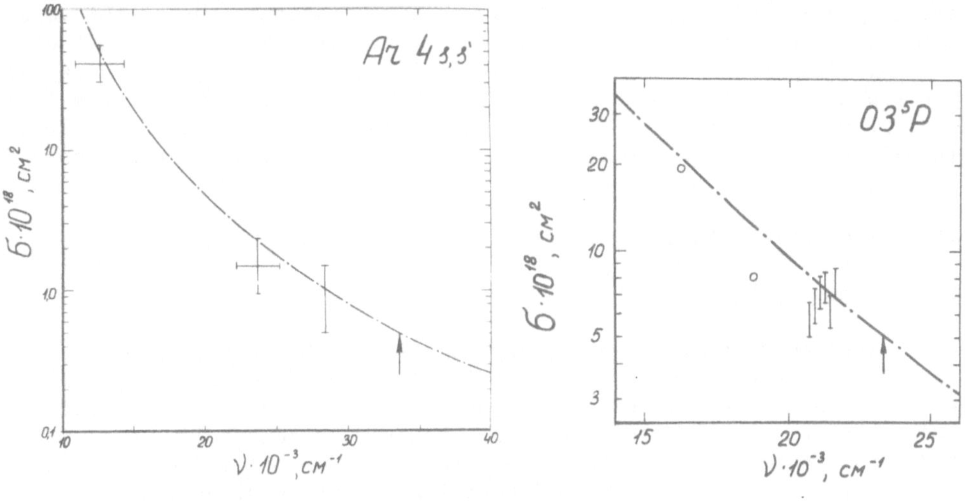

Fig. 2 Photoionization and photoexcitation cross-sections and ArI4s,s' OI$3p^5p$[19]; 1 - experimental points for photoexcitation cross-section; 2 - threshold values of photoionization cross-sections measured in Ref. 20; 3 - calculation according to quantum defect method; 4 - limit of isolated atom series.

There are many publications on so-called "optical" lowering of the ionization potential ΔE_{opt} (this term itself is not adequate). It was suggested that ΔE_{opt} be found by one of the two methods. First, on the basis of the last observed line of the series that makes this finding indefinitive. Second, on the basis of the comparison of measured continuum in the field of the spectrum determined mostly by the photoionization of high excited levels and electron-ion bremsstrahlung with the calculation according to the Unzold-Biberman-Norman formula. The quotient of their division was defined as exp $(\Delta E^{*}_{opt}/Kt)$. Evidently, in the second case ΔE_{opt} is an effective quantity markedly dependent on the specific features of the spectrum and calculation accuracy. In Debye plasma the value of exp $(\Delta E^{*}_{opt}/kT)$ does not differ from one noticeably, however, its consideration improves agreement between theory and experiment. The value ΔE_{opt} is found in these conditions close to the estimation according to Eq. (2.10).

Until recently, the above-mentioned WNA has had no strict proof, its application was based on common sense and non-contradictions of the comparison of calculation and experiment. In the literature there are known attempts to construct the "first principle" theory of bremsstrahlung and photorecombination continuum of non-ideal plasma and to obtain corrections to certain formulas, or as in the minimum, the criteria of applicability of the approximation of the WNA type.

The major portion of the first-principle theories deals with studying the bound-bound transitions, i.e., with the problem of line shift and broadening in a plasma. The theory of electron-ion photo-continuum of weakly-nonideal plasma was evolved ion Refs. 21 and 22. In Ref. 21 the theory of bremsstrahlung was constructed. In Ref. 22 the general quantum-mechanical expression for spectral radiation power is used for obtaining the expression for the intensity of recombination radiation of Debye plasma on the basis of the modified Kroll resolvent technique with those used in WNA; at the same time, some estimations made of the values of the effects out of consideration, namely, broadening and the shift of energy levels, alteration of the continuum density of states, dynamic screening of the fields of particles in a plasma. At the same time, within the framework of approximations used in Ref. 22, one failed to analyze the effects of rebuilding the higher energy levels which are characteristic of non-Debye plasma.

EFFECTS OF NON-IDEALITY UPON OPTICAL PROPERTIES OF PLASMA

Experiments in Non-Debye Plasma

Although a great number of experiments on the optics of plasma has been satisfactorily explained in WNA, however, there are several experimental results bearing on the optics of plasma which are not explained within the WNA framework. It is clear that the WNA-calculation yields an increase in the total photoabsorption cross-section at constant temperature as N_e increases owing to an additional contribution of quasicontinuums, because of the merging of high members of spectral series, to the continuous spectrum. In Ref. 23 as N_e increases up to 10^{19} cm^{-3}, there is observed a decrease in the photoabsorption cross-section in carbon-hydrogen and xenon plasma. In the experiments of Ref. 24, other than Inglis-Teller behavior of high members of spectral series was observed. It looks as though lines do not "disappear" from the spectrum as density grows, in other words, the continuum background does not form in their place.

Hypothesis of Non-Realization of High Atomic Levels

A hypothesis of proposed in Ref. 17 which explains the totality of experimental data in a unified manner on the basis of the idea about "non-realization" of part of high energy levels in a plasma as a result of the modification of the Coulomb potential of interaction of an optical electron with a parent ion to a short-range potential or as a results of the influence of the electric microfield of surrounding particles. The supposition of "non-realization" of part of high levels in a plasma was discussed earlier in literature in connection with the problem of cutting-off of the atomic partition function and briefly in connection with the optics of a plasma.

Two competing effects of non-ideality are dealt with in Ref. 17, namely, merging of the lines adjoined to the threshold resulting in an apparent shift of the photoionization threshold (let us denote the effective principal quantum number of the last observed line by ($\widetilde{n}_{\Delta\nu}$), and the non-realization of atomic energy levels ($\widetilde{n}_{\Delta E}$ is the principal quantum number of the last realizing level) under the action of strong microfields or due to reconstructing the effective potential of interaction of charged particles in a plasma from the Coulombic potential to a short-range one.

One can estimate $\widetilde{n}_{\Delta E}$ by equating the energy of the level to the lowering of an atomic potential barrier in a uniform electric field [see Eq. (2.10)]. The selection of $\widetilde{n}_{\Delta\nu}$ and $\widetilde{n}_{\Delta E}$ contains some uncertainty. Crossing of curves $\widetilde{n}_{\Delta\nu}$ and $\widetilde{n}_{\Delta E}$ takes place at densities between 10^{17} and 10^{18} cm^{-3} (for atoms). At greater densities of charge N_i one may observe peculiarities in the spectra near photoionization thresholds which are caused by the non-realization of atomic levels under the influence of plasma microfields.

In case of an effective cut-off Coulombic potential the value of $\widetilde{n}_{\Delta E}$ is determined by the limitation of the number of bound states in the short-range potential and is found by solving the Schredinger equation for an appropriate potential. The origin of short-range interaction may be associated with a screening action of both charged and neutral surrounding particles. Generally speaking, the concrete type of an effective short-range potential can be found by solving the many-body problem. Some published estimates of $\widetilde{n}_{\Delta E}$ are obtained for simplified model potentials - Debye and cut-off Coulomb, but their use in the problem of ENOP is not strictly substantiated.

According to Ref. 17 strongly perturbed levels belonging to the energy range determined from the formula (for a fixed value of microfield F):

$$\Delta E = 2e\sqrt{eF} \tag{3.1}$$

and from Eq. (2.10) (after Holtsmark averaging) and (2.9) (estimate according to the merging of levels) are assumed to be non-realizable and therefore not involved both in photon absorption and radiation. This picture features a vanishingly low density of states in the range of ΔE near the zero energy point. The two observable consequences are discussed in Ref. 17: "relative brightening" of dense plasma, i.e., a decrease in the total photoabsorption cross-section in those spectral regions where the energy levels belonging to ΔE interval must mostly contribute to the cross-section; and the appearance of frequency intervals of ΔE size free from spectral lines, i.e., "transparency windows" in the regions of the spectrum adjoining the photoionization thresholds from the long wave side.

Relative Brightening of Dense Plasma

One can readily estimate the relative brightening of hydrogen recombination continuum in such purposely simplified approach. Changing the upper limit of integration in so-called Unzold-Biberman-Norman integral formula for bound-free transitions from zero to -ΔE brings about the following estimate of a brightening degree ($h\nu > \Delta E$):

$$k_e/k_o \simeq (e^{h\nu/kT} - e^{\Delta E/kT})/(e^{h\nu/kT} - 1) \tag{3.2}$$

A more sophisticated approach within the framework of the same model presupposes the calculation of an absorption coefficient for a fixed value of the electric field F and subsequent averaging. In Ref. 17 the totality of experimental data on Xenon plasma radiation for frequency $\nu \simeq 22000$ cm^{-1} with the charge density changing from 10^{16} up to 10^{19} cm^{-3} were obtained and the behavior of the argon plasma absorption coefficient for the same frequency at high densities was predicted.

Figure 3 gives the results of experiments handling in the form of dependence ξ versus N_i. At low N_i , according to WNA, the calculated value is in good agreement with the experimental one. The curves in Fig. 3 are calculated with the assumption that the levels are not realized in the range of energies ΔE, where ΔE was determined by Eq. (3.1) or by as twice as much value.

The subsequent experiments 25-28 have corroborated with observations in Ref. 17, their results are generally in good agreement with predictions of relative brightening according to Ref. 17, It is noteworthy that the performed estimates of relative brightening are based on averaging the realization probability of a particular atomic level in a plasma with microfield distribution function, and on a respective decrease in the contribution of this level to the total absorption coefficient. In Ref. 29 the authors explained experimental results on relative brightening of argon plasma on the basis of "confined atom" approximation. The energy levels of Ar atom in the atomic cell were calculated by means of a self-consistent field method. The cell size was dependent on plasma density. Photoionization

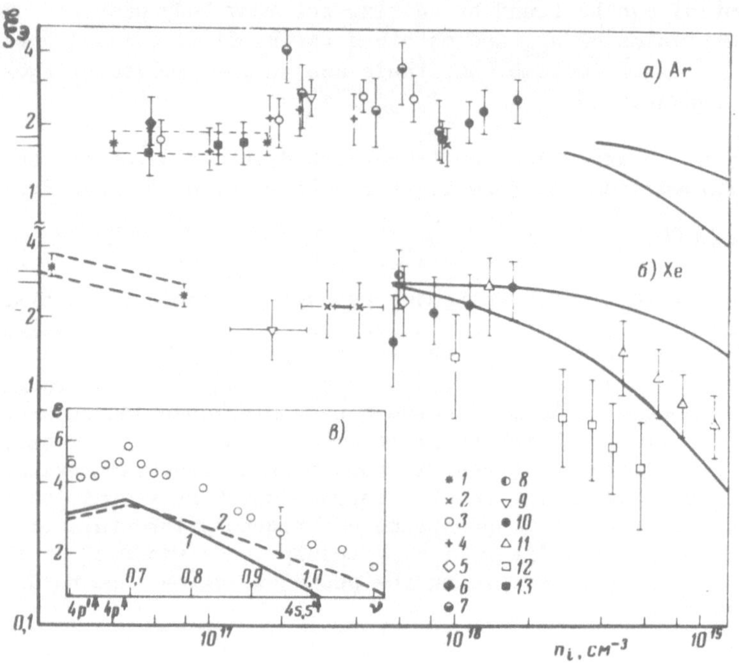

Fig. 3. Relationship between ξ-factor and charge density in a plasma:[17] a - AR; $\nu \approx 222000\ cm^{-1}$; b - Xe, ν - 22000-23000 cm^{-1}, $\xi_e = k_e/k_0$, where k_e is an experimental value of absorption coefficient, k_0 is a calculated value with $\xi = 1$. 1-13 are experimental data.

cross-sections of atomic levels were assumed to be independent of plasma parameters. Afterwards, more consistent calculation[30] was carried out of the total photoionization cross-section of the argon atom for experimental conditions of Ref. 25 on the basis of the modified Hartree-Fock-Slater model.

In Refs. 27 and 31, optical properties of air plasma were investigated at pressures from 40 to 150 atm and in a temperature range from 17000 to 20000 K with charge densities as high as $10^{19}\ cm^{-3}$.

At high pressures the measured spectral absorption coefficients were found to be lower than those calculated in WNA within a wide spectral interval. The calculation performed in Ref. 31 in accordance with the approach in Ref. 17 with due account of a relative brightening of dense plasma yielded the results which coincided with the experiment. It turned out that in the conditions of the experiment of Refs. 27 and 31 a group of 3d levels of oxygen and nitrogen atoms whose unperturbed ionization energy of about 1.5 eV, is strongly affected by plasma microfields.

It should be noted that within the limits of microfield description of non-ideality effects upon optical properties of plasma bremsstrahlung is usually calculated on the basis of the Kramers formula with the Biberman-Norman correction factor ξ_{ff}, which takes into account a non-hydrogenic nature of the absorption of radiation in the field of a complex ion. ξ_{ff}-factor is assumed to be independent of density, nor is an increase in the number of electrons taken into account which are involved in bremsstrahlung at the expense of those which in an ideal-gas approximation would belong to bound states in the range of energies from 0 to $-\Delta E$. For a high frequency part of the spectrum, these assumptions look natural because , generally speaking, only a small portion of such radiation can be produced by low-energy electrons. References 21, 32-35 are devoted to studying the electron-ion bremsstrahlung of non-ideal plasma. References 21 and 32 show that in Debye plasma the corrections to ideal-gas bremsstrahlung are proportional to degrees of factor a_0/r_D and can be neglected. Theoretical elaboration of bremsstrahlung of non-Debye plasma feature a considerable inconsistency in estimates (see e.g., Ref. 33) and is aimed at taking into account not so much non-ideality, but rather non-hydrogenic nature of the spectrum.

The reconstruction of a low-energy electron spectrum in a dense plasma, the violation of a pair nature of electron-ion scattering, an increase in the share of curvilinear sections of trajectories should inevitably change the low-frequency part of the bremsstrahlung spectrum. However, experimental data on bremsstrahlung of dense plasma are not available today. It would be useful to see the results of molecular dynamics and Monte-Carlo calculations for realistic plasma models.

In Ref. 36 on the basis of hypothesis of "non-realization" the estimates were carried out of the effect of relative brightening on one of the integral optical properties - Rosseland mean radiation length l_R for non-ideal caesium plasma. The basic contribution to l_R at $T \simeq 10^4$ K is conditioned by photoionization of Cs atoms. At high pressures and temperatures, Cs atoms are found to be strongly influenced by plasma microfields leading to a decrease in the absorption coefficient for photon energies of the order of $h\nu \simeq (3\text{-}5)$ kT and, correspondingly, to an increase in l_R. The estimates obtained show that density effects in the examined case bring about a relative brightening of caesium plasma noticeable even in an integral optical property. This conclusion qualitatively agrees with the results of experimental determination of radiative heat conductivity of dense caesium plasma.[37] The ranges of values of l_R must markedly increase as pressure grows within above limits. The effect of brightening on the coefficient of radiative heat conductivity of non-ideal plasma of alkali metals Li, Na, K and Cs was experimentally observed later in Ref. 38.

Near-Threshold "Transparency Windows"

The effect of brightening of dense plasma is defined by non-realization of initial states. By contrast, near threshold "transparency windows" are defined by non-realization of final states with unperturbed initial states and correspond to the disappearance of the high members of spectral series (without merging because of broadening) with an increase in plasma density. In so doing, in the value averaged along the spectrum - spectral density of oscillator strengths df/dE there must be observed a devication a deviation from law for the isolated atom dependence for the region $\Delta E/kT \simeq c \cdot \gamma$, where γ is non-ideality (strong coupling) parameter, and $C \simeq 3\text{-}4$ is numerical coefficient weakly dependent on γ.

In the region of ΔE the conditions of applicability of spectroscopic stability principle (SSP) are violated. In actual fact, SSP is applicable, if the perturbation of the final and initial state of a radiating system is insignificant,[18] and this perturbation can be described by unitarity transformation. In the case under consideration the upper state is shifted to the continuum under the action of a microfield and this transition cannot be described by unitarity transformation. The distribution of df/dE in the near-threshold region must differ from the unperturbed one.

The above considerations bearing on the appearance of near-threshold "transparency windows" in dense plasma are experimentally corrobrated.

Lithium. Analyzed in Ref. 36 are Yaakobi's data[39] on the radiation of dense lithium plasma with charge density of $1.7\ 10^{17} - 3.10^{18}\ cm^{-3}$

formed during an electrical explosion of wire. The time radiation spectrum measured integral in time contains lines of principal series in absorption on an intense continuum background. As the spectral lines approach the photoionization threshold, they merge and disappear, and a peak of intensity is observed in the near-threshold, they merge and disappear, and a peak of intensity is observed in the near-threshold region of the spectrum which goes beyond the framework of measuring error. This peak may be due to a dramatic decrease in the optical depth caused by a near-threshold transparency window.

Argon. Experimental results on continuous radiation of argon plasma[24] allow an interpretation on the basis of ideas about the non-realization of high atomic levels in a plasma. Shown by pointers in Fig. 2 is an unperturbed postion of thresholds of 4 s,s' and 4 p,p' states. The absolute values of the peaks of intensity in these regions correspond to the photoionization cross-sections of these states. Adjoining spectral lines are not observed in the experiment.

Mercury. In Ref. 40 the radiation of dense mercury plasma is measured. Spectral attention is paid to the spectra near photorecombination thresholds of $6p^1P_1^0$ nd $6p^3P_2^0$ levels. Series of spectral lines are traced with a view to determine the last line still discernible over continuum. In a less dense case ($N_i = 5.10^{15}$ cm^{-3}), the merging of spectral lines is observed in the near-threshold regions of the spectrum leading to an apparent shift of the photorecombination thresholds toward low=frequency side. With greater densities of charges ($N_i = 4.10^{17}$ cm^{-3}) a qualitatively new effect is confirmed,namely, the spectral lines disappear in the near-threshold region without overlapping, and the thresholds are found to be actually unshifted. As a result, a gap of radiation intensity develops near the photorecombination threshold compared with the WNA calculation.

Air. In Refs. 27 and 31 jth line spectrum of atoms and first ions of nitrogen and oxygen is investigated in addition to measurements of the total radiation continuum of air plasma. At maximum density ($N_i = 1.2 \cdot 10^{19}$ cm^{-3}) the only lines observed in spectral series np-s 3p-3s and 3d-3p transitions. In the Balmer series of hydrogen which is presented as a small admixture, it is only the H (3d-2P)-line that is observed.

Hydrogen. Hydrogen is the most attractive chemical element from the viewpoint of verification of theory because its atomic parameters are known as accurate and the near-threshold spectral regions are free from overlapping of the lines belonging to the spectral series different from the lines under consideration. There is known a number of papers, where measurements are carried out near the Balmer threshold. However, because of experimental difficulties, the values of charge densities achieved in experiments do not exceed, as a rule, $N_i \simeq 10^{17}$ cm^{-3}. The authors of most well-known papers[41,42] of this kind assume that they are in good agreement with the WNA calculation. Systematic studies of the radiation of hydrogen plasma at $N_i > 10^{17}$ cm^{-3} have only begun. Individual results are given in Refs. 43-46. In Ref. 43 there are not observed any anomalies in a relative distribution of spectral absorption coefficients near the Balmer series limit.

Figures 4 and 5 show the examples of experimental data from Refs. 45 and 46, displaying an anomalous behavior of an absorption coefficient near the series limit. They can be regarded as experimental confirmation of near-threshold "transparency windows."

It should be noted that the observation of disappearance of the high members of spectral series in the measured spectrum, as the plasma density grows, is still not enough to attest to the presence of a near-threshold "transparency window." It will be recalled that in Ref. 17 in order to single out the subject of discussion the case was considered where a "transparency window" appears in the absence of line merging. A similar situation - a decrease in intensity and the disappearance of non-merged lines - was observed in mercury and air dense plasmas[24,27,31] as well as in the spectrum of helium-like ions of silicon in laser plasma produced in the "Janus" facility.[47] The merging of lines to quasicontinuum seems to be more characteristic for hydrogen. This merging is accompanied by the suppression of this quasicontinuum. This case is obviously realized in Figs. 4 and 5. As distinct from this, a decrease in the line intensity due

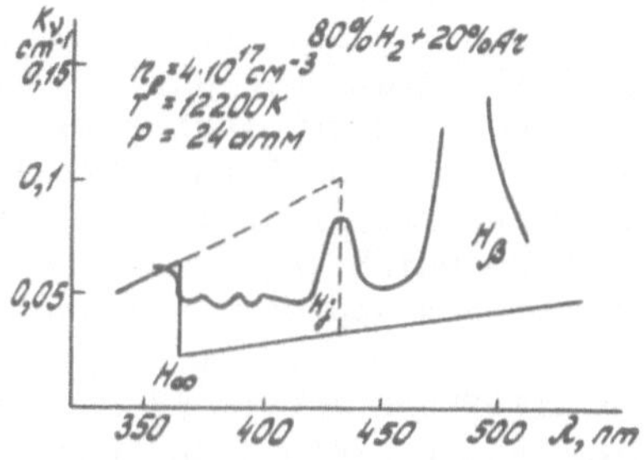

Fig. 4 Absorption coefficient of hydrogen-argon plasma. Solid thick line - experiment, Ref. 45. Solid thin line - calculation without Balmer series lines. Dotted line shows Inglis-Teller shift.

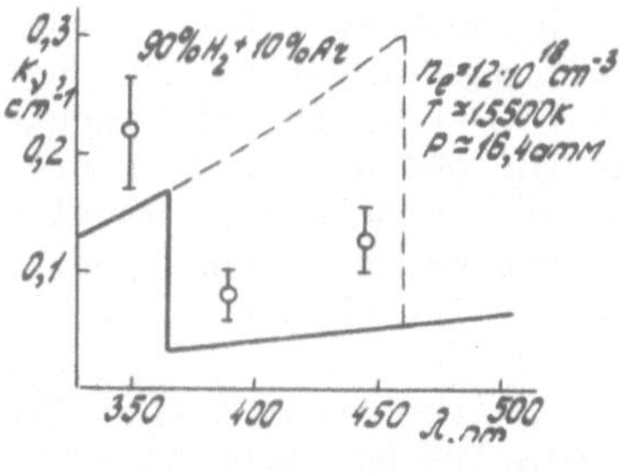

Fig. 5 Absorption coefficient of hydrogen-argon plasma near Balmer limit. Points - experiment, Ref. 46. Calculations - same as in Fig. 4.

to ionization (non-realization) of part of levels in the plasma microfield under the condition of applicability of the principle of spectroscopic stability should be compensated by a respective increase in the photo-recombination background (continuation of the photoionization cross-section toward a long-wave side). In keeping with Refs. 17 and 36, such compensation does not occur in dense plasma due to non-fulfillment of SSP. To calculate this effect in the approximation of Ref. 17 any addition to continuous spectrum is completely neglected. This approach, however, does not contain the "key principle" enabling one to define more precisely this description of the "window."

On Models Describing the ENOP

The published attempts of the ENOP-estimation for dense plasma are based on a microfield approach, on the simulation of the ENOP by means of simple central-symmetric potentials of the Debye-type and some modifications of the cut-off Coulomb one, on numerical calculation of simplified equations of a self-consistent field of Hartree-Fock or Hartree-Slater for an "average atom" in a cell whose dimensions are found by way of variation to obtain the minimum free energy or use is made of a different relationship between the cell size and plasma parameters.

It is the reaction of the atom on a uniform electric field that is usually considered in a microfield approach. Then, this reaction if averaged in view of the distribution of the plasma electric microfields. Solved in a similar manner are many problems of the optics of weakly non-ideal plasma, primarily those which are associated with line broadening. For recombination continuum such approach was used, for example, in Ref. 48. An analysis of merits and demerits of concrete papers is not our task. A disadvantage common to all these papers is a failure to take into account the reconstruction of high energy levels and the adjoining continuum as well as a change in the appropriate transition probabilities which may introduce a pronounced error int he results when this approach is made in the case of dense plasma, where the region of energy spectrum affected by a strong perturbing action of microfields is markedly expanded.

Different model approaches making use of effect cutoff of the Coulombic potential tail, i.e., effective short-range interaction, bring about such a reconstruction of the upper part of the energy spectrum when the number of levels in the perturbed region becomes limited. As the energy size of this region increases, which in one way or another is associated with a grown in the density of particles - neutral or charged, the number of levels drops as low as zero, and the "gap" appears between the last realizing level and the continuum. A consistent calculation of transition probabilities in models of this kind faces difficulties in constructing the wave function of the continuum. The exception is models of the "confined atom of hydrogen" type, for which accurate numerical and analytical quasiclassical solutions have been obtained. Reference 49 shows that for a wide class of central-symmetric Coulomb potentials with an effective short-range interaction the spectral density of oscillator strength, averaged on the energy level splitting or resonances in the photoionization cross-section, is a smooth energy function. The sophistication of the model through introduction of a collisional or microfield broadening of the spectral lines results in the emergence of a near-threshold window in the "observable" spectrum with a size close to ΔE (see Eq. 2.9) when there is less than one level in interval ΔE. Models with an effectively cutoff Coulomb tail are suitable for the description of a relative brightening, and the "window" effect manifests itself at densities higher than in experiments. It seems that centralsymmetric models with the cutoff Coulomb potential fail to describe an important property of states strongly perturbed by plasma mcirofields. The electron that belongs to these states is in the field of two or more ions. It is necessary to find more realistic models in order to describe the influence of non-ideality on the optical properties of non-ideal plasma.

REFERENCES

1. G. Kalman, ed.: "Strongly Coupled Plasma" 656:1978, N.Y.-London, Plenum Press.
2. W. Ebeling, W. Kreft, D. Kremp., "Theory of Bound States and Ionization Equilibrium in Plasma and Solid," 262:1979, Moscow, Mir Publ.
3. L. P. Kudrin, Statistical Physics of Plasma, 469:1974, Moscow, Atomizdat Publ.,
4. V. E. Fortov, I. T. Yakubox, The Physics of Non-Ideal Plasma, Chernogolovka, 263:1984.
5. N. M. Kuznetsov, Thermodynamic Functions and Shock Adiabats of Air at High Temperatures Mashinost. Publ., 1965, 462 pp.
6. H. R. Griem, Plasma Spectroscopy, McGraw-Hill, N.Y., 1964.
7. M. Lamoureaux, R. H. Pratt, in: Radiative Properties of Hot Dense Matter, p. 241, World Sci. Singapore, 1985.
8. L. M. Biberman, G. E. Norman, Usp. fiz. nauk 91:193 (1967).
9. G. A. Kobzev. IVTAN preprint No. 1 - 112, Moscow, 1983, 73 pp.
10. L. G. Dyachkov, G. A. Kobzev, Teplofiz. Vis. Temp. 14:681 (1976).
11. L. G. Dyachkov, G. A. Kobzev, G. E. Norman, Zhur. exp. teor. fiz. 65:1399 (1973).
12. L. D. Dyachkov, G. A. Kobzev, In: Khimiya plazmy, vol 8, 1981, 122 pp.
13. O. A. Golubev, L. G. Dyachkov, G. A. Kobzev, JQSRT, 20:175 (1978).
14. V. E. Bespalov, L. G. Dyachkov, G. A. Kobzev, V. E. Fortov, Teplofiz. Vis. Temp. 17:266 (1979).
15. D. R. Inglis, E. Teller, Astrophys. J. 90:439 (1939).
16. Yu. K. Kurilenkov. In: Proc. 4th All-Union Conf. "Dynamics of Radiative Gas," Vol. 1, Moscow, 1981, 140 pp.
17. G. A. Kobzev. Yu. K. Kurilenkov, G. E. Norman, Teplofiz. Vis. Temp. 15:193 (1977).
18. I. B. Levinson, A. A. Nikitin. The Manual on Theoretical Calculation of Line Intensity in Atomic Spectra, Leningrad, 1962.
19. G. A. Kobzev, G. E. Norman, K. N. Seryakov, Teplofiz. Vis. Temp. A, 473 (1966).
20. G. Z. Boldt. Physik, 154, 319 (1959); E. I. Asinovsky, V. M. Batenin, Teplofiz Vis. Temp. 3:530 (1965).
21. A. I. Alekseyev, M. A. Troitski, In: Application of Methods of Quantum Field Theory for Many-Body Problems, Gosatomizdat Publ., Moscow, 1963, 83 pp.
22. M. D. Ginzburg, L. E. Pargamanik, Ukr. phiz. zhurn. 23:69 (1978); 24:749 (1979).
23. N. N. Ogurtsova, I. I. Podmoshenski, V. M. Shelemina, Opt. Spectr., 16, 949 (1964); S. I. Andreyev, V. E. Gavrilov, Opt. Spectr. 26:666 (1969); S. I. Andreyev, Opt. Spectr. 38:432 (1975).
24. V. M. Batenin, P. V. Minayev, In: "Khimya plazmy," Vol. 2, 1975, 199 p, Teplofiz vis. temp. 9:676 (1971).
25. V. E Bespalov, V. E. Fortov. Zhurn., Techn. Fiz. (Pisma), 4:445 (1978).
26. V. A. Volkov, S. I. Titarov, B. K. Tkachenko, Teplofiz. vis. temp. 16:411 (1978); Phyzika plazmy 6:1115 (1980).
27. S. I. Andreyev, T. V. Gavrilova, Opt Spectr. 49:469 (1980).
28. B. G. Zhukov, V. G. Maslennikov, G. K. Tumakayev, Zhur. Tekch. Fiz. 51:2194 (1981).
29. V. E. Bespalov, V. K. Gryaznov, V. E. Fortov, JETP 76:140 (1979).
30. A. F. Nikiforov, N. Yu. Orlov, V. B. Uvarov, In: Problems of Nuclear Science and Engineering, Vol. 4, 1979, 36 p.
31. T. V. Gavrilova, Zhur. Tekch. Fiz. 49:652 (1979).
32. R. P. Johnson, JQSRT 7:815 (1967).

33. B. F. Rozsnyai, JQSRT 22:337 (1979); J. Glasser, J. Chapelle, Plamsa Phys. 21:477 (1979); M. Lamoureaux, Feng I. J. et al., JQSRT 27:477, Tekch. Fiz. 51:2980 (1981).
34. A. A. Valuyev, Yu. K. Kurilenkov, Teplofiz. Vis. Temp. 18:897 (1980).
35. G. E. Norman, Teplofiz. Vis. Temp. 17:453 (1979).
36. G. A. Kobzev, Yu. K. Kurilenkov, Teplofiz. vis. temp. 15:415 (1977).
37. P. P. Kulik, E. K. Rozanov, V. A. Ryaby, Teplofiz. vis. temp. 15:415 (1977).
38. P. P. Kulik, et al., In: Proc. XV ICPIG, V. 1, Minsk, 1981, p. 349.
39. B. Yaakobi, In: Exploding Wires, 4:87, New York, Plenum Press, 1968.
40. Yu. K. Kurilenkov, P. V. Minayev, JETP 74:87 , New York, Plenum Press, 1968.
41. K. Behringer, Z. Physik 246:333 (1971).
42. W. Wiese, D. E. Kelleher, D. R. Paquette, Phys. Rev. 6:1132 (1972).
43. K. Gunter, R. Radtke, Proc. XV ICPIG, V. 1., Minsk, 1981, p. 355.
44. V. E. Gavrilov, T. V. Gavrilova, T. N. Fyodorova, Opt. Spectr. 59:518 (1985).
45. A. A. Kon'kov, Teplofiz. vis. temp. 17:678 (1979).
46. S. I. Titarov, Thesis, Dolgroprudny, 1981.
47. J. C. Weisheit, C. B. Tarter, et al., JQSRT 16:649 (1976).
48. V. Vujnovic, J.Q.S.R.T. 10:929 (1970); V. Ts. Gurovich, V. S. Engel'sht, fiz. sb. 19:505 (1979); J. Davis, V. L. Jacobs, Phys. Rev. A 12:2017 (1975); V. G. Sevastyanenko, Preparint of ITAM, Sib. Br. USSR Ac. Sci., No. 30 (1980).
49. L. G. Dyachkov, G. A. Kobzev, P. M..Pankratov, Teplofiz. vis. temp. 23, I (1985).

SIMULATION STUDIES OF ION DYNAMIC EFFECTS

ON DENSE PLASMA LINE SHAPES

E. L. Pollock

Lawrence Livermore National Laboratory
University of California
Livermore, CA 94550

INTRODUCTION

Computer simulations have been widely used in studying dense plasma properties[1] including the local field properties important in spectral line broadening calculations.[2] We will review here a more recent use of simulation, possibly less familiar to this audience, where the time dependent ionic microfield generated by computer simulation of a plasma is used directly as a time dependent external potential for the evolution of the electronic structure of an ion. This permits calculation of the dipole correlation function and thus line shapes with the inclusion of ion dynamic effects.

Some of the first line shape results calculated this way, by Stamm and co-workers,[3] were questioned on: the usual concerns of numerical and statistical accuracy since this depends on how long an ion microfield time series is used for computing the dipole correlation function; the importance of omitted effects such as fine structure splitting, natural lifetimes, and Doppler broadening; the significance of the results to density diagnostics which depend on the line wings where dynamic effects are least important.[4] Some of these questions have been settled. Subsequent calculations have established the statistical accuracy of the original work.[5] The method has been extended to include effects such as natural lifetimes and fine structure splitting which dominates some lines.[6] Questions on the importance of ion dynamics to specific line widths or radiation redistribution problems are still being decided. The correct inclusion of ion dynamic effects is, nonetheless, crucial in many instances and the method discussed here, although more time consuming than the usual line shape calculation, is the best available way of doing this. Results obtained this way are already being used to benchmark quicker, more analytic methods.[7]

We begin by repeating the formulae used in line shape calculations and from these give a discussion of when ion dynamic effects should be important. As an illustration, some now classic experimental results (although not for a dense plasma) are presented. We then use simulation results as a further illustration. Although done for realistic plasma conditions these are intended for demonstration rather than experimental comparison since in order to isolate ion dynamic effects complicating, but theoretically straightforward effects such as fine structure splitting and Doppler broadening are not included. Finally we discuss Stark-Doppler coupling.

LINE SHAPE FORMALISM

The starting point in most line shape studies is the expression for the power spectrum of dipolar radiation

$$P(\omega) = \frac{4\hbar\omega^4}{3c^3} S(\omega) \tag{1}$$

where the line shape function

$$S(\omega) = \frac{\mathrm{Re}}{\pi} \int_0^\infty e^{i\omega t} \langle e^{ik\cdot r(t)} \vec{d}(t) \cdot \vec{d}(0) \rangle \quad . \tag{2}$$

In the following $\vec{d}$ refers to the dipole operator for a single radiator as coherence effects are not considered. The <> denotes an ensemble average over plasma conditions and the Doppler term, $e^{ik\cdot r(t)}$, will be considered only in the last section when we discuss its correlation with the electronic dipole terms.

Although more general cases can be treated with somewhat more effort, for radiative transitions to the ground state the expression for the dipole correlation function may be simplified to

$$C(t) \equiv \langle \vec{d}(t) \cdot \vec{d}(0) \rangle = \sum_u \sum_{u'} \rho_u e^{i\omega_{ul} t} \, \vec{d}_{ul} \cdot \vec{d}_{lu'} \langle U_{uu'}(t) \rangle \tag{3}$$

where u and u' denote upper states for the particular line and l the ground state and the $\vec{d}_{ul}$ etc. are dipole matrix elements. The time development operator for the emitter U(t) satisfies

$$i \frac{\partial U(t)}{\partial t} = [\vec{d} \cdot (\vec{E}_{ion}(t) + \vec{E}_{elec}(t))] \, U(t) \qquad (4)$$

assuming only dipolar coupling between the emitter electronic states and the total microfield of the plasma due to both electrons and other ions. Presently available time dependent plasma simulations give no information about the electronic component of the microfield. The main approximation in these calculations is to treat this electronic microfield component separately using the impact approximation which leads to

$$i \frac{\partial U}{\partial t} = (\vec{d} \cdot \vec{E}_{ion}(t) + \phi) \, U(t). \qquad (5)$$

For hydrogenic emitters evaluations of the electron collision operator ϕ are available.[8]

It remains to average the time development operator U(t) by solving the above equation for a representative ensemble of ion microfields. This is then used in equations 2 and 3 to give the line shapes. For the results we show below these were produced by standard molecular dynamics computations using systems of a hundred or so particles and periodic boundary conditions. Other methods have been used for weakly coupled plasmas where the use of periodic boundary conditions would require very large systems.

QUALITATIVE DISCUSSION OF DYNAMIC EFFECTS

Perhaps the fullest discussion of the effects of fluctuations on line shapes appears in the NMR literature.[9] Here we only want to indicate the time scales involved by considering a simplified model for equation 5

$$i \frac{\partial U(t)}{\partial t} = \vec{\mu} \cdot \vec{F}(t) \, U(t) \qquad (6)$$

where μ is a scalar rather than an operator. The averaged solution to this equation

$$\langle U(t) \rangle = \langle e^{\int_0^t \vec{\mu} \cdot \vec{F}(s) \, ds} \rangle \qquad (7)$$

simplifies when $\vec{F}(s)$ is almost constant during the times t where U(t) is non-negligible. The averaging then reduces to an average over initial field values. In this quasi-static limit

$$\langle U(t)\rangle_{Q.S.} = \int P(F)e^{-i\vec{\mu}\cdot\vec{F}t}\,dF \qquad (8)$$

A rough upper limit for the time scale over which the field is constant is the field autocorrelation time. If the quasi-static $\langle U(t)\rangle$ is negligible for times larger than this then dynamic effects are not important. Applying this to a dense plasma we may use the inverse of the ion plasma frequency as an estimate for the field autocorrelation time. Using a Gaussian as an approximation to the distribution of the field along $\vec{\mu}$ equation 8 gives

$$\langle U(t = \omega_{ion}^{-1})\rangle_{Q.S.} \simeq e^{-\frac{\mu^2\langle F^2\rangle}{6^2\hbar^2}\omega_{ion}^{-2}} \qquad (9)$$

If the exponent in this equation is large then $\langle U(t)\rangle$ is small for times such that the ion microfield changes significantly and the line in question is not strongly affected by ion dynamics.

The dipole moment μ for a hydrogenic emitter increases as the square of the upper level principal quantum number (n) and decreases with the nuclear charge Z. $\langle F^2\rangle$ may be estimated as $Z^2_{perturber}/r^4_{ion}$ where r_{ion} is the ion sphere radius for the perturbing plasma. Combining these with the expression for the plasma frequency shows that this exponent (R) scales as

$$R \sim \frac{n^4 M_{ion}}{Z^2}\rho_{ion}^{1/3}. \qquad (10)$$

Ion dynamic effects are thus reduced for the higher members of the Lyman series, because the dipole matrix elements increase. They also drop with ion density since the increase in the field strength magnitude more than compensates for the increasing plasma frequency. Although there is no explicit temperature dependence in this expression R will decrease with temperature since the ionization state Z increases. Not surprisingly, ion dynamic effects are predicted to be largest for the Lyman-α or β lines of highly ionized emitters in low mass perturbers.

The above formula provides only a qualitative estimate. Before turning to simulation results we show an early experimental indication of the importance of ion dynamic effects. This experiment is not strictly relevant to the above discussion, since the time dependence of the field is produced by the motion of the emitter rather than by the motion of the perturbers as we assumed above for a dense plasma, but the results are still instructive.

EXPERIMENTAL OBSERVATION OF ION DYNAMIC EFFECTS

One of the earliest observations of ion dynamic effects were for the Lyman spectra of hydrogen or deuterium impurities in an arc plasma of singly ionized argon.[10,11] The measured line spectra (crosses in fig. 1) differed considerably from theoretical predictions which assumed static ions (solid line in fig. 1). For example the Lyman-α width was over twice theoretical estimates and the central dip in Lyman-β markedly

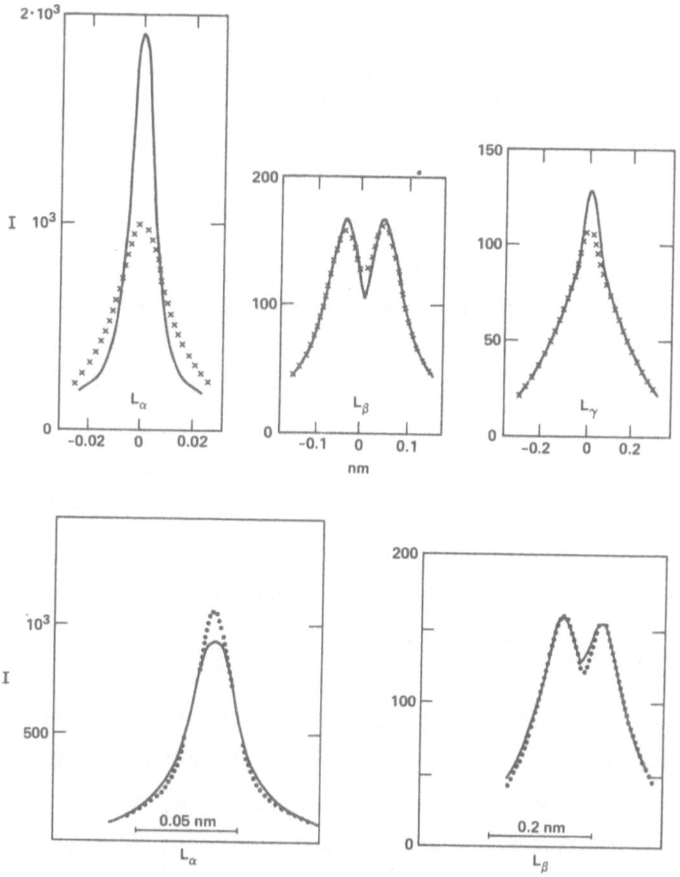

Fig. 1. The top graph shows the experimental (xxx) and theoretically predicted (---), for static ions, L_α, L_β, L_γ lines shapes for hydrogen in singly ionized Ar at T = 15,500 °K and $n_e = 2 \times 10^{17}$ cm^{-3}. The lower graph shows the decreased emitter motion effect when deuterium (•) is substituted for hydrogen (-). This figure is reproduced from references 10 and 11.

reduced. It was convincingly shown that these differences were due to ion motion when deuterium was substituted for hydrogen. As mentioned these effects are due to the fluctuating field caused by emitter motion and relate more to the Stark-Doppler coupling discussed in the last section than to the effects of perturber motion discussed above. These effects could have been anticipated. The measured half-width for the Lyman-α line shown in fig. 1 is ~ $2 \times 10^{12} \mathrm{sec}^{-1}$ corresponding to times for the dipole correlation function of ~ $.5 \times 10^{-12}$ sec^{-1}. From the plasma conditions of a temperature of 15,000 °K and an ion density of 9.3×10^{17} cm^{-3} the hydrogen emitter at mean thermal velocity traverses an ion sphere radius in ~ 10^{-12} sec so the ion field can not be treated as constant for the relevant dipole correlation times. Recent calculations,[12] some of them using computer simulation, now agree well with the spectra in fig. 1.

COMPARISON OF STATIC AND DYNAMIC LYMAN SPECTRA

The central object to be obtained from simulation for use in spectral studies is the time history of the local field (other things such as field gradients could also be obtained for studying quadrupolar effects) and the emitter trajectory, for use in Doppler broadening. Figure 2 shows a short section of the microfield time history for a moderately coupled plasma of hydrogenic argon ions. The three field components and the magnitude are displayed as functions of $\omega_{ion}t$ along with the distributions and field correlation obtained from much longer time histories. A noticeable feature here is the strong collision near the start of the time series, $\omega_{ion}t \sim .6$. An ensemble of such time histories is then used in equation 5 to compute the averaged time evolution matrix.

In terms of the dipole correlation function the main effect of ion microfield fluctuations is to reduce correlations for times beyond an inverse ion plasma frequency. As a simple example, for a Lyman-α line the correlation, in the quasi-static approximation, at a given field strength (also assuming the electron broadening is the same for all components) is

$$C(F,t) = \frac{e^{\phi t}}{3} \{2 + \cos (\frac{D_\alpha tF}{\hbar})\}$$

where $D_\alpha = (200 \mid d_z \mid 210)$. If this is averaged with an assumed Gaussian distribution for the field component distribution it becomes

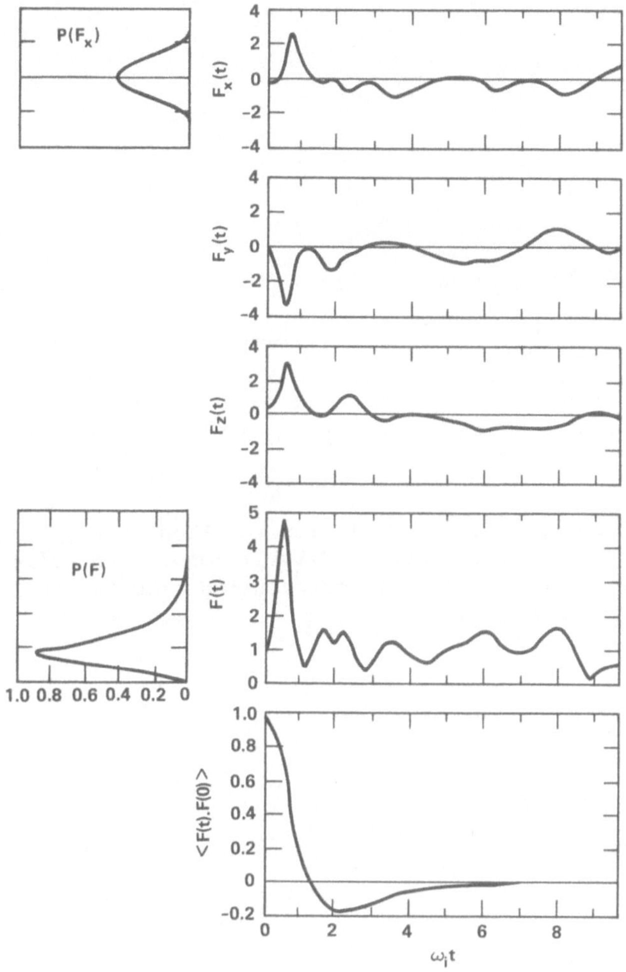

Fig. 2. Typical ion microfield time series from a simulated Ar^{+17} plasma at $T = 10^7$ °K and $n_e = 1.5 \times 10^{23} m^{-3}$. All field values are in units of Z/r_{ion}^2. The distributions for a field component, $P(F_x)$, and the field magnitude, $P(F)$, are shown to the side and the field autocorrelation function below. The time scales are $\omega_i t$.

$$C(t) = \frac{e^{\phi t}}{3} \{2 + [1 - (\omega_s^2 t)^2] \, e^{-\omega_s^2 t^2/2}\}$$

where $\omega_s \equiv D_\alpha \sqrt{<F^2>/3}$. If the electron broadening is weak the effect of ion dynamics on the unshifted component can be dramatic as shown in fig. 3 where this simplified quasi-static result is compared to a typical case including ion dynamics. The effect for Lyman-β, fig. 3b, is similar but smaller.

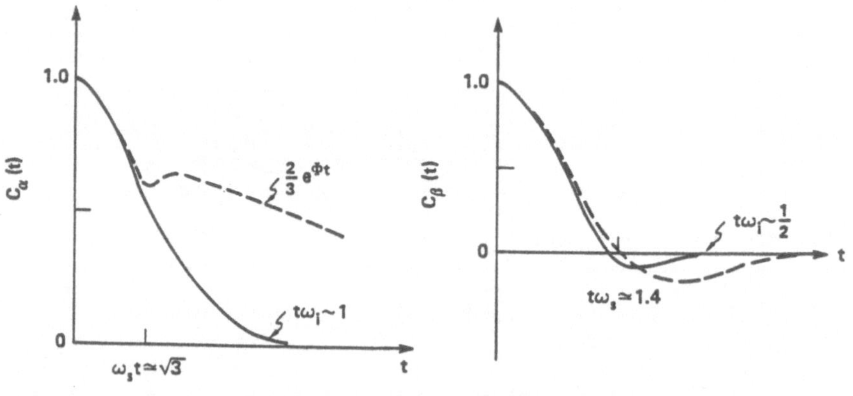

Fig. 3. Schematic dipole correlation functions for L_α and L_β. The dashed lines are for static ions. The solid lines are typical of the reduced correlation time due to ion motion.

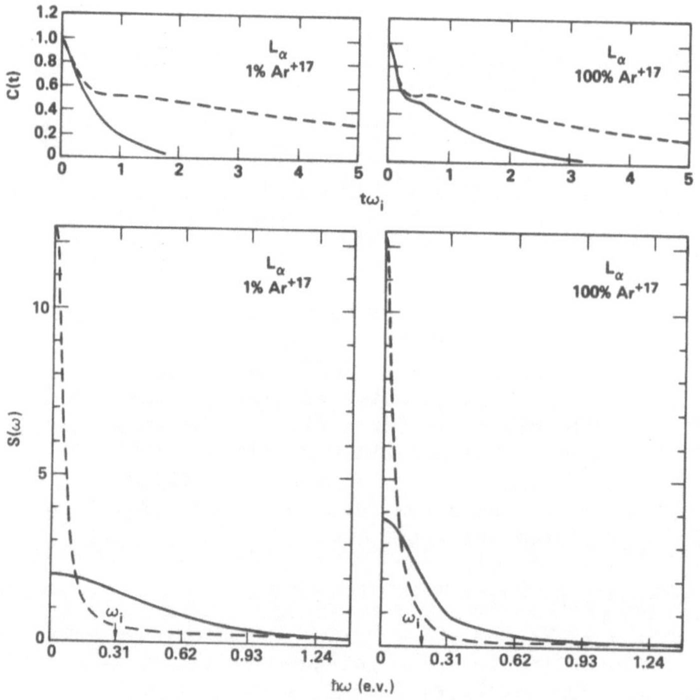

Fig. 4. Comparison of static (- - -) and dynamic (——) dipole L_α dipole correlation functions and line shapes for 1% Ar^{+17} in H^{+1} and 100% Ar^{+17} at $T = 10^7$ °K and $n_e = 1.5 \times 10^{23} cm^{-3}$.

Turning now to some numerical results, fig. 4 shows the correlation functions and line shapes for a plasma of hydrogenic argon in hydrogen ion perturbers and for pure hydrogenic argon at the same temperatures (10^7 °K) and electronic density (1.5×10^{23} cm^3). The difference between the static approximation (dashed line) and the results including dynamics (solid line) is similar to the qualitative comparison in fig. 3. The difference is larger for the hydrogen ion perturbers since the ion microfield due to these light mass perturber fluctuates more rapidly. The same trend is shown in fig. 5 for the Lyman-β line shapes. We caution again that the omitted fine structure splitting is important for these lines. (This is included in ref. 6.)

The effect of temperature, perturber mass, electron density and other variations on dynamic effects has also been studied.[5]

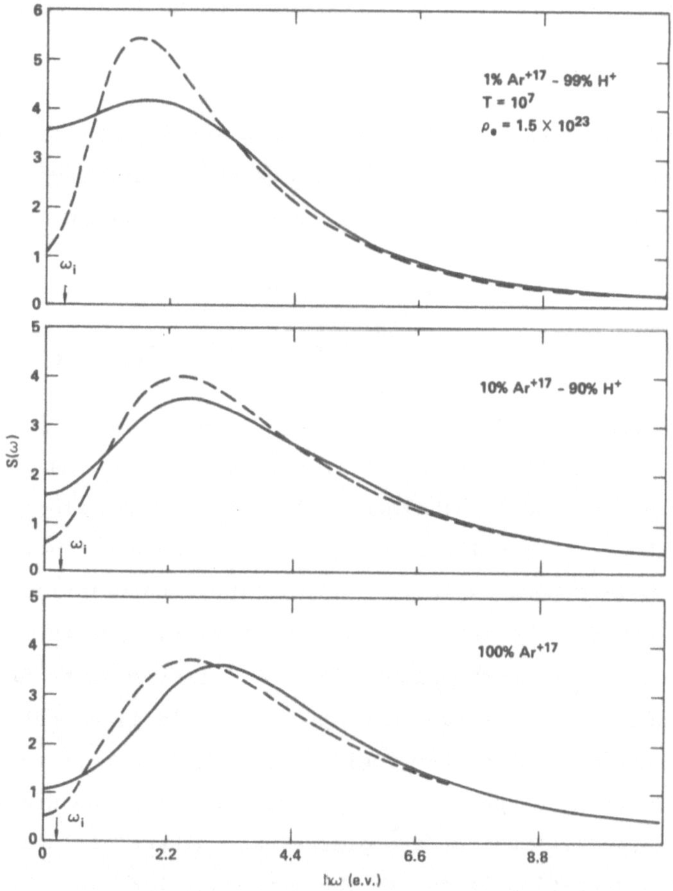

Fig. 5. Comparison of static (- - -) and dynamic (——) L_β line shapes at $T = 10^7$ °K and $n_e = 1.5 \times 10^{23}$ cm^{-3}.

So far we have only considered the dipole correlation $<\vec{d}(t) \cdot \vec{d}(0)>$ rather than $<e^{ik\cdot r(t)}\vec{d}(t) \cdot \vec{d}(0)>$ which includes Doppler effect. It is usually assumed that the Doppler term is independent of the dipole correlation and the final line shape comes from convoluting the two together. For this the free particle limit of

$$<e^{i\vec{k}\cdot\vec{r}(t)}> \rightarrow \begin{cases} e^{-k^2t^2/2\beta m} & \text{free particle limit} \\ e^{-3k^2\,Dt} & \text{diffusion limit} \end{cases} \quad (11)$$

is used. The diffusion limit is valid only for times $t\omega_{ion} \sim 10$ and is probably never relevant for dense plasma spectral line broadening.

Figure 6 shows these two limits for an Al^{+12} plasma at $T = 2.7 \times 10^6$ °K and electron density of 4×20^{21} cm^{-3} so the coupling parameter $\Gamma \sim 1$. The time limit beyond which the Lyman-α dipole correlation function is negligible, t_α, is seen to be in the free particle (dashed line) regime.

This does not mean that Stark-Doppler coupling is negligible. The experiments discussed in section 4 being a dramatic counterexample! For times sufficiently small that the initial velocity is unchanged we can write

$$<e^{ik\cdot r(t)}\vec{d}(t) \cdot \vec{d}(0)> = \int e^{ik\cdot v_o t} <\vec{d}(t) \cdot \vec{d}(0) \mid v_o> P(v_o)dv_o \quad (12)$$

where a dipole correlation function conditional on the (presumed constant) initial velocity of the emitter has been introduced which is integrated over the (Maxwellian) distribution of initial velocities. Intuitively, faster moving emitters will see more rapidly fluctuating fields and thus have a more rapidly decaying dipole correlation. Figure 7 shows this effect for the same plasma as in fig. 6. For this case the more rapidly moving ions (dashed line) do show a faster decay of the dipole correlation than the slower moving ones (solid line) but the effect is slight since most of the field variation is due to the motion of the perturbing ions as can be seen by comparing with the static ion case (dot dashed line). For the arc plasma case discussed in section 4 most of the field variation is due to the motion of the light emitter in the heavy perturbers and a plot similar to fig. 7 would show much larger variation presumably.

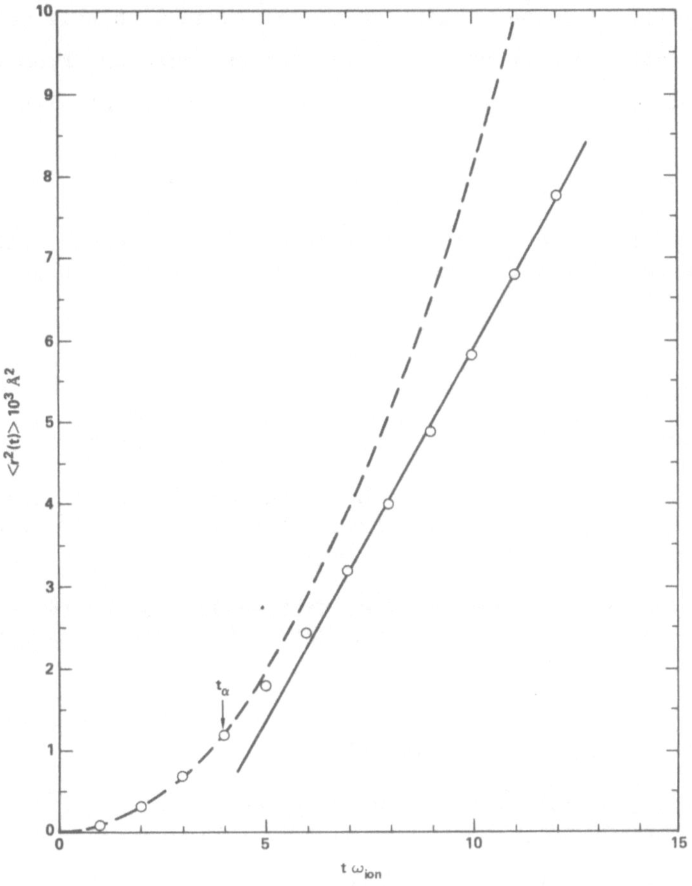

Fig. 6. Mean squared displacement of Al^{+12} ions (o) at $T = 2.7 \times 10^6$ °K and $n_e = 4 \times 10^{21}$ cm^{-3}. The dashed line is for free particles and the solid line is the diffusion limit. The time t_α indicates where the L_α dipole correlation function for this system become negligible.

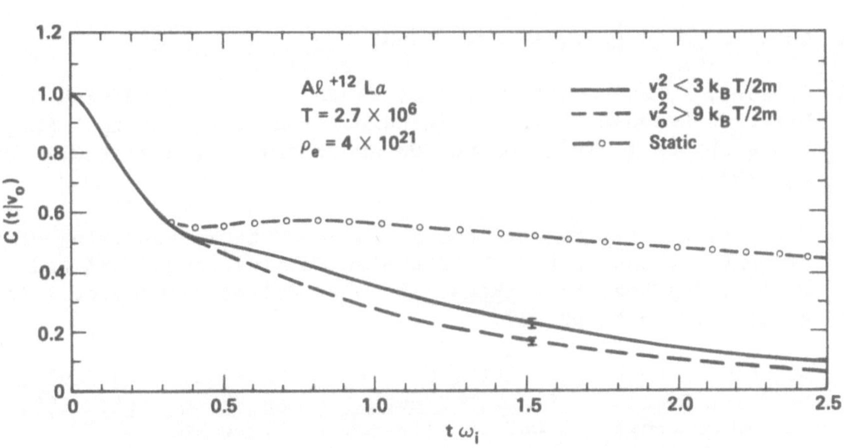

Fig. 7. The L_α dipole correlation function for slow (——) and fast (- - -) Al^{+12} emitters, (same conditions as fig. 6), compared to the static ion approximation (-o-).

We hope this review has given some indication of the types of plasma spectra questions which can be addressed with this method. Most of the specific results used as illustrations here are from research done in collaboration with Roland Stamm and Carlos Iglesias.

ACKNOWLEDGEMENTS

Work performed under the auspices of the U.S. Department of Energy by the Lawrence Livermore National Laboratory under contract number W-7405-ENG-48.

REFERENCES

1. M. Baus and J. P. Hansen, Physics Reports 59: (1980).

2. E. L. Pollock and J. C. Weisheit, "Local Fields in Strongly Coupled Plasmas," Spectral Line Shapes III, 1985, Walter de Gruyter & Co., Berlin-New York, pg. 181.

3. R. Stamm, Y. Botzanowski, V.P. Kaftandjian, B. Talin, and E. W. Smith, Phys. Rev. Lett. 52: 2217 (1984); Phys. Rev. Lett. 54: 2170 (1985).

4. J. P. Apruzese, P. C. Kepple, J. Davis, H. R. Griem, and R. Cauble, Phys. Rev. Lett. 54: 2167 (1985).

5. R. Stamm, B. Talin, E. L. Pollock, and C. A. Iglesias, Phys. Rev. A, to appear.

6. R. Stamm, B. Talin, E. L. Pollock, and C.A. Iglesias, "Line Broadening of Hydrogenic Ions in Dense Plasmas," Spectral Line Shapes IV, to appear.

7. D. B. Boercker, J. W. Dufty, and C. A. Iglesias, work in progress.

8. H. R. Griem, M. Blaha, and P. C. Kepple, Phys. Rev. A 19: 2421 (1979).

9. R. Kubo, J. Math Phys. 4: 174 (1963).

10. M. Geisler, K. Grützmacher, and B. Wende, "Stark Broadening of the Hydrogen Resonance Line L_γ in Comparison to L_α and L_β," Spectral Lines Shapes I, 1981, Walter de Gruyter & Co., Berlin-New York, pg. 103.

11. M. Geisler, K. Grützmacher, and B. Wende, "Stark Broadening of the Hydrogen L_α and L_β for the Atom-Ion Combinations H-Ar^+ and Ar-D^+", Spectral Line Shapes II, 1983 Walter de Gruyter & Co., Berlin-New York, pg. 37.

12. J. Seidel, "Effects of Emitter Motion on the Plasma Broadening of Hydrogen Resonance Lines," Spectral Lines Shapes III, 1985, Walter de Gruyter & Co., Berlin-New York, pg. 69.

ELECTRIC MICROFIELD DISTRIBUTIONS IN STRONGLY COUPLED PLASMAS FROM INTEGRAL EQUATION SOLUTIONS

F. Lado

Department of Physics
North Carolina State University
Raleigh, NC 27695-8202, U.S.A.

Ions radiating from a dense plasma reveal important information about plasma conditions through the detailed structure of their spectral lineshapes (Griem, 1974). This coupling of lineshape and plasma conditions is brought about by the local electric microfield distribution at the radiating ion due to all other charges in the plasma, a quantity first studied by Holtsmark (1919) by neglecting all correlations between plasma particles. More recently, Iglesias (1983) has shown that the calculation of the electric microfield distribution in a plasma is equivalent to the determination of the structure of a fictitious "fluid" whose intermolecular potential has both real and imaginary parts. This new formulation of the problem is the point of departure for the work presented below, which is based on the application of a straightforward generalization of the standard methods of liquid state theory coupled with a novel representation of the generalized bridge function.

The model to be studied is the one-component-plasma (OCP), consisting of N+1 ions, labeled 0 through N, each of charge e and all contained in a volume V at temperature T; a uniform background of opposite charge serves to neutralize the collection. The goal is to determine the distribution of electric field magnitudes at ion 0 due to the other N charges and the background. For a particular configuration, the field at $\vec{r}_0$ is

$$\vec{E}(\vec{r}^{N+1}) = \sum_{j=1}^{N} \frac{e}{r_{j0}^2}\hat{r}_{j0} - e\rho \int d\vec{r}\, \frac{\vec{r}_0-\vec{r}}{|\vec{r}_0-\vec{r}|^3}, \tag{1}$$

where $\rho = N/V$ and the integral term is the background contribution. The probability density for finding the value $\vec{\varepsilon}$ for $\vec{E}(\vec{r}^{N+1})$ is then

$$W(\varepsilon) = (2\pi^2\varepsilon)^{-1} \int_0^\infty dK\, KT(K)\sin(K\varepsilon), \tag{2}$$

where

$$T(K) = \langle \exp[i\vec{K}\cdot\vec{E}(\vec{r}^{N+1})]\rangle \tag{3}$$

is the characteristic function corresponding to ε. The angular brackets denote a canonical ensemble average with the OCP potential.

Iglesias (1983) has shown Eq. (3) to be formally equivalent to

$$\ln T(K) = ie\rho \int_0^K d\lambda \int d\vec{r}\, [g(\vec{r},\vec{\lambda})-1]\hat{K}\cdot\hat{r}/r^2, \tag{4}$$

where $g(\vec{r}_{01},\vec{\lambda})$ is the generalized pair distribution function (PDF) for a fluid with "potential energy" given by

$$U(\vec{r}^{N+1},\vec{\lambda}) = U_{OCP}(\vec{r}^{N+1}) - i\vec{\lambda}\cdot\vec{E}(\vec{r}^{N+1})/\beta. \tag{5}$$

Here, $\beta = (k_BT)^{-1}$ and $\hat{r}$ and $\hat{K}$ are unit vectors in the direction of $\vec{r}$ and $\vec{K}$ respectively, with $\vec{\lambda} = \lambda\hat{K}$. This reformulation of the problem has already been used to motivate an empirical but remarkably successful approximation dubbed APEX (Iglesias, Lebowitz, and MacGowan, 1983).

As noted by Iglesias (1983), the generalized PDF can be studied with the appropiate generalizations of the standard equations of liquid state theory: a generalized Ornstein-Zernike (OZ) equation

$$h(\vec{r}_{01},\vec{K}) = C(\vec{r}_{01},\vec{K}) + \rho \int d\vec{r}_2\, h(\vec{r}_{02},\vec{K})C(r_{21}) \tag{6}$$

and a generalized closure relation

$$C(\vec{r},\vec{K}) = h(\vec{r},\vec{K}) - \ln\{g(\vec{r},\vec{K})\exp[\beta\phi(r)-ie\vec{K}\cdot\hat{r}/r^2]\} + B(\vec{r},\vec{K}). \tag{7}$$

In these equations, $h = g-1$, $\phi(r)$ is the Coulomb potential, and $B(\vec{r},\vec{K})$ is the generalized bridge function for the "fluid", for which an approximation must be supplied. This is discussed below.

It is first necessary to simplifly Equations (6) and (7) before they can be numerically manipulated. In extracting the angular dependence of complex functions such as $g(\vec{r},\vec{K})$, we first note from its definition that it satisfies

$$g^*(\vec{r},\vec{K}) = g(\vec{r},-\vec{K}), \tag{8}$$

so that expansion in Legendre polynomials $P_\ell(x)$ automatically produces a separation of real and imaginary parts,

$$g(\vec{r},\vec{K}) = g(r,K,\cos\theta) = \sum_{\ell\ \text{even}} g_\ell(r,K)P_\ell(\cos\theta) + i \sum_{\ell\ \text{odd}} g_\ell(r,K)P_\ell(\cos\theta)$$

$$= \sum_{\ell=0}^{\infty} i^\sigma g_\ell(r,K)P_\ell(\cos\theta), \tag{9}$$

where for brevity in Eq. (9) and below we are using $\sigma = 0$ for ℓ even and $\sigma = 1$ for ℓ odd. Other generalized functions of $\vec{r}$ and $\vec{K}$ have similar expansions.

As for simple fluids, it is convenient to deconvolute Eq. (6) using Fourier transforms. The transform of a generalized function such as $C(\vec{r},\vec{K})$,

$$\tilde{C}(\vec{k},\vec{K}) = \int d\vec{r}\, C(\vec{r},\vec{K})e^{i\vec{k}\cdot\vec{r}}, \tag{10}$$

can be readily performed by orienting the z axis of the coordinate frame along $\vec{K}$, representing $C(\vec{r},\vec{K})$ as in Eq. (9), and using the Rayleigh expansion (Gray and Gubbins, 1984)

$$e^{i\vec{k}\cdot\vec{r}} = 4\pi \sum_{\ell,m} i^\ell j_\ell(kr)Y_{\ell m}(\theta,\phi)Y^*_{\ell m}(\theta',\phi'), \tag{11}$$

where the primed and unprimed angles specify the orientations of $\vec{k}$ and $\vec{r}$ respectively. Inserting (11) into (10), we then find

$$\tilde{C}(\vec{k},\vec{K}) = \sum_{\ell=0}^{\infty} \tilde{C}_\ell(k,K)P_\ell(\cos\theta'), \tag{12}$$

with the transform coefficients given by

$$\tilde{C}_\ell(k,K) = 4\pi i^{\ell+\sigma} \int_0^\infty dr\ r^2 C_\ell(r,K) j_\ell(kr), \tag{13}$$

where $j_\ell(x)$ is the spherical Bessel function of order ℓ. Note that the $\tilde{C}_\ell$ are real for all ℓ.

These ingredients suffice to define an iterative algorithm for the solution of the coupled equations (6) and (7) in terms of $S(\vec{r},\vec{K}) = h(\vec{r},\vec{K})-C(\vec{r},\vec{K})$ as the unknown:

i. Using the current (finite) set of coefficients $S_\ell(r,K)$, form

$$S(\vec{r},\vec{K}) = \sum_\ell i^\sigma S_\ell(r,K)P_\ell(\cos\theta) \tag{14}$$

and similarly for $B(\vec{r},\vec{K})$. (See below.) Then numerically evaluate the Legendre inversion integral of the closure equation (7) for

$$i^\sigma h_\ell(r,K) = (\ell+\tfrac{1}{2}) \int_{-1}^{1} d(\cos\theta)\{\exp[-\beta\phi(r)+ie\frac{\vec{K}\cdot\hat{r}}{r^2}+S(\vec{r},\vec{K})+B(\vec{r},\vec{K})]-1\}P_\ell(\cos\theta) \tag{15}$$

to get the DCF coefficients

$$C_\ell(r,K) = h_\ell(r,K) - S_\ell(r,K). \tag{16}$$

ii. Fourier transform these using Eq. (13) to yield the transform coefficients $\tilde{C}_\ell(k,K)$.

iii. Use the transform of the OZ equation to get

$$\tilde{S}_\ell(k,K) = \tilde{C}_\ell(k,K)\tilde{C}(k)/[1-\rho\tilde{C}(k)]. \tag{17}$$

iv. Invert these to find the new set of coefficients

$$S_\ell(r,K) = (2\pi^2 i^\ell)^{-1} \int_0^\infty dk\ k^2\tilde{S}_\ell(k,K)j_\ell(kr). \tag{18}$$

This completes an iteration. Here a test is made for self consistency of the input and output S_ℓ and the cycle is repeated until convergence is reached, whereupon Eq. (4) for T(K) is evaluated as

$$\ell n\ T(K) = -\tfrac{4}{3}\pi\rho e \int_0^K d\lambda \int_0^\infty dr\ h_1(r,\lambda), \tag{19}$$

the orthogonality of the Legendre polynomials having reduced $h(\vec{r},\vec{\lambda})$ in Eq. (4) to just the $\ell = 1$ coefficient.

Finally, we turn to the question of representing the generalized bridge function $B(\vec{r},\vec{K})$, about which essentially nothing is known. Trial calculations show that its omission, as in the Hypernetted-Chain equation, or approximation with no K dependence, leads to mediocre results. Guided by the exact solution of the Mean Spherical Model (MSM) for this problem (Lado, 1986), we have adopted for $B(\vec{r},\vec{K})$ the MSM analytic form of $g(\vec{r},\vec{K})$ and $C(\vec{r},\vec{K})$ and have thus used in this calculation the two-term expansion

$$B(\vec{r},\vec{K}) = B_{OCP}(r) + i\,\frac{K}{\beta e}\,B'_{OCP}(r)P_1(\cos\theta), \tag{20}$$

where the prime denotes differentiation and the OCP bridge function is itself approximated by that of hard spheres (Rosenfeld and Ashcroft, 1979; Lado, Foiles, and Ashcroft, 1983).

Correlation functions of imaginary fluids having some novelty interest, it seems worthwhile to comment briefly on their behavior before presenting the final results for the microfield distributions, though space here allows display of only one key feature. The leading real coefficient g_0 of $g(\vec{r},\vec{K})$ is similar to the PDF of simple fluids, with some additional oscillatory structure between the core region and the usual first peak. Higher coefficients are all oscillatory in the same range, going asymptotically to zero as g_0 goes to one. Imaginary terms show the same latter behavior. Both the magnitudes and frequencies of these oscillations increase rapidly with K, so that it becomes apparent that a sum of any practical number of terms would poorly reproduce the full $g(\vec{r},\vec{K})$. Note however that such a sum is not needed in the solution algorithm.

What is summed is the function $S(\vec{r},\vec{K})$ and here the story is quite different. We show in Fig. 1 the first three coefficients of both the real and imaginary parts of $S(\vec{r},\vec{K})$ for $\Gamma = \beta e^2/a = 10$ and $K\varepsilon_0 = 10$, where $\varepsilon_0 = e/a^2$ and a is the usual ion sphere radius. It is clear from these figures that both S_0 and S_1 are overwhelmingly the dominant terms of their respective sets; the expansion of Eq. (14) is very rapidly convergent. Further, halving K does not change this qualitative picture: S_0 decreases modestly only for small r while, significantly, S_1 roughly diminishes by half its magnitude. If we further note that S_1 has the form of the slope of S_0, we find in these features solid internal support for the $B(\vec{r},\vec{K})$ ansatz of Eq. (20).

The ultimate test of Eq. (20), which is the sole approximation in this calculation, must lie however in the final computed forms of the microfield distributions. These are shown for $\Gamma = 10$ and 100 in Fig. 2, along with the corresponding APEX results (Iglesias et al., 1983) and normative computer

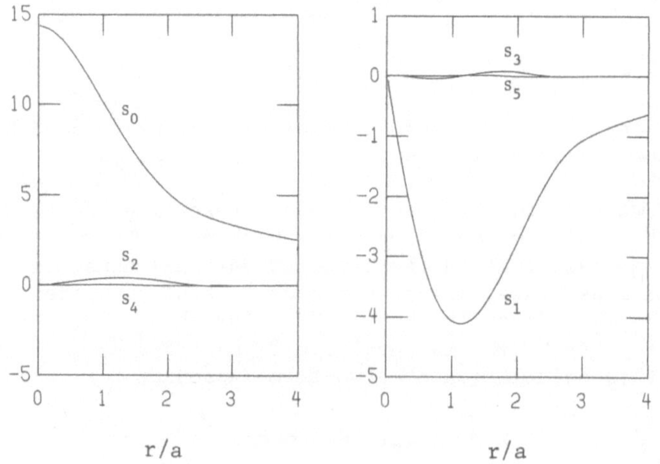

Figure 1. Coefficients of the real and imaginary parts of $S(\vec{r},\vec{K})$ for $\Gamma = 10$ and $K\varepsilon_0 = 10$.

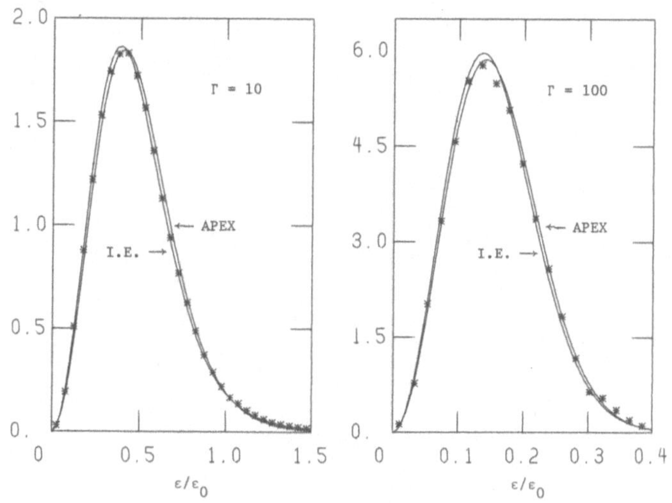

Figure 2. Electric microfield distributions $P(\varepsilon) = 4\pi\varepsilon^2 W(\varepsilon)$ at an ion for $\Gamma = 10$ and 100 obtained from the present work (I.E.), APEX, and computer simulation (stars).

simulation data (DeWitt, unpublished). It seems clear from these figures that the integral equation method initiated by Iglesias (1983), coupled with a realistic form for the generalized bridge function, is a fruitful approach to the calculation of the electric microfield distribution in a plasma.

This work was supported by the National Science Foundation under Grant No. CHE-84-02144.

REFERENCES

DeWitt, H. E., unpublished.
Gray, C. G., and Gubbins, K. E., 1984, "Theory of Molecular Fluids," Vol. 1, Clarendon, Oxford.
Griem, H. R., 1974, "Spectral Line Broadening by Plasmas," Academic, New York.
Holtsmark, J., 1919, Ann. Phys. (Leipzig), 58:577.
Iglesias, C. A., 1983, Phys. Rev. A, 27:2705.
Iglesias, C. A., Lebowitz, J. L., and MacGowan, D., 1983, Phys. Rev. A, 28:1667.
Lado, F., 1986, Phys. Rev. A, 34:(November 1 issue).
Lado, F., Foiles, S. M., and Ashcroft, N. W., 1983, Phys. Rev. A, 28:2374.
Rosenfeld, Y., and Ashcroft, N. W., 1979, Phys. Rev. A, 20:1208.

EFFECTS OF DIELECTRONIC SATELLITE BROADENING ON THE EMISSION SPECTRA FROM HOT PLASMAS

Abraham Goldberg and Balazs F. Rozsnyai

Lawrence Livermore National Laboratory
University of California
P. O. Box 808
Livermore, California 94550

OUTLINE OF THEORY

We recently presented[1] a model for the computation of spectral lines in hot plasmas based on the intermediate coupling scheme. In order to obtain meaningful photoabsorption cross-sections from which emission spectra can be calculated, the spectral lines must be endowed with realistic line-shape functions. In our present model we include the Doppler, electron impact and quasi-static ion-Stark broadening mechanisms in the computation of the line profiles. Assuming statistical independence for the above broadening processes, the resulting profiles are computed by convolution of the individual line-shape functions. In addition, we investigate the modification of the spectral profiles by the presence or absence of shifts caused by spectator electrons occupying high Rydberg levels. When a spectator electron is close to the radiating electron, the dielectronic satellite line is well separated from the principal line. On the other hand, a spectator electron in a high n level causes only a minor shift, and a distribution of spectator electrons over the high n levels causes an additional broadening of the lines that we call "dielectronic satellite broadening" (DSB). If the plasma is in local thermodynamic equilibrium (LTE), we assume that the average population p_i of a single- electron level ϵ_i is given by the Fermi statistics

$$p_i = w_i/\{\exp[(\epsilon_i-\mu/kT]+1\} \qquad (1)$$

where w_i, μ and kT stand for the statistical weight of the level ϵ_i, the Fermi level and the temperature of the plasma, respectively. Simple statistical argument shows that an absorption line associated with a one-electron transition $\epsilon_l-\epsilon_k$ will be broadened due to the statistical fluctuations in the occupancies of the spectator electrons by

$$\Delta^2{}_{lk} = \Sigma_j \quad (V_{lj}-V_{kj})^2 p_j(1-p_j/w_j) \qquad (2)$$

where V_{lj} and V_{kj} stand for the interaction energy between a spectator electron in level j and the active electron in the initial and final states designated by l and k, respectively. In Eqn. 2 Δ stands for the second moment of the level shifts due to the spectator electrons, and presently we approximate the distribution of shifts with a Gaussian profile. The summation in Eqn. (2) goes over to those sparsely populated states which are not included explicitly in the many-electron configurations contributing to the spectral lines. In the present case the latter include all the Rydberg levels with average occupancy of .1 electron or less.

RESULTS OF CALCULATIONS

The experimental spectra and theoretical computations are summarized in Fig. 1. Fig. 1.a shows the measured emission spectrum of a laser produced bromine plasma of Bailey et al.[2] The free-electron density and temperature were estimated as 5×10^{21} cm^{-3} and 480 eV, respectively, and the plasma was mainly neon- and fluorine-like and optically thin. Figs. 1b and 1c show our calculated emission spectra without and with the effect of dielectronic satellite broadening, respectively. The emission spectrum was obtained from a simple solution of the radiative transfer equation

$$I(\nu) = B(\nu)\{1-\exp[-\sigma(\nu)\rho L]\} \qquad (3)$$

where $I(\nu)$ is the intensity of the emerging radiation, $B(\nu)$ is the Planck function, $\sigma(\nu)$ is the frequency-dependent photoabsorption cross section, ρ is the matter density and L is the average distance inside of the plasma material through which the photons must pass before emerging and reaching the detector. The photoabsorption cross section from which Fig. 1b was obtained using Eqn.(3) is shown in Fig. 1d. We should mention that in order to mimic non-LTE conditions with our essentially LTE "average atom" (AA) model, as described in Ref. 1 and in references given there, we reduced our LTE temperature from 480 eV to 270 eV so as to obtain the neon- and fluorine-like configurations with the highest probabilities. The matter density corresponding to the experimental conditions was 2.6×10^{-2} g/cc and the plasma thickness L was estimated as 2.5×10^{-3} cm. The computation of the photoabsorption cross section was done by considering one-electron dipole excitations from a number of "parent", absorbing states with definite J values and parity, into a set of permissible J' "daughter" states of opposite parity. The list of parent configurations and their probabilities together with the number of parent J states is given in Table I. From the 54 parent states we computed explicitly all the possible 2-3, 2-4 and 2-5 transitions, where the numbers stand for the principal quantum numbers, and obtained 10632 spectral lines, which were then supplied with the line-shape functions obtained from Doppler, Lorentz and Stark broadenings. The inclusion of the dielectronic satellite broadening entailed adding the variance square given by Eqn. (2) to the square of the Doppler width. The remaining lines together with the bound-free (photoionization) and free-free (inverse bremsstrahlung) were computed as described in Ref. 1 and in references given there.

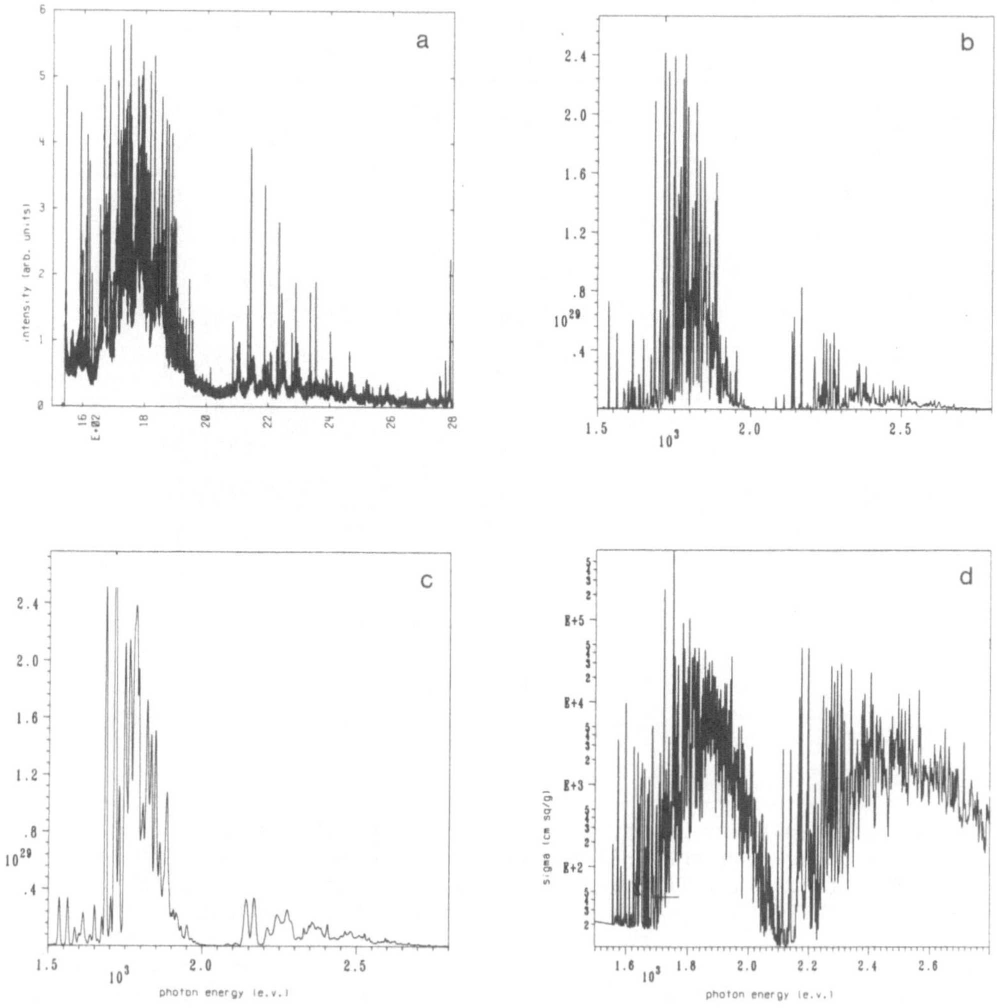

Figure 1. Experimental measurement of emission intensity from a laser produced bromine plasma (a), theoretical prediction for the emission spectrum without the effect of dielectronic satellite broadening (b), with the inclusion of the effect dielectronic satellite broadening (c), and the frequency photoabsorption cross section from which Fig. 1b was produced (d).

Table I. Parent configurations, number of J states and probabilities for bromine plasma at kT=270 eV and at ρ=2.6[-2] g/cm^3. The numbers in square brackets are exponents of 10.

Configuration	No. of J States	Σ_j P(J,α)
[Na]	1	8.504[-3]
[He]$2s^2 2p^5 3s^1$	4	9.011[-3]
[Ne]	1	3.095[-1]
[He]$2s^1 2p^6$	1	6.086[-2]
[He]$2s^2 2p^4 3s^1$	8	3.951[-3]
[F]	2	3.269[-1]
[He]$2s^1 2p^5$	4	6.427[-2]
[O]	5	1.438[-1]
[He]$2s^1 2p^4$	8	2.828[-2]
[N]	5	3.375[-2]
[He]$2s^1 2p^3$	10	6.635[-3]
[C]	5	4.455[-3]

CONCLUSION

The point of this report is shown in the results of two calculations, identical apart from the omission or inclusion of the dielectronic satellite broadening, illustrated in Figs. 1b and 1c. Since under LTE conditions the upper Rydberg states are populated according to the Fermi or Boltzmann statistics, the effect of dielectronic satellite broadening must be present. As is evident from the figures, the omission of the dielectronic satellite broadeneing mimics the experimental conditions, Fig. 1a, much better, as is expected since the experimental plasma conditions were likely to be far from LTE. We wish to point out that the detectibility of the dielectronic satellite broadening is an additional way to obtain information about plasma conditions.

ACKNOWLEDGEMENTS

Work performed under the auspices of the U.S. Department of Energy by the Lawrence Livermore National Laboratory under contract number W-7405-ENG-48.

REFERENCES

1. A. Goldberg, B. F. Rozsnyai and P. Thompson, Phys. Rev. A 34:421, (1986).
2. J. Bailey, R. E. Stewart, J. D. Kilkenney, R. S. Walling, T. Phillips, R. J. Fortner and R. W. Lee, J. Phys. (in press).

INFLUENCE OF EFFECTIVE INTERIONIC POTENTIALS ON THE LOW FREQUENCY ELECTRIC MICROFIELD DISTRIBUTIONS IN DENSE SEMI-CLASSICAL HYDROGEN PLASMAS

R. Mazighi,* J. P. Hansen,* and B. Bernu**

*Laboratoire de Physique Théorique des Liquides†
Université Pierre et Marie Curie
75252 Paris Cedex 05, France

**Chimie-Physique II,†† C.P. 231
Université Libre de Bruxelles
B-1050 Brussels, Belgium

1. INTRODUCTION

The low frequency electric microfield distribution, at a test particle (radiator) immersed in a plasma, has been extensively studied in view of its application to the theory of spectral line, due to ion-radiator collisions, in plasma spectroscopy[1].

Such a distribution can be calculated on the basis of an adiabatic approximation : the total field, acting on the radiator, is considered to be the sum of contributions due to the ions statically screened by the electrons. This amounts to reducing the original multicomponent ion-electron system to a purely ionic system characterized by effective interionic potentials.

For high temperatures and sufficiently low electron densities, this reduction is implicitly carried out by replacing the bare Coulomb potential between ions by the screened Debye Hückel (DH) potential[2] :

$$v_{\alpha\beta}(r) = Z_\alpha Z_\beta \frac{e^2}{r} \exp[-k_D r] \tag{1}$$

where Z_α is the valence of ionic species α, $\lambda_D = k_D^{-1} = (4\pi\rho_e e^2/k_B T)^{-1/2}$ is the electron screening length and ρ_e the electron density.

---

†Unité associée au C.N.R.S.
††Association Euratom-Etat Belge

At lower temperatures and higher densities, electron correlations and degeneracy effects come into play. These have recently been explicitly included in a calculation of electric microfield distribution of partially degenerate plasmas within a linear screening approximation[3,4] whereby the Fourier transform of the effective ion-ion potentials is taken of the form :

$$\hat{v}_{\alpha\beta}(k) = \frac{4\pi Z_\alpha Z_\beta e^2}{k^2 \epsilon_e(k)} \tag{2}$$

where $\epsilon_e(k)$ is the static dielectric function of the interacting electron gas.

The procedure described in this work is not limited to the linear screening approximation embodied in Eq.(2) and we show how the reduction from an ion-electron plasma to an effective ion plasma can be carried out systematically for strongly coupled but weakly degenerate plasmas. The resulting effective ion-ion potentials are then compared to the DH form (1) and used to calculate the second moment of the microfield distribution function which is a basic ingredient in any calculation of the latter quantity[5]. Other effective proton-proton potentials recently published in the literature[6] also indicate strong deviations from Debye like behaviour.

In section 2 a semi classical model for a multicomponent plasma is described. The reduction procedure is presented in section 3 with the closure approximations to extract the effective ion-ion potential. We assume, in section 4, the exponential approximation suggested by Iglesias[5], and we apply the simple APEX procedure[5,2] to work out the microfield distribution and discuss the results.

2. A SEMI CLASSICAL MODEL

Let us consider a fully ionized multicomponent plasma, confined in a volume Ω, consisting of point like ions and electrons having masses m_α, charges $Z_\alpha e$ and number densities $\rho_\alpha = N_\alpha/\Omega$ respectively ; e denotes the elementary charge. The system is globally neutral and assumed to be in thermal equilibrium at the temperature T.

A thermodynamic state of the plasma may be characterized by the two dimensionless quantities : the coupling parameter $\Gamma = e^2/k_B T a_e$, where $a_e = (3/4\pi\rho_e)^{1/3}$ is the electron sphere radius, and the density parameter $r_s = a_e/a_0$, $a_0 = \hbar^2/(m_e e^2)$ being the Bohr radius.

A strongly coupled state is characterized by $\Gamma \gtrsim 1$.

In order to ensure full ionization of the plasma, we restrict ourselves to temperatures of the order of (or higher than) the ionization temperature T_i ; we also assume the temperature sufficiently high enough to make the electron de Broglie wave-length $\lambda_e = \hbar/(2\pi\, m_e k_B T)^{1/2}$ shorter than a_e.

When the temperature is lowered and becomes of the order of the electron Fermi temperature T_F symmetry effects, due to the Pauli exclusion principle, can be included in the (effective) electron-electron interaction.

As has already been suggested[7], such a weakly degenerate plasma may be described by classical statistical mechanics provided that electrons and ions interact via the following effective pair potentials deduced from the two body Slater sum[8] :

$$v_{\alpha\beta}(r) = v_{\alpha\beta}^{(d)}(r) + \delta_{\alpha e}\,\delta_{\beta e}\,v_{ee}^{(s)}(r) \tag{3a}$$

$$v_{\alpha\beta}^{(d)}(r) = Z_\alpha Z_\beta \frac{e^2}{r}\,(1 - \exp(- r/\lambda_{\alpha\beta})) \tag{3b}$$

$$v_{ee}^{(s)}(r) = k_B T \ln 2 \exp(- \frac{(r/\lambda_{ee})^2}{\pi \ln 2}) \tag{3c}$$

where $\lambda_{\alpha\beta}^2$ is the sum of the squares of the partial de Broglie thermal wave-lengths. The last term is an average over the spin states. The range of validity of this model is sketched in Fig. 1.

For the special case of an impurity imbedded in a hydrogen plasma which shall be considered in the following, we reserve the indices 0 for the impurity of charge $Z_0 e$ $(Z_0 > 0)$, 1 for the protons and 2 for the electrons ; we have then $Z_1 = - Z_2 = 1$ which implies $\rho_1 = \rho_2$.

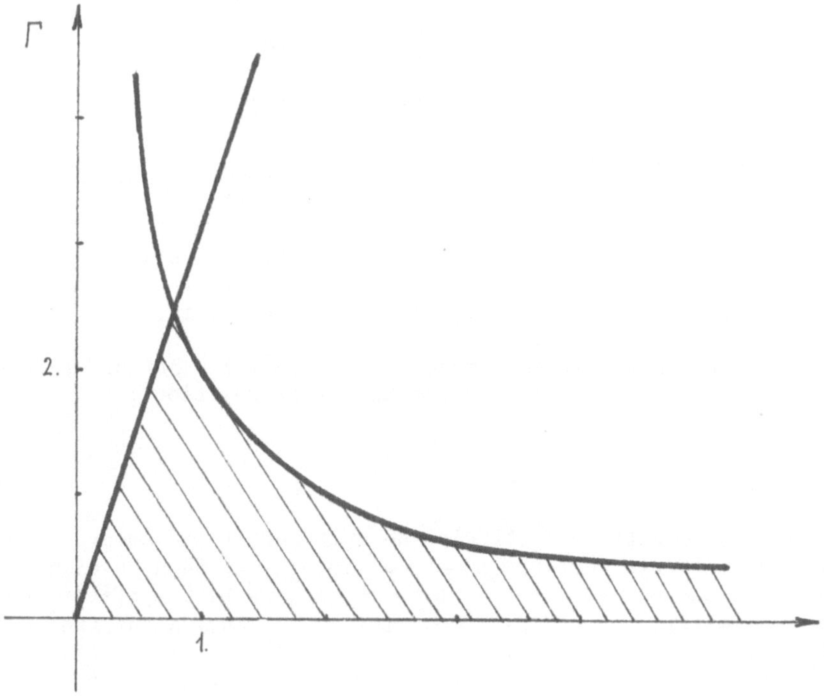

Fig. 1. Validity domain of the model (shaded area), delimited by the two curves : $\Gamma = \pi r_s$ and $\Gamma = 2/r_s$ which correspond to weak degeneracy and total ionization conditions respectively.

3. EFFECTIVE INTERIONIC PAIR POTENTIALS

3.1. The reduction procedure

For the take of clarity, we first restrict our study to the two-component case where the radiator is of the same species ($Z_0 = Z_1$) as the plasma ions. The generalization to radiating impurities of different charge is straightforward and shall be considered later.

The set of potentials (3) can be used in conjunction with the methods of classical statistical mechanics to compute the three pair distribution functions $g_{\alpha\beta}(r)$. The associated correlation functions $h_{\alpha\beta}(r) = g_{\alpha\beta}(r)-1$ and direct correlation functions $c_{\alpha\beta}(r)$ are related via the set of three Ornstein-Zernike relations (OZ) which read in Fourier space[9] :

$$\hat{h}_{\alpha\beta}(k) = \hat{c}_{\alpha\beta}(k) + \sum_{\gamma=1}^{2} \rho_\gamma \, \hat{c}_{\alpha\gamma}(k) \, \hat{h}_{\gamma\beta}(k) \tag{4}$$

The electron component can be formally eliminated by expressing the ion-ion correlation function h_{11} in terms of an effective direct correlation function via the OZ relation for an underlying one-component ion plasma, characterized by identical static ion-ion correlations :

$$\hat{h}(k) \equiv \hat{h}_{11}(k) = \hat{c}_{eff}(k) + \rho_1 \, \hat{c}_{eff}(k) \, \hat{h}(k) \tag{5}$$

A similar procedure has been used by Adelman[10] to eliminate the solvent in a microscopic description of electrolyte solutions. Comparison between Eqs. (4) and (5) yields the following expression for $\hat{c}_{eff}(k)$:

$$\hat{c}_{eff}(k) = \hat{c}_{11}(k) + \frac{\rho_2 \, [\hat{c}_{12}(k)]^2}{[1-\rho_2 \, \hat{c}_{22}(k)]} \tag{6}$$

Now, in a one-component description, a knowledge of h or c_{eff} uniquely determines an effective interionic pair potential[11].

Let us now consider the hydrogen plasma with an impurity as an infinite dilution limit of the three-component fluid made up of ρ_0 impurities, ρ_1 protons and ρ_2 electrons per unit volume.

As in the two-component case, one can write a set of OZ relations which can be summarized by the following matrix relation :

$$\underline{\underline{H}}^{(3)} = \underline{\underline{C}}^{(3)} \, (\underline{\underline{I}} - \underline{\underline{\rho}} \, \underline{\underline{C}}^{(3)})^{-1} \tag{7}$$

Here $\underline{\underline{I}}$ and $\underline{\underline{\rho}}$ denote respectively the unit matrix and numerical density matrix the elements of which are $\rho_{\alpha\beta} = \rho_\alpha \, \delta_{\alpha\beta}$. When we eliminate the electron component, we are left with a two by two effective direct correlation matrix : $\underline{\underline{C}}^{(2)}{}_{eff}$ describing an equivalent two-component system (without electrons). The elements of $\underline{\underline{C}}^{(2)}{}_{eff}$ are functions of the $\hat{c}_{\alpha\beta}$ elements of $\underline{\underline{C}}^{(3)}$only. The corresponding OZ matrix relation takes now the form :

$$\underline{\underline{H}}^{(2)} = \begin{bmatrix} h_{00} & h_{01} \\ h_{10} & h_{11} \end{bmatrix} = \underline{\underline{C}}^{(2)}_{eff} \, (\underline{\underline{I}} - \underline{\underline{\rho}} \, \underline{\underline{C}}^{(2)}_{eff})^{-1} \tag{8}$$

In the limit $\rho_0 \to 0$, $\underline{\underline{H}}^{(2)}$ reads :

$$\underline{\underline{H}}^{(2)} = \frac{1}{(1-\rho_1\hat{c}_{11}^{eff})} \begin{bmatrix} (1-\rho_1\hat{c}_{11}^{eff})\,\hat{c}_{11}^{eff} + \rho_1(\hat{c}_{01}^{eff})^2 & \hat{c}_{01}^{eff} \\ \hat{c}_{10}^{eff} & \hat{c}_{11}^{eff} \end{bmatrix} \tag{9}$$

On the other hand $\hat{h}_{01}$ and $\hat{h}_{11}$ are related to the three-component direct correlation functions via (7) ; by identification, one finally gets :

$$\hat{c}_{01}^{eff}(k) = \hat{c}_{01}(k) + \frac{\rho_1\,\hat{c}_{21}(k)\,\hat{c}_{02}(k)}{[1 - \rho_2\,\hat{c}_{22}(k)]} \tag{10}$$

relation which evidently reduces to Eq.(6) when $Z_o = 1$. Once the effective direct correlation functions $\hat{c}_{01}^{eff}$ and $\hat{c}_{11}^{eff}$ are calculated, an appropriate closure relation allows to define effective potentials for impurity-proton and proton-proton pairs.

3.2. The weak coupling limit ($\Gamma \ll 1$)

The simplest closure one can use is of mean field (MF) or Debye Hückel (DH) type :

$$c_{MF}^{eff}(r) = - v^{eff}(r)/k_BT \tag{11}$$

This leads to an analytic expression for the effective potential, e.g. for a proton-impurity pair :

$$\hat{v}_{01}^{eff}(k) = \hat{v}_{01}(k) - \frac{\rho_2\,\hat{v}_{12}(k)\,\hat{v}_{02}(k)/k_BT}{[1 + \rho_2\,\hat{v}_{22}(k)/k_BT]} \tag{12}$$

In the special case of purely coulombic interactions $v_{01}(r) = Z_o\,Z_1/r$, the resulting effective potential is precisely the DH potential given by Eq.(1).

3.3. The case of intermediate and strong coupling ($\Gamma > 0.1$)

In the framework of the HNC approximation, which is known to be a successfull method for treating coulombic systems[9], we use the following closure relations, to supplement the OZ relations :

$$g_{\alpha\beta}(r) = \exp\left[- \frac{v_{\alpha\beta}(r)}{k_BT} + h_{\alpha\beta}(r) - c_{\alpha\beta}(r) \right] \tag{13}$$

Expressing once more that ionic correlations must be the same in the effective ionic fluid and in the initial ion-electron fluid, we arrive at the following relation for the effective pair potentials :

$$\frac{v_{\alpha 1}^{eff}(r)}{k_B T} = - c_{\alpha 1}^{eff}(r) + \frac{v_{\alpha 1}(r)}{k_B T} + c_{\alpha 1}(r) \qquad (14)$$

Eq.(14) is closely related to a similar result recently obtained by Chihara[12].

Here, the calculation of $v_{\alpha 1}^{eff}(r)$ requires a numerical solution of the coupled HNC equations i.e. the OZ relations and Eq.(13) ; then we finally have in Fourier space :

$$\hat{v}_{\alpha 1}^{eff} = \hat{v}_{\alpha 1} - \frac{\rho_2 \hat{c}_{12}\, \hat{c}_{\alpha 2}\, k_B T}{[1 - \rho_2\, \hat{c}_{22}]}, \quad \alpha = 0,1 \qquad (15)$$

As a comparison between the two types of closure we recall that the difference between the corresponding effective ion-ion potentials is never very large ; in the case $Z_0 = 1$ and when the full potential given by Eq.(3) is used, the difference $v_{MF}^{eff}(r) - v_{HNC}^{eff}(r)$ is well fitted by a single decreasing exponential[13].

On the other hand, in the strong coupling limit, we found that $\hat{v}_{11}^{eff}(k)$ differs considerably from the DH form (1) at small and intermediate wavenumbers ; this is essentially due to the contribution of the electron symmetry term (3c) which will significantly affect the microfield distribution.

In the following section we shall use the HNC approximation (for the closure and the calculation of the pair correlation functions) to carry out the microfield calculations over an extensive range of Coulomb coupling ($0.01 \leq \Gamma < 2.$).

4. HNC CALCULATION OF THE MICROFIELD DISTRIBUTION

4.1. Low frequency component

The total potential energy of the system V is assumed to be a sum of effective pairwise interactions :

$$V = \sum_{j=1}^{N} \sum_{\substack{i=0 \\ j>i}}^{N} v^{eff}(i,j) \qquad (16)$$

$v^{eff}(i,j)$ means one of the effective potential derived in the last section.

Since the low frequency component is usually defined as the sum of the static electric fields due to the ions, at the radiator (located at $\vec{r}_0$), we have the following superposition of screened ion electric fields :

$$\vec{E} = - \frac{1}{(Z_0 e)} \vec{\nabla}_0 \sum_{i \neq 0}^{N} v^{eff}(r_{io}) \qquad (17)$$

The mean square electric microfield is easily expressed is terms of the impurity-ion pair correlation function calculated in the HNC approximation :

$$\langle \vec{E}^2 \rangle = \frac{\rho_1 k_B T}{(Z_0 e)} \int d\vec{r}\, g_{01}^{HNC}(r)\, \vec{\nabla}_0^2\, v_{01}^{eff}(r) \qquad (18)$$

At this point, we follow Iglesias et al.[2], assume that the exponential approximation can be applied and use the APEX procedure to approximate T(k) which is the Fourier transform of the probability density $W(\vec{\varepsilon})$ of finding an electric field $\vec{\varepsilon}$ equal to $\vec{E}$ at $\vec{r}_0$. The distribution function for the modulus of the electric field is then given by :

$$P(\varepsilon) = \frac{2\varepsilon}{\pi} \int_0^\infty kT(k) \sin k\varepsilon\, dk \qquad (19)$$

$T(k) = \langle \exp(i\vec{k}.\vec{\varepsilon}) \rangle$ is approximated by :

$$\exp\left\{ 4\pi\rho_1 \int_0^\infty dr\, r^2 g_{01}^{HNC}(r) \left[\frac{\sin k\, \varepsilon^*(r,\alpha)}{k\, \varepsilon^*(r,\alpha)} - 1 \right] \frac{\varepsilon(r)}{\varepsilon^*(r,\alpha)} \right\} \qquad (20)$$

where $\vec{\varepsilon}(r)$ is the electron-screened ion field at $\vec{r}_0$ and $\vec{\varepsilon}^*(r,\alpha)$ is the effective electric field, proper to the exponential approximation ; the parametrized form of the latter field is a DH one :

$$\varepsilon^*(r,\alpha) = \frac{Ze}{r^2} (1 + \alpha r)\, e^{-\alpha r} \qquad (21)$$

α is an inverse screening length to be determined from the sum rule equation (18) as suggested by APEX. The numerical results are limited to cases were Z_0 is set equal to one.

4.2. Numerical results

We have calculated the mean square microfield from the effective ion-ion potential obtained via Eq.(15) and from the DH form (1), using the respective HNC results for the ion-ion correlation functions. Representative results, relative to the one-component plasma value $\langle \vec{E}^2 \rangle_0 = 4\pi\rho_1 k_B T$, are given in table 1. The effective potential (15) and the DH form (1) lead to significantly different mean square microfields for strong coupling ($\Gamma > 1$) and at high density. For a fixed value of Γ, the results obtained with the potential (15) approach the DH results when the density is lowered, i.e. when degeneracy effects decrease.

At this point we can conclude that the DH model, which ignores quantum effects and is restricted to the linear screening regime, tends to underestimate the mean square microfield when the coupling is strong, while for weaker coupling ($\Gamma < 0.1$) the deviations from the unscreened one-component plasma results are at any rate negligible.

This situation is also reflected, as we shall see in the following, in the microfield distribution functions when calculated with the more realistic effective potential derived from Eq.(15).

The low frequency electric microfield distribution functions have been evaluated for the three values of the coupling parameter $\Gamma = 0.1$, $\Gamma = 1.$ and $\Gamma = 2.$ which is the strongest coupling we can study within our model (see Fig.1). We introduce the reduced field strength $\grave{E} = E/E_0$, where $E_0 = e/a_e^2$, and systematically compare the DH one-component results with those of the reduced two-component plasma (RTCP) using Eq.(15).

Table 1. Mean square microfield, divided by its one-component plasma value $\langle \vec{E}^2 \rangle_0 = 4\pi\rho_1 k_B T$, as calculated from the effective potential (15) and from the DH potential (1), for various values of Γ and r_s. The necessary pair distribution functions are calculated in the HNC approximation for potentials ; the results obtained with the DH potential are independente of r_s.

Γ	DH Potential	HNC Effective Potential						
		r_s=4.0	2.0	1.0	0.8	0.4	0.2	0.1
2.0	0.56			0.82				
1.5	0.64		0.75	0.83	0.85			
1.0	0.74		0.79	0.86	0.87			
0.5	0.86	0.83	0.87	0.90	0.91	0.94		
0.1	1.00	0.98	0.98	0.98	0.98	0.98	0.96	0.93
0.05	1.00	1.00	1.00	0.99	0.99	0.97	0.94	0.88

Figure 2 displays the low frequency distributions at $\Gamma = 0.1$; we see that we have, as expected, good agreement between the two models at any value of r_s.

The differences, between the two models, begin to show up at $\Gamma \sim 1$. (see Fig. 3). We find that the DH curve overestimates the probability density, with a peak shifted towards small fields.

When Γ is further increased to the value $\Gamma = 2$. the maximum height of the DH curves increases by a factor 2 (see Fig. 4) while the variation is only of the order of 20% in the RTCP case. Hence, we conclude that, in the strong coupling limit, the DH model yields a much too narrow distribution shifted towards low field values and this may be related to the fact that correlations between ions are neglected by this model.

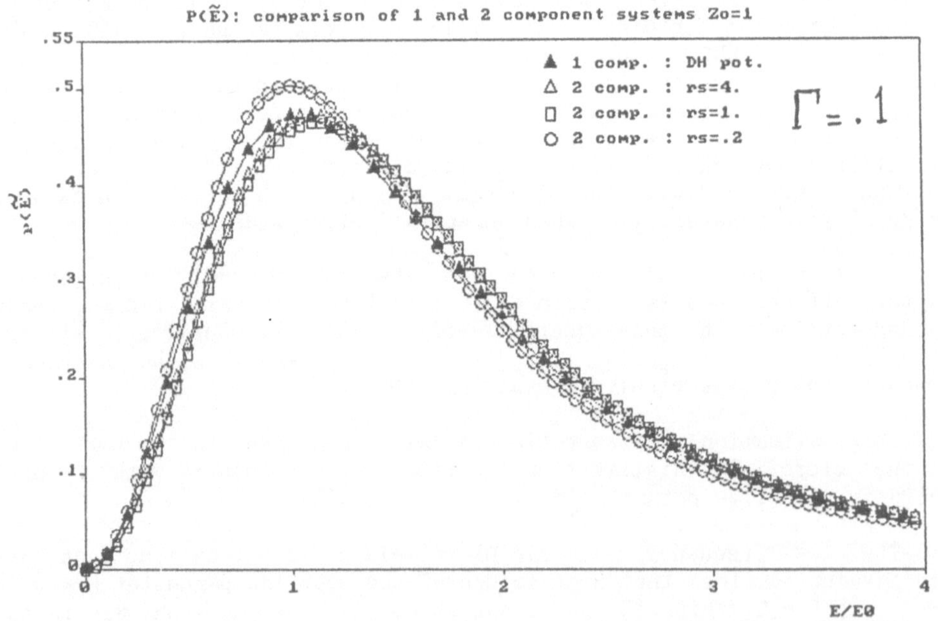

Fig. 2. Comparison of $P(\check{E})$ curves for a hydrogen plasma at a H^+ ion. Black triangles refer to the DH distribution while the other curves are for the RTCP model.

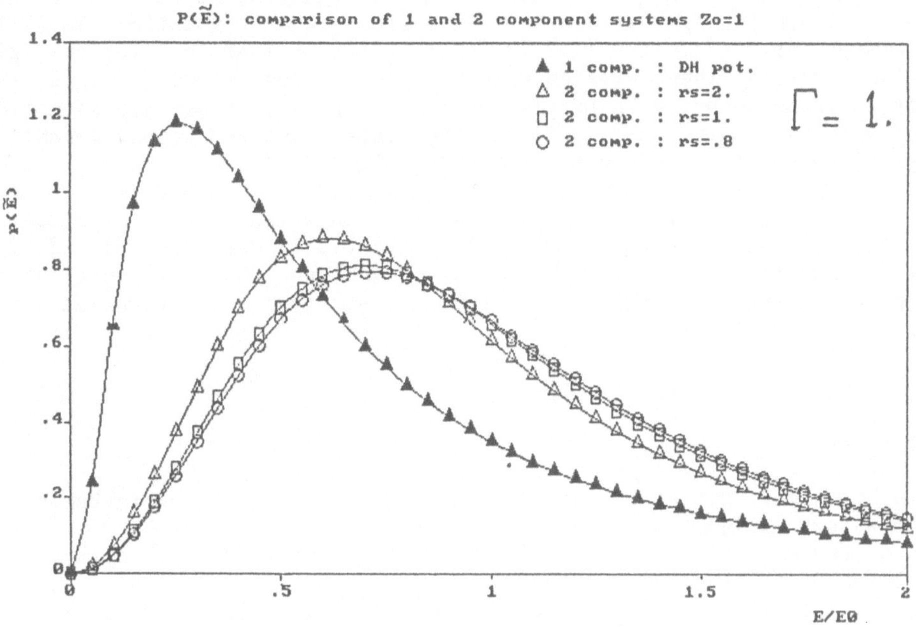

Fig. 3. Same as Fig. 2 but for Γ = 1. and r_s = 0.8, 1. and 2.

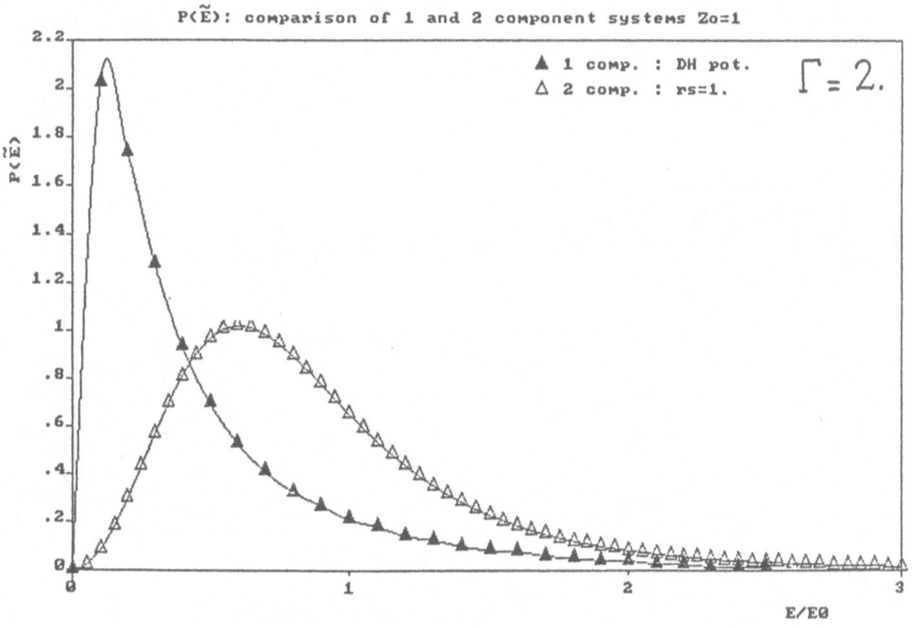

Fig. 4. Same as Fig. 3 but Γ = 2. and r_s 1.

CONCLUSION

In this comparative study, based on the estimation of the second moment of the low frequency microfield distribution, we have shown that effective ion-ion potentials obtained from HNC approximation (Eq.(15)) yield to very different microfield distribution functions for strongly coupled systems when compared with the predictions of the DH model. The differences between the respective distributions illustrate the effect of electronic screening and reflect the fact that the second moment is underestimated by the DH model in the case $Z_0 = Z_1$.

We are presently investigating the importance of the symmetry effects, between electrons, on the probability density $P(\vec{E})$ and also considering the more complicated case of a higher charge impurity still along the lines of the work of Iglesias et al.[2] to approximate the function T(k).

REFERENCES

1. H.R. Griem, "Spectral Line Broadening by Plasmas", Academic Press, New York (1974).
2. For recent work, see C.A. Iglesias, H.E. De Witt, J.L. Lebowitz, D.Mac Gowan, and W.B. Hubbard, Phys. Rev.A 31 : 1968 (1985).
3. C.A. Iglesias, and C.F. Hooper, Phys. Rev.A 25 : 1632 (1982).
4. X.Z. Yan, and S. Ichimaru, Phys. Rev.A 34 : 2167 (1986).
5. C.A. Iglesias, J.L. Lebowitz, and D.Mac Gowan, Phys. Rev.A 28 : 1667 (1983). See also H.C. Andersen, and D. Chandler, J. Chem. Phys. 57 : 1918 (1972).
6. M.W.C. Dharma-Wardana, F. Perrot, and G.C. Aers, Phys. Rev.A 28 : 344 (1983) ; M.W.C. Dharma-Wardana, and G.C. Aers, Phys. Rev.Lett. 56 : 1211 (1986) ; F.J. Rogers, Phys. Rev.A 29 : 868 (1984).
7. See for example B. Bernu, J.P. Hansen, and R. Mazighi, Phys. Lett. 100A : 28 (1984), and references therein.
8. H. Minoo, M.M. Gombert, and C. Deutsch, Phys. Rev.A 23 : 924 (1981).
9. See e.g. J.P. Hansen, in : "Laser-Plasma Interaction", R. Balian, and J.C. Adam, Ed., North Holland, Amsterdam (1982).
10. S.A. Adelman, J. Chem. Phys. 64 : 724 (1976).
11. N.D. Mermin, Phys. Rev.A 137 : 1442 (1965).
12. J. Chihara, Phys. Rev.A 33 : 2575 (1986).
13. B. Bernu, J.P. Hansen, and R. Mazighi, Europhys. Lett., 1 : 267 (1986).

STARK EFFECT IN DENSE, CORRELATED PLASMAS

Yves Vitel and Maurice Skowronek

Laboratoire des Plasmas denses, Tour 12 E5
Université P. et M. Curie, 4 place Jussieu
F-75252 PARIS CEDEX 05 France

INTRODUCTION

The plasmas produced in flashtubes are convenient to study an eventual effect of the non-ideality as they may reach an interaction parameter of the order of 0.2 where such effects are soon visible/1/. In the following, we will describe briefly the experimental set-up, giving some insight on the way used to solve some technical difficulties. Then we describe the density and the temperature measurements, using independent methods. As a matter of confirmation, we report results on H broadening and shift. Finally, we give the FWHM of some lines of ArII, KrII and XeII. This allows us to conclude on the Stark effect in dense correlated plasmas.

EXPERIMENTAL SET-UP

The plasma source is a cylindrical quartz flashtube designed by the Verre et Quartz Company. These flastubes are filled with a noble gas at initial pressures between 50 and 600Torrs. In the case of the H_α studies, an argon-hydrogen mixture in the proportion 97%-3% is used under 100Torrs. The electrical discharge is triggered by means of an auxiliary electrode, placed outside the tube, along a bulb generatrix. A simmer supply maintains a low current of about 600mA through the lamp during 100ms, before the main discharge is triggered. The effect of this thin channel of plasma is to decrease the lamp resistance and to place the start of the main discharge on the tube axis; then, the arc grows symmetrically from the axis. The simmer method allows the damping and even the vanishing of hydrodynamic perturbations which were previously described/2/. The pulse shaping network of the main discharge involves 5 LC cells forming a delay line where C=400μF and the inductance L is continuously adjustable between 10μH and 380μH. The impedance of the line and that of the discharge are well matched. The current pulse duration is about 2.5ms, with a constant plateau of 1.2ms, defined with a precision better than 2%; then the intensity is about 1kA.

Measurements of the plasma radiation. The spectrum is analyzed by means of an optical multichannel analyzer (OMA II) with an intensified silicon photodiode array detector placed in the focal plane of a spectrograph. The gated intensifier allows the record of the whole spectrum during 1us at the time corresponding to the best filling of the tube by the plasma. The optical magnification is 4. Two spectrographs are used in this study. The first one is of the Czerny-Turner type with a 250mm focal length and a 1200g/mm grating. It allows a record of a 41nm spectral range with a 0.06nm resolution in the first order. The second one is of the Ebert-Fastie type with a 1150mm focal length and a 1200g/mm grating. It may record a line profile over a 11nm spectral range with a 0.015nm resolution in the first order.

DETERMINATION OF THE ELECTRON DENSITY PROFILES

The continuum radiation varies as $n_e^2/\sqrt{T}$. Then, the electron density is mainly determined from the continuum emitted by the plasma during the quasi-stationary phase of the main discharge. The continuum is calibrated in absolute value against a standard tungsten ribbon lamp and a carbon arc in the visible and in the near IR, and against a standard argon arc in the spectral range 200nm-350nm. The electron density profiles $n_e(r)$ are calculated from the radial distribution of the emission coefficient, obtained through an Abel inversion in a spectral range where the plasma is optically thin, near 380nm. The spatial resolution is about 40um.

The $n_e(r)$ profiles are also determined from laser interferometry of the Ashby-Jephcott type/3/ at two wavelengths, 632.8nm and 3.39μm, at the output of an He-Ne laser. The plasma is placed inside the laser cavity; it modulates the laser output depending on the plasma refractive index. The number of fringes is found to be the same during the creation phase and the extinction phase of the plasma. Using the relative density distribution previously determined, it is now possible to calculate the electron density along the axis $n_e(0)$.

We have displayed on table 1 the electron densities on the tube axis $n_e(0)$ obtained by the two methods for different electrical energies in argon tubes. The agreement appears to be good. That means that the ξ-factor, which was taken independent of the electron density as in /4/, has a correct value. No plasma effect can be seen on the continuum at our electron densities.

DETERMINATION OF THE TEMPERATURE PROFILES

The radial electron temperature profiles $T_e(r)$ are obtained independently from the neutral lines, optically thick at their center, and by applying the thermodynamic equilibrium equations. The error on $T_e(r)$ is estimated to be less than 3%, mainly due to the absolute calibration of the neutral lines intensities. This method may be affected by the self-absorption of the neutral lines, which would lead to an underestimate of $T_e(0)$. In order to check the validity of our method, we have used special L-shaped flashtubes having the same inner diameter, the same interelectrode distance and which

Table 1. Electron densities obtained by two different methods, absolute value of the continuum and interferometry, for various values of the discharge energy

Initial pressure (in Torrs)	100	200	200	200	200	200	400	400	400	400
Energy (in kJoules)	0.8	1.0	1.2	1.4	1.7	1.96	1.0	1.2	1.4	1.7
$n_e(0)$ continuum (in $10^{17}cm^{-3}$)	5.9	9.6	11	12	13.6	15.1	10.1	11.5	13.1	14.6
$n_e(0)$ interferom. (in $10^{17}cm^{-3}$)	6.2	8.8	10	10.9	12.7	14.1	9.7	11.1	12.1	13.6

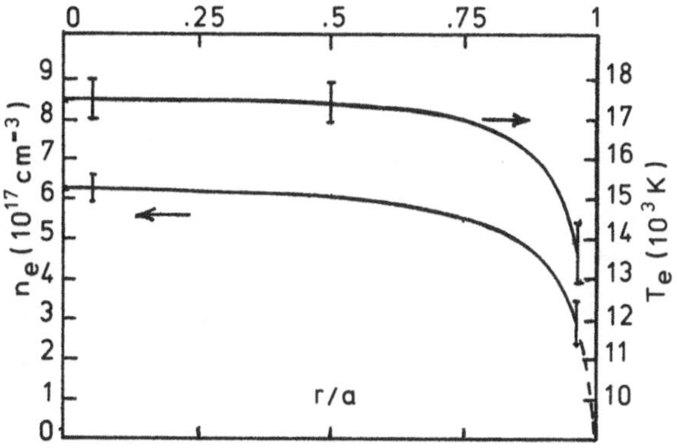

Fig. 1. Typical electron density and temperature profiles: argon 100Torrs, 1000Joules

enable end-on measurements at any point of the tube diameter, through a good quality quartz window. Using this device, the light emission is observed through a long optical path. The optical thickness becomes very high for wavelengths above 700nm and the plasma radiates like a blackbody at the local temperature. The different temperature profiles $T_e(r)$ obtained under various electrical conditions are in good agreement with those obtained by means of the side-on measurements of the neutral lines, mentioned above. We report on figure 1 typical $n_e(r/a)$ and $T_e(r/a)$ profiles obtained in our flashtubes of inner radius a. The main characteristic of these recorded profiles is that they are practically constant over the two thirds of the flashtube radius.

STUDY OF THE H_α LINE PROFILE

In this experiment, the H_α profiles are measured over the range of electron densities $6\times10^{17}cm^{-3}$ to $10^{18}cm^{-3}$ and over temperatures between 16000K and 19000K. In these conditions, the linear Stark broadening is the most important. As the temperature and density profiles, $T_e(r)$ and $n_e(r)$, are nearly flat over the two thirds of the radius, we make the simplifying assumption that the broadening and the shift of H are mainly due to the constant part of the profiles. We report on table 2 the width (FWHM) obtained for different discharge conditions in a tube at an initial pressure of 100Torrs. We estimate the relative error on the FWHM less than 15% due to the simplifying assumption. The experimental profile lays between the two theoretical ones /5,6/. This was also observed at lower electron densities (near $10^{17}cm^{-3}$) by Wiese et al./7,8/. A computer simulation /9/ of the Stark broadening gives a good agreement with experimental FWHM within 10^{16}–$10^{17}cm^{-3}$.

Table 2. Comparison of measured H_α FWHM with calculated values of $n_e(0)$ and $T_e(0)$

Discharge energy (in Joules)	800	1000	1200	1440	1700	1960
(in nm)	3.2	3.5	3.7	3.9	4.1	4.4
$n_e(0)$ (in $10^{17}cm^{-3}$)	6.1	6.2	7.4	8.2	9.4	10
$T_e(0)$ (in 10^4K)	1.65	1.74	1.77	1.8	1.85	1.87

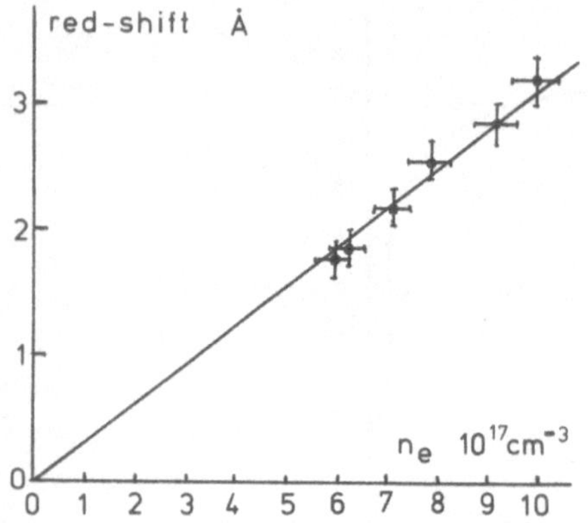

Fig. 2. Measured H_α red-shifts vs. the electron density

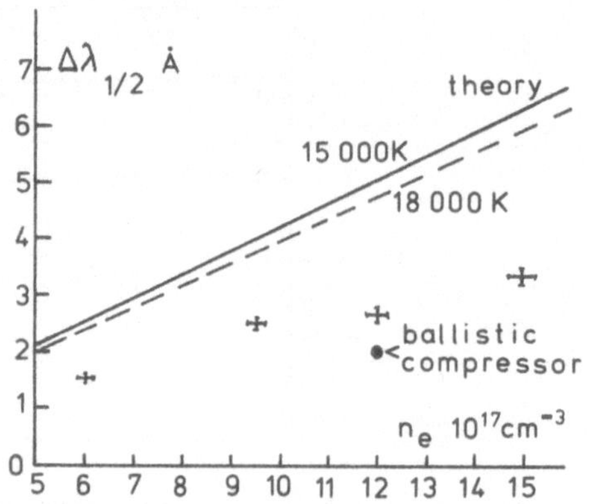

Fig. 3. Measured and calculated FWHM of ArII lines vs. electron density

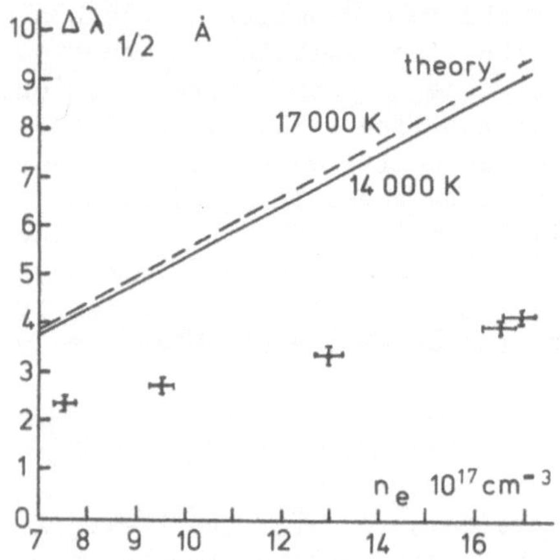

Fig. 4. Measured and calculated FWHM of KrII line vs. electron density

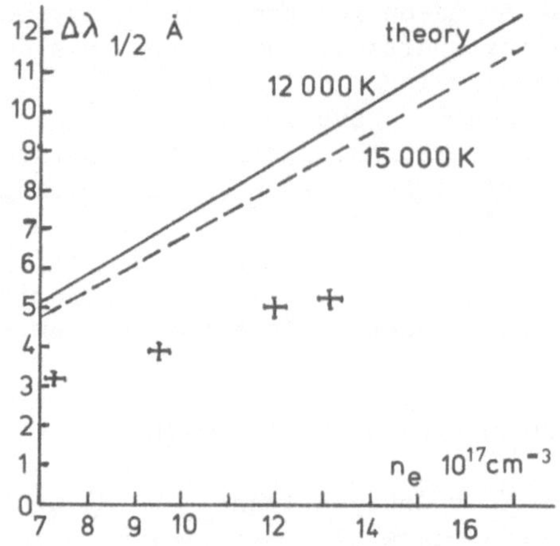

Fig. 5. Measured and calculated FWHM of XeII line vs. electron density

We have reported on table 3 the theoretical values of $\Delta\lambda_{1/2}$ extrapolated from Seidel's work/9/ and our experimental results for increasing electron density along the axis $n_e(0)$. The agreement is very good between them. The high resolution spectrograph is used to measure lineshifts. No asymmetry in the H_α profile is observed within the 0.02nm accuracy. The measured red-shifts are displayed on figure 2 vs. the electron density $n_e(0)$. They are found almost linear in density as expected from theories /10,11/.

Table 3. Calculated H_α FWHM from /9/ compared with our experimental results for various values of the axis electron density

$n_e(0)$ (in 10^{17}cm^{-3})	6.1	6.2	7.4	8.2	9.4	10
$\Delta\lambda_{1/2}$ theory (in nm)	3.1	3.2	3.6	3.8	4.2	4.3
$\Delta\lambda_{1/2}$ experim. (in nm)	3.2	3.5	3.7	3.9	4.1	4.4

The spectral lines are recorded in one shot, over 5nm, using an O.M.A. with a gate duration of 1us and a high resolution spectrograph. The linewidth and shift of some ArII, KrII and XeII lines are measured on the fitted Lorentzian profiles, using the measured values of n_e and T_e. Assuming flat $n_e(r)$ profiles over the diameter of the tube does not increase the error level higher than 5%. We have displayed on figure 3 the variation of the measured FWHM versus the electron density for the lines ArII 480.6nm and ArII 484.78nm. We have also drawn the calculated FWHM for 15000K and 18000K. A measurement using a ballistic compressor /12/ is also given in agreement with our results. The experimental values at a density of $1.5 \times 10^{18} cm^{-3}$ are about 50% lower than the theoretical ones. The same features are seen with the KrII 473.9nm line on figure 4, and with the XeII 529.22nm line on figure 5. These results, showing a saturation of the Stark broadening, seem to confirm the relative importance of the elastic long-range collisions in the broadening mechanism along with more efficient screening due to the non-ideality of the plasma.

CONCLUSION

The electron density and the temperature profiles in plasmas produced by flashtubes are carefully measured. A good check of these measurements is provided by the agreement between the continuum absolute values and the Hofsaess ξ-factor. On the other hand, the measurements of the linear Stark effect on H_α profiles, which is mainly due to the ionic part, are in agreement with theories and experiments of other authors. Then, the strong departure of the profile measurements (found Lorentzian) and the corresponding FWHM of some ionic noble gas lines from the theory shows an evidence of the non-ideal effect in coupled plasmas. Indeed, these lines are specially sensitive to the electron part of the Stark effect.

REFERENCES

1. A.A. Bakeev and R.E. Rovinskii, High Temp. 8:1121(1970); M.M. Popovic, S.S. Popovic and S.M. Vukovic, Fizika 6:29 (1974); R. Radtke and K. Günther, J.Phys.D:Appl.Phys. 9:1131 (1976)
2. Y. Vitel, M. Skowronek, K. Benisty and M.M. Popovic, J.Phys.D:Appl.Phys. 12:1125 (1979)
3. D.E.T.F. Ashby, D.F. Jephcott, A. Malein and F.A. Raynor, J.Appl.Phys. 36:29 (1965)
4. D. Hofsaess, J.Q.S.R.T. 19:339 (1979)
5. P. Kepple and H.R. Griem, Phys.Rev. 173:317 (1968)
6. C.R. Vidal, J. Cooper and E.W. Smith, Astrophys.J.Suppl. 25:37 (1973)
7. W.L. Wiese, D.E. Kelleher and D.R. Paquette, Phys.Rev.A 6:1132 (1972)
8. W.L. Wiese, D.E. Kelleher and V. Helbig, Phys.Rev.A 11:1854 (1975)
9. J. Seidel, Spectral line Shapes, 8th Int. Conf. (Virginia, USA), 1986, de Gruyter ed., New York A8 (to be published 1987)
10. T.L. Pittman and D.E. Kelleher, Spectral line shapes, 5th Int. Conf. (Berlin,FRG), 1980, de Gruyter ed., New York, B 165 (1981)
11. H.R. Griem, Phys.Rev.A 28:1596 (1983)
12. D.D. Burgess and J. Cooper, Proc.Phys.Soc. 86:1333 (1965)

RESONANT ABSORPTION IN DENSE CESIUM PLASMA

Jean Larour, Jean Rous and Maurice Skowronek

Laboratoire des Plasmas denses, Tour 12 E5
Université P. et M. Curie, 4 place Jussieu
F-75252 PARIS CEDEX 05 France

INTRODUCTION

Alkali metal vapours are well known to be easily ionizable and many applications of this property have been developed. Numerous theoretical and experimental works are also available, for example on the PVT data or on the transport properties /1/. Recent results /2/ have shown the peculiar behavior of cesium in the alkali metals family. Despite its high chemical reactivity which generates serious experimental troubles, the cesium is an ideal medium to study dense plasma effects. Its ionization is enhanced by a very low potential E(Cs)=3.89eV, and by a high atomic polarizability $\alpha=65\text{A}^3$ /3/, i.e. 13 times the corresponding value for a mercury atom. A high atomic density can be reached quite easily considering the low critical coordinates T_c=1924K, P_c=92bars and N_c=0.38g.cm^{-3} /2/. A great effort has been undertaken to describe the cesium vapour, especially its anomalous electrical conductivity approaching the metallic regime /4/. The vapour composition, taking into account the positive cluster ions, plays a significant role in the electron transport /4,5/. So we have developed an experimental procedure sensitive to the species present in the vapour. After a description of the experimental set-up, we present absorption profiles recorded around the first resonance doublet lines. Difficulties arise to connect these data with a vapour composition. Mainly, the broadening mechanisms (Stark, Van der Waals and resonance) seem to be weakened by a density effect.

EXPERIMENTAL SET-UP

We have used a classical transmission method to measure the absorption coefficient and the broadening and shift of the two resonance lines:

852.1nm ($6^2S_{1/2} - 6^2P_{3/2}$) and 894.3nm ($6^2S_{1/2} - 6^2P_{1/2}$).

The optical cell and the high pressure vessel have been described elsewhere /6/. They are designed to maintain a very thin slab of homogeneous cesium vapour (l=10µm) under a static argon overpressure and at a well defined temperature. The fluid is always in its vapour phase, along the coexistence curve, from T=900°C, P_{total}=9 bars up to T=1500°C and P_{total}=110 bars. We took a great care to degassing and purification in order to obtain reproducible results.

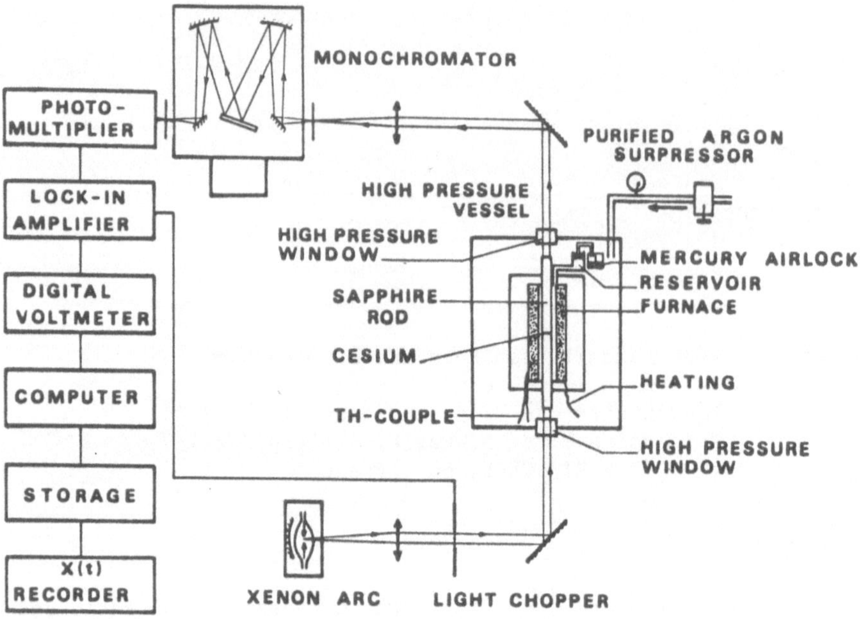

Fig. 1. Scheme of the experimental set-up

The incoming light I_0 at the output of a 150Watt xenon arc lamp is modulated at 110 Hz and then passes through the sapphire-cesium cell. The outgoing light I is analyzed by a motorized monochromator and recorded on a precision lock-in amplifier. This procedure is necessary to discriminate the transmitted light from the bright continuum blackbody radiation from the central part of the furnace.

As we deal with optically thick lines, we perform a complete numerical treatment of the profiles, or we consider quantities unsensitive to the instrument function, for example the total fractional absorption A:

$$A = \int_{-\infty}^{+\infty} (1 - I_0 / I) \, d\lambda$$

We follow the pioneering work of Chen and Phelps /7/ and we assume a theoretical Lorentzian profile for the spectral absorption coefficient $k(\lambda)$. After a convolution by the instrument function $G(\lambda)$, which is typically a Gaussian function with FWHM=1.2nm, we adjust the calculated profile to the experimental one:

$$k'(\lambda) = (1 / l) \ln (I_0 / I).$$

The *half-width* γ (HWHM) and the shift are taken on the calculated profile. The precision of these results can be estimated on the order of 10% to 20%.

RESULTS

We present now selected results on spectral absorption. The more precise ones are obtained around the strongest resonance line of the doublet. Records are not noisy and reproducible. Our discussion is based on these data. On the other hand, the second resonance line of the doublet can give mainly qualitative results, fairly agreeing with the previous ones.

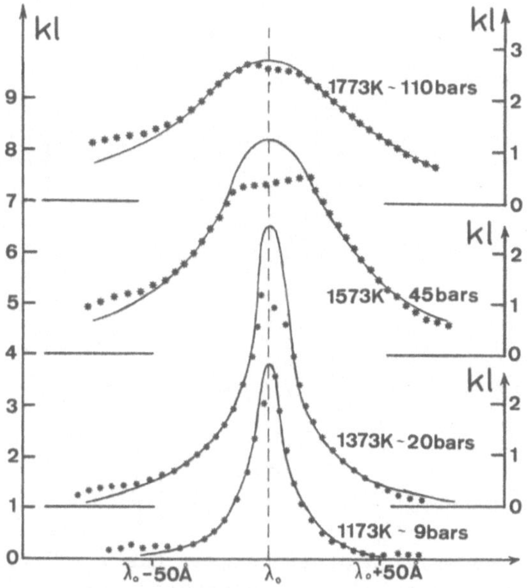

Fig. 2. Experimental (✸) and best Lorentzian (———) profiles near the $\lambda_0 = 852.1$nm resonance line for various temperatures and total pressures $P_{total} = P_{Cs} + P_{Ar}$. The slab thickness is l=10µm.

Typical experimental absorption profiles are displayed on figure 2, in the case of the strongest Cs 852.1nm line. Apart from the center of the line, affected by the resonance radiation transfer, and the extreme blue wing which is perturbated by a probable molecular band, the fit to a Lorentzian profile is satisfactory. No lineshift is noticeable on the records.

Similar results are obtained around the other resonance line (figure 3) but a greater noise on the signal makes it difficult to ensure a good fit. The spectral absorbance is quite the same as the previous one but a slight red-shift can be detected at high densities.

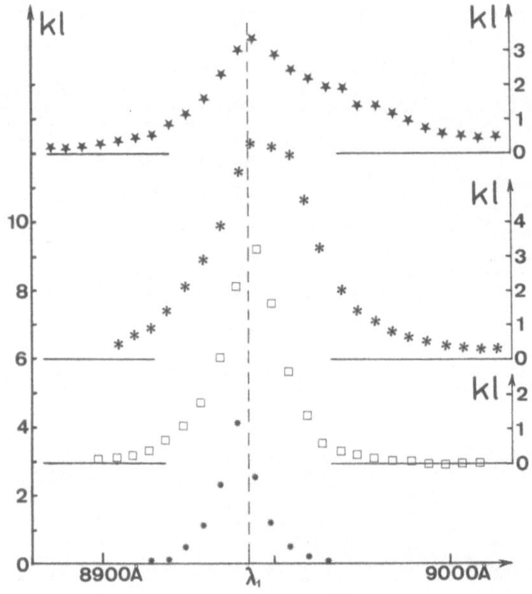

Fig. 3. Experimental absorption profiles near the λ_1=894.3nm line for various temperatures and total pressures. (✱) 1173K, 13.5bars (□) 1373K, 20bars (∗) 1573K, 45bars (✶) 1773K, 110bars

The table 1 summarizes the experimental conditions, temperature, pressure, total density, electrical density deduced from conductivity data /8/ assuming a non metallic state and a mobility regime transport. Absorption data, linewidth γ , extinction coefficient $k(\lambda_o)$ and total fractional absorption are also indicated.

Table 1. Main experimental conditions and data for the λ_0=852.1nm line

T	K	1173	1273	1373	1473	1573	1673	1773
P_{total}	bars	9	13.5	20	29	45	60	110
N_{Cs}	$g.cm^{-3}$	.009	.013	.021	.033	.049	.075	0.10
σ	$(\Omega.cm)^{-1}$	2.2	2.8	5.6	10	20	50	110
n_e	$10^{19}cm^{-3}$	0.04	0.09	0.4	1.4	3	9.1	20
γ	nm	0.8	1.25	1.4	2.1	3.6	6.4	4.8
$k(\lambda_0)$	μm^{-1}	0.5	0.55	0.67	0.58	0.42	0.47	0.28
A	nm	4.2	6.7	8.1	11.4	17.3	32.0	19.0

DISCUSSION

The main feature of these results is a very low value of the line broadening, by more than one order of magnitude, compared with an extrapolation of low density data /7/. Considering the data of the table 1, it is also impossible to match the linewidth with the sum of the three contributions, resonance by Cs atoms in the fundamental state, Van der Waals by Ar atoms and non-resonant Cs, Stark by the electrons. It seems necessary to consider a reduction of the number of Cs atoms in the fundamental state /9/ and a weakening of electrostatic interactions /10/. This hypothesis may be coherent with the formation of clusters and positive cluster ions /4,5/, favoured by a resonant excitation /11/, and with a continuous enhancement of the dielectric constant /12/.

REFERENCES

1. V.A. Alekseev and I.T. Iakubov, Phys.Reports 96:1 (1983)
2. S. Jüngst, B. Knuth and F. Hensel, Phys.Rev.Letters 55:2160 (1985)
3. W. Geerstma, J.Phys.C 18:2461 (1985)
4. J.P. Hernandez, Phys.Rev.A 31:932 (1985), Phys.Rev.A 34:1316 (1986)
5. A. Likal'ter, High Temp. 21:186 (1983), High Temp. 19:544 (1981)
6. J. Rous and A. Leycuras, Physica B 139 & 140:928 (1986)
7. C.L. Chen and A.V. Phelps, Phys.Rev. 173:62 (1968)
8. H. Renkert, F. Hensel and E.U. Franck, Ber.Bunsenges.Phys.Chem. 75:507 (1971)
9. M. Skowronek, J. Rous and J. Larour, XVIIth Int. Conf. on Phenomena in Ionized Gases, Bakos and Zsuzsa Sörlei ed. (Budapest), F-10:363 (1985)
10. Y. Vitel and M. Skowronek, this conference (to be published)
11. D.H. Pollock and A.O. Jensen, J.Appl.Phys. 36:3184 (1965)
12. W. Hefner and F. Hensel, Phys.Rev.Letters 48:1026 (1982)

CHAPTER XI

INTEGRAL EQUATIONS

ONSAGER-THOMAS-FERMI "ATOMS" AND "MOLECULES": "CHEMISTRY" OF CORRELATIONS IN DENSE PLASMAS

Yaakov Rosenfeld

Nuclear Research Center-Negev
P. O. Box 9001
Beer-Sheva, Israel

Following Widom,[1] the m-body correlation function $g^{(m)}$ $(r_1,r_2 \ldots r_m)$ of a fluid can be expressed through the free energy change upon fixing the positions of m fluid particles in the appropriate configuration to form an m-interaction-site molecule. As a special case, the zero-separation theorem[2] relates the r=0 value of the plasma pair-screening potential, $H(r) = \ell n(g(r)\exp(\beta\phi(r))$, to the thermodynamics of plasma mixtures.[3] This relation is the starting point for calculating enhancement factors of nuclear reaction rates.[4] It played a key role in the study of the short range behavior of the bridge function, notably their universal characteristics.[5] The application of Widom's relation for calculating the complete pair correlation function g(r), not to mention higher order correlation functions, has been out of reach for existing theories[6] for the thermodynamics of molecular fluids. A new theory for the statistical thermodynamics of interacting charged particles[7-10] is, however, of the required accuracy and simplicity to enable such a calculation. This physically transparent theory is applied here to a molecular fluid composed of clusters of positive ions in a uniform neutralizing background charge density. We calculate the m-particle screening potentials in classical plasmas. The results reported below represent the first accurate calculation of fluid many body correlation functions from a theory for the thermodynamics of molecular fluids.

To simplify the presentation we consider specifically the 3-D one component plasma (OCP),[4,11,12] containing N positive point ions of charge Qe, at temperature $k_BT=\beta^{-1}$, in a uniform neutralizing background of volume ν (eventually $N,\nu \rightarrow \infty$, $n=N/\nu$), characterized by the coupling parameter $\Gamma = \beta(Qe)^2/a$. The Wigner-Seitz radius, $a = (3/4\pi n)^{1/3}$, serves as the unit of length in this paper: $\beta\phi(r)=\Gamma/r$.

For pair correlations, the OCP potential of mean-force, $w(r) = -\ln g(r)$, is given by[13]

$$w(r) = -\beta \left[F_0^{ex}(N \bullet) - F_1^{ex}(\bullet\!\underset{r}{-}\!\bullet,\ N\text{-}2 \bullet) \right] \tag{1}$$

$F^{ex}{}_0(N\bullet)$ is the excess free energy of the OCP containing point ions. $F^{ex}{}_1$ is the excess free energy of an OCP in which one pair of ions is kept at fixed separation r. That is, $F^{ex}{}_1$ corresponds to an infinitely dilute solution of the 2-site-point-charge molecule ($\bullet\!\underset{r}{-}\!\bullet$) in a fluid of 1-site point charge ions, together in a uniform background. $F^{ex}{}_1$ contains the electrostatic interaction between different point charges in the molecule, i.e., $(Qe)^2/r$, so that the screening potential, $H(r) = -w(r) + \Gamma/r$, is finite for r=0.

From the exact diagrammatic expression one obtains[4,5] $H(r) = h(r)-c(r)-B(r)$, where $h(r) = g(r)-1$. $c(r)$, the direct correlation function is related to h(r) by the Ornstein-Zernike (OZ) equation, which (in k-space) takes the form $\tilde{c}(k)=\tilde{h}(k)/(1+n\tilde{c}(k))$. B(r), the bridge function, is expressed in terms of graphs with h(r)-bonds and at least triply connected field points. B(r)=0 defines the hypernetted-chain (HNC) approximation. The modified-HNC theory corrects it by employing (e.g.) the hard sphere bridge functions and the ansatz of universality. The HNC integral equation,[4,5,11,14] obtained from the HNC-closure $H(r)=h(r)-c(r)$ and the OZ relation, can be also derived variationally from a free energy functional. This HNC free energy is of about 1% accuracy for one- and multi-component plasmas in 2 and 3 dimensions, and it is much more accurate than the (short range) HNC-g(r).

Our approach is to iterate on the HNC approximation by using it not via its closure relations, but rather via its prediction for the free energies $F^{ex}{}_0$, $F^{ex}{}_1$, which is relatively more accurate. Yet exact solutions for the HNC equation for complex charge clusters, associated with high order correlations, is not within our reach. We need more insight in order to perform the calculations.

Our theory implements[7-10] the Onsager charge smearing optimization[15,16] into the variational free energy functional of the HNC theory or the closely related[14] mean spherical approximation (MSA). This leads to an approximate physically intuitive solution of the HNC integral equation for the structure. It yields, however, very accurate results for the HNC free energy for all $o < \Gamma < \infty$. Moreover, the leading $\Gamma \to \infty$ HNC results for c(r) and free energy in our theory, are the exact $\Gamma \to \infty$ HNC results. These are calculated as interactions between smeared charges at distance r, and as self energies of the Onsager "atoms" and "molecules" respectively. These self energies give an exact lower bound to the true potential energy of the system, which is a very tight bound when $\Gamma >> 1$.

In strong coupling ($\Gamma >> 1$) the free energies ${F^{ex}}_0$, ${F^{ex}}_1$ are dominated by the corresponding Madelung potential energy terms, for which the HNC-Onsager results are[10,17]:

$$F_0^{ex} \ (N \ \bullet) \xrightarrow[\Gamma \to \infty]{\text{HNC}} Nu \ (\odot) \tag{2}$$

$F_1^{ex}(\underset{r}{\bullet\!\!-\!\!\!-\!\!\bullet}, \ N\text{-}2 \ \bullet) \xrightarrow[\Gamma \to \infty]{\text{HNC}} (N\text{-}2)u(\odot) + u$ () (3)

u ($\odot$) denotes the self energy of an "Onsager atom" (the ion-sphere Thomas-Fermi "atom") consisting of a point charge at the center of a neutralizing sphere (of radius a, for the OCP) having the background charge density β u ($\odot$) = - 0.9Γ provides a tight bound to the bcc lattice energy -0895929Γ. u denotes the self energy of the "Onsager-molecule" composed of the pair of ions at distance r and a uniform neutralizing charge cloud of the background charge density. The shape of the molecule is determined by the ("isolation"-) condition: the electric potential and field vanish on its surface, i.e, there are no induced surface charges on the molecule if placed in an infinite neutral conductor. From (1)-(3) we obtain:

$w(r) \xrightarrow[\Gamma \to \infty]{\text{HNC-F}^{ex}} \quad \beta u$ () $- \ 2\beta u \ (\odot)$ (4)

The "Onsager molecule" concept is readily extended for higher order correlation functions. For e.g. the triplet correlations, $w^{(3)} (\underset{\sim}{r}_1,\underset{\sim}{r}_2,\underset{\sim}{r}_3) = - \ell n \ g^{(3)} (\underset{\sim}{r}_1,\underset{\sim}{r}_2,\underset{\sim}{r}_3)$. we get

$w^{(3)}(\underset{\sim}{r}_1,\underset{\sim}{r}_2,\underset{\sim}{r}_3) \xrightarrow[\Gamma \to \infty]{\text{HNC-F}^{ex}} \beta u$ () $- \ 3\beta u(\odot)$ (5a)

The triplet screening potentials

$$H^{(3)}(\underset{\sim}{r}_1,\underset{\sim}{r}_2,\underset{\sim}{r}_3) \ = \ -w^{(3)}(\underset{\sim}{r}_1,\underset{\sim}{r}_2,\underset{\sim}{r}_3) + \Gamma(\underset{\sim}{r}_{12}^{-1} + \underset{\sim}{r}_{13}^{-1} + \underset{\sim}{r}_{23}^{-1} \tag{5b}$$

are finite for r_{12} and/or r_{13} and/or $r_{23} = 0$.

A most important property of the Onsager-molecules as defined above is their "dissociation" whenever any molecular point charge is of distance larger than 2a for all others: e.g.

u (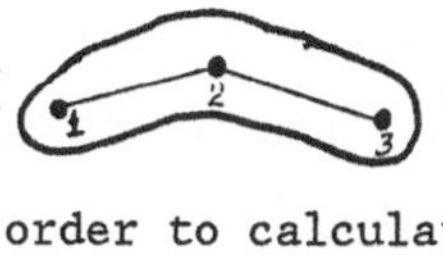 : $r_{13},r_{23} \geq 2a$) = u () + u () (6)

In order to calculate the self energies for complex geometries we seek a simple and accurate approximation that has the dissociation property

(6). It is given via the "convolution-ball-smearing"[8] by which every molecular point charge is uniformly smeared within a sphere of some radius d_i. The number of independent such parameters d_i depends on the symmetry of the molecule, e.g. we need only one parameter for a symmetric polytop. Note, however, that with this "ball smearing" the resulting Onsager lower bound for the energy approximates well the self energy of the Onsager molecule, but is is not the self energy of the resulting dumbbell configuration. u_d () remains, however, a useful symbol for this energy bound. For the "diatomic" 2-charge molecule we obtain[18]:

βu (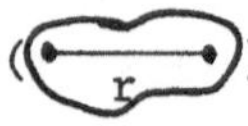) $\simeq -\Gamma\ [\ 2\frac{3}{10}d^2 + 2\frac{3}{5}d^{-1} + \Psi_d(r)\ ] + \Gamma/r = \beta u_d($  $)$ (7)

subject to $\partial u_d/\partial d = 0$ which gives d(r). $\Psi_d(r) = d^{-1}\Psi_1(r/d)$ is the electrostatic interaction between two uniformly charged spheres of radius d, unit total charge, and separation r, given for d=1 by[10]:

$$\Psi_1(r) = \begin{Bmatrix} \frac{6}{5} - \frac{1}{2}r^2 + \frac{3}{16}r^3 - \frac{1}{160}r^5, & r \leq 2 \\ r^{-1} & , r \geq 2 \end{Bmatrix} = -\lim_{\Gamma\to\infty} c_{HNC}(r)/\Gamma \qquad (8)$$

Using the optimized d(r) from (7), the defining z=r2d. $f(z \geq 1) = 0$, $f(z \leq 1) = 1-5z^2 + 5z^3 - z^5$, we obtain the following parametric expression for the screening potential representing the HNC result for the free energy difference:

$$\lim_{\substack{\Gamma\to\infty \\ \text{HNC-}F^{ex}}} H(r)/\Gamma \simeq H_d(z)/\Gamma = \frac{9}{5}\left\{[1 + \frac{1}{3}f(z) + \frac{5}{9}\Psi_1(2z)]\ /\ (1 + f(z))^{1/3} - 1\right\} \qquad (9a)$$

$$r = 2z(1 + f(z))^{1/3} \qquad (9b)$$

Without optimization, namely using d(r)=1, we obtain the HNC-closure result:

$$H_{d=1}(r)/\Gamma = \Psi_q(r) = \lim_{\substack{\Gamma\to\infty \\ \text{HNC-closure}}} H(r)/\Gamma = \lim_{\Gamma\to\infty} (h_{HNC}(r) - c_{HNC}(r))/\Gamma \qquad (10)$$

$H_{d(r)}$ (r) is in excellent agreement with both the simulations and modified-HNC[5] calculations for the strongly coupled fluid OCP (see Fig. 1). Comparison of (9) and (10) shows clearly that the physical meaning of the bridge diagrams contributions (i.e. those missing the the HNC closure) is provided by the Onsager-molecule concept. Note that $H_{d(r)}(r) = H_{d=1}(r) = \Gamma/r$ for $r \geq 2$ (the "dissociation" property), and that both functions are continuous with 3 continuous derivatives. That is why both functions are nearly the same already for $r \geq 1$. These features explain why the available simulation date, unguided by a suitable theory, led to different interpretations of the scaling properties of the screening potentials for multicomponent plasmas, and of their "linear" behavior for $0.5 < r < 1.5$.[19] Note also that for small r, $H_{d(r)} = \Gamma \frac{9}{10}(2^{5/3}-2) - \frac{\Gamma}{4} r^2 + \ldots$, in agreement with the well known ion-sphere result.

The "dissociation" property plays a key role for all values of Γ. The convolution-ball-smearing that lead to (7) and (9), when incorporated into the Debye-Huckel (DH) or MSA free energy functional, provides an effective interpolation between the weak ($\Gamma \ll 1$) and strong ($\Gamma \ll 1$) coupling regions which are well represented by the DH ($d_1 \sim 1$) and Onsager ($d_1 \sim 1$) lower bounds,[7,8] respectively. The OCP excess free energy is thus characterized by one smearing parameter, $d_0(r)$, which varies from $d_0=1$ for $\Gamma=\infty$ to $d_0=0$ for $\Gamma=0$, and represents the size of the effective hard core radius. The molecular smearing radius, $d_1(\Gamma, d_0, (\Gamma))$ is obtained by optimizing the free energy functional approximating $F^{ex}{}_1$.

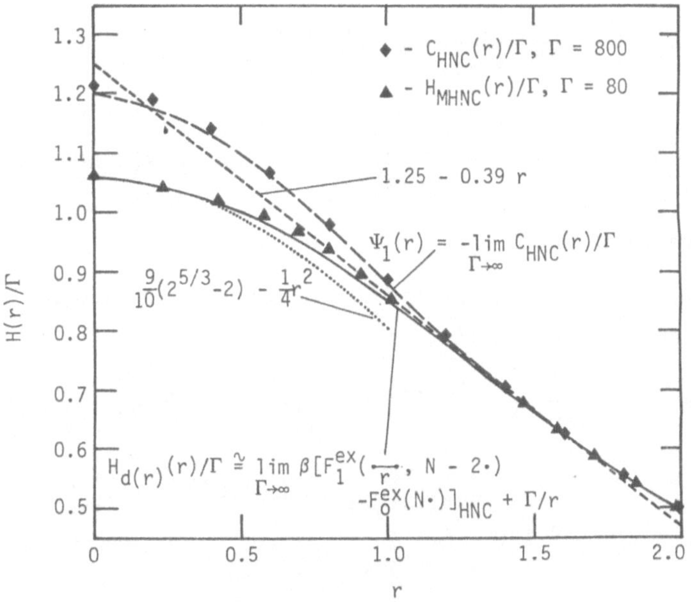

Fig. 1 Pair-screening potential, H(r), of the strongly coupled OCP. The line, 1.25-0.39r, fits well the simulation results in the range $0.5 < r < 1.8$ for all $\Gamma > 5$ (Fig. 7 in Ref. 4). The squares represent the HNC $\Gamma = 800$ results[20] for $c(r)/\Gamma$. The circles represent the $\Gamma = 80$ results for $H(r)/\Gamma$ as obtained by the Modified-HNC equation.[5] The theoretical predictions $\Psi_1(r)$ and $H_{d(r)}(r)$ are described in the text.

Using (as above) a as our unit of length,, $na^3 = 3/4\pi$, we thus write

$$\frac{\beta F_0^{ex}}{N} \simeq \mathscr{F}(d_0,\Gamma) = -\Gamma \left(\frac{3}{10}d_0^2 + \frac{3}{5}d_0^{-1} \right)$$

$$+ \frac{(2\pi)^{-3}}{2(3/4\pi)} \int d\underset{\sim}{k} \; \ell n \left[1 + \frac{3}{4\pi} \Gamma \tilde{\Psi}_{d_0}(k) \right] \tag{11}$$

which should be optimized, $\partial\mathscr{F}/\partial d_0|_\Gamma = 0$, to obtain $d_0(\Gamma)$. $\tilde{\Psi}_{d_0}(k)$, the Fourier - transform (FT) of $\Psi_{d_0}(r)$, is given by $\tilde{\Psi}_{d_0}(k) = d_0^2 \tilde{\Psi}_1(kd_0)$, where $\tilde{\Psi}_1(k)$ is the FT of $\Psi_1(r)$, given by

$$\tilde{\Psi}_1(k) = \frac{4\pi}{k^2} \tilde{\rho}_1(k) \tag{12}$$

where

$$\tilde{\rho}_1(k) = \frac{3(\sin k - k \cos k)}{k^3} \tag{13}$$

Using the fact that

$$\lim_{\Gamma\to\infty} (2\pi)^{-3} \int \frac{\tilde{\Psi}_1(k) \, d\underset{\sim}{k}}{1 + \frac{3}{4\pi}\Gamma\tilde{\Psi}_1(k)} = \frac{\sqrt{3}}{6}\Gamma^{-1/2} \tag{14}$$

We find that $\mathscr{F}(d_0(\Gamma),\Gamma) = -0.9\Gamma + \frac{\sqrt{3}}{6}\Gamma^{1/2} + \ldots$ and the HNC excess free energy $(\beta F_0^{ex}/N)_{HNC} \simeq -0.9\Gamma + 2\frac{\sqrt{3}}{6}\Gamma^{1/2} + \ldots$, tightly bracket the simulation OCP data for $\Gamma \gg 1$.

Similarly, the free energy difference in (1) is given by the following[18] generalization of (7):

$$H_{d_0,d_1}(r) = 2 \left[\frac{3}{10} d_1^2 + \frac{3}{5} d_1^{-1} \right] + \Psi_{d_1}(r) - 2 \left[\frac{3}{10} d_0^2 + \frac{3}{5} d_0^{-1} \right]$$

$$- (2\pi)^{-3} \int d\underset{\sim}{k} \frac{\Psi_{d_1}(k) \left[1 + \frac{\sin kr}{kr} \right] - \Psi_{d_0}(k)}{1 + \frac{3}{4\pi}\Gamma\tilde{\Psi}_{d_0}(k)} \tag{15}$$

in which we use the above $d_0(\Gamma)$, and optimize, $\partial H_{d_0,d_1}/\partial d_1|_\Gamma = 0$, to obtain $d_1(r,\Gamma)$. We find that for large Γ,

$$H_{d_0(\Gamma),d_1(r,\Gamma)}(r) = H_{Eq.\ (9)}(r) + \alpha(r)\Gamma^{1/2} + \ldots \tag{16}$$

where $|\alpha(r)| \ll H_{Eq.\ (9)}(r)/\Gamma$, and in particular, $H_{Eq.\ (9)}(r=0) = \frac{9}{10}(2^{5/3} - 2) \simeq 1.06$, $\alpha(r=0) \simeq 0.06$, explaining why the simulation results line in Fig. 1 is so weakly dependent on Γ. It should be noted that for $r \geq d_0(\Gamma)$ the "molecule" dissociates, i.e. the approximation $d_1(\Gamma) = d_0(\Gamma)$ is valid.

Recalling (8) we denote $C^{(\infty)}_{HNC} = -\Gamma\Psi_1(r)$, and using the Ornstein-Zernike (OZ) equation

$$\tilde{h}(k) = \tilde{c}(k)/[1 - \frac{3}{4\pi}\tilde{c}(k)] \tag{17}$$

we define

$$\tilde{h}^{(\infty)}_{HNC}(k) = -\Gamma\tilde{\Psi}_1(k)/[1 + \frac{3}{4\pi}\Gamma\tilde{\Psi}_1(k)] \tag{18}$$

to obtain the following leading $\Gamma \to \infty$ expression

$$H_{d_0=d_1=1}(r) = -C^{(\infty)}_{HNC}(r) + (2\pi)^{-3}\int d\underset{\sim}{k}\ \tilde{h}^{(\infty)}_{HNC}(k)\ \frac{\sin kr}{kr} =$$

$$h^{(\infty)}_{HNC}(r) - C^{(\infty)}_{HNC}(r) \tag{19}$$

i.e. as a generalization of (10) we now get an HNC closure form. Indeed, replacing (in the above steps) $-\Gamma\Psi_1(r)$ by $C_{HNC}(r)$, which is a good approximation, we obtain the full HNC closure:

$$H_{d_0(\Gamma),d_1(r,\Gamma)}(r \gtrsim d_0(\Gamma)) \simeq h_{HNC}(r) - C_{HNC}(r) \quad . \tag{20}$$

Comparison with (8) and (10) shows that $h_{HNC}(r)$ should be an entropic contribution for $F^{ex}{}_1$ at large Γ. As also revealed by numerical solutions[20] of the HNC equation for the OCP,

$h_{HNC}(r) \sim$ excess entropy $\sim \Gamma^{1/2}$.

Fuller analysis leads to the following general result for the screening potentials:

$$H^{(m)}_{HNCF^{ex}}(\underset{\sim}{r}_1, \underset{\sim}{r}_2\ \ldots,\ \underset{\sim}{r}_m:\ r_{ij} \gtrsim 2d_0(\Gamma) \text{ or } \Gamma \lesssim 1) \simeq$$

$$\sum_{\substack{j=1 \\ i>j}}^{m} [\, h_{HNC}(r_{ij}) - c_{HNC}(r_{ij})] \tag{21}$$

This is the Kirkwood superposition approximation[21] (KSA) combined with the HNC closure for the OCP. The short range behavior of the bridge functions and the cross-over to KSA-HNC are natural consequences of the Onsager-molecules and their dissociation property.

It should be emphasized that the crucial term sin kr/kr in (15) and (19), leading to the HNC closure, is not "forced" into the expression. It results[18] from the orientation average of the one-pair-string form-factor (with $\underset{\sim}{r} = \underset{\sim}{r}_1 - \underset{\sim}{r}_2$)

$$\frac{1}{4\pi}\int d\,\Omega \left| e^{i\underset{\sim}{k}\cdot\underset{\sim}{r}_1} + e^{i\underset{\sim}{k}\cdot\underset{\sim}{r}_2}\right|^2 = 2\,(1 + \frac{\sin kr}{kr})\ ,$$

and is a natural consequence of the theory[18] for $F^{ex}{}_1$ ($\bullet\!\overset{}{\underset{r}{—}}\!\bullet$, N-2$\bullet$) when the separation vector r can have any orientation in space with equal probability.

Our result Eq. (19) demands special attention. Define:

$$H(r) = -\,C_0(r) + \Delta H(r)$$

$$C(r) = c_0(r) + \Delta C(r)$$

and consider the HNC equation for ΔH and ΔC in the following form:

$$\Delta\tilde{H}(k) = \frac{\tilde{C}_0(k) + \frac{3}{4\pi}\,[\tilde{C}_0(k) + \Delta\tilde{C}(k)]\,\Delta\tilde{C}(k)}{1 - \frac{3}{4\pi}\,[\tilde{C}_0(k) + \Delta\tilde{C}(k)} \tag{22a}$$

$$h(r) = \exp\,[\,-\,(\,\frac{\Gamma}{r} + C_0(r)) + \Delta H(r)] - 1 \tag{22b}$$

$$\Delta C(r) = h\,(r) - \Delta H(r) \tag{22c}$$

For example, with the choice of $C_0(r) = -\frac{\Gamma}{r}\mathrm{erf}(1.08r)$, these equations constitute the elementary cycle in the iterative solution of Ng in Ref. 20. Recall[22] that such an iterative solution of Eqs. (22) correspond to graphs-summation to all orders. Indeed, for weak coupling ($\Gamma \ll 1$) one starts with $C_0(r) = -\Gamma/r = -\beta\phi_c(r)$, which is the Coulomb-"bond" and $(\Delta C)_0 = 0$, then the first iteration builds the Debye-bond, $\Delta H_i(r) = \beta\phi_D(r) = -\frac{\Gamma}{r}e^{-xr}$, from (22a), and the Meeron-bond,

$$\Delta C_1(r) = +\,\beta\phi_m(r) = \exp\,[\,-\frac{\Gamma}{r}\,e^{-xr}] + \frac{\Gamma}{r}e^{-xr} - 1, \qquad x = (3\Gamma)^{1/2} \tag{22d}$$

from (22b) - (22c). Continuing with this iteration loop for all non-elementary prototype graphs in the diagrammation expansion of H(r) are formally constructed.[22] For strong coupling ($\Gamma \gg 1$), however, the natural starting point for the iterations is the Onsager-Coulomb (OC) bond, $C_0(r) = C^{(\infty)}{}_{HNC}(r) = -\beta\phi_{OC}(r)$. If the solution of the HNC equation (22a-22d) indeed represents the sum of the appropriate infinite number of diagrams, then it is independent of $C_0(r)$ which is just a formal device chosen to speed up the convergence of the iteration procedure. The choice $CO(r) = C^{(\infty)}{}_{HNC}(r)$ is, however, crucial for obtaining a solution in the limit $\Gamma \to \infty$, and for understanding the solution in general for $\Gamma >> 1$. Starting with the OC-"bond", $C_0(r) = -\Gamma\Psi_1(r)$, and $(\Delta C)_0=0$, then the first iteration builds the Onsager-Debye (OD) "bond"

$$\Delta H_1(r) \equiv -\beta\phi_{OD}(r) = (2\pi)^{-3} \int d\underset{\sim}{k} \frac{-\Gamma\tilde{\Psi}_1(k)\frac{\sin kr}{kr}}{1 + \frac{3}{4\pi}\Gamma\tilde{\Psi}_1(k)} \tag{23}$$

from Eq. (22a), and the Onsager-Meeron (OM) "bond" (from 22c-22d),

$$\Delta C_1(r) \equiv \beta\phi_{OM}(r) = \exp\left[- (\frac{\Gamma}{r} -\Gamma\Psi_1(r)) + \Delta H_1(r)\right] - \Delta H_1(r) -1 \tag{24}$$

Note that (23) is nothing but our expression (19), which is the strong coupling analog of the Debye-Huckel (DH) total pair correlation function,

$$h_{Onsager}(r) = \beta\phi_{OM}(r) + \beta\phi_{OD}(r) \equiv h_1(r) = \Delta C_1(r) + \Delta H_1(r) =$$

$$= \exp\left[- \frac{\Gamma}{r} + h^{(\infty)}_{HNC}(r) - C^{(\infty)}_{HNC}(r) \right] -1$$

Both $h_1(r) = h_{Onsager}(r)$ and the second iteration result $h_2(r)$, along with the complete $\Gamma = 1000$, 7000 solution of the HNC equation by Ng,[20] are featured in Fig. 2. Note that $h_{Onsager}(r)$ already contains the split second peak (for $\Gamma = 7000$) which is associated with the formation of a glassy state. Note also in Fig. 3 that both $S_0(k) = [1 - \frac{3}{4\pi}C_0(k)]^{-1}$ and $S_1(k) = [1 - \frac{3}{4\pi}C_1(k)]^{-1}$ feature correct peak positions and qualitative trends of the structure factor $S(k)$. As Γ increases more peaks, starting from the first and in consecutive order, are identified with location near the corresponding zeroes of $\rho_1(k)$, i.e., $k_1 = 4.49\ldots$, $k_2 = 7.725\ldots$, $k_3 = 10.90\ldots$, etc. This trend, appearing already in $S_0(k)$ and $s_1(k)$ for the OCP, is not limited to the Coulomb potential and exists also in the simulation data for simple (e.g. inverse power) potentials. A comprehensive discussion of these features will be given elsewhere.

In conclusion we mention that our results provide the rational for solving[23] the Thomas-Fermi confined molecule problem in order to calculate enhancement factors for nuclear reaction rates, and other short range correlation effects in dense matter. Our method, concepts and qualitative results are valid for a general D-dimensional multicomponent

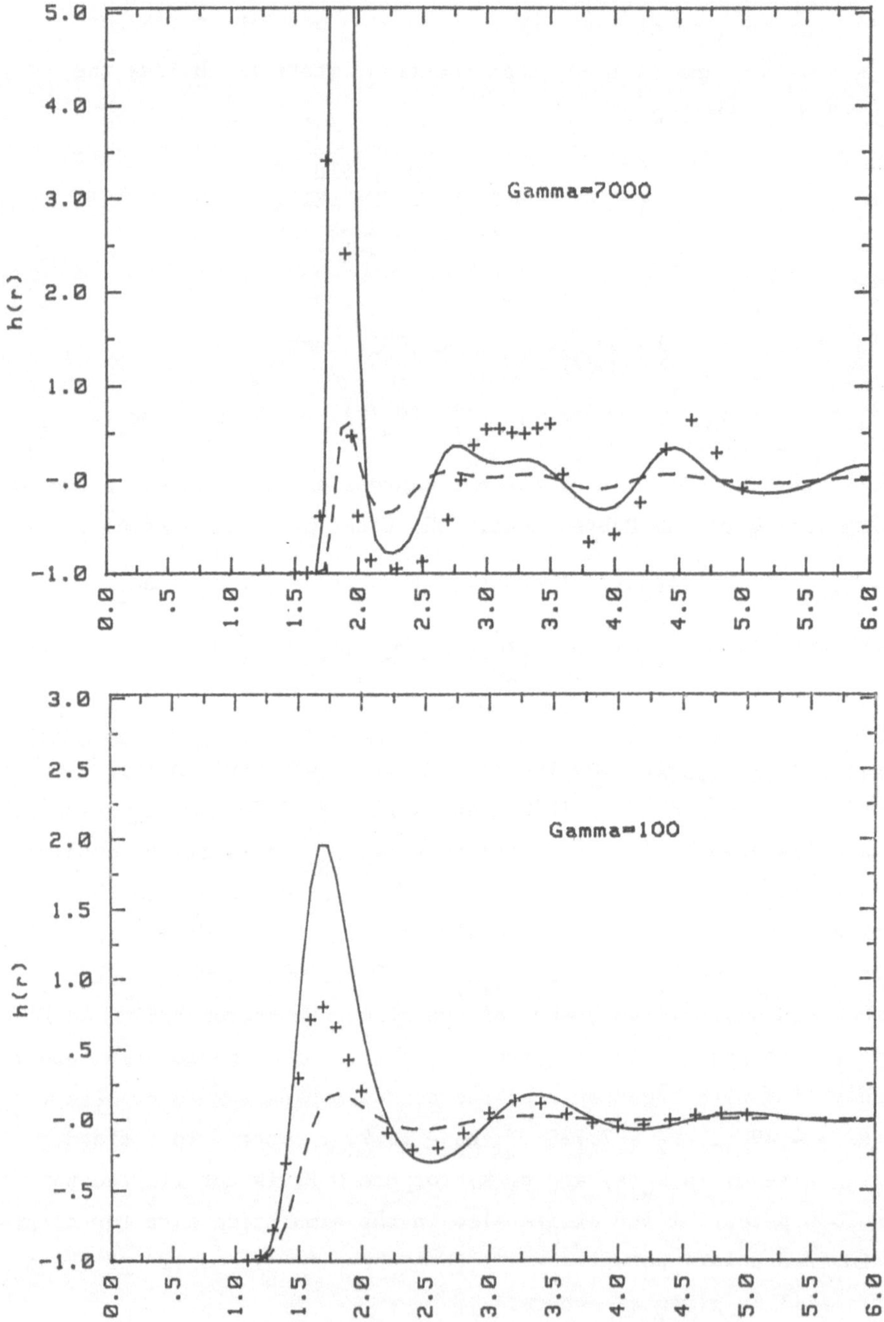

Fig. 2 OCP pair correlation function $h(r) = g(r)-1$, for Γ = 100,7000. Crosses, dotted line, and full line, represent the complete HNC results,[20] its first iteration, $h_1(r)$, and its second iteration, $h_2(r)$, respectively (see text).

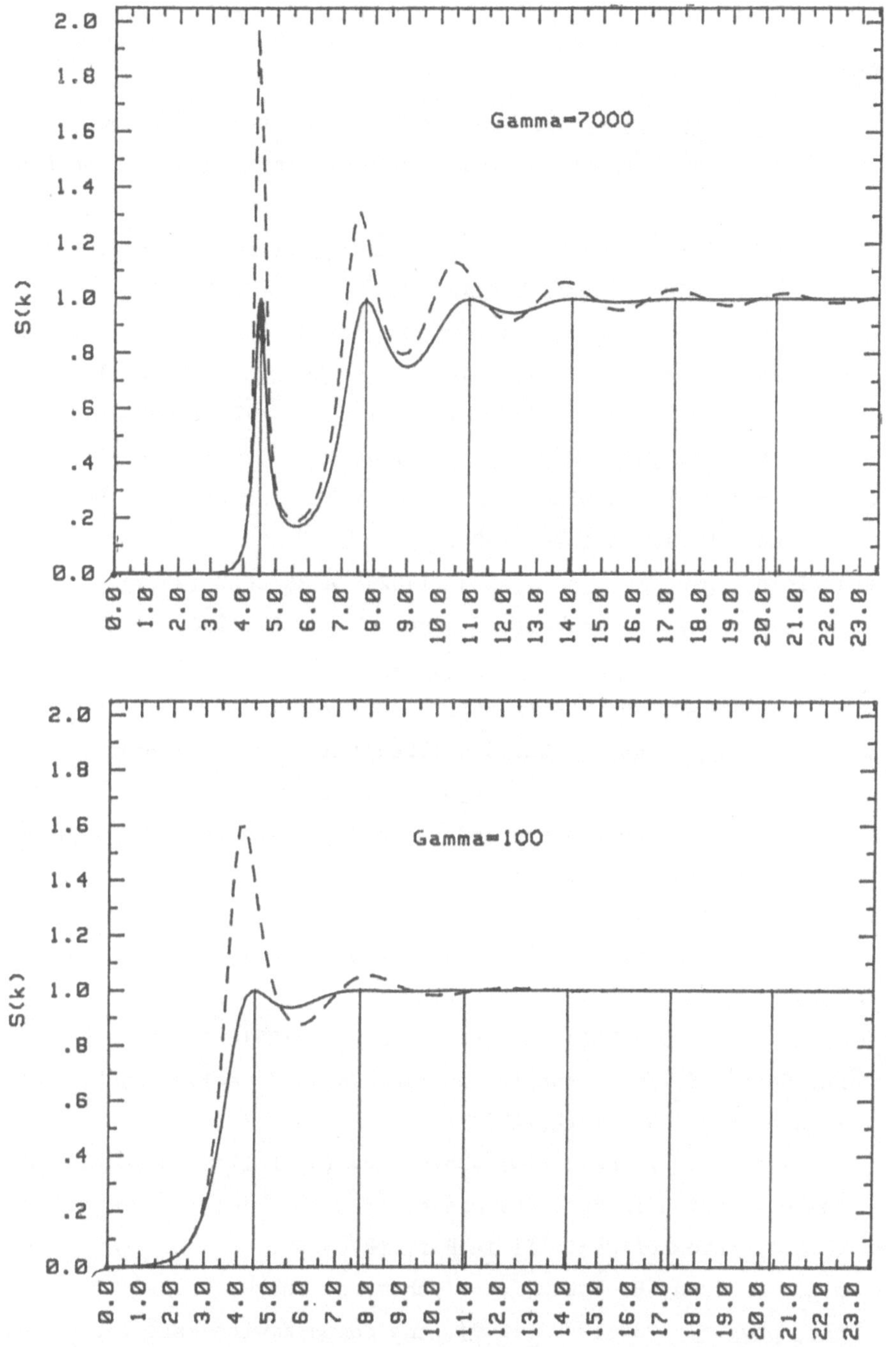

Fig. 3 OCP structure factor for Γ=100,7000 from the zero's order ($S_0(k)$, full line) and first order $S_1(k)$, dotted line) iterations of the HNC-Onsager scheme (see text). Vertical lines denote the positions of the zeroes of $\rho_1(k)$.

plasma, in which the charges may be associated with any Green's function potentials, the Coulomb and screened-Coulomb (Yukawa) being the most important cases.

REFERENCES

1. B. Windom, J. Chem. Phys. 39:2808 (1963). See also J. S. Rowlinson and B. Windom, "Molecular Theory of Capilarity", Clarendon-Press, Oxford 1982.
2. W. G. Hoover and J. C. Poirer, J. Chem. Phys. 37:1041 (1962).
3. B. Jancovici, J. Stat. Phys. 17:357 (1977).
4. See, e.g., the review by S. Ichimaru, Rev. Mod. Phys. 54:1017 (1982).
5. (a) Y. Rosenfeld and N. W. Ashcroft, Phys. Rev. A 20:1208 (1979);
   (b) Y. Rosenfeld, J. Phys. (Paris), Colloq. 41:c-77 (1980);
   (c) Y. Rosenfeld, Phys. Rev. Lett. 44:146 (1980).
6. C. G. Gray and K. E. Gubbins, "Theory of Molecular Fluids", Volume I, Clarendon Press, Oxford, 1984.
7. Y. Rosenfeld, Phys. Rev. A 25:1206 (1982): A 26:3622 (1982).
8. Y. Rosenfeld and W. M. Gelbart, J. Chem. Phys. 81:4574 (1984).
9. Y. Rosenfeld and L. Blum, J. Phys. Chem. 89:5149 (1985): J. Chem. Phys. 85:1556 (1986).
10. Y. Rosenfeld, Phys. Rev. A 32:1834 (1985): A 33:2025 (1986).
11. M. Baus and J. P. Hansen, Phys. Rep. 59:1 (1980).
12. H. E. DeWitt, in "Strongly Coupled Plasmas", edited by G. Kalman, Plenum Press, New York, 1977.
13. From direct application of Widom's relation.
14. Y. Rosenfeld, J. Stat. Phys. 37:215 (1984).
15. L. Onsager, J. Phys. Chem. 43:189 (1939).
16. E. H. Lieb and H. Narnhofer, J. Stat. Phys. 12:2916 (1975): Ph. Choquard, in "Strongly Coupled Plasmas", edited by G. Kalman, Plenum Press, New York, 1977.
17. Eq. (2) is derived in Ref. (10) above: Eq. (3) follows directly from the discussion in Ref. (9) above.
18. Direct application of Sec. III in Ref. (8) above.
19. See, e.g., N. Itoh, H. Totsuji, S. Ichimaru, and H. E. DeWitt, Astrophys. J. 234:1079 (1979), and compare with Ref. (5c) above.
20. K. C. Ng, J. Chem. Phys. 61:2680 (1974): F. J. Rogers and H. E. DeWitt (unpublished): see also Ref. (10) above.

21. J. G. Kirkwood, J. Chem. Phys. 3:300 (1935): compare with F. J. Pinski and C. E. Campbell, Phys. Rev. A 33:4232 (1986).

22. Appendix B in Ref. (5a) above, and references therein.

23. M. Friedman, A. Rabinovitch, Y. Rosenfeld and R. Thieberger "Thomas-Fermi Equation with Non-Spherical Boundary Conditions", J. Comp. Phys., in print.

QUANTAL HYPERNETTED CHAIN EQUATION

APPLIED TO LIQUID METALLIC HYDROGEN

Junzo Chihara

Department of Physics
Japan Atomic Energy Research Institute
Tokai-mura, Ibaraki 319-11, Japan

INTRODUCTION

In classical liquids, integral equation methods have been shown to give successful results in many cases. However, there is no standard integral equation for quantum liquids at finite temperature. Therefore, when a liquid metal is taken as a mixture of ions and electrons, it is necessary to set up an integral equation which can be applied to quantum and classical liquids in a unified manner, since the electrons constitute a quantum liquid and the ions behave as classical particles. In this situation, it is important to notice that the density functional theory can treat a quantum liquid at arbitrary temperature including a classical liquid.

QUANTAL HYPERNETTED CHAIN EQUATION

Let us consider a liquid metal (or a plasma) as a mixture of electrons and ions with densities, n_0^e and n_0^I, respectively: the ions can be regarded as forming a classical fluid and the electrons are assumed to constitute a quantum fluid. When external potentials $U_i(r)$, acting on ions (i=I) and electrons (i=e), are imposed on this system, the density distributions $n_i(r|U_I,U_e)$ of this inhomogeneous system can be determined by the functional derivative of the thermodynamic potential Ω with respect to $\gamma_i(r)\equiv\mu_i-U_i(r)$ at fixed temperature T and volume V (for example, Chihara 1978b)[1]

$$\left.\frac{\delta\Omega}{\delta\gamma_i(r)}\right|_{T,V} = -n_i(r|U_I,U_e) \; . \tag{1}$$

Here, μ_i denotes the chemical potential of i species. The natural variables of the thermodynamic potential Ω for this inhomogeneous system are T, V and $\gamma_i(r)$. The independent variable $\gamma_i(r)$ is replaced by $n_i(r)$ with the use of the Legendre transformation of Ω (Callen 1960), which introduces the intrinsic Helmholtz free energy:

$$\mathcal{F}\equiv \Omega - \sum_j\int\frac{\delta\Omega}{\delta\gamma_j(r)}\gamma_j(r)dr = \Omega + \sum_j\int n_j(r)\gamma_j(r)dr \; . \tag{2}$$

Then, $\gamma_i(r)$ is obtained from $\mathcal{F}[n_e,n_I]$

$$\frac{\delta \mathcal{F}[n_I, n_e]}{\delta n_i(\mathbf{r})} = \mu_i - U_i(r) = \gamma_i(r) \ . \tag{3}$$

In order to describe the inhomogeneous mixture, we take the noninteracting ion-electron mixture as a reference system, and introduce such effective potentials $U_i^{eff}(r)$, that induce the density distributions $n_i^0(r|U_i^{eff})$ in the noninteracting system so as to become exactly equal to the density distributions $n_i(r|U_I, U_e)$ in the real system:

$$n_i(r|U_I, U_e) = n_i^0(r|U_i^{eff}) \ . \tag{4}$$

Here,

$$n_I^0(r|U) \equiv n_0^I \exp\{-\beta U(r)\} \tag{5}$$

and

$$n_e^0(r|U) \equiv \sum_\ell f(\varepsilon_\ell)|\phi_\ell(r)|^2 \tag{6}$$

with

$$f(\varepsilon) \equiv [\exp\{\beta(\varepsilon - \mu_e^0)\}+1]^{-1}$$

which is calculated by solving the wave equation for an electron

$$\{ -\frac{\hbar^2}{2m}\nabla^2 + U(r) \}\phi_\ell(r) = \varepsilon_\ell \phi_\ell(r) \tag{7}$$

with μ_e^0, the chemical potential of a noninteracting electron gas. Similarly to (3), for the noninteracting system under the external potential $U_i^{eff}(r)$, there follows

$$\frac{\delta \mathcal{F}_0[n_I, n_e]}{\delta n_i(\mathbf{r})} = \frac{\delta \mathcal{F}_0[n_I^0, n_e^0]}{\delta n_i^0(\mathbf{r})} = \mu_i^0 - U_i^{eff}(r) \ , \tag{8}$$

since $U_i^{eff}(r)$ are defined so as to induce the same density distributions of the interacting system. Here, $\mathcal{F}_0$ is the intrinsic Helmholtz free energy of the noninteracting system, and can be written in the explicit form

$$\mathcal{F}_0 \equiv \frac{1}{\beta}\int n_I(r)[\ln(n_I(r)\lambda^3)-1]d\mathbf{r} + T_s[n_e] - TS_s[n_e] \tag{9}$$

where

$$T_s[n_e] \equiv \sum_i f(\varepsilon_i)\langle\phi_i| \frac{-\hbar\nabla^2}{2m}|\phi_i\rangle \tag{10}$$

$$S_s[n_e] = \sum_\ell\{f(\varepsilon_\ell)\ln f(\varepsilon_\ell)+[1-f(\varepsilon_\ell)]\ln[1-f(\varepsilon_\ell)]\} \tag{11}$$

with $\lambda \equiv (h\beta/m)^{1/2}$. From (3) and (8), the effective external potential defined by (4) can be expressed in terms of the interaction part of the intrinsic Helmholtz free energy

$$U_i^{eff}(r) = U_i(r) + \frac{\delta \mathcal{F}_{int}}{\delta n_i(\mathbf{r})} - \mu_i^{int} \ , \tag{12}$$

where

$$\mathcal{F}_{int} \equiv \mathcal{F} - \mathcal{F}_0 \tag{13}$$

with μ_i^{int}, the interaction part of the chemical potential of i species.

At this stage, we notice the fact that the density distributions $n_i(r|I)/n_0^I$ around a fixed ion in the liquid metal become identical with the radial distribution functions (RDF's) $g_{iI}(r)$ concerning an ion, if the ions can be treated as classical particles. Since a fixed ion in the liquid metal causes external potentials $U_i(r) \equiv v_{iI}(r)$ [with $v_{ij}(r)$: the interparticle interactions] acting on ions and electrons, Eqs.(5), (6) and (12) turn out to give exact expressions for the RDF's $g_{ij}(r)$ in the forms.

$$g_{II}(r) = n_I(r|I)/n_0^I = \exp\{ -\beta v_{II}^{eff}(r) \} \tag{14}$$

$$g_{eI}(r) = n_e(r|I)/n_0^e = n_e^0(r|v_{eI}^{eff})/n_0^e \, . \tag{15}$$

Therefore, the effective interactions $v_{ij}^{eff}(r)$ involved in (14) and (15) are given by (12) as

$$v_{ij}^{eff}(r) = v_{ij}(r) + \frac{\delta \mathcal{F}_{int}}{\delta n_i(\mathbf{r}|j)} - \mu_i^{int} \, . \tag{16}$$

It is important to note that we can define the direct correlation functions (DCF's) in quantum mixtures on the basis of $\mathcal{F}_{int}$ (Chihara 1984b):

$$C_{ij}(|\mathbf{r}-\mathbf{r}'|) \equiv -\beta \frac{\delta^2 \mathcal{F}_{int}}{\delta n_i(\mathbf{r}) \delta n_j(\mathbf{r}')}\bigg|_{U=0} \tag{17}$$

$$= \frac{\delta[\beta U_j(r) - \beta U_j^{eff}(r)]}{\delta n_i(r)}\bigg|_{U=0} \, . \tag{18}$$

Then, the effective interaction given by (12) can be rewritten

$$v_{ij}^{eff}(r) = v_{ij}(r) + \sum_\ell \int \frac{\delta^2 \mathcal{F}_{int}}{\delta n_i(\mathbf{r}) \delta n_\ell(\mathbf{r}')}\bigg|_0 \delta n_\ell(\mathbf{r}'|j) d\mathbf{r}' - B_{ij}(r)/\beta$$

$$= v_{ij}(r) - \Gamma_{ij}(r)/\beta - B_{ij}(r)/\beta \tag{19}$$

with

$$\Gamma_{ij}(r) \equiv \sum_\ell \int C_{i\ell}(|\mathbf{r}-\mathbf{r}'|)\{n_\ell(r'|j) - n_0^\ell\} d\mathbf{r}' \, , \tag{20}$$

in terms of the bridge functions $B_{ij}(r)$ and the DCF's. The Fourier transforms of (17) can be written in the matrix form

$$\sqrt{n}\,C(Q)\sqrt{n} = (\chi_Q^0)^{-1} - (\chi_Q)^{-1} \, , \tag{21}$$

in terms of the density response functions, $\chi_Q \equiv \|\chi_Q^{ij}\|$ and $\chi_Q^0 \equiv \|\chi_Q^{0i}\delta_{ij}\|$, of the interacting and noninteracting systems, respectively, and $n \equiv \|n_0^i \delta_{ij}\|$. Note that this relation can be derived from (18) and the linear response formula.

Inversely, Eq.(21) leads to expressions of the partial structure factors $S_{II}(Q)$, $S_{eI}(Q)$ and the density response function of electrons χ_Q^{ee} by the DCF's $C_{ij}(Q)$ and the density response function $\chi_Q^0 \equiv \chi_Q^{0e}$ of the non-interacting system in the forms:

$$S_{II}(Q) = \{ 1 - n_0^I C_{ee}(Q)\chi_Q^0 \}/D(Q) \tag{22}$$

$$S_{eI}(Q) = (n_0^e n_0^I)^{1/2} C_{eI}(Q)\chi_Q^0/D(Q) = \frac{\rho(Q)}{\sqrt{Z}} S_{II}(Q) \tag{23}$$

$$\chi_Q^{ee} = \{ 1-n_0^I C_{II}(Q) \}/D(Q) = \frac{|\rho(Q)|^2}{Z} S_{II}(Q) + \frac{\chi_Q^0}{1-n_0^e C_{ee}(Q)\chi_Q^0} \tag{24}$$

with

$$\rho(Q) \equiv n_0^e C_{eI}(Q)\chi_Q^0/[1-n_0 C_{ee}(Q)\chi_Q^0] \tag{24a}$$

$$D(Q) \equiv \{1-n_0^I C_{II}(Q)\}\{1-n_0^e C_{ee}(Q)\chi_Q^0\} - n_0^e n_0^I |C_{eI}(Q)|^2 \chi_Q^0 \quad . \tag{24b}$$

Here, we have used the fact that the density response functions χ_Q^{iI} concerning ions become identical with the structure factors $S_{II}(Q)$ and $\chi_Q^{0I}=1$, since the ions in a liquid metal can be treated as classical particles. The Ornstein-Zernike (OZ) relations for the mixture are obtained by the inverse Fourier transforms of the above equations as

$$g_{II}(r) - 1 = C_{II}(r) + \Gamma_{II}(r) \tag{25}$$

$$g_{eI}(r) - 1 = \hat{B}\cdot C_{eI}(r) + \hat{B}\cdot\Gamma_{eI}(r) \tag{26}$$

$$n_e(r|e)/n_0^e - 1 = \hat{B}\cdot C_{ee}(r) + \hat{B}\cdot\Gamma_{ee}(r) \tag{27}$$

where $\hat{B}$ denotes an operator defined by

$$\mathfrak{F}_Q[\ \hat{B}^\alpha\cdot f(r)\ ] \equiv (\chi_Q^0)^\alpha\cdot\mathfrak{F}_Q[f(r)] = (\chi_Q^0)^\alpha\cdot\int e^{i\mathbf{Q}\cdot\mathbf{r}} f(r)d\mathbf{r}$$

for an arbitrary real number α. Here, it should be remarked that the OZ relation for the electron density distribution $n_e(r|e)$ around a "fixed" electron in the mixture, is derived on the basis of the relation:

$$\mathfrak{F}_Q[n_e(r|e) - n_0^e\ ] = \chi_Q^{ee}/\chi_Q^0 - 1\ , \tag{28}$$

which has been derived from a certain ansatz (Chihara 1978a, 79, 83). In the conventional approach, a liquid metal is considered as a quasi-one component fluid interacting via an effective interaction $v^{eff}(r)$, in which the presence of electrons in a liquid metal is taken into account only. We can prove that this effective potential can be exactly represented as

$$\beta v^{eff}(Q) = \beta v_{II}(Q) - |C_{eI}(Q)|^2 \frac{n_0^e\chi_Q^0}{1-n_0^e C_{ee}(Q)\chi_Q^0} \tag{29}$$

in terms of the DCF's of the two component model, if the bridge function of the quasi-one component system is chosen to be equal with the bridge function $B_{II}(r)$ of the two component model (Chihara 1986b). In the above, it should be noted that Eqs. (14)∿(26) are formal, but exact expressions, provided that the ions constitute a classical fluid.

At this point, we introduce the hypernetted chain (HNC) approximation, which means the bridge functions in (19) are neglected. This approximation together with (28) leads to the quantal HNC (QHNC) equation, which can be written with the aid of the OZ relations (25)∿(27) as follows:

$$C_{II}(r) = \exp\{ -\beta v_{II}(r)+\Gamma_{II}(r) \} - 1 - \Gamma_{II}(r) \tag{30}$$

$$C_{eI}(r) = \hat{B}^{-1}\cdot[n_e^0(r|v_{eI}-\Gamma_{eI}/\beta)/n_0^e - 1] - \Gamma_{eI}(r) \tag{31}$$

$$C_{ee}(r) = \hat{B}^{-1}\cdot[n_e^0(r|v_{ee}-\Gamma_{ee}/\beta)/n_0^e - 1] - \Gamma_{ee}(r) \ . \tag{32}$$

In the HNC approximation, the DCF's $C_{ij}(r)$ are all determined self-consistently from (30) to (32). Here, it should be emphasized that the QHNC equation can be applied to an electron-ion mixture from the low-temperature region where the electrons are in a degenerate state to the high-temperature region where the electrons become a classical fluid. Also, it is interesting to note that we can derive many kinds of integral equations for the RDF's in a liquid metal by introducing further approximations to (30)∿(32) (Chihara 1986a). For example, if the electron-electron local-field correction $G_{ee}(Q)$ is approximated as that of the jellium model where the ions in a liquid metal are replaced by the background of uniform positive charge:

$$C_{ee}(Q) \equiv -\beta v_{ee}(Q)\{1-G_{ee}(Q)\} \simeq -\beta v_{ee}(Q)\{1-G^{jell}(Q)\} \ , \tag{33}$$

then, we need not to use (32). With the use of several approximations in addition to (33), the QHNC equation is shown to lead to the integral equations of Dharma-wardana and Perrot (DWP) (1982) and of Ichimaru et al (1985).

APPLICATION TO THE ELECTRON-PROTON SYSTEM

We have applied the QHNC equation to the electron-proton system: that is, liquid metallic hydrogens (LMH) or perfectly ionized hydrogen plasmas. The temperature and density $n_0 \equiv n_0^e = n_0^I$ of this system can be specified by the plasma parameter Γ and the Wigner-Seitz radius $r_s = a/a_B$ with $a \equiv (3/4\pi n_0)^{1/3}$ and the Bohr radius a_B. Features of the proton-electron system can be represented by Fig. 1. In the region between two lines: $T=E_F$ and $T=E_F^p$ (E_F and E_F^p are the Fermi temperatures of the electrons and protons, respectively.), the protons constitute a classical liquid and the electrons are taken as in a degenerate state. In the calculation, the electrons are assumed either to be at absolute zero temperature in the liquid metallic region (Chihara 1984a), or to form a classical fluid in the plasma region (Chihara 1978c).

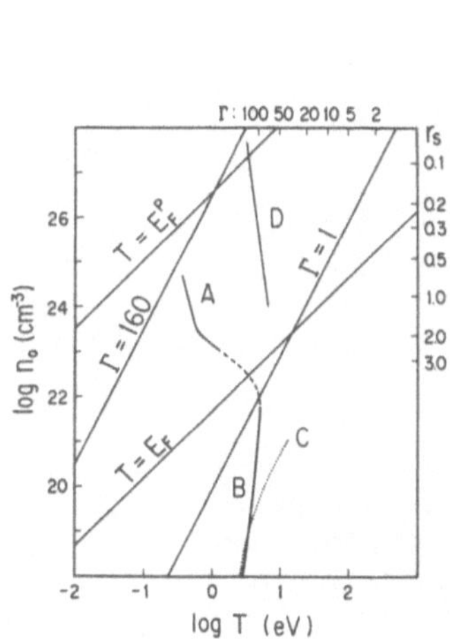

Fig. 1 Features of the proton-electron system. In the region right to the line AB, hydrogens can be treated as a proton-electron mixture; a metallic liquid and a perfectly ionized plasma. The curve C denotes 99% ionization calculated by the Saha equation. Near the line D, the first-order phase transition occurs in LMH.

In the conventional pseudopotential theory, to which the approach of Ichimaru et al (1985)is equivalent, the effective ion-ion interaction in a liquid metal modeled as the one-component system is dependent only on the electron density r_s, and independent of the temperature of the system. In LMH such as $r_s=1$, due to the strong electron-proton interaction, this standard approach breaks down as shown by Fig. 2, which indicates that the effective proton-proton interactions calculated by QHNC equation exhibit strong dependence on Γ. Three effective potentials obtained from the QHNC and the DWP equations and the screened Coulomb (used by Ichimaru et al) yield quite different RDF's with each other, as shown for the case of $\Gamma=10$ and $r_s=1$ in Fig. 3. Therefore, we cannot applied the conventional pseudopotential theory (the screened Coulomb) to LMH, generally.

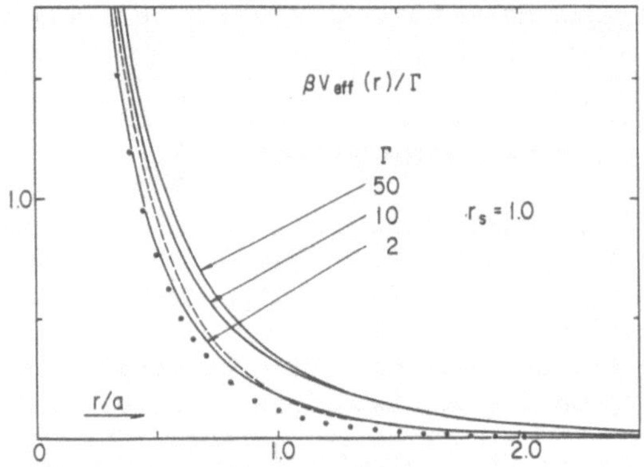

Fig. 2 Effective proton-proton interactions in the quasi-one component model of LMH for $r_s=1.0$. Full circles denote the interaction calculated by DWP equation at $\Gamma=10$. The screened Coulomb interaction used by Ichimaru et al is displayed by the broken curve, which is independent of Γ.

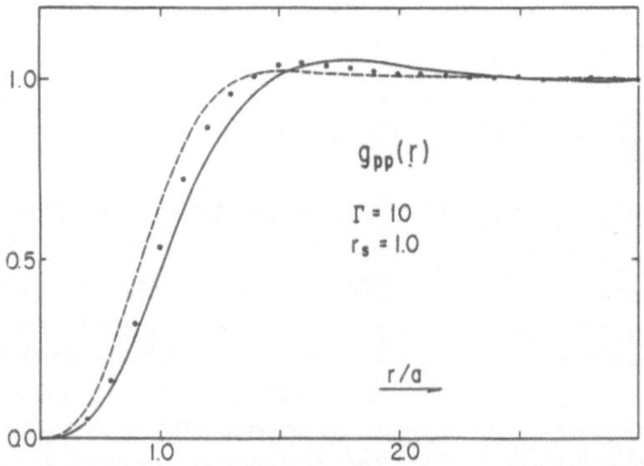

Fig. 3 The radial distribution functions between protons calculated with the use of QHNC, screened Coulomb and DWP interactions, which are denoted by the full and broken curves and full circles, respectively.

In the treatment of Ichimaru et al (which is identical to the conventional pseudopotential theory) and DWP, the exchange-correlation effect is approximated as that of the jellium model where the influence of the protons on the electrons is neglected. Furthermore, Ichimaru et al used the linear response formula to evaluate the RDF g_{ep}, and DWP neglected some terms which represent the electron-proton correlation effect in (19) (Chihara 1986a). These approximations bring about the differences from the QHNC results in Figs. 2 and 3.

Figure 4 shows the RDF's obtained from the QHNC equation for LMH varying r_s at fixed Γ=20. At high density region such as r_s=0.1, there is no significant difference in g_{pp} between LMH and the one-component plasma model. Naturally, with increasing r_s, the proton-proton correlation becomes week, because the screening reduces the proton-proton interaction. Contrary to g_{pp}, the RDF g_{ep} grows up with the increase of r_s. In varying parameter r_s across r_s=0.3 at fixed Γ=20, the RDF's change abruptly. Also, it should be noticed that the QHNC equation has two sets of solutions at r_s=0.3 and Γ=20 as shown in Fig. 5: the one shown by broken curves is continuous to the solution in the high temperature region (with smaller r_s) and the other drawn by full curves continues to the solution in the low temperature region. This fact indicates that some kind of the first order phase transition takes place across r_s=0.3 in LMH at Γ=20. This figure shows that this phase transition causes not so great variation to g_{pp} and g_{ep} in contrast with $n_e(r|e)$, which exhibits strong correlations and reflects the structure of g_{pp} in the low temperature phase. Therefore, this phase transition may be regarded as a result of change in the electronic structure in LMH. A similar phase transition is observed across the line D in Fig. 1: the phase on the right-hand side region of this line will be called the high-temperature phase and the phase on the other side, the low-temperature phase.

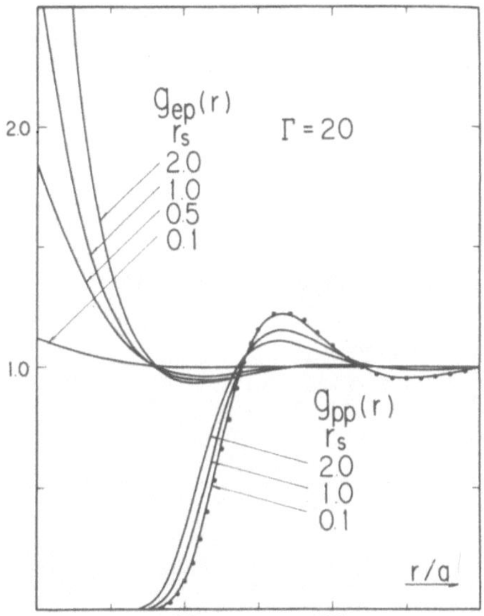

Fig. 4 The RDF's, g_{pp} and g_{ep}, at Γ=20 varying r_s from 0.1 to 20. The full circles are g_{pp} of the one-component plasma model.

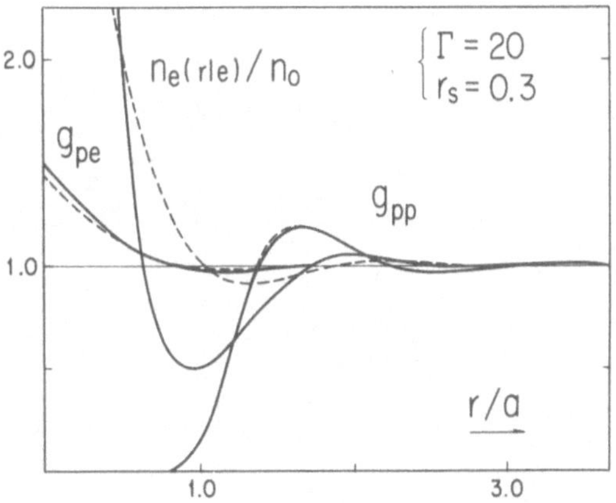

Fig. 5 Two sets of solutions for the QHNC equation at Γ=20 and r_s=0.3. The full curves denote solutions belonging to the low temperature phase, the broken curves, the high temperature phase.

This transition is clearly seen by tracing the change in the local-field correction $G_{ee}(Q)$ for electrons in the two component model by varying Γ from unity to twenty at r_s=0.5 as displayed in Fig. 6. As Γ increases from unity, the local-field correction deviates a lot from that of the jellium model, grows up and changes abruptly to have a negative part near origin, when Γ varies from 11 to 12, where some kind of phase transition occurs.

Next, we proceed to the plasma region, where the electrons behave as classical particles (for details, see Chihara 1978c). When the temperature is reduced at fixed density $\rho=6\text{x}10^{20}$ electrons/cm^3, the proton-proton RDF grows up near the origin as shown in Fig. 7, and the numerical solution can not be obtained due to the divergence in the iterative process near line B in Fig. 1. The peak in g_{pp}, which appears when the temperature decreases to $5.4\text{x}10^4$ K, may be regarded as indicating the tendency to form a hydrogen molecule. Also, a similar divergence is found in the liquid metallic region along the line A. The QHNC equation can provide us the bound energy levels of hydrogen in the plasma state. At low density region such as 10^{18} electrons/cm^3, there is no significant change from a free atom. A higher density, the bound energy levels become shallow and rise up as the temperature increases as shown in Fig. 8. As the temperature approaches $5.4\text{x}10^4$ K, where the molecular formation begins at $\rho=6\text{x}10^{20}$/cm^3, the levels rise up with larger variations.

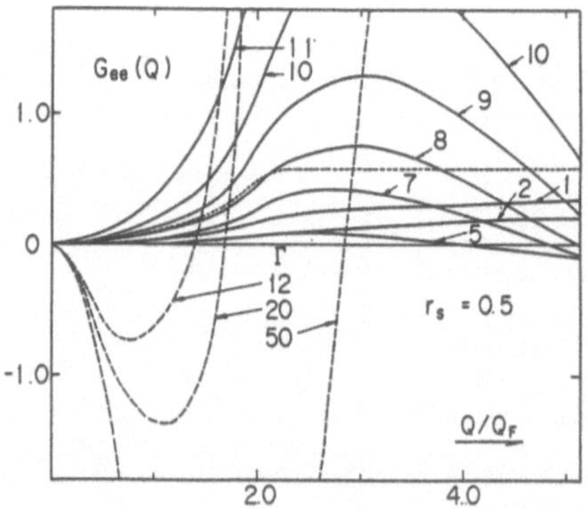

Fig. 6 The dependence of the local-field correction $G_{ee}(Q)$ on the plasma parameter Γ (shown on curves), which specified the proton configuration. The broken and full curves denote $G_{ee}(Q)$ of the LMH in the low and high temperature phases, respectively. The dotted curves is the $G^{jell}(Q)$ of the Hubbard form in the jellium model. At fixed r_s=0.5, the phase transition occurs between Γ=11 and Γ=12, where $G_{ee}(Q)$ exhibits an abrupt change.

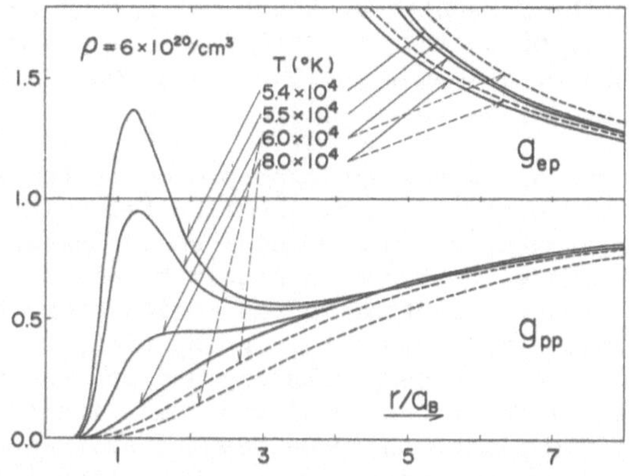

Fig. 7 The radial distribution functions, g_{pp} and g_{ep}, calculated by the QHNC (full curves) in a perfectly ionized hydrogen plasma varying the temperature at a fixed density $6x10^{20}/cm^3$. The broken curves denote the RDF's calculated by the nonlinear Debye-Hückel theory with the use of the wave-equation.

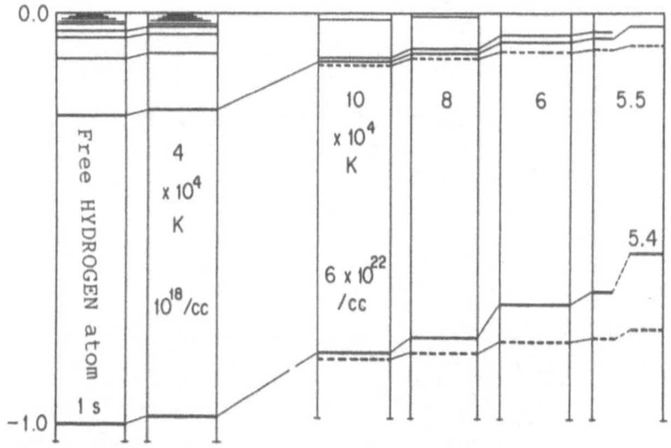

Fig. 8 The bound electronic energy levels in a hydrogen plasma, calculated by QHNC and Debye-Hückel potentials, which are denoted by full lines and broken lines, respectively. The levels of a free hydrogen atom are shown for the comparison. The bound levels rise up with larger variations, when the temperature is reduced to 5.4 K where hydrogen molecules begin to form.

CONCLUSION

The QHNC equation is superior to the integral equations of DWP and Ichimaru et al in that the influence of ions on the exchange-correlation effect can be taken account by treating the ions and electrons in the system on equal footing, besides the effect of electron-proton correlation (Chihara 1986a). Furthermore, their equations cannot be applied to a plasma state where the electrons become a classical fluid, since g_{ee} must be identical with g_{pp} there. On the other hand, the QHNC equation can be applied both to the liquid metallic state and to the plasma state in a unified manner. Results, when applied to LMH, are summarized in Fig. 1. On the right-hand side of the line B in Fig. 1, the electron-proton system forms a plasma state where hydrogens are perfectly ionized. The line Γ=160 in Fig. 1 gives a rough criterion that the hydrogen system becomes a solid phase. Therefore, hydrogens may be considered to become a liquid metallic state where electrons have no bound level around a proton, in the region above the line A surrounded by the lines, Γ=160 and $T=E_F$. In addition, there is two phases in liquid metallic state: the low temperature phase and the high temperature phase. A set of integral equations, (30) and (31) with (33) based on the jellium model, can yield almost the same results as the QHNC equation in the high-temperature phase region, but shows no phase transition, because of the exchange-correlation effect being fixed as that of the jellium model. Here, it should be noticed that the high temperature phase contains the plasma state, because there is no abrupt change in RDF's between the plasma and liquid metallic region.

Moreover, the QHNC equation has been extended to treat liquid metals or plasmas with core electrons: the average ionic charge, electronic bound energy levels, and the electron-ion interaction can be determined self-consistently by this equation (Chihara 1985).

REFERENCES

Callen, H. B., 1960, "Thermodynamics" John Wiley, New York

Chihara, J., 1978a, Prog. Theor. Phys. 59, 76.

----- 1978b, Prog. Theor. Phys. 59, 1085.

----- 1978c, Prog. Theor. Phys. 60, 1640 .

----- 1979, Prog. Theor. Phys. 62, 1533.

----- 1983, Prog. Theor. Phys. 70, 331.

----- 1984a, Prog. Theor. Phys. 72, 940.

----- 1984b, J. Phys. C 17, 1633.

----- 1985, J. Phys. C 18, 3103.

----- 1986a, Phys. Rev. A 33, 2575.

----- 1986b, J. Phys. C 19, 1665.

Dharma-wardana M. W. C. and Perrot F., 1982, Phys. Rev. A 26, 2096.

Ichimaru S., Mitake S., Tanaka S. and, Yan X.-Z., 1985, Phys. Rev. A 32, 1768.

COMPUTATION OF AN IMPROVED INTEGRAL EQUATION BY NON LINEAR RESUMMATION OF THE FIRST GRAPHS OF THE BRIDGE FUNCTION

Michel Lavaud and Jean-Marc Victor*

G.R.E.M.I. (UA 831) UFR Faculté des Sciences
45067 Orléans Cédex 2, France

* L.P.T.L., T 16, Université P. et M. Curie, 4 Place Jussieu
75230 Paris Cédex 05, France

INTRODUCTION

There have been recently many attempts to derive a "universal" self-consistent approximate integral equation for the correlation function h(r), which would be reliable for an extensive range of fluid systems and physical conditions[1]. Various closures have been proposed. They give results that are generally in good agreement with numerical experiments, for a wide variety of systems. It seems that, up to now, the closure of Zerah and Hansen,[1] that interpolates beween HNC and soft-core MSA, has the widest range of applicability.

All these approximate integral equations are very useful, because they give a simple and fast way to compute approximations of thermodynamic and structural properties of dense systems. However, the approximations involved are not very clear, and the only means to choose the best one for a given system, is to compare the results with numerical experiments. Insofar as all the closures can be considered as derived from particular approximations of the bridge function B(r), it seems interesting to study the properties of this function from first principles, and to compute universal approximations to it.

In this article, we make a first step in this direction. We first recall the explicit development of B(r) as a sum of 2-graphs (i.e. graphs with 2 root-points) with lines h(r). Then, we propose a new resummation method for n-graphs. We describe it on the simple case of the fourth virial coefficient B_4, and we apply it to B_5 and to the first few terms of B(r). Finally, we compute simple and accurate approximations for these coefficients, from their exact resummations.

DEVELOPMENT OF B(r) AS A SUM OF 2-GRAPHS

An exact integral equation for the correlation function h(r) has been given by several authors[2] :

$$- \beta w_{12} = - \beta u_{12} + B_{12} + \rho \int h_{13} \, [h_{32} + \beta(w_{32} - u_{32}) + B_{32}] \, d\vec{r}_3 \qquad (1)$$

u(r) is the interaction potential, w(r) the potential of mean force defined by $1 + h(r) = \exp\ -[\beta w(r)]$, and B(r) is the bridge function. w_{ij} stands for $w(r_{ij})$.

B(r) has the explicit development[2] :

$$B(r) = \sum_{m=2}^{\infty} \frac{\rho^m}{m!} \beta'_{m+2}(r) \qquad (2)$$

where $\beta'_{m+2}(r)$ is the sum of basic[2] 2-graphs with m field-points and lines h(r). One has :

$$\beta'_4(r) = \delta_5(r) = \int h_{13} \, h_{32} \, h_{14} \, h_{42} \, h_{34} \, d\vec{r}_3 \, d\vec{r}_4 \qquad (3)$$

$$\beta'_5(r) = 6\varepsilon 7c(r) + 6\varepsilon 7d(r) + 3\varepsilon 8a(r) + 6\varepsilon 8b(r) + \varepsilon 9(r) \qquad (4)$$

where the usual notation of 2-graphs is used[3]. We include $\beta'_5(r)$ in the development for two reasons. First, from results on PY2 equation[4], $\beta'_4(r)$ might be insufficient to obtain accurate results at liquid densities, for certain systems. Secondly, all the 2-graphs of $\beta'_5(r)$ are computable by known techniques[3].

DESCRIPTION OF OUR RESUMMATION METHOD

At low densities, it is possible to compute $\beta'_5(r)$ with a reasonable accuracy by computing independently the five 2-graphs of the development[3]. At high densities, this is no longer possible because h(r) has several oscillations. It is necessary to resum $\beta'_5(r)$ before computing it, to eleminate extensive cancellations among contributing 2-graphs and get sufficient accuracy with reasonable computer time.

Let C be any given linear combination of n-graphs with k field-points and identical lines h :

$$C = \sum_{i=1}^{p} a_i \, \gamma_i \qquad (5)$$

The principle of the method is to identify C to another linear combination of n-graphs with lines $X_L + h$:

$$C = \sum_{i=1}^{q} \alpha_i \, \Gamma_i \, (\{X_L\}) \qquad (6)$$

The real numbers α_i and X_L are solutions of a set of non linear (polynomial) equations.

RESUMMATION OF B_4

We have[5] :

$$- 8 \, B_4 = 3 \, D_4 + 6 \, D_5 + D_6 \qquad (7)$$

By identifying this development to the 2-graph :

$$\alpha \int h_{12}\, h_{23}\, h_{34}\, h_{41}\, (X_1 + h_{13})(X_2 + h_{24})\, d\vec{r}_2\, d\vec{r}_3\, d\vec{r}_4 \qquad (8)$$

$$= \alpha\, [X_1\, X_2\, D_4 + (X_1 + X_2)\, D_5 + D_6]$$

we find that α, X_1 and X_2 satisfy the set of equations :

$$\alpha\, X_1\, X_2 = 3$$

$$\alpha(X_1 + X_2) = 6$$

$$\alpha = 1$$

The solution is :

$$\alpha = 1, \quad X_1 = 3 + \sqrt{6}, \quad X_2 = 3 - \sqrt{6} \qquad (9)$$

From the exact resummation (8)-(9), we construct an approximation of B_4 by noticing that $X_1 + h \sim X_1$ in the fluid region. This gives :

$$-\, 8\, B_4 \sim 3\, D_4 + (3 + \sqrt{6})\, D_5$$

We find that it approaches exact values to less than 20 % for hard spheres, Lennard-Jones and square-well potential, with any width of well.

RESUMMATION OF B_5

B_5 is a sum of ten 1-graphs[5]. We have to solve a system of 10 equations with 11 unknowns. We find an infinite set of resummations with two 1-graphs, depending on a parameter t :

$$B_5 = \alpha_1\, \Gamma_1\, (X_1, \ldots X_5) + \alpha_2\, \Gamma_2\, (X_6, \ldots X_9)$$

We reduce it to 1 equation with 2 unknowns by using computer algebraic languages Macsyma and Reduce.

We find two manifolds of real resummations, for $1 < t < 1.182$ and $3.88 < t < + \infty$, and two simple approximations :

$- \, 30\, B_5 \sim 15.2$ 

$- \, 30\, B_5 \sim 322.2$ 

that are accurate respectively for short range and long-range systems. Full and dotted lines represent respectively the Mayer and Boltzmann functions[5].

RESUMMATION OF $\beta_5'(r)$

We have to solve a system of 4 equations with 5 unknowns.

We find an infinite set of resummations with two 2-graphs, depending on a parameter t. They are real for any t :

$$\beta_5' = \alpha_1\, \Gamma_1\, (X_1, X_2) + \alpha_2\, \Gamma_2\, (X_3, X_4) \qquad (11)$$

with

$$
\begin{cases}
\alpha_1 = 1/t, \quad \alpha_2 = 1 - 1/t & (12) \\
X_{1,2} = 1 + 2t \pm \sqrt{4\,t^2 - 2\,t + 1} & \\
 & \\
X_3 = 3t/(t-1), \quad X_4 = 2 & (13) \\
\Gamma_1 = \text{[graph]} \qquad \Gamma_2 = \text{[graph]} &
\end{cases}
$$

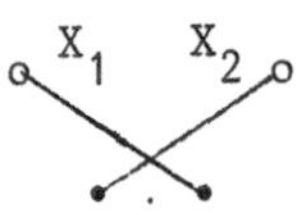

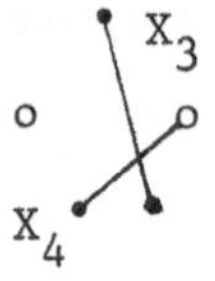

In (13), a line $\circ\overset{X_L}{\text{———}}\circ$ between points i and j represents the function $X_L + h_{ij}$, and no line represents the function h_{ij}.

An interesting resummation is obtained for $t = \dfrac{1 + \sqrt{33}}{4}$. We find :

$$
\begin{cases}
X_1 = X_3 = (9 + \sqrt{33})/2 \sim 7.37 \\
X_2 = (-\,3 + \sqrt{33})/2 \qquad \sim 1.37 \\
X_4 = 2 & (14) \\
\alpha_1 = (\sqrt{33} - 1)/8 \qquad \sim 0.59 \\
\alpha_2 = (7 - \sqrt{33})/8 \qquad \sim 0.41 \quad .
\end{cases}
$$

This resummation is exact. In the cases of gaussian gas and hard cubes at low density, where all the 2-graphs of $\beta_5'(r)$ are known analytically[5,6], we find that $\alpha_1\,\Gamma_1$ and $\alpha_2\,\Gamma_2$ have the same sign for all r. This means that cancellations between 2-graphs of opposite sign in (4) are completely removed.

Further, we find that the simple approximation obtained from the exact resummation (11)-(14) by taking $X_1 + h \sim X_1$ and $X_3 + h \sim X_3$ approaches the exact values of $\beta_5'(r)$ to less than 10% for all r.

CONCLUSION

We have found an exact resummation of $\beta_5'(r)$ that is a good candidate for computing accurate values of this coefficient, and a simplified approximation to it. The approximate integral equations obtained by keeping only $\beta_4'(r)$ and $\beta_5'(r)$ or its approximation in B(r) are universal because the resummation is independent of the potential, and the approximation 7.4 + h(r) $\sim$ 7.4 is valid for realistic systems.

These equations seem to be soluble by conjugate use of the Barker et al. method[7] of computing 2-graphs, and the Gillan-Zerah method[8] of solving equations of type (1). They are nevertheless much more difficult to solve than other approximate equations[1]. To simplify them, one would need accurate approximations for the basic 2-graphs with 4 and 5 points. We are trying to do this by combining the asymptotic developments of 2-graphs we found for large values of r[9], to the upper-bounds obtained at intermediate distances[10] and to transformation of small r values of these 2-graphs as small r values of simpler ones with lines $[h(r)]^k$, k integer.

REFERENCES

1. For an up-to-date review, see for example G. Zerah and J.P. Hansen, J. Chem. Phys. 84:2336 (1986).

2. J.M.J. Van Leeuwen, J. Groeneveld and J. de Boer, Physica, 25:792 (1959) ; E.Meeron, J. Math. Phys., 1:192 (1960) ; T. Morita and K. Hiroike, Prog. Theor. Phys., 23:1003 (1960).
3. See e.g. S. Kim, D. Henderson and L. Oden, Trans. Farad. Soc., 65:2308 (1969).
4. L. Verlet and D. Levesque, Physica, 36:254 (1967).
5. See for example G.E. Uhlenbeck and G.W. Ford, Theory of linear graphs, in "Studies in Statistical Mechanics", Vol.I, North Holland (1962).
6. W.G. Hoover and A.G. de Rocco, J. Chem. Phys., 36:3141 (1962).
7. J.A. Barker and J.J. Monaghan, J. Chem. Phys., 36:2564 (1962).
8. M.J. Gillan, Mol. Phys. 38:1781 (1979) ; G. Zerah, J. Comput. Phys. (in press).
9. M. Lavaud, Phys. Lett., 63A:76 (1977) ; J. Stat. Phys., 19:429 (1982).
10. M. Lavaud, Phys. Lett., 62A:295 (1977) ; J. Stat. Phys., 27:593 (1982) ; J. Stat. Phys., 27:57 (1982).

LECTURERS

A. Alastuey	Laboratoire de Physique Théorique et Hautes Energies Université de Paris XI
B. Alder	University of California Lawrence Livermore National Laboratory
M. Baus	Chimie Physique II Université Libre de Bruxelles
D. B. Boercker	University of California Lawrence Livermore National Laboratory
D. Bollé	Instituut voor Theoretische Fysica Universiteit Leuven
L. Brewer	Time & Frequency Division National Burearu of Standards
J. Chihara	Department of Physics Japan Atomic Energy Research Institute
C. Deutsch	Plasma Physics Laboratory Universite de Paris XI
C. Dharma-wardana	National Research Council of Canada
J. Dufty	Department of Physics University of Florida
W. Ebeling	Sektion Physik der Humboldt Universitat zu Berum
S. Eliezer	Plasma Physics Department Soreq Nuclear Research Center

G. Fontaine — Department of Physiqué
Unviersité de Montréal

V. Fortov — Institute of High Temperatures
USSR Academy of Sciences

K. Golden — Department of Electrical & Computer Engineering
Northeastern University

J.-P. Hansen — Laboratoire de Physique Théorique des Liquides
Université Pierre et Marie Curie

F. Hensel — Institute of Physical Chemistry
Phillips Universitat of Marburg

W. Hubbard — Lunar & Planetary Laboratory
University of Arizona

S. Ichimaru — Department of Physics
University of Tokyo

N. Itoh — Department of Physics
Sophia University

H. Iyetomi — Department of Physics
University of Tokyo

B. Jancovici — Laboratoire de Physique Théorique
et Hautes Energies
Université de Paris Sud

G. Kalman — Department of Physics
Boston College

G. Kobzev — Institute of High Temperatures
USSR Academy of Sciences

W. Kohn — Department of Physics
University of California, Santa Barbara

W. Kraeft — Sektion Physik/Electronnik
der Ernst-Moritz Arndt-Universität

O. Landen — University of California
Lawrence Livermore National Laboratory

P. Martin — Institut de Physique Théorique, Ecole Polytechnique, Fédérale de Lausanna

A. Mostovych — U. S. Naval Research Laboratory

T. O'Neill — Department of Physics, University of California, San Diego

F. Perrot — Centre d'Etudes de Limeil-Valenton

E. Pollock — University of California, Lawrence Livermore National Laboratory

M. Popović — Institute of Physics, Boegrad, Yugoslavia

S. Popović — Institute of Physics, Boegrad, Yugoslavia

Y. Rosenfeld — Nuclear Research Center, Beer-Sheva, Israel

E. Schatzman — Observatoire de Paris, Universite de Nice

G. Senatore — Instituto di Fisica Teorica, Universita di Trieste

J. Shaner — Los Alamos National Laboratory

L. Suttorp — Institute of Theoretical Physics, University of Amsterdam

S. Tanaka — Department of Engineering Sciences, Tohoku University

H. Totsuji — Department of Electronics, Okayama University

G. Zerah — Centre d'Etudes de Limeil-Valenton

INDEX